SEMI-LAGRANGIAN ADVECTION METHODS AND THEIR APPLICATIONS IN GEOSCIENCE

SEMI-LAGRANGIAN ADVECTION METHODS AND THEIR APPLICATIONS IN GEOSCIENCE

STEVEN J. FLETCHER
Research Scientist III
Cooperative Institute for Research in the Atmosphere (CIRA)
Colorado State University
Fort Collins, CO, United States

ELSEVIER

Elsevier
Radarweg 29, PO Box 211, 1000 AE Amsterdam, Netherlands
The Boulevard, Langford Lane, Kidlington, Oxford OX5 1GB, United Kingdom
50 Hampshire Street, 5th Floor, Cambridge, MA 02139, United States

Library of Congress Cataloging-in-Publication Data
A catalog record for this book is available from the Library of Congress

British Library Cataloguing-in-Publication Data
A catalogue record for this book is available from the British Library

ISBN: 978-0-12-817222-3

For information on all Elsevier publications
visit our website at https://www.elsevier.com/books-and-journals

Publisher: Candice Janco
Acquisition Editor: Amy Shapiro
Editorial Project Manager: Ruby Smith
Production Project Manager: Vignesh Tamil
Designer: Matthew Limbert

Typeset by VTeX

Contents

8. Advection with nonconstant velocities

9. Nonzero forcings

10. Semi-Lagrangian methods for two-dimensional problems

11. Semi-Lagrangian methods for three-dimensional problems

12. Semi-Lagrangian methods on a sphere

13. Shape-preserving and mass-conserving semi-Lagrangian approaches

14. Tangent linear modeling and adjoints of semi-Lagrangian methods

15. Applications of semi-Lagrangian methods in the geosciences

16. Solutions to select exercises

CHAPTER

1

Introduction

Advection plays a vital part in many different forms of geophysical modeling that can affect everybody, every day in some form. Advection, in its linear or nonlinear form, affects the weather and ocean forecasts, river sediment as well as chemical transport, along with the solar wind forecast, modeling of magma flows, and hydrothermal transport. An inaccurate forecast in any of these situations could lead to quite catastrophic, and life threatening, results.

Moisture transport for an on shore flow, if miscalculated, could lead to a misleading forecast of an extreme rain event, which in turn could lead to a flash flood warning not being issued. The onset of the tornado season in the United States' Midwest is partially dependent on the advection of the warm moist air from the Gulf of Mexico into the continent. An over-, or under-, prediction of this advection could lead to very different outcomes which could indicate severe, or not severe, weather.

As we saw in 2010, the eruption of Eyjafjallajökull in Iceland led to the shutdown of the North Atlantic air space for over a week because the numerical weather prediction centers did not have good enough models for the transport of the ash. Since then there have been major developments in this area, particularly in transport as a form of advection, where semi-Lagrangian approaches have been developed.

The statement above indicates that advection is also referred to as a transport problem in atmospheric chemistry, sediment transport in rivers and ocean modeling. If we go below the Earth's surface, when we are concerned with temperature, advection is often referred to as convection modeling. Advection also occurs in space through the ionosphere as well as solar wind interactions with the magnetosphere.

Approximations of advection in the early numerical weather predictions were considered in the Eulerian form which is associated with modeling the flow as it passes a point. On the other hand, the Lagrangian formulation deals **with the flow**. Both approaches have advantages and disadvantages. In the early numerical weather prediction, the modeling of advection in the Eulerian form was either done using a finite difference or finite element formulation. The problems that were observed in those early attempts were associated with the **Courant–Friedrichs–Lewy**, or **CFL condition**, which is associated with the **stability** of the numerical scheme. Effectively, this conditions places a restriction on the size of the time step, as well as the grid or element size.

It was shown early on in numerical weather prediction that this condition could lead to very unstable forecasts if it was not met, but this restricted the size of the time step, which in turn prevented running longer forecasts as the computational resources were not available to

Semi-Lagrangian Advection Methods and Their Applications in Geoscience
https://doi.org/10.1016/B978-0-12-817222-3.00005-0

run at such a fine temporal resolution. There was also the problem of damping with certain Eulerian-based finite difference schemes.

If we were to consider the Lagrangian framework, we would start with a set of particles, follow them in time, and then apply the numerical approximation to the differential operators there. The problem with this approach is that the particles could become too far apart, which is not practical to achieve viable approximations. We show a copy of Fig. 2 from Welander [198] in Fig. 1.1, which illustrates this problem.

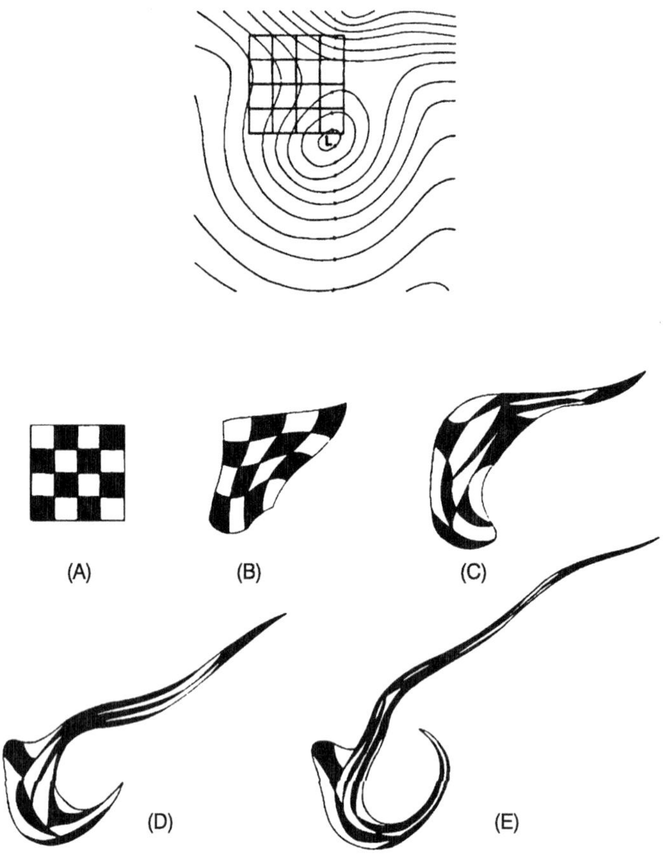

FIGURE 1.1 Copy of the Lagrangian deformation plot, Fig. 2 from Welander, 1955: Studies on the general development of motion in a two-dimensional ideal fluid, *Tellus*, **17**, 141–156. https://www.tandfonline.com/doi/abs/10.3402/tellusa.v7i2.8797. https://creativecommons.org/licenses/by/4.0/.

The advantage of the Lagrangian approach is that the value of the particle following the trajectory is constant. However, we cannot keep track of all these particles, but we would like some form of a technique that utilizes this. Here come semi-Lagrangian approaches. The basis of semi-Lagrangian approaches is that we have a fixed Eulerian grid with known values of the field at these grid points at time t^n and we know the location of the grid points at t^{n+1}, but we do not know what is the value of the field there.

The Lagrangian form of the advection equation, which you will see throughout the book, is given by

$$\frac{D\psi}{Dt} = 0,$$

subject to the kinematic equation

$$\frac{D\mathbf{x}}{Dt} = \mathbf{u},$$

where **u** is referred to as the advection velocity.

The semi-Lagrangian approach amounts to saying that the second equation above tells us how far the particle has traveled in one or two time steps without forcing. Whereas the first equation, when integrated with respect to time, tells us that the value of the field at t^{n-1}, or t^n, depending on whether one uses a two or three time level discretization for the time derivative, is the value of the tracer at the **arrival point** at t^{n+1}. The problem we have is that we are not guaranteed that the **departure point** will be at a grid point at t^n or t^{n-1}. The value of ψ can be found through interpolation but we have to be careful which order, and what properties, we wish the interpolation polynomial to satisfy. This is the finite difference point of view; there is also a similar view with the finite element approach where it not a point but an element to deal with. Finally, there are the flux form and finite volume approaches that conserve properties of the field between the time levels.

Given this motivation, we now move on to briefly summarize what we have in each chapter to address the situation described above. We finish with applications of semi-Lagrangian approaches in the geosciences, followed by solutions to select exercises.

In Chapter 2 we start by showing derivations of the advection/transport equations and show the characteristics based approach which will enable us to move on to the semi-Lagrangian approaches. We will present different explicit and implicit Eulerian-based finite difference schemes and demostrate the performance of these schemes with a smooth Gaussian bell advection, as well as with a discontinuous step function.

Chapter 3 covers the different properties of the Eulerian-based finite difference scheme, where we shall introduce different forms of stability, convergence, and consistency. We will look at dispersion and diffusion properties of the different schemes so that we know their likely performance.

Given the behavior of the Eulerian scheme, we move on to the history of the ideas towards adopting a semi-Lagrangian approach in Chapter 4; we should note here that the semi-Lagrangian approach was referred to initially as quasi-Lagrangian, but the schemes are also referred to as **Eulerian–Lagrangian** approach, and in some disciplines as a **particle trajectory tracking**.

In Chapter 5 we will introduce the semi-Lagrangian approach for the linear one-dimensional scalar advection and show its performance with different order of interpolation for the advection of the Gaussian bell curve and the step function.

Given that we require interpolation polynomials, we shall introduce different forms of interpolation polynomials in Chapter 6, and show their performances in reconstructing the bell curve and step function with different number of interpolation points to illustrate how well the approximations improve with the number of points.

The next stage of the introduction of the semi-Lagrangian approaches is to consider how to determine if there are restrictions on the time and space steps of the different schemes; this is accomplished in Chapter 7.

The properties and techniques to perform the semi-Lagrangian integration so far have been intended for the constant velocity and zero forcing one-dimensional scalar problem. In Chapter 8 we will introduce the techniques that are required to be able to solve the case where the trajectory velocity is no longer constant, which implies that we have to estimate the velocity along the trajectory, but also to evaluate the velocity at the departure point in an implicit iterative solution. We will consider either two or three time level schemes and look at the properties of both of these approaches.

In Chapter 9 we will introduce the situation where we have a forcing term on the right-hand side of the advection equations. There are many different forms of advection problems that have nonzero forcing terms, and given the dynamical situation that is being considered, the approach to deal with these situations could lead to nonphysical modes being excited in the numerical solution. A widely used approach to deal with this situation is referred to as the semi-implicit semi-Lagrangian (SISL) method, where nonlinear terms are treated as explicit, while the linear terms are treated implicitly. We shall introduce this technique along with other techniques that have been developed over the years to stabilize and increase the efficiently of the semi-Lagrangian integration when a forcing term is present.

We move on to the two-dimensional problem in Chapter 10. In this chapter we shall introduce the techniques of bivariate interpolation to estimate the value of the field at the departure point, along with the stability analysis in two dimensions. We shall also introduce different staggered horizontal and vertical grids that are quite often used for numerical modeling. We shall also look at two-dimensional approximations for nonconstant velocities, as well as for nonzero forcing terms where the departure point now is in an area, element, or cell for the finite difference, finite elements, or finite volume/flux form approaches. We shall also present some test cases that are frequently used to analyze the performance of new developments in semi-Lagrangian theory. Finally, we shall show an example of semi-Lagrangian advection with the Eady model which describes baroclinic instability in an x–z plane approximation of the atmosphere.

The next step is to introduce the third dimension for the advection problem, which we do in Chapter 11. In this chapter we shall extend the ideas from the previous chapters to the three-dimensional problem. We consider the case that we need to interpolate to a departure point in three dimensions. We shall introduce some different vertical coordinate systems that are used in operational numerical weather prediction systems, as well as in ocean modeling. We shall consider finite difference, finite element, and finite volume approaches, all as an extension of the flux-form from the last chapter on chemical transport. We revisit the assumptions of only needing two iterations to find the departure point, as well as being inconsistent in the order of the velocity interpolation, compared to the fields' interpolation to the departure point.

As we do not live on a cube, we need to look at how we can extend the technique to spherical coordinates, and we do this in Chapter 12. Here we shall introduce semi-Lagrangian development on the sphere. This chapter comprises of two parts: the first introduces how the vector calculus transfers to spherical coordinates, projections, numerical grids, as well as introduces the theory of spectral methods. The second part describes how semi-Lagrangian theory is applied to different models in spherical coordinates and multiple dimensions, where we will present finite difference and finite volume approaches, as well as semi-Lagrangian methods with spectral methods.

One of the problems that has been noticed in the standard finite difference approach to the semi-Lagrangian methods is that they may not preserve the shape of the object being advected; it may also be required that the *mass* is conserved. In Chapter 13 we will introduce different techniques that have been developed to ensure that the semi-Lagrangian scheme can preserve the shape of the body being advected, to avoid the Gibbs phenomena of under- and overshoots. We shall introduce the cascade interpolation methods between Eulerian and Lagrangian grids, along with different finite volume approaches that enable the conservation of mass either locally of globally. We will also introduce SLICE which is a finite volume based scheme utilizing the cascade interpolation. The chapter finishes by introducing the Vlasov family of nonlinear PDEs and the HWENO approach to solve them.

In the last theoretical chapter, Chapter 14, we shall introduce the concept of the tangent linear model, along with adjoints. We shall also introduce the notion of differentiating the code to obtain the adjoint rather than deriving it analytically. We will also introduce the perturbation forecast model which is seen as an alternative to the tangent linear approximation. Afterwards we will apply all these techniques to semi-Lagrangian methods, where we will see that there is an important property in the derivation that puts a limit on the size of the variation in the departure point and the interpolation.

The penultimate chapter of the book, Chapter 15, will preset different applications of semi-Lagrangian, Eulerian–Lagrangian, and quasi-Lagrangian approaches for solving advection, convection, transport, Navier–Stokes, and Vlasov–Maxwell equations. We will show applications in numerical weather and ocean prediction, hydrology glacier movement, volcanic ash transport, air pollution, ocean ridge hydrothermal models, to name a few, to show how far fetching semi-Lagrangian methods are used. The final chapeter is solutions to select exercises.

That being said, we now move on to learn about semi-Lagrangian advection methods and their applications in geosciences.

CHAPTER

2

Eulerian modeling of advection problems

As we shall mention again soon, there are different ways of describing the movement of a particle. It can be with respect to the particle itself, or it can be with respect to the particles going past a certain fixed point. It is the latter description that the next two chapters are concerned with. In this chapter we shall present different numerical approximations to the one-dimensional constant-velocity advection equation. For some of the numerical schemes, we shall present plots of their performance with *advecting* a Gaussian bell curve and a discontinuous step function around a periodic domain. We shall see some behaviors of the schemes that we shall quantify and verify in the next chapter. We start this chapter now with a brief summary of the one-, two-, and three-dimensional general Eulerian forms of the advection equation, followed by a couple of brief derivations of the advection equation.

2.1 Continuous form of the advection equation

There are two approaches that can be used to describe the process of advection: Eulerian, where we have a volume element that is fixed in space in a set frame of reference, or Lagrangian, where the surface of the volume element is **co-moving** with the fluid, in the fluid's frame of reference. Thus in Eulerian modeling of advection we are considering the problem of the flow past a point, and as such the continuous one-dimensional general linear advection model is given by the partial differential equation

$$\frac{\partial \psi (x,t)}{\partial t} + u (x,t) \frac{\partial \psi (x,t)}{\partial x} = f (x,t), \tag{2.1}$$

where $\psi (x,t)$ is the dependent variable, t is time, x is the spatial coordinate, $u (x,t)$ is the **advection velocity**, which can be a function of both space and time, and finally, $f (x,t)$ is referred to as the **forcing term**. Note that the forcing term can be zero, linear, or nonlinear. The advection velocity can also be constant, linear, or nonlinear.

If we now think that there is a field that is two-dimensional, meaning that it can move in both horizontal directions, or in one horizontal and the vertical directions, then there is an associated version of the advection equation in two dimensions. For example, if we consider the horizontal case, then the associated two-dimensional partial differential equation for this

Semi-Lagrangian Advection Methods and Their Applications in Geoscience
https://doi.org/10.1016/B978-0-12-817222-3.00006-2

situation is given by

$$\frac{\partial \psi (x, y, t)}{\partial t} + u (x, y, t) \frac{\partial \psi (x, y, t)}{\partial x} + v (x, y, t) \frac{\partial \psi (x, y, t)}{\partial y} = f (x, y, t), \tag{2.2}$$

where $v (x, y, t)$ is the advection velocity in the y direction, but it can be a function of the other spatial coordinate. Finally, if we are considering a three-dimensional field then it is possible that it can be advected with respect to all three spatial coordinates in a Cartesian formulation, or any other coordinate system, and the associated partial differential equation for this situation is given by

$$\begin{aligned}\frac{\partial \psi (x, y, z, t)}{\partial t} &+ u (x, y, z, t) \frac{\partial \psi (x, y, z, t)}{\partial x} + v (x, y, z, t) \frac{\partial \psi (x, y, z, t)}{\partial y} \\ &+ w (x, y, z, t) \frac{\psi (x, y, z, t)}{\partial z} = f (x, y, z, t),\end{aligned} \tag{2.3}$$

where $w (x, y, z, t)$ is the advection velocity in the vertical direction. Note that all three advecting velocities can be functions of all three spatial dimensions, along with time.

For the multidimensional equations above, (2.2) and (2.3), it is possible to simplify these equations using vector-differential notation as

$$\frac{\partial \psi (\mathbf{x}, t)}{\partial t} + \mathbf{u} \cdot \nabla \psi (\mathbf{x}, t) = f (\mathbf{x}, t), \tag{2.4}$$

where $\mathbf{x} \equiv \begin{pmatrix} x \\ y \\ z \end{pmatrix}$, $\mathbf{u} \equiv \begin{pmatrix} u \\ v \\ w \end{pmatrix}$, "$\cdot$" is the vector dot product operator, and ∇ is the spatial gradient operator given by $\nabla^T \equiv \begin{pmatrix} \frac{\partial}{\partial x} & \frac{\partial}{\partial y} & \frac{\partial}{\partial z} \end{pmatrix}$.

We now consider how to derive the one-dimensional advection equation that we presented above. We shall present two different approaches: one based on a dynamical description of the situation, and one that uses a more mathematical argument.

2.1.1 Derivation of the one-dimensional Eulerian advection equation

As just mentioned, in this section we shall present two different approaches to derive the one-dimensional advection equation. We should note here that the advection equation has many different names, it is quite often referred to as the transport equation, but it is also referred to as the convection equation, along with being associated with the wave equation. The first derivation that we consider here is a dynamics-based approach employing the ideas of mass conservation, continuity, and incompressibility.

Mass conservation derivation of the advection equation

At the beginning of this chapter we referred to a volume element, which we will now denote as dV as shown in Fig. 2.1, where $P (x, y, z)$ is the centroid of the volume element. As we are considering the Eulerian formulation, we have that the sides of the volume element are fixed in space. Therefore, we wish to consider the situation where we have a flow in and out of the volume element through the sides.

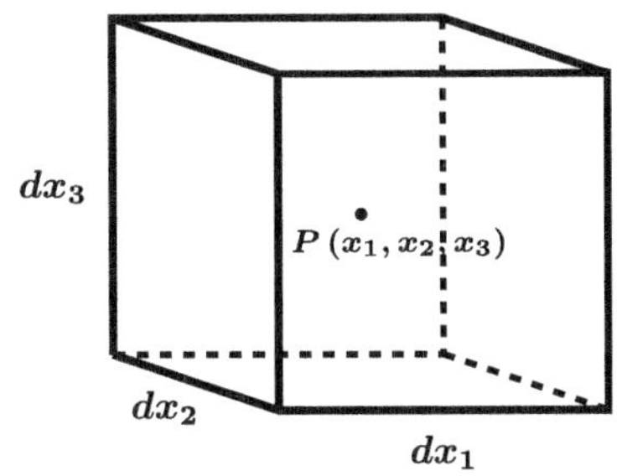

FIGURE 2.1 Schematic of a generic volume element, dV.

If we denote the mass density at the point $P(x, y, z)$ as $\rho(x, y, z)$, which represents the mass divided by the volume, then we assume that it is an average, and nearly uniform mass density in all of dV. Thus the total mass, M, that is within dv is given by the integral quantity

$$M = \int \rho dV \equiv \int \rho dx\, dy\, dz. \tag{2.5}$$

The next assumption that we make is that there are no sources or sinks of mass within dV. As such we have that $\frac{dM}{dt}$ represents the rate at which mass enters or exits the volume through the surface dS.

When we are considering the three-dimensional volume, there are multiple surfaces of the volume, as such we have a side of the volume denoted by $d\mathbf{S}$, and the unit normal vector denoted by $\mathbf{n}$, where it is assumed that the normal vector points "out" of the volume element. The next important quantity that we have to introduce is the **flux** of the mass, which represents mass per unit area, per unit time, that we have passing through a surface, and which is given by $\rho\mathbf{v}$, where $\mathbf{v}$ is the fluid velocity.

We now require that the mass per unit time that is flowing through the side of the volume, $d\mathbf{S}$, be given by $\rho\mathbf{v}\cdot d\mathbf{S} \equiv \rho\mathbf{v}\cdot\mathbf{n}dS$, and thus the total rate of flow of mass out of the volume element dV is given by the sum of the mass per unit time flowing through each of the surfaces/faces of the volume element. Thus in continuous form this is given by the surface integrals

$$\sum \rho\mathbf{v}\cdot d\mathbf{S} \equiv \oint_S \rho\mathbf{v}\cdot d\mathbf{S} = \oint_S \rho\mathbf{v}\cdot\mathbf{n}dS. \tag{2.6}$$

As mentioned before, we have $\frac{dM}{dt}$ representing the rate at which mass is entering or exiting the volume element through the surface, and as such we require (2.6) to be equal to

$-\frac{dM}{dt}$, which implies that

$$\frac{dM}{dt} = \frac{d}{dt}\int_v \rho dV = -\oint_S \rho \mathbf{v} \cdot \mathbf{n} dS. \tag{2.7}$$

As we are considering a fixed Eulerian surface, we are able to bring the total time derivative $\frac{d}{dt}$ inside the volume integral as a partial derivative, which then gives us

$$\int_v \frac{\partial \rho}{\partial t} dV = -\oint_S \rho \mathbf{v} \cdot \mathbf{n} dS. \tag{2.8}$$

While the expression in (2.8) may look quite intimidating, we now introduce a theorem that will help us here. This theorem is referred to as **Gauss' Theorem** and is stated below.

Theorem 2.1 (Gauss' Theorem). *Let V be a region in space with boundary ∂V. Then the volume integral of the divergence $\nabla \cdot F$ of F over V and the surface integral of F over the boundary ∂V of V are related by*

$$\int_V (\nabla \cdot F) dV = \oint_{\partial V} F \cdot dS. \tag{2.9}$$

Gauss' Theorem is quite often referred to as the **divergence theorem**, where the theorem is the mathematical expression for the physical situation that, in the absence of the creation or destruction of matter, the density within a region of space can change only by having it flow in or out from the region through its boundary, which is the situation that we have here.

Upon applying Gauss' theorem to the situation above, we obtain

$$\oint_S \rho \mathbf{v} \cdot \mathbf{n} dS = \int_v \nabla \cdot (\rho \mathbf{v}) dV. \tag{2.10}$$

Therefore, we now have

$$\int_V \left(\frac{\partial \rho}{\partial t} + \nabla \cdot (\rho \mathbf{v}) \right) dV = 0. \tag{2.11}$$

We require the expression in (2.11) to hold for all arbitrary shaped volumes, and thus the only way that the integral equation in (2.11) can be satisfied is if the integrand is equal to zero. This then implies that

$$\frac{\partial \rho}{\partial t} = -\nabla \cdot (\rho \mathbf{v}). \tag{2.12}$$

Eq. (2.12) is referred to as the **continuity equation**, where it is describing the **conservation of mass** within the Eulerian frame of reference. We can see that this is almost the advection equation that we presented earlier, but we have to make one more assumption, namely that the fluid is incompressible, which implies that the density within the volume is constant. Another way of interpreting this assumption is that the divergence of the flow velocity is zero. Mathematically the non-divergence of the velocity field is equivalent to $\nabla \cdot \mathbf{v} = 0$. If we now expand the term $\nabla \cdot (\rho \mathbf{v})$ we have

$$\nabla \cdot (\rho \mathbf{v}) \equiv \rho \nabla \cdot \mathbf{v} + \mathbf{v} \cdot \nabla \rho,$$

where we can see that we require the divergence of the velocity. Using the incompressibility assumption results in the continuity equation for an in compressible fluid as

$$\frac{\partial \rho}{\partial t} + \mathbf{v} \cdot \nabla \rho = 0, \tag{2.13}$$

which is the vectorial form of the three-dimensional advection equation that we presented in (2.4), but where there is no external forcing term.

Taylor series expansion derivation of the one-dimensional advection equation

This derivation of the advection equation is for the one-dimension situation, which starts by considering the one-dimensional drift of an incompressible fluid. Therefore, let $\psi(x, t)$ represent a particle density that only changes due to advection processes so that

$$\psi(x, t + \Delta t) = \psi(x - u\Delta t, t). \tag{2.14}$$

If we now assume that Δt is sufficiently small so that we can apply a Taylor series expansion to either side of (2.14), then we obtain, to first order,

$$\psi(x, t) + \Delta t \frac{\partial \psi(x, t)}{\partial t} \approx \psi(x, t) - u\Delta t \frac{\partial \psi(x, t)}{\partial x}, \tag{2.15}$$

which simplifies to

$$\frac{\partial \psi}{\partial t} + u \frac{\partial \psi}{\partial x} = 0, \tag{2.16}$$

where u is a nonzero constant velocity.

As an aside here, we should note that as we are dealing with a partial differential equation. There are three different types that a partial differential equation can be: hyperbolic, parabolic, or elliptical, which represent many different physical and geophysical processes. The partial differential equation in (2.16) is hyperbolic. Thus the unique solution of (2.16) is determined by an initial condition, denoted by u_0, such that $u_0 \equiv u(x, 0)$ and

$$u(x, t) = u_0(x - ct). \tag{2.17}$$

One approach to solve the problem above is by the **Method of Characteristics**.

2.1.2 Methods of characteristics

The first thing to notice here is that the advection equation in (2.16) can be factorized as

$$\left(\frac{\partial}{\partial t} + u \frac{\partial}{\partial x}\right) \psi = 0. \tag{2.18}$$

The next step is to transform x, t to new variables r, s such that the derivative transforms as

$$\frac{\partial}{\partial r} = \frac{\partial}{\partial t} + u\frac{\partial}{\partial x}. \tag{2.19}$$

If it is possible to introduce this change of variable, then we would have that $\frac{\partial \psi}{\partial r} = 0$, where this equation is easier to solve. Therefore, if we let $x = x(r, s)$ and $t = t(r, s)$, then, applying the chain rule, we obtain

$$\frac{\partial}{\partial r} = \frac{\partial x}{\partial r}\frac{\partial}{\partial x} + \frac{\partial t}{\partial r}\frac{\partial}{\partial t}. \tag{2.20}$$

If we now compare (2.20) with (2.18), we see that $\frac{\partial x}{\partial r} = u$ and $\frac{\partial t}{\partial r} = 1$. Integrating these two equations results in $x = ur + q(s)$ and $t = r + p(s)$. We now use the s dependence by applying the property that the initial conditions specify the solution along the x-axis. To make the application of the initial conditions in the transformed space, we introduce the condition that the x-axis at $t = 0$ transforms onto the s-axis ($r = 0$); specifically, we need $r = 0$, implying that $t = 0$ and $x = s$. Thus by setting $r = 0$ and $t = 0$, we have $q(s) = s$ and $p(s) = 0$, implying that

$$x = ur + s, \qquad t = r. \tag{2.21}$$

The next step is to invert the transformation above, which results in $r = t$ and $s = x - ut$, which then enable us to write (2.18) in the form $\frac{\partial \psi}{\partial r} = 0$, implying that $\psi = \psi(s) = \psi(x - ut)$. Given the initial conditions, we therefore can conclude that the solution of the advection problem is

$$\psi(x, t) = \psi_0(x - ut). \tag{2.22}$$

Therefore, the expression in (2.22) defines a traveling wave that moves with speed u, and it is also seen that the solution is constant along the lines of the form $x - ut = \beta$ where β is a constant. These lines are referred to as the **characteristics** for the partial differential equation, and the method that was applied here to find the solution is known as the **method of characteristics**.

In Fig. 2.2 we have an illustration of the appearance of the characteristics for the advection equation, were we see that the characteristics are defined by their value as they cross the x-axis, but then have that value along the whole path.

To help illustrate this approach, we shall consider the example of a step function, which we shall revisit many times throughout the book in different forms. The initial conditions that define the step function shown in Fig. 2.3 are given by

$$\psi_0(x) = \begin{cases} 1, & 0 \leq x \leq 1, \\ 0, & x > 1. \end{cases} \tag{2.23}$$

Thus from (2.22) the solution is

$$\psi(x, t) = \begin{cases} 1, & 0 \leq x - ut \leq 1, \\ 0, & x - ut > 1, \end{cases} \tag{2.24}$$

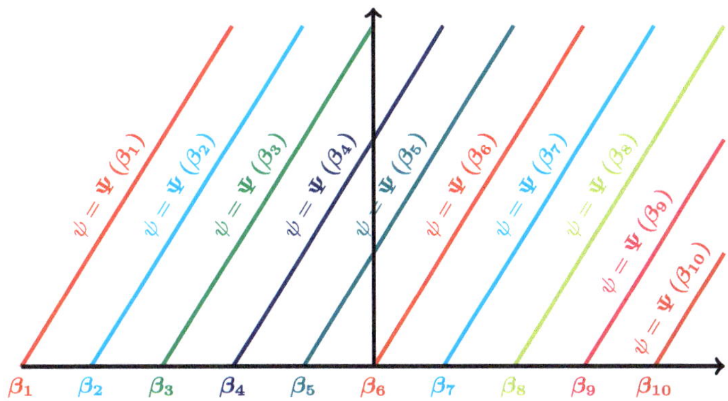

FIGURE 2.2 Example of the constant characteristics for the one-dimensional linear, constant velocity, advection partial differential equation, with initial conditions $\psi(x,0) = \Psi(x)$.

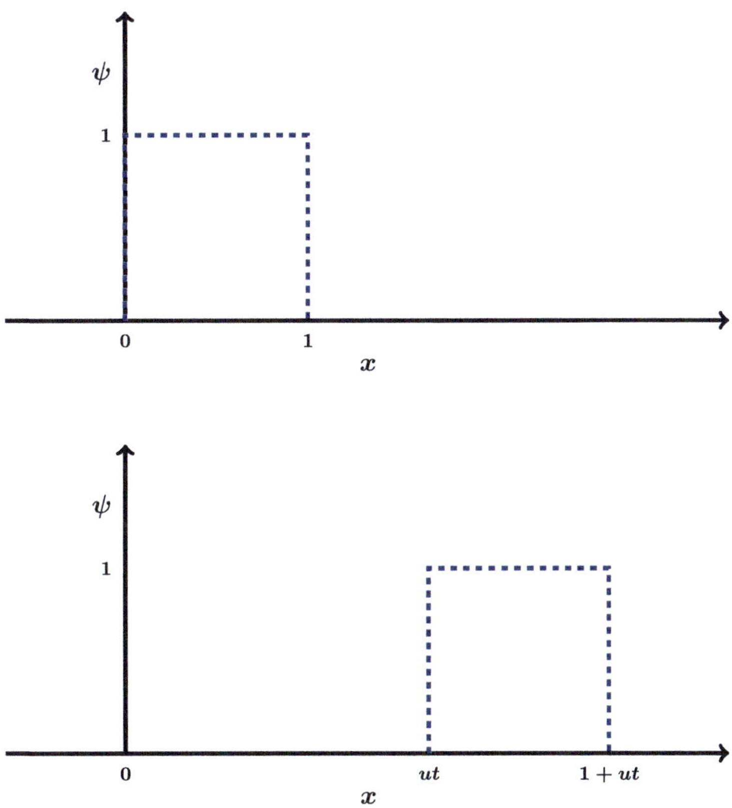

FIGURE 2.3 Schematic of advecting a step function starting at initial time in the top plot, and then after a time of 1 in the lower plot.

or equivalently,

$$\psi_0(x) = \begin{cases} 1, & ut \leq x \leq 1 + ut, \\ 0, & x < ut,\ x > 1 + ut. \end{cases} \tag{2.25}$$

In Fig. 2.3 we have presented both the initial step function, as mentioned above, and the step function at a future time t, where the solution is the original step function that has been moved to the interval $ut \leq x \leq 1 + ut$.

There are a few properties of the linear constant velocity equation that are useful when assessing the performance of the numerical approximation to it:

1. The initial profile is preserved; this is true for jumps and discontinuities.
2. The information of the initial profile travels at speed u, and the associated direction of the travel is one-way only.
3. The solution at a given spatial location $\widehat{x}$ and at time $\widehat{t}$ is entirely determined by the values of the initial conditions at $x = \widehat{x} - u\widehat{t}$. We have provided a schematic of this situation in Fig. 2.4.

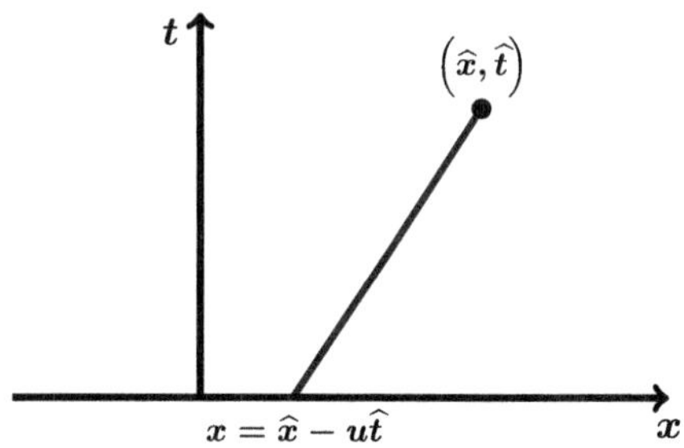

FIGURE 2.4 Schematic of back-tracing of the characteristics to the initial condition from the point $(\widehat{x}, \widehat{t})$.

Given the continuous Eulerian approach for advection, we now consider several different numerical approximation to the one-dimensional linear advection equation with constant velocity. The approaches that we shall show in the next section are referred to a **finite differencing** scheme, where we will present explicit and implicit schemes, and a combination of the two, which can be multistep algorithms referred to as predictor–corrector methods.

2.2 Finite difference approximations to the Eulerian formulation of the advection equation

In this section we shall introduce many different approximations to the linear, zero-forcing version of the one-dimensional advection equation. The first thing that we have to introduce here is the concept of the numerical grid for both time and space.

In Fig. 2.5 we have presented a uniform distributed set of grid points, where for this setup we have equal spacing in time and space, although we should note that spacing in the temporal direction is not always the same size as that in the spatial direction. We can see that there is a distance between the points in both time and space. The distance between the grid points in the x-direction is denoted by Δx, whereas the distance between the time grid points is denoted by Δt, and they are referred to as the grid spacing and time step, respectively.

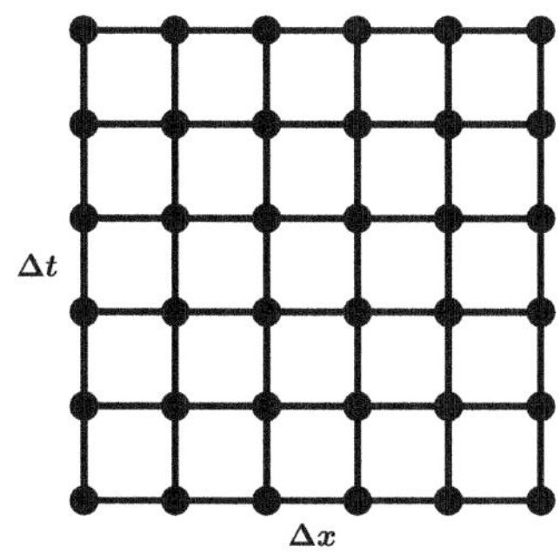

FIGURE 2.5 Illustration of a numerical grid.

If we start with the spatial direction, then we introduce the index i for the x direction, while for the temporal direction with have the index n. Thus at point (i, n) we have a numerical approximation to the continuous tracer, $\psi(x, t)$, as $\psi(x_i, t^n)$, which is usually denoted by ψ_i^n, for $i = 1, 2, \ldots, I$ and $n = 0, 1, \ldots, N$. Please note that the superscript n here does not represent a power of the function, it is simply the temporal index. Therefore, if we consider the x direction derivative, then we could use the following approximation:

$$\frac{\partial \psi}{\partial x} \approx \frac{\psi(x_{i+1}, t^n) - \psi(x_i, t^n)}{\Delta x} \equiv \frac{\psi_{i+1}^n - \psi_i^n}{\Delta x}, \tag{2.26}$$

where the approximation in (2.26) is referred to as an **upwind** scheme.

It is also possible to define an upwind scheme for the temporal derivative for the numerical approximation to the time component as well, which is given by

$$\frac{\partial \psi}{\partial t} \approx \frac{\psi\left(x_i, t^{n+1}\right) - \psi(x_i, t^n)}{\Delta t}. \tag{2.27}$$

The approaches presented here are examples of what is referred to as **finite difference** approximations to the gradients of the field in the partial differential equation. Given the concept of the grid points introduced above, we shall now present a series of different explicit numerical and implicit numerical schemes, and the schemes that combine explicit and implicit features for the Eulerian formulation of the one-dimensional advection equation.

2.2.1 Upwind forward Euler

Depending on the sign of the advective velocity, we can define two different versions of this scheme. If $u > 0$, then the version of the **forward Euler** approximation is based upon the upwind approximation to the spatial derivative presented above combined with the temporal approximation above as well, which results in

$$\frac{\psi_i^{n+1} - \psi_i^n}{\Delta t} = -u\frac{\psi_i^n - \psi_{i-1}^n}{\Delta x}. \tag{2.28}$$

To implement this numerical scheme presented in (2.28), we rearrange (2.28) as

$$\psi_i^{n+1} = (1 - C)\,\psi_i^n + C\psi_{i-1}^n, \tag{2.29}$$

where $C \equiv u\frac{\Delta x}{\Delta t}$ is referred to as the **Courant number**, which will become important in the next chapter.

In Fig. 2.6 we have presented the stencil, which is the set of points that are used in this approximation, for the forward Euler, using an upwind approach, scheme. It should be noted that forward Euler refers to the forward approximation of the temporal component. For all of the stencil plots that we show in this chapter, the green circles are those that are used to create that finite difference scheme.

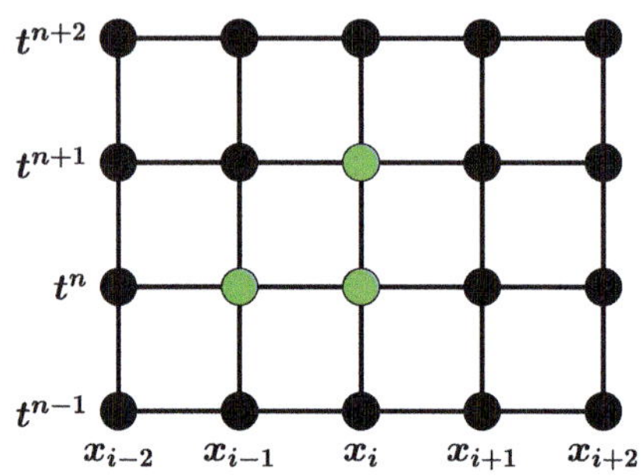

FIGURE 2.6 Illustration of the stencil for the upwind forward Euler finite difference scheme.

If the advective velocity is $u < 0$, then we can define a downwind approximation to the spatial derivative, and if we use the forward Euler approximation for the time component, then we obtain

$$\frac{\psi_i^{n+1} - \psi_i^n}{\Delta t} = -u\frac{\psi_{i+1}^n - \psi_i^n}{\Delta x}. \tag{2.30}$$

To implement this numerical scheme presented in (2.30), we rearrange it as

$$\psi_i^{n+1} = (1 + C)\,\psi_i^n - C\psi_{i+1}^n. \tag{2.31}$$

Notes: (1) The reason why the scheme in (2.30) is not referred to as the backward Euler is because that name belongs to the implicit versions of the time derivative that we shall present later. (2) Eqs. (2.28) and (2.30) are referred to as the **difference equations** for the continuous partial differential equation.

Throughout this chapter we shall assess the performance of some of the advection schemes with the advection of a step function, as well as a Gaussian bell curve. Before that we investigate some properties of each scheme in the next chapter. As we saw earlier, the characteristics tell us that the shapes should simply be moving along, without any distortions, with advection speed u, without slowing down or speeding up.

Upwind forward Euler with the bell curve

For the first set of plots for this test case, we shall consider an advection speed of $u = 1$, and spatial and temporal grid spacing of $\Delta x = \Delta t = 0.1$ units. The bell curve is defined by

$$\psi(x,0) = \exp\left\{-\frac{1}{2}\left(\frac{x-\mu}{\sigma}\right)^2\right\}, \tag{2.32}$$

where the parameter μ determines where the bell is initially centered, and σ determines the spread of the curve. We should note that this is similar to the definition of the Gaussian distribution, except that we do not have the normalizing factor of $\dfrac{1}{\sqrt{2\pi\sigma}}$.

For the demonstrations that we present in this chapter, we use $\mu = 5$, implying that the bell will initially be centered at $x = 5$, and $\sigma = 0.5$. The initial profile is presented in Fig. 2.7.

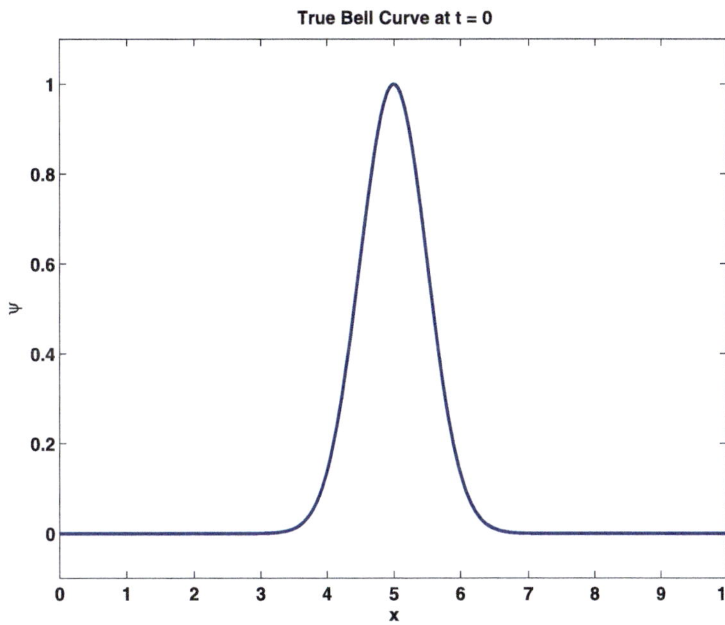

FIGURE 2.7 Plot of the true bell curve for the situation when $\mu = 5$ and $\sigma = 0.5$ at $t = 0$.

Given the information presented here we know that the Courant number for this configuration is $C \equiv u\dfrac{\Delta t}{\Delta x} = 1$. We are using a spatial domain of $x \in [0, 10]$, and a temporal domain of $0 \leq t \leq 10$. This means that there are 101 x grid points and 101 t points for this configuration. We are using a periodicity condition as the boundary conditions for either end of the domain, and with the advection speed of 1, this then means that the bell curve can take one complete circuit of the domain in this time.

We are able to plot the true solution at every time step as we know that it is given by

$$\psi(x,0) = \exp\left\{-\frac{1}{2}\left(\frac{x-(n-1)\,v\Delta t-\mu}{\sigma}\right)^2\right\}, \tag{2.33}$$

where n is the index for the specific time step that the numerical mode is at. This will enable us to quantify the *error* associated with each numerical scheme.

In Fig. 2.8 we have plotted both the true and numerical solutions for the configuration shown at every 10 time steps. We should point out that there is no error in this plot, simply since, when $C = 1$, the upwind forward Euler approximates the continuous bell function very well.

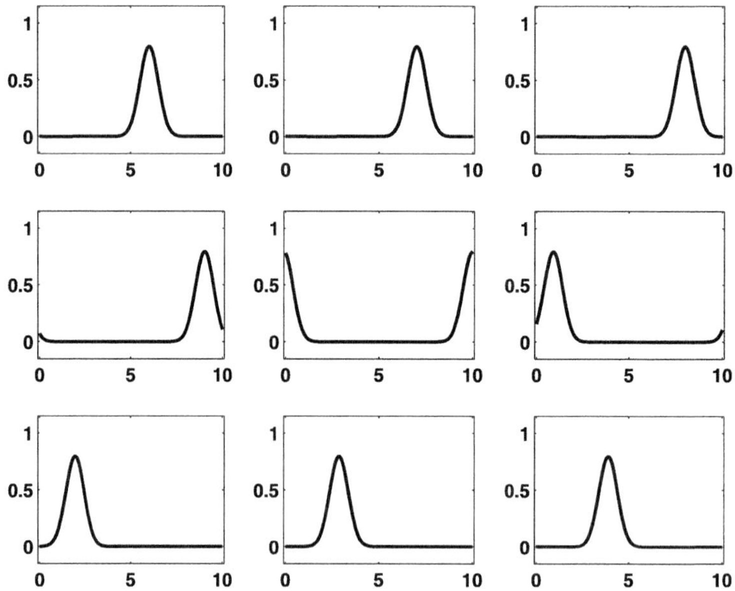

FIGURE 2.8 Plot of the true bell curve, in red, and the numerical solution from the upwind forward Euler scheme for $tn = 10\Delta t, 20\Delta t, \ldots, 90\Delta t$ steps.

We will now look at the effects of changing the spatial step size. In Fig. 2.9 we have presented a similar plot to that in Fig. 2.8, but now we have the situation where $C = 1.25$, which arises keeping $\Delta t = 0.1$ but having increased the resolution of the x direction to $\Delta x = 0.08$. While this change may not seem like a big deal, we can see from Fig. 2.9 that we are no longer maintaining the structure of the bell curves. We can clearly see some damping of the curve within 20 time steps, and that, by the time we arrive at 70 times steps (panel 7), we can clearly see that the solution is starting to display some feature behind the bell curve, and by 90 times steps we can see distortions both behind the wave and also traveling further behind the curve. This could indicate a relationship between C, Δt, and Δx.

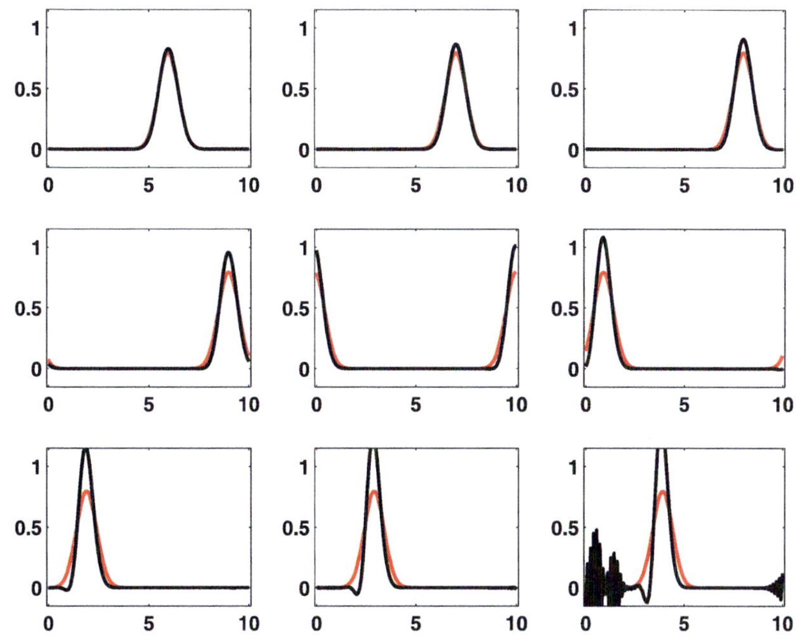

FIGURE 2.9 Plot of the upwind forward Euler scheme for the bell curve when the Courant number is $C = 1.25$.

The next set of plots that we present for this situation with this numerical scheme are for the case when we have $C = 0.5$. Here we are trying to save some storage space and also speed up the numerical integration. We have kept $\Delta t = 0.1$, but now $\Delta x = 0.2$. The first feature to notice in Fig. 2.10 is that the upwind forward Euler scheme show what is referred to as *damping* of the bell curve as time goes on; on a positive note, we can see that the scheme is not *out of phase* with the true solution. Therefore, by reducing the number of spatial grid points, we have activated another feature of the numerical scheme that was not present when we had a *higher spatial resolution*.

As mentioned earlier, this upwind forward Euler is supposed to be used when $u > 0$, and we now present a plot of the affects of applying this scheme when $u < 0$. In Fig. 2.11 we have plotted the results at 10 and 20 times steps of applying the upwind forward Euler scheme when $u < 0$. We see that after 10 times steps, the numerical solutions is starting to go out of phase with the true solution, and is also introducing an *amplitude error*, as well as overshoots. However, only 10 time steps later, we see that the numerical solution is nothing like the true solution and is starting to blow up. There was no point in producing one of the sets of 9 plots for this case as the scheme became completely unstable after a few time steps.

We now move on to the discontinuous case of the step function with the same setup used here to see the impact that discontinuities have on the performance of the upwind forward Euler scheme.

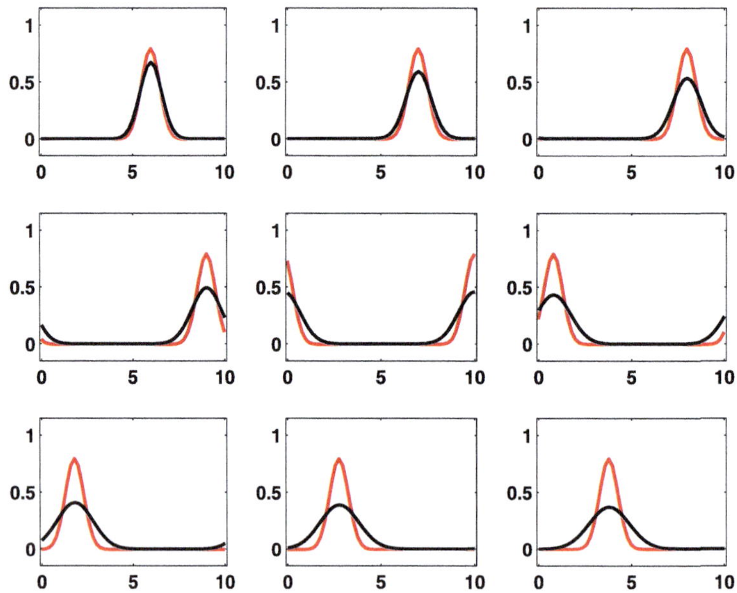

FIGURE 2.10 Plot of the true bell curve, in red, and the numerical solution from the upwind forward Euler scheme for $tn = 10\Delta t, 20\Delta t, \ldots, 90\Delta t$ steps, for $\Delta t = 0.1$, $\Delta x = 0.2$, $u = 1$, which implies a Courant number of 0.5.

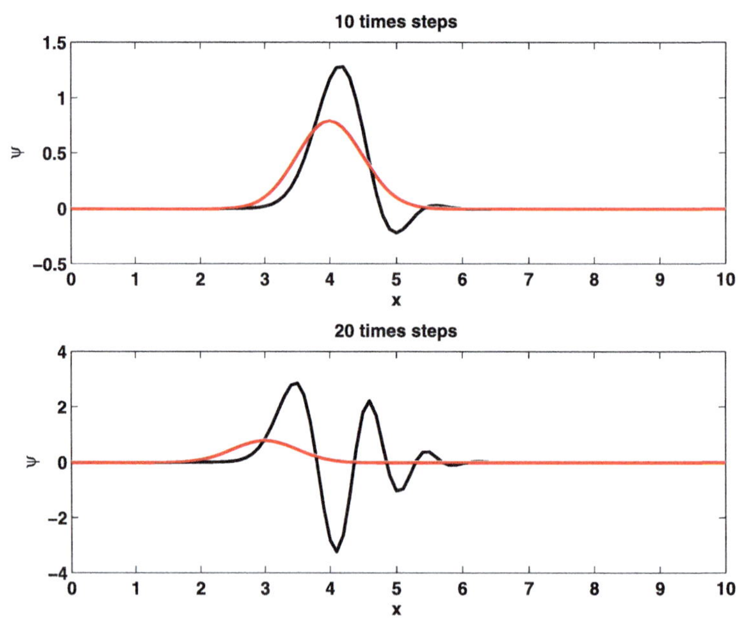

FIGURE 2.11 Plot of the performance of the upwind forward Euler finite difference scheme when we have a negative advection velocity.

Upwind forward Euler with the step function

We now move on to assess the performance of the upwind forward Euler scheme when applied to a step function, where the step function is defined as

$$\psi(x,0) \equiv \begin{cases} 0, & 0 \leq x < 3, \\ 1, & 3 \leq x \leq 7, \\ 0, & 7 < x \leq 10. \end{cases} \tag{2.34}$$

We are going to present four sets of results here, where each plot has 9 panels representing the results at ten time step intervals. The results are for $C = 1, 0.75, 0.5$, and 0.25, which is achieved by keeping $\Delta t = 0.1$ but using $\Delta x = 0.1, 0.1333, 0.2$, and 0.4, respectively. We can see for the case when the Courant number is equal to 1, Fig. 2.12, that the upwind forward Euler scheme does really well at approximating this discontinuous function at all time stages.

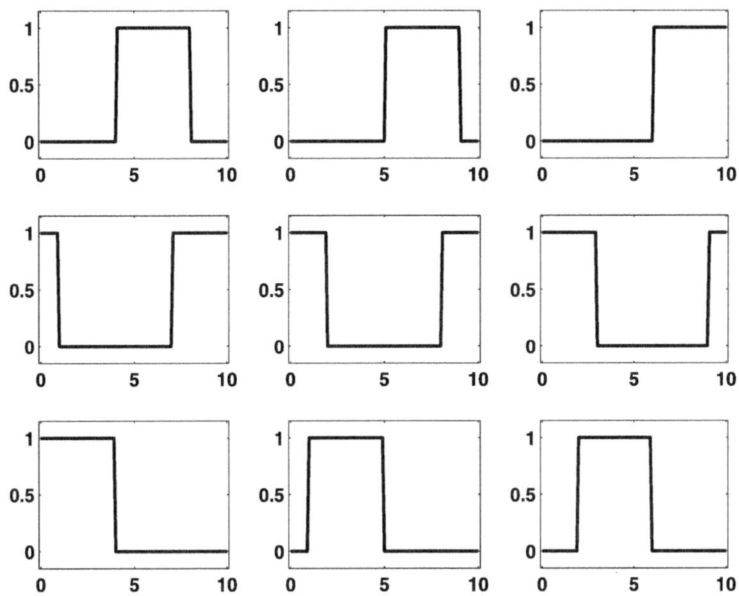

FIGURE 2.12 Plot of the true step function, in red, and the numerical solution from the upwind forward Euler scheme for $tn = 10\Delta t, 20\Delta t, \ldots, 90\Delta t$ steps, for $\Delta t = 0.1$, $\Delta x = 0.1$, $u = 1$, which implies a Courant number of 1.

However, when considering the case where $C = 0.75$, Fig. 2.13, we can clearly see that we are no longer capturing the sharp features of the discontinuities of the step function, rather the scheme is smoothing out these features. On a positive note, we can see that the scheme neither goes out of phase with the true solution nor damps it, but rather just smoothes it.

When we increase the spatial step size to $\Delta x = 0.2$, which results in a decrease in the spatial resolution, we see from Fig. 2.14 that while the solution stays in phase with the true solution,

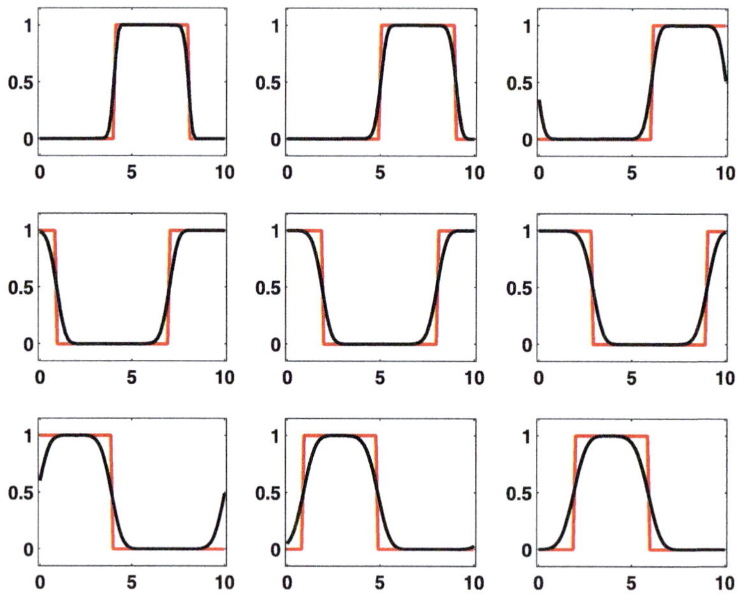

FIGURE 2.13 Plot of the true step function, in red, and the numerical solution from the upwind forward Euler scheme for $tn = 10\Delta t, 20\Delta t, \ldots, 90\Delta t$ steps, for $\Delta t = 0.1$, $\Delta x = 0.1333$, $u = 1$, which implies a Courant number of 0.75.

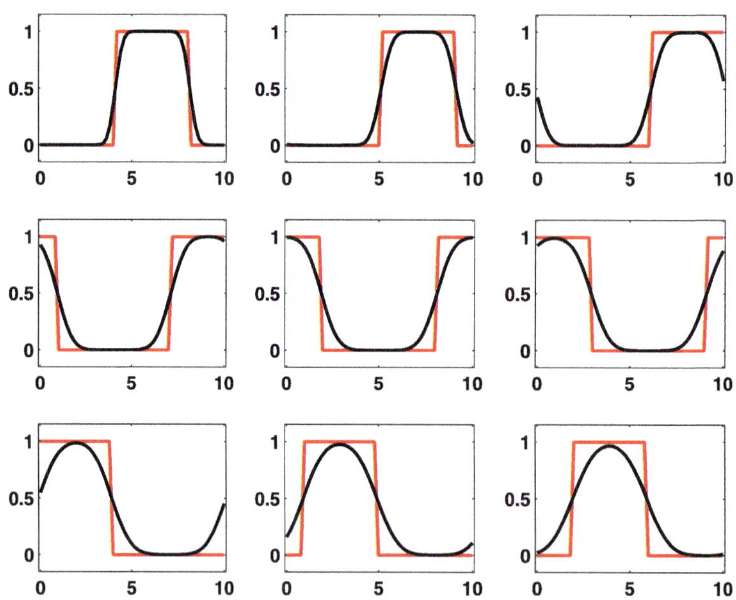

FIGURE 2.14 Plot of the true step function, in red, and the numerical solution from the upwind forward Euler scheme for $tn = 10\Delta t, 20\Delta t, \ldots, 90\Delta t$ steps, for $\Delta t = 0.1$, $\Delta x = 0.2$, $u = 1$, which implies a Courant number of 0.5.

and has not introduced any spurious minima or maxima, the smoothing of the discontinuities is more prevalent and that, by the time we are at 90 time steps, the numerical solution looks more like a wave than a step function.

The last situation that we look at for this test case is where $C = 0.25$, Fig. 2.15. This time we are using a spatial step size of $\Delta x = 0.4$. As we have seen from the results for the other three test cases for the upwind forward Euler solution with the step function, our solution still does not go out of phase; however, the smoothing effect is still there, but now we appear to have a damping affect on the solution which is not a very desirable result that we want to see from our numerical scheme.

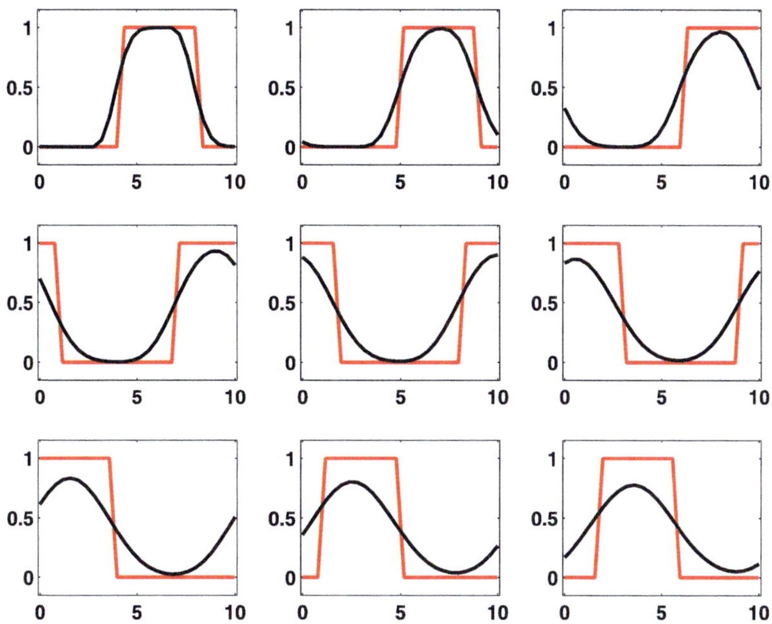

FIGURE 2.15 Plot of the true step function, in red, and the numerical solution from the upwind forward Euler scheme for $tn = 10\Delta t, 20\Delta t, \ldots, 90\Delta t$ steps, for $\Delta t = 0.1$, $\Delta x = 0.4$, $u = 1$, which implies a Courant number of 0.25.

2.2.2 Forward-time, centered-space, FTCS

We now consider what is referred to as a **central difference** for the spatial derivative combined with a forward Euler approximation for the temporal derivative. This numerical scheme is referred to as the **forward-time, centered-space** scheme and is defined as

$$\frac{\psi_i^{n+1} - \psi_i^n}{\Delta t} + u\frac{\psi_{i+1}^n - \psi_{i-1}^n}{2\Delta x} = 0. \tag{2.35}$$

The stencil for the forward-time, centered-space scheme is shown in Fig. 2.16, where we can see that we are using the points on either side of ψ_i^n to form the approximation to $\frac{\partial \psi}{\partial x}$.

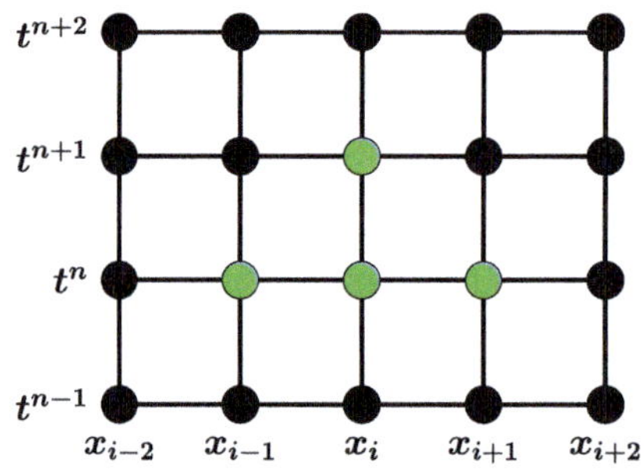

FIGURE 2.16 Illustration of the stencil for the forward-time-centered-space (FTCS) finite difference scheme.

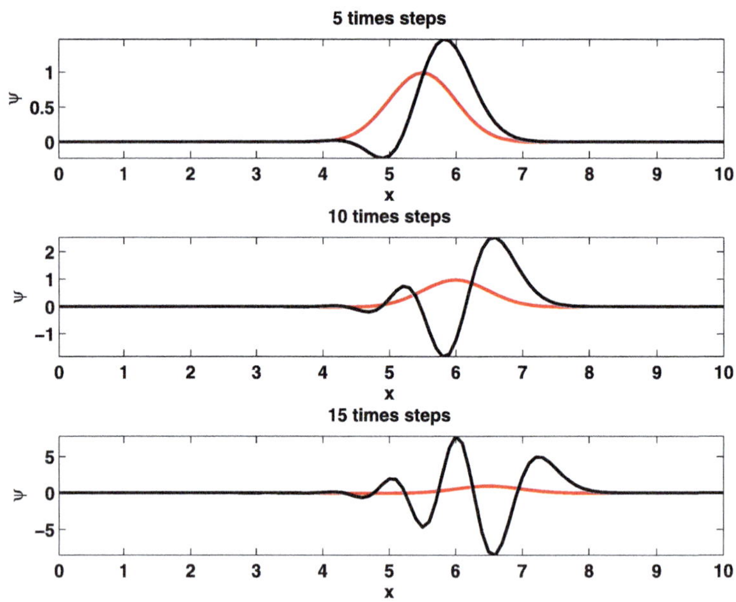

FIGURE 2.17 Plot of the true step function, in red, and the numerical solution from the forward-time-centered-space scheme at $t^n = 5\Delta t$, $10\Delta t$, and $15\Delta t$ to illustrate the instability of this finite difference scheme.

To implement (2.35), we again rearrange to separate out ψ_j^{n+1} so that

$$\psi_i^{n+1} = \psi_i^n + \frac{C}{2}\psi_{i-1}^n - \frac{C}{2}\psi_{i+1}^n. \tag{2.36}$$

We shall only present one figure for the bell curve case where we are using $u = 1$, $\Delta t = 1$, and $\Delta x = 1$, giving us a Courant number of $c = 1$, which from the upwind forward Euler scheme appeared to be a quite optimal choice for these parameters. We do not present the nine panels in Fig. 2.17 due to the same property mentioned in the situation for the upwind forward Euler case when the panels become all black. We do, however, present the numerical solution in black and the true state in red at $t^n = 5, 10, 15$, to show that the scheme becomes unstable quite quickly at this resolution, as we have seen when using an upwind scheme, which is less sophisticated than the central difference scheme presented here. We can clearly see that even by the fifth time step, the scheme is already starting to have problems, and they are just continuing to grow; there is an amplitude error as well as a phase error. By the time we are at $t^n = 15$, the numerical solution looks nothing like the true state and it can clearly be seen to be growing more unstable, which is not a good thing. This scheme should be avoided. The instability can be reduced by increasing the resolution, but there is nothing that can be done with this scheme to make it stable.

2.2.3 Lax–Friedrichs scheme

We are now moving to discuss the schemes where we are using different approximations of certain terms, or even values at points, which is this case for the Lax–Friedrichs scheme. For reasons that we shall explain in the next chapter, there are methods that utilize the upwind feature in time of the forward-time, centered-space scheme, but not with what it brings for this scheme. The Lax–Friedrichs scheme is considered as a forward-time, centered-space scheme with an artificial viscosity term that is divided by two. The difference equation for this scheme is given by

$$\frac{\psi_i^{n+1} - \frac{1}{2}\left(\psi_{i+1}^n + \psi_{i-1}^n\right)}{\Delta t} + u\frac{\psi_{i+1}^n - \psi_{i-1}^n}{2\Delta x} = 0. \tag{2.37}$$

Therefore, rearranging to separate out ψ_i^{n+1} and expressing the difference equation in terms of the Courant number gives

$$\psi_i^{n+1} = \frac{1}{2}(1 + C)\,\psi_{i-1}^n + \frac{1}{2}(1 - C)\,\psi_{i+1}^n. \tag{2.38}$$

We have plotted the stencil for the Lax–Friedrichs scheme in Fig. 2.18, where we can see that it is different compared to the stencil for the forward-time, centered-space, in that the central point ψ_i^n is not used for this scheme.

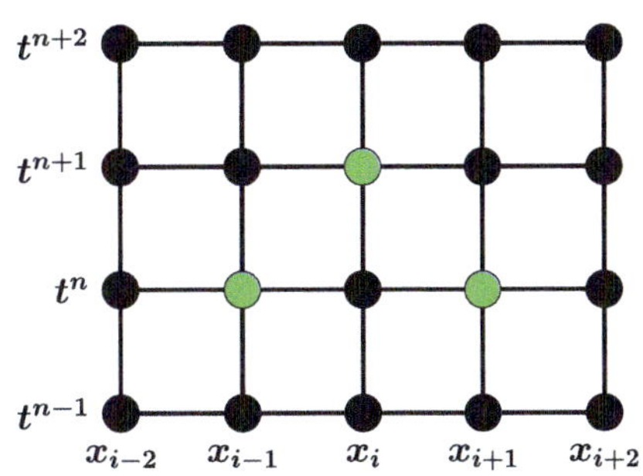

FIGURE 2.18 Illustration of the stencil for the Lax–Friedrichs finite difference scheme.

We have implemented the Lax–Friedrichs scheme for both the bell curve and the step function for Courant numbers $C = 0.75, 0.5$, and 0.25. Taking the bell curve first with $C = 0.75$,

Fig. 2.19, we can see that the numerical scheme is damping the amplitude of the bell, as well as introducing a slight phase error. The damping and slight phase error are already evident after only 10 time steps, and that by the time $t^n = 90\Delta t$ we see that the amplitude is almost half of what the true solution is, but we should also note that what initially appeared as a phase error is, in fact, an expansion of the curve as the peak of the numerical solution is at the same point as the true solution for all of the plots here.

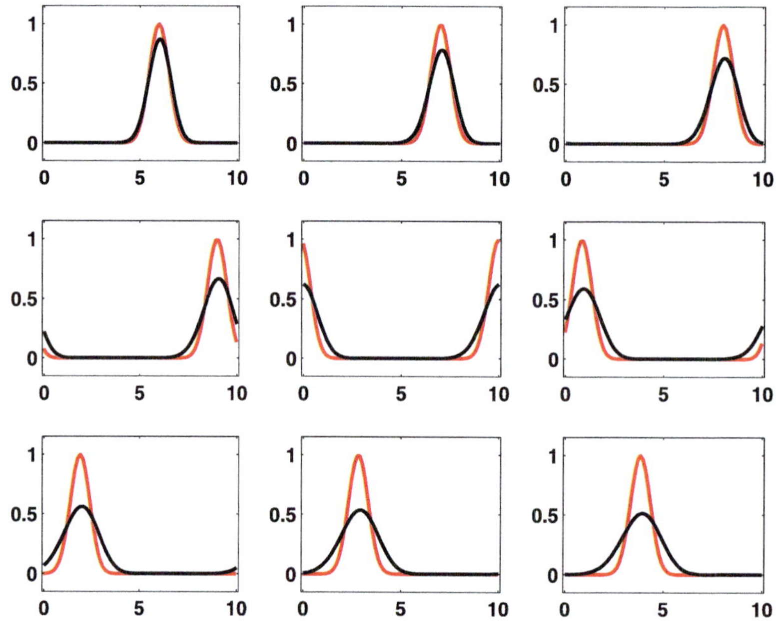

FIGURE 2.19 Plot of the true bell curve, in red, and the numerical solution from the Lax–Friedrichs finite difference scheme for $tn = 10\Delta t, 20\Delta t, \ldots, 90\Delta t$ steps, for $\Delta t = 0.1$, $\Delta x = 0.1333$, $u = 1$, which implies a Courant number of 0.75.

When we consider the results for $C = 0.5$, Fig. 2.20, we can clearly see the damping and the spreading out of the numerical solution within the first 10 times steps, and that this diffusion of the numerical solution along the x-axis is more prevalent in $t^n = 10\Delta t$ panel (top left). We should note that there does not appear to be a phase error and that the scheme is producing heavy diffusion.

In Fig. 2.21 we have provided the results for the case $C = 0.25$. It is quite clear that at this coarse resolution in the x direction the Lax–Friedrichs scheme is completely dominated by the numerical diffusion where within 40 time steps the bell has been completely flatten, but the peak, or what is left of it, is located where the peak of the true solution is. We should make a note here that, when $C = 1$, the Lax–Friedrichs scheme did reproduce the true state quite well, and there was no indication of any diffusion.

Given the performance of the Lax–Friedrichs scheme with the bell curve, it should be quite interesting to see how it does with the step function. The results for this scheme applied to

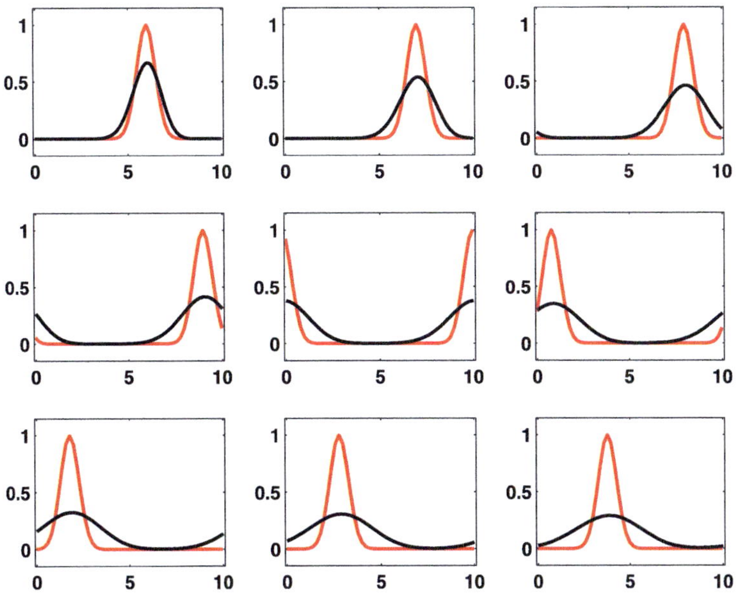

FIGURE 2.20 Plot of the true bell curve, in red, and the numerical solution from the Lax–Friedrichs finite difference scheme for $tn = 10\Delta t, 20\Delta t, \ldots, 90\Delta t$ steps, for $\Delta t = 0.1$, $\Delta x = 0.2$, $u = 1$, which implies a Courant number of 0.5.

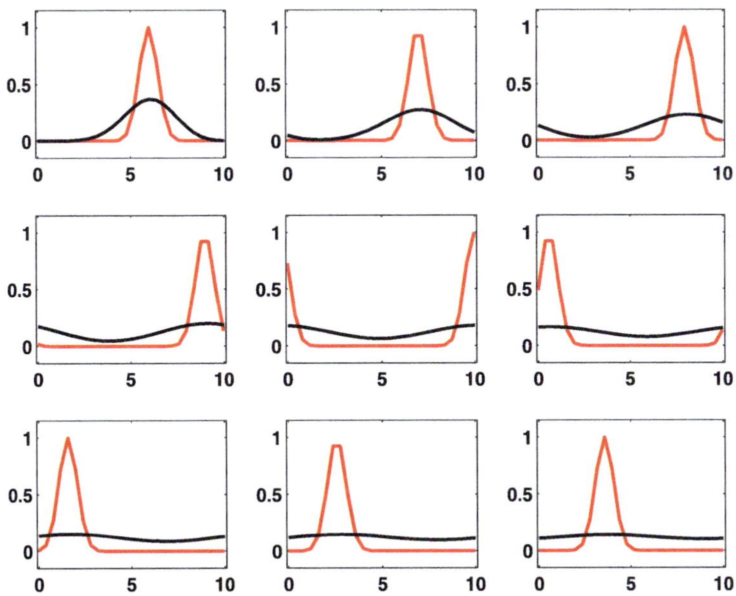

FIGURE 2.21 Plot of the true bell curve, in red, and the numerical solution from the Lax–Friedrichs finite difference scheme for $tn = 10\Delta t, 20\Delta t, \ldots, 90\Delta t$ steps, for $\Delta t = 0.1$, $\Delta x = 0.4$, $u = 1$, which implies a Courant number of 0.25.

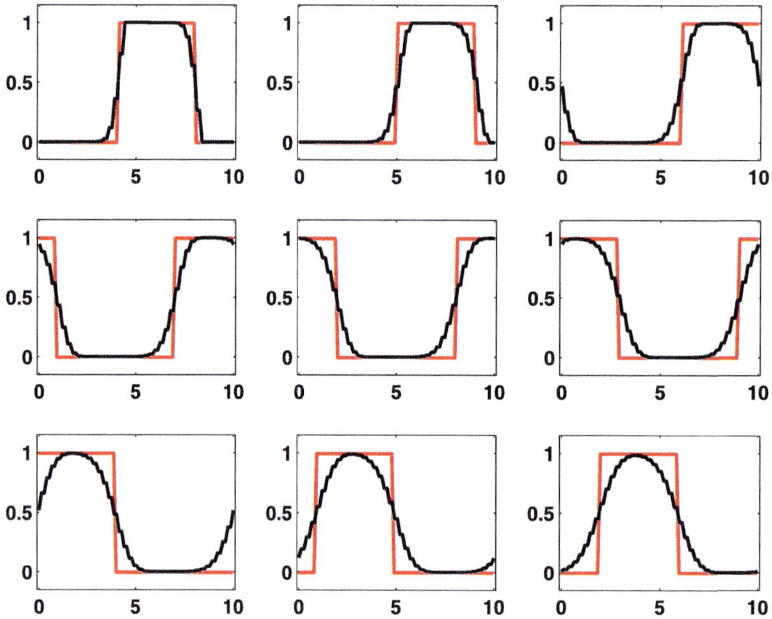

FIGURE 2.22 Plot of the true step function, in red, and the numerical solution from the Lax–Friedrichs finite difference scheme for $tn = 10\Delta t, 20\Delta t, \ldots, 90\Delta t$ steps, for $\Delta t = 0.1$, $\Delta x = 0.1333$, $u = 1$, which implies a Courant number of 0.75.

the step function with $C = 0.75$ are shown in Fig. 2.22. We can see that for this case the scheme is not damping the numerical solution, nor is it out of phase with the true state. In fact, the Lax–Friedrichs scheme appears to be smoothing out the function, similar to the behavior we have seen with other schemes, but it is not diffusing the step profile.

However, the same cannot be said for when $C = 0.5$, Fig. 2.23. While we still do not have a phase error with this scheme, we are now starting to see the numerical diffusion associated with this scheme. We can quite clearly see that the numerical diffusion is not as strong as for the bell case with the same Courant number, but that it is still present in the scheme.

That being said, when considering the results for the case $C = 0.25$, Fig. 2.24, we can clearly see that the diffusion is almost as strong here as it was for the bell curve. While the solution is not completely flattened, it is quite clear that the Lax–Friedrichs scheme suffers from a severe diffusion affect from the numerical computations if the Courant number is too low, but is able to maintain the correct speed and as such does not have a phase error, just a very large amplitude error.

As we have seen from the plots presented in this section, the Lax–Friedrichs schemes suffers from severe diffusion, but it is not unstable, which was the case for the forward-time-centered-space scheme, to which this scheme is quite similar. We now consider yet another scheme that was developed by Lax to see if we can remove some of the problems caused here, but still only using the forward Euler for the time derivative.

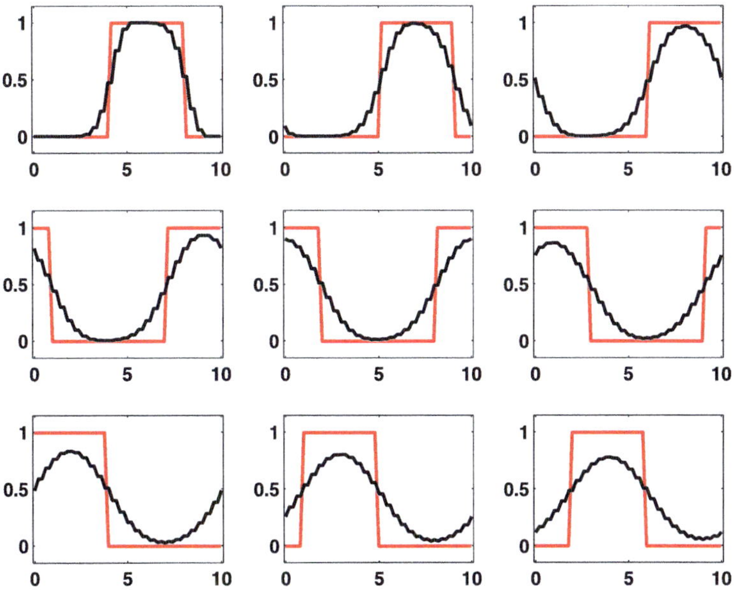

FIGURE 2.23 Plot of the true step function, in red, and the numerical solution from the Lax–Friedrichs finite difference scheme for $tn = 10\Delta t, 20\Delta t, \ldots, 90\Delta t$ steps, for $\Delta t = 0.1$, $\Delta x = 0.2$, $u = 1$, which implies a Courant number of 0.5.

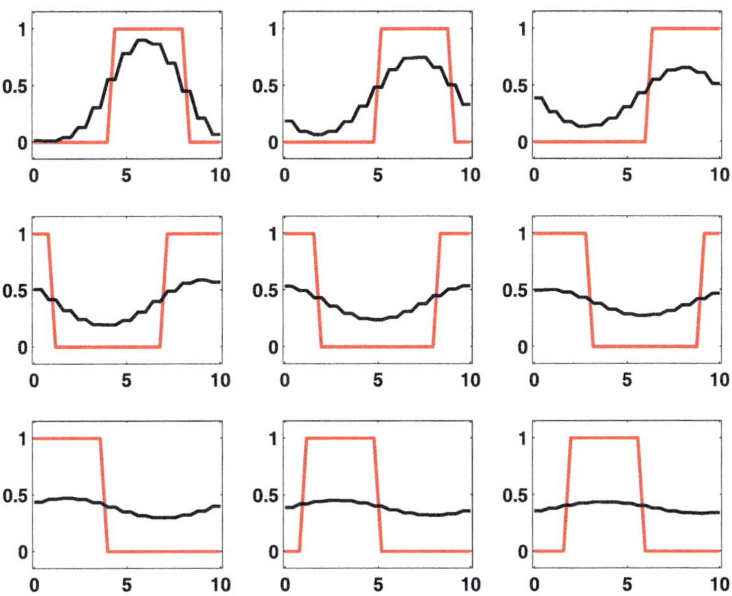

FIGURE 2.24 Plot of the true step function, in red, and the numerical solution from the Lax–Friedrichs finite difference scheme for $tn = 10\Delta t, 20\Delta t, \ldots, 90\Delta t$ steps, for $\Delta t = 0.1$, $\Delta x = 0.4$, $u = 1$, which implies a Courant number of 0.25.

2.2.4 Lax–Wendroff scheme

The methods that we have presented so far to approximate the one-dimensional advection equation have been straightforward finite difference approximation to the temporal and spatial derivatives, but there are other techniques that can be used to derive numerical approximations to the advection equation. One such technique is to consider a Taylor series expansion of the field u with respect to the time component such that

$$\psi\left(x,t+\Delta t\right)=\psi\left(x,t\right)+\Delta t\psi'\left(x,t\right)+\frac{\Delta t^2}{2}\psi''\left(x,t\right)+\cdots, \tag{2.39}$$

where $\psi'\equiv\frac{\partial\psi}{\partial t}$ and $\psi''\equiv\frac{\partial^2\psi}{\partial t^2}$. As we have seen many times in this chapter, the partial differential equation that we are solving is $u_t=-cu_x$, so if we differentiate this equation with respect to t then we obtain

$$\psi''=-u\psi'_x=-u\frac{\partial}{\partial t}\psi_x=-u\left(-u\psi_x\right)_x=u^2\psi_{xx}. \tag{2.40}$$

This now means that we have an expression for the second time derivative in terms of the second spatial derivative. This then enables us to write (2.39), using (2.40), as

$$\psi\left(x,t+\Delta t\right)=\psi\left(x,t\right)-u\Delta tu\psi_x\left(x,t\right)+\frac{\Delta t^2}{2}\psi_{xx}\left(x,t\right)+\cdots. \tag{2.41}$$

We now apply second-order central differencing to the two sets of spatial derivatives in (2.41), yielding

$$\psi_{xx}\approx\frac{\psi_{i+1}^n-2\psi_i^n+\psi_{i-1}^n}{\Delta x^2}. \tag{2.42}$$

Thus the numerical approximation to (2.39) is given by

$$\psi_i^{n+1}=\psi_i^n-\frac{u\Delta t}{2\Delta x}\left(\psi_{i+1}^n-\psi_{i-1}^n\right)+\frac{u^2\Delta t^2}{2\Delta x^2}\left(\psi_{i+1}^n-2\psi_i^n+\psi_{i-1}^n\right). \tag{2.43}$$

The difference equation in (2.43) is referred to as the **Lax–Wendroff** scheme for the advection equation. An interesting feature to notice about (2.43) is that the first two terms are those for the forward-time, centered-space, which we have shown to be unconditionally unstable. We shall now investigate the impact the second-order term in (2.43) has on the stability of this scheme. The second-order term in (2.43) is the definition of **diffusion** and as such the approximation in (2.43) is called the **advection–diffusion** equation,

$$\psi\left(x,t+\Delta t\right)=\psi\left(x,t\right)-u\Delta t\psi_x\left(x,t\right)+u^2\frac{1}{2}\left(\Delta t\right)^2\psi_{xx}\left(x,t\right)+\cdots. \tag{2.44}$$

The difference equation for the Lax–Wendroff scheme comes about by considering central differences in the space component for the first and second-order spatial derivatives in (2.44) given by

$$\psi_{xx} \approx \frac{\psi_{j+1}^n - 2\psi_j^n + \psi_{j-1}^n}{(\Delta x)^2}. \tag{2.45}$$

Thus applying (2.45) to (2.44) results in

$$\psi_j^{n+1} = \psi_j^n - \frac{u\Delta t}{2\Delta x}\left(\psi_{j+1}^n - \psi_{j-1}^n\right) + \frac{u^2(\Delta t)^2}{2(\Delta x)^2}\left(\psi_{j+1}^n - 2\psi_j^n + \psi_{j-1}^n\right). \tag{2.46}$$

As we can see from (2.46), the Lax–Wendroff scheme is an explicit approximation to the advection equation. We should note that we have what appears to be a forward-time, centered-space scheme combined with an extra term. The extra term, $\frac{\partial^2\psi}{\partial x^2}$, represents a **diffusion**. We shall explain the importance of this extra term in the next chapter in relation to the forward-time, centered space scheme. Now if we rearrange (2.46) to consolidate terms and to express the difference equation in terms the Courant number, we obtain

$$\psi_j^{n+1} = \left(1 - C^2\right)\psi_j^n + (C+1)\frac{C}{2}\psi_{j-1}^n + (C-1)\frac{C}{2}\psi_{j+1}^n. \tag{2.47}$$

The stencil for this scheme is presented in Fig. 2.25 where we can see that it uses the same points as the forward-time, centered-space scheme.

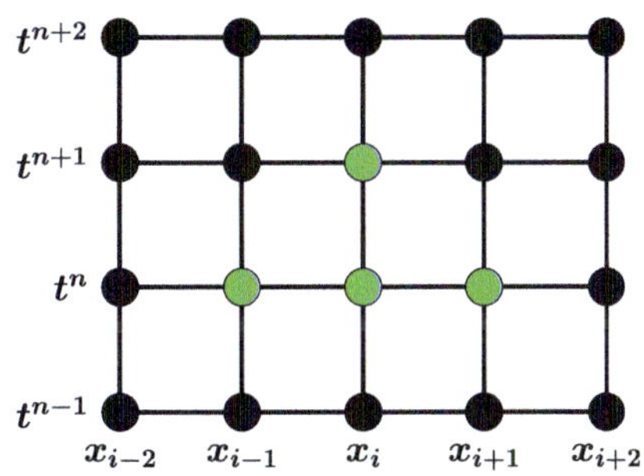

FIGURE 2.25 Illustration of the stencil for the Lax–Wendroff finite difference scheme.

As with the previous schemes, we will present some results of using the Lax–Wendroff scheme with the bell curve and the step function. We shall again present these for the case where we have the Courant number $C = 0.75, 0.5$, and 0.25. We do not show the results for $C = 1$ as they are very similar to those from the upwind forward Euler, but we do see some interesting features of this scheme for other Courant numbers.

We start with $C = 0.75$, where these results are presented in Fig. 2.26. Here $u = 1$, $\Delta t = 0.1$, and $\Delta x = 0.1333$ again. As before, the red line marks the true solution and the black shows the numerical solution from the Lax–Wendroff scheme. The first three top panels as before represent the solutions at $t^n = 10, 20$, and 30, respectively. We can start to see some red line emerging from the black in the third panel, which indicates a possible phase error, but it appears not to be too bad. However, when we arrive at the bottom left panel, $t^n = 70$, we see

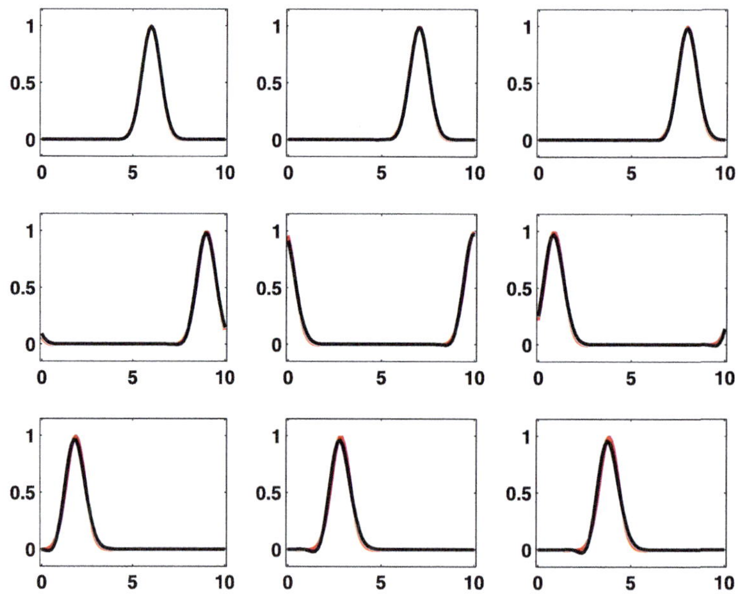

FIGURE 2.26 Plot of the true bell curve, in red, and the numerical solution from the Lax–Wendroff finite difference scheme for $tn = 10\Delta t, 20\Delta t, \ldots, 90\Delta t$ steps, for $\Delta t = 0.1$, $\Delta x = 0.1333$, $u = 1$, which implies a Courant number of 0.75.

that there appears to be some form of undershoot as the scheme starts to come into the bell curve from the left. By the time we arrive at $t^n = 90$, the bottom right panel, we can definitely see more red showing through, but also that there appears to be a phase error as well as an amplitude error, along with the *wiggles* that appear to be forming downwind.

In Fig. 2.27 we have the results of applying the Lax–Wendroff scheme to the bell curve when we have a Courant number of 0.5, where $u = 1$, $\Delta t = 0.1$, and $\Delta x = 0.2$. We can clearly see the red appearing from behind the black within the first 10 time steps; this becomes more prevalent as time goes on and the scheme is developing a phase error, as well as an amplitude error, indicative of damping, but also the wiggles downstream are starting to become quite obvious. So the Lax–Wendroff suffers from properties that the upwind Euler did not appear to have trouble with.

When we reduce the spatial size to $\Delta x = 0.4$, given $u = 1$ and $\Delta t = 0.1$, which implies a Courant number of 0.25, we see from Fig. 2.28 that the lower resolution does not improve the performance of the Lax–Wendroff scheme. We can quite clearly see red lines in all of the 9 plots, where the wiggle is present at $t^n = 10$, and that the numerical solution goes out of phase with the true solution very quickly. We can quite clearly see that the downstream wiggles are growing faster than what we have seen in the other plots, but also that the damping is more severe in these plots. By the time we are at $t^n = 90$, we can clearly see that the numerical scheme does not look like the true solution at all.

Given the problems that the Lax–Wendroff scheme had with the bell curve, we must be quite curious to see what happens when it is applied to the discontinuous step function test case. We present results for the same three values of the Courant number as for the bell curve

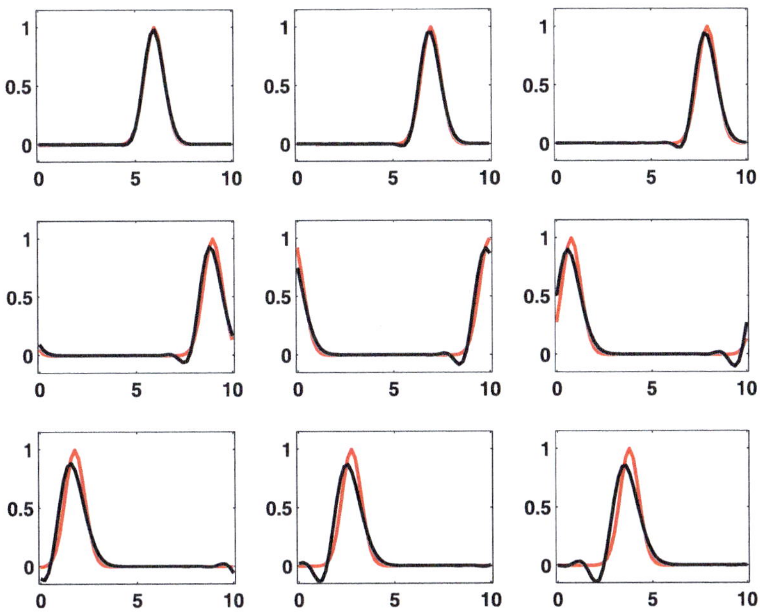

FIGURE 2.27 Plot of the true bell curve, in red, and the numerical solution from the Lax–Wendroff finite difference scheme for $tn = 10\Delta t, 20\Delta t, \ldots, 90\Delta t$ steps, for $\Delta t = 0.1$, $\Delta x = 0.2$, $u = 1$, which implies a Courant number of 0.5.

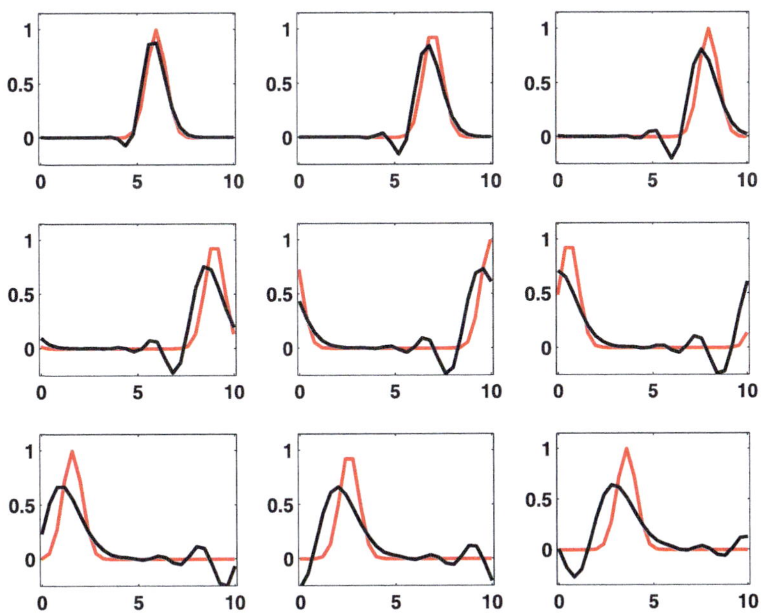

FIGURE 2.28 Plot of the true bell curve, in red, and the numerical solution from the Lax–Wendroff finite difference scheme for $tn = 10\Delta t, 20\Delta t, \ldots, 90\Delta t$ steps, for $\Delta t = 0.1$, $\Delta x = 0.4$, $u = 1$, which implies a Courant number of 0.25.

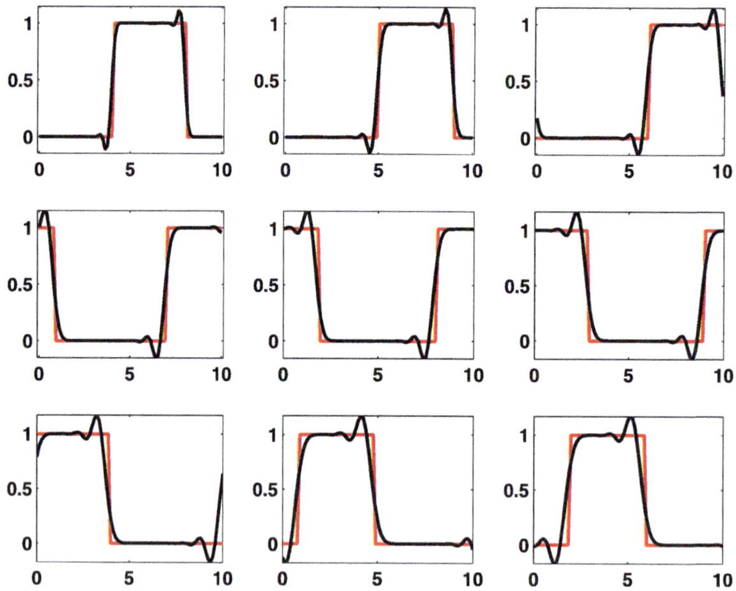

FIGURE 2.29 Plot of the true step function, in red, and the numerical solution from the Lax–Wendroff finite difference scheme for $tn = 10\Delta t, 20\Delta t, \ldots, 90\Delta t$ steps, for $\Delta t = 0.1$, $\Delta x = 0.1333$, $u = 1$, which implies a Courant number of 0.75.

results. We start with $C = 0.75$, where these results are presented in Fig. 2.29. We can clearly see that the scheme is having trouble at the discontinuity points, and that it is producing under- and overshoots at these points. We can see that the wiggles are downstream of both discontinuities, that they are growing as time carries on, but also that the numerical scheme is trying to smooth out the steps. On a positive note, it does appear that the numerical is staying more in phase with the step function, but is introducing errors downstream.

When we reduce the Courant number to 0.5 for this test case with the Lax–Wendroff scheme, Fig. 2.30, we see that the over- and undershoots are larger and at an earlier time step than for the Courant number equal to 0.75. While the numerical solution stays in phase with the true solution, we can clearly see that the under- and overshoots are growing a lot faster than in Fig. 2.29, and that they are distorting the solutions quite a lot. Even the smoothing of the function does not appear to remove these over- and undershoots and their effects downstream from the two discontinuity points.

The last plot that we consider for the Lax–Wendroff scheme is for the step function with a Courant number of 0.25, Fig. 2.31. We can clearly see that the over- and undershoots at the discontinuity points are growing faster than in the two previous sets of results for this scheme. While the numerical solution does stay in phase, sort of, with the true state, the downstream wiggles have overwhelmed the scheme, and as such the numerical solution has an amplitude error, as well as looks nothing like the true solution at all.

All of the schemes that we have considered so far consist of using the forward Euler algorithm for the time component, and then using different mathematical tricks to find ways to stabilize the schemes. Given that the advection equation is a function of both time and

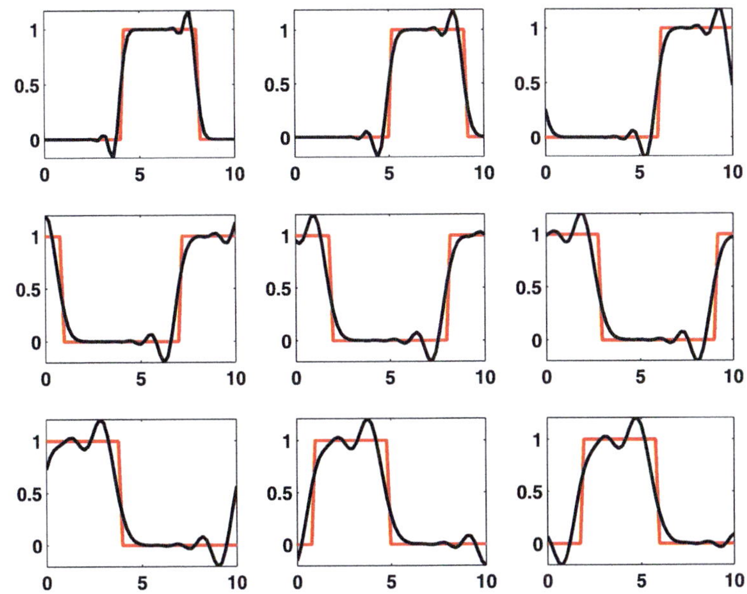

FIGURE 2.30 Plot of the true step function, in red, and the numerical solution from the Lax–Wendroff finite difference scheme for $tn = 10\Delta t, 20\Delta t, \ldots, 90\Delta t$ steps, for $\Delta t = 0.1$, $\Delta x = 0.2$, $u = 1$, which implies a Courant number of 0.5.

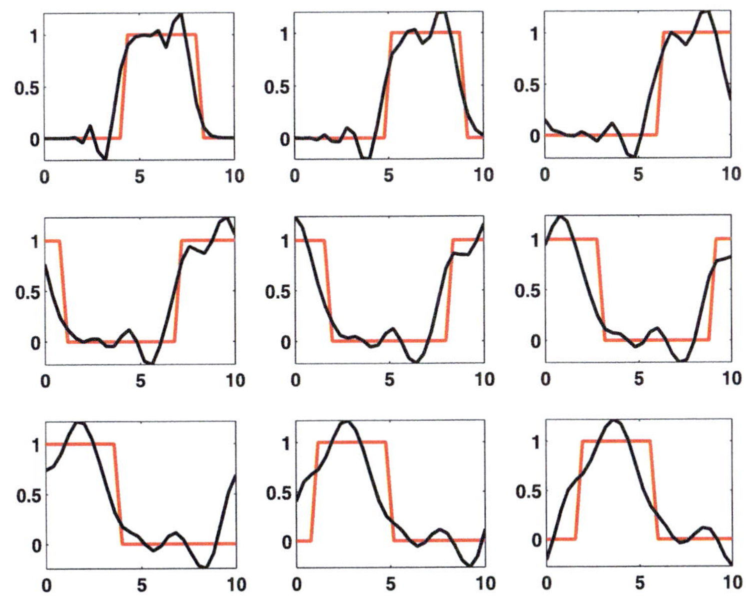

FIGURE 2.31 Plot of the true step function, in red, and the numerical solution from the Lax–Wendroff finite difference scheme for $tn = 10\Delta t, 20\Delta t, \ldots, 90\Delta t$ steps, for $\Delta t = 0.1$, $\Delta x = 0.4$, $u = 1$, which implies a Courant number of 0.25.

space, we now turn our focus to schemes to try improve the representation of the temporal component.

2.2.5 Leap-frog, centered-time–centered-space (CTCS)

As just mentioned, so far all of the schemes that have been presented use an upwind approximation to the time derivative in the advection equation. We saw in the last subsection a few numerical schemes that had a centered approximation to the spatial derivative. Therefore, if we were to use a centered-time, and a centered-space, approach, then we are *leap-frogging* over the central points, but the derivatives are valid at these points. Thus this scheme is quite often referred to as the **leap-frog scheme**, and has been, and still is, quite widely used in many geophysical modeling applications in some form. This scheme is also referred to as **centered-time–centered-space** scheme. The difference equation for this scheme is given by

$$\frac{\psi_i^{n+1}-\psi_i^{n-1}}{2\Delta t}+u\frac{\psi_{i+1}^n-\psi_{i-1}^n}{2\Delta x}=0. \tag{2.48}$$

The stencil for the leap-frog scheme in (2.48) is shown in Fig. 2.32, where we can see that the points that we are using are different again in some parts to the previously presented schemes.

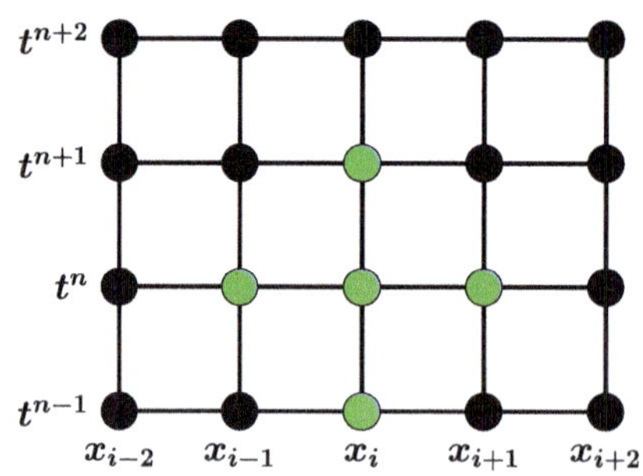

FIGURE 2.32 Illustration of the stencil for the centered-time–centered-space (CTCS), or leap-frog, finite difference scheme.

The implementation of the difference equation in (2.48) is given by

$$\psi_i^{n+1}=\psi_i^{n-1}+C\psi_{i-1}^n-C\psi_{i+1}^n. \tag{2.49}$$

An important feature to note here is that at the initial time we do not have a ψ_i^{-1} to start the leap-frog scheme, and we cannot introduce what are referred to as ghost points before the starting time. There are many different approaches that can be used here to step forward in time for the first time step. One candidate could be the forward-time, centered-space scheme, but we will mention a note of caution here using this scheme. As we mentioned earlier, there is some form of instability associated with this scheme. Another approach could be to use the Lax–Wendroff scheme that we introduced in the last subsection. To show the impact that the two just mentioned schemes have on the leap-frog scheme if they are used to provide that first step, we present the next six figures for both the bell curve and the step function, and for Courant number equal to 1 and 0.5.

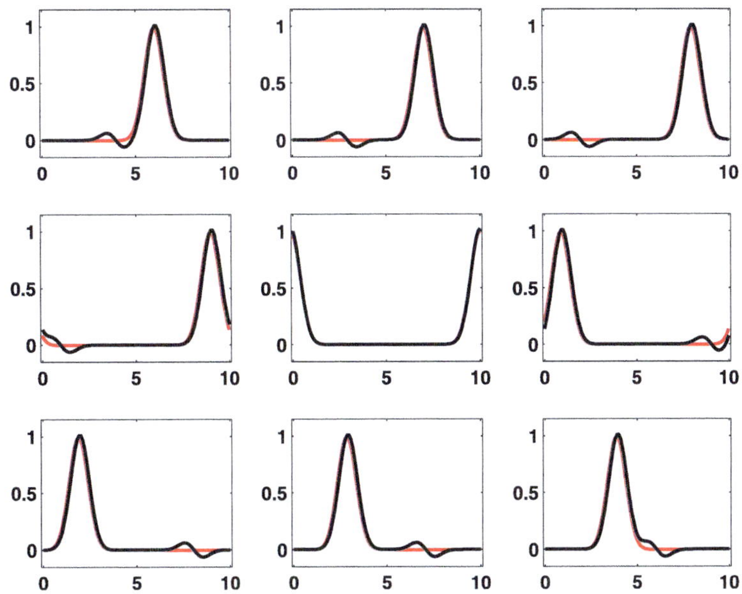

FIGURE 2.33 Plot of the true bell curve, in red, and the numerical solution from the centered-time–centered-space finite difference scheme initialized by the FTCS scheme for $tn = 10\Delta t, 20\Delta t, \ldots, 90\Delta t$ steps, for $\Delta t = 0.1$, $\Delta x = 0.1$, $u = 1$, which implies a Courant number of 1.

We have already alluded to the fact that when the Courant number is equal to 1, the forward Euler scheme is quite accurate. We have seen that both the forward-time, centered-space scheme and the Lax–Wendroff use this approach for the temporal derivative approximation, but if we compare the results from Fig. 2.33 and Fig. 2.34, then we can see that even at this Courant number the forward-time, centered space scheme introduces a wiggle downstream of the start of the curve, and while the leap-frog scheme does not damp it out, it also does not excite along the way, but as we can see when it approaches $t^n = 90$, there could be some spurious interactions that could cause the scheme to create something that is not physical. When we look at the Lax–Wendroff scheme results in Fig. 2.34, we can clearly see that there are neither wiggles nor phase errors. We should point out that the latter is also true for the Lax–Wendroff scheme initialized forward-time–centered-space scheme.

However, when we consider the case when we have a Courant number equal to 0.5, we see that for both initialization approaches in the leap-frog scheme, Figs. 2.35 and 2.36, for the forward-time–centered-space and Lax–Wendroff, respectively, we see that, while the Lax–Wendroff initialized scheme is quite good in the first panel, there is a little red patch showing through. We see that the forward-time–centered-space initialization has produced a spurious wave downwind, but as time goes on the leap-frog scheme does advect the wave backwards. We should also point out that in the Lax–Wendroff initialized scheme the leap-frog scheme

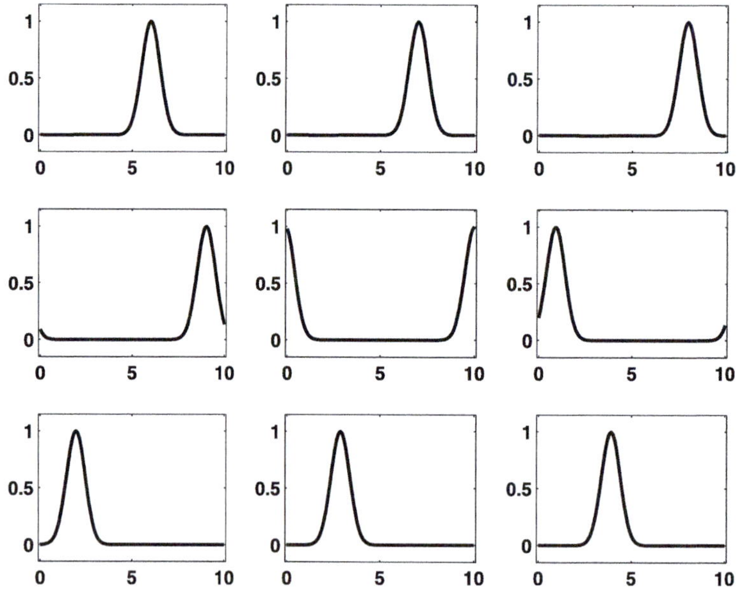

FIGURE 2.34 Plot of the true bell curve, in red, and the numerical solution from the centered-time–centered-space finite difference scheme initialized by the Lax–Wendroff scheme for $tn = 10\Delta t, 20\Delta t, \ldots, 90\Delta t$ steps, for $\Delta t = 0.1$, $\Delta x = 0.1$, $u = 1$, which implies a Courant number of 1.

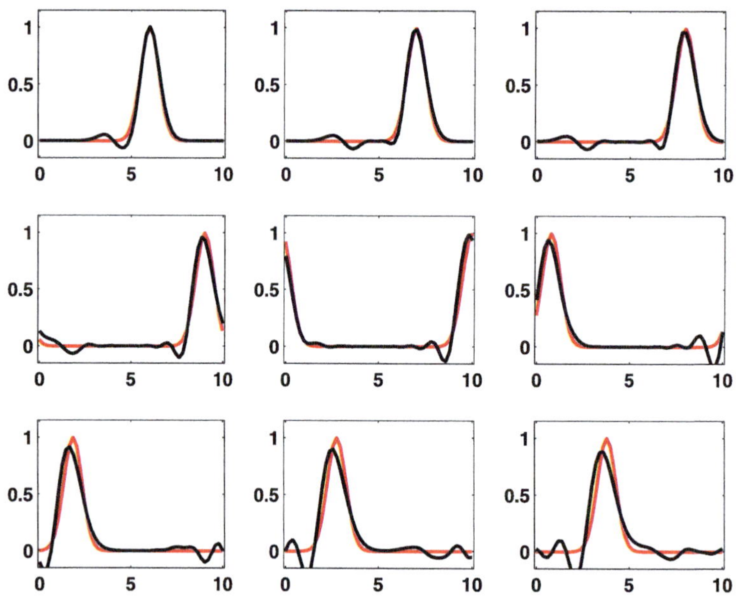

FIGURE 2.35 Plot of the true bell curve, in red, and the numerical solution from the centered-time–centered-space finite difference scheme initialized by the FTCS scheme for $tn = 10\Delta t, 20\Delta t, \ldots, 90\Delta t$ steps, for $\Delta t = 0.2$, $\Delta x = 0.1$, $u = 1$, which implies a Courant number of 0.5.

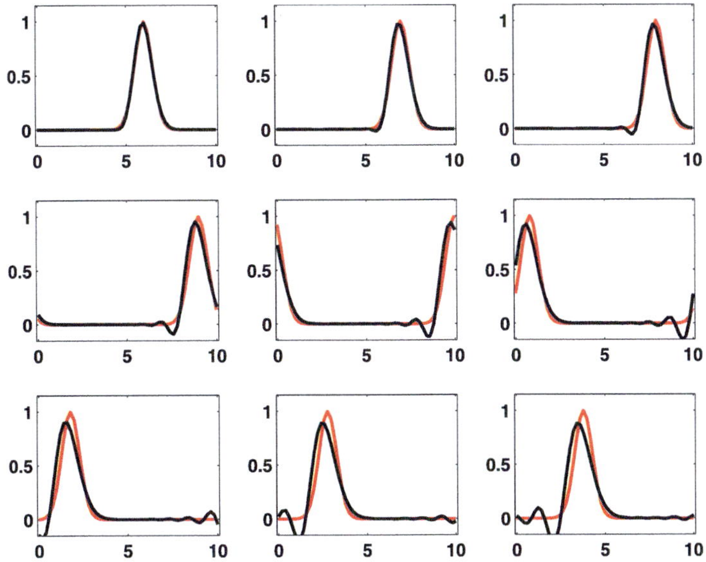

FIGURE 2.36 Plot of the true bell curve, in red, and the numerical solution from the centered-time–centered-space finite difference scheme initialized by the Lax–Wendroff scheme for $tn = 10\Delta t, 20\Delta t, \ldots, 90\Delta t$ steps, for $\Delta t = 0.2$, $\Delta x = 0.1$, $u = 1$, which implies a Courant number of 0.5.

itself initializes a spurious wiggle downstream of the curve which then creates more wiggles behind.

We now move on to the application of the two different initialization approaches for the leap-frog scheme with the step function. We are only going to show the results for the case where the Courant number is equal to 0.9 because as you will see in Figs. 2.37 and 2.38, if we have this instability straightaway with these schemes for this high Courant number, then it will be even worse for the lower Courant numbers. On a positive note, neither of the initialized schemes go out of phase with the step function, and while both produce wiggles initially, the leap-frog scheme itself has problems with this form of discontinuity. We do see that the wiggles produced by the Lax–Wendroff initialization are nowhere as bad as those from the forward-time–centered-space approach, and in general this type of approach using the centered time discretization may not be as advantageous compared to the upwind forward Euler in this case.

2.2.6 Linear multistep methods: Adams–Bashforth schemes

So far in this section we have presented methods that utilize either one or two time levels to approximate the time derivative, but it is possible to use information about the x derivative at previous time levels to build a better numerical approximation to this derivative. The approaches that use more than one time level to approximate either of the derivatives are referred to as the **multistep methods**, and when the relationship is linear between the different time levels, the schemes are referred to as the **linear multistep methods, or LMM.** One such

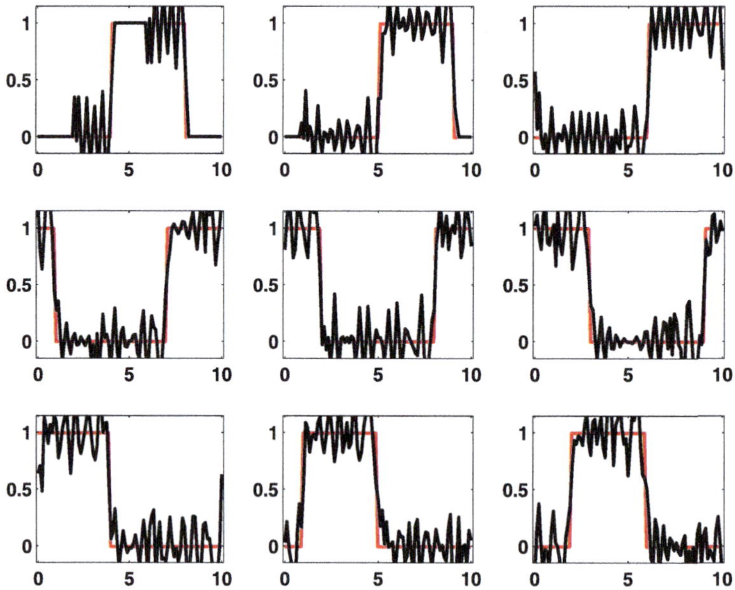

FIGURE 2.37 Plot of the true step function, in red, and the numerical solution from the centered-time–centered-space finite difference scheme initialized by the FTCS scheme for $tn = 10\Delta t, 20\Delta t, \ldots, 90\Delta t$ steps, for $\Delta t = 0.1111$, $\Delta x = 0.1$, $u = 1$, which implies a Courant number of 0.9.

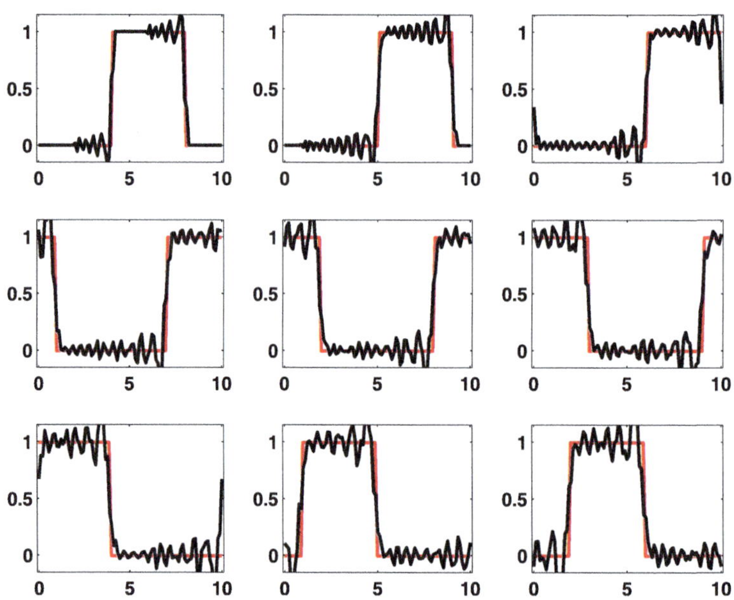

FIGURE 2.38 Plot of the true step function, in red, and the numerical solution from the centered-time–centered-space finite difference scheme initialized by the Lax–Wendroff scheme for $tn = 10\Delta t, 20\Delta t, \ldots, 90\Delta t$ steps, for $\Delta t = 0.1111$, $\Delta x = 0.1$, $u = 1$, which implies a Courant number of 0.9.

set of linear multistep methods are the **Adams–Bashforth schemes**. We start by integrating the general initial value problem

$$\begin{cases} \dfrac{d\psi}{dt} = f(\psi, t) \quad 0 \leq t \leq T, \\ x(0) = x_0. \end{cases} \tag{2.50}$$

From the method of characteristics, we know that we can obtain an exact solution to (2.50) at time $t + \Delta t$ through integration, namely

$$\psi(t + \Delta t) = \psi(t) + \int_t^{t+\Delta t} f(s, \psi(s))\, ds. \tag{2.51}$$

The next step is to consider numerical quadrature approximations to the integral in (2.51). It can be shown that some of the numerical schemes that we have presented so far are equivalent to different quadrature approximations to the integral. If we were to consider a second-order approximation, at time $t_2 \equiv t_0 + 2\Delta t$, and were to use the fact that we already know $f(t_1 \psi(t_1)) = f_1$, and at the initial time $f(t_0, \psi(t_0)) = f_0$, then we could write the solution at t_2 in integral form as

$$\psi(t_2) = \psi(t_1) + \int_{t_1}^{t_2} f(\tau, \psi(\tau))\, d\tau. \tag{2.52}$$

We approximate the integrand by a linear interpolant, using the data that we have at t_0 and t_1, which implies

$$f(t, \psi(t)) = P_1(t) + \xi_2, \tag{2.53}$$

where P_1 represents a linear polynomial interpolation function, and ξ_2 represents the error from the interpolation method. We should note that we are approximating this function in time.

The methods that we describe here use what is referred to as the Lagrange interpolation formula. We shall go into more details about this polynomial interpolation method in Chapter 6. The linear Lagrange polynomial interpolation is given by

$$P_1(t) \equiv \frac{t - t_1}{t_0 - t_1} f_0 + \frac{t - t_0}{t_1 - t_0} f_1 = \frac{t - \Delta t}{-\Delta t} f_0 + \frac{t}{\Delta t} f_1. \tag{2.54}$$

The next step is to integrate (2.54) between Δt and $2\Delta t$, which implies that

$$\begin{aligned} \int_{\Delta t}^{2\Delta t} f(\tau, \psi(\tau))\, d\tau &\approx \int_{\Delta t}^{2\Delta t} \frac{t - \Delta t}{-\Delta t} f_0 dt + \int_{\Delta t}^{2\Delta t} \frac{t}{\Delta t} f_1 dt \approx \left[\frac{\frac{t^2}{2} - t\Delta t}{-\Delta t} f_0 \right]_{\Delta t}^{2\Delta t} + \left[\frac{\frac{t^2}{2}}{\Delta t} \right]_{\Delta t}^{2\Delta t}, \\ &\approx \left(\frac{\frac{4(\Delta t)^2}{2} - 2(\Delta t)^2 - \frac{(\Delta t)^2}{2} + \Delta t}{-\Delta t} \right) f_0 + \left(\frac{\frac{4(\Delta t)^2}{2} - \frac{(\Delta t)^2}{2}}{\Delta t} \right) f_1, \\ &\approx -\frac{\Delta t}{2} f_0 + \frac{3\Delta t}{2} f_1, \end{aligned} \tag{2.55}$$

where (2.55) is now combined with the other part of the equation in (2.51) to give what is referred to as the second-order Adams–Bashforth scheme given by

$$\psi^2 = \psi^1 + \frac{\Delta t}{2}\left(3f^1 - f^0\right). \tag{2.56}$$

The derivation above is applicable to any time step and is generalizable such that the general expression for the second-order Adams–Bashforth scheme is given by

$$\psi^{n+1} = \psi^n + \frac{\Delta t}{2}\left(3f^n - f^{n-1}\right). \tag{2.57}$$

The expression in (2.57) is for a general function f, but for the linear advection equation that we are considering, we have $f = -u\frac{\partial \psi}{\partial x}$. The numerical approximation to the linear advection equation using the second-order Adams–Bashforth scheme is therefore given by

$$\psi_i^{n+1} = \psi_i^n - \frac{3u\Delta t}{4\Delta x}\left(\psi_{i+1}^n - \psi_{i-1}^n\right) + \frac{v\Delta t}{4\Delta x}\left(\psi_{i+1}^{n-1} - \psi_{i-1}^{n-1}\right). \tag{2.58}$$

The difference equation in (2.58) is referred to as a two-step method as it uses two different time levels behind the one that we are seeking the values for. This can be seen by the stencil associated with (2.58) which is presented in Fig. 2.39, and as we saw with the leap-frog schemes, there are two steps with respect to the time derivative approximation.

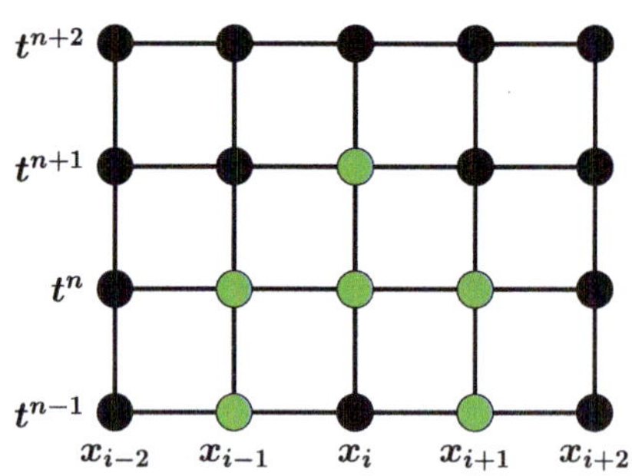

FIGURE 2.39 Illustration of the stencil for the two level Adams–Bashforth scheme.

There are more advanced versions of the Adams–Bashforth scheme, and the next formulation that we derive here is for the third-order Adams–Bashforth scheme, where now we are using three different time levels to approximate the function at the point $t = t_3$, given the known function values are t_2, t_1, and t_0. This then means that we are fitting a quadratic polynomial, denoted by $P_2(t)$. We will use the Lagrange interpolant polynomial, which for the quadratic case is given by

$$P_2(t) = \frac{\left(t - t^1\right)\left(t - t^2\right)}{\left(t^0 - t^1\right)\left(t^0 - t^2\right)}\psi_{\delta 2x}^0 + \frac{\left(t - t^0\right)\left(t - t^2\right)}{\left(t^1 - t^0\right)\left(t^1 - t^2\right)}\psi_{\delta 2x}^1 + \frac{\left(t - t^0\right)\left(t - t^1\right)}{\left(t^2 - t^0\right)\left(t^2 - t^1\right)}\psi_{\delta 2x}^2, \tag{2.59}$$

where $\psi_{\delta 2x}^n$ is referring to the central difference approximation to the x-derivative of ψ evaluated at time $t = t^n$. As with the derivation of the second-order Adams–Bashforth scheme, we now need to integrate (2.59) with respect to t, but this time the limits of integration are from $t = 2\Delta t$ to $t = 3\Delta t$. Thus we have

$$\int_{2\Delta t}^{3\Delta t} \left(\frac{\left(t-t^1\right)\left(t-t^2\right)}{\left(t^0-t^1\right)\left(t^0-t^2\right)} \psi_{\delta 2x}^0 + \frac{\left(t-t^0\right)\left(t-t^2\right)}{\left(t^1-t^0\right)\left(t^1-t^2\right)} \psi_{\delta 2x}^1 + \frac{\left(t-t^0\right)\left(t-t^1\right)}{\left(t^2-t^0\right)\left(t^2-t^1\right)} \psi_{\delta 2x}^2 \right) dt. \qquad (2.60)$$

Substituting $t^2 = 2\Delta t$, $t^1 = \Delta t$, and $t^0 = 0$ into (2.60) results in the following factors of $\psi_{2\delta x}^0$, $\psi_{2\delta x}^1$ and $\psi_{2\delta x}^2$ that have to be integrated respectively:

$$\begin{aligned} f^0 &: \quad \frac{(t-\Delta t)(t-2\Delta t)}{(-\Delta t)(-2\Delta t)} = \frac{t^2 - 3t(\Delta t)^2 + 2(\Delta t)^2}{2(\Delta t)^2}, \\ f^1 &: \quad \frac{t(t-2\Delta t)}{\Delta t(-\Delta t)} = \frac{t^2 - 2t\Delta t}{-(\Delta t)^2}, \\ f^2 &: \quad \frac{t(t-\Delta t)}{(2\Delta t)(\Delta t)} + \frac{t^2 - t\Delta t}{2(\Delta t)^2}. \end{aligned} \qquad (2.61)$$

Integrating the factors in (2.61), we then obtain

$$\begin{aligned} \psi_{2\delta x}^0 &: \quad \left[\frac{1}{2(\Delta t)^2} \left(\frac{t^3}{3} - \frac{3(\Delta t)t^2}{2} + 2(\Delta t)^2 t \right) \right]_{2\Delta t}^{3\Delta t} = \frac{5\Delta t}{12}, \\ \psi_{2\delta x}^1 &: \quad \left[-\frac{1}{(\Delta t)^2} \left(\frac{t^3}{3} - (\Delta t)t^2 \right) \right]_{2\Delta t}^{3\Delta t} = -\frac{4\Delta t}{3}, \\ \psi_{2\delta x}^2 &: \quad \left[\frac{1}{2(\Delta t)^2} \left(\frac{t^3}{3} - \frac{(\Delta t)t^2}{2} \right) \right]_{2\Delta t}^{3\Delta t} = \frac{23\Delta t}{12}. \end{aligned} \qquad (2.62)$$

This then means that the third-order Adams–Bashforth difference equation for the one-dimensional advection equation at any time step is given by

$$\psi_i^{n+1} = \psi_i^n - \frac{u\Delta t}{24\Delta x} \left(23\left(\psi_{i+1}^n - \psi_{i-1}^n\right) - 16\left(\psi_{i+1}^{n-1} - \psi_{i-1}^{n-1}\right) + 5\left(\psi_{i+1}^{n-2} - \psi_{i-1}^{n-2}\right) \right). \qquad (2.63)$$

In Fig. 2.40 we have presented the stencil for the third-order Adams–Bashforth scheme.

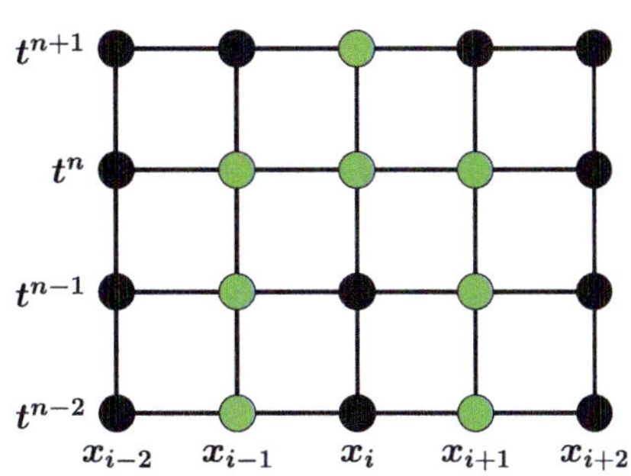

FIGURE 2.40 Illustration of the stencil for the three level Adams–Bashforth scheme.

It is possible to define a general expression for the kth time step Adams–Bashforth scheme for any f, but here we shall do it for our case, which is $-u\delta_{2x}$. This then results in

$$\psi_i^{n+1} = \psi_i^n - \frac{u\Delta t}{2\Delta x}\sum_{j=0}^{k-1}\beta_j\left(\psi_{i+1}^{n-j} - \psi_{i-1}^{n-j}\right), \tag{2.64}$$

where the β terms are the integrated terms from the jth order Lagrange polynomial between $(k-1)\Delta t$ and $k\Delta t$.

2.2.7 Explicit Runge–Kutta methods

The Runge–Kutta methods are based upon forming a one step nonlinear approximation to the differential equation, instead of a linear method, which the multistep methods are. To obtain this nonlinear approximation, we start by expanding y and y^{n+1} in Taylor series as

$$y(t_n + h) = y(t_n) + hy'(t_n) + \frac{h^2}{2!}y''(t) + \cdots, \tag{2.65}$$

$$y^{n+1} = y^n + hf^n + \frac{h^2}{2!}\left(\frac{\partial f}{\partial t} + f\frac{\partial f}{\partial y}\right) + \frac{h^3}{3!}\cdots, \tag{2.66}$$

where $y' = f(t, y)$ and $y'' = \dfrac{\partial f}{\partial t} + \dfrac{\partial f}{\partial y}\dfrac{dy}{dt}$.

Therefore, we have

$$y^{n+1} = y^n + h\tilde{f}^n, \tag{2.67}$$

where $\tilde{f}^n$ is an estimate of the right-hand side of the differential equation. The general form of a Runge–Kutta approximation is given by

$$y^{n+1} = y^n + h\sum_{i=1}^{s} b_i k_i, \tag{2.68}$$

where $k_i = f\left(t_n + c_i h, y + a_{ij}k_{i-1}\right)$ and expands as

$$\begin{aligned}
k_1 &= f\left(t_n, y^n\right),\\
k_2 &= f\left(t_n + c_2 h, y^n + h\left(a_{21}, k_1\right)\right),\\
k_3 &= f\left(t_n + c_3 h, y^n + h\left(a_{31}k_1 + a_{32}k_2\right)\right),\\
&\vdots\\
k_s &= f\left(t_n + c_s h, y^n + h\left(a_{s1}k_1 + a_{s2}k_2 + \cdots + a_{ss-1}k_{s-1}\right)\right).
\end{aligned} \tag{2.69}$$

We can see from (2.69) that the Runge–Kutta schemes are based upon linear combinations of the slopes of the function. The matrix $\left[a_{ij}\right]$ is referred to as the **Runge–Kutta matrix**, while the b_is are referred to as the **weights** and the c_is as the **nodes**. There is a popular tabulation approach to visualize the coefficients, referred to as the **Butcher tableau**, which for the general case above in (2.69) is

$$\begin{array}{c|ccccc} 0 & & & & & \\ c_2 & a_{21} & & & & \\ c_3 & a_{31} & a_{32} & & & \\ \vdots & \vdots & \vdots & \ddots & & \\ c_s & a_{s1} & a_{s2} & \cdots & a_{ss-1} & \\ \hline & b_1 & b_2 & \cdots & b_{s-1} & b_s \end{array}, \tag{2.70}$$

which can also be displayed as $\begin{array}{c|c} \boldsymbol{c} & \mathbf{A} \\ \hline & \boldsymbol{b}^T \end{array}$.

The Runge–Kutta general scheme presented above is referred to as the *explicit* Runge–Kutta method, as the k_nth term for $n = 1, 2, \ldots, s$ depends on all of the previous k_is for $i = 1, 2, \ldots, n-1$. The explicit Runge–Kutta scheme, as we can see from (2.70), implies that **A** is a strictly lower-triangular array, which means that all the nonzero values are below the diagonal entries. There is a class of Runge–Kutta methods that are called **semi-implicit** Runge–Kutta schemes in that the array **A** is a lower-triangular matrix, i.e., there are values on the diagonal but not in the upper-triangular part of the matrix.

We have already presented the simplest Runge–Kutta method, which corresponds to a one-stage Runge–Kutta approach, which is the upwind forward Euler scheme. A commonly used second-order method is the **midpoint method**, given by

$$y^{n+1} = y^n + hf\left(t_n + \frac{1}{2}h, y^n + \frac{1}{2}hf\left(t_n, y^n\right)\right), \tag{2.71}$$

where the associated Butcher tableau is

$$\begin{array}{c|cc} 0 & & \\ \frac{1}{2} & \frac{1}{2} & \\ \hline & 0 & 1 \end{array}.$$

A more commonly used version of the Runge–Kutta scheme is the fourth-order version, which we derive below for the general case and then apply it to the one-dimensional advection equation.

Derivation of the general explicit fourth-order Runge–Kutta method

When we considered the second-order Runge–Kutta schemes, we saw that the condition for the scheme to be consistent was that it matched the different order of derivatives in the Taylor series expansion of $y(t+h)$. Thus, the starting point for the derivation of the fourth-order Runge–Kutta scheme is to evaluate the general formula

$$y^{n+1} = y^n + h\sum_{i=1}^{s} b_i k_i, \qquad k_i = f\left(t_n + c_i h, y^n + h\sum_{j=1}^{s} a_{ij} k_j\right),$$

where the fourth-order scheme is going to be evaluated at the starting point, the midpoint, and the end point of any time interval $[t_n, t_n + h]$, implying that we use the following values for the coefficients of a_{ij} and c_i:

$$c_1 = 0,\ c_2 = \frac{1}{2},\ c_3 = \frac{1}{2},\ c_4 = 1,\ a_{21} = \frac{1}{2},\ a_{32} = \frac{1}{2},\ a_{43} = 1,$$

where $a_{ij} = 0$ for the other values of i and j. To make things easier with the derivation, as you can see from the $k_i = f\left(t_n + c_i h, y^n + h\sum_{j=1}^{s} a_{ij}k_j\right)$ term, there are going to be up to 4 terms in each k_i, so to simplify we introduce the following notation:

$$\begin{aligned}
y_1^{n+1} &\equiv y^n + hf\left(t_n, y^n\right), \\
y_1^{n+\frac{1}{2}} &\equiv \frac{y^n + y_1^{n+1}}{2}, \\
y_2^{n+1} &\equiv y^n + hf\left(t_n + \frac{h}{2}, y_1^{n+\frac{1}{2}}\right), \\
y_2^{n+\frac{1}{2}} &\equiv \frac{y^n + y_2^{n+1}}{2}, \\
y_3^{n+1} &\equiv y^n + hf\left(t_n + \frac{h}{2}, y_2^{n+\frac{1}{2}}\right),
\end{aligned}$$

which leads to the following expressions for the k_is:

$$\begin{aligned}
k_1 &\equiv f\left(t_n, y^n\right), \\
k_2 &\equiv f\left(t_n + \frac{h}{2}, y_1^{n+\frac{1}{2}}\right), \\
k_3 &\equiv f\left(t_n + \frac{h}{2}, y_2^{n+\frac{1}{2}}\right), \\
k_4 &\equiv f\left(t_n + h, y_3^{n+1}\right).
\end{aligned}$$

Next we have to derive formulas for the k_is in terms of differential operators, which are given by:

$$\begin{aligned}
k_2 &= f\left(t_n + \frac{h}{2}, y_1^{n+\frac{1}{2}}\right) = f\left(t + \frac{h}{2}, y^n + \frac{h}{2}k_1\right) \\
&= f\left(t_n, y^n\right) + \frac{h}{2}\frac{\mathrm{d}}{\mathrm{d}t} f\left(t_n, y^n\right), \qquad (2.72) \\
k_3 &= f\left(t_n + \frac{h}{2}, y_2^{n+\frac{1}{2}}\right) = f\left(t_n + \frac{h}{2}, y^n + \frac{h}{2} f\left(t_n + \frac{h}{2}, y^n + \frac{h}{2}k_1\right)\right) \\
&= f\left(t_n, y^n\right) + \frac{h}{2}\frac{\mathrm{d}}{\mathrm{d}t}\left(f\left(t_n, y^n\right)\right) + \frac{h^2}{4}\frac{\mathrm{d}^2}{\mathrm{d}t^2}\left(f\left(t_n, y^n\right)\right), \qquad (2.73) \\
k_4 &= f\left(t_n + h, y_3^{n+1}\right) = f\left(t_n + h, y^n + hf\left(t_n + \frac{h}{2}, y^n + \frac{k_2}{2}\right)\right)
\end{aligned}$$

$$
\begin{aligned}
&= f\left(t_n+h, y^n+hf\left(t_n+\frac{h}{2}, y^n+\frac{h}{2}f\left(t_n+\frac{h}{2}, y^n+\frac{h}{2}f\left(t_n, y^n\right)\right)\right)\right) \\
&= f\left(t_n, y^n\right)+h\frac{\mathrm{d}}{\mathrm{d}t}f\left(t_n, y^n\right)+\frac{h^2}{2}\frac{\mathrm{d}^2}{\mathrm{d}t^2}f\left(t_n, y^n\right)+\frac{h^3}{4}\frac{\mathrm{d}^3}{\mathrm{d}t^3}f\left(t_n, y^n\right). \qquad (2.74)
\end{aligned}
$$

Now we substitute (2.72)–(2.74) into the general expression for the Runge–Kutta methods to obtain the values for the b_is to complete the expression for the fourth-order Runge–Kutta scheme. Thus we start with

$$
\begin{aligned}
y_{n+1} &= y^n+h\left(b_1 f\left(t_n, y^n\right)+b_2\left(f\left(t_n, y^n\right)+\frac{h}{2}\frac{\mathrm{d}}{\mathrm{d}t}f\left(t_n, y^n\right)\right)\right. \\
&+ b_3\left(f\left(t_n, y^n\right)+\frac{h}{2}\frac{\mathrm{d}}{\mathrm{d}t}f\left(t_n, y^n\right)+\frac{h^2}{4}\frac{\mathrm{d}^2}{\mathrm{d}t^2}f\left(t_n, y^n\right)\right) \\
&+ \left. b_4\left(f\left(t_n, y^n\right)+h\frac{\mathrm{d}}{\mathrm{d}t}f\left(t_n, y^n\right)+\frac{h^2}{2}\frac{\mathrm{d}^2}{\mathrm{d}t^2}f\left(t_n, y^n\right)+\frac{h^3}{4}\frac{\mathrm{d}^3}{\mathrm{d}t^3}f\left(t_n, y^n\right)\right)\right) \\
&+ \mathcal{O}\left(h^5\right). \qquad (2.75)
\end{aligned}
$$

Collecting all similar terms of the powers of h and powers of the derivatives of f results in

$$
\begin{array}{r|l}
hf\left(t_n, y^n\right) & b_1+b_2+b_3+b_4, \\
h^2\frac{\mathrm{d}}{\mathrm{d}t}f\left(t_n, y^n\right) & \frac{b_2}{2}+\frac{b_3}{2}+b_4, \\
h^3\frac{\mathrm{d}^2}{\mathrm{d}t^2}f\left(t_n, y^n\right) & \frac{b_3}{4}+\frac{b_4}{2}, \\
h^4\frac{\mathrm{d}^3}{\mathrm{d}t^3}f\left(t_n, y^n\right) & \frac{b_4}{4}.
\end{array} \qquad (2.76)
$$

To ascertain what the values of the b_is should be, we consider the Taylor series expansion of $y\left(t_n+h\right)$ to fourth order of h about $y\left(t_n\right)$, where we use the relationship that $\frac{\mathrm{d}}{\mathrm{d}t}y\left(t_n\right) \equiv f\left(t_n, y^n\right)$, which is

$$
\begin{aligned}
y\left(t_n+h\right) &= y\left(t_n\right)+h\frac{\mathrm{d}}{\mathrm{d}t}y\left(t_n\right)+\frac{h^2}{2}\frac{\mathrm{d}^2}{\mathrm{d}t^2}y\left(t_n\right)+\frac{h^3}{6}\frac{\mathrm{d}^3}{\mathrm{d}t^3}y\left(t_n\right)+\frac{h^4}{24}\frac{\mathrm{d}^4}{\mathrm{d}t^4}y\left(t_n\right)+\mathcal{O}\left(h^5\right) \\
&\equiv y\left(t_n\right)+hf\left(t_n, y^n\right)+\frac{h^2}{2}\frac{\mathrm{d}}{\mathrm{d}t}f\left(t_n, y^n\right)+\frac{h^3}{6}\frac{\mathrm{d}^2}{\mathrm{d}t^2}f\left(t_n, y^n\right)+\frac{h^4}{24}\frac{\mathrm{d}^3}{\mathrm{d}t^3}f\left(t_n, y^n\right) \\
&+ \mathcal{O}\left(h^5\right). \qquad (2.77)
\end{aligned}
$$

Therefore, equating (2.75) with (2.77) enables us to obtain four simultaneous equations for four unknowns:

$$
\begin{aligned}
b_1+b_2+b_3+b_4 &= 1, \\
\frac{b_2}{2}+\frac{b_3}{2}+b_4 &= \frac{1}{2},
\end{aligned}
$$

$$
\begin{aligned}
\frac{b_3}{4} + \frac{b_4}{2} &= \frac{1}{6}, \\
\frac{b_4}{4} &= \frac{1}{24}.
\end{aligned}
$$

Thus we have $b_1 = \frac{1}{6}$, $b_2 = \frac{1}{3}$, $b_3 = \frac{1}{3}$, and $b_4 = \frac{1}{6}$, and so the final Butcher tableau is

$$
\begin{array}{c|cccc}
0 & & & & \\
\frac{1}{2} & \frac{1}{2} & & & \\
\frac{1}{2} & 0 & \frac{1}{2} & & \\
1 & 0 & 0 & 1 & \\
\hline
 & \frac{1}{6} & \frac{1}{3} & \frac{1}{3} & \frac{1}{6}
\end{array}.
$$

Therefore, collecting all the terms we present the steps that are involved in programming the fourth-order Runge–Kutta scheme for the one-dimensional advection equation that we are considering in this chapter. The first step is to calculate the right-hand side of the differential equation at the half-way, midpoints, and endpoints in time:

- First, we calculate the term $\widetilde{\psi}_i^{n+\frac{1}{2}} = \psi_i^n + \frac{\Delta t}{2} f\left(t^n, \psi_i^n\right)$.
- Next, given our estimate of ψ at the half-way time step, we now evaluate what the forward upwind would be from this new state backward, which is given by $\widehat{\psi}_i^{n+\frac{1}{2}} \psi_n + \frac{\Delta t}{2} f\left(t^{n+\frac{1}{2}}, \widetilde{\psi}_i^{n+\frac{1}{2}}\right)$.
- Now given these updated states, we now perform the equivalent of an integration forward to the next full time level, which is given by $\widetilde{\psi}_i^{n+1} \psi_i^n + \Delta t f\left(t^n, \widehat{\psi}_i^{n+\frac{1}{2}}\right)$ *(midpoint rule)*.
- Finally, we effectively calculate the new value as a the weighted sum of f evaluated at the three values above, as well as including $f\left(t^n, \psi_i^n\right)$ which we already know. Thus we have

$$
\psi_i^{n+1} = \psi_i^n + \frac{\Delta t}{2}\left[f\left(t^n, \psi_n\right) + 2f\left(t^{n+\frac{1}{2}}, \widetilde{\psi}_i^{n+\frac{1}{2}}\right) + 2f\left(t^{n+\frac{1}{2}}, \widehat{\psi}_i^{n+\frac{1}{2}}\right) + f\left(t^{n+1}, \widetilde{\psi}_i^{n+1}\right)\right]. \tag{2.78}
$$

The actual expression for the fourth order Runge–Kutta method applied to the linear advection case would have the centered difference scheme applied to the f term at each of the different times steps involved in this scheme. Therefore, the algorithm for the advection equation would look as follows:

$$
\begin{aligned}
\widetilde{\psi}_i^{n+\frac{1}{2}} &= \psi_i^n - \frac{u\Delta t}{4\Delta x}\left(\psi_{i+1}^n - \psi_{i-1}^n\right), \\
\widehat{\psi}_i^{n+\frac{1}{2}} &= \psi_i^n - \frac{u\Delta t}{4\Delta x}\left(\widetilde{\psi}_{i+1}^{n+\frac{1}{2}} - \widetilde{\psi}_{i-1}^{n+\frac{1}{2}}\right), \\
\widetilde{\psi}_i^{n+1} &= \psi_i^n - \frac{u\Delta t}{2\Delta x}\left(\widehat{\psi}_{i+1}^{n+\frac{1}{2}} - \widehat{\psi}_{i-1}^{n+\frac{1}{2}}\right).
\end{aligned}
$$

We now have values for ψ at the two midpoints and at either end of the time interval, which means that we can now evaluate f at all four points. Thus applying a central difference again for f, we have

$$\psi_i^{n+1} = \psi_i^n - \frac{u\Delta t}{6}\left(\delta_{2x}^n\left(\psi_i^n\right) + 2\delta_{2x}\left(\widetilde{\psi}_i^{n+\frac{1}{2}}\right) + 2\delta_{2x}\left(\widehat{\psi}_i^{n+\frac{1}{2}}\right) + \delta_{2x}\left(\widetilde{\psi}_i^{n+1}\right)\right), \tag{2.79}$$

where δ_{2x} represents the spatial central difference operator applied to the different ψs defined above.

The schemes that we have presented in this section are classified as explicit schemes, as we do not have any terms with the time index of $n+1$ on the right-hand side of any of the presented difference equations. In the next section we shall present schemes where this is not the case, and such schemes will be referred to as **implicit** schemes.

2.3 Implicit schemes

As just mentioned, the implicit schemes involve some sort of function evaluation at the future time step $n+1$, rather than at the current time step. Given this loose definition of what implicit finite difference schemes entail, we start with the simplest, as we did with the explicit schemes, which is a backward Euler scheme.

2.3.1 Implicit, or backward Euler, scheme

This scheme can be considered as a backward-time, centered space approach. For the advection equation that we are considering, the difference equation of the backward Euler scheme is given by

$$\frac{\psi_j^{n+1} - \psi_j^n}{\Delta t} = -u\frac{\psi_{j+1}^{n+1} - \psi_{j-1}^{n+1}}{2\Delta x}. \tag{2.80}$$

To implement this numerical scheme presented in (2.80), we rearrange (2.80) as before, but now we have to have all of the terms with the time index $n+1$ on the left-hand side of the difference equation, which results in

$$\psi_i^{n+1} - \frac{C}{2}\psi_{i-1}^{n+1} + \frac{C}{2}\psi_{i+1}^{n+1} + \psi_i^n. \tag{2.81}$$

Therefore, we have a system of linear equations that forms a matrix–vector equation of the form $\mathbf{A}\boldsymbol{\Psi}^{n+1} = \boldsymbol{\Psi}^n$, which in component form for this difference equation is

$$\underbrace{\begin{pmatrix} 1 & \frac{C}{2} & 0 & 0 & \cdots & -\frac{C}{2} \\ -\frac{C}{2} & 1 & \frac{C}{2} & 0 & \cdots & 0 \\ 0 & -\frac{C}{2} & 1 & \frac{C}{2} & \cdots & 0 \\ \vdots & \ddots & \ddots & \ddots & \ddots & \vdots \\ \vdots & \ddots & \ddots & \ddots & \ddots & \vdots \\ \frac{C}{2} & 0 & 0 & \cdots & -\frac{C}{2} & 1 \end{pmatrix}}_{\mathbf{A}} \underbrace{\begin{pmatrix} \psi_1^{n+1} \\ \psi_2^{n+1} \\ \psi_3^{n+1} \\ \vdots \\ \psi_{I-1}^{n+1} \\ \psi_I^{n+1} \end{pmatrix}}_{\mathbf{\Psi}^{n+1}} = \underbrace{\begin{pmatrix} \psi_1^{n} \\ \psi_2^{n} \\ \psi_3^{n} \\ \vdots \\ \psi_{I-1}^{n} \\ \psi_I^{n} \end{pmatrix}}_{\mathbf{\Psi}^{n}},$$

that has be inverted to find the value at the next time step.

The stencils for the backward Euler approximation is presented in Fig. 2.41.

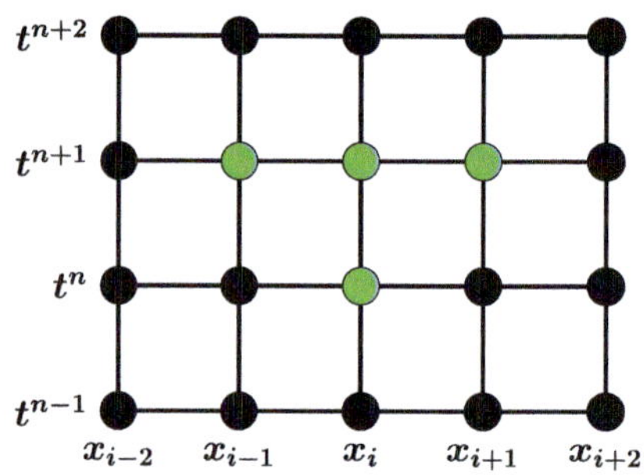

FIGURE 2.41 Illustration of the stencil for the implicit backward Euler finite difference scheme.

As with the explicit finite difference methods that were presented in the last subsection, we have applied the backward Euler to the bell curve and the step function for Courant number of 1 and 10 (No this is not a mistake, we have results for Courant number equal to 10, where we are using $\Delta x = 0.01$ with $\Delta t = 0.1$!).

In Fig. 2.42, we have the results with the bell curve using the backward Euler scheme. We can clearly see that it is dampening the numerical solution, while keeping it in phase. The surprising feature that we see when we increase the spatial resolution to $\Delta x = 0.01$ is that we obtain very similar results to the case when $\Delta x = 0.1$, see Fig. 2.43. There appears to be some sort of dampening of the solution, no matter what value we take for Δx, but we are not seeing the instability that the upwind forward Euler scheme displayed.

Moving on to the step function is where we see some interesting results using this scheme. The results for the two cases are presented in Figs. 2.44 and Fig. 2.45, where we can clearly see that there are wiggles that are very dissimilar to those seen with the other schemes. It does appear in Fig. 2.44 that the scheme is eventually able to smooth out these wiggles, yet it is also smoothing out the discontinuous function that we are approximating to the point where at $t^n = 90$ it appears that we are really advecting the bell curve again. When we compare the results here to those in Fig. 2.45, we see that the wiggles have been smoothed out, the numerical solution is in phase with the true state, but it does appear more like the bell curve than a step function.

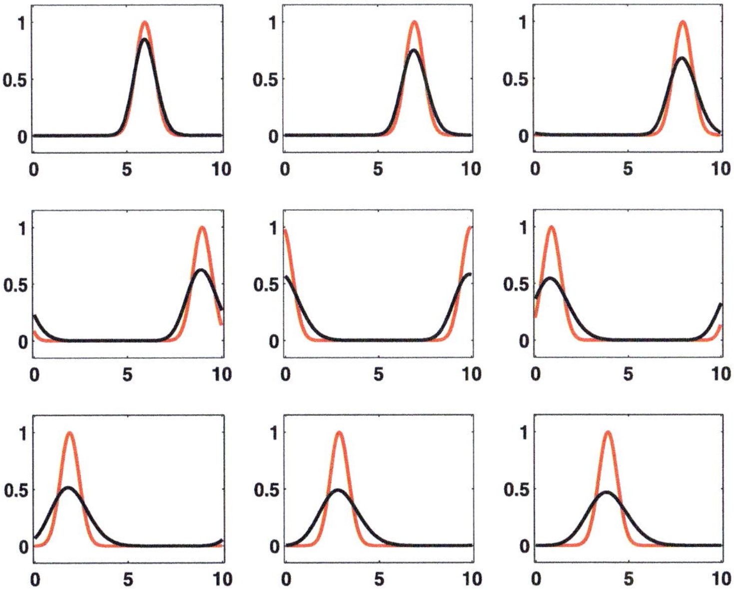

FIGURE 2.42 Plot of the true bell curve, in red, and the numerical solution from the implicit backward Euler finite difference scheme for $tn = 10\Delta t, 20\Delta t, \ldots, 90\Delta t$ steps, for $\Delta t = 0.1$, $\Delta x = 0.1$, $u = 1$, which implies a Courant number of 1.

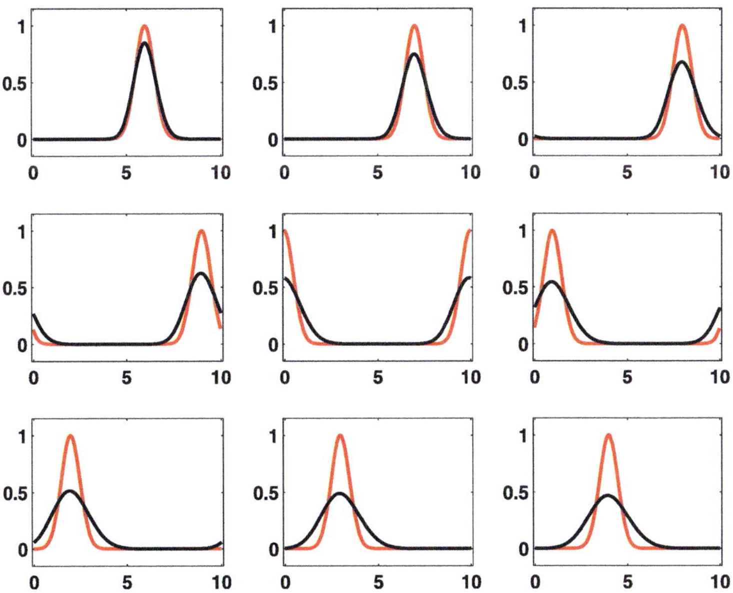

FIGURE 2.43 Plot of the true bell curve, in red, and the numerical solution from the implicit backward Euler finite difference scheme for $tn = 10\Delta t, 20\Delta t, \ldots, 90\Delta t$ steps, for $\Delta t = 0.1$, $\Delta x = 0.01$, $u = 1$, which implies a Courant number of 10.

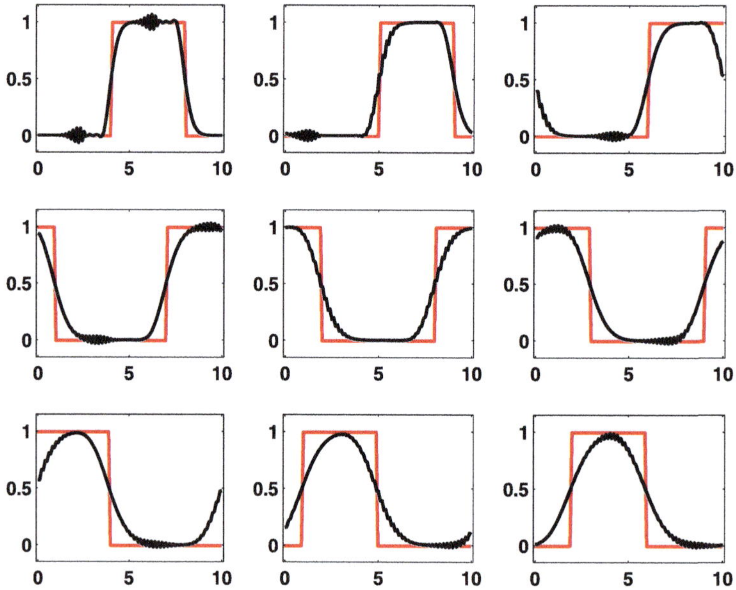

FIGURE 2.44 Plot of the true step function, in red, and the numerical solution from the implicit backward Euler finite difference scheme for $tn = 10\Delta t, 20\Delta t, \ldots, 90\Delta t$ steps, for $\Delta t = 0.2$, $\Delta x = 0.1$, $u = 1$, which implies a Courant number of 1.

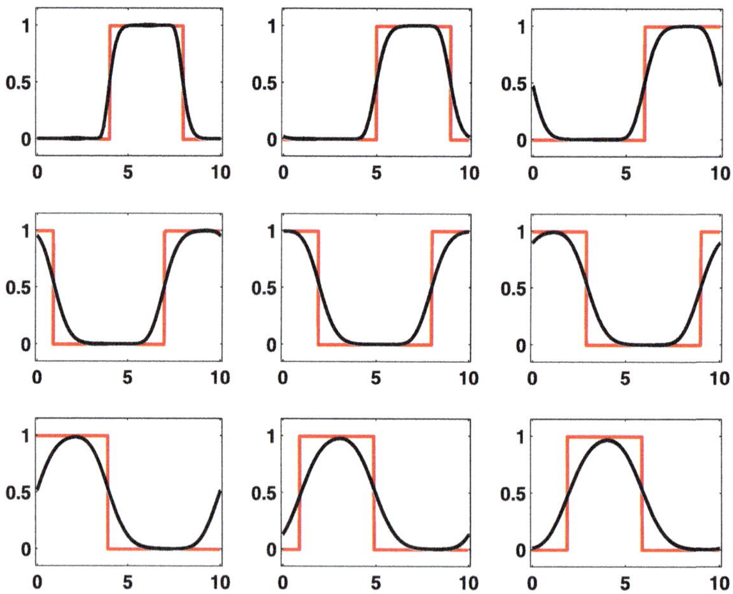

FIGURE 2.45 Plot of the true step function, in red, and the numerical solution from the implicit backward Euler finite difference scheme for $tn = 10\Delta t, 20\Delta t, \ldots, 90\Delta t$ steps, for $\Delta t = 0.2$, $\Delta x = 0.01$, $u = 1$, which implies a Courant number of 10.

2.3.2 Crank–Nicolson scheme

The Crank–Nicolson (CN) scheme is more associated with the heat equation, which we shall not go into details here, however, see [61] for more details, while the original theory of the heat equation application can be found in [41]. The main idea of the CN scheme is to take an average of the spatial gradient at time $t = t^n$ and $t = t^{n1}$. Thus the continuous form of the CN scheme to the advection equation is given by

$$\frac{\psi_i^{n+1} - \psi_i^n}{\Delta t} = -\frac{u}{2}\left(\left.\frac{\partial \psi}{\partial x}\right|_i^n + \left.\frac{\partial \psi}{\partial x}\right|_i^{n+1} \right), \tag{2.82}$$

where $\left.\frac{\partial \psi}{\partial x}\right|_i^n$ is referring to the central difference approximation to the first spatial derivative at time n, centered at x_i. Approximating the continuous derivatives in (2.82) with central differences as mentioned results in

$$\frac{\psi_i^{n+1} - \psi_i^n}{\Delta t} = -\frac{u}{2}\left(\frac{\psi_{i+1}^n - \psi_{i-1}^n}{2\Delta x} + \frac{\psi_{i+1}^{n+1} - \psi_{i-1}^{n+1}}{2\Delta x} \right). \tag{2.83}$$

Now collecting terms and recognizing the Courant number where it occurs, results in

$$\psi_i^{n+1} + \frac{C}{4}\psi_{i+1}^{n+1} - \frac{C}{4}\psi_{i-1}^{n+1} = \psi_i^n - \frac{C}{4}\psi_{i+1}^n + \frac{C}{4}\psi_{i-1}^n. \tag{2.84}$$

We have plotted the stencil that the C–N scheme uses in Fig. 2.46.

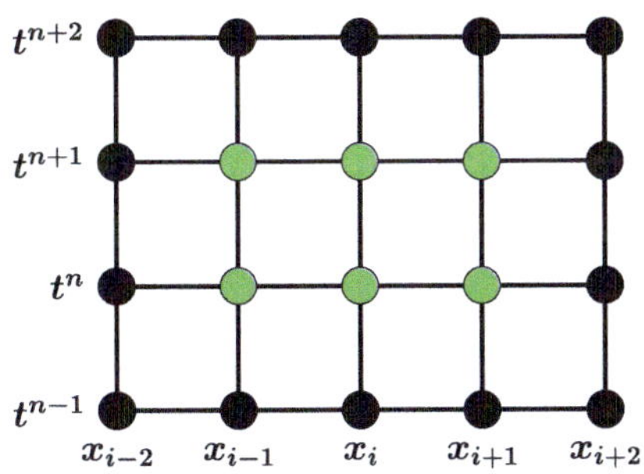

FIGURE 2.46 Illustration of the stencil for the Crank–Nicolson finite difference scheme.

As with the backward Euler approach, this implicit difference equation results in a matrix–vector problem to invert. The CN approach results in a different tridiagonal matrix to that of the backward Euler, but also we now are solving the matrix–vector equation of the form

$\mathbf{A}\boldsymbol{\Psi}^{n+1} = \mathbf{B}\boldsymbol{\Psi}^{n}$, where $\mathbf{A}$ and $\mathbf{B}$ are given by

$$\mathbf{A} = \begin{pmatrix} 1 & \frac{C}{4} & 0 & 0 & \cdots & -\frac{C}{4} \\ -\frac{C}{4} & 1 & \frac{C}{4} & 0 & \cdots & 0 \\ 0 & -\frac{C}{4} & 1 & \frac{C}{4} & \ddots & 0 \\ \vdots & \ddots & \ddots & \ddots & \ddots & \vdots \\ \vdots & \ddots & \ddots & \ddots & \ddots & \vdots \\ \frac{C}{4} & 0 & 0 & \cdots & -\frac{C}{4} & 1 \end{pmatrix}, \quad \mathbf{B} = \begin{pmatrix} 1 & -\frac{C}{4} & 0 & 0 & \cdots & \frac{C}{4} \\ \frac{C}{4} & 1 & -\frac{C}{4} & 0 & \cdots & 0 \\ 0 & \frac{C}{4} & 1 & -\frac{C}{4} & \ddots & 0 \\ \vdots & \ddots & \ddots & \ddots & \ddots & \vdots \\ \vdots & \ddots & \ddots & \ddots & \ddots & \vdots \\ -\frac{C}{4} & 0 & 0 & \cdots & \frac{C}{4} & 1 \end{pmatrix}.$$

2.3.3 Box scheme

The next scheme that we consider is referred to as the **box scheme**, which uses a Crank–Nicholson type approach, but for both first-order derivatives. However, this approach uses upwind approximations for the derivatives, which are given by

$$\begin{aligned} \psi_t &\approx \frac{1}{2}\frac{\left(\psi_j^{n+1} - \psi_j^n\right)}{\Delta t} + \frac{1}{2}\frac{\left(\psi_{j+1}^{n+1} - \psi_{j+1}^n\right)}{\Delta t}, \\ \psi_x &\approx \frac{1}{2}\frac{\left(\psi_{j+1}^{n} - \psi_j^n\right)}{\Delta x} + \frac{1}{2}\frac{\left(\psi_{j+1}^{n+1} - \psi_{j}^{n+1}\right)}{\Delta x}. \end{aligned}$$

Therefore, applying the expressions above as a numerical approximation to (2.16), we have

$$\frac{1}{2}\frac{\psi_j^{n+1} - \psi_j^n}{\Delta t} + \frac{1}{2}\frac{\psi_{j+1}^{n+1} - \psi_{j+1}^n}{\Delta x} = -\frac{u}{2}\left(\frac{\psi_{j+1}^n - \psi_j^n}{\Delta x} + \frac{\psi_{j+1}^{n+1} - \psi_j^{n+1}}{\Delta x}\right). \tag{2.85}$$

As before, collecting like terms and identifying the Courant number results in the implicit difference equation

$$(1 - C)\,\psi_j^{n+1} + (1 - C)\,\psi_{j+1}^{n+1} = (1 + C)\,\psi_j^n + (1 - C)\,\psi_{j+1}^n. \tag{2.86}$$

While the box scheme uses some of the techniques from the Crank–Nicholson scheme, it only results in an upper bidiagonal matrix to invert, which is given by

$$\mathbf{A} = \begin{pmatrix} (1-C) & (1+C) & 0 & \cdots & \cdots & 0 \\ 0 & (1-C) & (1+C) & 0 & \cdots & 0 \\ 0 & 0 & (1-C) & (1+C) & \ddots & 0 \\ \vdots & \vdots & \ddots & \ddots & \ddots & \vdots \\ \vdots & \vdots & \ddots & \ddots & (1-C) & (1+C) \\ (1+C) & 0 & \cdots & \cdots & 0 & (1-C) \end{pmatrix}. \tag{2.87}$$

We should note that if $u < 0$ then we would adapt the scheme above to be a downwind approximation to the spatial derivatives, and not an upwind as presented here. The stencil

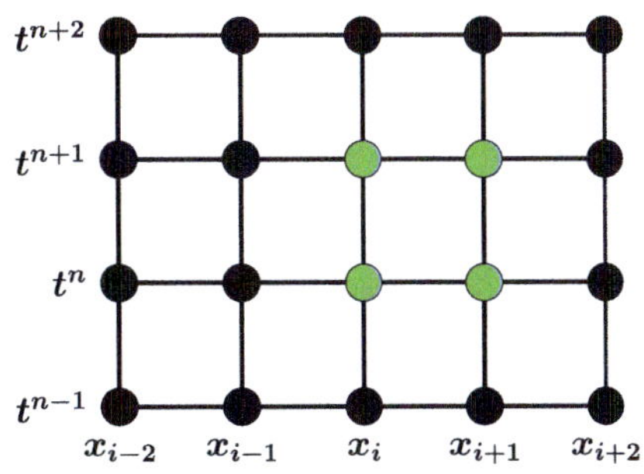

FIGURE 2.47 Illustration of the stencil for the box finite difference scheme.

for the box scheme is shown in Fig. 2.47 where we can see the subtle differences compared to the stencil for the Crank–Nicholson scheme.

2.3.4 Adams–Moulton methods

As the first name in the title of this subsection suggests, we are now going to consider a family of methods that are based on the same arguments as for the Adams–Bashforth schemes, but must be implicit in some way as they are in the implicit part of the chapter. The Adams–Moulton (AM) schemes appear by fitting an interpolant to f through t^n to t^{n+1}, instead of t^n to t^{n-1}. If we consider the second-order AM approximation using Lagrange interpolants again, then we have that

$$P_2(t) = f\left(t^{n+1}, \psi^{n+1}\right)\frac{t-t^n}{t^{n+1}-t^n} + f\left(t^n, \psi^n\right)\frac{t-t^{n+1}}{t^n-t^{n+1}}. \tag{2.88}$$

We know that the difference between two time levels is Δt, so substituting this into (2.88) gives us

$$P_2(t) = f\left(t^{n+1}, \psi^{n+1}\right)\frac{t-t^n}{\Delta t} + f\left(t^n, \psi^n\right)\frac{t^{n+1}-t}{\Delta t}. \tag{2.89}$$

Integrating (2.89) between $t = t^n$ and $t = t^{n+1} = t^n + \Delta t$, we have

$$\begin{aligned}
\int_{t^n}^{t+\Delta t} P(t)\,dt &= \int_{t^n}^{t^n+\Delta t} f\left(t^{n+1}, \psi^{n+1}\right)\frac{t-t^n}{\Delta t} + f\left(t^n, \psi^n\right)\frac{t^{n+1}-t}{\Delta t}dt \\
&= \left[\frac{\left(t^n+\Delta t\right)t - \frac{t^2}{2}}{\Delta t}\right]_{t^n}^{t^n+\Delta t} f\left(t^{n+1}, \psi^{n+1}\right) + \left[\frac{\frac{t^2}{2} - tt^n}{\Delta 2}\right]_{t^n}^{t^n+\Delta t} f\left(t^n, \psi^n\right) \\
&= \frac{\Delta t}{2} f\left(t^{n+1}, \psi^{n+1}\right) + \frac{\Delta t}{2} f\left(t^n, \psi^n\right).
\end{aligned} \tag{2.90}$$

Thus the second-order Adams–Moulton scheme's implicit difference equation is given by

$$\psi_i^{n+1} = \psi_i^n + \Delta t \left[\frac{1}{2} f\left(t^{n+1}, \psi^{n+1}\right) + \frac{1}{2} f\left(t^n, \psi^n\right) \right]. \tag{2.91}$$

If we were to substitute $\frac{\partial \psi}{\partial x}$ and apply a central difference to the derivative at both time levels, then we would see that we have the **Crank–Nicholson (trapezoidal)** scheme. If we were to apply an upwind approximation for the spatial derivative at both time levels, then we would have the box scheme.

We now consider how to derive the third-order Adams–Moulton scheme by interpolating between t^{n-1}, t^n, and t^{n+1}. The starting point is to determine the Lagrange polynomials for each term, which can easily be shown to be

$$\begin{aligned} P_3(t) &= \frac{\left(t - t^{n+1}\right)\left(t - t^n\right)}{\left(t^{n-1} - t^{n+1}\right)\left(t^{n-1} - t^n\right)} f\left(t^{n-1}, \psi^{n-1}\right) + \frac{\left(t - t^{n+1}\right)\left(t - t^{n-1}\right)}{\left(t^n - t^{n+1}\right)\left(t^n - t^{n-1}\right)} f\left(t^n, \psi^n\right) \\ &+ \frac{\left(t - t^n\right)\left(t - t^{n-1}\right)}{\left(t^{n+1} - t^n\right)\left(t^{n+1} - t^{n-1}\right)} f\left(t^{n+1}, \psi^{n+1}\right). \end{aligned} \tag{2.92}$$

Upon integrating the three terms in the polynomial in (2.92), lots of cancelations will occur, and the final expressions for the third-order AM scheme can be shown to be

$$\psi_i^{n+1} = \psi^n + \frac{\Delta t}{12}\left(5f\left(t^{n+1}, \psi^{n+1}\right) + 8f\left(t^n, \psi^n\right) - f\left(t^{n-1}, \psi^{n-1}\right)\right). \tag{2.93}$$

The stencil for (2.93) with a central difference applied to the spatial derivatives is shown in Fig. 2.48, where we can see that the time steps appear to be leap-frogging over the current time, and we are using information from the time step before the current time step, along with that where we are seeking the value at. Also note here that the matrix equation now becomes $\mathbf{A}\boldsymbol{\Psi}^{n+1} = \mathbf{B}\boldsymbol{\Psi}^n + \mathbf{D}\boldsymbol{\Psi}^{n-1}$, as we are using the information about the field at the previous time step as well. The matrices **A** and **B** will be full diagonal matrices, which means that there are values in all three components, the matrix **D** does not have any entries on the diagonal, but does on the off-diagonals. The entries for these three matrices are

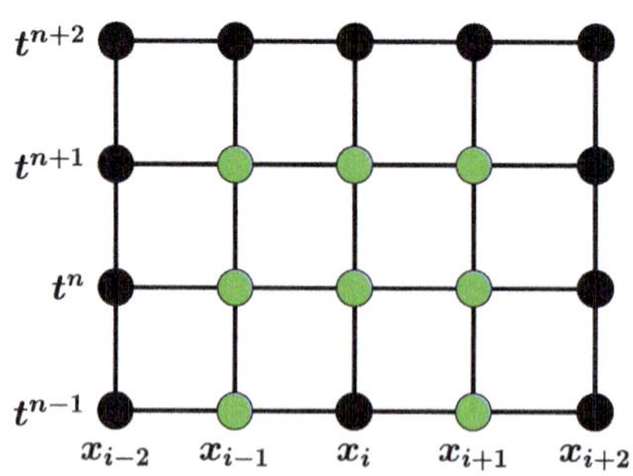

FIGURE 2.48 Illustration of the stencil for the three-level Adams–Moulton finite difference scheme.

$$
\begin{aligned}
a_{i,i} &= 1, & b_{i,i} &= 1, & d_{i,i} &= 0, \\
a_{i-1,i} &= -\frac{5C}{24}, & b_{i-1,i} &= \frac{C}{3}, & d_{i-1,i} &= -\frac{C}{24}, \\
a_{i+1,i} &= \frac{5C}{24}, & b_{i+1,i} &= -\frac{C}{3}, & d_{i+1,i} &= \frac{C}{24}.
\end{aligned} \tag{2.94}
$$

Exercise 2.2. *Verify that the expression in* (2.93) *is the result of integrating* (2.92) *between* t^n *and* $t^n + \Delta t$.

2.3.5 Backward differentiation formula

The last of the family of implicit methods that we consider here are referred to as the **backward differentiation formula (BDF)**, and, as their name suggests, we will be applying differentiation rather than integration. The derivation is similar to the Adams–Bashforth and Adams–Moulton schemes, but now we apply a polynomial approximation to $\psi(x,t)$ rather than to $\psi_x(x,t)$.

If we consider the first-order approach first, then the simplest polynomial that we can apply is $P_1(t) = At + B$. Therefore, substituting t^n and t^{n+1}, we have that

$$\psi^n = At^n + B, \tag{2.95a}$$

$$\psi^{n+1} = At^{n+1} + B. \tag{2.95b}$$

We now use the condition that the first time derivative of the polynomial must satisfy the original differential equation at time t^{n+1}, which then means that $P_1'\left(t^{n+1}\right) = f\left(t^{n+1}, \psi^{n+1}\right)$, thus $P_1' = A = f\left(t^{n+1}, \psi^{n+1}\right)$. If we now subtract (2.95a) from (2.95b), then we have

$$
\begin{aligned}
A\left(t^{n+1} - t^n\right) &= \psi^{n+1} - \psi^n, \\
A\Delta t &= \psi^{n+1} - \psi^n \Rightarrow A = \frac{\psi^{n+1} - \psi^n}{\Delta t}.
\end{aligned}
$$

We now have two expressions for A, and therefore they must be equal. Thus equating them and rearranging results in

$$\psi^{n+1} = \psi^n + \Delta t f\left(t^{n+1}, \psi^{n+1}\right),$$

which is the backward Euler difference equation. For the advection application, we would apply a central difference for the spatial derivative that would be the function f above, at time level t^{n+1}.

For the second-order methods, which are quite often referred to as BDF2, we again consider the second-order Lagrange interpolation polynomial, but now applied to ψ, between t^n and t^{n+1} which yields

$$P_2(t) = \frac{(t-t^n)(t-t^{n+1})}{(t^{n-1}-t^n)(t^{n-1}-t^{n+1})}\psi^{n-1} + \frac{(t-t^{n+1})(t-t^{n-1})}{(t^n-t^{n+1})(t^n-t^{n-1})}\psi^n + \frac{(t-t^n)(t-t^{n-1})}{(t^{n+1}-t^n)(t^{n+1}-t^{n-1})}\psi^{n+1}. \tag{2.96}$$

The next step is to differentiate (2.96) with respect to t, which produces

$$\frac{dP_2(t)}{dt} = \frac{2t-t^n-t^{n+1}}{(-\Delta t)(-2\Delta t)}\psi^{n-1} + \frac{2t-t^{n+1}-t^{n-1}}{(-\Delta t)(\Delta t)}\psi^n + \frac{2t-t^n-t^{n-1}}{(\Delta t)(2\Delta t)}\psi^{n+1}. \tag{2.97}$$

Now we have to evaluate the derivative in (2.97) at $t = t^{n+1}$, which results in

$$\left.\frac{dP_2(t)}{dt}\right|_{t=t^{n+1}} = \frac{1}{2\Delta t}\psi^{n-1} - \frac{2}{\Delta t}\psi^n + \frac{3}{2\Delta t}\psi^{n+1}. \tag{2.98}$$

The final step is to set (2.98) equal to $f\left(t^{n+1}, \psi^{n+1}\right)$ and rearrange to obtain

$$\psi^{n+1} = \frac{4}{3}\psi^n - \frac{1}{3}\psi^{n-1} + \frac{2\Delta t}{3} f\left(t^{n+1}, \psi^{n+1}\right). \tag{2.99}$$

We have presented the stencil for the second-order backward differentiation formula, which has been combined with a central difference approximation for the spatial derivative in the advection problem in Fig. 2.49, where we can see that it is quite different from the other implicit methods where we are not using points on either side of the central point except at $t = t^{n+1}$. Therefore we again have a tridiagonal matrix to invert to be able to implement this implicit finite difference method.

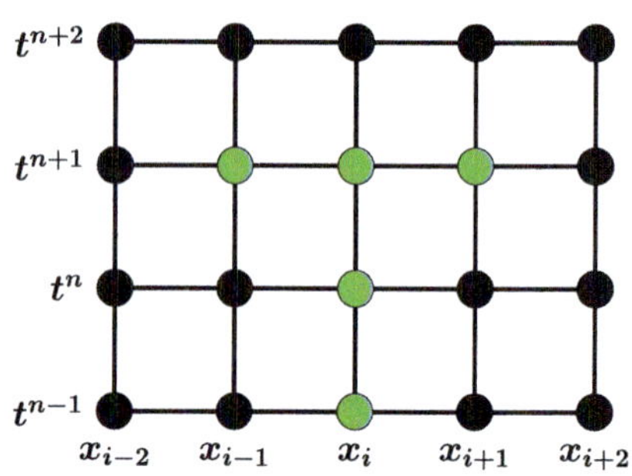

FIGURE 2.49 Illustration of the stencil for the three level backward differentiation formula finite difference scheme.

Exercise 2.3. *By using the third-order Lagrange polynomial in the interval* $t \in \left[t^{n-2}, t^{n+1}\right]$ *show that the difference equation for the third-order backward differentiation formula (BDF3) is*

$$\psi^{n+1} = \frac{2}{11}\psi^{n-2} - \frac{9}{11}\psi^{n-1} + \frac{18}{11}\psi^n + \frac{6\Delta t}{11} f\left(t^{n+1}, \psi^{n+1}\right).$$

Exercise 2.4. *By using the fourth-order Lagrange polynomial in the interval* $t \in \left[t^{n-3}, t^{n+1}\right]$ *show that the difference equation for the fourth-order backward differentiation formula (BDF4) is*

$$\psi^{n+1} = -\frac{3}{25}\psi^{n-3} + \frac{16}{25}\psi^{n-2} - \frac{36}{25}\psi^{n-1} + \frac{48}{25}\psi^{n} + \frac{12\Delta t}{25} f\left(t^{n+1}, \psi^{n+1}\right).$$

Given the multistep methods that we have presented, namely Adams–Bashforth, Adams–Moulton, and the backward differentiation formulas, it is possible to write a general expression relating the parameters α_k and β_k for a polynomial of order s, namely

$$\sum_{k=0}^{s} \alpha^k \psi^{n+1-k} = \sum_{k=0}^{s} \beta^k f^{n+1-k}, \tag{2.100}$$

where for the Adams–Bashforth approaches we have $\alpha^k = 0$, for $k > 1$ with $\beta^0 = 0$, whereas for the Adams–Moulton methods we have $\alpha^k = 0$ for $k > 1$, but now $\beta^0 \neq 0$, and finally, for the backward differentiation formula we have $\beta^k = 0$ for $k > 0$.

2.3.6 Implicit Runge–Kutta methods

We have seen in this section that it is possible to derive implicit schemes by either integrating or differentiating polynomial approximations to either the function f or the field ψ. Another family of implicit schemes are the so-called **implicit Runge–Kutta** schemes. There are many different approaches for deriving implicit Runge–Kutta methods, but we shall consider two different approaches for what are referred to as one-stage, second-order schemes, where one approach is based upon the midpoint integration technique, while the other is based upon the trapezoidal approach, and then present one approach for the **collocation based** implicit Runge–Kutta schemes.

Before we move on to the derivations of some of the implicit Runge–Kutta difference equations, we recall the definition of the general s-stage Runge–Kutta difference equation, which is

$$\psi^{n+1} = \psi^n + \Delta t \sum_{i=1}^{s} b_i k_i, \qquad k_i = f\left(t^n + c_i \Delta t, \psi^n + \Delta t \sum_{j=1}^{s} a_{ij} k_j\right), \tag{2.101}$$

where each of the methods to be presented will be summarized in terms of the a_{ij}, c_i, and b_i in (2.101), and for the more complicated methods, we will have their associated Butcher tableau as well.

Implicit midpoint derivation

The midpoint approximation is a slight refinement of the forward Euler method that we saw earlier, where the derivative is approximated by

$$\psi'\left(t^n\right) \approx \frac{\psi\left(t^n + \Delta t\right) - \psi\left(t^n\right)}{\Delta t}, \tag{2.102}$$

and remember that $\psi' = f(t^n, \psi^n)$. The midpoint approximation replaces (2.102) with the expression

$$\psi'\left(t^n + \frac{\Delta t}{2}\right) \approx \frac{\psi(t^n + \Delta t) - \psi(t^n)}{h}, \tag{2.103}$$

where upon rearranging (2.103) we obtain

$$\psi(t^n + \Delta t) \approx \psi(t^n) + h\psi'\left(t^n + \frac{\Delta t}{2}\right) = \psi(t^n) + \Delta t f\left(t^n + \frac{\Delta t}{2}, \psi\left(t^n + \frac{\Delta t}{2}\right)\right). \tag{2.104}$$

Obviously, we do not know $\psi^{n+\frac{1}{2}}$, and a particular choice of approximation would define either an explicit or an implicit scheme. The explicit scheme appears by considering the Taylor series expansion of $\psi\left(t^n + \frac{\Delta t}{2}\right)$ about t^n and using the fact that $\psi' = f$.

To obtain the implicit midpoint difference equation, we use a line segment from ψ^n to ψ^{n+1}, and then take the midpoint along that line as

$$\psi\left(t^n + \frac{\Delta t}{2}\right) \approx \frac{1}{2}\left(\psi(t^n) + \psi(t^n + \Delta t)\right). \tag{2.105}$$

Now recalling (2.103) and using (2.105), we arrive at

$$\frac{\psi(t^n + \Delta t) - \psi(t^n)}{\Delta t} \approx \psi'\left(t^n + \frac{\Delta t}{2}\right) \approx k = f\left(t^n + \frac{\Delta t}{2}, \frac{1}{2}\left(\psi(t^n) + \psi(t^n + \Delta t)\right)\right). \tag{2.106}$$

Finally, by rearranging the left-hand side of (2.106), we obtain $\psi(t^n + \Delta t) = \psi(t^n) + \Delta t k$, where, upon substituting this expression in to the right-hand side of (2.106), we obtain the implicit midpoint difference equation which is also an implicit second-order Runge–Kutta method as

$$\psi^{n+1} = \psi^n + \Delta t k, \tag{2.107a}$$

$$k = f\left(t^n + \frac{\Delta t}{2}, \psi^n + \frac{\Delta t}{2}k\right). \tag{2.107b}$$

Comparing (2.107a) and (2.107b) with the general expression for an implicit Runge–Kutta difference equation in (2.101), we have

$$c_1 = \frac{1}{2}, \qquad a_{11} = \frac{1}{2}, \qquad b_1 = 1. \tag{2.108}$$

Implicit trapezoidal derivation

This approach is actually a collocation based method, which we shall study in more details soon, but here the starting point is to define two collocation points, where we have $c_1 = t^n$ and $c_1 = t^n + \Delta t$, and now we are going to fit a second-order polynomial $p(t) = \alpha(t - t^n)^2 + \beta(t - t^n) + \gamma$ that satisfies

$$p(t^n) = \psi^n,$$

$$\begin{aligned} p'\left(t^n\right) &= f\left(t^n, p\left(t^n\right)\right), \\ p'\left(t^n + \Delta t\right) &= f\left(t^n + \Delta t, p\left(t^n + \Delta t\right)\right). \end{aligned}$$

If we take the first condition above then we see that $\gamma = \psi^n$. Now if we differentiate the polynomial and evaluate it at $t = t^n$, then we have that $\beta = f(t^n, p(t^n))$, and finally by using the third condition above, along with the value for β and the fact that $t^n + \Delta t - t^n = \Delta t$, yields the expression for α as $\alpha = \frac{1}{2\Delta t}(f(t^n + \Delta t, p(t^n + \Delta t)) - f(t^n, p(t^n)))$. Thus we have an implicit difference equation given by

$$\psi^{n+1} = p\left(t^n + \Delta t\right) = \psi^n + \frac{\Delta t}{2}\left(f\left(t^n + \Delta t, \psi^{n+1}\right) - f\left(t^n, \psi^n\right)\right). \tag{2.109}$$

Now comparing the expression in (2.109) with the general expression for the implicit Runge–Kutta difference equation in (2.101), we have

$$c_0 = 0 \qquad c_1 = 1, \qquad a_{11} = 1, \qquad b_1 = \frac{1}{2}, \qquad b_2 = \frac{1}{2}. \tag{2.110}$$

Collocated implicit Runge–Kutta schemes

As just mentioned, the implicit trapezoidal scheme is an implicit collocated based Runge–Kutta scheme, but it is possible to build more advanced implicit schemes using more accurate forms of quadrature rules to integrate the ODE. One of the very well known collocated schemes involves the quadrature rule of Gauss–Legendre. The collocated part in the name here is referring to the locations of the time steps between t^n and t^{n+1} we are going to evaluate the function f and the field ψ at, such that the accuracy of the integration is improved.

There are several different quadrature rules that have been discovered and applied to the implicit Runge–Kutta scheme, where a more advance and detailed explanation of these schemes can be found in [27].

The Legendre polynomials are the polynomial solutions $P_n(x)$ to Legendre's differential equation

$$\frac{d}{dx}\left[\left(1 - x^2\right)\frac{dP_n(x))}{dx}\right] + n(n+1)P_n(x) = 0, \tag{2.111}$$

with $n \geq 0$ and with the convention $P_n(1) = 1$. The polynomials $P_n(x)$ form a sequence of orthogonal polynomials of degree n. The Legendre polynomials can also be derived by using **Rodrigues' formula** given by

$$P_n(x) = \frac{1}{2^n n!}\frac{d^n}{dx^n}\left(x^2 - 1\right)^n. \tag{2.112}$$

However, for the application to get the implicit Runge–Kutta formula, we require the equation for the shifted Legendre polynomials, which for order s is given by

$$P_s(2t - 1) = \frac{1}{s!}\frac{d^s}{dt^s}\left(t^s(t-1)^n\right). \tag{2.113}$$

If we now consider a two step approach, then we require $s=2$ and also the roots of the polynomial

$$P_2(2t-1)=\frac{1}{2!}\frac{d^2}{dt^2}\left(t^2(t-1)^2\right)\equiv\frac{1}{2}\frac{d^2}{dt^2}\left(t^4-2t^3+t^2\right).$$

Upon differentiating the expression on the far right-hand side, we need to find the roots of the equation

$$6t^2-6t+1=0.$$

Applying the standard quadratic equation solver yields the following two roots:

$$c_1=\frac{1}{2}-\frac{\sqrt{3}}{6},\qquad c_2=\frac{1}{2}+\frac{\sqrt{3}}{6},\tag{2.114}$$

which are now our collocation points in time, and form the left-hand side of the Butcher tableau for this method.

Now that we have the collocation points in time, we introduce the following definition and theorem to enable us to find the a_{ij}s and the b_is.

Definition 2.5. Let $c_1, c_2, \ldots, c_s$ be distinct real numbers in $[0, 1]$. The **collocation polynomial** $p(t)$ is a polynomial of degree s satisfying

$$p(t_0)=\psi_0,\tag{2.115a}$$

$$p(t_0+c_i\Delta t)=f(t_0+c_i\Delta t, p(t_0+c_i\Delta t)),\quad i=1,2,\ldots,s,\tag{2.115b}$$

and the numerical solution of the associated **collocation method** is defined by $\psi_1 = p(t_0+\Delta t)$.

Theorem 2.6. *The collocation method in the definition above is equivalent to the s-stage Runge–Kutta method with coefficients*

$$a_{i,j}=\int_0^{c_i} l_j dt,\tag{2.116a}$$

$$b_i=\int_0^1 l_i(t)\,dt,\tag{2.116b}$$

$$l_i=\prod_{k\neq i}\frac{t-c_k}{c_i-c_k}.\tag{2.116c}$$

Returning to the c_is that we derived earlier, we now can calculate the associated a_{ij}s and b_is, but also need to find the two Lagrange polynomials, which are given by

$$l_1=\frac{t-c_2}{c_1-c_2}=\frac{t-c_2}{-\frac{\sqrt{3}}{3}},\qquad l_2=\frac{t-c_1}{c_2-c_1}=\frac{t-c_1}{\frac{\sqrt{3}}{3}}.$$

Therefore, it can be shown that

$$b_1 = \frac{1}{2}, \qquad b_2 = \frac{1}{2}. \tag{2.117}$$

There are four a_{ij}s that we are seeking:

$$a_{11} = \int_0^{c_1} l_1 \mathrm{d}t, \quad a_{12} = \int_0^{c_1} l_2 \mathrm{d}t, \tag{2.118}$$

$$a_{21} = \int_0^{c_2} l_1 \mathrm{d}t, \quad a_{12} = \int_0^{c_2} l_2 \mathrm{d}t, \tag{2.119}$$

which can be shown to be equal to

$$a_{11} = \frac{1}{4}, \qquad a_{12} = \frac{1}{4} - \frac{\sqrt{3}}{6}, \tag{2.120}$$

$$a_{21} = \frac{1}{4} + \frac{\sqrt{3}}{6}, \quad a_{22} = \frac{1}{4}. \tag{2.121}$$

Thus the associated Butcher tableau for this scheme is given by

$$\begin{array}{c|cc}
\frac{1}{2} - \frac{\sqrt{3}}{6} & \frac{1}{4} & \frac{1}{4} - \frac{\sqrt{3}}{6} \\
\frac{1}{2} + \frac{\sqrt{3}}{6} & \frac{1}{4} + \frac{\sqrt{3}}{6} & \frac{1}{4} \\
\hline
 & \frac{1}{2} & \frac{1}{2}
\end{array}. \tag{2.122}$$

Expanding the stages from the Butcher tableau above results in the two stage implicit Runge–Kutta scheme

$$\psi^{n+1} = \psi^n + \frac{\Delta t}{2}\left(k_1 + k_2\right), \tag{2.123a}$$

$$k_1 = f\left(t^n + \left(\frac{1}{2} - \frac{\sqrt{3}}{6}\right)\Delta t, \psi^n + \Delta t\left(\frac{1}{4}k_1 + \left(\frac{1}{4} - \frac{\sqrt{3}}{6}\right)k_2\right)\right), \tag{2.123b}$$

$$k_2 = f\left(t^n + \left(\frac{1}{2} + \frac{\sqrt{3}}{6}\right)\Delta t, \psi^n + \Delta t\left(\left(\frac{1}{4} + \frac{\sqrt{3}}{6}\right)k_1 + \frac{1}{4}k_2\right)\right). \tag{2.123c}$$

This is quite obviously an implicit scheme that is presented in (2.123a), (2.123b), and (2.123c), but as we shall see in the next chapter, there are some very important properties of these schemes that are desirable, even if the schemes are difficult to evaluate.

There are other quadrature rules that have been used to derive implicit Runge–Kutta schemes and the reader is referred to the excellent book by Professor John Butcher titled *"Numerical Methods for Ordinary Differential Equations"*, [27] is the most up-to-date version.

2.3.7 Diagonally implicit Runge–Kutta schemes (DIRK)

In this subsection we shall summarize the very well written paper by Roger Alexander from 1977 [2]. We start by stating the problem that we wish to solve, which is to find an approximation to the solution $\psi : [0, T] \to \mathbb{R}^m$ of an initial value problem

$$\psi' = f(\psi, t), \quad \psi(0) = \psi_0, \tag{2.124}$$

in which the function f is defined as $f : [0, T] \times \mathbb{R}^m \to \mathbb{R}^m$. As we have mentioned before, the objective of Runge–Kutta methods is to move from t^n to $t^{n+1} = t^n + \Delta t$, by approximating the integral equation

$$\psi\left(t^{n+1}\right) = \psi\left(t^n\right) + \int_{t^n}^{t^{n+1}} f(\psi(t), t)\, \mathrm{d}t. \tag{2.125}$$

Runge–Kutta methods, as we have just seen, select quadrature points, $c_1, c_2, \ldots, c_q$, and the associated weights, $b_1, b_2, \ldots, b_q$, and use the quadrature rule

$$\psi\left(t^{n+1}\right) = \psi\left(t^n\right) + \Delta t \sum_{i=1}^{q} b_i f\left(t^n + c_i \Delta t, \psi\left(t^n + c_i \Delta t\right)\right). \tag{2.126}$$

We now introduce the notation $t_i^n \equiv t^n + c_i \Delta t$, as we have multiple time locations involved in the Runge–Kutta methods. We assume that we already have a value for ψ^n to $\psi(t^n)$, but to be able to use (2.126) we require values for ψ_i^n to use for $\psi\left(t_i^n\right)$. These values are also computed applying a quadrature rule on the same time nodes and are defined as

$$\psi_i^n = \psi^n + \Delta t \sum_{j=1}^{q} a_{ij} f\left(t_i^n, \psi_i^n\right), \quad i = 1, 2, \ldots, q. \tag{2.127}$$

As we have seen earlier, this is a set of implicit equations that have to be solved and used in (2.126) for the next value of ψ,

$$\psi^{n+1} = \psi^n + \Delta t \sum_{i=1}^{q} b_i f\left(t_i^n, \psi_i^n\right). \tag{2.128}$$

Here we make the remark that implicit Runge–Kutta methods were introduced by John Butcher in 1963 [23], and further developed over a series of papers [24–26].

The implicit Runge–Kutta schemes that were introduced by Butcher in [24,25] have a problem in that they require us to able to integrate a system of m differential equations, we need to solve mq simultaneous implicit, and in general probably highly nonlinear, equations at each time step. A way to overcome this is to use a lower-triangular matrix for the a_{ij}s in the Butcher tableau. Thus Eq. (2.127) may be solved in q successive stages with only an m-dimensional system to be solved at each stage. Butcher and Alexander refer to such schemes as **semi-implicit**.

There were several investigations of the semi-implicit Runge–Kutta methods in the late 1960s, and for a set of references to these studies, please see [2]. However, in [2] the author introduces a Newton type iteration to solve linear systems at each stage with a coefficient matrix of

$$I - \Delta t a_{ii} \frac{\partial f}{\partial \psi}. \tag{2.129}$$

Alexander then states that if all the diagonal entries in **A**, i.e., a_{ii}, are equal, then the hope is to repeatedly use a stored lower–upper LU-factorization of a single matrix **M**. When the semi-implicit Runge–Kutta scheme has this property, it is stated in [2] that these schemes will be referred to as diagonally implicit Runge–Kutta, or DIRK, formulas. There are some very nice properties that the DIRK numerical approximations have, and we shall go into more details about them in the next chapter. However, to finish this subsection, we shall present the Butcher tableau for two different versions of DIRK for a two- and a three-step approach that are from [2]:

$$q=2 \qquad \begin{array}{c|cc} \frac{1}{2}+\frac{1}{2\sqrt{3}} & \frac{1}{2}+\frac{1}{2\sqrt{3}} & 0 \\ \frac{1}{2}-\frac{1}{2\sqrt{3}} & -\frac{1}{\sqrt{3}} & \frac{1}{2}+\frac{1}{2\sqrt{3}} \\ \hline & \frac{1}{2} & \frac{1}{2} \end{array}, \tag{2.130}$$

$$q=3 \qquad \begin{array}{c|ccc} \frac{(1+\alpha)}{2} & \frac{(1+\alpha)}{2} & 0 & 0 \\ \frac{1}{2} & -\frac{\alpha}{2} & \frac{(1+\alpha)}{2} & 0 \\ \frac{(1-\alpha)}{2} & 1+\alpha & -(1+2\alpha) & \frac{(1+\alpha)}{2} \\ \hline & \frac{1}{6\alpha^2} & 1-\frac{1}{3\alpha^2} & \frac{1}{6\alpha^2} \end{array}, \tag{2.131}$$

where $\alpha \equiv \dfrac{2\cos\left(\frac{\pi}{18}\right)}{\sqrt{3}}$.

2.4 Predictor–corrector methods

One idea behind the predictor–corrector methods is to use a suitable combination of an explicit and implicit technique to obtain a new method with better convergence characteristics; we shall go into more details about the convergence property in the next chapter.

As we have seen, the implicit methods require values of the function f or the field ψ at a time after t^n, where we do not have values at these times explicitly. We do, however, have explicit methods that can create an *estimate* of the field ψ^{n+1}, but it may not be as *accurate* as solving an implicit method; however, it does avoid both the matrix inversion or the iterative solver for the implicit Runge–Kutta methods.

The first predictor–corrector method we present uses the forward Euler method to create a value for ψ at t^{n+1} which we shall denote $\widetilde{\psi}^{n+1}$ and then uses this value to evaluate the trapezoidal difference equation, which is also the second-order Adams–Moulton scheme. Therefore we would have

$$\widetilde{\psi}_i^{n+1} = \psi_i^n + \Delta t f\left(t^n, \psi^n\right), \tag{2.132}$$

$$\psi_i^{n+1} = \psi_i^n + \frac{1}{2}\Delta t\left(f\left(t^{n+1}, \widetilde{\psi}^{n+1}\right) + f\left(t^n, \psi^n\right)\right), \tag{2.133}$$

where the first step above is the **predictor** step which is effectively creating a *first guess* of the field at the next time step, and then the second step is creating a *correction* to the explicit step; however, we should note here that this scheme is an explicit scheme as we do not have a matrix to invert due to the fact that we have a value for $\widetilde{\psi}^{n+1}$. Also note that we have seen that the forward Euler scheme does have some undesirable properties, and as such we would like to consider some more *accurate* explicit scheme to create the estimate of $\widetilde{\psi}^{n+1}$. The predictor–corrector method presented above is sometimes referred to as the **Euler–Trapezoidal** method.

2.4.1 Adams–Bashforth–Adams–Moulton predictor–corrector

As just mentioned, we have seen that using the forward Euler scheme is not very practical, and as such we require a more accurate method to create the *predictor* step. One such approach is to combine the Adams–Bashforth and Adams–Moulton schemes. These schemes are multistep approaches as they require values of ψ at previous times steps, but as we shall show in the next chapter, these schemes are quite accurate.

There are several versions of the Adams' based methods we could use here, but the scheme we present here uses AB4 for the predictor step and then AM4 for the corrector step, which is given by

$$\widetilde{\psi}^{n+1} = \psi^n + \frac{\Delta t}{24}\left(-9f^{n-3} + 37f^{n-2} - 59f^{n-1} + 55f^n\right),$$

$$\widetilde{f}^{n+1} = f\left(\widetilde{\psi}^{n+1}, t^{n+1}\right),$$

$$\psi^{n+1} = \psi^n + \frac{\Delta t}{24}\left(f^{n-2} - 5f^{n-1} + 19f^n + 9\widetilde{f}^{n+1}\right).$$

2.5 Summary

In this chapter we have presented different numerical approximation to the Eulerian formulation of the advection equation, in the form of explicit, implicit, semi-implicit, and combination of implicit–explicit predictor–corrector methods. We have illustrated for some of the methods their performance with advecting a bell curve and a step function around a periodic domain, but we have not mathematically quantified the features that we see in the performance of these schemes.

We now move on to quantify in the next chapter how good a numerical approximation to the advection equation is, work out if there are restrictions on the step sizes or not, and if the approximations are converging to the true solution. These properties are referred to as **Truncation Error**, **Stability**, **Convergence**, and **Consistency**. We will also discuss how to quantify **dissipation, dispersive**, and **phase errors**.

CHAPTER

3

Stability, consistency, and convergence of Eulerian advection based numerical methods

In the last chapter we introduced a series of explicit and implicit numerical approximations to the one-dimensional advection equation. We investigated a few of the finite difference schemes through advecting a bell curve and a step function around a periodic domain, where we saw some different behaviors for different schemes, for different choices of the time step size, as well as the spatial grid spacing. In this chapter we present the theory to quantify the features we saw and to determine how good the approximations to the advection equation in the different numerical schemes are, discuss why we observed different *wiggles* with some of them, and determine why the forward-time–centered-space scheme performed the way it did. The first feature we quantify is that of how good an approximation is for a numerical scheme, and to do that we have to look at what is referred to as the **truncation error**, which we introduce now.

3.1 Truncation error

We start by considering an easier ordinary differential equation and move up to the advection partial differential equation. This simpler ordinary differential equation is given by

$$\frac{d\psi}{dt} = \lambda\psi, \quad \psi(0) = 1, \tag{3.1}$$

which has the solution $\psi(t) = e^{\lambda t}$. The differential equation in (3.1) has different characteristics as shown in Fig. 3.1, where we see that if $\lambda > 0$, then the solution is growing, while if $\lambda = 0$, then the solution is constant, and finally, if $\lambda < 0$, then the solution is decaying.

An equivalent numerical approximation to (3.1) could be

$$\frac{\psi^{n+1} - \psi^{n}}{\Delta t} = \lambda\psi^{n}. \tag{3.2}$$

Semi-Lagrangian Advection Methods and Their Applications in Geoscience
https://doi.org/10.1016/B978-0-12-817222-3.00007-4

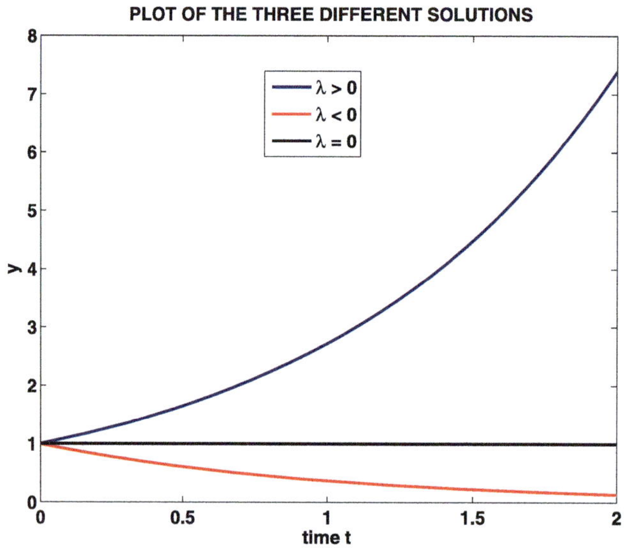

FIGURE 3.1 Plot of the three possible solutions to the differential equation $\psi' = \lambda\psi$.

Given the equations in (3.1) and (3.2), we now consider how to define the **local and global truncation errors**. If we consider the difference equation in (3.2), then we can rewrite it as

$$\psi^{n+1} = (1+\lambda\Delta)\,\psi^n, \tag{3.3}$$

where upon expanding (3.3) as a sequence starting at $t=0$ we have, by induction at the end,

$$\begin{aligned}
\psi^1 &= (1+\lambda\Delta t)\,\psi^0,\\
\psi^2 &= (1+\lambda\Delta t)\,\psi^1 \equiv (1+\lambda\Delta t)(1+\lambda\Delta t)\,\psi^0 = (1+\lambda\Delta t)^2\,\psi^0,\\
&\vdots\\
\psi^n &= (1+\lambda\Delta t)\,\psi^{n-1} \equiv (1+\lambda\Delta t)(1+\lambda\Delta t)\,\psi^{n-2} = \cdots = (1+\lambda\Delta t)^n\,\psi_0,
\end{aligned} \tag{3.4}$$

where the superscript on the ψs is the time index, while that on the $(1+\lambda\Delta t)$ is referring to the power. However, if $\lambda\Delta t < -1$, then the function defined in (3.4) will oscillate as the term will be negative and as such will change sign between the even and odd numbered powers. The aim of the numerical approximation is to satisfy the condition that if we take the limit as $\Delta t \to 0$, then the approximation in (3.4) tends to the solution of (3.1).

The equation that is defined in (3.1) is a linear scalar problem and has the analytical solution $e^{\lambda t}$ as mentioned earlier. The difference equation given in (3.2) is the forward Euler scheme for the time component that we introduced in the last chapter. However, we have to answer the following question: How accurate is the numerical approximation?

When we are considering the accuracy of a numerical scheme, we quantify this through analyzing two types of error: local truncation error and global truncation error. If we consider the Euler method just presented, we have that $e^{\lambda t^{n+1}} = e^{\lambda(t^n+\Delta t)} = e^{\lambda\Delta t}e^{\lambda t^n}$, where the latter relationship here comes by assuming that at the previous time step the estimate of the true state is almost correct and as such we have that $\psi^n \equiv e^{\lambda t^n}$. If we now consider the difference between the true solution $\psi\left(t^{n+1}\right)$ and the numerical approximation valid at the same time, ψ^{n+1}, then

$$\begin{aligned}\tau_{Lfe} \equiv \psi\left(t^{n+1}\right) - \psi^{n+1} &= e^{\lambda t^{n+1}} - (1+\lambda\Delta t)\,\psi^n = e^{\lambda\Delta t}\psi^n - (1+\lambda\Delta t)\,\psi^n \\ &= \left(e^{\lambda\Delta t} - (1+\lambda\Delta t)\right)\psi^n. \end{aligned} \tag{3.5}$$

Taking the Taylor series expansion of $e^{\lambda\Delta t}$, and substituting this expression into (3.5), we can show that the first two terms of the expansion cancel with the two terms from the numerical approximation. Therefore, the error caused by this approximation is $\frac{(\lambda\Delta t)^2}{2!}$. This is referred to as the **local truncation error** and is referring to **the error that is incurred in one time step**. Given the remaining term just mentioned, we would then state that the forward Euler scheme has a local truncation error of order $(\Delta t)^2$, and is denoted by what is referred to as the big O notation, therefore mathematically we would state that $\tau_{Lfe} \sim O\left((\Delta t)^2\right)$.

To recap, the general expression for the local truncation error, τ_L, is defined as

$$\tau_L \equiv \psi\left(t^{n+1}\right) - \psi^{n+1}. \tag{3.6}$$

We next consider how to derive a definition for the **global truncation error**, that is to say, how much the local truncation error is compounded up to the nth time step. Therefore, if we now let

$$L(\psi) \equiv \frac{d\psi}{dt} - \lambda\psi = 0, \tag{3.7a}$$

$$L_{\Delta t}\left(\psi^n\right) \equiv \frac{\psi^{n+1} - \psi^n}{\Delta t} - \lambda\psi^n = 0, \tag{3.7b}$$

then the **global truncation error**, τ, is defined as

$$\tau \equiv L_{\Delta t}(\psi) - L_{\Delta t}\left(\psi^n\right). \tag{3.8}$$

That is to say, we substitute the true state ψ into the finite difference equation and then expand the true state evaluated at the different grid and spacial points about ψ_i^n with respect the increments Δt and Δx.

Returning to the forward Euler approximation to the test ordinary differential equation, we have

$$\tau_{fe} = \frac{\psi\left(x_i, t^{n+1}\right) - \psi\left(x_i, t^n\right)}{\Delta t} - \lambda\psi\left(x_i, t^n\right). \tag{3.9}$$

As we just mentioned, if we now expand the term $\psi\left(x_i, t^n + \Delta t\right)$ about t^n, then we obtain

$$\begin{aligned}\tau_{fe} &= \frac{\psi_i^n + \Delta t \psi_i^{n'} + \frac{(\Delta t)^2}{2!}\psi^{n''} + \cdots - \psi_i^n}{\Delta t} - \lambda \psi_i^n, \\ &= \psi_i^{n'} - \lambda \psi_i^n + \frac{\Delta t}{2!}\psi_i^{n''} + O\left((\Delta t)^2\right) = \frac{\Delta t}{2!}\psi_i^{n''} + O\left((\Delta t)^2\right),\end{aligned} \tag{3.10}$$

which then tells us that the global truncation error for the forward Euler scheme for this differential equation is of order $O(\Delta t)$. We should note that the reason why the first terms have disappeared above is because they represent the original differential equation evaluated at the true state, which is equal to zero.

To be able to consider more difficult functions for the right-hand side of the differential equation, we have to be certain that: first, there exist a solution to the differential equation, and second, it is unique. If we consider the general form of the first order differential equation as

$$\frac{d\psi}{dt} = f(\psi, t), \quad \psi(0) = \psi_0, \tag{3.11}$$

then the important questions to consider here is how can we know that the differential equation in (3.11) has a solution? And is this solution unique? The answer depends upon f. To address these questions, we consider the **Lipschitz condition** which enables us to label the function as *Lipschitz continuous* in ψ over some range t and ψ. This then leads to the definition of the Lipschitz condition as:

Definition 3.1 (Lipschitz condition). If ψ and $\widehat{\psi}$ are any two values then there exists a number L such that

$$\left|\frac{f(\psi, t) - f(\widehat{\psi}, t)}{\psi - \widehat{\psi}}\right| \leq L, \tag{3.12}$$

for $\psi \neq \widehat{\psi}$.

If f is differentiable with respect to ψ and this derivative $f_\psi \equiv \frac{\partial f}{\partial \psi}$ is bounded, then we can take

$$L = \max_{\psi} \left| f_\psi(\psi, t) \right|, \tag{3.13}$$

because

$$f(\psi, t) = f(\widehat{\psi}, t) + f_\psi(\widetilde{\psi}, t)(\psi - \widehat{\psi}), \tag{3.14}$$

for some value $\widetilde{\psi}$ between ψ and $\widehat{\psi}$.

3.1.1 Consistency

Given the expression that we have for the truncation error, we now need to consider whether or not the numerical scheme has the property of what is referred to as being **consistent** with the partial differential equation. The property of consistency is quantified through the following definition:

Definition 3.2. A numerical scheme $L_{\Delta t}\left(\psi_i^n\right) = 0$ for a given partial differential equation $L\left(\psi\left(x,t\right)\right)$ with a time step Δt and grid spacing Δx is said to be **consistent** if the truncation error

$$\tau_{\Delta t}\left(\Delta x, \Delta t\right) \equiv L\left(\psi\right) - L_{\Delta t}\left(\psi_i^n\right) \to 0,$$

satisfies

$$\lim_{\Delta x, \Delta t \to 0} \tau_{\Delta t} = 0. \tag{3.15}$$

If the truncation error of the numerical scheme satisfies the condition above, then that numerical scheme is said to be consistent of order (p, q) provided

$$\tau_{\Delta t}\left(\Delta x, \Delta t\right) = O\left(\left(\Delta x\right)^p + \left(\Delta t\right)^q\right),$$

indicating the order of the first nondisappearing terms in the truncation error, if we were considering the advection equation. If we return to the forward Euler approximation to the test differential equation, we see that it is consistent as the remaining terms in the global truncation error are multiples of Δt and its higher powers and will tend to zero as $\Delta t \to 0$, and that the leading order term in the global truncation error contains Δt, which means that this scheme is first-order in time, $O\left(\Delta t\right)$.

3.1.2 Truncation errors and consistency analysis of the linear multistep methods

We start by recalling from the last chapter that the general form of a linear multistep method is given by

$$\frac{\alpha_k \psi^{n+k} + \alpha_{k-1}\psi^{n+k-1} + \cdots + \alpha_0 \psi^n}{\Delta t} = \beta_k f^{n+k} + \beta_{k-1} f^{n+k-1} + \cdots + \beta_0 f^n. \tag{3.16}$$

- If $k = 1$ then

  $$\frac{\alpha_1 \psi^{n+1} + \alpha_0 \psi^n}{h} = \beta_1 f^{n+1} + \beta_0 f^n,$$

  where we can have

  $$\begin{aligned} \alpha_1 &= 1, \alpha_0 = -1, & \beta_1 &= 0, \beta_0 = 1, \\ \alpha_1 &= 1, \alpha_0 = -1, & \beta_1 &= \frac{1}{2}, \beta_0 = -\frac{1}{2}, \end{aligned}$$

  which then results in the following difference equations:

  $$\begin{aligned} \frac{\psi^{n+1} - \psi^n}{\Delta t} &= f^n, \\ \frac{\psi^{n+1} - \psi^n}{\Delta t} &= \frac{1}{2}\left(f^{n+1} - f^n\right). \end{aligned}$$

- If $k=2$ then

$$\alpha_2=1,\quad \alpha_1=0,\quad \alpha_0=-1,\quad \beta_2=0,\quad \beta_1=2,\quad \beta_0=0,$$

which results in a **midpoint** approximation

$$\frac{\psi^{n+2}-\psi^n}{2\Delta t}=f^{n+1}.$$

Therefore, the question here is how do we choose the values for the αs and βs to obtain the maximum accuracy while avoiding instability? The way to address this question is to use the truncation error, which as we already know is a measure of accuracy.

We now consider the following three schemes:

$$\frac{\psi^{n+1}-\psi^n}{\Delta t} = f^{n+1}, \tag{3.17a}$$

$$\frac{\psi^{n+1}-\psi^n}{\Delta t} = \frac{1}{2}\left(f^n+f^{n+1}\right), \tag{3.17b}$$

$$\frac{\psi^{n+2}-\psi^n}{\Delta t} = f^{n+1}, \tag{3.17c}$$

and for each approximate solution of $\psi'=f(\psi,t)$ we will check the consistency conditions, determine the order of accuracy, and find a bound on the truncation error in terms of a bound on the modulus of a derivative of ψ.

We start with the general expansion of the truncation error given by

$$\tau_n=\frac{C_0}{\Delta t}+C_1\psi+C_2\Delta t\psi'+C_3(\Delta t)^2\psi''+\cdots. \tag{3.18}$$

We can see that the truncation error τ_n can tend to zero only if C_0 and C_1 are equal to zero, and as we know this is the consistency condition. If we now have $C_2=C_3=C_4=0$, up to $C_p=0$ but $C_{p+1}\neq 0$, then

$$\tau_n=C_{p+1}(\Delta t)^p\psi^{(p+1)}+C_{p+2}(\Delta t)^{p+1}\psi^{(p+2)}+\cdots=C_{p+1}(\Delta t)^p\psi^{(p+1)}+O\left((\Delta t)^{p+1}\right). \tag{3.19}$$

Therefore, the first remaining value of the power of Δt, in this case $(\Delta t)^p$, defines the **order of accuracy**. We should recall that the p-coefficient term is defined as

$$C_p=\sum_{i=1}^{k}\left(\frac{i^p}{p!}\alpha_i-\frac{i^{(p-1)}}{(p-1)!}\beta_i\right). \tag{3.20}$$

Finally, we address how to determine the **bound on the truncation error**. We start with the Taylor series expansion of $\psi(t^n+\Delta t)$ with a remainder term as

$$\psi\left(t^n+\Delta t\right)=\psi\left(t^n\right)+\frac{(\Delta t)^p}{p!}\psi^{(p)}\left(t^n+\theta h\right)\quad(0<\theta<1). \tag{3.21}$$

Expanding the truncation error term yields

$$\tau_n = C_{p+1}(\Delta t)^p \psi^{(p+1)} = \sum_{i=0}^{k}\left(\frac{i^{p+1}}{(p+1)!}\alpha_i \psi^{p+1}(\xi_i) - \frac{i^p}{p!}\beta_i \psi^{p+1}(\eta_i)\right)(\Delta t)^p, \tag{3.22}$$

where $0 < \xi_i < 1$ and $0 < \eta_i < 1$. Next taking the modulus of (3.22) results in

$$|\tau_n| \leq \sum_{i=0}^{k}\left(\frac{i^{p+1}}{(p+1)!}|\alpha_i|\left|\psi^{p+1}(\xi_i)\right| - \frac{i^p}{p!}|\beta_i|\left|\psi^{p+1}(\eta_i)\right|\right)(\Delta t)^p. \tag{3.23}$$

Now suppose that $\left|\psi^{(p+1)}(\theta)\right| \leq M_{p+1}$, then we have

$$|\tau_n| \leq \sum_{i=0}^{k}\left(\frac{i^{p+1}}{(p+1)!}|\alpha_i| - \frac{i^p}{p!}|\beta_i|\right)M_{p+1}(\Delta t)^p. \tag{3.24}$$

Therefore, knowing k, α_i, β_i, and p, and assuming that M_{p+1} exists, we can work out a bound on the truncation error. To make the linear multistep scheme formula unique, we have to normalize, which is achieved through dividing by $\sum_{i=1}^{k}\beta_i$.

Recalling the definition of the truncation error, we have

$$L(\psi) = \frac{d\psi}{dt} - f(\psi, t) = 0, \tag{3.25}$$

$$L_{\Delta t}(\psi^n) = \sum_{i=0}^{k}\frac{\alpha_i \psi^{n+i}}{\Delta t} - \sum_{i=1}^{k}\beta_i f^{n+i}. \tag{3.26}$$

We do not require $L_{\Delta t}(\psi)$ and $L(\psi^n)$ to be equal to zero, but we do require them to be quite small. As such we shall use the second expression as a measure of error. Therefore, we now put the exact solution into (3.26), which results in

$$\begin{aligned} L_{\Delta t}(\Delta t) &= \sum_{i=0}^{k}\frac{\alpha_i \psi\left(t^{n+1}\right)}{\Delta t} - \sum_{i=1}^{k}\beta_i f\left(\psi\left(t^n\right), t^n\right) \\ &= \sum_{i=0}^{k}\frac{\alpha_i \psi\left(t^{n+1}\right)}{\Delta t} - \sum_{i=0}^{k}\left(\beta_i \frac{d\psi}{dt}\left(t^{n+1}\right)\right). \end{aligned} \tag{3.27}$$

We should note that

$$\psi\left(t^{n+1}\right) = \psi\left(t^{n+i\Delta t}\right), \tag{3.28}$$

$$\frac{d\psi}{dt}\left(t^{n+1}\right) = \frac{d\psi}{dt}\left(t^{n+i\Delta t}\right). \tag{3.29}$$

Expanding (3.28) and (3.29) via a Taylor series results in

$$\psi\left(t^{n+1}\right) = \psi(t_n) + i\Delta t\psi'(t_n) + \frac{(i\Delta t)^2}{2!}\psi''\left(t^n\right) + \cdots, \tag{3.30}$$

$$\psi'\left(t^{n+1}\right) = \psi'\left(t^{n}\right) + i\Delta t \psi''\left(t^{n}\right) + \frac{(i\Delta t)^2}{2!}\psi'''\left(t^{n}\right) + \cdots. \tag{3.31}$$

Therefore, the truncation error is defined as

$$\tau = \frac{1}{\Delta t}C_{-1} + C_0 + \Delta t C_1 + (\Delta t)^2 C_2 + \cdots, \tag{3.32}$$

therefore, collecting the terms of the same powers of Δt when (3.30) and (3.31) are substituted into (3.27) results in

$$C_{-1} = \sum_{i=0}^{k} \alpha_i \psi \qquad = \sum_{i=0}^{k} \alpha_i \psi, \tag{3.33a}$$

$$C_0 = \sum_{i=0}^{k} i\alpha_i \psi' - \sum_{i=1}^{k} \beta_i \psi' \qquad = \sum_{i=0}^{k} \left(i\alpha_i - \beta_i\right)\psi', \tag{3.33b}$$

$$C_1 = \sum_{i=0}^{k} \frac{i^2}{2!}\alpha_i \psi'' - \sum_{i=1}^{k} i\beta_i \psi'' \qquad = \sum_{i=0}^{k} \left(\frac{i^2}{2!}\alpha_i - i\beta_i\right)\psi''. \tag{3.33c}$$

Given the expression in (3.33a) to (3.33c), we can state that the pth coefficient of the truncation error is

$$C_p = \sum_{i=0}^{k} \left(\frac{i^{p+1}}{(p+1)!}\alpha_i - \frac{i^p}{p!}\beta_i\right)\psi^{(p+1)}, \tag{3.34}$$

where the superscript $p+1$ of ψ refers to the $(p+1)$th derivative of ψ. Finally, as we let $\Delta t \to 0$, we desire the truncation error to also tend to zero, however, at the moment we have one coefficient C_{-1} that is dependent on h^{-1} and one coefficient, C_0, that is not dependent on Δt, both of which do not tend to zero as Δt tends to zero. This then gives us two necessary conditions for the truncation error to tend to zero, which is that C_{-1} and C_0 must both be equal to zero, and thus the α_is and β_is must have these values to ensure that these two conditions are satisfied, and are the **consistency conditions** for the linear multistep methods.

This then gives rise to the following theorem that enables us to determine if the linear multistep scheme is consistent and also obtain the order of the scheme.

Theorem 3.3. *A general multistep method given by*

$$\sum_{i=0}^{s} \alpha \psi^{(n+1)-i} = \sum_{i=0}^{s} \beta_i f\left(\psi^{(n+1)-i}, t^{(n+1)-i}\right) \tag{3.35}$$

is of order $p \geq 1$ if and only if

$$\sum_{i=0}^{s} = 0, \tag{3.36a}$$

$$\sum_{i=0}^{s} \frac{i^k}{k!}\alpha_{s-i} = \sum_{i=0}^{s} \frac{i^{k-1}}{(k-1)!}\beta_{s-i}, \quad k = 1, 2, \ldots, p, \tag{3.36b}$$

$$\sum_{i=0}^{s} \frac{i^{p+1}}{(p+1)!}\alpha_i \neq \sum_{i=0}^{s} \frac{i^p}{p!}\beta_i. \tag{3.36c}$$

It should be noted that condition (3.36a) is the consistency condition for all multistep methods.

3.2 Dispersion and dissipation errors

The truncation error does not just inform us of the order of the scheme, the leading order of the derivative that remains in the truncation terms enables us to determine other properties of the numerical schemes. The first of these is referred to as **dispersion error**. It occurs when the leading terms in τ have an **odd order** derivative. These errors are characterized by **oscillations** in the numerical solution, which most of the time come in the form of moving waves. The reason that this error is referred to as dispersion error is because waves of different wavelengths prorogate at different speeds, where the wave speed is a function of the wavenumber, k, in the numerical approximation, which causes dispersion of waves. Given the partial differential equation we are considering in this chapter, all of the Fourier components that describe the one-dimensional advection equation, see [47] for more details, should move at the same speed, u.

The next error that we introduce is referred to as the **dissipation error**. It occurs when the leading order term in τ has an **even order** derivative. This type of error is characterized by a loss of wave amplitude. The effect is also referred to as **artificial viscosity**, but this property is implicit in the numerical solution.

To derive the two errors that were just mentioned, we require the **modified equation**, which is the partial differential equation that is actually solved by the finite difference scheme which is applied to the original partial differential equation. If we recall the difference equation of the upstream forward Euler scheme for the advection equation, then we have

$$\frac{\psi_i^{n+1} - \psi_i^n}{\Delta t} + u\frac{\psi_i^n - \psi_{i-1}^n}{\Delta x} = 0. \tag{3.37}$$

Substituting the true solution $\psi(x_i, t^n)$ into (3.37) and expanding as a Taylor series about this point gives

$$\frac{\partial \psi}{\partial t} + u\frac{\partial \psi}{\partial x} = -\frac{\Delta t}{2}\psi'' + \frac{u\Delta x}{2}\psi_{xx} - \frac{(\Delta t)^2}{6}\psi''' - \frac{u(\Delta x)^2}{6}\psi_{xxx} + O\left((\Delta x)^3 + (\Delta t)^3\right). \tag{3.38}$$

The next step is to replace the ψ'' term, and in general all of the time derivatives, with a spatial derivative equivalent. Therefore, we start by taking the time derivative of (3.38), which results in

$$\psi'' + u\psi'_x = -\frac{\Delta t}{2}\psi''' + \frac{u\Delta x}{2}\psi'_{xx} - \frac{(\Delta t)^2}{6}\psi'''' - \frac{c\,(\Delta x)^2}{6}\psi'_{xxx} + \cdots, \tag{3.39}$$

and then take the x derivative of (3.38) and multiply the result by $-u$, which yields

$$-u\psi'_x - u^2\psi_{xx} = u\frac{\Delta t}{2}\psi''_x - \frac{u^2\Delta x}{2}\psi_{xxx} + u\frac{(\Delta t)^2}{6}\psi'''_x + \frac{u^2(\Delta x)^2}{6}\psi_{xxxx} + \cdots. \tag{3.40}$$

If we now add (3.39) and (3.40) together, then we obtain

$$\psi'' = u^2\psi_{xx} + \Delta t\left(\frac{-\psi'''}{2} + \frac{u}{2}\psi''_x + O(\Delta t)\right) + \Delta x\left(\frac{u}{2}\psi'_{xx} - \frac{u^2}{2}\psi_{xxx} + O(\Delta x)\right). \tag{3.41}$$

It can quite easily be shown that the equivalent expression for the remaining third-order temporal derivatives in (3.41) can be replaced by

$$\begin{aligned}\psi''' &= -u^3\psi_{xxx} + O(\Delta x + \Delta t),\\ \psi''_x &= u^2\psi_{xxx} + O(\Delta x + \Delta t),\\ \psi'_{xx} &= -u\psi_{xxx} + O(\Delta x + \Delta t).\end{aligned} \tag{3.42}$$

Thus combining (3.37) with (3.41) and (3.42), we obtain the **modified equation** of the upwind forward Euler finite difference scheme for the advection equation as

$$\psi' + u\psi_x = \frac{u\Delta t}{2}(1 - C)\psi_{xx} - \frac{u(\Delta x)^2}{6}\left(2C^2 - 3C + 1\right)\psi_{xxx} + O(\Delta x + \Delta t). \tag{3.43}$$

We can see from (3.43) that the leading term is of the form $K\frac{\partial^2\psi}{\partial x^2}$, which, for $1 - C > 0$, represents a dissipation process, and therefore the dominant error is dissipative in nature. Another feature to note here is that when $C = 1$, the scheme is exact, but when $C \neq 1$, the leading term in the truncation error causes severe damping of the numerical solution, as we saw in the last chapter.

3.3 Amplitude and phase errors

The next set of errors that are important with respect to quantifying the performance of the different numerical schemes are the **phase and amplitude errors**. To assess them, we return again to the general initial value problem

$$\frac{d\psi}{dt} = F(\psi, t), \quad \psi(0) = \psi_0, \tag{3.44}$$

which has a unique solution, provided function F is sufficiently smooth. If we consider the case of $F(\psi, t) \equiv \nu\psi \equiv (\lambda + i\omega)\psi$, where λ and ω are real numbers, then we have the ordinary differential equation

$$\frac{d\psi}{dt} = \nu\psi \equiv (\lambda + i\omega)\,\psi, \tag{3.45}$$

with the solution

$$\psi(t) = \psi_0 e^{\lambda t} e^{i\omega t}, \tag{3.46}$$

where the real component of ν is λ, which determines the rate of change of the magnitude (modulus) of ψ, while the imaginary part of ν is ω, which quantifies the rate of change of the phase of the solution ψ.

For the case $\nu = i\omega$, the associated differential equation is referred to as the **oscillation equation**, which is given by

$$\frac{d\psi}{dt} = i\omega\psi, \tag{3.47}$$

which is a model for many nondissipative dynamical systems. If we now integrate (3.47) over a time interval Δt, then we obtain the solution

$$\psi(t + \Delta t) = \psi\left(t^0\right) \equiv \psi\left(t^0\right)\xi_c, \tag{3.48}$$

where ξ_c is the exact amplification factor, which we can see from (3.48) is a complex number, with modulus equal to 1. We can now define the amplitude and phase errors as functions of the numerical resolution that we are using, through evaluating the modulus and argument of the numerical solution $\psi^n = |\xi|\, e^{i\theta}$, respectively, where

$$|\xi| = \sqrt{Re(\xi)^2 + Im(\xi)^2}, \qquad \text{and} \qquad \theta = \arctan\frac{Im(\xi)}{Re(\xi)}. \tag{3.49}$$

The phase error of a numerical scheme can be characterized by considering the relative phase change, which is defined as $R\dfrac{\theta}{\omega\Delta t}$ and represents the ratio of the phase advancement that is produced by one time integration of the numerical scheme to the change in phase that is experienced by the true state in the same time interval. Therefore we have the following three situations:

1. If $R > 1$, then the method is accelerating.
2. If $R = 1$, then the method is traveling at the same speed as the true state.
3. If $R < 1$, then the method is decelerating.

We now focus on the modulus of the amplification factor to quantify the amplitude error from the numerical solution, which is achieved by comparing the modulus of ξ with that of ξ_c, which we know has the value of unity. Therefore we have the following three situations:

1. If $|\xi| < 1$, then the scheme is **damping** the numerical solution.
2. If $|\xi| = 1$, then the scheme is referred to as being **neutral**.
3. If $|\xi| > 1$, then the scheme is **amplifying** the numerical solution.

If we now consider the advection equation

$$\frac{\partial \psi}{\partial t}+u\frac{\partial \psi}{\partial x}=0, \tag{3.50}$$

then the solution to (3.50) can be found by considering a Fourier series, as the solution is wavelike, and as such a solution for each individual Fourier mode can be sought as a traveling wave given by

$$\psi(x,t)=e^{i(kx-\omega t)}, \tag{3.51}$$

where here k is equal to 2π divided by the wavelength, and is the wavenumber that we referred to earlier, while ω here is equal to 2π divided by the period, and referred to as the **frequency**. If we substitute (3.51) into (3.50) then we see that there can only be a solution if the **dispersion relation** $\omega=uk$ is satisfied. Therefore, the wave travels with constant amplitude at a **phase speed**, which is given by $\dfrac{\omega}{k}=u$.

For us to be able to assess these errors for the numerical schemes, we now introduce what is referred to as the **differential–difference** equation which, if we consider the upwind scheme, is given by

$$\frac{d\psi_j}{dt}+u\frac{\psi_j-\psi_{j-1}}{\Delta x}=0, \tag{3.52}$$

and if we now substitute an individual wavelike solution of the form

$$\psi_j(t)=e^{i(kj\Delta x-\omega_{uw}t)}, \tag{3.53}$$

where ω_{uw} is the frequency associated with the upwind spatial discretization, into (3.52), then we obtain

$$-\omega_{uw}i\psi_j=-u\frac{\left(1-e^{-ik\Delta x}\right)}{\Delta x}\psi_j. \tag{3.54}$$

We now use the trigonometric identity

$$e^{-ik\Delta x}\equiv \cos k\Delta x-i\sin k\Delta x,$$

which gives us the dispersion relation for the frequency associated with the upwind discretization as

$$\omega_{uw}=\frac{u}{i\Delta x}\left(1-e^{-k\Delta x}\right)=\frac{u}{\Delta x}\left(\sin k\Delta x+i\left(\cos k\Delta x-1\right)\right), \tag{3.55}$$

where we have also used the fact that $\dfrac{1}{i}\equiv\dfrac{-ii}{i}=-i$.

If we take the real and imaginary components separately and consider the real component first, then we have $\omega_{uw}^{r}=u\dfrac{\sin k\Delta x}{\Delta x}$, and the associated phase speed is given by

$$u_{uw}\equiv\frac{\omega_{uw}}{k}=u\frac{\sin k\Delta x}{k\Delta x}, \tag{3.56}$$

but we have to notice here that the phase speed is a function of the wave number k, and as such these waves are dispersive, see [47] for other properties of this term.

As we have a imaginary component in the numerical frequency for the upwind scheme, this implies that there are amplitude errors introduced by this scheme. Thus the amplitude of the differential–difference solution will grow, or decay, at the rate

$$\exp\left\{-\frac{u}{\Delta x}\left(1-\cos k\Delta x\right)t\right\}.$$

3.4 Stability

Given the definitions above for the ordinary differential equation in (3.44), we now move on to the study of the **stability** of a numerical scheme. We shall be presenting the mathematical expression and technique for determining many different forms of stability that enable us to understand how well the finite difference scheme can perform. We start with a general definition of stability which is given by:

Definition 3.4 (Stability). A numerical scheme $L\left(\psi_i^n\right)=0$ for a time evolving partial differential equation on the time domain $[0, T]$ is said to be stable if the discrete solution satisfies

$$\max_{i=1,2,\ldots,I}\left|\psi_i^n\right|\leq C, \quad \forall n=1,2,\ldots,N, \tag{3.57}$$

where $C>0$ is constant independent of the grid size, Δx and Δt.

This then leads to the following theorem for convergence, which is referred to as the Lax equivalence theorem.

Theorem 3.5 (Convergence, Lax equivalence theorem). *If the original partial differential equation is well-posed, then the discrete solution of the numerical scheme converges to the solution of the partial differential equation when $\Delta x, \Delta t\rightarrow 0$ if and only if the scheme is consistent and stable.*

We have already presented how to prove if a finite difference scheme is consistent with the partial differential equation that we are approximating, so we now move onto the first technique to either determine if a finite difference scheme is stable for all choices of Δt and Δx, or if there are restrictions on the grid and time steps.

3.4.1 Courant–Friedrichs–Lewy condition

The next condition that we consider is the **Courant, Friedrichs, and Lewy condition,** which is more commonly know as the **CFL** condition, named after its discoverers, Richard Courant, Kurt Friedrichs, and Hans Lewy, who presented this condition in 1928. It is based upon what are referred to as the **domain of dependence** and the **numerical domain of dependence**. The *true* domain of dependence is that which is found through tracing the characteristics backwards in time, whereas the numerical domain of dependence is that of the all of the previous spatial grid points used at each time level to approximate the original partial differential equation. The CFL condition is expressed as follows: **the true domain of dependence must be contained within the numerical domain of dependence**. We should make a note here that the CFL condition is only a necessary one, and not necessarily sufficient. We shall see this with the forward-time, centered-space scheme later. In Fig. 3.2 we have an example of schemes where the CFL condition is met and one where it is not, to help illustrate the condition. The actual situation presented in Fig. 3.2 is for the upwind forward Euler scheme with positive velocity on the left and negative velocity on the right, which is consistent with the numerical behavior we saw with this scheme under these situations.

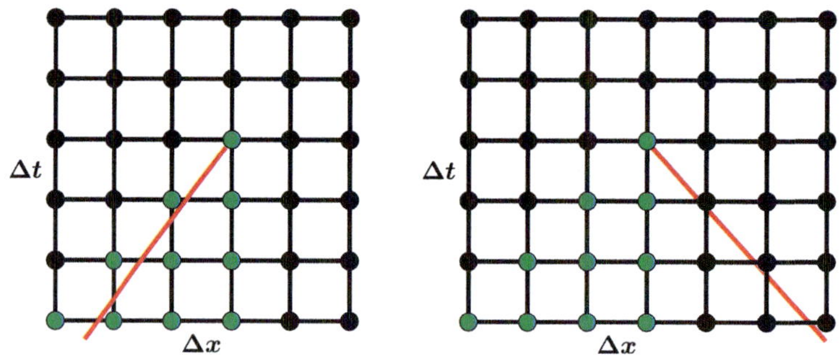

FIGURE 3.2 Two examples to illustrate the Courant–Friedrichs–Lewy (CFL) condition, where on the left we have the numerical domain of dependence capturing the analytical domain of dependence as the characteristic, the red line, is within the green dots which are showing which points feed into the answer at the t^{n+1} level. In the right plot we have an illustration where the numerical scheme is not capturing the true trajectory as it uses the points in the wrong direction.

As we can see from Fig. 3.2, the CFL condition is a restriction on the size of both the spatial and temporal grid steps. We see that if the time step is too big, given the grid spacing, then it is possible for the numerical domain to not contain the true domain of dependence. Therefore, the CFL condition is a condition on the ratio of the time step to the grid spacing, which also depends on the advection velocity u. Thus the CFL condition can be expressed mathematically for the linear advection equation as

$$u\frac{\Delta t}{\Delta x} \leq 1,$$

in which the quantity on the left we have already identified as the Courant number.

3.4.2 Von Neumann stability analysis

We now introduce a method that was discovered in the 1950s by John von Neumann, which enables us to quantify the **stability** of the numerical scheme. The von Neumann stability analysis is based upon applying a trial solution into the difference equation and determining if the solution grows. The solution is in the form of a Fourier node given by

$$\psi(x,t) = z(t)\,e^{ikx},$$

where i is referring to the imaginary number $i = \sqrt{-1}$, and the term $z(t)$ is capturing the growth factor. At the moment this is just one Fourier component of the entire solution, and as such, when applied to the numerical grid, we have

$$\psi_j^n = z^n e^{ikj\Delta x} = \xi^n e^{ikj\Delta x}, \tag{3.58}$$

where ξ is referred to as the **amplification factor**, defined as $z^n = \xi z^{n-1}$, but the superscript of ξ in (3.58) is actually a power. The von Neumann stability is the process of inserting the trial function above into the difference equation and examining whether or not $|\xi|$ lies between -1 and $+1$. This condition on the amplitude factor is to ensure that the numerical scheme does not grow between each time step artificially.

3.4.3 Multistep method stability

Another form of stability is for the multistep numerical schemes, where we are evaluating over many different time levels. We consider the continuous case for an ordinary differential of the form

$$\psi' = f(\psi, t), \qquad \psi(0) = \psi_0,$$

where the functions on either side are nonlinear, in general. A linear multistep scheme involves $\psi_0, \psi_1, \psi_2, \psi_3, \ldots,$ and the truncation error is the error in the differential equation; however, we need the error in the solution.

We now consider the boundedness of the operator which *solves* the difference scheme for the following linear ordinary differential equation problem:

$$\psi' = \lambda\psi, \qquad \psi(0) = \psi_0,$$

where λ is a constant. Therefore, as presented earlier, the exact solution to this problem is $\psi = \psi_0 e^{\lambda t}$.

If we now consider the general expression for a difference equation that solves the differential equation above with $\lambda = 0$, given by

$$\alpha_k \psi^{n+k} + \alpha_{k-1}\psi^{n+k-1} + \cdots + \alpha_0\psi^0 = 0, \tag{3.59}$$

as we wish to solve for ψ_n, and if we know $\psi^0, \psi^1, \ldots, \psi^{k-1}$, then the formula above gives the value for ψ^k. Therefore, if one has $\psi^1, \psi^2 \ldots, \psi^k$, then it is possible to obtain ψ^{k+1}, and so on, which is therefore generating a sequence. If we take the first step, $k = 1$, we have

$$\alpha_1\psi^{n+1} + \alpha_0\psi^n = 0 \Longrightarrow \psi^{n+1} = -\frac{\alpha_0}{\alpha_1}\psi^n, \ \alpha_1 \neq 0. \tag{3.60}$$

Therefore, given the first value of ψ, ψ^0, it is possible to generate ψ^1, and thus ψ^2, and as such we can define a sequence that is dependent on the previous values up to ψ^n as

$$\psi^1 = -\frac{\alpha_0}{\alpha_1}\psi^0,$$

$$\begin{aligned}\psi^2 &= -\frac{\alpha_0}{\alpha_1}\psi^1 \equiv \left(-\frac{\alpha_0}{\alpha_1}\right)^2 \psi^0, \\ &\vdots \\ \psi^n &= \left(-\frac{\alpha_0}{\alpha_1}\right)^n \psi^0.\end{aligned} \tag{3.61}$$

If we consider the case $k=2$, then we will have

$$\alpha_2\psi^{n+2} + \alpha_1\psi^{n+1} + \alpha_0\psi^n = 0. \tag{3.62}$$

Before we continue deriving properties of the difference equations, we briefly consider the following differential equation theory. Given the general form of an ordinary differential equation

$$\alpha_k \frac{d^{n+k}\psi}{dt^{n+k}} + \alpha_{k-1}\frac{d^{n+k-1}\psi}{dt^{n+k-1}} + \cdots + \alpha_0 \frac{d^n\psi}{dt^n} = 0, \tag{3.63}$$

the solution to (3.63) is of the form $\psi = e^{mt}$. As such we can form the auxiliary equation as

$$\alpha_k m^{n+k} + \alpha m^{n+k-1} + \cdots + \alpha_0 m^n = 0, \tag{3.64}$$

thus, if m is a root of (3.64), then e^{mt} is a solution. Given this information, we return to the difference equation and substitute $\psi^n = z^n$, which results in

$$\rho(z) = \alpha_k z^k + \alpha_{k-1} z^{k-1} + \cdots + \alpha_0 = 0. \tag{3.65}$$

Therefore, the exist up to k roots, $z_1, z_2, \ldots, z_k$, and as such there are $z_1^n, z_2^n, \ldots, z_k^n$ solutions to the difference equation (3.59), so we have

$$y^n = A_1 z_1^n + A_2 z_2^n + \cdots + A_k z_k^n. \tag{3.66}$$

As an example let us consider the difference equation

$$\psi^{n+2} - 3\psi^{n+1} + 2\psi^n = 0, \tag{3.67}$$

and if we try the solution of the form $\psi^n = z^n$, then (3.67) becomes

$$z^{n+2} - 3z^{n+1} + 2z^n = 0. \tag{3.68}$$

Factoring out z^n from (3.68) results in the following quadratic equation

$$\rho(z) = z^2 - 3z + 2 = 0 \equiv (z-2)(z-1) \Longrightarrow z_1 = 1, z_2 = 2, \tag{3.69}$$

which implies that $\psi^n = A_1(1)^n + A_2 2^n = A_1 + A_2 2^n$. If we consider the differential equation from (3.1) with $\lambda = 0$, then the exact solution is $\psi = \psi_0$, and therefore $\psi^n = A_1 z_1^n + A_2 z_2^n + \cdots + A_k z_k^n$ is an approximate solution to ψ_0. However, given the values of z that we have found, we now introduce another definition for **stability** to quantify the effects of the values of z.

Definition 3.6 (Stability). **1.** If $|z_i| < 1$, then $|z_i|^n$ decays when $n \to \infty$, and therefore the numerical scheme is said to be **stable**.

2. If $|z_i| = 1$, then $|z_i|^n$ is always equal to one, and therefore the numerical scheme is said to be **neutrally stable**.

3. If $|z_i| > 1$ then $|z_i|^n \to \infty$ as $n \to \infty$, and therefore the numerical scheme is said to be **numerically unstable**.

Therefore, given the definition above, the example scheme in (3.67) is numerically unstable. If we now take a general p-step method in the form

$$\psi_{n+1} = \sum_{i=0}^{p} \alpha_{p-i} \psi^{n-i} + \sum_{i=-1}^{p} \beta_{p-i} f\left(t^{n-i}, \psi^{n-i}\right), \tag{3.70}$$

then we can form what is referred to as the **stability polynomial** by considering the homogenous version of (3.70) which is where we assume that $f(t, \psi) = 0$, and so the true solution is of the form $\psi^n = r^n$. Thus the stability polynomial is given by

$$\phi(r) = r^{p+1} - \sum_{i=0}^{p} \alpha_{p-i} r^{p-i}. \tag{3.71}$$

Therefore, the exact solution may be written as a linear combination $\psi^k = \sum_{i=0}^{p} \gamma_i r_i^k$.

We now introduce an important definition, which is referred to as the **root condition**:

Definition 3.7. The stability polynomial satisfies the root condition if and only if all the roots of $\phi(r)$ are such that

$$|r_i| \leq 1, \quad 0 \leq i \leq p, \tag{3.72}$$

and if each root r_i such that $|r_i| = 1$ is simple. It is said that the strong root condition is satisfied if and only if

$$r_0 = 1, \quad |r_i| < 1, \quad i = 1, 2, \ldots, p. \tag{3.73}$$

Given the root condition above, we now use it to define what is referred to as **zero stability**.

Definition 3.8. A multistep method is said to be **zero-stable** if the numerical solution it generates remains bounded as $\Delta t \to 0$ for a finite final time T. That is, we require that all solutions to

$$\sum_{i=0}^{p} \alpha_i \psi^{n+1-i} = 0 \tag{3.74}$$

remain bounded as $n \to \infty$. This is achieved if the root condition above is satisfied.

We again consider the ordinary differential equation, $\psi' = \lambda\psi$, where we are applying the general multistep numerical approximation from (3.70), which results in the equation

$$\sum_{i=1}^{p} \left(\alpha_i \psi^{n-i} - \lambda \Delta t \beta_i \psi^{n-1}\right) = 0. \tag{3.75}$$

If we now assume a solution of the form $\psi^n = c\phi^{(n)}$ and substitute this solution into (3.75), then we obtain

$$\sum_{i=0}^{p} \left(\alpha_i c\phi^{(n-i)} - \lambda \Delta t \beta_i c\phi^{(n-i)} \right) = 0, \tag{3.76}$$

where the subscripts in the parentheses are referring to powers in (3.76). If we now divide (3.76) by $c\phi^{(p-k)}$, then we obtain

$$\rho(\phi) - \lambda \Delta t \sigma(\phi) = 0, \tag{3.77}$$

where $\rho(\phi)$ is defined as

$$\rho(\phi) \equiv \sum_{i=0}^{p} \alpha_i \phi^{(p-i)} \tag{3.78}$$

and is referred to as both the stability and the first characteristic polynomial, and where $\sigma(\phi)$, defined as

$$\sigma(\phi) \equiv \sum_{i=0}^{p} \beta_i \phi^{(p-i)}, \tag{3.79}$$

is referred to as the **secondary characteristic polynomial**.

We notice that one way to see if a linear multistep scheme is consistent with the partial differential equation is to check if $\rho(1) = 1$ and $\rho'(1) - \sigma(1) = 0$. We can see that there are p solutions to (3.78), but only one of them is approximating the true state, the remaining $p-1$ are referred to either **parasitic** or **computational modes**. If the roots are distinct, then the solution is of the form

$$\psi^n = \sum_{i=1}^{p} c_i \phi_i^{(n)}. \tag{3.80}$$

If, however, the roots of the characteristic polynomials are not distinct, then we would have to build solutions similar to those used in control theory, see [61] for more details on control theory and eigenvalue solutions.

The exact solution for the ordinary differential equation that we are considering here is an exponential function that is dependent on the parameter λ. When we have $\lambda = 0$, this requires $|\phi_i| \leq 1$, which is the zero stability condition; when we have $\lambda < 0$, we require $|\phi_i| \leq 1$, which is referred to as the **absolute stability** condition; finally, when we have $\lambda > 0$, then we can allow for $|\phi_i| > 1$ but not by too much as we are approximating a growing exponential function, so we must restrain the numerical solution from becoming unstable.

Another stability definition that we have to introduce to quantify a property of the linear multistep methods is referred to a **A-stability**.

Definition 3.9. A linear multistep method is said to be **A-stable** if its stability region is the whole left half of the complex plane.

The **stability regions, or stability domains,** are the constraints that we have seen on the step sizes so that the van Neumann analysis stays stable. For the multistep methods, the stability region is defined through the ratio of the first characteristic equation to the second characteristic equation, given by

$$\mu = \frac{\rho(\phi)}{\sigma(\phi)}. \tag{3.81}$$

A way to analyze the stability regions is to transform the problem into polar coordinates $\phi = e^{i\theta}$ with $0 \leq \theta \leq 2\pi$. We shall plot some of the associated stability domains in the next subsection for the different Adams–Bashforth, Adams–Moulton, and the backward differentiation formula schemes.

Finally, we can come to the **Dahlquist theorem**, which is:

Theorem 3.10. *A linear multistep method is convergent **if and only if** it is zero-stable and consistent.*

However, given the theorem above, we now present two properties that restrict the highest order of the multistep scheme to achieve convergence. These are referred to as the first and second Dahlquist barriers, which are given by

Theorem 3.11 (Dahlquist first barrier). *A zero-stable and linear p-step method cannot obtain an order of convergence greater than $p+1$ if p is odd, and $p+2$ if p is even. If the method is explicit then it cannot attain an order greater than p.*

Theorem 3.12 (Dahlquist second barrier). *There are **no explicit A-stable** and linear multistep methods. The implicit multistep methods only have convergence order of at most 2.*

A final definition of stability that we consider here is that of **L-stability**.

Definition 3.13. A finite difference scheme is said to be L-stable if it is A-stable, and when applied to the scalar test differential equation $\psi' = \lambda\psi$, where λ is complex but its real component is less than zero, then its associated amplification factor $\xi(\lambda\Delta t)$ must tend to zero as $\Delta t\lambda \rightarrow \infty$.

This property is implying that there are a limited number of implicit methods, as we know that there are no explicit methods that are A-stable, and from the Dahlquist second barrier we can only have up to second-order implicit schemes that can be A-stable, which infers that there are only a few schemes that can be L-stable.

3.5 Quantifying the properties of the explicit finite difference schemes

In this section we shall look at the different explicit finite difference schemes that were introduced in the last chapter to quantify the performances that we saw in the plots with the advection of the bell curve and the step function. We start with the upwind forward Euler scheme.

3.5.1 Upwind forward Euler scheme

Recalling the upwind forward Euler difference equation for the 1D advection equation, we have

$$\psi_i^{n+1} = \psi_i^n - u\frac{\Delta t}{\Delta x}\left(\psi_{i+1}^n - \psi_i^n\right).$$

Now if we substitute the true state into the difference equation above, then we obtain

$$\psi\left(x_i, t^{n+1}\right) = \psi\left(x_i, t^n\right) - u\frac{\Delta t}{\Delta x}\left(\psi\left(x_i + \Delta x, t^n\right) - \psi\left(x_i, t^n\right)\right). \tag{3.82}$$

Next we expand the true state about the point $\psi\left(x_i, t^n\right)$ into a Taylor series for each term in (3.82), and obtain the following for the truncation error:

$$\begin{aligned}\tau_{UFE} &= \psi_i^n + \Delta t\psi_i^{n'} + \frac{(\Delta t)^2}{2}\psi_i^{n''} + O\left((\Delta t)^3\right) - \psi_i^n \\ &+ u\frac{\Delta t}{\Delta x}\left(\psi_i^n + \Delta x\psi_x^n + \frac{(\Delta x)^2}{2}\psi_{i,xx}^n + O\left((\Delta x)^2\right) - \psi_i^n\right) \\ &= \Delta t\left(\psi_i^{n'} + u\psi_{i,x}^n\right) + \frac{(\Delta t)^2}{2}\psi_i^{n,''} + u\frac{\Delta t}{\Delta x}(\Delta x)^2\psi_{i,xx}^n + O\left((\Delta t)^3\right) \\ &+ O\left((\Delta x)^3\right).\end{aligned} \tag{3.83}$$

The first term in (3.83) is the original partial differential equation, and as such this term is zero. If we now divide throughout by Δt and Δx, we see that the remaining terms are multiplied by Δt and Δx, which then means that this scheme is consistent with the partial differential equation to first order in time and space. We can see that it is consistent as all the remaining terms in the truncation error are multiplied by Δt and Δx in some form and as such these tend to zero as the time step and the grid size tend to zero.

Exercise 3.14. *Show that the local truncation error for the upwind forward Euler scheme for the advection equation is* $O\left((\Delta t)^2, (\Delta x)^2\right)$.

Now we check the stability of the upwind forward Euler scheme by using the von Neumann stability criterion. We recall that we substitute the first Fourier node of a Fourier expansion into the difference equation for ψ, which yields

$$z^{n+1}e^{ikj\Delta x} = z^n e^{ijk\Delta x} - u\frac{\Delta t}{\Delta x}\left(z^n e^{ik(j+1)\Delta x} - z^n e^{ikj\Delta x}\right). \tag{3.84}$$

If we now factorize out $z^n e^{ikj\Delta x}$ in (3.84), then we obtain

$$z^{n+1}e^{ikj\Delta x} = z^n e^{ikj\Delta x}\left(1 - Ce^{ik\Delta x} + C\right). \tag{3.85}$$

If we recall that we are looking for the amplification factor for this scheme, which is the factor multiplying z^n, which in (3.85) is $\xi = \left(1 + C - Ce^{ik\Delta x}\right)$. We now use the trigonometric identity

$$e^{ijk\Delta x} \equiv \cos k\Delta x + i\sin k\Delta x,$$

which then enables us to write the amplification factor from (3.85) as

$$\xi = 1 + C - C\cos k\Delta x - i\sin k\Delta x = 1 + C\left(1 - \cos k\Delta x\right) - Ci\sin k\Delta x. \tag{3.86}$$

The next step is to find the magnitude of ξ which appears by taking the sum of the separate squares of the real and complex components in (3.86), resulting in

$$\begin{aligned} \left|\xi^2\right| &= \left(1 + C\left(1 - \cos k\Delta x\right)\right)^2 + C^2\sin^2 k\Delta x \\ &= 1 + 2C\left(1 - \cos k\Delta x\right) + C^2\left(1 - \cos k\Delta x\right)^2 + C^2\sin^2 k\Delta x \\ &= 1 + 2C\left(1 - \cos k\Delta x\right) + C^2\left(1 - 2\cos k\Delta x + \cos^2 k\Delta x\right) + C^2\sin^2 k\Delta x \\ &= 1 + 2C\left(1 - \cos k\Delta x\right)\left(1 + C\right). \end{aligned} \tag{3.87}$$

We now require the amplification factor to be less than 1, if we wish for the numerical scheme to be stable. Rearranging (3.87), we have the inequality

$$2C\left(1 - \cos k\Delta x\right)\left(1 + C\right) < 0, \tag{3.88}$$

where we know that $-1 \leq \cos k\Delta \leq 1$, $\forall k$. Therefore, we have to look at the term $1 + C$ which we require to be greater than zero, implying that $C < -1 \Rightarrow u\frac{\Delta t}{\Delta x} < -1$. We now recall for the upwind scheme that $u < 0$, therefore the condition becomes $|u|\frac{\Delta t}{\Delta x} < 1$, which is the CFL condition for the upwind forward Euler numerical scheme that is stable as long as the numerics do not transport information in a single time step Δt more than one single cell Δx.

Given the application of the von Neumann stability analysis, and definition of the dispersion and dissipation, along with amplitude and phase errors, we now consider how to combine the discrete solution to find these errors for the full discrete approximation of the advection equation. Thus we have

$$\psi_j^n = e^{i\left(kj\Delta x - \omega n\Delta t\right)}. \tag{3.89}$$

We now substitute the numerical solution (3.89) into the upwind forward Euler scheme,

$$\frac{\psi_j^{n+1} - \psi_j^n}{\Delta t} - u\frac{\psi_j^n - \psi_{j-1}^n}{\Delta x}, \tag{3.90}$$

where we obtain

$$e^{-i\omega\Delta t} - 1 = C\left(e^{-ik\Delta x} - 1\right). \tag{3.91}$$

If now we consider the real and the imaginary components of the frequency as $\left(\omega^r + i\omega^i\right)$, then the numerical solution becomes

$$\psi_j^n = e^{\omega^i n\Delta t} e^{i\left(kj\Delta x - \omega^r n\Delta t\right)} \equiv |\xi|^n e^{i\left(kj\Delta x - \omega^r n\Delta t\right)}, \tag{3.92}$$

and we can express (3.91) in terms of the real and imaginary components and equate them on either side, which gives the following set of equations:

$$|\xi|\cos\omega^r\Delta - 1 = C\left(\cos k\Delta x - 1\right), \tag{3.93a}$$

$$|\xi|\sin\omega^r\Delta t = C\sin k\Delta x. \tag{3.93b}$$

An interesting feature to notice here is that determining the imaginary component of ω is equivalent to the von Neumann analysis that we just presented, as ω^i determines the amplification factor, which then governs the rate of numerical dissipation. Therefore, we now focus on the real component ω^r which gives us information about the phase speed error.

Next we square both sides of the equalities in (3.93a) and (3.93b) and add these together, which yields

$$\left|\xi_{ufe}\right|^2 = 1 - 2C\left(1 - C\right)\left(1 - \cos kx\right), \tag{3.94}$$

which tells us that the scheme is stable, as we saw with the von Neumann stability analysis earlier, but now we can see that it is also damping, except for when $C = 1$, which is consistent with what we saw in the previous chapter in the advection of the bell curve and the step function.

We can now obtain the discrete-dispersion relationship for the upwind forward Euler scheme ω^r_{ufe} through dividing (3.93b) by (3.93a), and taking the arctangent of both sides, which results in

$$\omega^r_{ufe} = \frac{1}{\Delta t}\arctan\left(\frac{c\sin k\Delta x}{1 + C\left(\cos k\Delta x - 1\right)}\right). \tag{3.95}$$

The next step is to form the phase speed $c_{ufe} \equiv \dfrac{\omega^r_{ufe}}{k}$ and to approximate the expression in terms of the Courant number to see if the scheme has a phase error.

We shall use the following series approximations for the different functions to achieve this:

$$\begin{aligned}
\frac{1}{1-x} &= \sum_{n=0}^{\infty} x^n, \\
\sin x &= x - \frac{x^3}{6} + \frac{x^5}{120} + \cdots, \\
\cos x &= 1 - \frac{x}{2} + \frac{x^4}{24} + \cdots, \\
\arctan x &= x - \frac{x}{3} + \frac{x^5}{5} + \cdots.
\end{aligned}$$

It can therefore be shown that the phase speed for the upwind forward Euler scheme is approximately

$$c_{ufe} \approx u\left(1 - \frac{\left(k\Delta x\right)^2}{6}\left(1 - C\right)\left(1 - 2C\right)\right). \tag{3.96}$$

Thus we can see from (3.96) that the upwind forward Euler scheme will be decelerating for $C < \frac{1}{2}$, but is accelerating for $C > \frac{1}{2}$, and that the phase speed is almost zero when $C = 1$ or $C = \frac{1}{2}$.

3.5.2 Forward-time–centered-space scheme

Recalling the forward-time, centered-space scheme's difference equation

$$\frac{\psi_i^{n+1} - \psi_i^n}{\Delta t} + u\frac{\psi_{i+1}^n - \psi_{i-1}^n}{\Delta 2x} = 0,$$

to form the global truncation error for this scheme, we substitute $\psi\left(x_i, t^n\right)$ into the difference equations above and expand about this point, which results in

$$\begin{aligned}
\tau_{FTCS} &= \psi\left(x_i, t^{n+1}\right) - \psi\left(x_i, t^n\right) - u\frac{\Delta t}{2\Delta x}\left(\psi\left(x_{i+1}, t^n\right) - \psi\left(x_{i-1}, t^n\right)\right) \\
&= \psi\left(x_i, t^n + \Delta t\right) - \psi\left(x_i, t^n\right) - u\frac{\Delta t}{2\Delta x}\left(\psi\left(x_i + \Delta x, t^n\right) - \psi\left(x_i - \Delta x, t^n\right)\right) \\
&= \psi_i^n + \Delta t\psi_i^{n'} + \frac{(\Delta t)^2}{2}\psi_i^{n''} - \psi_i^n + u\frac{\Delta t}{2\Delta x}\left(\psi_i^n + \Delta x\psi_{i,x}^n + \frac{(\Delta x)^2}{2}\psi_{i,xx}^n\right. \\
&+ \left.\frac{(\Delta x)^3}{3!}\psi_{i,xxx}^n - \psi_i^n + \Delta x\psi_{i,x}^n - \frac{(\Delta x)^2}{2}\psi_{i,xx} + \frac{(\Delta x)^3}{3!}\right) \\
&= \Delta t\left(\psi_i^{n'} + u\psi_{i,x}^n\right) + \frac{(\Delta t)^2}{2}\psi_i^{n''} + O\left((\Delta t)^3\right) \\
&+ u\frac{\Delta t}{2\Delta x}\left(\frac{2(\Delta x)^3}{3!}\psi_{i,xxx}^n + O\left((\Delta x)^5\right)\right).
\end{aligned} \tag{3.97}$$

We know that the first term in (3.97) is equal to zero as it is the original differential equation and we are evaluating that ordinary differential equation at the true state ψ, therefore the remaining terms tell us that the scheme is first order in time and second order in space. That is to say,

$$\tau_{FTCS} \sim O(\Delta t) + O\left((\Delta x)^2\right). \tag{3.98}$$

This then means that if we increase the number of grid points in the x direction, then we obtain a rate of convergence to the true state of $(\Delta x)^2$.

Exercise 3.15. *Show that the local truncation error for the forward-time–centered-space is* $O\left((\Delta t)^2, (\Delta x)^2\right)$.

However, we shall see that just because a scheme is **consistent** with the ordinary differential equation does not mean that it will converge to the true solution. We now consider the von Neumann stability analysis for this scheme to identify whether or not it is stable. We have seen some evidence in the last chapter that this scheme may not be stable.

As introduced before, we substitute the trial solution $\psi_j^n = z^n e^{ikj\Delta x}$ into the difference equation for this scheme, which results in

$$z^{n+1}e^{ikj\Delta x} = z^{n+1}e^{ikj\Delta x} - \frac{C}{2}\left(z^n e^{ik(j+1)\Delta x} - z^n e^{ik(j-1)\Delta x}\right).$$

Factorizing as before, we obtain

$$z^{n+1}e^{ikj\Delta x} = z^{n+1}e^{ikj\Delta x}\left(1 - \frac{C}{2}\left(e^{ijk\Delta x} - e^{-ijk\Delta x}\right)\right). \tag{3.99}$$

Thus we can identify the amplification factor in (3.99) for the forward-time–centered-space scheme as

$$\xi_{FTCS} = 1 - \frac{C}{2}\left(e^{ijk\Delta x} - e^{-ijk\Delta x}\right) \equiv 1 - Ci\sin k\Delta x. \tag{3.100}$$

Now if we form the modulus of ξ_{FTCS} in (3.100), we have

$$\left|\xi^2\right| = 1 + C^2\sin^2 k\Delta x,$$

which can only lie on the unit circle, but cannot lie inside of the unit circle, if $k\Delta x = m\pi$. Therefore, the forward-time–centered-space scheme is **unconditionally unstable**.

3.5.3 Lax–Friedrichs scheme

As we did with the previous numerical methods, we start by recalling the difference equation for this scheme, which is given by

$$\frac{\psi_i^{n+1} - \frac{1}{2}\left(\psi_{i+1}^n + \psi_{i-1}^n\right)}{\Delta t} + \frac{u}{2\Delta x}\left(\psi_{i+1}^n - \psi_{i-1}^n\right) = 0.$$

As before we put the true state $\psi(x_i, t^n)$ into the difference equation for this scheme, which results in

$$\begin{aligned} \tau_{LF} &= \frac{\psi(x_i, t^n + \Delta t) - \frac{1}{2}\left(\psi(x_i + \Delta x, t^n) + \psi(x_i - \Delta x, t^n)\right)}{\Delta t} \\ &+ \frac{u}{2\Delta x}\left(\psi\left(x_i + \Delta x, t^n\right) - \psi\left(x_i - \Delta x, t^n\right)\right). \end{aligned}$$

Expanding the true state in a Taylor series, we obtain

$$\begin{aligned} \tau_{LF} &= \psi_i^n + \Delta t\psi_i^{n'} + \frac{(\Delta t)^2}{2}\psi_i^{n''} + O\left((\Delta t)^3\right) \\ &- \frac{\psi_i^n + \Delta x\psi_{i,x}^n + \frac{(\Delta x)^2}{2}\psi_{i,xx}^n + O\left((\Delta x)^3\right) + \psi_i^n - \Delta x\psi_{i,x}^n + \frac{(\Delta x)^2}{2}\psi_{i,xx}^n + O\left((\Delta x)^3\right)}{2} \end{aligned}$$

$$
\begin{aligned}
&+ \frac{u\Delta t}{2\Delta x}\left(\psi_i^n + \Delta x \psi_{i,x}^n + \frac{(\Delta x)^2}{2}\psi_{i,xx}^n + \frac{(\Delta x)^3}{3!}\psi_{i,xxx}^n + O\left((\Delta x)^5\right)\right. \\
&- \left.\psi_i^n + \Delta x \psi_{i,x}^n - \frac{(\Delta x)^2}{2}\psi_{i,xx}^n + \frac{(\Delta x)^3}{3!}\psi_{i,xxx}^n + O\left((\Delta x)^5\right)\right) \\
&= \Delta t\left(\psi_i^{n} + u\psi_{i,x}^n\right) - \frac{(\Delta t)^2}{2} + O\left((\Delta x)^3\right) + \frac{u\Delta t}{2\Delta x}\left(\frac{(\Delta x)^3}{3!}\psi_{i,xxx}^n + O\left((\Delta x)^5\right)\right).
\end{aligned}
$$

Therefore the Lax–Friedrichs scheme is only first order in time, but is second order in space. The reason for showing this scheme will become obvious now as we consider its stability analysis. One can see that it has some of the structure of the forward-time–centered-space scheme.

Exercise 3.16. *Show that local truncation error for the Lax–Friedrichs scheme is* $O\left((\Delta t)^2, (\Delta x)^3\right)$.

Applying the von Neumann stability analysis to the difference equation for the Lax–Friedrichs scheme, we have

$$
\begin{aligned}
z^{n+1} &= \frac{1}{2}\left(z^n e^{ik(j+1)\Delta x} + z^n e^{ik(j-1)\Delta x}\right) - \frac{C}{2}\left(z^n e^{ik(j+1)\Delta x} - z^n e^{ik(j-1)\Delta x}\right) \\
&= z^n e^{ikj\Delta x}\left(\frac{1}{2}\left(e^{ik\Delta x} + e^{-ik\Delta x}\right) - \frac{C}{2}\left(e^{ik\Delta x} - e^{-ik\Delta x}\right)\right) \\
&= z^n e^{ikj\Delta x}\left(\cos k\Delta x - iC\sin k\Delta x\right). \qquad (3.101)
\end{aligned}
$$

Therefore, from (3.101) we can see that the amplification factor for the Lax–Friedrichs scheme is

$$
\xi_{LaxF} = \cos k\Delta x - iC\sin k\Delta x. \qquad (3.102)
$$

Forming the modulus of (3.102), we obtain

$$
\left|\xi^2\right| = \cos^2 k\Delta x + C^2\sin^2 k\Delta x \equiv 1 - \left(1 - C^2\right)\sin^2 k\Delta x. \qquad (3.103)
$$

We require (3.103) to be less than or equal to 1, which can only occur if the term $1 - C^2 > 0$, which implies that $u\frac{\Delta t}{\Delta x} \leq 1$. The important point about the stability analysis for this scheme is that, by adding the artificial viscosity to the forward-time–centered-space scheme, we have been able to stabilize the scheme. We have shown that it is consistent with the differential equation as well, as there are neither $(\Delta t)^{-1}$ nor $(\Delta t)^0$ terms in the truncation error.

If we now look at the truncation error and replace ψ_i'' with $u^2\psi_{i,xx}$, then we can show that the modified equation that this scheme solves for is

$$
\psi_i^{n'} + u\psi_{i,xx}^n = \frac{(\Delta x)^2}{2\Delta t}\left(1 - C^2\right)\psi_{i,xx}^n, \qquad (3.104)
$$

where, due to the even order derivative on the right-hand side of (3.104), we see that this scheme is dissipative, and as such we would see some damping of the wave amplitude.

Exercise 3.17. *Verify that the modified equation for the Lax–Friedrichs scheme is* (3.104).

3.5.4 Lax–Wendroff scheme

Recalling the difference equation for the Lax–Wendroff scheme, we have

$$\psi_i^{n+1} = \psi_i^n - \frac{c\Delta t}{2\Delta x}\left(\psi_{i+1}^n - \psi_{i-1}^n\right) + \frac{c^2(\Delta t)^2}{2(\Delta x)^2}\left(\psi_{i+1}^n - 2\psi_i^n + \psi_{i-1}^n\right) = 0.$$

The global truncation error for this scheme is given by

$$\begin{aligned}
\tau_{LW} &= \frac{\psi\left(x_i, t^{n+1}\right) - \psi\left(x_i, t^n\right)}{\Delta t} + \frac{u}{2\Delta x}\left(\psi\left(x_i + \Delta x, t^n\right) - \psi\left(x_i - \Delta x, t^n\right)\right) \\
&- u^2\frac{\Delta t}{(\Delta x)^2}\left(\psi\left(x_i + \Delta x, t^n\right) - 2\psi\left(x_i, t^n\right) + \psi\left(x_i - \Delta x, t^n\right)\right) \\
&= \left(\psi_i^{n'} + u\psi_{i,x}^n\right) + \frac{\Delta t}{2}\left(\psi_i^{n''} - c^2\psi_{i,xx}^n\right) + \frac{(\Delta t)^2}{3!}\psi_i^{n'''} + c\frac{(\Delta x)^2}{2}\psi_{i,xxx}^n \\
&- \frac{c^2\Delta t(\Delta x)^2}{4!}\psi_{i,xxxx}^n.
\end{aligned} \tag{3.105}$$

At first glance, it may appear that the Lax–Wendroff scheme should only be first order in time, but, because we have replaced temporal derivatives with a spatial one, we know that the term multiplying Δt in (3.105) is actually the second order version of the advection/wave equation, and so both the first and second term in (3.105) are zero, and as such the Lax–Wendroff scheme is second order in time and space. Whilst we have the $\Delta t(\Delta x)^2$ as well in (3.105), we can see from the consistency analysis that the second-order spatial term dominates the first-order temporal term as they tend to zero.

Exercise 3.18. *Show that the local truncation error for the Lax–Wendroff scheme is* $O\left((\Delta t)^3, (\Delta x)^3\right)$.

To determine if there are any constraints on the spatial and temporal step sizes, we again consider the Fourier stability of (2.46) by substituting $\psi_j^n = z^n e^{ikj\Delta x}$ into (2.46), which yields

$$\begin{aligned}
z^{n+1}e^{ikj\Delta x} &= z^n e^{ikj\Delta x} - \frac{C}{2}\left(z^n e^{ik(j+1)\Delta x} - z^n e^{ik(j-1)\Delta x}\right) \\
&+ \frac{C^2}{2}\left(z^n e^{ik(j+1)\Delta x} - 2z^n e^{ikj\Delta x} + z^n e^{ik(j-1)\Delta x}\right).
\end{aligned} \tag{3.106}$$

Factorizing as before, we can see that the amplification factor is

$$\xi_{LaxW} = 1 - C^2(1 - \cos k\Delta x) - Ci\sin k\Delta x. \tag{3.107}$$

The amplification factor in (3.107) lies on an ellipse centered at $\left(1 - C^2, 0\right)$ with semiaxes $\left(C^2, -C\right)$, which lies inside the unit circle if $|C| \leq 1$.

It can be shown that the modified equation for the Lax–Wendroff scheme is

$$\psi' + u\psi_x = -\frac{1}{6}u\left(\Delta x\right)^2\left(1 - C^2\right)\psi_{xxx}, \tag{3.108}$$

which implies that the Lax–Wendroff scheme is dispersive, and it can be shown through the group velocity analysis that the scheme will emit oscillations behind the traveling wave.

Exercise 3.19. *Show that the leading term in the modified equation for the Lax–Wendroff scheme is* (3.108).

Exercise 3.20. *Determine if the Lax–Wendroff scheme has any amplitude or phase errors.*

3.5.5 Leap-frog, centered-time–centered-space scheme

The next scheme that we presented in the last chapter after the Lax–Wendroff scheme was the centered-time–centered-space, or leap-frog scheme. If we recall the difference equation for this scheme, then we have

$$\frac{\psi_i^{n+1} - \psi_i^{n-1}}{\Delta t} + u\frac{\psi_{i+1}^n - \psi_{i-1}^n}{2\Delta x} = 0,$$

and, to form the global truncation error for this scheme, we substitute $\psi\left(x_i, t^n\right)$ into the difference equations above and expand about this point, which results in

$$\begin{aligned}
\tau_{CTCS} &= \psi\left(x_i, t^{n+1}\right) - \psi\left(x_i, t^n\right) - u\frac{\Delta t}{2\Delta x}\left(\psi\left(x_{i+1}, t^n\right) - \psi\left(x_{i-1}, t^n\right)\right) \\
&= \psi\left(x_i, t^n + \Delta t\right) - \psi\left(x_i, t^n\right) - u\frac{\Delta t}{2\Delta x}\left(\psi\left(x_i + \Delta x, t^n\right) - \psi\left(x_i - \Delta x, t^n\right)\right) \\
&= \psi_i^n + \Delta t\psi_i^{n'} + \frac{(\Delta t)^2}{2}\psi_i^{n''} + \frac{(\Delta t)^3}{3!}\psi_i^{'''} + O\left((\Delta t)^4\right) - \psi_i^n + \Delta t\psi_i^{n'} \\
&- \frac{(\Delta t)^2}{2}\psi_i^{n''} + \frac{(\Delta t)^3}{3!}\psi_i^{n'''} + O\left((\Delta t)^4\right) \\
&+ u\frac{2\Delta t}{2\Delta x}\left(\psi_i^n + \Delta x\psi_{i,x}^n + \frac{(\Delta x)^2}{2}\psi_{i,xx}^n + \frac{(\Delta x)^3}{3!}\psi_{i,xxx}^n\right. \\
&- \left.\psi_i^n + \Delta x\psi_{i,x}^n - \frac{(\Delta x)^2}{2}\psi_{i,xx} + \frac{(\Delta x)^3}{3!}\right) \\
&= 2\Delta t\left(\psi_i^{n'} + u\psi_{i,x}^n\right) + \frac{(\Delta t)^3}{3!}\psi_i^{n'''} + O\left((\Delta t)^4\right) \\
&+ u\frac{2\Delta t}{2\Delta x}\left(\frac{(\Delta x)^3}{3!}\psi_{i,xxx}^n + O\left((\Delta x)^5\right)\right).
\end{aligned} \tag{3.109}$$

Therefore, the leap-frog scheme has a global truncation error of

$$\tau_{CTCS} \sim O\left((\Delta t)^2\right) + O\left((\Delta x)^2\right),$$

and there we can see that this scheme is second order in time and space.

Exercise 3.21. *Show that the local truncation error for the centered-time–centered-space scheme is* $O\left((\Delta t)^3, (\Delta x)^3\right)$.

Exercise 3.22. *Derive the modified equation for this scheme and determine if it is dispersive or dissipative.*

Exercise 3.23. *Determine if this scheme has any amplitude and/or phase errors.*

If we perform the full numerical-dispersion analysis on the leap-frog scheme, we obtain

$$\sin \omega \Delta t = C \sin k \Delta x. \tag{3.110}$$

If $C < 1$, then ω is a real quantity, and as such the scheme is neutral, which implies that the scheme is propagating at the correct speed. Forming the differential-discrete equation for the second-order centered difference for the spatial component shows that this scheme is dispersive.

If we now apply the von Neumann stability analysis to the leap-frog scheme, we have

$$z^{n+1} e^{ikj\Delta x} = z^{n-1} e^{ikj\Delta x} - C\left(z^n e^{ik(j+1)\Delta x} - e^{ik(j-1)}\right). \tag{3.111}$$

This is the first of our multistep numerical schemes, and we see that the characteristic equation is given by

$$z^2 - 1 + 2iCz \sin k\Delta x = 0, \tag{3.112}$$

which has solutions

$$z(k) = -iC \sin k\Delta x \pm \left(1 - C^2 \sin^2 k\Delta x\right)^{\frac{1}{2}}. \tag{3.113}$$

From the definition earlier, we require the roots z_1, z_2 to have real parts that are less than or equal to 1. We see from (3.113) that this is only possible if $|C| \leq 1$, which means that $C\frac{\Delta t}{\Delta x} \leq 1$. However, is should be noted here that one of the two roots in (3.113) is associated with the computational mode, which for this case is undamped and can grow for features of a certain lengthscale [48]. One way to control this undamped mode is to introduce what is referred to as the Robert–Asselin–Williams (RAW) filter [163,5,201,202]. We shall not go into details here about this filter, but the reader is referred to the references just cited.

3.6 Linear multistep methods

In this section we shall look at some of the Adams–Bashforth schemes to determine their order and stability domains.

3.6.1 Stability of Adams–Bashforth 2 scheme

If we recall the difference equation for the Adams–Bashforth 2 scheme, we have

$$\psi^{n+1} - \psi^n = \Delta t \left(\frac{3}{2} f\left(\psi^n, t^n\right) - \frac{1}{2} f\left(\psi^{n-1}, t^{n-1}\right) \right).$$

Thus comparing to the general formula in (3.35), we have

$$\alpha_0 = 1, \quad \alpha_1 = -1, \quad \alpha_2 = 0, \quad \beta_0 = 0, \quad \beta_1 = \frac{3}{2}, \quad \beta_2 = -\frac{1}{2}.$$

Therefore, checking condition (3.36a) first to show that this scheme is consistent with the ordinary differential equation, we have

$$\sum_{i=0}^{2} \alpha_i = \alpha_0 + \alpha_1 + \alpha_2 = 1 - 1 + 0 = 0,$$

and thus the Adams–Bashforth 2 scheme is consistent. Next we determine the order of the scheme. Moving on to condition (3.36b), we start with $k = 1$, which gives us

$$\sum_{i=0}^{2} \frac{i^k}{k!} \alpha_{s-i} = \sum_{i=1}^{2} \frac{i^{k-1}}{(k-1)!} \beta_{s-i},$$
$$\frac{0^1}{1!}\alpha_2 + \frac{1^1}{1!}\alpha_1 + \frac{2^1}{1!}\alpha_0 = \frac{0^0}{0!}\beta_2 + \frac{1^0}{0!}\beta_1 + \frac{2^0}{0!}\beta_0,$$
$$1 = 1,$$

and from condition (3.36c) we can say that this scheme is more than first order in time. Now we consider the case $k = 2$, which implies

$$\sum_{i=1}^{2} \frac{i^2}{2!} \alpha_{s-i} = \sum_{i=0}^{2} \frac{i^1}{1!} \beta_{s-i},$$
$$\frac{0^2}{2!}\alpha_2 + \frac{1^2}{2!}\alpha_1 + \frac{2^2}{2!}\alpha_0 = \frac{0^1}{1!}\beta_2 + \frac{1^1}{1!}\beta_1 + \frac{2^1}{1!}7\beta_0,$$
$$\frac{3}{2} = \frac{3}{2}.$$

If we now consider the case $k = 3$, we have

$$\sum_{i=1}^{2} \frac{i^3}{3!} \alpha_{s-i} = \sum_{i=0}^{2} \frac{i^2}{2!} \beta_{s-i},$$
$$\frac{0^3}{3!}\alpha_2 + \frac{1}{3!}\alpha_1 + \frac{2^3}{3!}\alpha_0 = \frac{0}{2!}\beta_2 + \frac{1^2}{2!}\beta_0 + \frac{2^2}{2!}\beta_0$$
$$\frac{7}{6} \neq \frac{3}{4}.$$

Therefore, by evaluating condition (3.36c), we see that $p = 2$, and thus this scheme is second order in time.

We now consider the von Neumann stability for the second-order Adams–Bashforth scheme, where we have

$$
\begin{aligned}
z^{n+2}e^{ikj\Delta x} &= z^{n+1}e^{ikj\Delta x} - Ce^{ijk\Delta x}\left[\frac{3z^{n+1}}{2}\left(e^{ik\Delta x} - e^{-ik\Delta x}\right) - \frac{z^n}{2}\left(e^{ik\Delta x} - e^{-ik\Delta x}\right)\right], \\
\Longrightarrow \xi_{AB2} &= z^2 - z\left(1 - \frac{3Ci}{2}\sin k\Delta x\right) + \frac{Ci}{2}\sin k\Delta x, \\
\Longrightarrow \xi_{AB2} &= \frac{1}{2}\left[1 - \frac{3i}{2}C\sin k\Delta x \pm \sqrt{1 - iC\sin k\Delta x - \frac{9C^2}{4}\sin^2 k\Delta x}\,\right]. \qquad (3.114)
\end{aligned}
$$

We can see that this method produces both a physical and a computational mode. It is shown in [47] that the computational mode is heavily damped, which is a desirable feature of the two step schemes. However, it is also shown that the physical mode is slightly unstable, yet is not dominating the performance of the scheme. It should be noted as well that we have used a second-order central difference for the spatial derivatives here, it is mentioned in [47] that if a fourth-order central difference is applied to the spatial derivative, then the instability of the physical mode can be damped further for more accurate results, which avoids applying the RAW filter, which is needed with the leap-frog scheme.

If we now consider the zero stability condition for linear multistep methods, we can see that for the Adams–Bashforth 2 scheme we have the first characteristic polynomial given by $\phi^{p+1} - \phi^p = 0$, which means that there are two roots, the primary root being $\psi_0 = 1$, as required for consistency, and then $\phi_1 = 0$, which means that the scheme is strongly stable. In fact, if we recall the derivation of the Adams–Bashforth schemes, we will remember that the multistep component of this model is dependent on different time evaluations of function f, and not the time derivative on the left-hand side, as such we can say that **all** Adams–Bashforth schemes are strongly stable.

If we now form the stability region, we see that the ratio of the two characteristic polynomials for the AB2 scheme is given by

$$
\mu = \lambda\Delta t = \frac{2\left(\phi^2 - \phi\right)}{3\phi - 1}. \qquad (3.115)
$$

Converting the region defined in (3.115) into polar coordinates results in the shaded domain in Fig. 3.3.

In Fig. 3.4 we have plotted the stability domains of the first to fifth order Adams–Bashforth schemes to illustrate that the size of the stability domains decreases when we increase the number of steps, but also that the region spills over into the positive real part of the domain.

Exercise 3.24. *Show that the Adams–Bashforth 3 scheme given by*

$$
\psi_i^{n+1} = \psi_i^n + \Delta t\left(\frac{23}{12}f_i^n - \frac{4}{3}f_i^{n-1} + \frac{5}{12}f_i^{n-2}\right)
$$

is third order in time.

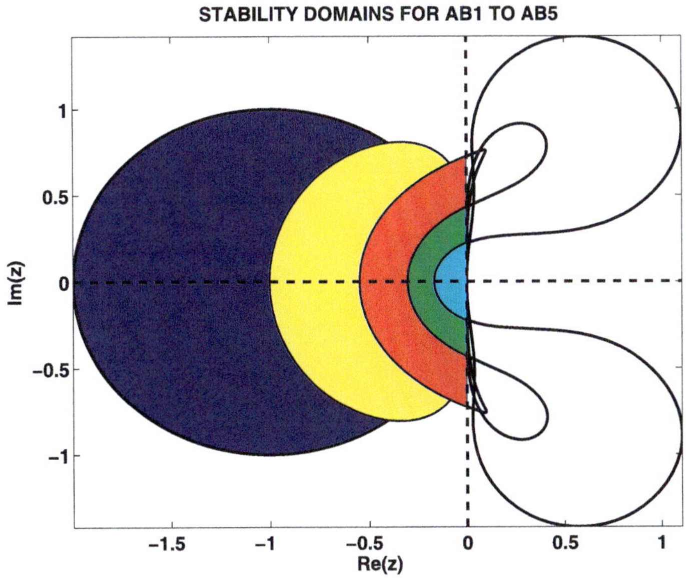

FIGURE 3.3 Plot of the complex plane where the values in the shaded shape are the values for which the second-order Adams–Bashforth finite difference scheme is stable.

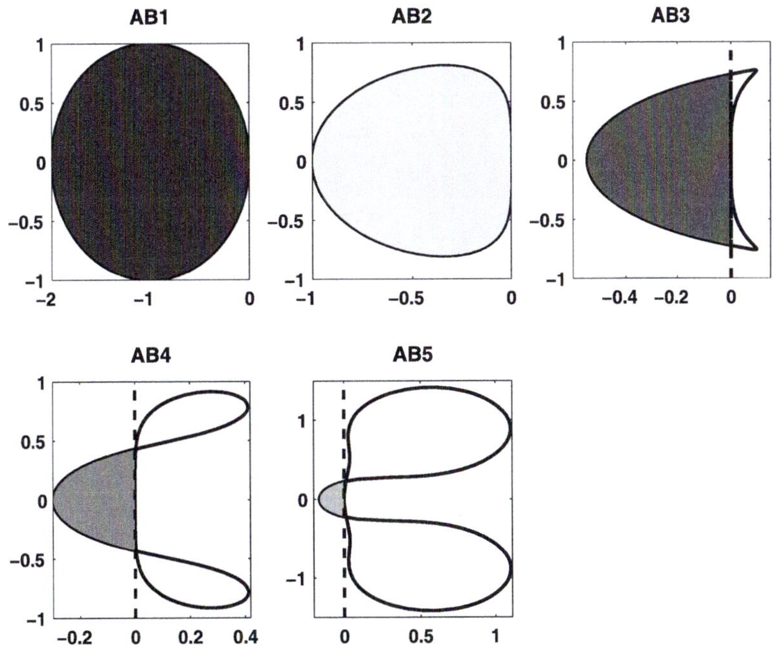

FIGURE 3.4 Plot of the stability domains for the first to the fifth order Adams–Bashforth schemes to illustrate that the stability domain becomes smaller, the higher the order of the finite difference scheme.

Exercise 3.25. *Find the stability domains for the third-, fourth-, and fifth-order Adams–Bashforth schemes.*

Exercise 3.26. *Show that the numerical approximation to the advection equation given by*

$$\psi_i^{n+1} = \psi_i^n \pm \frac{u\Delta t}{2\Delta x}\left(3\psi_i^n - 4\psi_{i\pm1}^n + \psi_{i\pm2}^n\right) + \frac{u^2(\Delta t)^2}{2(\Delta x)^2}\left(\psi_i^n - 2\psi_{i\pm1}^n + \psi_{i\pm2}^n\right) \tag{3.116}$$

is consistent with the differential equation, given u is positive or negative, and that its global truncation error shows it to be second order is space and time. Show that its CFL condition is $\Delta t \leq 2\frac{\Delta x}{|u|}$, *and show that the scheme is dispersive by finding its modified equation. Note that this scheme is referred to as the* ***beam-warming scheme****.*

3.7 Consistency and stability of explicit Runge–Kutta methods

As we have seen from the multistep linear scheme, the requirement for a scheme to be consistent is for the truncation error to tend to zero as $h \to 0$. Therefore, the truncation error for the general form of a Runge–Kutta scheme is

$$\begin{aligned}\tau_n &\equiv \frac{\psi\left(t^{n+1}\right) - \psi\left(t^n\right)}{\Delta t} - \sum_{i=1}^{s} b_i f\left(\psi^n + h\sum_{j=1}^{s} a_{ij}k_j, t^n + c_i\Delta t\right)\\ &= \left(1 - \sum_{i=1}^{s} b_i\right)\psi^{n'} + \frac{\Delta t}{2}\psi'' - \sum_{i=1}^{s} b_i\left(c_i\Delta t f_t + \Delta t\sum_{j=1}^{s} a_{ij} f f_\psi\right)\\ &+ O\left((\Delta t)^2\right),\end{aligned} \tag{3.117}$$

where $f_t \equiv \frac{\partial f}{\partial t}$ and $f_\psi \equiv \frac{\partial f}{\partial \psi}$. Therefore, the first-order condition for the general Runge–Kutta scheme to be consistent is $\sum_{i=1}^{s} b_i = 1$. This removes the term in (3.117) that is not dependent on Δt. To make the Runge–Kutta scheme consistent with the second-order derivative term $\psi'' = f_t + f f_\psi$, we require $\frac{1}{2} = \sum_{i=1}^{s} b_i c_i = \sum_{i=1}^{s} b_i \sum_{j=1}^{s} a_{ij}$.

To help illustrate these consistency requirements, let's consider the explicit case $s = 2$. Then there are two possible formulation for the Runge–Kutta methods which satisfy the consistency conditions $b_1 + b_2 = 1$ and $\frac{1}{2} = b_1c_1 + b_2c_2 = b_2a_{21}$, which are

$$\begin{array}{c|cc} 0 & 0 & 0 \\ \frac{1}{2} & \frac{1}{2} & 0 \\ \hline & 0 & 1 \end{array} \quad\Longrightarrow\quad \psi^{n+1} = \psi^n + \Delta t f\left(\psi^n + \frac{\Delta t}{2} f^n, t^n + \frac{\Delta t}{2}\right),$$

$$\begin{array}{c|cc} 0 & 0 & 0 \\ 1 & 1 & 0 \\ \hline & \frac{1}{2} & \frac{1}{2} \end{array} \quad \Longrightarrow \quad \psi^{n+1} = \psi^n + \frac{\Delta t}{2}\left(f^n + f\left(\psi^n + \Delta t f^n, t^n\right)\right),$$

where the first scheme above is the forward Euler method, and the second scheme is referred to as the improved Euler method.

We can clearly see from the consistency conditions that the two Runge–Kutta schemes above are consistent to the second order term in the Taylor series expansion of the differential equation.

If we now consider a Taylor series expansion truncated at the third-order term, then we have

$$\psi^{n+1} \approx \psi^n + \Delta t \psi^{n'} + \frac{(\Delta t)^2}{2!}\psi^{n''} + \frac{(\Delta t)^3}{3!}\psi^{n'''}. \tag{3.118}$$

For stability we substitute the solution $\psi' = \lambda\psi$ into (3.118) and obtain

$$\psi^{n+1} \approx \left(1 + \Delta t\lambda + \frac{(\Delta t)^2\lambda^2}{2!} + \frac{(\Delta t)^3\lambda^3}{3!}\right)\psi^n, \tag{3.119}$$

where the stability depends on the amplification factors, the term multiplying ψ^n in (3.119) lying in, or on, the unit circle. Therefore, we now have the **stability polynomial** for the Runge–Kutta methods, which is defined for third-order schemes as

$$\rho(z) = z - \left(1 + h\lambda + \frac{h^2\lambda^2}{2!} + \frac{h^3\lambda^3}{3!}\right). \tag{3.120}$$

Given the stability polynomial in (3.120), the stability condition for the third order schemes is given by

$$\left|1 + h\lambda + \frac{h^2\lambda^2}{2!} + \frac{h^3\lambda^3}{3!}\right| \leq 1. \tag{3.121}$$

In Fig. 3.5 we have plotted the stability regions for the Runge–Kutta schemes up to the fourth order. As we can see from Fig. 3.5, when $s = 1$, the stability region is in the negative part of the complex plane and is of the form of an ellipse; when we consider the second-order methods, we have that the stability region has stretched and includes a larger range of values. For the third-order Runge–Kutta methods, we have a larger region than for the two lower order schemes, but now there is some creepage into the positive values for the real number component. Finally, when we consider the stability region for the fourth-order methods, we see that it is possible for a larger group of the complex components of the roots of (3.121) to have a positive component.

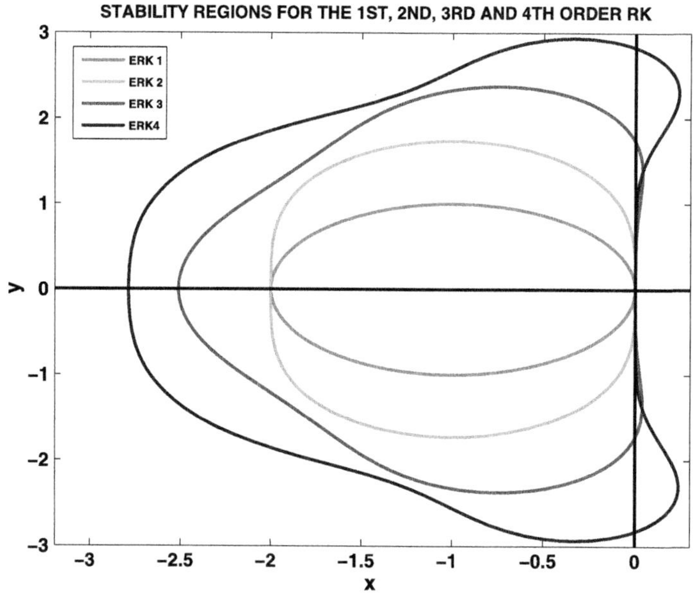

FIGURE 3.5 Plot of the stability domain for the Runge–Kutta methods up to fourth order.

Exercise 3.27. *Find the conditions that make the 3-stage explicit Runge–Kutta scheme*

$$\psi^{n+1} = \psi^n + \sum_{i=1}^{3} b_i k_i,$$
$$k_1 = f(\psi, t),$$
$$k_2 = f(\psi + \Delta t c_2 k_1, t + hc_2),$$
$$k_3 = f(y + \Delta t(c_3 - a_{32})k_1 + \Delta t a_{32} k_2 t + \Delta t c_3),$$

second order. Derive the associated stability polynomial for this scheme.

3.8 Implicit schemes

We now move on to consider a few of the implicit schemes that we introduced in the last chapter. The first one that we consider is the backward Euler scheme.

3.8.1 Backward Euler scheme

The next scheme we consider is the backward Euler scheme, where we recall that the difference equation for this scheme is given by

$$\frac{\psi_i^{n+1}-\psi_i^n}{\Delta t}+u\frac{\psi_{i+1}^{n+1}-\psi_{i-1}^{n+1}}{2\Delta x}=0.$$

If we now substitute the true state into the difference equation above and expand the Taylor series about $\psi\left(x_i, t^{n+1}\right)$ then we have

$$\begin{aligned}\tau_{BE} &= \psi_i^{n+1}-\psi_i^{n+1}+\Delta t\psi_i^{(n+1)'}-\frac{(\Delta t)^2}{2}\psi^{(n+1)''}+O\left((\Delta t)^3\right)\\ &+ \frac{u\Delta t}{2\Delta x}\left(\psi_i^{n+1}+\Delta x\psi_{i,x}^{n+1}+\frac{(\Delta x)^2}{2}\psi_{i,xx}^{n+1}+O\left((\Delta x)^3\right)-\psi_i^{n+1}\right.\\ &+ \left.\Delta x\psi_{i,x}^{n+1}-\frac{(\Delta x)^2}{2}\psi_{i,xx}^{n+1}+O\left((\Delta x)^3\right)\right). \end{aligned} \tag{3.122}$$

We can see the usual terms that cancel in (3.122), but because we expanded about $\psi\left(x_i, t^{n+1}\right)$, or rather we expanded $\psi\left(x_i, t^n\right)\equiv\psi\left(x_i, t^{n+1}-\Delta t\right)$, we can see that the global truncation error for the backward Euler is $O\left((\Delta t)^2\right)+O\left((\Delta x)^2\right)$. We can also see that the scheme is consistent with the advection equation.

We now move on to consider the stability properties of this scheme through the von Neumann analysis, thus applying the Fourier stability analysis with $\psi_j^n=z_ne^{ikj\Delta x}$, where it can easily be shown that

$$\left(1+\frac{c\Delta t}{\Delta x}\left(1-e^{-k\Delta x}\right)\right)z_{n+1}=z_n. \tag{3.123}$$

Therefore, the amplification factor is

$$z_{n+1}=\frac{1}{1+\dfrac{c\Delta t}{\Delta x}\left(1-e^{-ik\Delta x}\right)}z_n\equiv\left(\frac{1}{1+\nu-\nu e^{ik\Delta x}}\right)z_n. \tag{3.124}$$

As with the explicit upwind scheme, the amplitude factor for the implicit scheme also consists of a real and complex component. The real component is a circle centered at $1+\nu$ which is outside the unit circle, however, as we require the reciprocal of this circle, that reciprocal circle lies within the unit circle, which then makes the implicit upwind scheme **unconditionally stable**.

3.8.2 Crank–Nicolson scheme

Recalling the difference equation for the Crank–Nicolson scheme as

$$\frac{\psi_i^{n+1}-\psi_i^n}{\Delta t}=-\frac{u}{2}\left(\frac{\psi_{i+1}^n-\psi_{i-1}^n}{2\Delta x}+\frac{\psi_{i+1}^{n+1}-\psi_{i-1}^{n+1}}{2\Delta x}\right),$$

to form the truncation error, we write a Taylor series of ψ about the value $\psi(x_i, t^n)$, which yields

$$
\begin{aligned}
\tau_{CN} &= \frac{\psi_i^n + \Delta t \psi_i^{n'} + \frac{(\Delta t)^2}{2}\psi_i^{n''} + \frac{(\Delta t)^3}{3!}\psi_i^{n'''} + O\left((\Delta t)^4\right) - \psi_i^n}{\Delta t} \\
&\quad + \frac{u}{4\Delta x}\left(\left(\psi_i^n + \Delta x \psi_{i,x}^n + \frac{(\Delta x)^2}{2}\psi_{i,xx}^n + \frac{(\Delta x)^3}{3!}\psi_{i,xxx}^n + \cdots \right.\right. \\
&\quad \left. -\psi_i^n + \Delta x \psi_{i,x}^n - \frac{(\Delta x)^2}{2}\psi_{i,xx}^n + \frac{(\Delta x)^3}{3!}\psi_{i,xxx}^n - \cdots \right) \\
&\quad + \left(\psi_i^n + \Delta x \psi_{i,x}^n + \Delta t \psi_i^{n'} + \frac{(\Delta x)^2}{2}\psi_{i,xx}^n + \frac{(\Delta t)^2}{2}\psi_i^{n''} + \Delta t \Delta x \psi_{i,x}^{n'} \right. \\
&\quad + \frac{(\Delta x)^3}{3!}\psi_{i,xxx}^n + \cdots - \psi_i^n + \Delta x \psi_{i,x}^n - \Delta t \psi_i^{n'} + \frac{(\Delta x)^2}{2}\psi_{i,xx}^n - \frac{(\Delta t)^2}{2}\psi_i^{n''} \\
&\quad \left.\left. + \Delta x \Delta t \psi_{i,x}' + \frac{(\Delta x)^3}{3!}\psi_{i,xxx}^n + \cdots \right)\right).
\end{aligned} \tag{3.125}
$$

While at first it may appear that the Crank–Nicolson scheme may only be first order in time, we have to notice the cross-derivative term in (3.125), which can be shown to be $\psi' = -u\psi_x \Rightarrow \psi_x' = -u\psi_{xx}$, which means that the second-order mixed term can be rewritten in terms of the second-order spatial derivative of ψ, and as such the Crank–Nicolson scheme is second order both in time and space.

If we now apply the von Neumann stability analysis to the Crank–Nicolson scheme, and now recognize that the second-order spatial discretization produces $i \sin k\Delta x$ in the stability analysis, then we have

$$
z^{n+1}\left(1 + \frac{Ci}{2}\sin k\Delta x\right) = z^n\left(1 - \frac{Ci}{2}\sin k\Delta x\right), \tag{3.126}
$$

and so from (3.126) we can see that the amplification factor for this scheme is

$$
\xi_{CN} = \frac{\left(1 - \frac{Ci}{2}\sin k\Delta x\right)}{\left(1 + \frac{Ci}{2}\sin k\Delta x\right)}. \tag{3.127}
$$

It can be seen from (3.127) that there is no value of k that can make $|\xi_{CN}| > 1$, nor any value of C, therefore, as with the backward Euler scheme, the Crank–Nicolson scheme is unconditionally stable. It should be noted here that the modulus of the amplification factor is 1 for all wavenumbers, which implies that the scheme does not dampen the advective quantity. Following the same arguments that we have presented here, it is also possible to show that the box scheme that we presented in the last chapter is also second order in space and time, as well as being unconditionally stable.

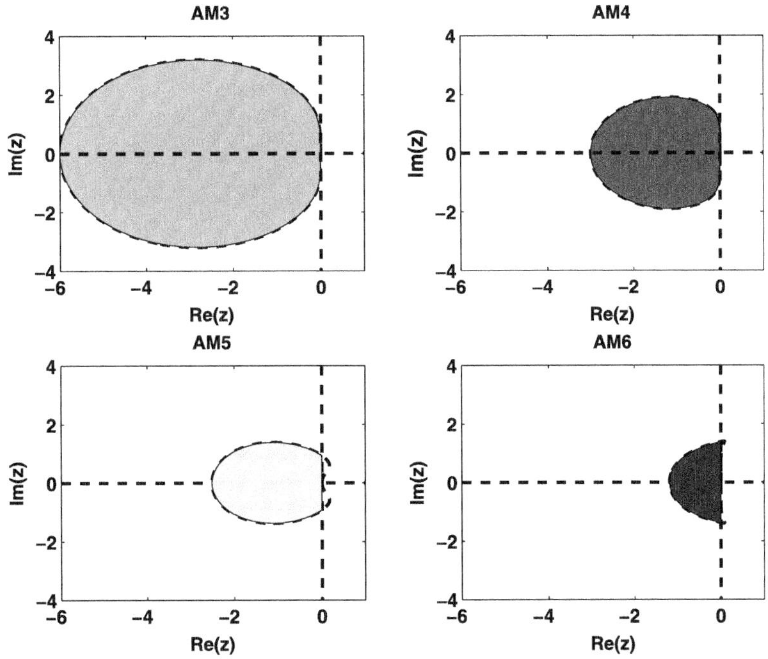

FIGURE 3.6 Plot of the stability domain for the Adams–Moulton schemes from third to sixth order.

Exercise 3.28. *Prove that the box numerical approximation to the one-dimensional advection equation,*

$$\frac{1}{2}\frac{\psi_j^{n+1}-\psi_j^n}{\Delta t}+\frac{1}{2}\frac{\psi_{j+1}^{n+1}-\psi_{j+1}^n}{\Delta x}=-\frac{u}{2}\left(\frac{\psi_{j+1}^n-\psi_j^n}{\Delta x}+\frac{\psi_{j+1}^{n+1}-\psi_j^{n+1}}{\Delta x}\right),$$

is second order in time and space, and prove that it is unconditionally stable by using von Neumann stability analysis.

While there were many other implicit schemes that were introduced in the last chapter, we should notice that the Adams–Moulton schemes and the backward differentiation methods are all multistep methods, and as such follow the conditions that we presented for the Adams–Bashforth schemes. However, we do have some plots of their stability domains when applied to the test differential equations, where we can see that both of the shaded unstable areas incur into to the negative half of the complex plane at order three and higher for the BDF schemes (Figs. 3.7 and 3.8), but that the stability domains for the Adams–Moulton schemes are decreasing in areas like the Adams–Bashforth schemes do, but not as quickly as we increase the order of the scheme (see Fig. 3.6). For the implicit Runge–Kutta schemes, we again would follow the analysis set out for the explicit versions, but the reader is referred to the great book by Butcher for a more detailed analysis of these schemes.

We will, however, now look over the properties of combined Adams–Bashforth and Adams–Moulton schemes, namely the predictor–corrector schemes.

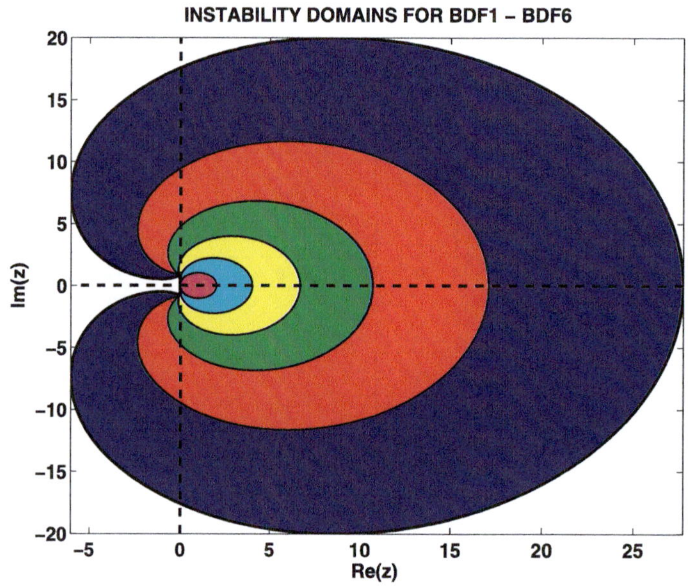

FIGURE 3.7 Plot of the overlapped instability domain for the backward differentiation formula from order 1 to 6.

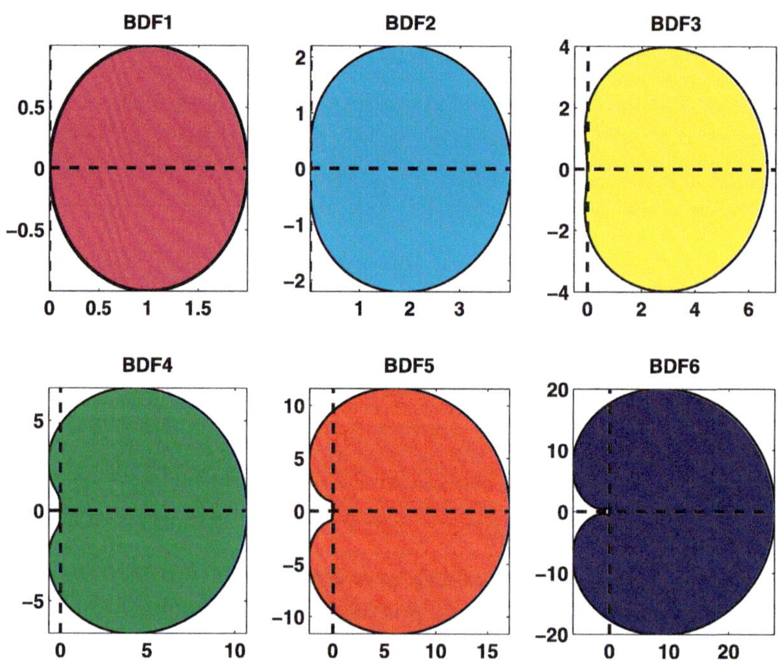

FIGURE 3.8 Plot of the instability domain for the backward differentiation formula from order 1 to 6 in separate plots.

3.9 Predictor–corrector methods

As we mentioned in the last chapter, it is quite often desirable to combine the explicit and implicit schemes into a predictor–corrector formulation, where one would use the explicit scheme to move the solution ψ to ψ^{n+1}, and then use this to evaluate the right-hand side of the partial differential equation at the *implicit* level, to use this in place of the implicit step in the implicit scheme, and then create a corrected estimate for ψ^{n+1}, but recalculate a new version of $f\left(\psi^{n+1}, t^{n+1}\right)$. In [27] it is stated that the scheme that we have just mentioned is referred to as *predict–evaluate–correct–evaluate*, which is quite often abbreviated to *PECE*. However, if we had not done the second evaluation step, then this method would be referred to as a *PEC*. In fact, if one reads the wider literature on the predictor–corrector methods, one could even form an iterative process to try and improve the estimates of both ψ^{n+1} and f^{n+1}. These schemes would be denoted as $P\left(EC\right)^k$ or $P\left(EC\right)^k E$ schemes.

Here we shall only analyze the second and third order Adams–Bashforth–Adams–Moulton predictor–corrector schemes, but it should be noted that there are many different combinations of different explicit and implicit schemes that can be developed.

First we recall that the second-order Adams–Bashforth scheme is given by

$$\psi_i^{n+1} = \psi_i^n - \frac{u\Delta t}{2}\left(3 * \psi_x^n - \psi_x^{n-1}\right), \tag{3.128}$$

and the equivalent order Adams–Moulton scheme, which is the trapezoidal rule, is given by

$$\psi_i^{n+1} = \psi_i^n - \frac{u\Delta t}{2}\left(\psi_x^n + \psi_x^{n+1}\right). \tag{3.129}$$

It can be shown that the local truncation errors for the two schemes above, when used in a predictor–corrector environment, are

$$\psi\left(t^{n+1}\right) - \widetilde{\psi}^{n+1} = \frac{5}{12}\psi'''\left(\zeta_{AB}\right), \tag{3.130a}$$

$$\psi\left(t^{n+1}\right) - \psi^{n+1} = -\frac{1}{12}\psi'''\left(\zeta_{AM}\right), \tag{3.130b}$$

where ζ_{AB} and ζ_{AM} are the error terms to evaluate the Taylor series exactly. If we now make the assumption that ψ''' does not change much in the interval between ζ_{AB} and ζ_{AM}, then taking the difference between (3.130b) and (3.130a), we obtain

$$\psi^{n+1} - \widetilde{\psi}^{n+1} \approx \frac{1}{12}\psi'''\left(\zeta\right). \tag{3.131}$$

Thus the error at each time step is

$$\left|\psi\left(t^{n+1}\right) - \psi^{n+1}\right| \approx \frac{1}{12}\psi'''\left(\zeta\right) \approx \frac{1}{6}\left|\psi^{n+1} - \widetilde{\psi}^{n+1}\right|. \tag{3.132}$$

Therefore, we can maintain the same order of accuracy for the predictor–corrector scheme as the original Adams–Bashforth and Adams–Moulton schemes, but the two approximations need to be somewhat close for the accuracy to be obtained. This comes back to the $PECE$, and how many times one does the EC steps in the center. The predictor–corrector scheme that we have presented here can be found in [92], where one can also find the full von Neumann stability analysis for this scheme, along with a plot of the stability domain. It is stated in Harrier et al. (1996) that by introducing the explicit Adams–Bashforth predictor into the implicit Adams–Moulton scheme, a large part of the stability domain of the implicit scheme is destroyed [70]. This was proven by Chase in 1962 [34].

3.10 Summary

In this chapter we have introduced the concepts of local and global truncation errors, consistency, together with many forms of stability: zero stability, absolute stability, A-stability, L-stability, along with convergence, and applied them to many of the explicit and implicit schemes introduced in the last chapter. We have seen that there are pros and cons for the Eulerian based schemes, that there are several problems of spurious modes for the multilevel schemes, and that controlling these computational or parasitic modes is quite complicated.

We have seen that there are restrictions on the ratio of the time step and the spatial grid spacing to ensure that the schemes are stable and consistent with the original differential equation. While we saw that these restrictions were somewhat removed for the implicit schemes, we also saw that we had to invert a matrix to find the solution. When the grid spacing is quite small to create a more accurate answer, we have a bigger matrix to invert, so in practical circumstances, an iterative scheme is used to approximate the matrix inversion, which introduces more errors into the problem. The reason for this restriction on the spacing is due to the Courant–Friedrichs–Lewy condition where the numerical domain of dependence has to contain the true domain of dependence of the original differential equation. We have also introduced different mechanism to determine dissipation, dispersive, amplitude, and phase errors of the numerical schemes to quantify the behavior we saw with some of the numerical schemes in the last chapter, where we can have damping or exciting of waves, as well as the numerical solution not traveling at the same speed as the true solution.

It is because of these just mentioned restrictions that people started to look at other ways to describe the flow of particles in the advection/transport situations, which is what lead to the introduction of the semi-Lagrangian approaches. In the next chapter we shall summarize the research undertaken trying to use the Lagrangian dynamical approach to advection, rather than the Eulerian through a graphical approach, and then describe a full attempt to use a pure Lagrangian approach to model advection, and finally, we will discuss what in the past was referred to as the *quasi-Lagrangian* approach, which formed the foundation of what is now referred to as **semi-Lagrangian methods**.

CHAPTER

4

History of semi-Lagrangian methods

As seen in the previous chapters on numerical modeling, we consider the advection, wave equation, in its Eulerian form, but we can also consider the flow in a Lagrangian formulation as well and then discretize this formulation of the advection equation. Before introducing the theory, we shall review the different developments of the approximations to the Lagrangian formulation that led to adopting parts of both the Eulerian and the Lagrangian description of motion.

We start with a summary of the introduction to the 1982 paper by Bates and MacDonald [14]; the latter author has published many papers in the development of semi-Lagrangian schemes for numerical weather prediction in both Cartesian and spherical coordinates, for multiple dimensions, as well as for different order of time discretizations, along with different approximations to the forcings. As Bates and MacDonald indicate, this approach was first published with an application in meteorology in 1952 [56]. In [56] a graphical method was applied to solve the barotropic vorticity equation with a time step of 24 hours. This paper will be summarized in the next section.

The next development of the theory that leads to the semi-Lagrangian approach came in a 1955 paper by Welander [198], where it is shown that if we consider a set of fluid particles that are initially regularly distributed, then they soon become greatly deformed and are then rendered unsuitable for numerical integration. The first indication of a semi-Lagrangian approach for advection comes in [200], referred there as quasi-Lagrangian advection, where the approach still focuses on the movement of particles, but now the authors focused on the set of particles that arrived at a regular set of grid point and then traced **backwards** over a single time interval, Δt, to their ***departure points***. It is not always the case that the departure points will coincide with a grid point, and so the values of the tracer, or dynamical quantities, at neighboring grid points are interpolated to the departure points. Therefore, the reason why this approach is referred to as *semi-Lagrangian* is that in the true Lagrangian setting the set of particles change at every time step, whereas in the semi-Lagrangian approach we are following the set of particles that arrive at a grid point.

Throughout the 1960s and 1970s the semi-Lagrangian approach was applied to many different numerical prediction models as well as analyses of the properties of the semi-Lagrangian schemes; a summary of these applications can be found in [14]. However, it is the property of the semi-Lagrangian schemes that was demonstrated in by Robert in 1981 [161], with the primitive equations that made the semi-Lagrangian schemes appealing to operational numerical weather, ocean, and hydrological prediction. The property that was demonstrated in [161] was that the semi-Lagrangian approach, when compared to the equivalent

Semi-Lagrangian Advection Methods and Their Applications in Geoscience
https://doi.org/10.1016/B978-0-12-817222-3.00008-6

Euler formulation of advection, could use significantly larger time steps than the Eulerian formulation and remain numerically stable. This was demonstrated in a divergent barotropic model with a consistent spatial resolution for the time period, but with a time step greater than or equal to an hour.

As indicated in [14], the underlying idea for the semi-Lagrangian approach in [161] is to find the value of the field/tracer at the departure point through interpolating the values of the surrounding grid points near the departure point to the departure point. This then implies that if the advecting winds are strong then the departure point could be many spatial grid intervals, Δx, upstream from the arrival grid point. In [14] the authors call an approach where the interpolation points for the departure points are many gird intervals away form the arrival point a **multiply-upstream** scheme.

The focus of the Bates and MacDonald 1982 paper is first to verify that the scheme proposed in [161] is stable, which we shall derive in a later chapter, and determine some damping properties of the semi-Lagrangian scheme, and second, to extend the semi-Lagrangian approach to the horizontal advection term of a multi-level primitive equation model, as well as show that it is possible to obtain a faster numerical integration of these equations than was possible with the current Eulerian formulation at the time.

An important advancement in semi-Lagrangian approaches comes from the Bates and MacDonald (1982) paper [14], where they introduce the two-dimensional version of a semi-Lagrangian approach. This then means that for a bilinear interpolation scheme they use 4 surrounding grid points that form a box around departure point to obtain the value of the dynamical quantity at the departure point. They also introduce the bi-quadratic interpolation scheme which requires 9 grid points, where 8 of the points form a box around the departure point and then the last grid point is the nearest one in the interior of the box.

Another important result from [14] was that the stability criterion for the biinterpolation schemes was equal to the product of the stability criteria in the two different spatial directions. Therefore, using an equally spaced grid/mesh to approximate the flow, one only needs to derive the stability criterion in one direction and then ensure that both directions satisfy that stability criterion.

In [117], titled "*Accuracy of Multiply-Upstream, Semi-Lagrangian Advection Schemes*", we are introduced to the bicubic and biquartic interpolation schemes, as well as their stability criteria. As mentioned when discussing the biquadratic interpolation scheme, we require 9 grid points to obtain an approximation to the value of the tracer at the departure point; for the bicubic interpolation approach we require 16 grid points to form the interpolation, and for the biquartic interpolation we need 25 grid points. As one can see, to obtain a more accurate value of the tracer at the departure point requires a quickly increasing number of surrounding grid points. Remember, this is still only in two dimensions.

MacDonald published another ground-breaking paper on semi-Lagrangian methods in 1986 [118], which became the foundation for most of the world's operational numerical weather prediction center's implementations of semi-Lagrangian advection. The MacDonald 1986 paper titled "*A Semi-Lagrangian and Semi-Implicit Two Time-Level Integration Scheme*" introduces the use of midpoints for the back trajectory calculations, but also the semi-implicit approximation to the forcing term.

In [127], titled "*Semi-Lagrangian integration of a gridpoint shallow water model on the sphere*", as the title suggests, the semi-Lagrangian approach was extended to spherical coordinates;

we shall go into more details about numerical modeling on a sphere in Chapter 12. As well as the work by MacDonald, there was also a lot of work being undertaken by Purser [151], as well as by Staniforth and Côté [176,67,186].

Now given this brief overview of the development/history of the semi-Lagrangian theory, we shall present brief summaries of the papers by Fjørtoft, Welander, Wiin-Nielsen, and Robert. We start with a review of the theory and thoughts from Ranger Fjørtoft's 1952 paper "*On a Numerical Method of Integrating the Barotropic Vorticity Equation*".

4.1 Fjørtoft (1952) paper

The abstract for this paper is quite short, but quite detailed. It states that a method is presented by means of which quasipermeant "velocity" fields may be constructed in which absolute vorticity is conserved if the atmosphere is treated barotropic. It then states that vorticity displacements may then be considered as constant for time intervals of 24 hours. These approximation are then going to be applied to Poisson and Helmholtz equations, and then a graphical procedure is introduced to derive the isolines of vorticity and to "integrate", where integration is the mathematical equivalent of the evolution of the isolines in the atmosphere in time.

An interesting statement that is made in the introduction of this paper is the following and is still true for all forms of operational prediction of any geophysical system:

> *The practical aspect of numerical weather forecasting problem is related to the time necessary for carrying out a numerical solution. For it is clear that a forecast, how good it else may be, is completely useless in daily weather forecasting if it cannot be completed ahead of the weather. From that reason one may prefer one numerical method for another even on the cost of* ***accuracy*** *if the* ***computation time*** *is reduced essentially. This will especially be the case if the additional errors introduce thereby are well within the limits or errors present for other reasons.*

Given this motivation we shall now summarize the derivation from [56].

4.1.1 Barotropic problem

In the barotropic model it is assumed that the motion at vertical levels close to the 500 mb level is only in the horizontal direction and is also nondivergent, which then implies that the vertical component of the absolute vorticity is conserved by the assumption of conservative forces. The model that Fjørtoft introduces is therefore on a plane with the assumption of periodicity around the Earth for the quantity α, which is derived from some atmospheric variable, and as such this quantity can be represented by a Fourier series. The series that Fjørtoft introduces for α is a finite Fourier series, as we are dealing with numerical quantities. Thus

$$\alpha = \sum_{m,n}^{L_N} \alpha_{m,n} \exp\left\{2\pi i \left(\frac{x}{\lambda_m} + \frac{y}{\omega_n}\right)\right\} + c, \tag{4.1}$$

where L_N indicates the shortest wavelength occurring in the series and c is the constant in the series.

The next step is to introduce the law of individual conservation of vorticity, which is given by

$$\frac{\partial \widehat{\eta}}{\partial t} = -\mathbf{v} \cdot \nabla \widehat{\eta}, \tag{4.2}$$

where $\hat{\eta}$ is vorticity, $\mathbf{v}$ is the velocity, and ∇ is the del-operator in the plane of motion. Given the assumption of nondivergence for the velocity field, velocity may be written in terms of a streamfunction, ψ, such that $\mathbf{v} \equiv -\nabla \psi \times \mathbf{k}$, where $\mathbf{k}$ is the unit vector perpendicular to the plane of motion. This then enables us to write vorticity as $\nabla^2 \psi = \widehat{\eta}$.

The next step that is described in [56] is to consider how to solve (4.2), given the periodicity from (4.1), and as such we consider the numerical approximation to $\nabla^2 \psi$ through the finite difference given by

$$\begin{aligned} \hat{\eta} &= \nabla^2 \psi \approx -\frac{4}{d^2} \left(\psi(x, y) - \overline{\psi}(x, y, d) \right), & (4.3) \\ \overline{\psi}(x, y, d) &\equiv \frac{1}{4} \left(\psi(x+d, y) + \psi(x-d, y) + \psi(x, y+d) + \psi(x, y-d) \right), & (4.4) \end{aligned}$$

with d being the grid spacing. An interesting point that Fjørtoft makes next is that even if $\nabla^2 \psi$ is approximated by (4.3), it may be that we do not necessarily require the solution for grid points only. Fjørtoft now makes the assumption that even if a numerical solution cannot yield a solution everywhere, the theory that is derived initially here is required to be valid for any x and y.

If we now substitute $\widehat{\eta} = -\frac{4}{d^2} \left(\psi(x, y) - \overline{\psi}(x, y, d) \right)$ into (4.2) and divide out the factor $\frac{4}{d^2}$, then the resulting new continuous problem to be solved, under condition (4.1), is

$$\frac{\partial \eta}{\partial t} = -\mathbf{v} \cdot \nabla \eta, \tag{4.5}$$

where $\nabla = \psi - \overline{\psi}$. Thus the numerical problem of solving (4.5) for a sufficiently small δt may be divided into the following two independent steps [56]:

1. Displacing properly the initial isolines $\eta = const$ over a time interval δt to find the new position of these lines and subtracting the increment $\delta \eta$ up to time δt.
2. Solving thereafter $\delta \psi - \delta \overline{\psi} = \delta \eta$ to find the corresponding increment in the streamfunction of the velocity field.

The next part of discussion in [56] is concerned with the consequences of what is referred to as the overbar operation, which has been used L times on the Fourier representation in (4.1). This is represented as

$$\overline{\alpha}^{\{L\}} = \sum_{m,n}^{L_N} A_{m,n}^{L} \alpha_{m,n} e^{2\pi i \left(\frac{x}{\lambda_m} + \frac{y}{\omega_n} \right)}, \tag{4.6}$$

where

$$A_{m,n} = \frac{1}{2}\left(\cos\frac{4d\pi}{2\lambda_m} + \cos\frac{4d\pi}{2\omega_n}\right). \tag{4.7}$$

It is then stated that if $\frac{4d}{\lambda_m}$ and $\frac{4d}{\omega_n}$ are not equal to an even number, then we require $\left|A_{m,n}\right| < 1$ so that the solution does not blow up, which is followed by a discussion of the damping affects of this approach for certain wavenumbers.

Eventually the paper arrives at the section that is titled *"The problem of time-integration"*, where we have seen some of these problems in the last two chapters.

4.1.2 The problem with time integration

The opening paragraphs of this section of [56] are about justifying Lagrangian integration:

In Lagrangian terms the time-integration of $\frac{\partial v}{\partial t} = -\mathbf{v}\cdot\nabla\eta$ may in theory be accomplished by repeated displacement of the isolines in velocity fields which are considered as constant for a sufficiently small time increment Δt. This Lagrangian formulation of the time integration problem will make sure that the fundamental assumption of conservation of vorticity is realized during the computation in the sense that no isolines $\eta = const$ then will be present at any later time which did not already exist in the beginning, nor will any isolines be lost.
The assumption of constant velocity fields during the partial displacements of the isolines of vorticity introduces in the general case an error in the distribution of η at a later time t. One may however keep this error within arbitrarily small limits by choosing a sufficiently small value for Δt. Decisive for how rapid a forecast can be computed for a period T is the ratio $\frac{\Delta t}{T}$. In the atmosphere the actual velocity field are usually changing so rapidly that one is forced to keep Δt less than 3–4 hours.

The remainder of the section in [56] is concerned with developing an expression for the displacement of the constant isolines. We start by introducing a fictitious field to the streamfunction and the velocity fields so that they can be written as

$$\psi \equiv \psi^{\star} + \psi^{\star\star}, \qquad \mathbf{v} = \mathbf{v}^{\star} + \mathbf{v}^{\star\star}, \tag{4.8}$$

where

$$\psi^{\star} = \psi - c\left(\psi - \overline{\psi}\right), \quad \psi^{\star\star} = c\left(\psi - \overline{\psi}\right), \tag{4.9}$$

and c is a constant. Next the velocity field $\mathbf{v} = \mathbf{v}^{\star} + \mathbf{v}^{\star\star}$ is substituted into the vorticity equation, yielding

$$\frac{\partial\left(\psi - \overline{\psi}\right)}{\partial t} = -v^{\star}\cdot\left(\psi - \overline{\psi}\right). \tag{4.10}$$

It should be noted here that the term $-\mathbf{v}^{\star\star}\cdot\nabla\eta$ drops out of the equation because $\mathbf{v}^{\star\star}$ is parallel to the isolines $\eta = \psi - \overline{\psi} = const$ and as such is ineffective in changing the distribution of the vorticity.

The next comment in the derivation in [56] focuses on how to displace the equiscalar line of vorticity in the fictitious $\mathbf{v}^\star$-fields as in the actual velocity field. The aim of this discussion here is to find if a useful displacement field exists as it would imply that it could be more easily derived from $\psi^\star = \overline{\psi}$ as the field $\overline{\psi}$ exists when $\eta_{t=0}$ is computed and, here comes the important point in [56], because at later time-steps the solution of $\Delta\psi - \overline{\Delta\psi} = \Delta\eta$ provides the change in the $\overline{\psi}$ field. The remainder of this section in [56] is a discussion on the displacement size and time step for the graphical approach to be practical and accurate.

To finish out this section on [56] we present copies of the two figures from that paper to illustrate the predicted changes in the z field, Fig. 4.1, which is height of the 500 mb from the surface. The lines in the chart of Fig. 4.1 are spaced at 40 m, whereas the chart in Fig. 4.2 shows the observed changes in the same field. It is claimed that the two charts match up quite well, and then there is the caveat about the fact what the situation in these charts were if there was a strong change in the mean field.

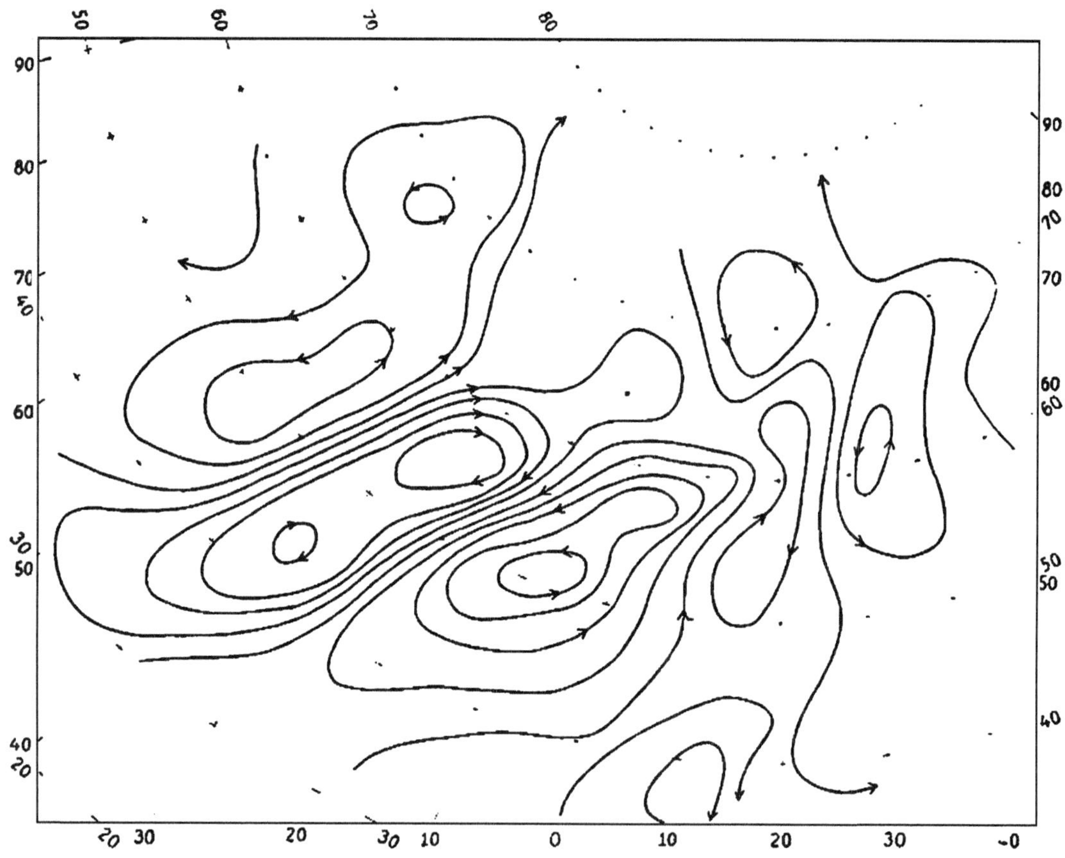

FIGURE 4.1 Copy of the plot of the forecast using the graphical method described in Fjørtoft, R., 1952: On a numerical method of integrating the barotropic vorticity equation, *Tellus*, **4**, 179–194.

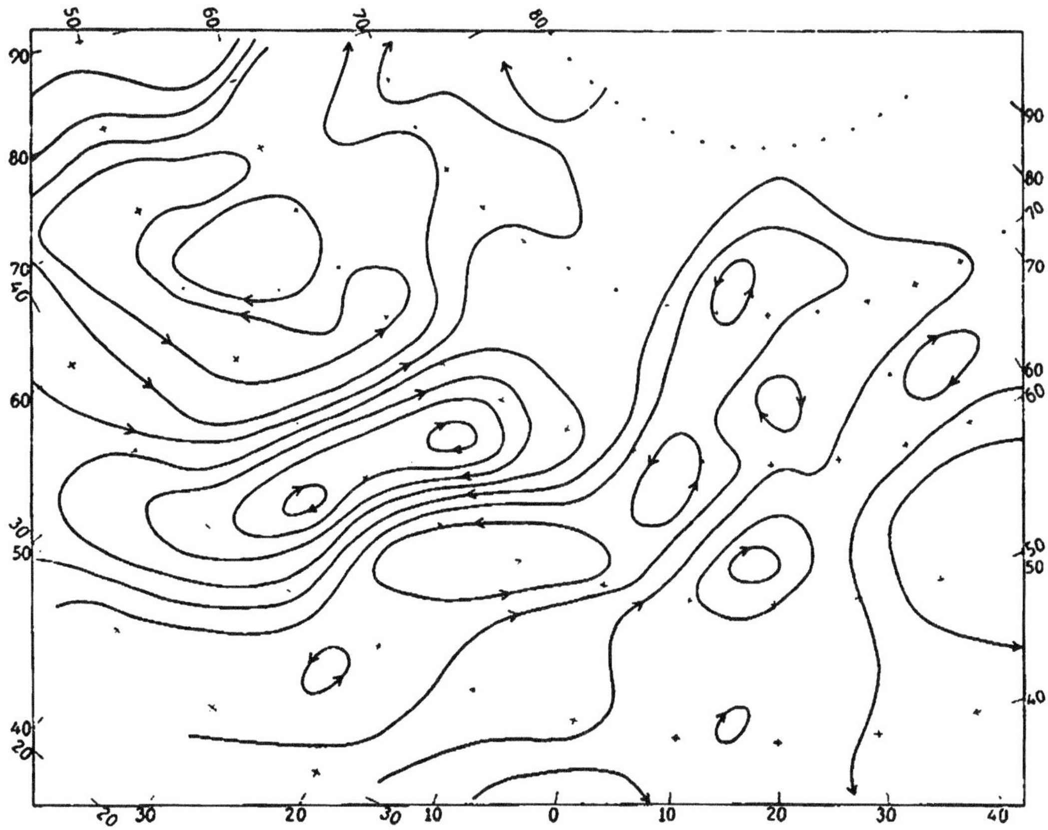

FIGURE 4.2 Copy of the plot of the observed field to compare the numerical forecast in the previous figure from Fjørtoft, R., 1952: On a numerical method of integrating the barotropic vorticity equation, *Tellus*, **4**, 179–194.

Obviously graphical methods are not used to predict the movement of the vorticity field in the atmosphere anymore, but the technique from [56] is similar to what we would consider as Lagrangian time integration.

We now move on to the next step in the development of the semi-Lagrangian methods, which is the paper by Welander, showing that it is not possible to computationally use the Lagrangian formulation to model advection in the atmosphere.

4.2 Welander (1955) paper

The abstract of this paper is as follows:

The purpose of this paper is to discuss qualitatively certain kinds of asymptotic motion in a two-dimensional, ideal fluid by help of methods of statistical mechanics. It is stressed that the final development of such a fluid cannot be adequately described by use of the ordinary equations of motion, but that a "coarse grain" representation should be used. In this representation, the development is characterized by the forming of a single, large vortex accompanied by a certain non-viscous dissipation. The final equilibrium is probably reached almost explosively after a finite time. Some experiments which are carried out seem to support this result. In the earlier stages of development we may expect to have some kind of a quasi-equilibrium motion. It is attempted to find conditions under which such a motion can exist, by studies of a point-vortex model.

The technique introduced in [198] is based upon classical statistical mechanics developed for conservative, Hamiltonian systems, where the equations of motion for such a system are given by

$$\left\{\begin{aligned} \frac{dp_i}{dt} &= -\frac{\partial W}{\partial q_i}, \\ \frac{dq_i}{dt} &= \frac{\partial w}{\partial p_i} \end{aligned}\right. \qquad (i = 1, 2, \ldots, n), \tag{4.11}$$

where $W(p_1, p_2, \ldots, p_n, q_1, q_2, \ldots, q_n)$ is the energy of the system, expressed in the coordinates $q_1, q_2, \ldots, q_n$ and the corresponding *generalized momenta*, $p_1, p_2, \ldots, p_n$. The instantaneous state of a system can be represented as a point, the representative point, in a phase space (or ensemble space) of $2n$ dimensions with the p_is and q_is as rectangular coordinates. The time development of the system is then represented by a certain trajectory in this space.

The next comment that Welander makes is that we suppose now that we have, in an actual case, only an incomplete knowledge of the initial state of the system. As a specific example, let us assume that we can control at the initial time only a certain function of the initial phase space coordinates, $F(p_1, p_2, \ldots, p_n, q_1, q_2, \ldots, q_n)$. If we carry out a number of similar experiments, under steady outer conditions and with a fixed value of F, say M, then the initial values p_{i0} and q_{i0} are in general found to vary from one experiment to the next due to uncontrollable disturbances in the system. However, carrying out a great number of experiments, it is possible to find a certain statistical distribution of initial values, and corresponding to this a certain distribution of points on the hypersurface $F(p_1, p_2, \ldots, p_n, q_1, q_2, \ldots, q_n) = \alpha$ in the phase space. The whole statistical structure of the experiment $F = \alpha$ is represented in the phase space by the bundle of trajectories starting at these points. In Fig. 4.3 we have a copy of the original figure of the trajectories from the hyperplane $F = \alpha$ from [198] to illustrate the fact that the initial points head off in different directions, which would result in a deformation of the original area.

An assumption of sorts is now made in [198] that for every part of the phase space there is a related universal a priori *possibility* distribution density of initial points of the system. To determine the density distribution just mentioned is one fundamental task in statistical mechanics.

We return to the Hamiltonian dynamics mentioned earlier, where now we introduce a velocity vector in phase space given by $\mathbf{v} = \left(\frac{dp_1}{dt}, \frac{dp_2}{dt}, \ldots, \frac{dp_n}{dt}, \frac{dq_1}{dt}, \frac{dq_2}{dt}, \ldots, \frac{dq_n}{dt}\right)$, and if

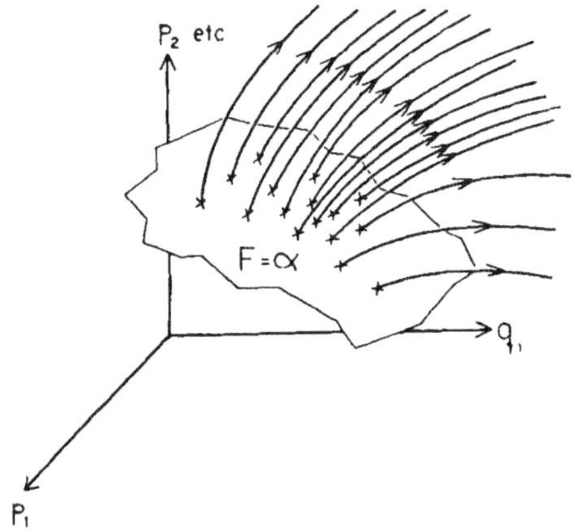

FIGURE 4.3 Copy of the hyperplane plot, Fig. 1 from Welander, P., 1955: Studies on the general development of motion in a two-dimensional ideal fluid, *Tellus*, **17**, 141–156. https://www.tandfonline.com/doi/abs/10.3402/tellusa.v7i2.8797. https://creativecommons.org/licenses/by/4.0/.

we compute the divergence of the velocity field above, then we have

$$\nabla \cdot \mathbf{v} = \sum_{i=1}^{n} \left[\frac{\partial}{\partial p_i} \left(\frac{dp_i}{dt} \right) + \frac{\partial}{\partial q_i} \left(\frac{dq_i}{dt} \right) \right] = \sum_{i=1}^{n} \left[\frac{\partial}{\partial p_i} \left(-\frac{dW}{dq_i} \right) + \frac{\partial}{\partial q_i} \left(\frac{dW}{dp_i} \right) \right] = 0, \tag{4.12}$$

which indicates that the flow of representative points in the phase space is an incompressible one. This is then implying that there is a uniform distribution of points throughout the entire phase space that is maintained permanently.

The next important part of the derivation in [198] involves what is referred to as the *H-Theorem* which states that a certain quantity,

$$H = \underbrace{\int \cdots \int}_{\text{all space}} P \log P \, dV, \tag{4.13}$$

where P is the probability density of the representative points and dV stands for the volume elements dp_i, dq_i for $i = 1, 2, \ldots, n$, will decrease in the run of time towards an absolute minimum. For molecular systems, H is proportional to the negative of the entropy, and the H-theorem then expresses the second law of thermodynamics.

The important property of the H-theorem is that it enables the determination of the state of statistical equilibrium of the system, which is the main task that is undertaken by statistical mechanics. Therefore, the minimum of H in variational form, see [61] for an introduction to calculus of variation which is the mathematical technique that is being used here, is given by

$$\delta H = \underbrace{\int \cdots \int}_{\text{all space}} (\log P + 1) \, \delta P \, dV = 0, \tag{4.14}$$

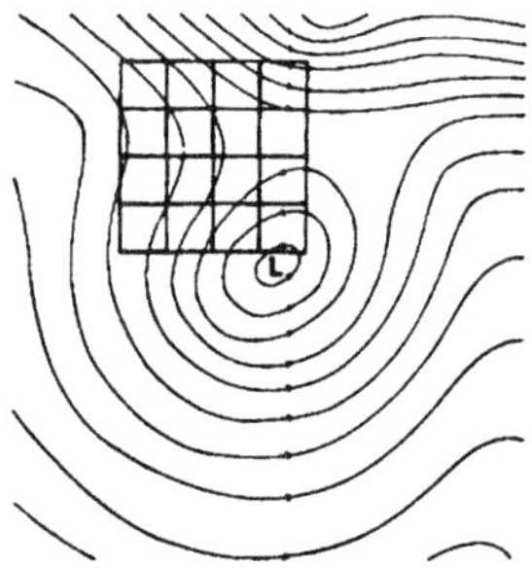

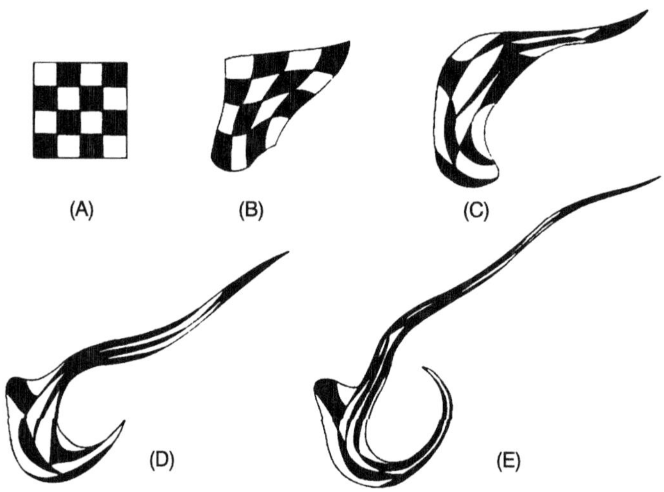

FIGURE 4.4 Copy of the Lagrangian deformation plot, Fig. 2 from Welander, P., 1955: Studies on the general development of motion in a two-dimensional ideal fluid, *Tellus*, **17**, 141–156. https://www.tandfonline.com/doi/abs/10.3402/tellusa.v7i2.8797. https://creativecommons.org/licenses/by/4.0/.

and Welander then introduces a subsidiary condition of

$$\underbrace{\int \cdots \int}_{\text{all space}} P\, dV = 1. \tag{4.15}$$

After a long discussion about different conditions that can be added to the problem, Welander moves on to the experiment where a fluid is initially divided into a number of surface elements, which are colored black and white (for instance, in a chess-board pattern). The color should follow the motion materially without any essential diffusion. When setting the fluid into motion, the elements are deformed, and after some time they are drawn out in the form of long and thin bands, as shown in Fig. 4.4. We now want to investigate the conservation of color density during the motion. If the black color is assigned the density value of 1 and the

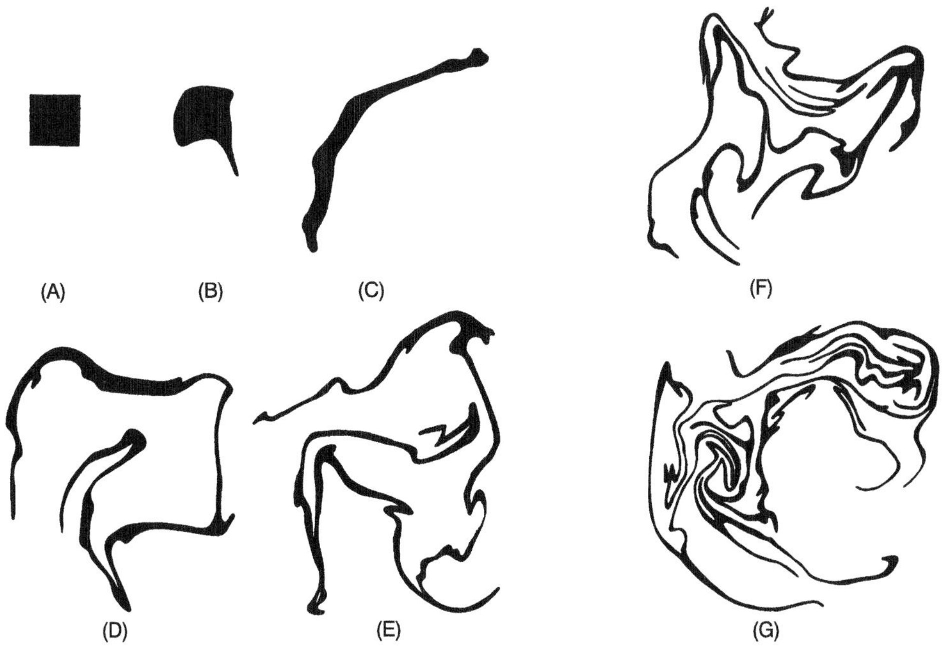

FIGURE 4.5 Copy of the first and second deformation plots, Fig. 3 from Welander, P., 1955: Studies on the general development of motion in a two-dimensional ideal fluid, *Tellus*, **17**, 141–156. https://www.tandfonline.com/doi/abs/10.3402/tellusa.v7i2.8797. https://creativecommons.org/licenses/by/4.0/.

white color is assigned the density value of 0, then the total color mass of the fluid is

$$\int_F \varrho \, dF = F_b, \tag{4.16}$$

where F is the total area of the fluid, F_b is the total area of the black elements, and ϱ is a density that represents the mean mass per unit volume. The other quantity that is introduced here is the total square density mass,

$$\int_F \varrho^2 dF = F_b, \tag{4.17}$$

were the two quantities in (4.16) and (4.17) are conservative because the area F_b is preserved during the motion. We also have the plots from Figs. 2–3 in [198]: Fig. 4.4 shows the results from the numerical integration of the barotropic vorticity model, while the results in Fig. 4.5 are for the physical experiment with a fluid.

It should now be obvious from the results in Figs. 4.4 and 4.5 that the Lagrangian approach, which is also referred to as Hamiltonian dynamics, would be quite difficult to undertake; here it could result in smaller time step restrictions than the Eulerian stability conditions, as we would not want the phase space to distort too much as it would make the numerical integration quite difficult. We have to remember though that in a global numerical weather prediction model, we are not just solving the advection equation. This equation is coupled to other equations that use its solution as a possible forcing. Therefore, we need something that is in-between the full Eulerian and Lagrangian approaches, thus we now consider the paper by Wiin-Nielsen titled *"On the application of trajectory methods in numerical forecasting"*. As mentioned in the introduction, what we now refer to as semi-Lagrangian methods was referred to as **quasi-Lagrangian** in [200].

4.3 Wiin-Nielsen (1959) paper

The abstract to this paper is:

On the basis of a comparison between the results so far obtained by the usual numerical methods in physical weather prediction and the graphical methods it is likely that certain defects in the present forecast methods are due to the Eulerian technique. The possibilities for the use of a quasi-Lagrangian technique are investigated, and it is found that such a technique is possible and becomes especially simple when combined with a space-smoothing which eliminates the shorter fast moving waves in the spectrum. The Lagrangian method of computation is described in the barotropic case and also for a simplified two-parameter baroclinic model.

As mentioned at the end of this quote, Wiin-Nielsen refers to this new technique as quasi-Lagrangian and not semi-Lagrangian, yet. The opening paragraph of this paper refers back to the work by Fjørtoft and says that up until [56] the development of the numerical methods used in numerical weather prediction had been in the Eulerian formulation and that the graphical method from [56] and the latter work in [58] were the first attempts to use the Lagrangian approach.

Back in the 1950s the barotropic vorticity equation model was in wide use as a numerical weather prediction model. As stated at the start of the second section of [200], the barotropic vorticity equation applied to an isobar surface expresses simply that the absolute vorticity is conserved along the path of a particle. An important fact that we have already seen in the Eulerian chapters is that this property is not explicitly used in the Eulerian integration models, where in that formulation we are always computing the local change in vorticity, in the barotropic model only due to the horizontal advection of absolute vorticity. Given this statement, Wiin-Nielsen then remarks that he wishes to use the property that the absolute vorticity is conserved during the trajectory of the particle.

Before we summarize the mathematics used to justify the quasi-Lagrangian approach, we present the motivation from the paper that exactly describes how we handle semi-Lagrangian methods nearly 60 years later:

When we want to follow a particle during its motion in the isobaric surface, it is not evident that a rectangular grid is the most convenient to use, because the successive positions of the particle in general will run between the grid points. The relaxation methods which are used to the solution of the finite difference form of the second order differential equations are, however, well suited for the rectangular grid. We shall in the first instant try to arrange the numerical procedure in such a way that we can still use the relaxation methods without major changes. This is obtained if we ask for the positions of the particles which after a certain time end up in the grid points of the rectangular grid.

Thus to start we let the Lagrangian coordinates be denoted by (x, y). Let further the Lagrangian coordinates coincide with the Cartesian coordinates for $t = 0$. These latter coordinates are denoted (a, b). The trajectory is then expressed by the two equations:

$$\frac{dx}{dt} = 0, \tag{4.18a}$$

$$\frac{dy}{dt} = 0. \tag{4.18b}$$

The equations in (4.18a) and (4.18b) are solved through their Eulerian formulation

$$\frac{\partial x}{\partial t} = -\mathbf{v} \cdot \nabla x, \tag{4.19a}$$

$$\frac{\partial y}{\partial t} = -\mathbf{v} \cdot \nabla y. \tag{4.19b}$$

It is now assumed that the field $\mathbf{v} = \mathbf{v}(x, y, t)$ is known, so that the solution of (4.19a) and (4.19b) will provide the coordinates for the particle which at $t = T$ end up in a certain given point. Wiin-Nielsen then makes a reference back to his work in 1957 where a series of experiments had shown that the Eulerian solution to the equations above provided trajectories with a good accuracy, given that the time integration is not extended too far. The reason for this statement is presented as a result of the initial grid being rectangular and the truncation error of ∇x and ∇y in the beginning being extremely small. Later, however, the initial rectangular grid is deformed to such an extent that the truncation error in the gradients of the coordinates field will play a larger and larger role besides the truncation error which always exists, when the wind vector is computed from a height field or from a streamfunction.

By computing trajectories from (4.19a) and (4.19b) in a stationary field, Wiin-Nielsen found that the truncation errors are quite small. A particle, which initially has a position on a certain isoline, should in a stationary field stay there, which in [200] it is stated that the tests show that it is very nearly so.

We now consider up to a time $t = T$ the particles that are at the points of a rectangular grid. From the solutions of (4.19a) and (4.19b) it is possible to know that the coordinates to $t = 0$ will in general not be at the grid points. In the barotropic case the problem is then to determine the absolute vorticity for the particle which by $t = T$ arrives at a grid point. The first step is to compute the relative vorticity. Let now A, B, C, and D to $t = 0$ be the positions of the particles which at $t = T$ occupy the grid points A_T, B_T, C_T, and D_T. We have a copy of the schematic from [200] showing the concept of where the points end up in Fig. 4.6.

The trajectories are denoted AA_T, BB_T, CC_T, and DD_T. Any trajectory can, as a curve, be characterized by the three parameters a, b, and t, where a and b are Lagrangian coordinates. A parametric representation of a trajectory can therefore be given in the form:

$$x = x(a, b, t), \quad y = y(a, b, t). \tag{4.20}$$

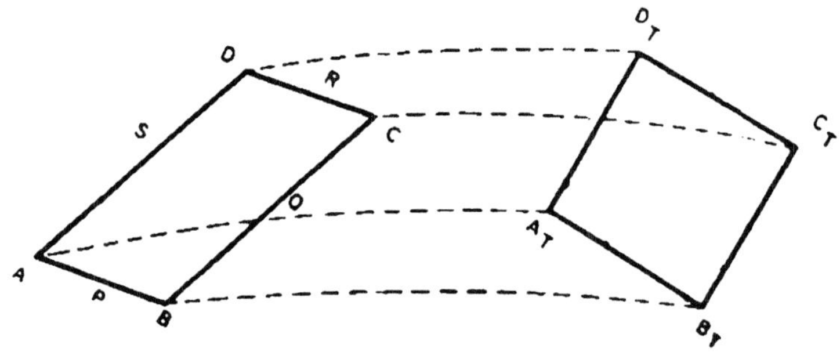

FIGURE 4.6 Schematic showing the initial positions of 4 points that arrive at the regular grid points, Fig. 1 from Wiin-Nielsen, A., 1959: On the application of trajectory methods in numerical forecasting, *Tellus*, **11**, 180–196. https://www.tandfonline.com/doi/abs/10.3402/tellusa.v11i2.9300. https://creativecommons.org/licenses/by/4.0/.

We now let the coordinates for A_T, B_T, C_T, and D_T be denoted by

$$A_T : (a_T, b_T), \quad B_T : (a_T + \delta a, b_T), \quad C_T : (a_T + \delta a, b_T + \delta b), \quad D_T : (a_T, b_T + \delta b).$$

Therefore, the particles that to $t = T$ are on the line through $A_T B_T$ are to $t = 0$ situated on the curve through AB, and as such the parametric representation for the curve is

$$x = x(a, b_T, 0), \quad y = y(a, b_T, 0).$$

The coordinates for A and B are

$$A : (x(a_T, b_T, 0), y(a_T, b_T, 0)), \qquad B : (x(a_T + \delta a, b_T, 0), y(a_T + \delta a, b_T, 0)).$$

Thus the associated infinitesimal vector AB is given by

$$\overrightarrow{AB} = \left(\frac{\partial x}{\partial a}\delta a, \frac{\partial y}{\partial a}\delta a\right).$$

By the same argument, the remaining three vectors can be shown to be:

$$\overrightarrow{BC} = \left(\frac{\partial x}{\partial a}\delta b, \frac{\partial y}{\partial b}\delta b\right),$$
$$\overrightarrow{CD} = -\left(\frac{\partial x}{\partial a}\delta a, \frac{\partial y}{\partial a}\delta a\right),$$
$$\overrightarrow{DA} = -\left(\frac{\partial x}{\partial b}\delta b, \frac{\partial y}{\partial b}\delta b\right).$$

This then implies that the coordinates of a particle are given by $(x(a, b, t), y(a, b, t))$, and the velocity for a particle is $\left(\frac{\partial x}{\partial t}, \frac{\partial y}{\partial t}\right)$.

The next part of the Wiin-Nielsen paper is concerned with linking these new coordinates back to the relative vorticity and involves some extensive mathematics which is not really important here. However, after deriving the link to the relative vorticity and defining an associated streamfunction, the important statements are slightly apart but start from

When $\psi^{(N)}$ now is assumed constant in time for some hours, we need not to recompute the streamfunction in every time step. We can then integrate the trajectory equations in the following first with:

$$x^{\tau+1} = x^{\tau-1} + 2\Delta t J\left(x^{\tau}, \psi^{(N)}\right),$$
$$y^{\tau+1} = y^{\tau-1} + 2\Delta t J\left(y^{\tau}, \psi^{(N)}\right).$$

The computation can be performed using these equations until we have the positions of the particles to time $t = T$ ($T \sim 12$ hours). We know the initial positions of the particles which to $t = T$ occupy grid points.

The operator J above is referring to the Jacobian operator, defined as

$$J(\alpha, \beta) = \frac{\partial \alpha}{\partial a}\frac{\partial \beta}{\partial b} - \frac{\partial x}{\partial b}\frac{\partial \beta}{\partial a}.$$

We can see that the trajectory equations allow us to develop the semi-Lagrangian theory later. An interesting point about the equations above is that they are a two-level approximation, this is because we are integrating from $\tau - 1$ to $\tau + 1$, but this comes from the Eulerian experience that the forward Euler, or single-level schemes, damp the numerical solution. It is then quite ironic that the next and final important sentence from [200] should then involve bilinear interpolation, which we shall investigate in later chapters and show why this may not have been a good choice of interpolation polynomial.

The rather complicated expression is clearly transformed into finite differences, if only the values of ψ are known in the initial positions of the particles. These positions will in general be between the grid points, but an interpolation between the known initial values in the grid points will probably be sufficient. The interpolation can be performed in the following way: Let us for simplicity denote the grid size by 1, and let $P = (p, q)$ (see a copy of figure from [200] in Fig. 4.7) be the point in which we

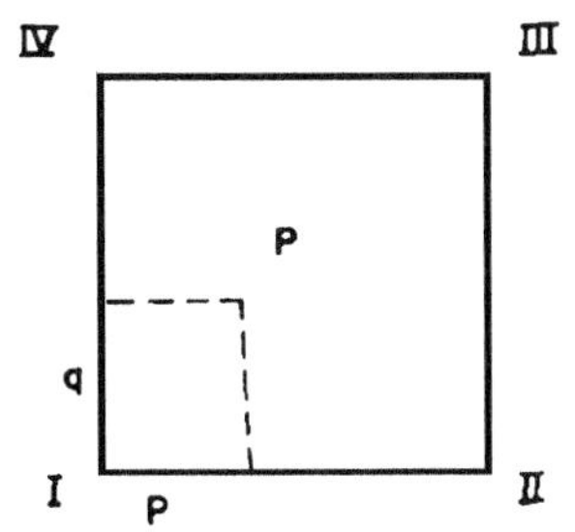

FIGURE 4.7 Copy of the interpolation grid box that Wiin-Nielsen suggests should be used to calculate the values at the departure point, Fig. 3 from Wiin-Nielsen, A., 1959: On the application of trajectory methods in numerical forecasting, *Tellus*, **11**, 180–196. https://www.tandfonline.com/doi/abs/10.3402/tellusa.v11i2.9300. https://creativecommons.org/licenses/by/4.0/.

want to interpolated values of the streamfunction. One finds then easily that

$$\psi_P = (1-p)(1-q)\psi_I + p(1-q)\psi_{II} + pq\psi_{III} + (1-p)q\psi_{IV} \tag{4.21}$$

And there we have it. In the quote above, Wiin-Nielsen has laid the foundations for what would eventually become semi-Lagrangian methods. For historical prospectus, it is claimed in the paper "*Numerical advection algorithms and their role in atmospheric transport and chemistry models*" by Richard Hood in (1987) [80], that the papers by Robert in 1981–1982 [161,162] are where the term semi-Lagrangian was first used. In fact, it is in the title of the 1982 paper. However, this is not true, the first use of semi-Lagrangian term is in a paper by Sawyer in 1963, titled "*A semi-Lagrangian method of solving the vorticity advection equation*" [167]. As we can see in the title, the semi-Lagrangian approach is only being used with the vorticity equations. It was not till Robert's paper in 1981 that it became possible for the semi-Lagrangian approach to be used with the primitive meteorological equations, or the shallow water equations model, and we shall summarize the justification for considering the semi-Lagrangian approach.

4.4 Robert's (1981) paper

The title of this Robert's paper is "*A stable numerical integration scheme for the primitive meteorological equations*". The abstract for this paper is

A stable numerical integration scheme is applied to the non-divergent barotropic vorticity equation. Integrations are performed with time steps ranging from 15 min to 4 h. The root-mean-square differences between the forecasts are calculated in order to measure the sensitivity of the predictions to the size of the time step. These experiments show that the truncation errors remain reasonably small with time steps as large as two hours.

This scheme is then associated with the semi-implicit algorithm in the integration of the shallow water equations. Integrations are carried out with large time steps and the resulting predictions are presented in order to demonstrate that the scheme is perfectly stable.

Ironically, the opening paragraph is still true today in that we keep outgrowing computational power availability as we try to improve the forecast, but we are also always trying to improve the algorithms used to make them more efficient:

The accurate integration of numerical models of the atmosphere requires a very large number of arithmetic operations on a computer. The fastest computers available today cannot easily integrate the models that scientists would like to use for general circulation studies. The models used for weather forecasting also consume a considerable amount of computing power. Fortunately, the speed of computers is rapidly increasing and it appears that they will soon be able to satisfy our needs. Scientists can wait until these supercomputers become available, but in the meanwhile, they can add their own contribution by designing more efficient computational algorithms.

Robert then mentioned advancements that had been made in the 1970s to improve the efficiency of the algorithms, and made a reference to a semi-implicit scheme that there was a problem with the size of the time step being too small to ensure stability. Therefore, Robert was looking for a new scheme to allow for a larger time step, yet ensuring the same accuracy. To analyze this feature, Robert looked at the truncation error of the approaches for the advection equation, where he stated that the advection equation without forcing is equivalent to saying that the variable ψ moves along the x-axis with speed u, and has the exact solution, as we have already seen, of

$$\psi = A\cos k\left(x - ut\right), \tag{4.22}$$

but that the finite difference approximation to the advection equation will have a solution of the form

$$\psi^{\star} = A\cos k\left(x - Ut\right), \tag{4.23}$$

where U is the speed that the numerical solution is traveling at. Next we define the error as the difference between (4.22) and (4.23), ε, given as

$$\varepsilon = \psi^{\star} - \psi = A\left(\cos k\left(x - Ut\right) - \cos k\left(x - ut\right)\right). \tag{4.24}$$

It is then stated in [161] that we can rewrite (4.24) in terms of sine functions as

$$\varepsilon = 2A\sin k\frac{\left(U-u\right)}{2}t\sin k\left(x - \frac{\left(U-u\right)}{2}\right)t. \tag{4.25}$$

The root-mean-square error E over one or many full wavelengths is given by

$$E = A\sqrt{2}\sin k\frac{\left(U-u\right)}{2}t. \tag{4.26}$$

If we now assume that we are only considering the error over a short integration, then the Taylor series approximation of the sine function enables us to write E as

$$E = \frac{\sqrt{2}}{2}kA\left(U-u\right)t. \tag{4.27}$$

Thus the error increases linearly with time, where the growth rate is proportional to the amplitude of the wave, as well as the error in the phase speed, $U-u$.

Now if we were to use a central difference for both the temporal and spatial derivatives then we would know that the phase speed U of the computation solution is given by the expression

$$\sin\left(kU\Delta t\right) = \frac{u\Delta t}{\delta x}\sin\left(k\Delta x\right), \tag{4.28}$$

assuming that we are modeling small angles. Then it is possible to expand the sine function in a Taylor series again, which results in

$$U = u\left(1 - \frac{1}{6}k^2\left(\left(\Delta x\right)^2 - u^2\left(\Delta t\right)^2\right)\right), \tag{4.29}$$

where substituting (4.29) into (4.27) results in

$$E = \frac{\sqrt{2}}{12} k^3 u A \left((\Delta x)^2 - u^2 (\Delta t)^2 \right). \tag{4.30}$$

If we wish for the error to remain small, then we must ensure that both Δx and Δt are as small as possible, and if we set Δt and Δx equal to zero, then the errors for the spacial and temporal differences are given by

$$E(\Delta x) = \frac{\sqrt{2}}{12} k^3 u A (\Delta x)^2, \tag{4.31a}$$

$$E(\Delta t) = \frac{\sqrt{2}}{12} k^3 u^3 A (\Delta t)^2, \tag{4.31b}$$

where, upon taking the ratio of (4.31b) to (4.31a), we obtain

$$\frac{E(\Delta t)}{E(\Delta x)} = \frac{u^2 (\Delta t)^2}{(\Delta x)^2} \equiv \frac{u^2}{u_0^2}, \quad \text{where} \quad u_0 = \frac{\Delta t}{\Delta x}. \tag{4.32}$$

After this derivation, Robert goes on to explain that the restriction of the CFL condition for meteorological modeling is based upon the speeds of the different waves in the atmosphere, stating that the average wave has a speed of $u = 15\ \text{ms}^{-1}$, but for the numerical integration to remain stable the constant u_0 must be larger than the largest phase speed encountered in the calculations. After some explanation about different waves speed, Robert shows that the ratio between the spatial and temporal errors must satisfy

$$E(\Delta t) = \frac{1}{700} E(\Delta x). \tag{4.33}$$

He then goes on to explain that the semi-implicit method that he had introduced eliminated the need to model the fastest wave, but still needed a ratio between the two errors of $\frac{1}{40}$.

It is stated in [161] that if the ideal wind speed were $u_0 = c = 15\ \text{ms}^{-1}$, for a grid size of 200 km, then the time step would be about approximately 3.7 hours. This then leads Robert to take up the work that was done by Sawyer in 1963 applied to the vorticity equation

$$\frac{\partial Q}{\partial t} + u \frac{\partial Q}{\partial x} + v \frac{\partial Q}{\partial y} = R, \tag{4.34}$$

where Q is the absolute vorticity, and R represents the remaining terms in the vorticity equation. Thus in the semi-Lagrangian framework, Robert discretizes (4.34) as

$$Q(x, y, t + \Delta t) = Q(x - 2a, y - 2b, t - \Delta t) + 2\Delta t R(x - a, y - a, t), \tag{4.35a}$$

$$a = \Delta t u (x - a, y - b, t), \tag{4.35b}$$

$$b = \Delta t v (x - a, y - b, t), \tag{4.35c}$$

where we can see that this is an implicit method and would require some iterations, where Robert says that experiments using between two and four iteration indicated no justification to go as high as four iterations, with two being sufficient, and where the upstream values of u, v, Q, and R are obtained through applying a bicubic interpolation.

The results presented in [161] for both the vorticity equation and the shallow water equations model are compared to the Eulerian formulation and were able to use a larger timestep in the semi-Lagrangian formulation, which is a great point about the schemes.

4.5 Summary

At the beginning of this chapter we provided a briefly summary from the famous Bates and McDonald 1982 paper which laid the foundations for semi-Lagrangian methods to become operationally viable in numerical weather prediction in models that were not dependent on the barotropic vorticity assumption. In that summary Bates and McDonald referenced three historic papers that initialized the development of the semi-Lagrangian theory. This development was essential to prove three points: (1) Eulerian based finite difference methods had too much damping and to many phase errors and restrictions on the time and spatial step sizes; (2) A purely Lagrangian based approach would end up with the original particles being in an infinitesimal small strip in a short finite time; and (3) There are good properties from both approaches that should be combined.

Given this historic overview in this chapter, we now move on to a fundamental component of semi-Lagrangian methods, which was alluded to in the summary of the Wiin-Nielsen paper, in the fact that we need to look at interpolation formulas to assess which of them could reduce the error introduced from finding what eventually became known as the departure point from the interpolation used in the horizontal directions.

CHAPTER

5

Semi-Lagrangian methods for linear advection problems

In this chapter we shall start by presenting an introduction to the Lagrangian formulation of the equations of motion which is where the advection equation is derived from for the Lagrangian framework. As we saw in the last chapter, the idea of using a complete Lagrangian formulation for modeling advection has problems due to the rapid deformation of the fluid's original set of particles, and so we will move onto the semi-Lagrangian formulation to utilize the advantages of the Eulerian and Lagrangian forms of dynamical modeling.

5.1 Derivation of the Lagrangian form for advection

What we have to mention first is that the field of mechanics that we are working with here is referred to as **kinematics**, which is the branch of mechanics dealing with quantities that involve space and time only. This field is concerned with variables such as displacement, acceleration, deformation, velocity, and rotation of fluid elements, but does not refer to the forces that are responsible for these motions. It is said in [94] that kinematics essentially describes the *appearance* of a motion.

As we have mentioned before, there are two ways of describing a fluid motion: in the Eulerian formulation we are monitoring a flow past fixed points, whereas in the Lagrangian formulation we are equivalently following the history of individual fluid particles. A visualization of the Lagrangian description of fluid motion in one-dimension is presented in Fig. 5.1. So in the Lagrangian formulation we require two independent variables, which are taken as time and a label for the fluid particles, as we saw in the last chapter. A convenient label that can be used for the fluid particles could be the position vector $\mathbf{x}_0$ at time $t = t_0$.

Given the description presented above, we have the situation that any flow variable F is expressed as $F(\mathbf{x}_0, t)$. If we now write the particle position in the form $\mathbf{x}(\mathbf{x}_0, t)$, then we can interpret this as the location at time t of a particle whose position was $\mathbf{x}_0$ at time $t = t_0$.

Using the same argument above in the Eulerian description, the model concentrates on what is happening at a spatial point, as mentioned many times before, $\mathbf{x}$, and so the independent variable is taken as $\mathbf{x}$ at time t, which implies that the flow variable is expressed as $F(\mathbf{x}, t)$. This then leads to the velocity and the acceleration of a fluid particle in a Lagrangian

Semi-Lagrangian Advection Methods and Their Applications in Geoscience
https://doi.org/10.1016/B978-0-12-817222-3.00009-8

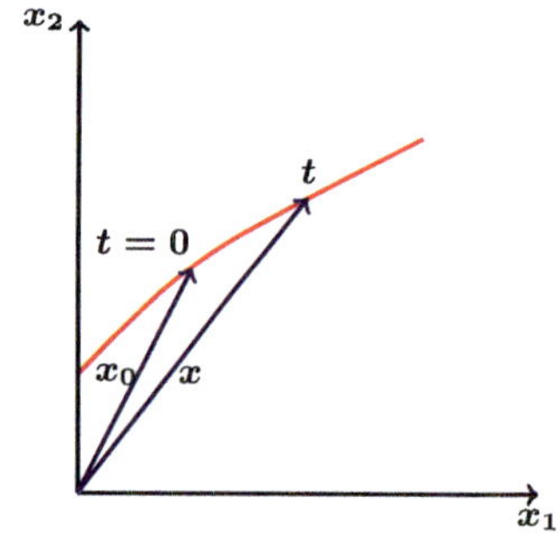

FIGURE 5.1 Lagrangian description of a fluid motion where the red line represents the particle's path.

description being expressed as partial time derivatives

$$u_i = \frac{\partial x_i}{\partial t}, \qquad a_i = \frac{\partial u_i}{\partial t} = \frac{\partial^2 x_i}{\partial t^2}, \tag{5.1}$$

as the particle identity is kept constant during the differentiation. What we have to notice here is that in the Eulerian framework the partial time derivative, $\frac{\partial}{\partial t}$, only represents the local rate of change at a point $\mathbf{x}$, and we have to also note here that it is not the total rate of change as seen by a fluid particle.

If we now let F be any field, which could involve temperature, velocity, or stress as examples, then if we consider the Eulerian coordinate in three dimensions, we have (x, y, z, t) and would like to calculate the rate of change of F at each point following a particle of fixed identity. We therefore seek a way to represent this concept, which is essentially a Lagrangian formulation, but in terms of the Eulerian framework.

To this end, if we now introduce arbitrary and independent increments, $\delta\mathbf{x}$ and δt, then the increment in $F(\mathbf{x}, t)$ is given by

$$\delta F = \frac{\partial F}{\partial t}\delta t + \sum_{i=1}^{3} \frac{\partial F}{\partial x_i}\delta x_i. \tag{5.2}$$

We now relax the arbitrary assumption on the spatial increments, and assume that these increments are actually those associated with following a particle of fixed identity. Given this new assumption, we now have that the increments $\delta\mathbf{x}$ and δt are no longer independent, and are related to the velocity component by the relations between the spatial components and the velocities through

$$\delta x_i = u_i \delta t, \qquad i = 1, 2, 3. \tag{5.3}$$

If we now substitute (5.3) into (5.2) and take the limit as $\delta t \to 0$, then

$$\frac{dF}{dt} = \frac{\partial F}{\partial t} + \sum_{i=1}^{3} u_i \frac{\partial F}{\partial x_i}. \tag{5.4}$$

What we have to recall when dealing with the Lagrangian framework is that we require the evaluated time derivative to be that which is following a particle. There is a standard notation used to represent this formulation, namely $\frac{D}{Dt}$ instead of $\frac{d}{dt}$. Thus we can write (5.4) as

$$\frac{DF}{Dt} = \frac{\partial F}{\partial t} + \sum_{i=1}^{3} u_i \frac{\partial F}{\partial x_i}. \tag{5.5}$$

The total rate of change given by $\frac{D}{Dt}$ is referred to by many names, but the most common one is the **material derivative**, which is to emphasize the fact that the derivative is taken following a fluid element. As we can see from (5.5), the material derivative comprises two components: $\frac{\partial F}{\partial t}$, which represents the local rate of change of F at a given point, where this derivative will be zero if we are in a steady flow, and the second components, $u_i \frac{\partial F}{\partial x_i}$, which is referred to the **advective derivative**, as it represents the change in F that is as a result of advection of the particle from one location to the another where the value of F is different.

Therefore, one approach to find the solution of (5.5) as an initial value problem would be to choose a regularly spaced distribution of fluid parcels, as suggested by Welander [198], assign values for ψ to each of the fluid parcels from the initial conditions provided, and then integrate the ordinary differential equations for the definition of the velocity and the advection equation in Lagrangian form to determine the location and the concentration of the tracer of each parcel as a function of time. However, as we have seen from [198], this can provide a problem in that the parcels become highly nonuniformly distributed in space, and as such the numerical approximation to $\psi(x, t)$ could become inaccurate in the regions where the fluid parcels are widely separated. We also have the other situation where there are areas with several fluid parcels, which are overrepresenting these areas.

If we now recall the derivation of the Eulerian formulation of the advection equation, then in the Lagrangian formulation, when there is no forcing term, we have

$$\frac{DF}{Dt} = 0. \tag{5.6}$$

Given the basic derivation of the Lagrangian version of the advection equation, we now combine the advantages of the Eulerian and Lagrangian frameworks of advection to form the semi-Lagrangian, Eulerian–Lagrangian, or quasi-Lagrangian model for advection.

5.2 Derivation of the semi-Lagrangian approach

The starting point for the derivation of the semi-Lagrangian numerical schemes is to describe the advection equation, as one would expect, in terms of the Lagrange derivative

$$\frac{D\psi}{Dt} = f(x, t), \tag{5.7}$$

which as we have seen has the equivalent Eulerian form given by

$$\frac{\partial \psi}{\partial t} + \bar{u}\frac{\partial \psi}{\partial x} = f(x,t), \tag{5.8}$$

where ψ is our field or tracer of interest, and at this point the function $f(x,t)$ is a linear forcing. We shall go into more details about how to deal with different forcing terms in Chapter 9.

We now have to consider how to discretize (5.7). We have seen in Chapter 2 that when we have a time component in the partial differential equation, we could discretize the scheme through an upwind scheme, a central difference approximation, or a multistep upwind scheme. The same approaches can be applied to the Lagrange derivative operator in (5.7). We shall consider the first-order upwind scheme approximation to the time derivative in (5.7) as

$$\frac{D\psi}{Dt} \approx \frac{\tilde{\psi}\left(x_i, t^{n+1}\right) - \tilde{\psi}\left(\tilde{x}_i^n, t^n\right)}{\Delta t} = f_i^n, \tag{5.9}$$

where $\tilde{\psi}$ is the approximation of the tracer/field ψ at location $\tilde{x}_j^n$, where $\tilde{x}_j^n$ is referred to as the **departure point** of a trajectory originating at time t^n that arrives at the grid point $\left(x_i, t^{n+1}\right)$. As we saw in Chapter 2, we know that the value of the field/tracer is constant along the characteristic which starts at the departure point at $t = t^n$ and arrives at the grid point x_i at $t = t^{n+1}$, where this grid point is referred to as the **arrival point** and so the value that we require at the arrival point is equal to the value of the tracer at the departure point. To find the departure point, we must travel backwards following the characteristics, where this movement backwards is referred to as the **backwards trajectory**.

The first situation that we shall consider is the case where the forcing term, the right-hand side function in (5.7), is equal to zero. We shall also consider the case where the advection wind is constant with respect to x and t, which are the situations that we considered in Chapters 2 and 3. Given these situations, the backward trajectory calculation to the departure points, $\tilde{x}_j^n$, is provided by

$$\tilde{x}_j^n = x_j^{n+1} - u\Delta t. \tag{5.10}$$

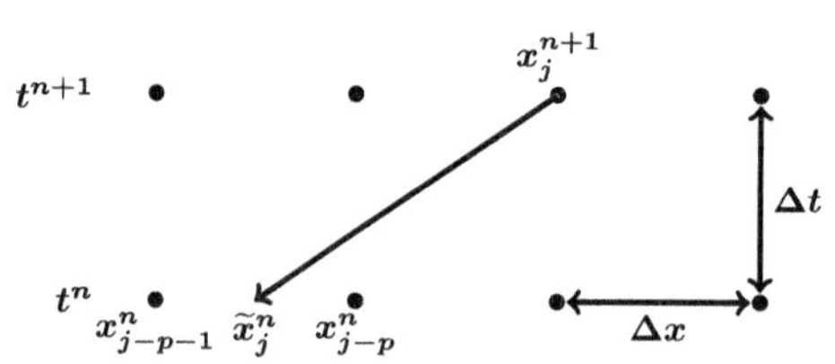

FIGURE 5.2 Schematic of the backward trajectory to find the departure point.

In Fig. 5.2 we have presented a schematic of the backward trajectory approach. An important feature to note about Fig. 5.2 is that it is likely that the departure point at time t^n does not coincide with a grid point that arrives at our grid point at t^{n+1}. When the departure point does not coincide with a grid point, we need to define a distance from the nearest grid point downwind to the direction of the characteristic arriving at time $t = t^n$, where we should note here that this is the situation that Wiin-Nielsen's mention may be needed.

Given our departure point $\tilde{x}_j^n$, which is not at a grid point, we have to use some form of interpolation polynomial, of some order, to obtain a value for ψ at the departure point, because when there is no forcing term, the value for ψ at the departure point becomes the value for ψ at our grid point at time t^{n+1}.

The first step in a semi-Lagrangian algorithm is to find the integer p of the grid point that is the nearest, downwind or upwind, to our departure point, i.e., p is the integer part of $u\frac{\Delta t}{\Delta x}$. Therefore, out departure point at t^n, $\tilde{x}_j^n$, will lie in the interval $x_{i-p-1}^n \leq \tilde{x}_i^n < x_{i-p}^n$, as shown in Fig. 5.2. Note that given the magnitude of the advection winds, the departure point could be in a region that is several grid points downwind of the arrival grid point. This is an important feature to note for programming a semi-Lagrangian based method.

The next step in the implementation of the semi-Lagrangian method is to calculate the distance from the departure point to the nearest grid point downwind of the departure point. This distance is quite often referred to as $a \equiv \tilde{x}_i^n - x_{i-p}^n$. When programming this step, it is quite often useful to run a loop over grid points with an "if" statement till one finds a grid point where $-1 \leq \dfrac{a}{\Delta x} < 0$, and then to note the index and set that to index to p. The final step is to create the weighting for the interpolation polynomial which is referred to as α and represents the ratio of the distance from departure point to the upwind grid point, divided by Δx, which is given by

$$\alpha = \frac{\widetilde{x}_i^n - x_{i-p}^n}{\Delta x}, \tag{5.11}$$

where $\alpha \in [-1, 0]$.

From this information it will be possible to calculate the weighting necessary for the interpolation polynomial, but now comes the important question: Which order of interpolation polynomial should we use? It could appear trivial, but first let's use a linear interpolation, which was suggested in [200], for the two-dimensional situation. This would then make the tracer's value at the departure point

$$\psi\left(\tilde{x}_{i-p}, t^n\right) = (1+\alpha)\,\psi_{i-p}^n - \alpha\psi_{i-p-1}^n. \tag{5.12}$$

However, it can be shown quite easily when implemented in a computer code that the linear interpolation semi-Lagrangian approach has a strong damping effect and is not shape preserving, which is a desired property for an advection scheme.

Now consider a quadratic polynomial based interpolation approach. As we saw above for the linear interpolation approach, we use the values of ψ at the grid points on either side of the departure point, weighted by their distance from the departure point. For the quadratic approach, we consider an interpolation that is upwind of the departure point, we shall still use the points on either side but now we include the value for ψ at x_{i-p+1}. A quadratic interpolation polynomial that could be used for the semi-Lagrangian approach is given by

$$\psi\left(\tilde{x}_{i-p}, t^n\right) = \frac{\alpha(\alpha+1)}{2}\psi_{i-p+1}^n + \left(1-\alpha^2\right)\psi_{i-p}^n + \frac{\alpha(\alpha-1)}{2}\psi_{i-p-1}^n. \tag{5.13}$$

A third possible choice for the interpolation polynomial could be a cubic interpolation polynomial, where now we have quite a few choices of points that can be used to interpolate the value of ψ to the departure point. Here we shall present an upwind scheme where use two points before the departure point and then two after the departure point. Given this choice, a possible cubic interpolation polynomial is given by

$$\begin{aligned}\tilde{\psi}\left(\tilde{x}_i^n, t^n\right) &= \frac{\alpha\left(1-\alpha^2\right)}{6}\psi_{i-p-2}^n + \frac{\alpha\left(\alpha-1\right)\left(\alpha+2\right)}{2}\psi_{i-p-1}^n + \frac{\left(1-\alpha^2\right)\left(2+\alpha\right)}{2}\psi_{i-p}^n \\ &+ \frac{\alpha\left(1+\alpha\right)\left(2+\alpha\right)}{6}\psi_{i-p+1}^n. \end{aligned} \tag{5.14}$$

To recap the grid points that are used for the interpolation to find the value of the tracer at the departure point given in Fig. 5.3, we have summarized the values for the three different interpolation polynomials that have been presented at each grid point for reference in Table 5.1. Note just that the interpolation polynomials presented here are referred to as the **Lagrange interpolation polynomials**. We shall go into more details about the derivation of these polynomial and a few other interpolation polynomials that have been used in a semi-Lagrange advection applications later.

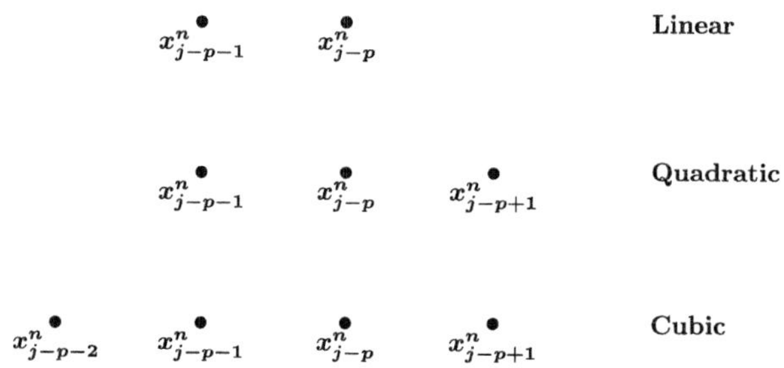

FIGURE 5.3 Schematic of the points that are used for the linear, quadratic, and cubic upwind versions of the Lagrange interpolation polynomial.

TABLE 5.1 Summary of the different coefficients of the Lagrange interpolation for the linear, quadratic, and cubic based semi-Lagrangian based advection schemes.

Order of polynomial	$i-p-2$	$i-p-1$	$i-p$	$i-p+1$
Linear		$-\alpha$	$1+\alpha$	
Quadratic		$\frac{\alpha\left(\alpha-1\right)}{2}$	$\left(1-\alpha^2\right)$	$\frac{\alpha\left(\alpha+1\right)}{2}$
Cubic	$\frac{\alpha\left(1-\alpha^2\right)}{6}$	$\frac{\alpha\left(\alpha-1\right)\left(2+\alpha\right)}{2}$	$\frac{\left(1-\alpha^2\right)\left(2+\alpha\right)}{2}$	$\frac{\alpha\left(1+\alpha\right)\left(2+\alpha\right)}{6}$

An algorithm to implement a semi-Lagrangian scheme is as follows:

Step 1: Determine the departure point at $t=t^n$, $\tilde{x}_i^n$, such that the tracer arrives at x_i^{n+1} through the backward trajectory. If the advection wind, u, is constant with respect to time and spacial direction of the advection, then backward calculation is simply $\tilde{x}_i^n = x_i^{n+1} - u\Delta t$.

Step 2: Determine if the departure point coincides with a grid point, recalling that this grid point may be several spacial steps away from the arrival grid point, and derive the number of grid points behind the arrival point, which is the integer component of $u\frac{\Delta t}{\Delta x}$.

Step 3a: If the departure point is at a grid point then the value of the tracer at this point, ψ_{i-p}^{n}, becomes the value for ψ_i^{n+1}.

Step 3b: If the departure point is not at a grid point then determine the nearest grid point that is upwind of the departure point, x_{i-p}, and calculate the distance from the departure point to the nearest upwind grid point, $a = \tilde{x}_i^n - x_{i-p}$.

Step 4: Calculate the interpolation weight, α, as $\alpha = \frac{a}{\Delta x}$.

Step 5: Select the order of the Lagrange interpolation polynomial, linear, quadratic, cubic, quartic, quintic, etc., or a different type of interpolation.

Step 6: Evaluate the interpolation, which becomes the value of the tracer at the arrival point,

$$\psi_i^{n+1} = f\left(\alpha, \psi_{i-p+\{1,2,\ldots,l\}}^{n}, \psi_{i-p-\{0,1,2,\ldots,k\}}^{n}\right).$$

Note that l and k may not be equal.

Step 7: Continue along the spatial grid repeating each step till the boundary is reached.

As we did with the Eulerian finite difference schemes, we shall apply these three different interpolation polynomials in the semi-Lagrangian framework to the advection of a Gaussian bell curve and a step function, to highlight their behaviors, but to also show how they are able to perform under conditions that would not have been possible with some of the Eulerian based schemes.

5.3 Semi-Lagrangian advection of the bell curve

In this section we shall present the results of using the linear, quadratic, and cubic Lagrange interpolation polynomials with the semi-Lagrangian framework with constant advection velocity and zero forcing. We shall present all of the results for $u = 1$, $\Delta x = 0.1$, and $\Delta x = 0.1$, which results in a Courant number of -1, meaning that we are exactly at the grid point one step size back, and $\alpha = 1$, and the case where $u = 1.5$, $\Delta x = 0.1$, and $\Delta t = 0.1$, but now we have a Courant number of 1.5, which means that we have moved beyond one grid point behind and are actually halfway between the two grid points, with $\alpha = -0.5$.

5.3.1 Semi-Lagrangian advection using linear Lagrange interpolation

In Fig. 5.4 we have plotted the solution using the linear interpolation formula for the first test case described above. The first thing to note here is that in theory the departure point is the value of the previous grid point to the arrival point at t^{n+1} at t^n. This then means that the term multiplying ψ_i^n is equal to zero, and we just have the value at ψ_{i-1}^n. We can see from the

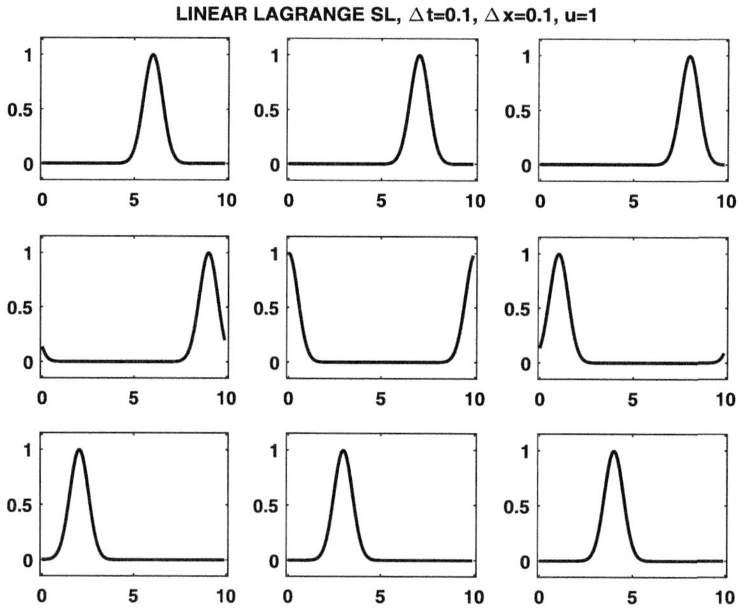

FIGURE 5.4 Plot of the bell curve being advected using a linear Lagrange interpolation polynomial to determine the value of ψ at the departure point. Here $\Delta t = \Delta x = 0.1$ and $u = 1$, which gives an α value of -1, and a Courant number of 1.

nine panels in Fig. 5.4 that we are able to track the solution almost exactly in time, and there appears to be no damping of the curve, nor are there any phase errors with this approach. This is consistent with what we saw using the upwind forward Euler approach when we have a Courant number of 1.

We now move on to consider the second choice of parameter that we mentioned above, using the same values for the spatial and temporal step sizes, but now taking a stronger/faster advection wind. With the advection wind being $u = 1.5$, this means checking for the nearest upwind integer grid point, since we are between two grid points, specifically we are halfway between the two points. The results for this formulation are presented in Fig. 5.5, where we can see that we are starting to see some damping effects from having to interpolate to the departure point. It should be noted that this is the worst case scenario for the interpolation. We shall go into more details about this effect in Chapter 7. We can see that there does not appear to be a phase error when using the linear Lagrange interpolation polynomial in the semi-Lagrangian framework, but that the damping that we saw from the upwind forward Euler finite difference approach scheme does appear to be present; however, this effect does not appear to be as severe as for the Eulerian finite difference scheme. Moreover, the important thing to notice here is that the Courant number is 1.5!!! We saw in Chapter 2 that the upwind forward Euler was unstable for Courant numbers greater than 1 as this would violate the CFL condition for that finite difference scheme, yet here we see, while slightly damped, that we do not have to increase the spatial or temporal resolution to cope with a advection speed of 1.5. This is consistent with the comments made by Robert in [161].

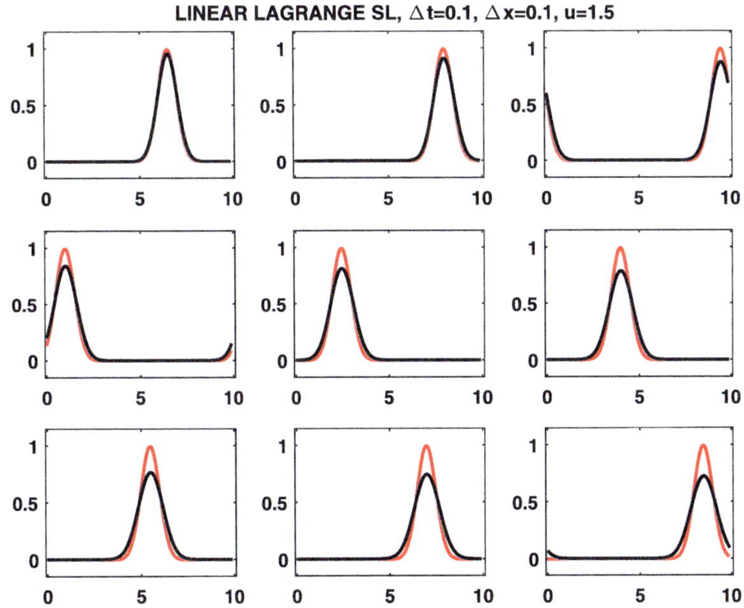

FIGURE 5.5 Plot of the bell curve being advected using a linear Lagrange interpolation polynomial to determine the value of ψ at the departure point. Here $\Delta t = \Delta x = 0.1$ and $u = 1.5$, which gives an α value of -0.5, and a Courant number of 1.5.

5.3.2 Quadratic Lagrange interpolation polynomial

We now move on to consider the results of using a quadratic Lagrange interpolation polynomial to estimate the value of the field at the departure point. We shall show results using the same configuration for the temporal and spatial step sizes, as well as the advection wind speed, that were presented earlier in this section. The results with the Courant number equal to 1 for the quadratic Lagrange interpolation polynomial are presented in Fig. 5.6.

As with the linear Lagrange polynomial results for this test case, we know that because the departure point coincides with a grid point that is shifted by one spatial grid point downstream from the arrival point at t^{n+1}, the quadratic polynomial is also able to capture the bell curve exactly for all time steps that are presented. This is a good sign that the interpolation polynomial will collapse to the exact value at the grid point, rather than giving weights to the values of the surrounding points when the departure point does match up with a grid point.

However, we now consider the second test case for the quadratic Lagrange interpolation polynomial where we have a Courant number of 1.5 and where α is such that the departure point in halfway between two grid points, Fig. 5.7.

The first obvious difference that we can see in Fig. 5.7 compared to the results for the linear Lagrange interpolation for this situation in Fig. 5.5 is that there does not appear to be as much, if any, damping of the bell curve, however, it does appear that in the quadratic case we see that the solution is slightly out of phase, appearing to be slightly behind the true solution.

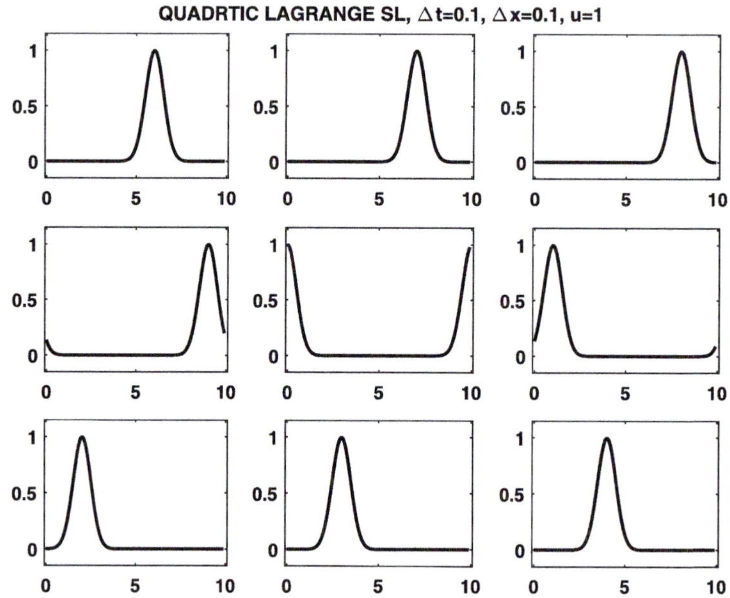

FIGURE 5.6 Plot of the bell curve being advected using a linear Lagrange interpolation polynomial to determine the value of ψ at the departure point. Here $\Delta t = \Delta x = 0.1$ and $u = 1$, which gives an α value of -1, and a Courant number of 1.

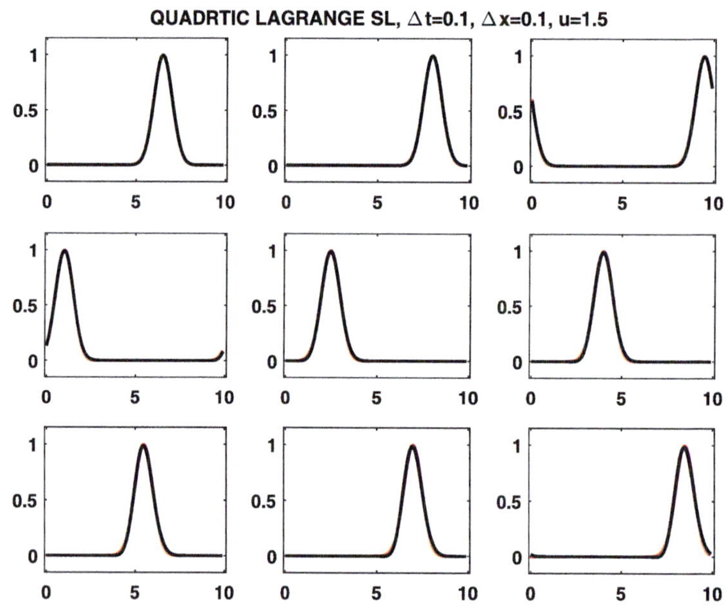

FIGURE 5.7 Plot of the bell curve being advected using a linear Lagrange interpolation polynomial to determine the value of ψ at the departure point. Here $\Delta t = \Delta x = 0.1$ and $u = 1.5$, which gives an α value of -0.5, and a Courant number of 1.5.

5.3.3 Cubic Lagrange interpolation polynomial

We now move on to consider the performance of the cubic Lagrange interpolation polynomial with the same two test cases presented for the linear and quadratic Lagrange interpolation polynomials. The results for the first set of parameters for the bell curve are presented in Fig. 5.8, where we can see that we are able to recreate the true solution quite accurately as would be expected when the departure point is a grid point at the previous time step. It should be noted here that this is the case no matter now many grid points behind the arrival point the departure point ends up being.

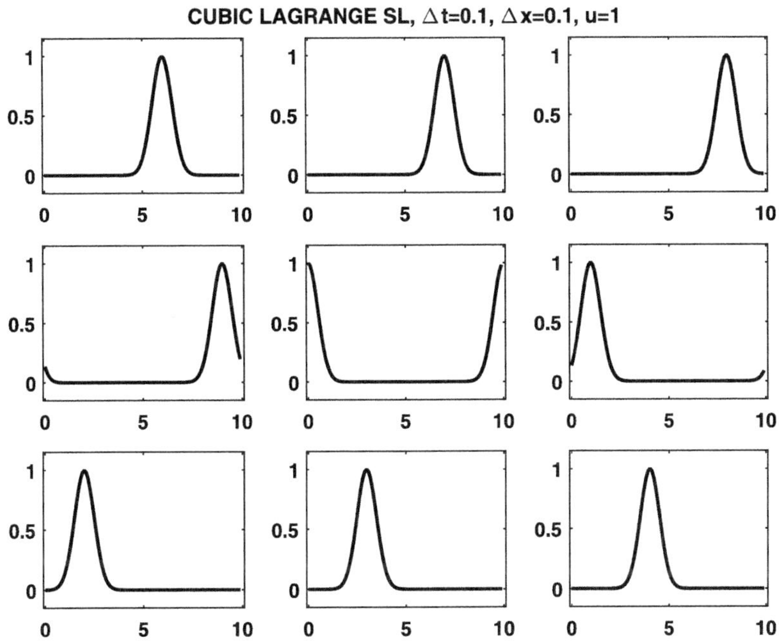

FIGURE 5.8 Plot of the bell curve being advected using a cubic Lagrange interpolation polynomial to determine the value of ψ at the departure point. Here $\Delta t = \Delta x = 0.1$ and $u = 1$, which gives an α value of 1, and a Courant number of 1.

If we now consider the second configuration for this test case using the cubic Lagrange interpolation polynomial, Fig. 5.9, we see that there is still a very slight damping from the scheme and that the numerical solution appears to be slightly out of phase again, but not as severely as when using the quadratic interpolation polynomial, which in itself was not that bad, but this is still a major improvement on the linear results; see Fig. 5.9. This is an important result from this order of interpolation polynomial, as it may not be obvious that we would need to consider a higher order polynomial to obtain a much higher order of accuracy above the cubic polynomial; we also have to take into account how long would it take to calculate a quartic polynomial, and whether the model could still be run in time for a forecast to be produced on time.

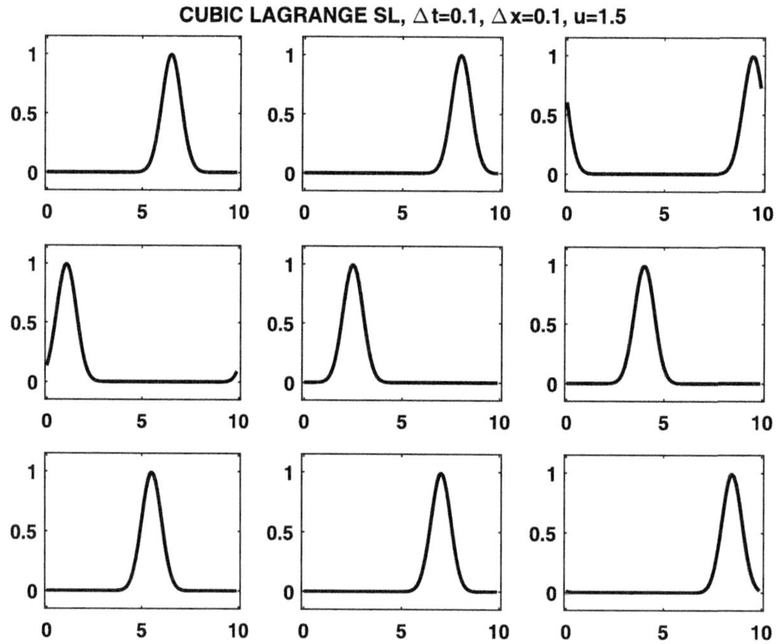

FIGURE 5.9 Plot of the bell curve being advected using a linear Lagrange interpolation polynomial to determine the value of ψ at the departure point. Here $\Delta t = \Delta x = 0.1$ and $u = 1.5$, which gives an α value of 0.5, and a Courant number of 1.5.

Another important feature to note here from all of the results that were presented in this section for all three order of polynomials is that we have been able to apply all three of them for a numerical situation where the Courant number is much higher than those which could produce stable solutions with most of the explicit Eulerian finite difference schemes, which is what makes the semi-Lagrangian method quite an interesting and appealing mechanism.

Given the results for the three different order Lagrange interpolation polynomials presented here with the continuous bell curve, we now move on to consider their performances with the discontinuous step function.

5.4 Semi-Lagrangian advection of the step function

As with the Eulerian finite difference schemes, we now consider how the three different Lagrange interpolation polynomials perform with the discontinuous step function. We have used the same parameters to define the step function as in Chapter 2, and as mentioned earlier, we shall be considering the performance of the three schemes for $\Delta t = \Delta x = 0.1$, with $u = 1$, which implies that we have a Courant number $C = 1$ and $\alpha = -1$; and for $\Delta x = \Delta t = 0.1$, where now the advection wind is $u = 1.5$, which results in a Courant number of $C = 1.5$ and a value of $\alpha = -0.5$.

5.4.1 Linear Lagrange interpolation polynomial

In Fig. 5.10 we have presented the results of using the linear Lagrange interpolation polynomial for the first set of parameters mentioned above. We can see that the linear interpolation does collapse to using the value at the next grid point downstream of the arrival point at t^{n+1}, at t^n. We can clearly see that there are neither phase nor amplitude errors. Another clear feature that is present in Fig. 5.10 is that there are no spurious wiggles as we saw with the Lax–Wendroff scheme in Chapter 2. We should note that the upwind forward Euler scheme could also recreate the step function almost exactly when we had a Courant number of 1.

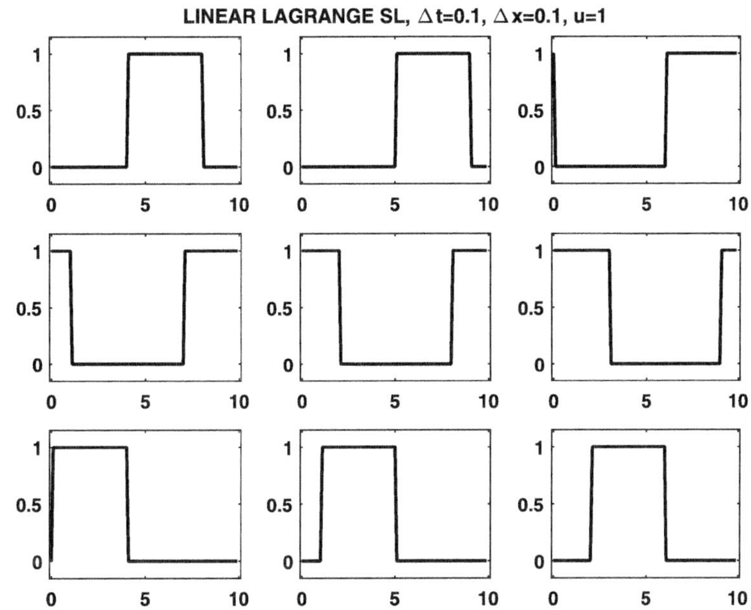

FIGURE 5.10 Plot of the step function being advected using a linear Lagrange interpolation polynomial to determine the value of ψ at the departure point. Here $\Delta t = \Delta x = 0.1$ and $u = 1$, which gives an α value of -1, and a Courant number of 1.

We now move on to consider the case with a Courant number of 1.5 and a linear Lagrange interpolation polynomial, where the results for this situation are presented in Fig. 5.11. We can clearly see that the scheme is stable for this Courant number, which would not be the case for most of the explicit Eulerian finite difference schemes presented in Chapter 2. Another positive feature to note here is that the linear Lagrange interpolation polynomial does not appear to be damping the numerical solution, yet it is trying to *smooth* out the discontinuities on either side of the step. The last positive feature to mention here is that the numerical solution using this interpolation polynomial has remained in phase with the true solution. We have not had to increase the resolution of either, or both, the spatial or temporal grids, which would not have been the case for most of the Eulerian finite difference schemes, to bring the Courant number to be less than or equal to 1.

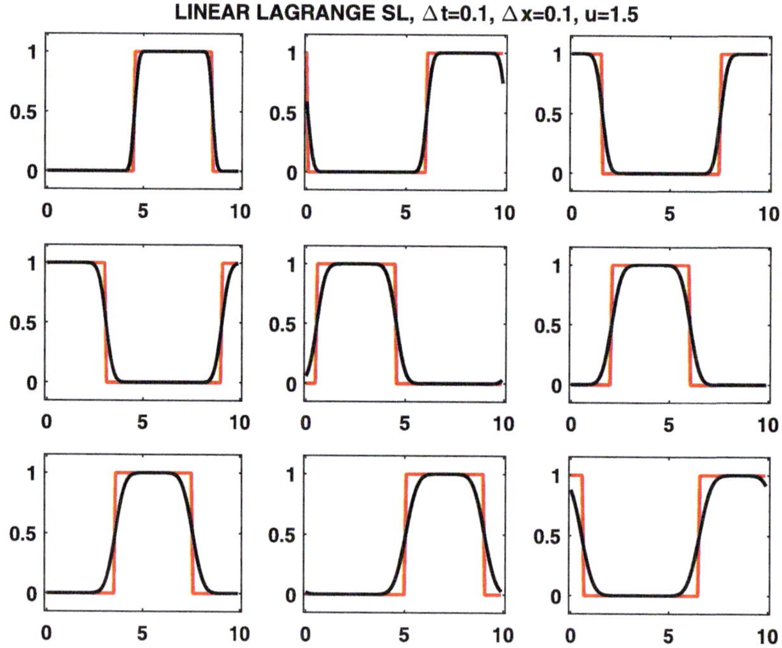

FIGURE 5.11 Plot of the step function being advected using a linear Lagrange interpolation polynomial to determine the value of ψ at the departure point. Here $\Delta t = \Delta x = 0.1$ and $u = 1.5$, which gives an α value of -0.5, and a Courant number of 1.5.

5.4.2 Quadratic Lagrange interpolation polynomial

As with the linear Lagrange interpolation polynomial, we shall consider the same two sets of the parameters for the spatial and temporal step sizes with the step function. The results for the Courant number equal to 1 are presented in Fig. 5.12, where we can again see that when the departure point is at a grid point at the previous time step, the semi-Lagrangian scheme, at least for the constant velocity and zero forcing case, can capture the true state quite well, when it is discontinuous like a step function.

The interesting case is that with the second set of parameters, as it involves both a Courant number greater than 1, but more importantly for the semi-Lagrangian approaches, a departure point in the center of two grid points, which is where we shall see the effects of the interpolation polynomials. The results using the quadratic Lagrange interpolation polynomial for this situation are shown in Fig. 5.13.

While it is quite clear from Fig. 5.13 that the quadratic Lagrange interpolation polynomial based semi-Lagrangian scheme does appear to be able to better capture the discontinuity than the linear Lagrange interpolation approach, in that it is not smoothing the numerical solution as much, we do, however, have a new issue where we are seeing wiggles at the boundaries of the discontinuities, and that as time goes on these wiggles start to propagate downstream of the discontinuities. Therefore, we have to be aware that this scheme produces spurious waves

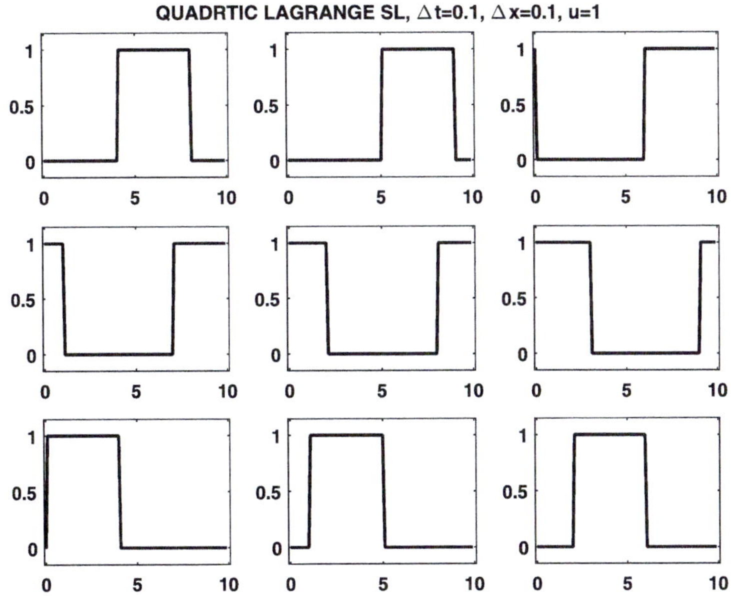

FIGURE 5.12 Plot of the step function being advected using a quadratic Lagrange interpolation polynomial to determine the value of ψ at the departure point. Here $\Delta t = \Delta x = 0.1$ and $u = 1$, which gives an α value of -1, and a Courant number of 1.

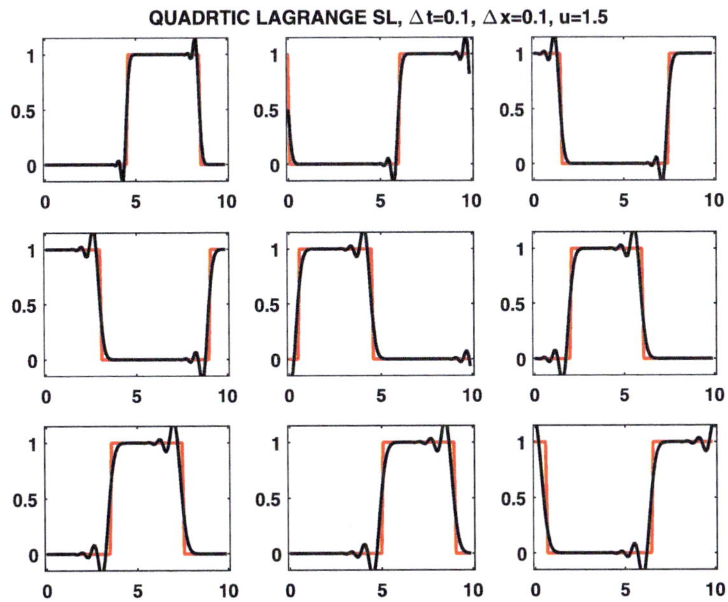

FIGURE 5.13 Plot of the step function being advected using a quadratic Lagrange interpolation polynomial to determine the value of ψ at the departure point. Here $\Delta t = \Delta x = 0.1$ and $u = 1.5$, which gives an α value of -0.5, and a Courant number of 1.5.

downstream of the discontinuities, which will feed back into other equations that could be coupled with the advection model.

5.4.3 Cubic Lagrange interpolation polynomial

We now move on to the last two sets of results in this chapter with the Lagrange interpolation polynomials. We can see in Fig. 5.14 that, when $\alpha = 1$, the interpolation scheme sets the departure point to the grid point that is downstream of the arrival point, and as such the scheme advects the function almost exactly. Note that we have used the value of α as mentioned in the last section for the cubic interpolation scheme.

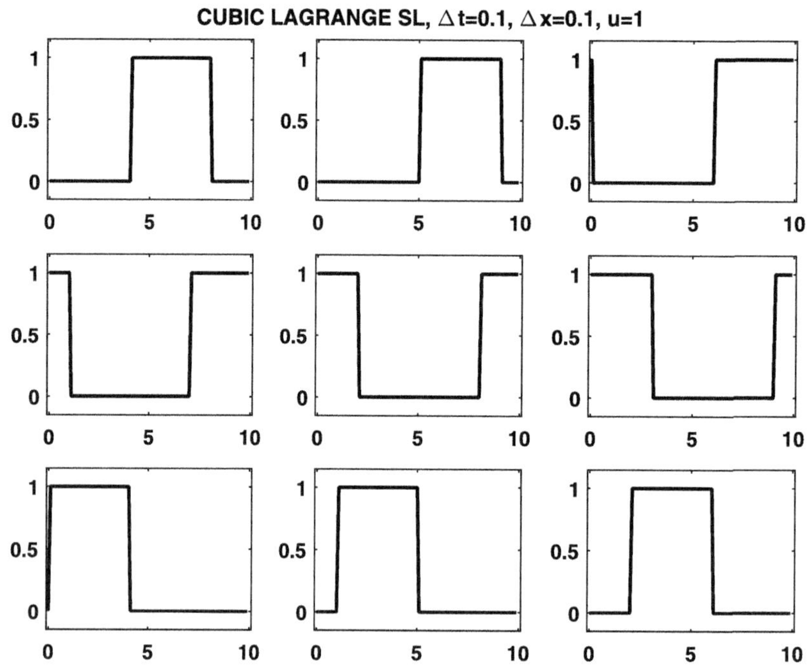

FIGURE 5.14 Plot of the step function being advected using a cubic Lagrange interpolation polynomial to determine the value of ψ at the departure point. Here $\Delta t = \Delta x = 0.1$ and $u = 1$, which gives an α value of -1, and a Courant number of 1.

As with the results from the linear and quadratic Lagrange interpolation results for $C = 1.5$, $\alpha = -0.5$, this is where we can see the affects of the different interpolation formulas. In Fig. 5.15 we have the results of using the cubic Lagrange interpolation polynomial, where we neither have the smoothing as with the linear interpolation nor the severe wiggles as we did with the quadratic interpolation, and nor do we see any wiggles propagating downstream of the discontinuities. But we do see that we have some over- and undershoots at the discontinuities which, while appearing not to be growing, are present in the solution which would then feed back into any other equations that are coupled to it.

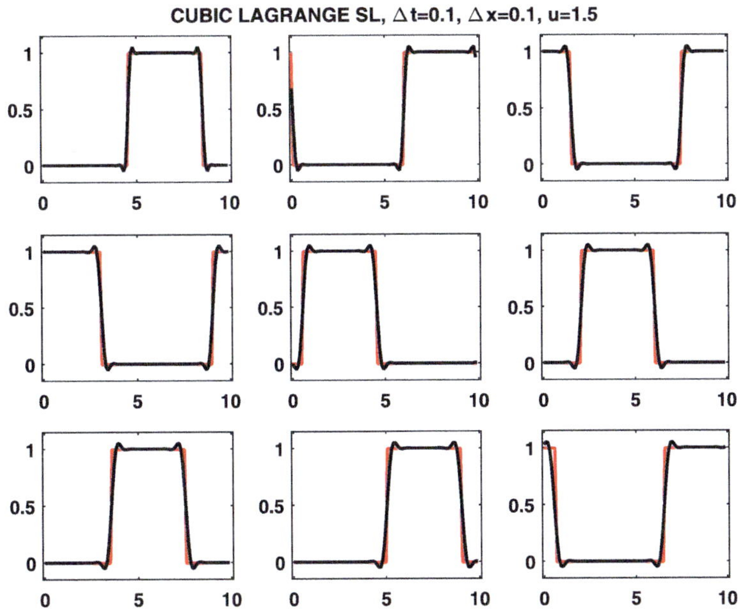

FIGURE 5.15 Plot of the step function being advected using a cubic Lagrange interpolation polynomial to determine the value of ψ at the departure point. Here $\Delta t = \Delta x = 0.1$ and $u = 1.5$, which gives an α value of -0.5, and a Courant number of 1.5.

5.5 Summary

In this chapter we have introduced a brief overview of the Lagrangian formulation of advection, and indicated that it may not be suitable for numerical modeling applications. We then moved on to explained how to find the departure points. When the departure point does not coincide with a grid point at the previous time step, we need to interpolate the values at the grid point for the field of interest to the departure points. We introduced linear, quadratic, and cubic interpolation polynomials that are part of the Lagrange family of interpolation polynomials. Given these three formulas, we tested them with the continuous bell curve and the step function, where we showed that we were able to recreate the true solution nearly exactly if the departure point does occur at a grid point. This was true for both the bell curve and the step function.

However, when we examined the performance of the different interpolation schemes where the departure point was halfway between two grid points, then we saw different behaviors for each scheme for each test case, and we have summarized these below:

- Linear Bell – Damping of the curve
- Linear Step – Smoothing of the discontinuities
- Quadratic Bell – Slight damping and phase error
- Quadratic Step – Wiggles downstream of the discontinuities that were growing and traveling behind

- Cubic Bell – Slight phase error and slight damping, not as severe as for the quadratic case
- Cubic Step – Over- and undershoots at either side of the discontinuities but did not cause any downwind effects

Given these results, we now move on to consider what different interpolations polynomials could be used in some different semi-Lagrangian schemes in different geophysical applications, as well as introduce how to derive the three forms of the Lagrange interpolation polynomials presented in this chapter.

CHAPTER

6

Interpolation methods

As seen in the previous chapter, we are seeking the values of the tracers at the grid points at the next time level, given the distance that the particle/field/flux has traveled in Δt. From the theory of characteristics we know that the value at the arrival point is equal to the point at the departure point, but it is not always the case that the departure point is at a grid point at the previous time step, and so we need to interpolate between grid points for the value at this departure point. In this chapter we shall introduce different interpolation methods that are commonly used in applications in the geosciences, applied to a simple one-dimensional problem and explain some of the pros and cons of each technique.

We start by introducing a series of definitions of properties that we wish for the interpolation schemes to have. The first property that would be extremely useful for the interpolation polynomials to possess is **monotonicity**. We first define what a monotonic function is and then how this definition applies to interpolation polynomials.

Definition 6.1. A function f that is defined on a subset S of the real numbers, $\mathbb{R}$, is said to be **monotonic** if and only if it is entirely increasing or decreasing over the subset. The function is referred to as **monotonically increasing** if for all the x_i and x_j that are in the subset, such that $x_i \leq x_j$, we have $f(x_i) \leq f(x_j)$, and as such the function f preserves the ordering of the values in the set S.

If we reverse the inequality above, then we would have a definition for **monotonically decreasing** function. By association, it we were to tighten the inequalities to greater than, or less than for the decreasing case, we would have the definition for a **strictly monotonically increasing (decreasing)** function.

This is an important property that we would like the interpolation polynomial to preserve, and that is the essence of the definition for a monotonic interpolation polynomial.

Definition 6.2. An interpolation polynomial is said to be a **monotonic interpolation** if it preserves the monotonicity of the underlying data.

As an example, linear interpolation preserves monotonicity, but not all higher order interpolation polynomials do. What does it mean if an interpolation polynomial is not monotonic? In short, it means that while the polynomial will fit the data well, it may do what is referred to as **overshoot** or **undershoot** between points or features that are being advected.

In Chapter 5 we have already introduced a set of polynomials that could be used for an interpolation approach. These polynomials are referred to as the **Lagrange interpolation poly-**

Semi-Lagrangian Advection Methods and Their Applications in Geoscience
https://doi.org/10.1016/B978-0-12-817222-3.00010-4

nomials, and we have also seen these in Chapter 2 when we derived the Adams–Bashforth and Adams–Moulton schemes. We now move to put these polynomials into a general framework.

6.1 Lagrange interpolation polynomials

The Lagrange interpolation polynomials play a vital part in some applications of the semi-Lagrangian schemes. These sets of interpolation polynomials are named after Joseph Louis Lagrange after he published them in 1795. However, they were first discovered in 1779 by Edward Wary and then rediscovered in 1783 by Leonhard Euler. The definition for a general Lagrange interpolation polynomial is given below.

Definition 6.3. The Lagrange interpolating polynomial $P(x)$ of degree less that or equal to $(n-1)$ passing through n points $(x_1, y_1 = f(x_1)), \ldots, (x_n, y_n = f(x_n))$ is given by

$$P(x) = \sum_{i=1}^{n} P_i(x), \text{ where } P_i(x) \equiv y_i \prod_{\substack{k=1 \\ k \neq i}}^{n} \frac{x - x_k}{x_i - x_k}. \tag{6.1}$$

We shall now apply the definition in (6.1) to derive the coefficients for the linear, quadratic, and cubic Lagrange interpolation polynomials that were stated in the last chapter. We shall drop the i component from the subscripts to derive the three polynomials set around the point from the arrival point index.

The first step in deriving the coefficients for the three different orders of polynomial is to identify the expressions for the distances from the departure point and the surrounding grid points in terms of α. We have defined α to be the distance from the nearest upwind grid point to the departure point, i.e., $\alpha = x_p - \tilde{x}_i$. Therefore, if we consider the grid points on either side of the departure point, recalling that the distance between each grid point has been normalized with respect to Δx, we have that $\tilde{x}_i - x_p = -\alpha$, $\tilde{x}_i - x_{p-1} = 1 - \alpha$, which occurs as a result of considering that the normalized distance between the x_{p-1} grid point to the x_p grid point is 1, and then we have to go back by α to arrive at the departure point. Following this same argument implies that the distance from $\tilde{x}_i$ to the x_{p-2} grid point is $2 - \alpha$.

Finally, considering the distance in the upwind direction from the departure point, we have that the distance $\tilde{x}_i - x_{p+1}$ is as follows: first, we have a distance $-\alpha$ from the departure point to the x_p grid point and then a further -1 between the x_p and x_{p+1} grid points. Therefore, we have that $\tilde{x}_i - x_{p+1} = -\alpha - 1 \equiv -(1 + \alpha)$.

Given all these distances, it is now possible to derive the coefficients of the for the linear, $P_1(\alpha)$, quadratic, $P_2(\alpha)$, and cubic, $P_3(\alpha)$, Lagrange interpolation polynomials as:

Linear: $\dfrac{\left(\tilde{x}_i^n - x_p\right)}{\left(x_{p-1} - x_p\right)}\phi_{p-1} + \dfrac{\left(\tilde{x}_j^n - x_{p-1}\right)}{\left(x_p - x_{p-1}\right)}\phi_p$, which is equivalent to $\dfrac{-\alpha}{-1}\psi_{p-1} + \dfrac{(1-\alpha)}{1}\psi_p$, resulting in $P_1(\alpha) \equiv \alpha\psi_p + (1-p)\psi_{p-1}$.

Quadratic: To fit a quadratic polynomial through the departure point, we require three points x_{p-1}, x_p, and x_{p+1}, which, upon substituting these points into (6.1), results in

$$\begin{aligned} P_2(\alpha) &= \frac{(\tilde{x}-x_p)(\tilde{x}-x_{p+1})}{(x_{p-1}-x_p)(x_{p-1}-x_{p+1})}\psi_{p-1} + \frac{(\tilde{x}-x_{p-1})(\tilde{x}-x_{p+1})}{(x_p-x_{p-1})(x_p-x_{p+1})}\psi_p \\ &+ \frac{(\tilde{x}-x_{p-1})(\tilde{x}-x_p)}{(x_{p+1}-x_{p-1})(x_{p+1}-x_p)}\psi_{p+1}, \end{aligned} \tag{6.2}$$

which is equivalent to

$$P_2(\alpha) = \frac{(\alpha^2+\alpha)}{2}\psi_{p-1} + \left(\alpha^2-1\right)\psi_p + \frac{(\alpha^2-\alpha)}{2}\psi_{p+1}. \tag{6.3}$$

Cubic: To fit a cubic Lagrange polynomial, we require four points, which have already been identified in the last chapter: we have two points downwind of the departure point, x_{p-2} and x_{p-1}, and then two grid points upwind, x_p and x_{p+1}. Therefore, the four coefficients of the cubic Lagrange polynomial $P_3(\alpha)$ are

$$\begin{aligned} C_1(\alpha) &= \frac{(\tilde{x}-x_{p-1})(\tilde{x}-x_p)(\tilde{x}-x_{p+1})}{(x_{p-2}-x_{p-1})(x_{p-2}-x_p)(x_{p-2}-x_{p+1})} \\ &= \frac{(1-\alpha)(-\alpha)(-\alpha-1)}{(-1)(-2)(-3)} \\ &= -\frac{\alpha(\alpha^2-1)}{6}, \end{aligned} \tag{6.4a}$$

$$\begin{aligned} C_2(\alpha) &= \frac{(\tilde{x}-x_{p-2})(\tilde{x}-x_p)(\tilde{x}-x_{p+1})}{(x_{p-1}-x_{p-2})(x_{p-1}-x_p)(x_{p-1}-x_{p+1})} \\ &= \frac{(2-\alpha)(-\alpha)(-\alpha-1)}{(1)(-1)(-2)} \\ &= \frac{\alpha(2-\alpha)(1+\alpha)}{2}, \end{aligned} \tag{6.4b}$$

$$\begin{aligned} C_3(\alpha) &= \frac{(\tilde{x}-x_{p-2})(\tilde{x}-x_{p-1})(\tilde{x}-x_{p+1})}{(x_p-x_{p-2})(x_p-x_{p-1})(x_p-x_{p+1})} \\ &= \frac{(2-\alpha)(1-\alpha)(-\alpha-1)}{(2)(1)(-1)} \\ &= \frac{(2-\alpha)(1-\alpha)(1+\alpha)}{2}, \end{aligned} \tag{6.4c}$$

$$\begin{aligned} C_4(\alpha) &= \frac{(\tilde{x}-x_{p-2})(\tilde{x}-x_{p-1})(\tilde{x}-x_p)}{(x_{p+1}-x_{p-2})(x_{p+1}-x_{p-1})(x_{p+1}-x_p)} \\ &= \frac{(2-\alpha)(1-\alpha)(-\alpha)}{(3)(2)(1)} \\ &= -\frac{\alpha(2-\alpha)(1+\alpha)}{6}. \end{aligned} \tag{6.4d}$$

Combining (6.4a)–(6.4d) with their respective tracer values at their grid points results in the cubic Lagrange interpolation polynomial as $P_3(\alpha) = C_1(\alpha)\psi_{p-2} + C_2(\alpha)\psi_{p-1} + C_3(\alpha)\psi_p + C_4(\alpha)\psi_{p+1}$, where substituting the expression for the C_is results in the expression for the explicit cubic interpolation scheme stated in (5.14).

Exercise 6.4. *Derive the distances for $\tilde{x}_i - x_{p-3}$, $\tilde{x}_i - x_{p+2}$, and $\tilde{x}_i - x_{p+3}$, and use these distances to determine the quartic Lagrange interpolation polynomial, $P_4(\alpha)$, for a central, upwind, and downwind based framework.*

6.1.1 Illustrations of the Lagrange polynomials

In this section we present the quadratic and cubic Lagrange polynomials to illustrate the features. The first plot that we can consider is that of the three quadratic polynomials which are given by

$$l_{-1} = \frac{x^2 + x}{2}, \qquad l_0 = 1 - x^2, \qquad l_1 = \frac{x^2 - x}{2},$$

where we have $x \in [-1, 1]$. The plots for these basis functions can be found in Fig. 6.1.

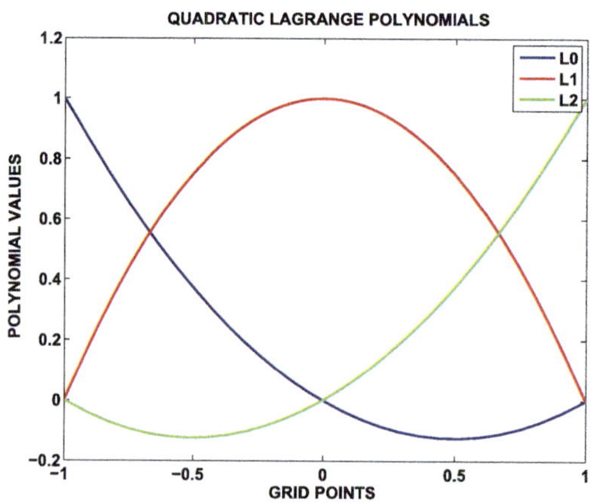

FIGURE 6.1 Plot of the quadratic Lagrange interpolation polynomials basis functions.

If one has ever worked with finite elements then the graphs in Fig. 6.1 should look quite familiar as they are quite often used as basis functions for a quadratic approximation with a triangular mesh. As we can see in Fig. 6.1, the functions are equal to 1 at the points where the interpolation point needs to be weighted and zero at the other points.

For the cubic Lagrange polynomials, we have

$$l_{-2} = -\frac{x\left(1 - x^2\right)}{6}, \quad l_{-1} = \frac{x(1 + x)(2 - x)}{2}, \quad l_0 = \frac{\left(1 - x^2\right)(2 - x)}{2}, \quad l_1 = -\frac{x(1 - x)(2 - x)}{6},$$

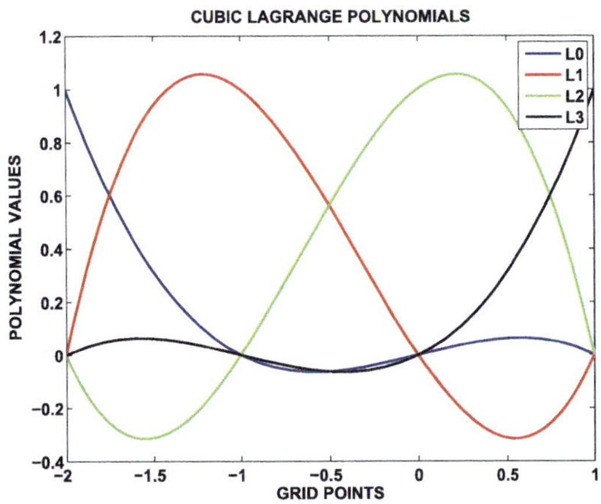

FIGURE 6.2 Plot of the cubic Lagrange interpolation polynomials basis functions.

where now $x \in [-2, 1]$. We have plotted these four cubic basis functions in Fig. 6.2. As with the quadratic basis functions, we can see that each polynomial is equal to 1 at the grid point that it is being used and zero at the other points. But a feature to notice here now is the *wiggles* in the graphs.

Now that we have seen the basis functions, we move on to consider how well these interpolation formulations, along with the Lagrange linear interpolation polynomial, work with a continuous and a discontinuous function.

6.1.2 Application of the Lagrange interpolation polynomials

As just mentioned, we are going to show the performance of the polynomials with the continuous and discontinuous functions that we introduced for the Eulerian finite difference schemes.

Continuous bell curve

We start with the continuous function, which is a bell, or normal/Gaussian, curve on the domain $x \in [0, 1]$, which is a periodic domain, and defined by

$$f(x) = \exp\left\{-\frac{1}{2}\left(\frac{x-\mu}{\sigma}\right)^2\right\}, \tag{6.5}$$

where μ is the mean, or the point where one wishes to center the curve, and σ is the standard deviation which determines the spread of the curve. For the figures that we presented with the bell curve, we have $\mu = 0.5$ and a standard deviation of $\sigma = 0.1$.

To be able to create an interpolation for this function, we require "observations" of the function/curve. For the illustrations, we have taken different number of observations, n, to show how the interpolation improves when there are more data to fit to.

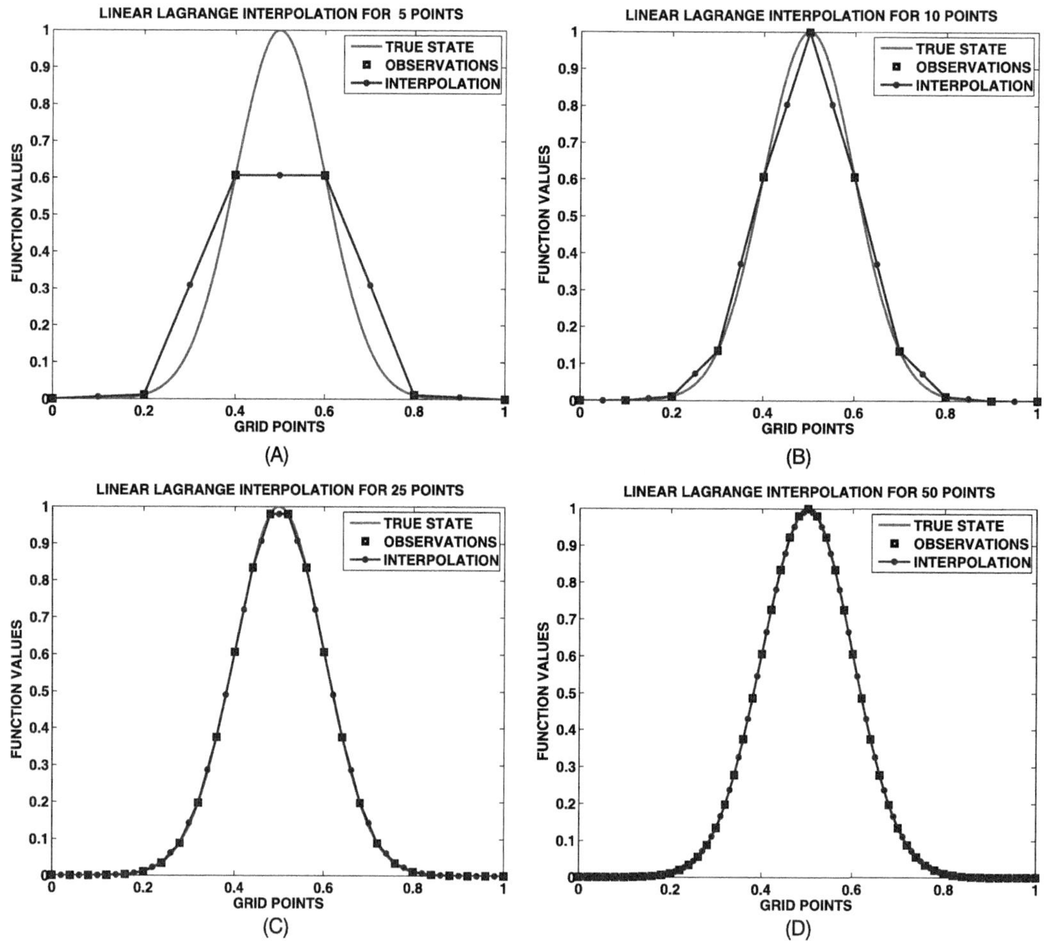

FIGURE 6.3 Plot of the linear Lagrange interpolation polynomial applied to the continuous bell curve for: (A) $n = 5$, (B) $n = 10$, (C) $n = 25$, and (D) $n = 50$.

The first approximation we consider is the linear Lagrange polynomial approximation, where we are considering how well we can approximate the curve with $n = 5, 10, 25$, and 50, and the results are shown in Fig. 6.3A–D, respectively. We can clearly see from Fig. 6.3A that when we only have 5 points to form a linear approximation, the maximum is grossly underestimated and that without a point near there it is not possible to capture this feature. The results with 10 data points are shown in Fig. 6.3B, and we can clearly see that while we do have a point at the maximum now, the curvature of the function is not captured with only 10 points. In Fig. 6.3C we have the results with 25 data points, and again because we do not have a point at the maximum of the function, the interpolation polynomial fits a straight line between the two points on either side, so the interpolation point in the center lies on this line. Finally, in Fig. 6.3D we have the results with 50 points. We can clearly see in this plot that we are able to capture the curve quite well and, by having a data point at the maximum, we are able to have some curvature there.

We now move on to consider the quadratic Lagrange polynomial approximation to the Gaussian curve with again $n = 5, 10, 25$, and 50. These results are shown in Fig. 6.4.

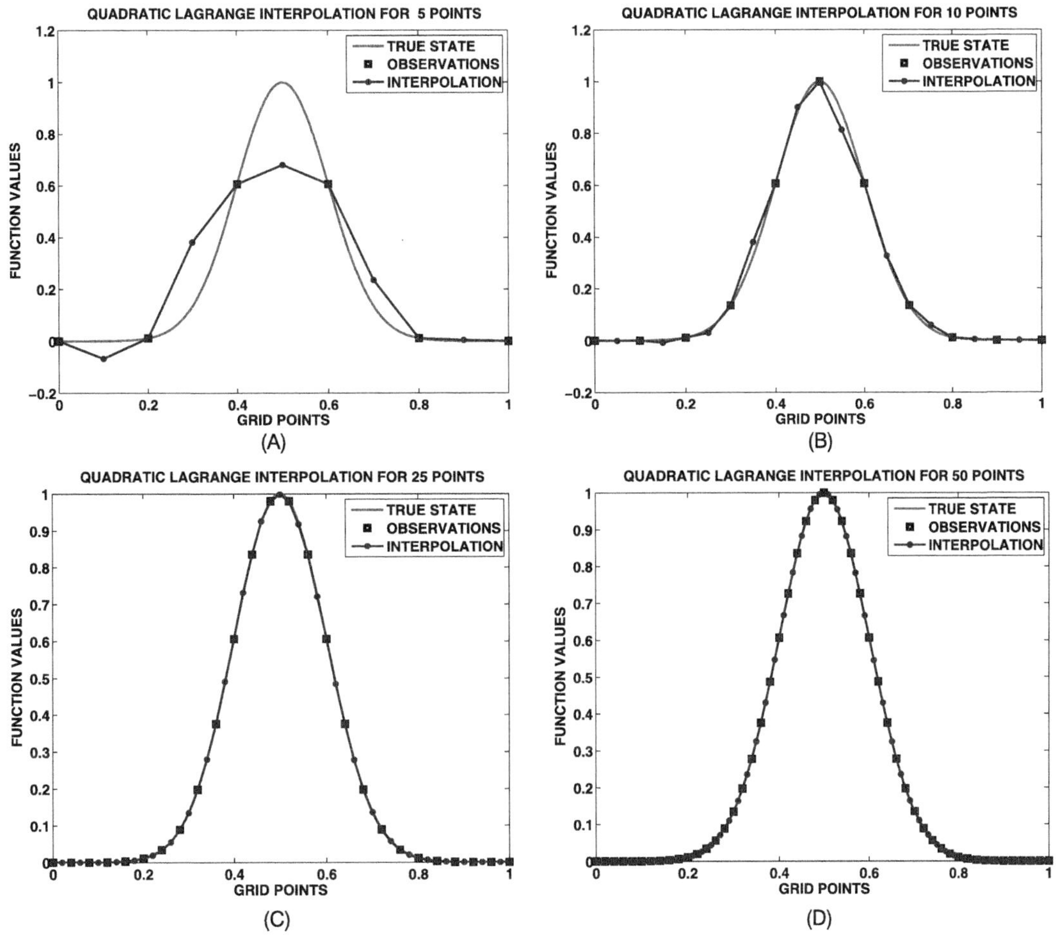

FIGURE 6.4 Plot of the quadratic Lagrange interpolation polynomial applied to the continuous bell curve for: (A) $n = 5$, (B) $n = 10$, (C) $n = 25$, and (D) $n = 50$.

The first feature to note here for the $n = 5$ case, Fig. 6.4A, is that we do not have a straight line between the two points on either side of the maximum of the curve and so the quadratic approximation does assign a slightly increased value here than in the linear case. In Fig. 6.4B we have the results with $n = 10$ and can see that on the left-hand side of the curve we are fitting quite well, but on the initial descent on the right-hand side of the maximum we have a different value than on the left. This is because the quadratic approach is not symmetric, as we are taking one point to the right of the interpolation point, the effect of which can be seen in Fig. 6.4B. However, when we increase the number of points to $n = 25$, the feature just described is not as present in Fig. 6.4C, and is quite clearly gone for the case of $n = 50$ in Fig. 6.4D.

We now move on to consider the cubic Lagrange polynomial interpolation. In Fig. 6.5A we have the results for $n = 5$. We can see that we do not have enough points to fill in the gaps with the interpolation, where it is clear that the maximum is underestimated. In Fig. 6.5B we have the results for $n = 10$ where we can see that we still do not have enough points to completely capture the curve, but that we are now closer to the maximum than with only five points. For the case $n = 25$ we can see in Fig. 6.5C that we are able to better fit the curve, and even though we do not have a point at the maximum, the interpolation is about to produce a reasonably good estimate. Finally, in Fig. 6.5D, when we have 50 points, we can see that we are able to quite well approximate the curve throughout the whole domain.

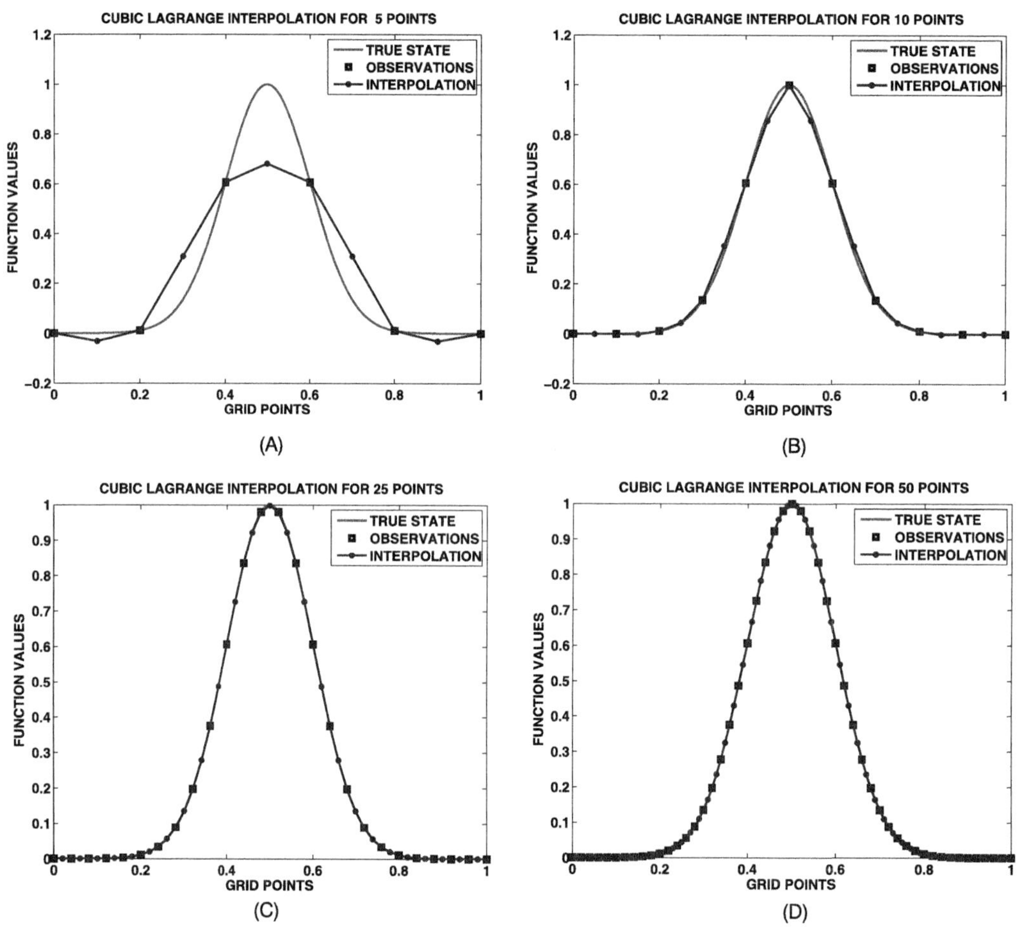

FIGURE 6.5 Plot of the cubic Lagrange interpolation polynomial applied to the continuous bell curve for: (A) $n = 5$, (B) $n = 10$, (C) $n = 25$, and (D) $n = 50$.

Discontinuous step function

We now consider a discontinuous function and illustrate problems that the Lagrange polynomials have with approximating a **step function**, which is defined as

$$
\begin{aligned}
f(x) &= 0, & 0 \le x < 0.3, \\
f(x) &= 1, & 0.3 \le x \le 0.7, \\
f(x) &= 0, & 0.7 < x \le 1.
\end{aligned}
\tag{6.6}
$$

We start again by considering the linear Lagrange interpolation, where the results for this situation can be found in Fig. 6.6.

The case when $n = 5$ is shown for the linear Lagrange interpolation in Fig. 6.6A. We can see that it can capture the horizontal straight parts, but is underestimating the values on the

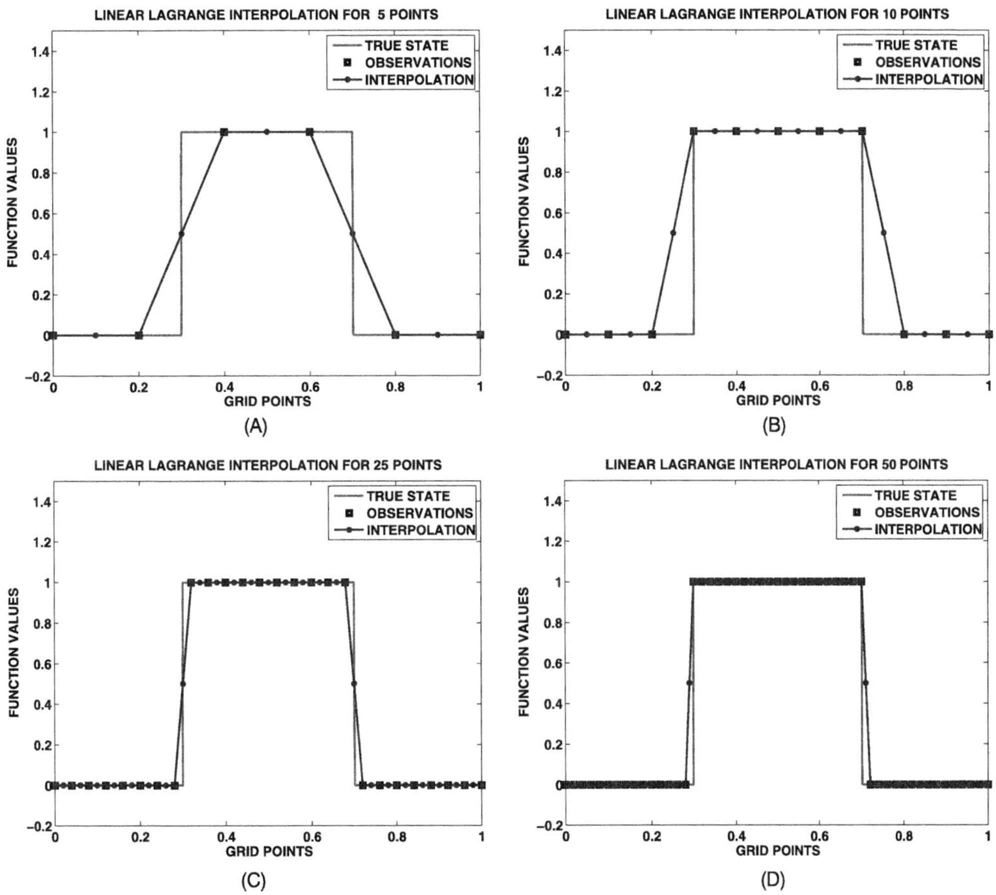

FIGURE 6.6 Plot of the linear Lagrange interpolation polynomial applied to the discontinuous step function for: (A) $n = 5$, (B) $n = 10$, (C) $n = 25$, and (D) $n = 50$.

edges of the step. In Fig. 6.6B we have the results with $n = 10$, where we can see that we are doing a better job of approximating the step, however, the approximation is not too good at the transition points between the two values of the step function. When we increase the number of points to 25 (Fig. 6.6C), we see that we are doing a very good job of approximating the discontinuous function and that the transition points are becoming better approximated. Finally, in Fig. 6.6D we have the results with $n = 50$ and can see that due to the distribution of the points, when we have an even number of sample points, we do not approximate the transition stages as well, but that the other parts of the step function are quite well approximated.

We now consider the results using a quadratic Lagrange polynomial interpolation for the step function. These results are presented in Fig. 6.7.

The results for $n = 5$ with the quadratic Lagrange polynomial interpolation are presented in Fig. 6.7A. An interesting feature to note here is what appears to be under- and overshoots

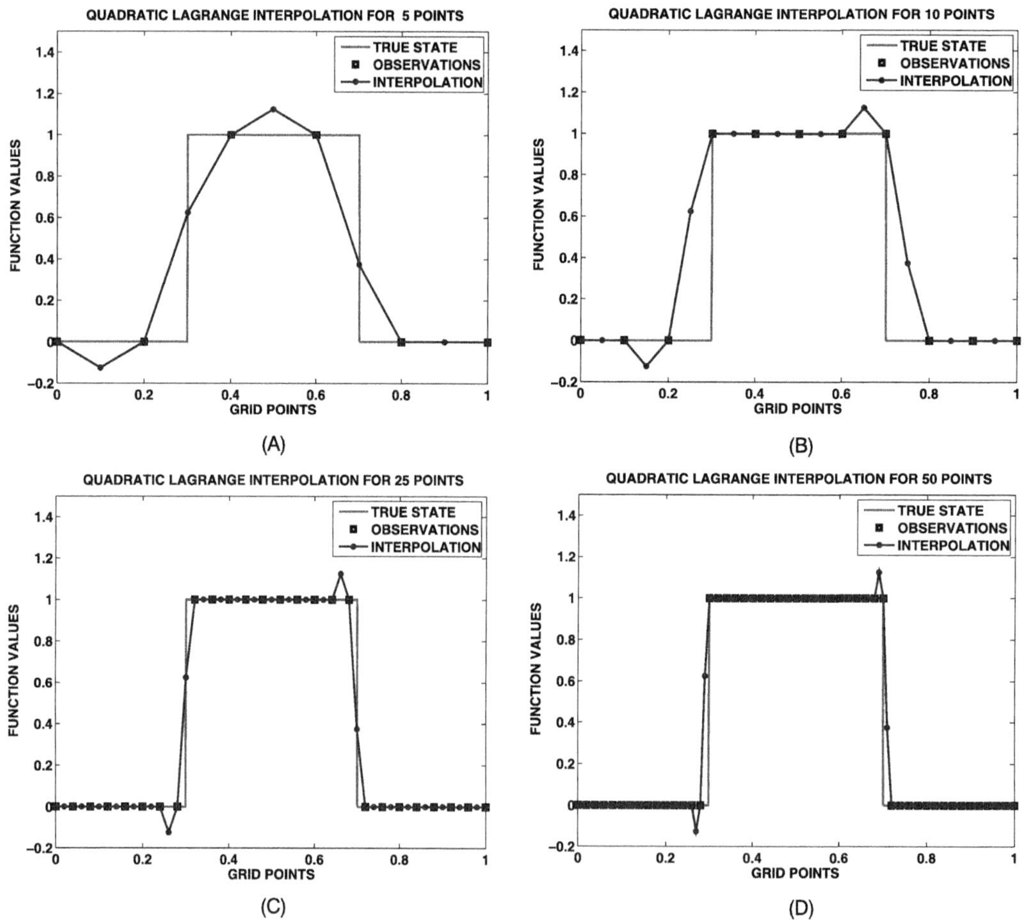

FIGURE 6.7 Plot of the quadratic Lagrange interpolation polynomial applied to the discontinuous step function for: (A) $n = 5$, (B) $n = 10$, (C) $n = 25$, and (D) $n = 50$.

of the function at both transition stages and on the top part of the step function. In Fig. 6.7B we have the results with $n = 10$, and can again see the lack of symmetry that was noticed in the continuous case earlier with the quadratic Lagrange approximations, but there appear to again be over- and undershoots with this approach. When we increase the number of points to $n = 25$, we see that the lack of symmetry still persists and the over- and undershoots at the transitions points are still present (Fig. 6.7C). Finally, in Fig. 6.7D we have the results for $n = 50$; they are improving, but the under- and overshoots are still present, which was not the case for the linear Lagrange interpolation.

We now go on to consider the results for the cubic Lagrange polynomial interpolation, which are show in Fig. 6.8. We start with the result in Fig. 6.8A for the case $n = 5$, where we can clearly see that we do not have sufficiently many points to capture nonlinear features, but we should also note what again appears to be over- and undershoots. When we increase the

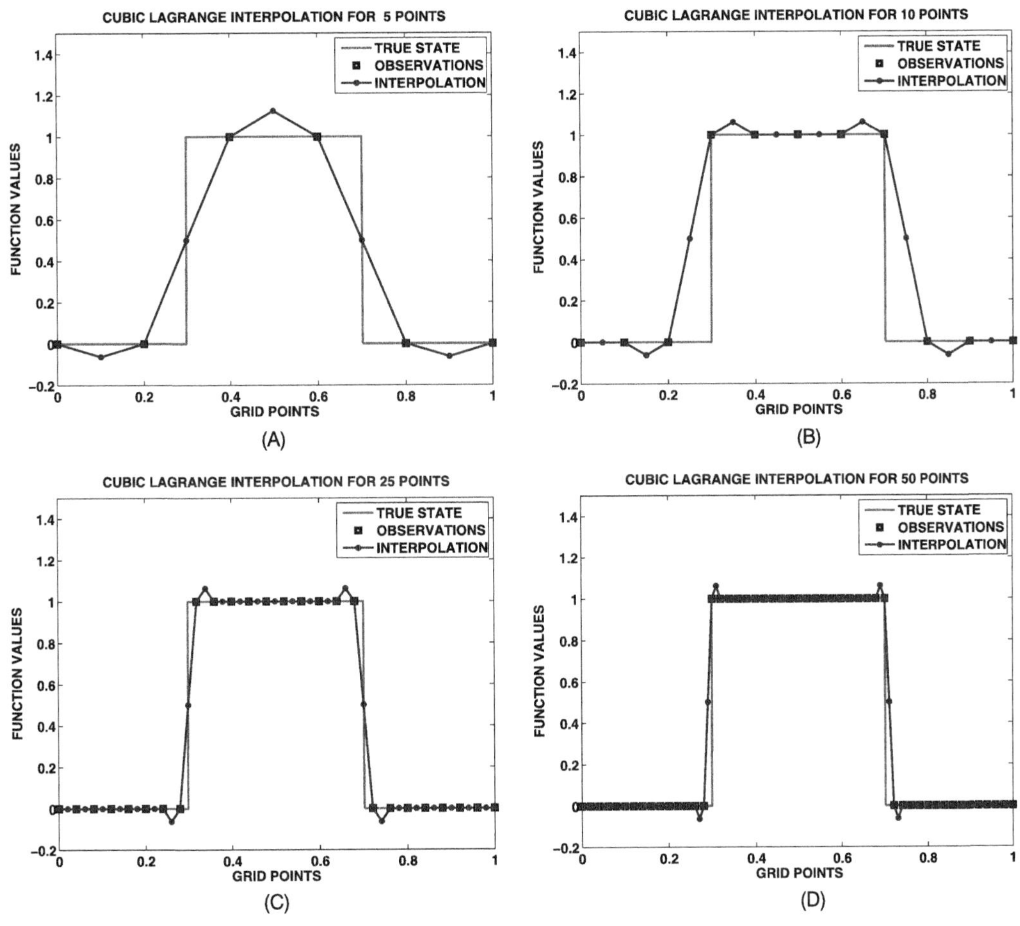

FIGURE 6.8 Plot of the cubic Lagrange interpolation polynomial applied to the discontinuous step function for: (A) $n = 5$, (B) $n = 10$, (C) $n = 25$, and (D) $n = 50$.

number of sample points to 10, we can see in Fig. 6.8B that we are able to approximate the step function, but the over- and undershoots are still present; however, these are not as prevalent as in Fig. 6.8A. In Fig. 6.8C we have increased the number of points to 25 as with the Gaussian curve case, but now we see that we are not fitting as well to the step function, and that the over- and undershoots are still present. Finally, in Fig. 6.8D we have increased the number of points to 50 and are fitting better to the step function, but the over- and undershoots are still present.

In this section we have presented the theory of the Lagrange interpolation polynomials and have illustrated the improvement in the accuracy of the interpolation with the increase of sample points for both a continuous bell curve case, and for a discontinuous step function case. We saw that the interpolations performed quite well in the continuous case, but for the discontinuous case we saw over- and undershoots for the higher order polynomials at the transition points. We now move on to consider the next set of interpolation polynomials which are referred to as the **Newton Divided Difference Polynomials**.

6.2 Newton divided difference interpolation polynomials

The basis of the Newton divided difference interpolation polynomials is that we would like to write the nth order polynomial, p_n, in terms of a p_{n-1} polynomial plus a correction term, $C(x)$, which is given by

$$p_n(x) = p_{n-1}(x) + C(x). \tag{6.7}$$

In general, $C(x)$ is a polynomial of degree n. When we have a function $f(x)$, the correction term becomes

$$C(x_i) = p_n(x_i) - p_{n-1}(x_i) = f(x_i) - f(x_i) = 0, \quad i = 0, \ldots, n-1. \tag{6.8}$$

Therefore, we have

$$C(x) = a_n(x - x_0)\cdots(x - x_{n-1}). \tag{6.9}$$

Given that $p_n(x_n) = f(x_n)$, we can obtain an expression for the a_ns as

$$a_n = \frac{f(x_n) - p_{n-1}(x_n)}{(x_n - x_0)\cdots(x_n - x_{n-1})}.$$

The coefficient a_n is called the *nth order Newton divided difference* and, when we have a function $f(x)$, it is defined as

$$a_n \equiv f(x_0, x_1, \ldots, n_n).$$

This then makes the polynomial

$$p(x_n) = p_{n-1}(x_n) + (x - x_0)(x - x_1)\cdots(x - x_{n-1}) f(x_0, x_1, \ldots, x_n). \tag{6.10}$$

The next stage in the derivation of the Newton divided difference polynomials is to consider the Lagrange formulation for the polynomial. We start by defining the polynomial $\Phi(x)$ as

$$\Phi(x) \equiv (x - x_0)(x - x_1)\cdots(x - x_{n-1}), \tag{6.11}$$

which implies that

$$\Phi'_n(x_i) \equiv (x_i - x_0)(x_i - x_1)\cdots(x_i - x_{i-1})(x_i - x_{i+1})\cdots(x_i - x_{n-1}). \tag{6.12}$$

The two polynomials in (6.11) and (6.12) should look quite familiar as they are the same polynomials as from the Lagrange polynomial approximation. Therefore, substituting (6.11) and (6.12) into (6.7), we obtain an expression for the polynomial as

$$p_n(x) = \sum_{j=0}^{n} \frac{\Phi_n(x)}{(x - x_j)\,\Phi'_n(x_j)} f(x_j). \tag{6.13}$$

The expression in (6.13) is for the case where x is not at what we refer to as grid points, but it is sometimes referred to as nodes [6].

We now return to addressing the factors a_n which are the coefficients of x^n in the polynomial $p_n(x)$. To obtain expressions for these coefficients, we use the Lagrange formula at each nth-degree term in (6.13) and we find that

$$f[x_0, x_1, \ldots, x_n] = \sum_{j=0}^{n} \frac{f(x_j)}{\Phi'_n(x_j)}, \tag{6.14}$$

where

$$f[x_0, x_1, \ldots, x_n] = \frac{f[x_1, \ldots, x_n] - f[x_0, \ldots, x_{n-1}]}{x_n - x_0}, \tag{6.15}$$

and it is because of (6.15) that the name associated with this interpolation formula is the **divided difference**. It is possible to write (6.15) in terms of the sum of two $(n-1)$th order polynomials as

$$p_n(x) = \frac{(x_n - x)\, p_{n-1}^{\{0:n-1\}}(x) + (x - x_0)\, p_{n-1}^{\{1:n\}}(x)}{x_n - x_0}, \tag{6.16}$$

where the polynomial $p_{n-1}^{\{0:n-1\}}(x)$ represents the polynomial of degree $n-1$ that interpolates $f(x)$ from the set of grid points $\{x_0, \ldots, x_{n-1}\}$ and $p_{n-1}^{\{1:n\}}$ represents another polynomial of degree $n-1$ that interpolates $f(x)$ again, but now from the set of grid points $\{x_1, \ldots, x_n\}$. Therefore, the terms in the formula given by (6.13), combined with (6.14) and (6.15) and applied to our even spaced numerical mesh, componentwise are:

$$\begin{aligned}
p_0(x) &= f(x_0), \\
p_1(x) &= f(x_0) + (x - x_0)\, f[x_0, x_1] \\
&\equiv p_0(x) + (x - x_0)\frac{f(x_1) - f(x_0)}{(x_1 - x_0)}
\end{aligned}$$

$$
\begin{aligned}
&\equiv p_0(x) + (x - x_0)\frac{f(x_1) - f(x_0)}{\Delta x}\\
&\approx p_0(x) + (x - x_0) f'(x),\\
p_2(x) &= f(x_0) + (x - x_0) f[x_0, x_1] + (x - x_0)(x - x_1) f[x_0, x_1, x_2]\\
&\equiv p_0(x) + (x - x_0)\frac{f(x_1) - f(x_0)}{(x_1 - x_0)}\\
&+ (x - x_0)(x - x_1)\frac{\frac{f(x_1) - f(x_0)}{(x_1 - x_0)} - \frac{f(x_2) - f(x_1)}{(x_2 - x_1)}}{(x_2 - x_0)}\\
&\equiv p_1(x) + (x - x_0)(x - x_1)\frac{\frac{f(x_1) - f(x_0)}{\Delta x} - \frac{f(x_2) - f(x_1)}{\Delta x}}{2\Delta x}\\
&\approx p_1(x) + (x - x_0)(x - x_1)\frac{f'(x_1) - f'(x_0)}{2\Delta x}\\
&\approx p_0 + (x - x_0) f'(x_0) + \frac{(x - x_0)(x - x_1)}{2} f''(x_0),\\
&\vdots\\
p_n(x) &= f(x_0) + (x - x_0) f[x_0, x_1] + (x - x_0)(x - x_1) f[x_0, x_1, x_2] + \cdots\\
&+ (x - x_0)(x - x_1)\cdots(x - x_{n-1}) f[x_0, x_1, \ldots, x_n]. \qquad (6.17)
\end{aligned}
$$

We can see from the derivation of the different order polynomials in (6.17) that these polynomials appear to be utilizing centered finite difference approximations to the different derivatives in a Taylor series expansion about the point x_0. The formula in (6.17) is referred to as the **Newton divided difference formula for the interpolation polynomial**.

At the moment the formula in (6.17) only approximates the value of the function $f(x)$ at a grid point. In the semi-Lagrangian advection approximation, we know that our departure point is not always at a grid point, otherwise there we would be no need to interpolate. Thus we have to consider the formulation of the Newton divided difference when $x \neq x_i$. If we now consider the point $\tilde{x}$, say, which is our departure point, then the polynomial approximation of our function at $x = \tilde{x}$ is

$$
\begin{aligned}
p_{n+1}(x) &= f(x_0) + (x - x_0) f[x_0, x_1] + \cdots + \left(\prod_{i=0}^{n-1}(x - x_i)\right) f[x_0, x_1, \ldots, x_n]\\
&+ \left(\prod_{i=0}^{n}(x - x_i)\right) f\left[x_0, x_1, \cdots, x_n, \tilde{x}\right], \qquad (6.18)
\end{aligned}
$$

where in (6.18) we have the polynomial approximation for the function $f(x)$ at the grid point x_0 plus the interpolation term to the point $\tilde{x}$. Therefore, since the polynomial $p_{n+1}(x)$ in (6.18) is now an approximation to $f(\tilde{x})$, this then implies that if we let $x = \tilde{x}$ in (6.18), we have an expression for the interpolation polynomial for $f(\tilde{x})$ as

$$f(\tilde{x}) = p_n(\tilde{x}) + \left(\prod_{i=0}^{n} (\tilde{x} - x_i)\right) f\left[x_0, x_1, \ldots, x_n, \tilde{x}\right]. \tag{6.19}$$

We now apply (6.19) to the interpolation to the departure point situation for the semi-Lagrangian schemes. If we consider the linear polynomial first, we have

$$\begin{aligned} p_N^{(1)}(\tilde{x}) &\equiv \psi_{i-p} + (\tilde{x} - x_{i-p})\left(\psi_{i-p} - \psi_{i-p-1}\right) = \psi_{i-p} - \alpha\left(\psi_{i-p} - \psi_{i-p-1}\right) \\ &= (1-\alpha)\psi_{i-p} + \alpha\psi_{i-p-1}. \end{aligned} \tag{6.20}$$

Thus, the linear approximation for the Newton polynomial in (6.20) is equivalent to the linear Lagrange polynomial.

We shall now see if the quadratic Lagrange polynomial and the Newton divided difference polynomial are also equivalent. We have to form the first order upwind approximations to the first derivative as

$$f\left[x_{i-p-1} - x_{i-p}\right] \equiv \frac{\psi_{i-p} - \psi_{i-p-1}}{\left(x_{i-p} - x_{i-p-1}\right)}, \quad f\left[x_{i-p+1}, x_{i-p}\right] \equiv \frac{\psi_{i-p+1} - \psi_{i-p}}{\left(x_{i-p+1} - x_{i-p}\right)}.$$

The next step is to form the second-order divided differences between the two first-order divided differences above; this then results in

$$f\left[x_{i-p-1}, x_{i-p}, x_{i-p+1}\right] = \frac{\dfrac{\psi_{i-p+1} - \psi_{i-p}}{\left(x_{i-p+1} - x_{i-p}\right)} - \dfrac{\psi_{i-p} - \psi_{i-p-1}}{\left(x_{i-p} - x_{i-p-1}\right)}}{(x_2 - x_1)}. \tag{6.21}$$

Substituting the distances between the grid points above and multiplying by $\left(\tilde{x} - x_{i-p-1}\right)\left(\tilde{x} - x_{i-p+1}\right)$ expressed in terms of the rational component of the Courant number results in

$$a_2 \equiv (1-\alpha)(-\alpha)\frac{\left(\psi_{i-p+1} - 2\psi_{i-p} + \psi_{i-p-1}\right)}{2}. \tag{6.22}$$

To obtain the quadratic Newton divided distance interpolation polynomial, we add (6.22) to (6.20), yielding

$$p_N^{(2)}(\tilde{x}) = (1-\alpha)\psi_{i-p} + \alpha\psi_{i-p-1} + (1-\alpha)(-\alpha)\frac{\left(\psi_{i-p+1} - 2\psi_{i-p} + \psi_{i-p-1}\right)}{2}.$$

Collecting the different factors of ψ results in

$$p_N^{(2)}(\tilde{x}) = \frac{\alpha + \alpha^2}{2}\psi_{i-p-1} + \left(1 - \alpha^2\right)\psi_{i-p} + \frac{\alpha^2 - \alpha}{2}\psi_{i-p+1}. \tag{6.23}$$

As with the linear Newton divided difference polynomial, the quadratic Newton interpolation polynomial can be shown to be equivalent to the quadratic Lagrange interpolation polynomial. We now consider the cubic Newton divided difference interpolation polynomial. We only have to evaluate the divided differences between two different numerical approximations to the second-order derivatives where we are now introducing another interpolation gird point at x_{i-p-2}, which results in

$$f\left[x_{i-p-2}, x_{i-p-1}, x_{i-p}, x_{i-p+1}\right] = \frac{\dfrac{\dfrac{\psi_{i-p+1}-\psi_{i-p}}{\Delta x}-\dfrac{\psi_{i-p}-\psi_{i-p-1}}{\Delta x}}{2\Delta x}-\dfrac{\dfrac{\psi_{i-p}-\psi_{i-p-1}}{\Delta x}-\dfrac{\psi_{i-p-1}-\psi_{i-p-2}}{\Delta x}}{2\Delta x}}{3\Delta x}. \tag{6.24}$$

The expression in (6.24) can be simplified for our case because Δx is equivalent to 1, as we have normalized the distances between the grid points. However, expanding the divided differences in (6.24) and simplifying results in

$$\begin{aligned} f\left[x_{i-p-2}, x_{i-p-1}, x_{i-p}, x_{i-p+1}\right] &= \frac{\dfrac{\psi_{i-p+1}-2\psi_{i-p}+\psi_{i-p-1}}{2\Delta x^2}-\dfrac{\psi_{i-p}-2\psi_{i-p-1}+\psi_{i-p-2}}{2\Delta x^2}}{3\Delta x} \\ &= \frac{1}{6\Delta x^3}\left(\psi_{i-p+1}-3\psi_{i-p}+3\psi_{i-p-1}-\psi_{i-p-2}\right) \\ &\equiv \frac{1}{6}\left(\psi_{i-p+1}-3\psi_{i-p}+3\psi_{i-p-1}-\psi_{i-p-2}\right). \end{aligned} \tag{6.25}$$

This then enables us to define the coefficient a_3 as

$$\begin{aligned} a_3 &= (-\alpha)(1-\alpha)(-\alpha-1)\frac{1}{6}\left(\psi_{i-p+1}-3\psi_{i-p}+3\psi_{i-p-1}-\psi_{i-p-2}\right) \\ &= \frac{\alpha\left(1-\alpha^2\right)}{6}\left(\psi_{i-p+1}-3\psi_{i-p}+3\psi_{i-p-1}-\psi_{i-p-2}\right). \end{aligned} \tag{6.26}$$

Thus the expression for the cubic Newton divided difference polynomial is (6.23) plus (6.26), which is

$$\begin{aligned} p_N^{(3)}(\tilde{x}) &= \frac{\alpha+\alpha^2}{2}\psi_{i-p-1}+\left(1-\alpha^2\right)\psi_{i-p}+\frac{\alpha^2-\alpha}{2}\psi_{i-p+1} \\ &+ \frac{\alpha\left(1-\alpha^2\right)}{6}\left(\psi_{i-p+1}-3\psi_{i-p}+3\psi_{i-p-1}-\psi_{i-p-2}\right). \end{aligned} \tag{6.27}$$

Collecting the different factors of the four ψ grid point values culminates in

$$\begin{aligned} p_N^{(3)} &= -\frac{\alpha-\alpha^3}{6}\psi_{i-p-2}+\left(\alpha+\frac{\alpha^2}{2}-\frac{\alpha^3}{2}\right)\psi_{i-p-1}+\left(1-\frac{\alpha}{2}-\frac{\alpha^2}{2}+\frac{\alpha^3}{2}\right)\psi_{i-p} \\ &+ \left(-\frac{\alpha}{3}+\frac{\alpha^2}{2}-\frac{\alpha^3}{6}\right)\psi_{i-p+1}. \end{aligned} \tag{6.28}$$

Once again we have that the Newton divided difference interpolation polynomial for this order of approximation is the same as that for the Lagrange interpolation. While so far the first three Newton polynomials match the Lagrange after some multiplying out parenthesis the reason to present the Newton based interpolation is because of its ease of coding; we are literally forming a finite difference approximation to the Taylor series of f. It is stated in many different sources that from a computational view the Newton method is computationally easier to expand, to include another data point to the original set.

To add another data point at the end of the current set of points is quite easy, and once the new data point is in the set, one simply builds up the new difference for that point, given all the divided differences from the original set. In comparison, for the Lagrange interpolation one would have to code the next order Lagrange polynomial formula, which would be quite lengthy compared to a simple change of a "do" or "for" loop, and then add an extra row to the array.

Exercise 6.5. *Derive the fourth-order Newton divided difference polynomial for the semi-Lagrange situation and show that it is equivalent to the relevant quartic Lagrange polynomial.*

The next set of interpolations polynomials that we consider are the **Hermite Interpolation Polynomials**, which are seen as a more advanced interpolation polynomial technique.

6.3 Hermite interpolating polynomials

As we saw in the derivations of the Lagrange and Newton divided difference interpolation polynomials, we are only using the function's values as constraints for the interpolation. However, there are several methods that enable us to obtain higher orders of accuracy from the interpolation formula. The first approach is to increase the number of points to interpolate through, however, with this approach we are using more and more points that are further away from the region where we wish to evaluate the interpolation.

The second approach to make the interpolation polynomial more accurate is to fit the polynomial to higher order derivative values of the function at the points surrounding the interpolation point. This then constrains the interpolation polynomial to match both $f\left(x_j\right)$ and $f'\left(x_j\right)$ for $j=0,1,\ldots,N$. One such set of interpolating polynomials that match these two conditions are referred to as **Hermite interpolating polynomials**, and we show their derivation below.

We start by stating the property that we require our interpolating polynomial $p\left(x\right)$ to satisfy, which is

$$p\left(x_j\right)=f_j \text{ and } p'\left(x_j\right)=f'_j, \text{ for } j=1,2,\ldots,N, \tag{6.29}$$

where the x_js are our grid points, and f_j and f'_j are the given values of our function and its first derivative at the grid points, respectively. The expression in (6.29) defines $2N$ constraints on the interpolation polynomial, and as such the highest degree the resulting interpolating polynomial can take is $2N-1$.

We now introduce the following notation to help shorten expressions in the derivation of the Hermite polynomial:

$$\chi_N\left(x\right) = \left(x-x_1\right)\left(x-x_2\right)\cdots\left(x-x_n\right), \tag{6.30a}$$

$$\begin{aligned} L_j(x) &= \frac{(x-x_1)\cdots(x-x_{j-1})(x-x_{j+1})\cdots(x-x_N)}{(x_j-x_1)\cdots(x_j-x_{j-1})(x_j-x_{j+1})\cdots(x_j-x_N)} \equiv \frac{\chi_n(x)}{(x-x_j)\chi_n'(x_j)}, &\quad (6.30b)\\ \tilde{h}_j(x) &= (x-x_j)\left[L_j(x)\right]^2, &\quad (6.30c)\\ h_j(x) &= \left(1-2L_j'(x_j)(x-x_j)\right)\left[L_j(x)\right)^2. &\quad (6.30d) \end{aligned}$$

Given the definitions above, we have the following two extra conditions for $j,k=1,\ldots,N$:

$$\begin{aligned} h_j(x_k) &= \tilde{h}_j(x_k)=0, \quad 1\le j,k\le N, &\quad (6.31a)\\ h_j(x_k) &= \tilde{h}_j'(x_j)=\begin{cases} 0, & j\neq k,\\ 1, & j=k. \end{cases} &\quad (6.31b) \end{aligned}$$

The interpolating polynomial that satisfies the conditions in (6.29) is defined as

$$H_N(x)=\sum_{j=1}^{N} y_j h_j(x)+\sum_{j=1}^{N} y_j'\tilde{h}_j(x). \tag{6.32}$$

Given the expressions above for the Hermite interpolation polynomial, the most frequently used order of the Hermite polynomial is the third. The definitions above imply that we require the values of the tracer at points on either side of the departure point, and the derivative of the tracer at these points. Therefore, we have

$$\begin{aligned} p(x_{i-p-1}) &\equiv \psi_{i-p-1}, & p'(x_{i-p-1}) &\equiv \psi_{i-p-1}',\\ p(x_{i-p}) &\equiv \psi_{i-p}, & p'(x_{i-p}) &\equiv \psi_{i-p}'. \end{aligned}$$

The next step is to form the Lagrange basis, L_1 and L_2, which for the semi-Lagrange advection will be L_{i-p-1} and L_{i-p}, respectively. Thus we have

$$\begin{aligned} L_{i-p-1} &= \frac{x-x_{i-p}}{x_{i-p-1}-x_{i-p}}, & L_{i-p} &= \frac{x-x_{i-p-1}}{x_{i-p}-x_{i-p-1}},\\ L_{i-p-1}' &= \frac{1}{x_{i-p-1}-x_{i-p}}, & L_{i-p}' &= \frac{1}{x_{i-p}-x_{i-p-1}}. \end{aligned}$$

Next forming the h_js and $\tilde{h}_j$s results in

$$\begin{aligned} h_{i-p-1} &= \left(1+\frac{2(x_{i-p-1}-x)}{x_{i-p}-x_{i-p-1}}\right)\left(\frac{x_{i-p}-x}{x_{i-p}-x_{i-p-1}}\right)^2,\\ h_{i-p} &= \left(1+\frac{2(x_{i-p}-x)}{x_{i-p}-x_{i-p-1}}\right)\left(\frac{x-x_{i-p-1}}{x_{i-p}-x_{i-p-1}}\right)^2,\\ \tilde{h}_{i-p-1} &= \frac{(x-x_{i-p-1})(x_{i-p}-x_{i-p})^2}{(x_{i-p}-x_{i-p-1})^2}, \end{aligned}$$

$$\tilde{h}_{i-p} = -\frac{\left(x - x_{i-p-1}\right)^2 \left(x_{i-p} - x\right)}{\left(x_{i-p} - x_{i-p-1}\right)^2}.$$

The final form of the cubic Hermite polynomial is given by

$$H_2(x) = h_{i-p-1}\psi_{i-p-1} + h_{i-p}\psi_{i-p} + \tilde{h}_{i-p-1}\psi'_{i-p-1} + \tilde{h}_{i-p}\psi'_{i-p}. \tag{6.33}$$

Substituting the distances from the grid points to the departure points, and the plus or minus 1 for the distance between the grid points themselves results in

$$\begin{aligned} H_2(x) &= \left(1 + 2\frac{1-\alpha}{1}\right)\left(\frac{\alpha}{1}\right)^2 \psi_{i-p-1} + \left(1 + 2\frac{\alpha}{1}\right)\left(\frac{1-\alpha}{1}\right)^2 \psi_{i-p} \\ &+ \frac{(1-\alpha)\alpha^2}{1^2}\psi'_{i-p-1} - \frac{(1-\alpha)^2\alpha}{1^2}\psi'_{i-p}. \end{aligned} \tag{6.34}$$

Expanding the brackets in (6.34) enables us to arrive at a simpler expression for the cubic Hermite interpolation polynomial as

$$\begin{aligned} H_2(x) &= \left(3\alpha^2 - 2\alpha^3\right)\psi_{i-p-1} + \left(1 - 3\alpha^2 + 2\alpha^3\right)\psi_{i-p} \\ &+ \left(\alpha^2 - \alpha^3\right)\psi'_{i-p-1} - \left(\alpha - 2\alpha^2 + \alpha^3\right)\psi'_{i-p}. \end{aligned} \tag{6.35}$$

However, it may be the case that the derivatives are not known at the grid points and so we require techniques that could be applied to approximate them. At this point it could appear that we have lost the advantage for the number of points needed to fit the interpolant compared to the Lagrange polynomial and the Newton divided difference polynomial. While this may be true for certain derivative estimates, as mentioned in [172], because of the flexibility of the Hermite interpolations, it is possible, as shown in [207], that the resulting semi-Lagrangian scheme preserves monotonicity.

It is possible to define many different finite difference approximations to the derivatives that are required for the Hermite interpolation polynomial, where these estimates of the derivatives could be linear or non-linear, but they do need to be *centered* at the grid points, and as we have just indicated. Following the notation from [172], we define the discrete gradient operator Δ_j which is acting on the interval $\left[x_j, x_{j+1}\right]$ as

$$\Delta_j = \frac{\psi_{j+1} - \psi_j}{x_{j+1} - x_j}. \tag{6.36}$$

From the definition in (6.36) it is possible to define approximations to the derivative at the points x_{j-1} and x_j through different combinations of the discrete gradient at nearby points.

In [172] there are three schemes that are listed, where one is a second-order approximation to the gradients and is calculated as an arithmetic mean from considering the discrete gradient at the points $j-1$ and j, i.e., $\psi'_j \approx \dfrac{\Delta_{j-1} + \Delta_j}{2}$. This scheme can be shown to be second order. Two others schemes that are presented in [172] are fourth order and taken from [85] and [146], and are respectively given by

$$\psi'_j \approx \frac{-\Delta_{j-2} + 7\Delta_{j-1} + 7\Delta_j - \Delta_{j+1}}{12},$$
$$\psi'_j \approx \frac{-3\Delta_{j-2} + 19\Delta_{j-1} + 19\Delta_j - 3\Delta_{j+1}}{32}.$$

We now move on to consider another set of more advanced interpolation polynomials, the **cubic splines**.

6.4 Cubic spline interpolation polynomials

The spline interpolation polynomials are part of a general class of interpolation methods that are referred to as *piecewise polynomial interpolations*. Besides being used for the interpolation of a value of a tracer to departure points in semi-Lagrangian modeling, piecewise polynomials, and especially the **spline** family, are used in many disciplines: data fitting, numerical integration, as well as differentiation, numerical solutions to integral and differential equations. However, for our use here we will focus of the piecewise polynomial theory as applied to interpolation problems.

We begin by consider a piecewise polynomial function that we shall denote as $p(x)$, which has an associated grid such that

$$-\infty < x_0 < x_1 < \cdots < x_N < \infty,$$

where the points indicated above are the bounds on the region where the piecewise polynomials are defined. The function $p(x)$ is a polynomial defined on each interval

$$(-\infty, x_0], [x_0, x_1], \ldots, [x_N, \infty). \tag{6.37}$$

It should be noted that in most cases the two intervals that contain either ∞ or $-\infty$ are usually not included in the fitting of the piecewise polynomials.

The order of the piecewise polynomial $p(x)$ is determined by being one order of magnitude higher than the largest magnitude of the piecewise polynomials over the intervals of consideration. An assumption that is made about $p(x)$ is that it does not have to be continuous, nor does its first derivative, but it is quite common for the piecewise polynomial to be continuous.

There are two possible classifications in general for piecewise polynomial interpolation problems; the first classification is referred to as a local problem, while the second classification is referred to as a global problem. For our purposes for the semi-Lagrangian advection, we consider spline functions which are associated with global problems.

Definition 6.6. A **spline function** $S(x)$ is said to be of order $m \geq 1$ on the interval

$$a = x_0 < x_1 < x_2 < \cdots < x_{N-1} < x_N = b,$$

if it satisfies the following two properties:

1. $S(x)$ is a polynomial of degree that is less than m on each of the subintervals $[x_{j-1}, x_j]$.
2. The kth derivative of the spline function, $S^{(k)}(x)$, is continuous on the full interval $[a, b]$ for all of the derivatives up to $m - 2$.

An interesting feature to remember about the derivatives of the spline functions of order m is that they too are spline functions but of order $m - 1$. An opposite but equivalent property of the spline functions is also true for the integral, which means that the antiderivative of a spline function of order $m - 1$ is a spline function but now of order m.

Among the most commonly used spline functions are the **cubic splines**, where m would be 4. There are many reasons why cubic splines are popular: first, they are smooth functions with which we fit data, but more importantly for the interpolation purposes, they do not have an oscillatory behavior that is common for higher order degree polynomials associated with interpolation, as we saw with the cubic Lagrange interpolation polynomial earlier.

The interpolation problem can be stated as follows: we wish to find a cubic spline, $S_4(x)$, such that $S(x_i) = y_i$ for $i = 1, 2, \ldots, N$, and thus the general form of the cubic spline is given by

$$S_4(x) \equiv a_j + b_j x + c_j x^2 + d_j x^3, \tag{6.38}$$

where we have $x_{j-1} \leq x \leq x_j$ for $j = 1, 2, \ldots, N$. The equation in (6.38) contains four unknowns for each spline, a_j, b_j, c_j, and d_j for a total of $4N$ unknowns over the whole interval, so we will require conditions on the splines to be able to determine these coefficients. The first two constraints/conditions for each spline are (1) that the spline must satisfy $S_4(x)$ and (2) the continuity of the kth derivative property state earlier.

The second constraint implies that the kth derivative must be equal from either direction at the point x_j for each of the splines at each grid point. This constraint then results in $4N - 2$ conditions, but we require $4N$ conditions. Thus there are two degrees of freedom in choosing the coefficients in (6.38). For us to be able to obtain a unique interpolation spline, we should expect that we will have to impose two more constraints to ensure this desired property.

We now consider how to define the cubic splines. We start by denoting the second derivative of the cubic splines $S_4''(x)$ as Q_j for $j = 1, 2, \ldots, N$. As we have chosen the spline to be a cubic polynomial, we know that its second derivative is a linear polynomial on the interval $[x_{j-1}, x_j]$. This then implies that

$$S''(x) = \frac{(x_{j+1} - x) Q_j + (x - x_j) Q_{j+1}}{\Delta x}, \qquad j = 0, 1, \ldots, N - 1, \tag{6.39}$$

where Δx is still our grid spacing. However, in general it could be that the nodes are not equally spaced and the denominator in (6.39) would be replaced with Δx_j, where $\Delta x_j \equiv x_{j+1} - x_j$.

Given the expression for the second derivative of the spline in (6.39), we have that this second derivative is continuous on the whole interval of consideration, $[x_0, x_N]$. Thus with this continuity property it is possible to integrate (6.39) twice to obtain

$$S(x) = \frac{(x_{j+1} - x)^3 Q_j + (x - x_j)^3 Q_{j+1}}{6\Delta x} + A(x_{j+1} - x) + B(x - x_j), \tag{6.40}$$

where A and B are arbitrary constants to be determined. To find them we use the condition that the spline must be equal to the known values of the function y at the two grid points x_j and x_{j+1}, which implies that $S_4(x_j) = y_j$ and $S_4(x_{j+1}) = y_{j+1}$. Substituting these conditions into (6.40) and rearranging results in the expressions for the two constants as

$$A = \frac{y_j}{\Delta x} - \frac{\Delta x Q_j}{6}, \qquad B = \frac{y_{j+1}}{\Delta x} - \frac{\Delta x Q_{j+1}}{6}.$$

Given the expression above for the two constants, the equation for the cubic splines is given by

$$\begin{aligned} S_4(x) &= \frac{(x_{j+1} - x)^3 Q_j + (x - x_j)^2 Q_{j+1}}{6\Delta x} + \frac{(x_{j+1} - x) y_j + (x - x_j) y_{j+1}}{\Delta x} \\ &- \frac{\Delta x \left((x_{j+1} - x) Q_i + (x - x_j) Q_{i+1} \right)}{6}, \end{aligned} \tag{6.41}$$

where $x_j \leq x \leq x_{j+1}$, for $0 \leq j \leq N-1$.

The expression for the cubic splines in (6.41) implies that the splines are continuous on the interval $[x_0, x_N]$, as well as satisfy the interpolating condition that the splines must be equal to the value of the function at the grid points.

We now focus our attention to determining the constants Q_j, for $j = 0, 1, \ldots, N$. We have the condition that the first derivatives of the splines, $S'(x)$, are required to be continuous at the interior grid points, $x_1, \ldots, x_{N-1}$. The continuity condition, just mentioned, is expressed mathematically as

$$\lim_{x \to x_j^+} S'(x) = \lim_{x \to x_j^-} S'(x), \quad \text{for } j = 1, 2, \ldots, N-1, \tag{6.42}$$

where the $^+$ superscript is referring to taking the limit from above the point x_j, and the $^-$ superscript refers to taking the limit from below the grid point x_j.

Given the new condition on the splines from (6.42), we now consider the first derivative of the cubic spline over two adjacent intervals. The first interval is $[x_j, x_{j+1}]$, which results in

$$S'(x) = -\frac{(x_{j+1} - x)^2 Q_j + (x - x_j)^2 Q_{j+1}}{2\Delta x} + \frac{y_{i+1} - y_i}{\Delta x} - \frac{(Q_{i+1} - Q_i)\Delta x}{6}, \tag{6.43}$$

while on the adjacent interval $[x_{j-1}, x_j]$ we have

$$S'(x) = -\frac{(x_j - x)^2 Q_{j-1} + (x - x_{j-1})^2 Q_j}{2\Delta x} + \frac{y_i - y_{i-1}}{\Delta x} - \frac{(Q_i - Q_{i-1})\Delta x}{6}. \tag{6.44}$$

Note that while we are deriving the cubic spline for our equally spaced grid, we should remember that if the data points are not equally spaced, then in (6.43) and (6.44) the terms Δx

should be replaced with Δx_j and Δx_{j-1}, respectively, to ensure that the correct weighting of the distances between the grid points is applied.

Therefore, taking the limit as $x \to x_j$ from above in (6.43) and from below in (6.44) and setting the two equations to be equal at that point results in the first terms in both equations tending to zero, which leaves only the terms involving y and Q as

$$\frac{\Delta x_{j-1}}{6} Q_{j-1} + \frac{\Delta x_j + \Delta x_{j-1}}{3} Q_j + \frac{\Delta x_j}{6} Q_{j+1} = \frac{y_{j+1} - y_j}{\Delta x_j} - \frac{y_j - y_{j-i}}{\Delta x_{j-1}}, \tag{6.45}$$

for $j = 1, 2, \ldots, N-1$, where we have included the subscripts on the Δx terms to give the full expression for an unequally spaced grid situation.

The formula in (6.45) results in $n-1$ equations for $N-1$ unknowns in the interior of the interval. However, we still require conditions at the end points of the interval to capture the last two remaining degrees of freedom associated with fitting cubic splines.

The more commonly used condition to capture the two missing degrees of freedom associated with the end points is based upon requiring the cubic spline, or any other order spline, to match not just the function values at x_0 and x_N but also its first derivative values at these grid points, that is to say,

$$S'(x_0) = y'(x_0) \text{ and } S'(x_N) = y'(x_N).$$

Substituting these conditions into (6.43) evaluated at $x_j = x_0$ and $x_j = x_N$ results in the two end conditions:

$$\frac{\Delta x_0}{3} Q_0 + \frac{\Delta x_0}{6} Q_1 = \frac{y_1 - y_0}{\Delta x_0} - y_0', \tag{6.46a}$$

$$\frac{\Delta x_{N-1}}{6} Q_{N-1} + \frac{\Delta x_{N-1}}{3} Q_N = y_n' - \frac{y_N - y_{N-1}}{\Delta x_{N-1}}. \tag{6.46b}$$

Combining the two end conditions above with the conditions for the interior points of the interval enables the problem of finding the Q_js be stated as the inversion of a matrix–vector equation $\mathbf{AQ} = \hat{\mathbf{y}}$ where

$$\hat{\mathbf{y}} = \begin{pmatrix} \frac{y_1 - y_0}{\Delta x_0} - y_0' \\ \frac{y_2 - y_1}{\Delta x_1} - \frac{y_1 - y_0}{\Delta x_0} \\ \vdots \\ \frac{y_N - y_{N-1}}{\Delta x_{N-1}} - \frac{y_{N-1} - y_{N-2}}{\Delta x_{N-2}} \\ y_N' - \frac{y_N - y_{N-1}}{\Delta x_{N-1}} \end{pmatrix}, \quad \mathbf{Q} = \begin{pmatrix} Q_0 \\ Q_1 \\ \vdots \\ Q_{N-1} \\ Q_N \end{pmatrix}, \tag{6.47}$$

$$\mathbf{A} = \begin{pmatrix} \frac{\Delta x_0}{3} & \frac{\Delta x_0}{6} & 0 & \cdots & \cdots & 0 \\ \frac{\Delta x_0}{6} & \frac{\Delta x_0 + \Delta x_1}{3} & \frac{\Delta x_1}{6} & 0 & \cdots & 0 \\ 0 & & \ddots & & & \vdots \\ \vdots & & & \ddots & & \vdots \\ 0 & & & \frac{\Delta x_{N-2}}{6} & \frac{\Delta x_{N-2} + \Delta x_{N-1}}{3} & \frac{\Delta x_{N-1}}{6} \\ 0 & \cdots & \cdots & \cdots & \frac{\Delta_{N-1}}{6} & \frac{\Delta x_{N-1}}{3} \end{pmatrix}.$$

The matrix $\mathbf{A}$ is tridiagonal, symmetric, positive definite, and diagonally dominant, which implies that the matrix is invertible and therefore there exists a unique solution to the matrix–vector problem.

We now address how to apply the cubic splines to the interpolation required in the semi-Lagrangian application. A good explanation of how to apply the cubic spline approach for a semi-Lagrangian approximation in a transport model can be found in [164], and we summarize that explanation here.

The starting point is the formula of a general interpolant, for both the cubic Lagrange and cubic spline, which is given by

$$\begin{aligned} \psi_i^{n+1} &= \alpha \psi_{i-p-1}^n + (1-\alpha)\psi_{i-p}^n + \frac{1}{6}\left(C^3 - C\right)(\Delta x)^2 \psi_{i-p-1}^{n''} \\ &+ \frac{1}{6}\left((1-\alpha)^3 - (1-\alpha)\right)(\Delta x)^2 \psi_{i-p}^{n''}, \end{aligned} \tag{6.48}$$

where $\psi_{i-p}^{n''}$ is a discrete approximation to the second derivative of ψ at the grid point in the subscript.

It is stated in [164] that the Lagrange polynomial and the cubic spline approaches differ in their definitions for the second derivative $\psi_{i-p}^{n''}$. For the Lagrange interpolation we have

$$\psi'' = \frac{\psi_{i-p-1} - 2\psi_{i-p} + \psi_{i-p+1}}{(\Delta x)^2}. \tag{6.49}$$

If we insert (6.49) into (6.48), then we obtain the cubic Lagrange interpolation formula from before. This is left as an exercise to verify this relationship. We move on to the equivalent expression for the cubic spline interpolation approach.

As we mentioned above, the cubic spline interpolant is derived from requiring that the first derivative of the interpolant be differentiable at the grid points. Given this condition, it is shown in [145] that our interpolation polynomial must satisfy

$$\frac{\psi_{i-p-1}''}{6} + \frac{2\psi_{i-p}''}{3} + \frac{\psi_{i-p+1}''}{6} = \frac{\psi_{i-p-1} - 2\psi_{i-p} + \psi_{i-p+1}}{(\Delta x)^2}. \tag{6.50}$$

If we now form the sum $\frac{1}{6}\psi_{i-p-1}^{n+1}+\frac{2}{3}\psi_{i-p}^{n+1}+\frac{1}{6}\psi_{i-p+1}^{n+1}$ and evaluate the right-hand side of (6.48) at the relative grid points, then we obtain

$$\begin{aligned}\frac{1}{6}\psi_{i-p-1}^{n+1} &= \frac{1}{6}\left(\alpha\psi_{i-p-2}^{n}+(1-\alpha)\psi_{i-p-1}^{n}+\frac{1}{6}\left(\alpha^3-\alpha\right)(\Delta x)^2\psi_{i-p-2}^{n''}\right.\\ &+ \left.\frac{1}{6}\left((1-\alpha)^3-(1-\alpha)\right)(\Delta x)^2\psi_{i-p-1}^{n''}\right), \end{aligned} \tag{6.51a}$$

$$\begin{aligned}\frac{2}{3}\psi_{i-p}^{n+1} &= \frac{2}{3}\left(\alpha\psi_{i-p-1}^{n}+(1-\alpha)\psi_{i-p}^{n}+\frac{1}{6}\left(\alpha^3-\alpha\right)(\Delta x)^2\psi_{i-p-1}^{n''}\right.\\ &+ \left.\frac{1}{6}\left((1-\alpha)^3-(1-\alpha)\right)(\Delta x)^2\psi_{i-p}^{n''}\right), \end{aligned} \tag{6.51b}$$

$$\begin{aligned}\frac{1}{6}\psi_{i-p+1}^{n+1} &= \frac{1}{6}\left(\alpha\psi_{i-p}^{n}+(1-\alpha)\psi_{i-p+1}^{n}+\frac{1}{6}\left(\alpha^3-\alpha\right)(\Delta x)^2\psi_{i-p}^{n''}\right.\\ &+ \left.\frac{1}{6}\left((1-\alpha)^3-(1-\alpha)\right)(\Delta x)^2\psi_{i-p+1}^{n''}\right). \end{aligned} \tag{6.51c}$$

Summing the three expressions in (6.51a), (6.51b), and (6.51c), we have expressions that are equivalent to the left-hand side of (6.50). We therefore have

$$\left(\alpha^3-\alpha\right)(\Delta x)^2\left(\frac{\psi_{i-p-2}^{n''}}{6}+\frac{2\psi_{i-p-1}^{n''}}{3}+\frac{\psi_{i-p}^{n''}}{6}\right) \equiv \left(\alpha^3-\alpha\right)\left(\psi_{i-p-2}^{n}-2\psi_{i-p-1}^{n}+\psi_{i-p}^{n}\right),$$

$$\begin{aligned}&\frac{1}{6}\left((1-\alpha)^3-(1-\alpha)\right)\left((\Delta x)^2\left(\frac{\psi_{i-p-1}^{n''}}{6}+\frac{2\psi_{i-p}^{n''}}{3}+\frac{\psi_{i-p+1}^{n''}}{6}\right)\right)\\ &\equiv \frac{1}{6}\left((1-\alpha)^3-(1-\alpha)\right)\left(\psi_{i-p-1}^{n}-2\psi_{i-p}^{n}+\psi_{i-p+1}^{n}\right).\end{aligned}$$

Substituting the expressions above into (6.51a)–(6.51c) and factorizing enables us to write the interpolation polynomial for the cubic spline in the form

$$\frac{1}{6}\psi_{i-p-1}^{n+1}+\frac{2}{3}\psi_{i-p}^{n+1}+\frac{1}{6}\psi_{i-p+1}^{n+1}=c_{-2}\psi_{i-p-2}^{n}+c_{-1}\psi_{i-p-1}^{n}+c_0\psi_{i-p}^{n}+c_1\psi_{i-p+1}^{n}. \tag{6.52}$$

The equivalent cubic Lagrange interpolation polynomial of (6.52) is

$$\psi_i^{n+1}=l_{-2}\psi_{i-p-2}^{n}+l_{-1}\psi_{i-p-1}^{n}+l_0\psi_{i-p}^{n}+l_1\psi_{i-p+1}^{n}. \tag{6.53}$$

Below we have put the definitions of the coefficients of the cubic spline and the cubic Lagrange interpolation polynomials side by side to illustrate the differences between the two approximations:

$$l_{-2} \equiv \frac{\alpha^3 - \alpha}{6}, \qquad c_{-2} \equiv \frac{\alpha^3}{6}, \tag{6.54a}$$

$$l_{-1} \equiv \frac{-\alpha^3 + \alpha^2 + 2\alpha}{2}, \qquad c_{-1} \equiv \alpha + \frac{(1-\alpha)^3 - 2\alpha^3}{6}, \tag{6.54b}$$

$$l_0 \equiv \frac{\alpha^3 - 2\alpha^2 - \alpha + 2}{2}, \qquad c_0 \equiv (1-\alpha) + \frac{\alpha^3 - 2(1-\alpha)^3}{6}, \tag{6.54c}$$

$$l_1 \equiv \frac{-\alpha^3 + 3\alpha^2 - 2\alpha}{6}, \qquad c_1 \equiv \frac{(1-\alpha)^3}{6}. \tag{6.54d}$$

It is clear from (6.54a)–(6.54d) that there are substantial differences between the coefficients for the cubic Lagrange and the cubic spline interpolation polynomials. There is also an important difference on the left-hand side of the interpolation polynomials as well. For the cubic spline interpolation polynomial, we have an implicit expression. As stated in [164], we have to solve a matrix equation for the cubic spline approach to obtain the values of the tracer at the departure point in the form $\mathbf{M}\boldsymbol{\psi}^{n+1} = \mathbf{S}\boldsymbol{\psi}^n$. Thus if we apply a cubic spline approach, we have to invert $\mathbf{M}$ to obtain an approximation to $\psi(\tilde{x}_i)$.

Exercise 6.7. *Verify that the terms for the Lagrange interpolation formula in* (6.54a)–(6.54d), *when substituted into* (6.48), *result in the equation for the cubic Lagrange interpolation formula given earlier.*

6.5 Summary

In this chapter we have introduced a series of interpolation polynomials that are quite often used in many different applications of semi-Lagrangian advection in the geosciences. We started with the Lagrange interpolation polynomial, which was introduced in the last chapter, but not defined or named. We presented the quadratic and cubic basis functions for this interpolation method, along with applications using a continuous bell curve and a discontinuous step function. We saw that, as the number of sample points/observations increases, the interpolation for the continuous case improved for the linear, quadratic, and cubic polynomials. However, when we applied the same order polynomials to the step function, we saw that the linear case was actually better at approximating the step than the higher order polynomials, which suffered from over- and undershoots.

We have also introduced the Newton divided difference polynomials, which are seen as a more computationally efficient way of implementing the Lagrange interpolation polynomials, especially when one wishes to increase the order of the polynomial. Next we introduced the Hermite interpolation polynomials which are a more robust approach than Lagrange interpolation, but require either the continuous derivative of the function at the grid points, or higher order finite difference approximations to this derivative.

Finally, the last set of interpolations that we introduced were the splines, where we focused on the application of the cubic splines, resulting in an implicit formulation that leads to a matrix inversion problem for all the interpolation points.

What we have to remember here is that in this chapter we have only introduced different interpolation polynomials that are commonly used in different applications of semi-

Lagrangian advection, but like with the Eulerian finite difference methods presented in Chapter 2, we have to now consider how stable these semi-Lagrangian schemes are with different interpolation polynomials, and also whether or not they are consistent with the underlying differential equation, which is what we intend to do in the next chapter.

CHAPTER

7

Stability and consistency analysis of semi-Lagrangian methods for the linear problem

In the last two chapters we have introduced the concepts of the semi-Lagrangian approach and shown that it can perform well with Courant numbers greater than 1, which was rarely the case for the explicit Eulerian finite difference schemes presented earlier. We have also introduced different interpolation polynomials that could be used to determine the value of the field/tracer at the departure point when it does not coincide with a grid point at the previous time level. In this chapter we will look at the stability, consistency, as well as the dispersion/dissipation phase and amplitude errors for some of the polynomials introduced in the last chapter when applied to the one-dimensional, constant velocity, zero forcing advection equation.

7.1 Stability of semi-Lagrangian schemes

In this chapter we shall be using the notation and expression from [172] to help explain the differences from standard von Neumann stability derivations that we showed for the Eulerian finite difference schemes in Chapter 5, as the departure points are not always at the spatial grid points, and so we have to take into account this in the spacial stability analysis. It is stated in [172] that this form of analysis is referred to as Fourier analysis, which describes how well a given interpolation scheme represents the amplitude and phase speed of an infinite, single frequency, progressive harmonic wave. Thus we start by defining the initial data for the advection equation as

$$\psi_0(x) = e^{ik\Delta x}, \tag{7.1}$$

where k is the wavenumber. Therefore, the analytical solution to the linear, constant velocity, zero forcing version of the advection equation is

$$\psi(x,t) = e^{ik(x-u\Delta t)}. \tag{7.2}$$

Semi-Lagrangian Advection Methods and Their Applications in Geoscience
https://doi.org/10.1016/B978-0-12-817222-3.00011-6

If we consider solutions that are separated by the time interval Δt, then they are related by

$$\psi(x, t+\Delta t) = e^{ik(x-u(t+\Delta t))} = e^{ik(x-ut-u\Delta t)} = e^{ik(x-ut)}e^{-iku\Delta t} = e^{-iku\Delta t}\psi(x,t). \quad (7.3)$$

For any set of initial data for the advection equation, the solutions at different times are related through the **evolution operator**, E, as

$$\psi(x, t+\Delta t) = \mathsf{E}(\Delta t)\psi(x,t) \equiv \psi(x - u\Delta t, t), \quad (7.4)$$

where we have used the fact that the term involving Δt is independent of t and as such acts upon the x component. The factor $e^{-iku\Delta t}$ in (7.3) is what is referred to as the **Fourier symbol** of the evolution operator. Differences in the modulus and the argument between the numerical and the analytic Fourier symbols represent errors in the amplitude and phase, respectively, of the numerically advected waves.

Given the derivation above, we know that the numerical solution depends upon four variables: the time step Δt, the spatial step size Δx, the wavenumber k, and the wave speed u. We also introduce two dimensionless parameters: the Courant number, which as related to our semi-Lagrangian parameters, is ν,

$$\nu = u\frac{\Delta t}{\Delta x},$$

but here it indicates the number of spatial grid points that have been crossed by a trajectory in one time step, and the wavenumber of the analytical solution, which is related to the wavelength λ through

$$k = \frac{2\pi}{\lambda}. \quad (7.5)$$

If we consider the ratio of the wavelength to the mesh, $\lambda\Delta x$, then a dimensionless wavenumber is defined by

$$\phi = \frac{2\pi}{\lambda\Delta x} \equiv k\Delta x. \quad (7.6)$$

It is now possible to define the analytical Fourier symbol, F, as

$$\mathsf{F} = e^{-iku\Delta t} = e^{-\phi u\frac{\Delta t}{\Delta x}} = e^{-\nu\phi}. \quad (7.7)$$

Next we determine equivalent properties of those presented for the Eulerian based finite difference schemes. We start by introducing the spatial shift operator S [172], which is defined as

$$\mathsf{S}\psi_p = \psi_{p+1}. \quad (7.8)$$

An important feature of the semi-Lagrangian schemes that we have to take note of here is that the finite difference stencil associated with the semi-Lagrangian schemes adapts to the trajectory of the flow; by this it is meant that for the Eulerian based finite differences we take the difference between points that are at the same location at every time step, which is not the case for the semi-Lagrangian based advection schemes.

To represent the mechanism that the stencils for the semi-Lagrangian based numerical schemes are flow dependent, we have to split the Courant number into its integer and noninteger components, where the integer component represents the nearest grid point upwind of the departure point, and the noninteger component represents the distance from the departure point to that nearest upwind grid point, which is represented by

$$\nu = \mathrm{p} + \alpha, \quad -1 \leq \alpha < 1, \quad \mathrm{p} \in \{0, 1, 2, \ldots\}, \tag{7.9}$$

where the integer p represents the number of grid points behind, or in front, of the nearest grid point at time $t = t^n$ to the departure point from ψ_i^{n+1}, and α is the ratio of the distance to the nearest upwind point of the grid space, and as a remainder is defined as

$$\alpha \equiv u\frac{\Delta t}{\Delta x} = \frac{x_{dp} - x_{i-p}}{\Delta x}. \tag{7.10}$$

Once we have an expression for the Fourier symbol, it is then possible to determine the phase error, ε_p, and the amplitude error, ε_a, through a similar argument that we presented in Chapter 3 as

$$\varepsilon_p = \frac{|\mathsf{F}_i|}{|\mathsf{F}|} = |\mathsf{F}_i|, \tag{7.11}$$

$$\varepsilon_a = \frac{\arg(\mathsf{F}_i)}{\arg(\mathsf{F})} = -\frac{\arg(\mathsf{F}_i)}{\nu\phi}. \tag{7.12}$$

Given all the definitions in this section, we now move on to apply these to the different interpolation polynomials from the last chapter.

7.2 Stability analysis of Lagrange interpolation polynomials

In this section we shall apply the definitions from the last section to the three different order Lagrange interpolation polynomials that we have introduced in the last two chapters. We shall start with the analysis of the linear polynomial.

7.2.1 Linear Lagrange interpolation polynomial

We start the stability analysis for the different Lagrange interpolations, and by association with the Newton divided difference interpolation polynomial, with the linear Lagrange interpolation. We mention here that the definition for α that is used in the papers we are partly summarizing in this chapter is the other way around to how we have presented, namely as

$$\alpha \equiv \frac{x_i - x_{dp}}{\Delta x}. \tag{7.13}$$

Therefore, with this in mind, we can write the linear Lagrange interpolation polynomial as

$$\psi_i^{n+1} = (1-\alpha)\,\psi_{i-p}^n + \alpha\psi_{i-p-1} = \mathsf{S}^{-\mathrm{p}}\left[(1-\alpha)+\alpha\mathsf{S}^{-1}\right]\psi_i^n,$$

where the shift operator, S^{-p}, indicates that we have to inversely shift p grid points from the nearest grid point to the departure point to the arrival point index at $t=n+1$. This then makes the numerical evolution operator for the linear interpolation semi-Lagrangian, E_1, as

$$\mathsf{E}_1 = \mathsf{S}^{-\mathrm{p}}\left[(1-\alpha)+\alpha\mathsf{S}^{-1}\right]. \tag{7.14}$$

The associated Fourier symbol, F_1, for the linear Lagrange interpolation polynomial scheme is derived by considering the effects the evolution operator, E_1, has on the discrete wavefunction, $e^{ij\varphi}$, where j is a spatial index, as

$$\mathsf{F}_1 \equiv \mathsf{E}_1 e^{ij\varphi} = e^{-ip\varphi}\left[(1-\alpha)+\alpha e^{-i\varphi}\right], \tag{7.15}$$

where we have substituted the number of spatial grid points moved backwards from the index for the arrival point to the index of the nearest grid point upwind of the departure point which is equivalent to −p. The −1 power in the second exponential term in (7.15) comes from the shift from the nearest point to the departure point upwind to the first grid point downwind of the departure point.

To start the stability analysis of the linear interpolation semi-Lagrangian scheme, we need to recall the amplification factor of the first-order upwind forward Euler finite difference scheme, which is

$$\left|\xi_{ufe}^2\right| = \left|\left(1-u\frac{\Delta t}{\Delta x}+u\frac{\Delta t}{\Delta x}e^{-i\varphi}\right)^2\right| \equiv 1-2C\,(1-C)\,(1-\cos\varphi)\,,$$

and has the CFL condition as $C\leq 1$, where we remind that C is the Courant number from the Eulerian finite difference schemes. What we are going to consider here now is the modulus of Fourier symbol for this interpolation scheme, where we have

$$|\mathsf{F}_1| = \left|e^{-ip\varphi}\left[(1-\alpha)+\alpha e^{-i\varphi}\right]\right|. \tag{7.16}$$

We can separate the two moduli in (7.16), and if we consider the rational component, $\left|\mathsf{F}_1^2\right|_r$, first, then we see that

$$\begin{aligned}\left|\mathsf{F}_1^2\right|_r &= (1-\alpha-\cos\varphi)^2+\alpha^2\sin^2\varphi = 1-2\alpha+\alpha^2+\alpha^2\cos^2\varphi-2\alpha\cos\varphi+2\alpha^2\sin^2\varphi \\ &= 1-2\alpha\,(1-\alpha)\,(1-\cos\varphi)\,. \end{aligned} \tag{7.17}$$

Therefore, we have the same amplification factor in (7.17) as for the upwind forward Euler finite difference scheme, but now we require the α term be less than 1. Here comes the important trick with the semi-Lagrangian schemes: because we have that the departure point is between the points $i-p$ and $i-p-1$, we have that α is always between 0 and 1, which is the condition for the amplification factor for the rational component of the Courant number to be stable, and therefore this is always satisfied.

We now turn our attention to the modulus of the integer component of the Courant number in (7.16), $\left|\mathsf{F}_1^2\right|_i$, which is $\left|\mathsf{F}_1^2\right|_i = \left|e^{-ip\varphi}\right|$. We can clearly see that we only have the complex component to deal with here. We now have to recall that when one is taking the modulus of a complex number, it is equivalent to computing the product of the complex number and its conjugate, which implies that $\left|e^{-ip\varphi}\right| = e^{-ip\varphi}e^{ip\varphi} = e^0 = 1$. Therefore, for all shifts, no matter how many grid points we consider, the integer component will always be stable, but we also have that α has been designed to be within $0 \leq \alpha \leq 1$, which means that the linear Lagrange interpolation polynomial when used for the linear, constant velocity, zero-forcing advection equation is **unconditionally stable**, which is an important property of semi-Lagrangian methods.

We have plotted the amplification factor of the linear Lagrange interpolation polynomial in Fig. 7.1, to highlight that: (1) the scheme is stable for all of the values of α that have been presented here to illustrate that this scheme is stable, but (2) there is heavy damping for the shortest wavelengths, where for the shortest resolvable wave at $\lambda = 2\Delta x$, there is complete extinction of this wave when $\alpha = 0.5$. However, the damping effect becomes less severe as the wavelength increases.

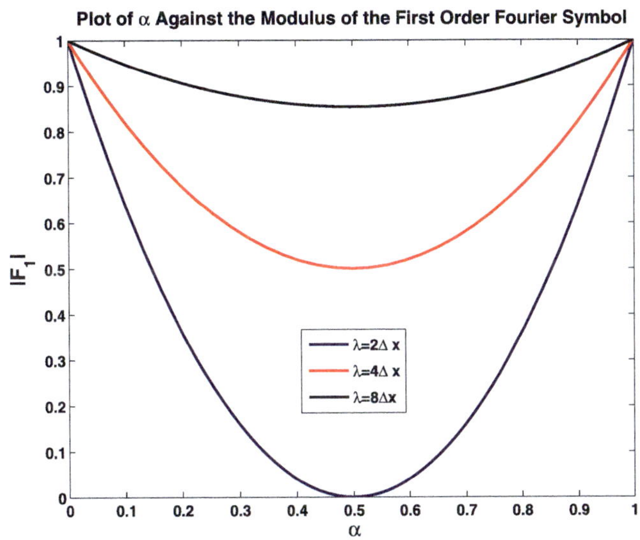

FIGURE 7.1 Plot of the amplification factor against the rational part of the Courant number for wavelengths $2\Delta x$, $4\Delta x$, and $8\Delta x$, respectively for the linear Lagrange interpolation polynomial.

We now move on to consider the phase errors for the linear Lagrange interpolation polynomial. This means that we are considering the Fourier symbol as

$$\mathsf{F}_1 = \left|\mathsf{F}_1\right| e^{-i\omega^\star \Delta t}, \tag{7.18}$$

and forming the complex argument of (7.18) to isolate $\omega^\star$, to then compare this expression to the analytical frequency, $ku \equiv \dfrac{k(p+\alpha)\Delta x}{\Delta t}$. If we now divide both sides of (7.18) by $|\mathsf{F}_1|$ and take the complex argument arg, then we obtain

$$\arg\left(e^{-i\omega^\star \Delta t}\right) = \arg(\mathsf{F}_1) - \arg(|\mathsf{F}_1|) = \arg(\mathsf{F}_1), \tag{7.19}$$

as the complex argument of a real number is zero which is what $|\mathsf{F}_1|$ is in (7.19). If we now use the trigonometric identity for the exponential of a complex number, we obtain

$$\begin{aligned}\arg\left(\cos\omega^\star\Delta t - i\sin\omega^\star\Delta t\right) &= \arg\left(\cos pk\Delta x - i\sin pk\Delta x\right)\\ &+ \arg\left(1-\alpha\left(1-\cos k\Delta x + i\sin k\Delta x\right)\right),\end{aligned} \tag{7.20}$$

where we have used the property that $\arg(ab) = \arg(a) + \arg(b)$. Recalling now that the definition of the complex argument is the $\tan^{-1}$ of the ratio of the imaginary to the real component of the complex number results in

$$-\tan^{-1}\left(\frac{\sin\omega^\star\Delta t}{\cos\omega^\star\Delta t}\right) = -\tan^{-1}\left(\frac{\sin pk\Delta x}{\cos pk\Delta x}\right) - \tan^{-1}\left(\frac{\alpha\sin k\Delta x}{1-\alpha\left(1-\cos k\Delta x\right)}\right), \tag{7.21}$$

which yields

$$\omega^\star\Delta t = pk\Delta x + \tan^{-1}\left(\frac{\alpha\sin k\Delta x}{1-\alpha\left(1-\cos k\Delta x\right)}\right). \tag{7.22}$$

The expression in (7.22) comes from [14], where if we now form the ratio of the numerical frequency above to the analytical frequency, which is referred to as the relative frequency R_1 in [14], then we obtain

$$R_1 \equiv \frac{\omega^\star}{\omega} = \frac{1}{(p+\alpha)k\Delta x}\left(pk\Delta x + \tan^{-1}\left(\frac{\alpha\sin k\Delta x}{1-\alpha\left(1-\cos k\Delta x\right)}\right)\right) \tag{7.23}$$

$$= \frac{p}{p+\alpha} + \frac{1}{(p+\alpha)k\Delta x}\tan^{-1}\left(\frac{\alpha\sin k\Delta x}{1-\alpha\left(1-\cos k\Delta x\right)}\right). \tag{7.24}$$

We have recreated a figure from [14] concerning the effects of the increasing wavelength in terms of grid spacing, with the decreasing of the phase error when the number of interval downstream of the arrival point increases, in Fig. 7.2. In [14] the authors state that in the long wave limit, as $k\Delta x \to 0$, $R_1 \to 1$, but also that if a departure point coincides with a grid point, R_1 is also equal to 1. We see from Fig. 7.2 that as the Courant number exceeds 1, the phase error is decreasing, that is to say, $R_1 \to 1$. It is also stated in [14] that this is a peculiar result for the constant velocity case because the departure points can be found exactly.

7.2.2 Stability analysis of the quadratic Lagrange interpolation

We now consider the stability of the quadratic Lagrange interpolation polynomial based semi-Lagrangian scheme, which as with the linear case starts with defining the quadratic

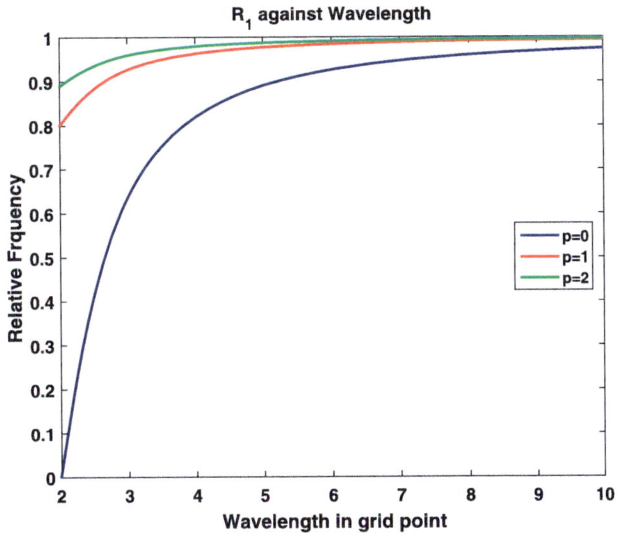

FIGURE 7.2 Plot of relative frequency as a function of wavelengths in terms of the grid length for the linear Lagrange interpolation polynomial for $\alpha = 0.25$.

interpolation semi-Lagrangian difference equation in terms of the shift operator, which is

$$\begin{aligned} \psi_i^{n+1} &= \frac{\alpha(\alpha+1)}{2}\psi_{i-p-1}^n + \left(1-\alpha^2\right)\psi_{i-p}^n + \frac{\alpha(\alpha-1)}{2}\psi_{i-p+1} \\ &\equiv \mathsf{S}^{-p}\left[\frac{\left(\alpha^2+\alpha\right)}{2}\mathsf{S}^{-1} + \left(1-\alpha^2\right) + \frac{\left(\alpha^2-1\right)}{2}\mathsf{S}^{+1}\right]\psi_i^n, \\ &\Rightarrow \mathsf{E}_2 \equiv \mathsf{S}^{-p}\left[\frac{\left(\alpha^2+\alpha\right)}{2} + \left(1-\alpha^2\right) + \frac{\left(\alpha^2-\alpha\right)}{2}\right]. \end{aligned} \tag{7.25}$$

Given the evolution operator for the quadratic Lagrange interpolation based semi-Lagrangian approach, E_2, we now move to define the Fourier symbol for this configuration of the semi-Lagrangian approach. As in the linear interpolation case, we apply the quadratic evolution operator to the discrete wavefunction $e^{ij\varphi}$, which results in

$$\mathsf{F}_2 = \mathsf{E}_2 e^{ij\varphi} \equiv e^{-ip\varphi}\left[\frac{\left(\alpha^2+\alpha\right)}{2}e^{-i\varphi} + \left(1-\alpha^2\right) + \frac{\left(\alpha^2-\alpha\right)}{2}e^{i\varphi}\right]. \tag{7.26}$$

The next step in the stability analysis for the quadratic interpolation based semi-Lagrangian scheme is to collect like terms for powers of α, which results in

$$\begin{aligned} \mathsf{F}_2 &= e^{-ip\varphi}\left(1-\alpha^2\left(1-\frac{1}{2}\left(e^{-i\varphi}+e^{i\varphi}\right)\right) + \frac{\alpha}{2}\left(e^{-i\varphi}-e^{i\varphi}\right)\right) \\ &\equiv e^{-ip\varphi}\left(1-\alpha^2\left(1-\cos\varphi\right) - i\alpha\sin\varphi\right). \end{aligned} \tag{7.27}$$

To compete the stability analysis for the quadratic interpolation semi-Lagrangian scheme, we need to consider the modulus and argument of (7.27) to see if they are bounded and if there are constraints on the step sizes.

Considering the modulus of F_2 first, we have

$$\begin{aligned}|\mathsf{F}_2| &= \left(1-\alpha^2(1-\cos\varphi)\right)^2+\alpha^2\sin^2\varphi\\ &= 1-2\alpha^2(1-\cos\varphi)+\alpha^4(1-\cos\varphi)^2+\alpha^2\left(1-\cos^2\varphi\right)\\ &= 1-\alpha^2+2\alpha^2\cos\varphi-\alpha^2\cos^2\varphi+\alpha^4(1-\cos\theta)^2\\ &= 1-\alpha^2\left((1-\cos\varphi)^2-\alpha^2(1-\cos\varphi)\right)\\ &= 1-\alpha^2\left(1-\alpha^2\right)(1-\cos\varphi)^2. \end{aligned} \tag{7.28}$$

To ensure stability, we require (7.28) to be less than or equal to 1, which implies that the following inequality:

$$-1\leq 1-\alpha^2\left(1-\alpha^2\right)\leq 1 \Longrightarrow -1\leq\alpha\leq 1 \tag{7.29}$$

must hold. The condition for the rational component of the Courant number is always satisfied due to the restriction that the distance between the departure point and the nearest upwind grid point has been normalized by the spatial step size, therefore, the quadratic interpolation semi-Lagrangian scheme is unconditionally stable, which is the same conclusion as for the linear interpolation, because we know that the integer part of the Fourier symbol has a modulus equal to 1 as it the case for the linear case.

In Fig. 7.3 we have plotted the modulus of the quadratic Lagrange interpolation polynomial Fourier symbol against the rational component of the Courant number for different wavelengths. This plot and Fig. 7.1 are recreations of those from [14], where they are explained as showing the scales that are damped by the different order interpolation schemes. In [14] they state that for the semi-Lagrangian scheme with linear interpolation there is heavy damping for the shortest wavelengths, where for the shortest resolvable wave at $\lambda = 2\Delta x$, there is complete extinction of this wave when $\alpha = 0.5$. However, the damping effect becomes less severe as the wavelength increases. If we consider the associated plot for the semi-Lagrangian scheme with quadratic interpolation, Fig. 7.3, then we see that there is still extinction of the short waves for the quadratic interpolation, but this does not occur till $\alpha \pm \frac{1}{\sqrt{2}}$ and quickly dissipates for larger wavelengths.

The relative frequency for the quadratic Lagrange interpolation polynomial can be shown to be

$$R_2 = \frac{p}{p+\alpha}+\left(\frac{1}{(p+\alpha)k\Delta x}\right)\tan^{-1}\left(\frac{\alpha\sin k\Delta x}{1-\alpha^2(1-\cos k\Delta x)}\right), \tag{7.30}$$

and we have plotted the relative frequency against wavelengths in terms of grid points in Fig. 7.4, where, as mentioned in [14], we again see that in the long wave limit $R_2 \rightarrow 1$, and

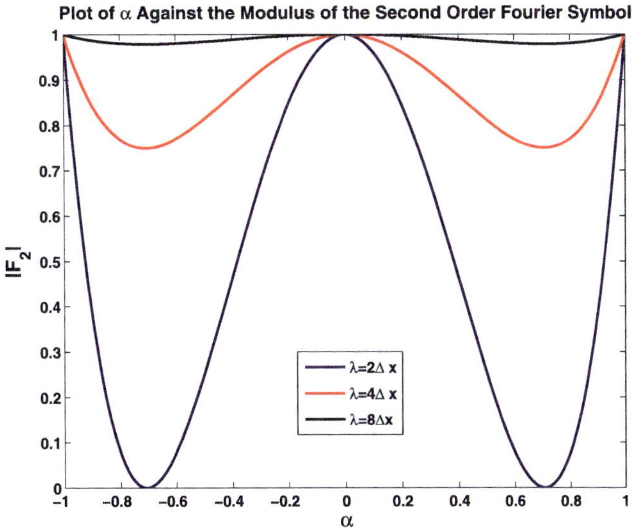

FIGURE 7.3 Plot of the amplification factor against the rational part of the Courant number for wavelengths $2\Delta x$, $4\Delta x$, and $8\Delta x$, respectively, for the quadratic Lagrange interpolation polynomial.

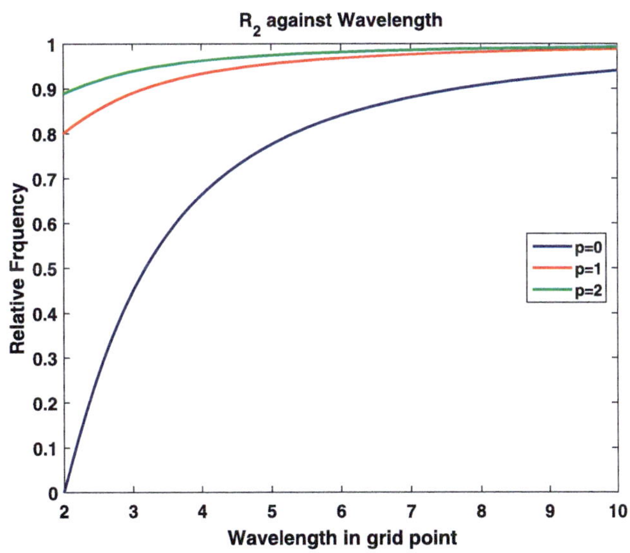

FIGURE 7.4 Plot of relative frequency as a function of wavelengths in terms of the grid length for the quadratic Lagrange interpolation polynomial for $\alpha = 0.25$.

we see again that if the departure point is at a grid point, then $R_2 = 1$ as for R_1. An important feature to note here, mentioned in [14], is that by going to a quadratic interpolation formula, we have been able to improve the amplitude error for most waves, but there appears to not be much of an reduction in the phase error. We can see the reason why by comparing R_2 with the expression for R_1 and seeing that the only difference is α^2 multiplying $1 - \cos k\Delta x$ in R_2 compared to α in R_1.

7.2.3 Stability analysis of the cubic Lagrange interpolation

If we now consider the cubic Lagrange interpolation polynomial based semi-Lagrangian scheme, upon multiplying out all of the brackets, then it is possible to write the polynomial in terms of powers of α as

$$\begin{aligned}\psi_i^{n+1} &= \psi_{i-p}^n - \frac{1}{2}\alpha\left(\psi_{i-p+1}^n - \psi_{i-p-1}^n\right) + \frac{\alpha^2}{2}\left(\psi_{i-p+1}^n - 2\psi_{i-p}^n + \psi_{i-p-1}^n\right)\\ &+ \frac{\alpha\left(1-\alpha^2\right)}{6}\left(\psi_{i-p+1}^n - 3\psi_{i-p}^n + 3\psi_{i-p-1}^n - \psi_{i-p-2}^n\right).\end{aligned} \tag{7.31}$$

Next we express (7.31) in terms of the shift operator as

$$\begin{aligned}\psi_i^{n+1} &= \mathsf{S}^{-p}\left[1 - \frac{1}{2}\alpha\left(\mathsf{S}^{+1} - \mathsf{S}^{-1}\right) + \frac{\alpha^2}{2}\left(\mathsf{S}^{+1} - 2 + \mathsf{S}^{-1}\right)\right.\\ &- \left.\frac{\alpha\left(1-\alpha^2\right)}{6}\left(\mathsf{S}^{+1} - 3 + 3\mathsf{S}^{-1} - \mathsf{S}^{-2}\right)\right]\psi_i^n.\end{aligned}$$

Therefore the evolution operator for the cubic Lagrange interpolation based semi-Lagrangian scheme, E_3, is

$$\begin{aligned}\mathsf{E}_3 &= \mathsf{S}^{-p}\left[1 - \frac{1}{2}\alpha\left(\mathsf{S}^{+1} - \mathsf{S}^{-1}\right) + \frac{\alpha^2}{2}\left(\mathsf{S}^{+1} - 2 + \mathsf{S}^{-1}\right)\right.\\ &- \left.\frac{\alpha\left(1-\alpha^2\right)}{6}\left(\mathsf{S}^{+1} - 3 + 3\mathsf{S}^{-1} - \mathsf{S}^{-2}\right)\right].\end{aligned}$$

Now we apply the third-order numerical evolution operator above to the Fourier node, $e^{ij\varphi}$, such that

$$\begin{aligned}\mathsf{F}_3 &\equiv \mathsf{E}_3 e^{ik\varphi} = e^{-ip\varphi}\left[1 - \frac{\alpha}{2}\left(e^{i\varphi} - e^{-i\varphi}\right) + \frac{\alpha^2}{2}\left(e^{i\varphi} - 2 + e^{-i\varphi}\right)\right.\\ &+ \left.\frac{\alpha\left(1-\alpha^2\right)}{6}\left(e^{i\varphi} - 3 + 3e^{-i\varphi} - e^{-2i\varphi}\right)\right].\end{aligned} \tag{7.32}$$

To obtain the amplification factor, we require $|\mathsf{F}_3|^2$, but first we will replace the exponentials in (7.32) with their relative sine and cosine equivalents, which then yields

$$\begin{aligned}
\mathsf{F}_3 &= e^{-ip\varphi}\left[1 - i\alpha\sin\varphi - \alpha^2(1-\cos\varphi)\right. \\
&+ \left.\frac{\alpha\left(1-\alpha^2\right)}{6}\left(4\cos\varphi - 2i\sin\varphi + 2 - 2\cos^2\varphi - 2i\cos\varphi\sin\varphi\right)\right] \\
&= e^{-ip\varphi}\left[1 - i\alpha\sin\varphi - \alpha^2(1-\cos\varphi) - \frac{\alpha\left(1-\alpha^2\right)}{3}\left((1-\cos\varphi)^2 + i\sin\varphi(1-\cos\varphi)\right)\right].
\end{aligned} \tag{7.33}$$

Collecting the real and imaginary parts results in

$$\operatorname{Re}(\mathsf{F}_3) = 1 - \alpha^2(1-\cos\varphi) - \frac{\alpha\left(1-\alpha^2\right)}{3}(1-\cos\varphi)^2, \tag{7.34}$$

$$\operatorname{Im}(\mathsf{F}_3) = -\alpha\sin\varphi - \frac{\alpha\left(1-\alpha^2\right)}{3}\sin\varphi(1-\cos\varphi). \tag{7.35}$$

The modulus of the Fourier symbol for the cubic interpolation is equivalent to the sum of the squares of (7.34) and (7.35). Therefore, squaring these two parts above results in

$$\begin{aligned}
|\mathsf{F}_3| &= 1 + \alpha^4(1-\cos\varphi)^2 + \frac{\alpha^2\left(1-\alpha^2\right)^2}{9}(1-\cos\varphi)^4 - 2\alpha^2(1-\cos\varphi) \\
&- \frac{2\alpha\left(1-\alpha^2\right)}{3}(1-\cos\varphi)^2 + \frac{2\alpha^3\left(1-\alpha^2\right)}{3}(1-\cos\varphi)^3 \\
&+ \alpha^2\sin^2\varphi + \frac{\alpha^2\left(1-\alpha^2\right)^2}{9}\sin^2\varphi(1-\cos\varphi)^2 + \frac{2\alpha^2\left(1-\alpha^2\right)}{3}\sin^2\varphi(1-\cos\varphi).
\end{aligned} \tag{7.36}$$

The expression in (7.36) looks quite daunting, and it is, but to help simplify things we shall expand the powers of $(1-\cos\varphi)$, which are

$$\begin{aligned}
C^2 = (1-c) &= 1 - 2c + c^2, \\
C^3 = (1-c)^3 &= 1 - 3c + 3c^2 - c^3, \\
C^4 = (1-c)^4 &= 1 - 4c + 6c^2 - 4c^3 + c^4,
\end{aligned}$$

where $c = \cos\varphi$. The reason for displaying these expansions is because the expression for the modulus of the third-order Fourier symbol is presented in [117], but in terms of $C = (1-\cos\varphi)$. The factorized expression for the amplification error in [117] is given in (7.37), and below we have expanded the brackets to ascertain which expressions we are looking for to verify that we have derived the correct amplification factor. Thus

$$\begin{aligned}
|\mathsf{F}_3| &= 1 - \alpha(2-\alpha)\left(1-\alpha^2\right)C^2\left(\frac{3 + 2C\alpha(1-\alpha)}{9}\right) \\
&= 1 - \left(2\alpha - \alpha^2 - 2\alpha^3 + \alpha^4\right)C^2\left(\frac{3 + 2C\left(\alpha - \alpha^2\right)}{9}\right)
\end{aligned} \tag{7.37}$$

$$
\begin{aligned}
&= 1-\frac{1}{9}\Big(6\alpha C^2-3\alpha^2C^2-6\alpha^3C^2+3\alpha^4C^2+4\alpha^2C^3-2\alpha^3C^3\\
&- 4\alpha^4C^3+2\alpha^5C^3-4\alpha^3C^3+2\alpha^4C^3+4\alpha^5C^3-2\alpha^6C^3\Big).
\end{aligned}
\tag{7.38}
$$

The next step is to collect all the factors of the different powers of α in (7.38), which results in:

$$
\begin{aligned}
\alpha: &\quad -\tfrac{6}{9}C^2,\\
\alpha^2: &\quad \tfrac{3}{9}C^2-\tfrac{4}{9}C^3,\\
\alpha^3: &\quad \tfrac{6}{9}C^2+\tfrac{2}{9}C^3+\tfrac{4}{9}C^3,\\
\alpha^4: &\quad -\tfrac{3}{9}C^2+\tfrac{4}{9}C^3-\tfrac{2}{9}C^3,\\
\alpha^5: &\quad -\tfrac{2}{9}C^3-\tfrac{4}{9}C^3,\\
\alpha^6: &\quad \tfrac{2}{9}C^3.
\end{aligned}
\tag{7.39}
$$

As we can see above, we do not have a C^4 term in the final expression from [117], yet we do have it in (7.36). This indicates that we have to find a way to eliminate the C^4 term. To do this, we have to *complete the cube* on two different terms in (7.36). The first term that we consider is the quartic term, and we find a way to express the quartic as a sum of cube expressions. This is achieved by

$$
\begin{array}{c|ccccc}
(1-c)^4 & 1 & -4c & 6c^2 & -4c^3 & c^4\\
(1-c)^3 & 1 & -3c & 3c^2 & -c^3 & \\
\hline
 & 0 & -c & 3c^2 & -3c^3 & -c^4
\end{array},
$$

where the last line in the polynomial division above is equivalent to $-c\,(1-c)^3$. Therefore, it is possible to write the quartic term as $(1-c)^4=C^3-cC^3$. The next term that we have to consider is $\left(1-c^2\right)(1-c)^2$:

$$
\begin{array}{c|ccccc}
\left(1-c^2\right)(1-c)^2 & 1 & -2c & & 2c^3 & -c^4\\
(1-c)^3 & 1 & -3c & 3c^2 & -c^3 & \\
\hline
 & 0 & c & -3c^2 & 3c^3 & -c^4
\end{array}.
$$

The last line in the array above is cC^3. Therefore, the cC^3 terms cancel, and we have matched the α^6 term.

We now consider the lower powers of α from (7.39). From (7.36) we already have the correct coefficient for α; however, for α^2 we have a term that is only a factor of C, yet all of the coefficients in (7.39) are in terms of C^2 or C^3 and none of them are in terms of $\sin^2\varphi$. Therefore, collecting all the factors of α^2, using the identity $\sin^2\varphi\equiv 1-\cos^2\varphi$, and putting all of the coefficients under a common denominator, we obtain

$$
\begin{aligned}
\alpha^2: \quad & \sin^2\varphi+\frac{2}{3}\alpha^2\sin^2\varphi\,(1-\cos\varphi)-2\,(1-\cos\varphi)\\
& 3-3c^2+2-2c^2-2c+2c^3-6+6c\\
& -1+4c-5c^2+2c^3.
\end{aligned}
\tag{7.40}
$$

The expression in (7.40) appears to be in the form of C^3, and if we complete the cube on (7.40) with $-2C^3$, we obtain

Eq. (7.40)	-1	$4c$	$-5c^3$	$2c^3$
$-2C^3$	-2	$6c$	$-6c^2$	$2c^3$.
	1	$-2c$	c^2	

Therefore, the factors of the α^2 terms combine to $-\frac{6}{9}C^3 + \frac{3}{9}C^2$, which gives us the C^2 factor for α^2 in (7.39). To obtain the factor of C^3, we should note now that what was C^4 is now in terms of $2C^3$ and, upon expanding the $\dfrac{2\alpha^2\left(1-\alpha^2\right)^2}{9}$ term, we have $\frac{2}{9}\alpha^2C^3$, which completes the factors of α^2 in (7.39).

Moving on to α^3, we first need to notice that the there are two factors of the C^3 term here, which when added together give a factor of $\frac{6}{9}C^3$. We can then see that both of the factors of α^3 are present in (7.36). Therefore, we have matched the α^3 factors. However, for the α^4 factors we have to do some rearranging and completing the cube again. Combining all of the current factors of α^4, substituting for $\sin^2\varphi$ again, and putting all the terms under a common denominator results in

$$
\begin{aligned}
\alpha^4: \qquad & 3\left(1-\cos\varphi\right)^2 - 2\left(1-\cos^2\varphi\left(\left(1-\cos\varphi\right)\right.\right. \\
& 3-6c+3c^2-2+2c+2c^2-2c^3 \\
& 1+4c+5c^2-2c^3 \\
\Longrightarrow \quad & -2C^3-C^2.
\end{aligned}
$$

Next putting the terms above under the other common denominator of 9 gives us the required factor of C^2 for α^4. Combining with the old C^4 term which also contains an α^4 of 4 results in the factor of C^3 when the two terms are added together in (7.39).

This now just leaves the α^5 factors to consider. There are no C^2s in the factors of α^5, so we are then left with two factors of C^3, which when added together require $-\frac{6}{9}C^3$, which, upon looking at (7.36), we can see is present. Thus we have verified the expression for the amplification factor of the cubic interpolation semi-Lagrangian scheme from [117]. Given the expression for the amplification factor in the factorized form, we can see that the restriction on the spatial and temporal step sizes is $0 \leq \alpha < 1$, which comes about because if $\alpha < 0$, then the $-\alpha$ terms make the second term in (7.37) positive and make the amplification greater than 1. If we consider the case $\alpha \geq 1$, then the amplification factor is either equal to 1, which is not a desirable property, or is greater than 1 since one of the terms involving α will be negative and as such change the sign of the second term in (7.37).

Given that we now know that the fractional part of the Courant number must satisfy $0 \leq \alpha < 1$, as we stated for the linear case, this condition is always satisfied because the distance between grid points has been normalized to 1, therefore the cubic interpolation semi-Lagrange scheme is also unconditionally stable, because we have the same shift operator as for the other two presented lower order Lagrange interpolation polynomials.

This may appear to be a lot of work to obtain the amplification factor, but we need this expression to ascertain if we have the damping effects that we have seen for the linear and

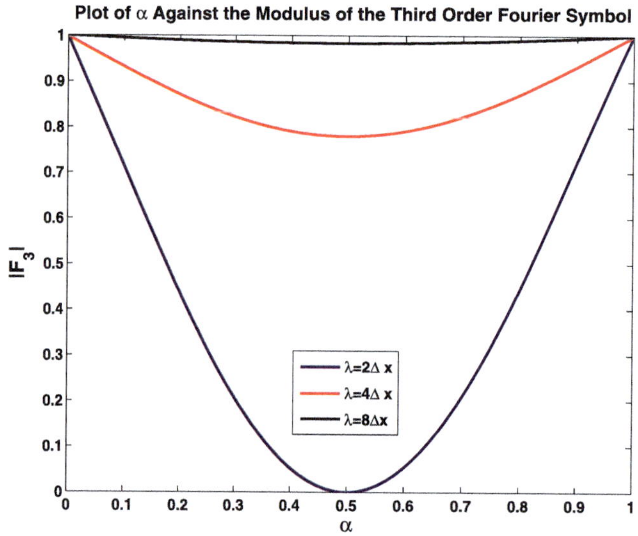

FIGURE 7.5 Plot of the amplification factor against the rational part of the Courant number for wavelengths $2\Delta x$, $4\Delta x$, and $8\Delta x$, respectively, for the cubic Lagrange interpolation polynomial.

quadratic interpolation semi-Lagrangian schemes. In Fig. 7.5 we have plotted the amplification factor for the cubic interpolation.

As we can see from Fig. 7.5, the short wavelengths are still being damped to the point that if the departure point is halfway between two points, we lose all information of the feature. However, when considering the longer wavelengths, we see that the cubic interpolation scheme does not damp as heavily as the linear interpolations. This improvement is quite significant, and is one of the reasons the cubic interpolation based scheme is used, while it does take more coding than the two lower order interpolations and does take longer to run, it is equivalent to a third order in space discretization.

In [117] the phase error, or relative frequency, for this scheme is given by

$$R_3(\alpha, k) = \frac{p}{\alpha} + \left(\frac{1}{\alpha k \Delta x}\right)\tan^{-1}\left(\frac{\alpha \sin k\Delta x\left(1 + \frac{c\left(1-\alpha^2\right)}{3}\right)}{1 - c\alpha^2 - \frac{c^2\alpha\left(1-\alpha^2\right)}{3}}\right), \tag{7.41}$$

where $c \equiv 1 - \cos k\Delta x$ as before.

For completeness, we have plotted the relative frequency for the cubic Lagrange interpolation polynomial in Fig. 7.6, where we can see that if we compare to Figs. 7.2 and 7.4, then there is a similar structure to that of the linear relative error plot, but it is approaching 1 faster for the cubic Lagrange interpolation than for the linear or quadratic. Therefore, there is some advantage of implementing a cubic interpolation over linear and quadratic with respect to reducing phase errors.

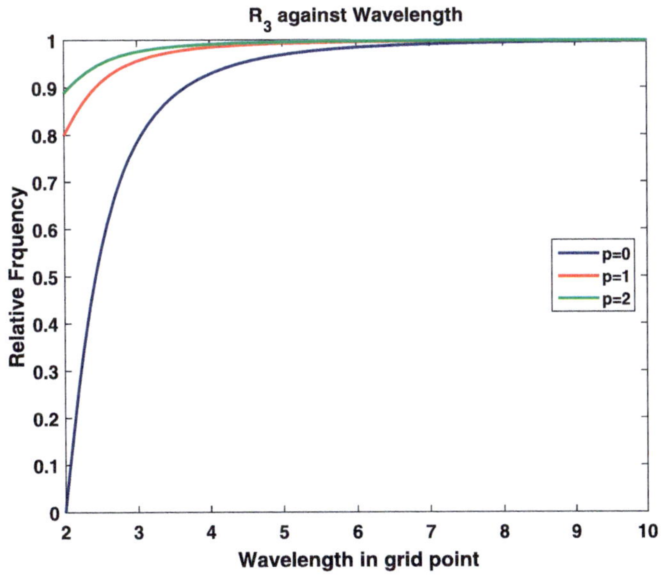

FIGURE 7.6 Plot of relative frequency as a function of wavelengths in terms of the grid length for the cubic Lagrange interpolation polynomial for $\alpha = 0.25$.

Exercise 7.1. *Verify that the relative frequency for the cubic Lagrange interpolation polynomial is* (7.41).

Exercise 7.2. *For the quartic Lagrange interpolation that uses the points*

$$\{i-p-2.i-p-1, i-p, i-p+1, i-p+2\},$$

derive the fourth-order evolution operator E_4 *and then the fourth-order Fourier symbol* F_4. *Show that the real and imaginary components of the Fourier symbol are*

$$\begin{aligned}\text{Re}(\mathsf{F}_4) &= 1 - C\alpha^2 - \frac{C^2\alpha^2\left(1-\alpha^2\right)}{6},\\ \text{Im}(\mathsf{F}_4) &= \alpha\sin\varphi\left(1+\frac{C\left(1-\alpha^2\right)}{3}\right).\end{aligned}$$

Taking the modulus of F_4, *show that the amplification factor can be factorized as*

$$|\mathsf{F}_4| = 1-\alpha^2\left(1-\alpha^2\right)\left(4-\alpha^2\right)C^3\frac{\left(C\left(1-\alpha^2\right)\right)}{36}. \tag{7.42}$$

Given the amplification factor in (7.42) *determine the values for* α *making the quartic Lagrange interpolation scheme stable and deduce the type of stability for this scheme.*

The expressions above as we have stated are from [117] where the authors introduce the extra constraint on α for the quadratic and quartic schemes, namely that the rational component of the Courant number, or rather the normalized (by Δx) distance from the departure point, cannot be more or less than 0.5. We can clearly see the reason for this from Fig. 7.3, where if we have $-0.5 \leq \alpha \leq 0.5$, then the scheme will not damp completely the smaller wavelengths and is, in fact, less damping on the longer wavelengths than the cubic interpolation [117].

7.3 Stability analysis of the cubic Hermite semi-Lagrangian interpolation scheme

As we have observed with the different order of Lagrange polynomial based interpolation schemes for the semi-Lagrangian advection case, there are two different ways to factorize the interpolation polynomial. The first is to collect factors of different powers of the Courant number, while the second approach is to collect the factors of the grid points. We are going to consider the latter case for a specific second-order version of the derivatives in a Hermite interpolation formulation.

The second-order approximation to the derivative scheme that we present here comes from [103], where the derivatives at the endpoints are approximated by $\frac{\psi_{i-p} - \psi_{i-p-2}}{2}$ for ψ'_{i-p-1} and by $\frac{\psi_{i-p+1} - \psi_{i-p-1}}{2}$ for ψ'_{i-p}. Therefore, the Hermite interpolation polynomial is going to be using the same points as the cubic Lagrange interpolation polynomial but here with different factors. We should note that this is equivalent to the arithmetic mean approximation to the derivatives mentioned in Chapter 6.3.

In [103] it is stated that the one-dimensional Hermite interpolation scheme acts on an interval $x \in (x_{i-1}, x_i)$ of length Δx, but where we have normalized the distance between the grid points in terms of the Courant number, this then gives the general form of the cubic Hermite interpolation formula as

$$\begin{aligned} \psi(\tilde{x}) &= (1-\alpha)^3 \left(2\psi_{i-p-1} - 2\psi_{i-p} + \psi'_{i-p-1} + \psi'_{i-p}\right) \\ &+ (1-\alpha)^2 \left(-3\psi_{i-p-1} + 3\psi_{i-p} - 2\psi'_{i-p-1} - \psi'_{i-p}\right) \\ &+ (1-\alpha)\psi'_{i-1} + \psi_{i-1}. \end{aligned} \tag{7.43}$$

Substituting the second-order approximations to the derivative terms at the endpoints into (7.43) and collecting all the factors of the four grid points results in

$$\psi_i^{n+1} = g_{-2}\psi_{i-p-2}^n + g_{-1}\psi_{i-p-1}^n + g_0\psi_{i-p}^n + g_{+1}\psi_{i-p+1}^n,$$

where the coefficients above are defined as

$$\begin{aligned} g_2 &\equiv -\frac{1}{2}(1-\alpha)^3 + (1-\alpha)^2 - \frac{1}{2}(1-\alpha), \\ g_{-1} &\equiv \frac{3}{2}(1-\alpha)^3 - \frac{5}{2}(1-\alpha)^2 + 1, \end{aligned}$$

$$g_0 \equiv -\frac{3}{2}(1-\alpha)^3 + 2(1-\alpha)^2 + \frac{1}{2}(1-\alpha),$$
$$g_{+1} \equiv \frac{1}{2}(1-\alpha)^3 - \frac{1}{2}(1-\alpha)^2.$$

To perform the von Neumann stability analysis for (7.43), we begin by identifying the evolution operator,

$$\mathsf{E}_{cH} = \mathsf{S}_{-p}\left[g_{-2}\mathsf{S}^{-2} + g_{-1}\mathsf{S}^{-1} + g_0 + g_1\mathsf{S}^{+1}\right],$$

which would then make the Fourier symbol for the second-order derivative based cubic Hermite interpolation,

$$\mathsf{F}_{cH} = \mathsf{E}_{cH}a_n e^{ik\varphi} \equiv e^{-pi\varphi}\left[g_{-2}e^{-2i\varphi} + g_{-1}e^{-i\varphi} + g_0 + g_{+1}e^{i\varphi}\right]. \tag{7.44}$$

If we simply wish to examine the amplification factor for the cubic Hermite interpolation, it is possible to write a short piece of code to run through different values of α between 0 and 1 in the modulus of (7.44) to see where the amplification factor goes above 1. In Fig. 7.7A we have plotted the amplification factor against the Courant number for different wavelengths to see which scales are damped with this scheme, while in Fig. 7.7B we have plotted the amplification factor against wavelengths for different Courant numbers to see which wavelengths are damped due to the distance of the departure point to the nearest grid point.

We can see from Fig. 7.7A that the cubic Hermite approximations is stable for Courant numbers between 0 and 1, which means that it is again unconditionally stable; we can see this for the 3 wavelength amplification curves that have been plotted. As with the cubic Lagrange interpolation, we see that for wavelengths $2\Delta x$, if the Courant number is $\alpha = 0.5$, which corresponds to the departure point being in the center of the interval, then again this feature would be completely damped. However, the structure of the curve associated with $k = 2\Delta x$ for the other values of the Courant number is not as severe.

If we now analyze the response to the amplification for the $k = 4\Delta x$ wavelength, then we see that there is a drastic improvement over the performance of the Hermite polynomial for the $k = 2\Delta x$ case. While there is still some damping when the departure point is in the center of the interval, the wave feature will not be too severely damped. Looking at the amplification factor for the $k = 8\Delta x$ case, we see that there is almost no damping at this wavelength at all.

As mentioned earlier, in Fig. 7.7B we have plotted the amplification factors by Courant number against the wavelength, which is another useful way to identify where the interpolation schemes are not themselves, introducing too much of an error into the numerical scheme. We see that for $\alpha = 0.5$, as the wavelength reaches π, the scheme damps this feature to zero. As the Courant numbers moves away from $\alpha = 0.5$, the damping effect reduces quite quickly, suggesting that if the departure point is away from the center part of the interval, then the interpolation is drastically improved.

If, however, we wish to determine the stability analytically for this version of the cubic Hermite interpolation, then we would start by expanding the coefficients in (7.44) into powers of α and, by collecting all the factors of each of the four powers of α, would obtain

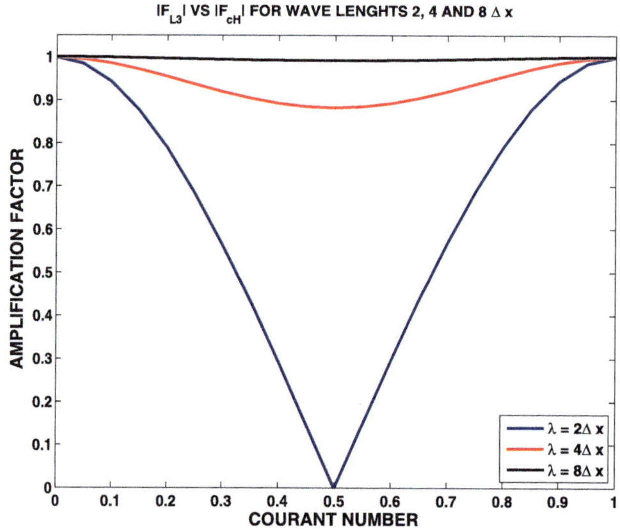

(A)

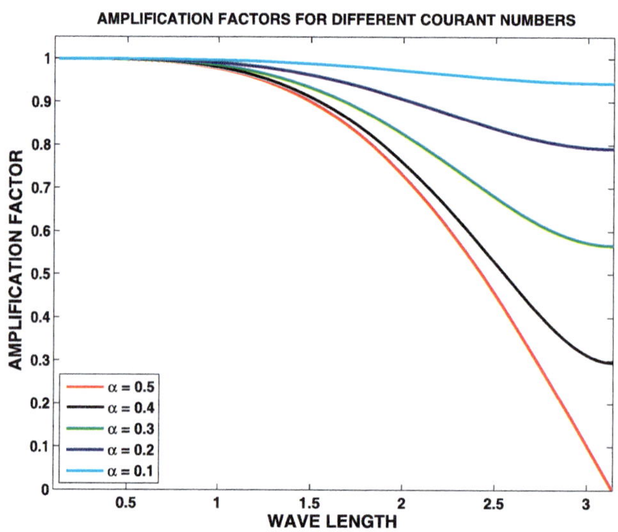

(B)

FIGURE 7.7 (A) Plot of the amplification factor against Courant number for the cubic Hermite interpolation scheme for different wavelengths, (B) plot of amplification factors by Courant number against the wavelength for the cubic Hermite interpolation scheme.

$$\alpha^3 : \frac{e^{-2i\varphi}}{2} - \frac{3e^{-i\varphi}}{2} + \frac{3}{2} - \frac{e^{i\varphi}}{2}, \qquad (1-C)^2 + i\sin\varphi\,(1-C),$$

$$\alpha^2 : -\frac{e^{-2i\varphi}}{2} + 2e^{-i\varphi} - \frac{5}{2} + e^{i\varphi}, \qquad -(1-C)^2 - i\sin\varphi\,(1-C) - (1-C),$$

$$\alpha^1 : \frac{e^{-i\varphi}}{2} - \frac{e^{i\varphi}}{2}, \qquad\qquad -i\sin\varphi,$$
$$\alpha^0 : 1, \qquad\qquad 1.$$

Combining all of the information above into the real and imaginary parts of the Fourier symbol results in

$$\mathrm{Re}\left(\mathsf{F}_{cH}\right) = 1 - \left((1-C) + (1-C)^2\right)\alpha^2 + (1-C)^2\alpha^3, \tag{7.45a}$$
$$\mathrm{Im}\left(\mathsf{F}_{cH}\right) = -i\sin\varphi\alpha - i\sin\varphi\,(1-C)\,\alpha^2 + i\sin\varphi\,(1-C)\,\alpha^3. \tag{7.45b}$$

Therefore, the amplification factor of the cubic Hermite interpolation is the modulus of the sum of squares of (7.45a) and (7.45b).

As mentioned before the derivation above, we now do a comparison of the amplification factors of the cubic Lagrange interpolation with the cubic Hermite interpolation with the arithmetic mean/central difference approximation for the gradient terms. In Fig. 7.8 we have presented the amplification factors for the two schemes in the same way as in Fig. 7.7.

As in Fig. 7.7A we have plotted the amplification factor by Courant number for different wavenumbers for the comparison between Lagrange interpolation and the Hermite polynomial with the central difference approximation in Fig. 7.8A. The first feature to note here is that while both the Lagrange and Hermite polynomials do damp out the wavenumber $2\Delta x$ feature, the Hermite interpolation that we are considering in this stability analysis does so at quite a slower rate than the cubic Lagrange interpolation polynomial. For the Lagrange interpolation, we see that the damping effect starts immediately for the departure points that are not at a grid point, but for the Hermite interpolation we see that the damping is less severe for all values of the Courant number except when the departure point is in the center of the interval.

Moving on to the results for wavenumber $k = 4\Delta x$, we see that while both schemes have drastically reduced the damping feature seen for the $2\Delta x$ case, there is still some damping but it is clearly smaller using the Hermite interpolation than the Lagrange interpolation. For the wavenumber $8\Delta x$ case, we again see a reduction in the damping effect but there is still a slight improvement at this number with the Hermite interpolation over using the Lagrange interpolation.

In Fig. 7.8B we have again plotted the amplification factor by wavelength for the $\alpha = 0.1$, $0.2, 0.3, 0.4$, and 0.5, for both the cubic Lagrange polynomial and our choice for the cubic Hermite polynomial. We can see that the Hermite polynomial damping effect is always smaller than that of the Lagrange interpolation. Another striking feature of Fig. 7.7B is the difference between the amplification factors of the two schemes as the Courant number decreases towards zero. The amplification factor for the Hermite interpolation when $\alpha = 0.4$ is 2.5 units larger than that of the equivalent Lagrange interpolation. The difference in the amplification factors between the two interpolations grows even more when $\alpha = 0.3$. However, the difference between the two schemes does start to decrease for $\alpha = 0.2$; we see that the version of the Hermite interpolation that we consider in this section has less of a damping effect than the Lagrangian interpolating.

Exercise 7.3. *Derive the relative frequency to determine the damping features of the cubic Hermite interpolation polynomial.*

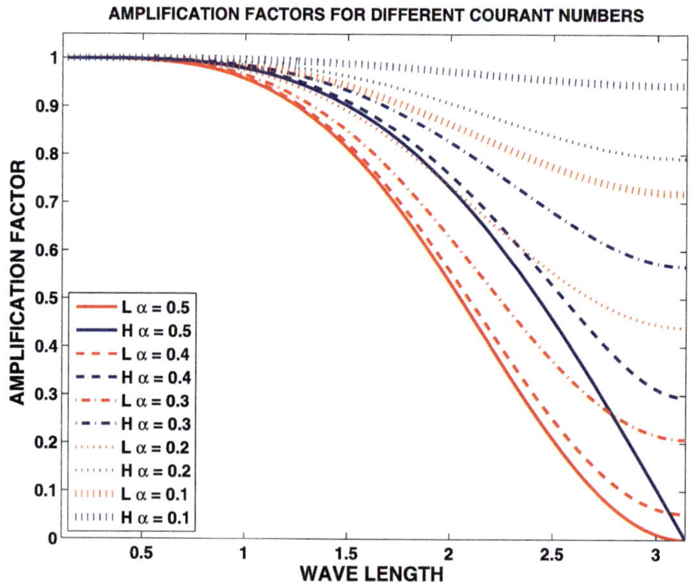

(A)

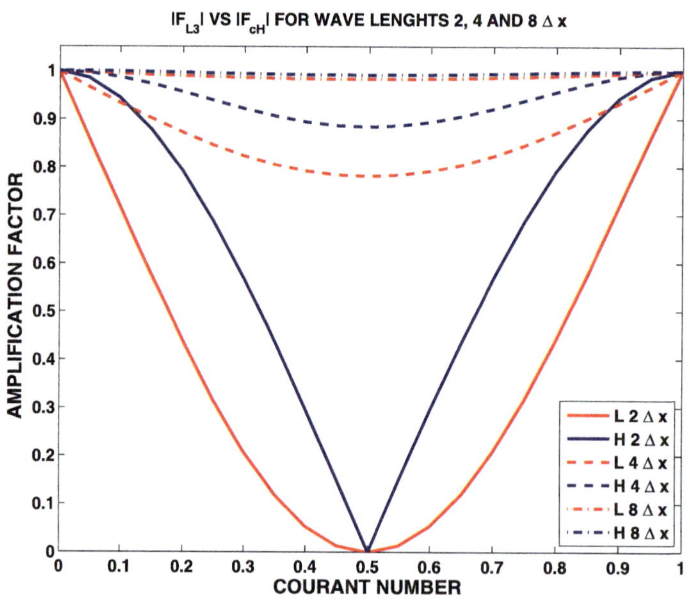

(B)

FIGURE 7.8 (A) Plot of the amplification factor against Courant number for the cubic Hermite and Lagrange interpolation schemes for different wavelengths, (B) plot of amplification factors by Courant number against the wavelength for the cubic Hermite and Lagrange interpolation scheme.

7.4 Stability analysis of the cubic spline semi-Lagrangian interpolation scheme

The stability for the cubic spline interpolation polynomial for the semi-Lagrangian advection problem is stated in [164]. The derivation of the von Neumann stability for the cubic spline starts by writing the associated polynomial in terms of the shift operator as

$$\frac{1}{6}\psi_{i-1}^{n+1} + \frac{2}{3}\psi_i^{n+1} + \frac{1}{6}\psi_{i+1}^{n+1} = \mathsf{S}^{-p}\left[\mathsf{S}^{-2}c_{-2} + \mathsf{S}^{-1}c_{-1} + c_0 + \mathsf{S}^{+1}c_1\right],$$

where the right-hand side of the equation above is at time n and would be our evolution operator for the cubic spline. Introducing the Fourier node $\psi_i^n = a_n e^{ik\varphi}$, where this time the time component is included into the Fourier node due to the implicit nature of the cubic spline, we have Fourier symbol as

$$\mathsf{F}_{cs}^n = \mathsf{E}_{cs}^n a_n e^{ij\phi} \equiv e^{-ip\varphi}\left(e^{-2i\varphi}c_{-2} + e^{-i\varphi}c_{-1} + c_0 + e^{i\varphi}c_1\right), \tag{7.46a}$$

$$\mathsf{F}_{cs}^{n+1} = a_{n+1}\left(\frac{1}{6}e^{-i\varphi} + \frac{2}{3} + \frac{1}{6}e^{i\varphi}\right) \equiv a_{n+1}\left(\frac{2}{3} + \frac{1}{3}\cos\varphi\right). \tag{7.46b}$$

Equating the two expressions in (7.46a) and (7.46b) and rearranging to have the time components on the left-hand side results in

$$\frac{a_{n+1}}{a_n} = \frac{c_{-2}e^{-2i\varphi} + c_{-1}e^{-i\varphi} + c_0 + c_1 e^{i\varphi}}{\frac{2}{3} + \frac{1}{3}\cos\varphi}. \tag{7.47}$$

The expression in (7.47) is referred to as the *symbol of the scheme* in [164], which is equivalent to our Fourier symbol terminology used for the other presented interpolation based semi-Lagrangian advection schemes.

To determine the amplification factor for the cubic spline based semi-Lagrangian advection, we need to take the modulus of (7.47). If we now expand the c_i coefficients in (7.47) and collect the factors for each of the powers of α, then we have

$$\alpha^0 : \frac{1}{6}\left(4 + e^{i\varphi} + e^{i\varphi}\right) \equiv \frac{1}{3}\left(2 + \cos\varphi\right), \tag{7.48a}$$

$$\alpha^1 : \frac{1}{6}\left(-3e^{-i\varphi} + 3e^{i\varphi}\right) \equiv -i\sin\varphi, \tag{7.48b}$$

$$\alpha^2 : \frac{1}{6}\left(-6 + 3e^{i\varphi} + 3e^{-i\varphi}\right) \equiv -\left(1 - \cos\varphi\right), \tag{7.48c}$$

$$\alpha^3 : \frac{1}{6}\left(3 - e^{i\varphi} - 3e^{-i\varphi} + e^{-2i\varphi}\right) \equiv \frac{1}{3}\left(\left(1 - \cos\varphi\right)^2 + i\left(1 - \cos\varphi\right)\sin\varphi\right). \tag{7.48d}$$

Combining the real and imaginary components from (7.48a)–(7.48d) results in

$$\mathrm{Re}\left(\mathsf{F}_{cs}\right) = \frac{1}{3}\left(2 + \cos\varphi\right) - \left(1 - \cos\varphi\right)\alpha^2 + 2\left(1 - \cos\varphi\right)^2\alpha^3, \tag{7.49a}$$

$$\mathrm{Im}\left(\mathsf{F}_{cs}\right) = -\sin\varphi + 2\left(1 - \cos\varphi\right)\sin\varphi, \tag{7.49b}$$

which we can see is slightly different compared to the real and imaginary components of the Fourier symbol for the cubic Lagrange interpolation polynomial. To obtain the amplification factor, we need to take the modulus of (7.49a) with (7.49b) divided by (7.46b); however, we should note that the first term in (7.49a) is the same as that in (7.46b). The modulus of a quotient of complex numbers is equal to the quotient of the moduli, therefore for the amplification factor for the cubic spline is given by

$$|\mathsf{F}_{cs}| \equiv \left| \frac{\mathrm{Re}\left(\mathsf{F}_{cs}\right) + i\mathrm{Im}\left(\mathsf{F}_{cs}\right)}{\frac{2}{3} + \frac{1}{3}\cos\varphi} \right| \equiv \frac{\left|\mathrm{Re}\left(\mathsf{F}_{cs}\right) + i\mathrm{Im}\left(\mathsf{F}_{cs}\right)\right|}{\left|\frac{2}{3} + \frac{1}{3}\cos\varphi\right|}. \tag{7.50}$$

It is left as an exercise to finish the derivation of the amplification factor for a cubic spline based semi-Lagrangian advection scheme, but there is a detailed analysis of the properties of the cubic splines in [164], and we shall provide a brief summary of the findings for this scheme.

We have already seen that the cubic Lagrange interpolation scheme completely damps waves that are of wavelength $2\Delta x$ when the rational component of the Courant number, α, is equal to 0.5. In [164] the authors reiterate this finding from [117] but compare it to the properties of the cubic spline approach. We have recreated Fig. 5A from [164] in Fig. 7.9, and also recreated Fig. 4A from that paper to illustrate the differences in the amplification factors for the cubic spline and cubic Lagrange interpolation. As mentioned in [164], the cubic spline's amplification factor when $\alpha = 0.5$ appears to not have as much of a damping effect on the shorter wavelengths compared to the cubic Lagrange interpolation approach. It is clear from Fig. 7.9B that the cubic spline approach has a very sharp decline in its amplification factor to the complete damping of the wave, while the cubic Lagrange interpolation based scheme covers a larger range of wavelengths that are damped in some form.

To finish the stability section of the semi-Lagrangian schemes, we have sets of figures associated with amplification factors of the introduced in this chapter schemes. We have amplification factors for the linear, quadratic, and cubic Lagrange interpolations, along with the cubic Hermite polynomial and the cubic spline, for wavenumbers $2\Delta x$, $4\Delta x$, and $8\Delta x$ in Fig. 7.10, A, B, and C, respectively. Taking the smallest wavenumber case first, it is clear that the Lagrange interpolation formulas do not perform as well as the cubic Hermite and spline polynomials, and while it might appear that one of the schemes is missing, this is not true. For the $2\Delta x$ case, it appears that the cubic Hermite and cubic spline based interpolation schemes have the same amplification factor.

When we consider wavenumber $k = 4\Delta x$, we see that all of the schemes have reduced the damping effects from wavenumber $2\Delta x$, but the linear scheme still has a strong damping effect. We should also notice that the scale on the y axis now starts at 0.5 and not 0; the reason for this scale is to highlight the difference between the other schemes, where we can clearly see that the Hermite and spline approaches have drastically smaller damping effects than any of the Lagrange interpolations that we have considered. However, we should also note that the damping effect associated with the cubic spline approach is still a lot smaller than that with the Hermite polynomial.

Finally, in Fig. 7.10C, which is for wavenumber $k = 8\Delta x$, we have again reduced the scale on the y axis so as to highlight the differences between the difference schemes. Therefore, we

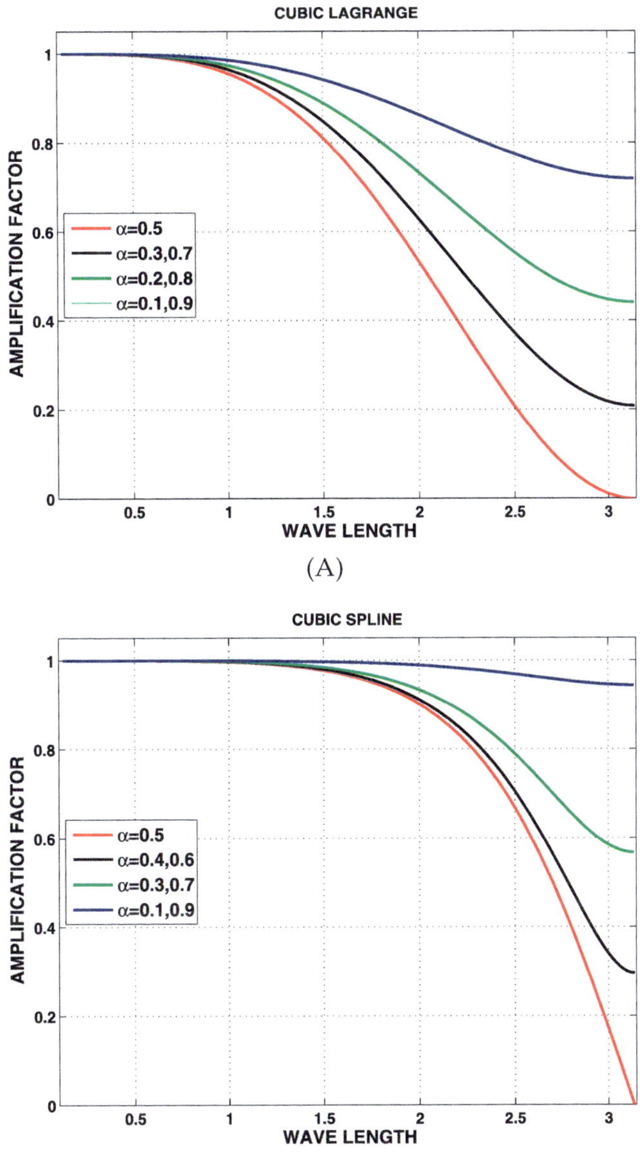

FIGURE 7.9 Recreation of the stability plot from Riishøjgaard et al. (1998) where (A) is for the cubic Lagrange interpolation and (B) is for the cubic spline interpolation.

can clearly see that nearly all of the schemes, except for the linear Lagrange polynomial case, which is still not that bad, have nearly eliminated their damping effects. However, as with the other cases, we see that the cubic spline approach still appears to be best out of the methods that we have introduced.

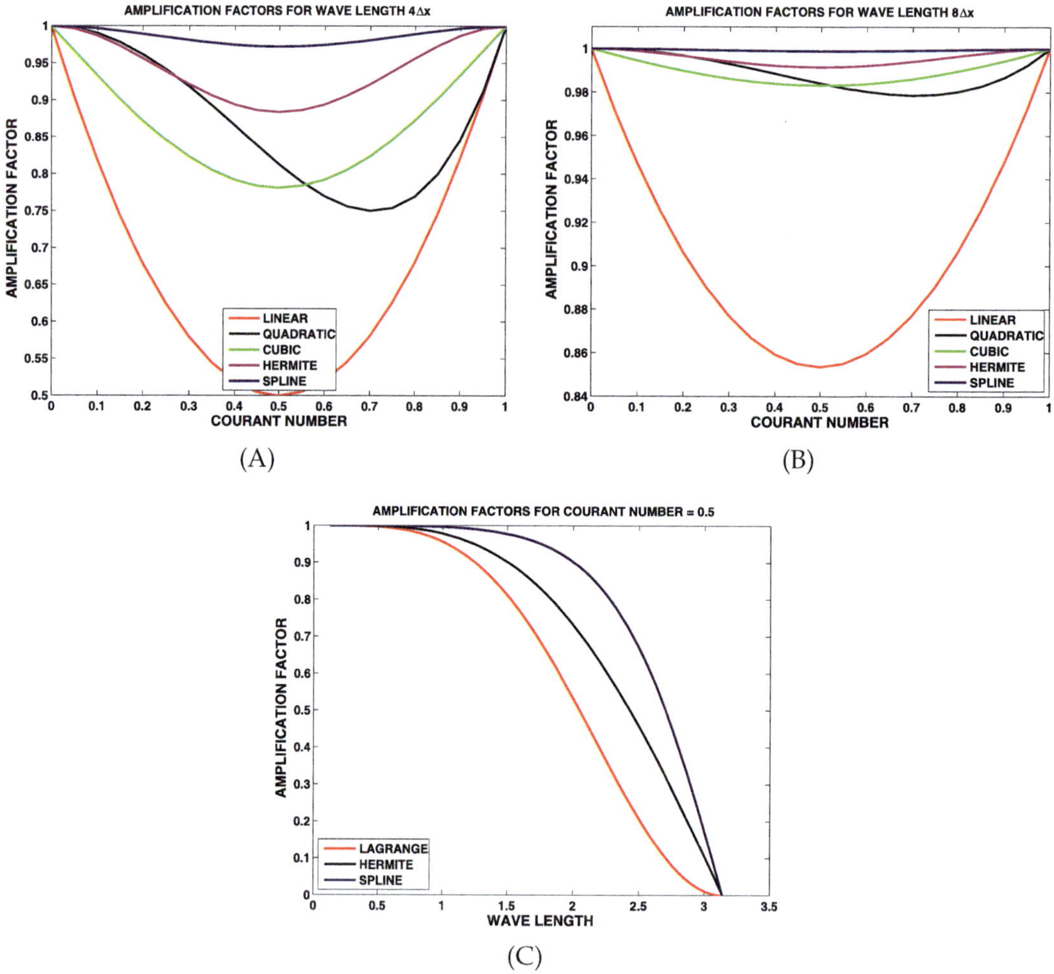

FIGURE 7.10 Amplification factors for the linear, quadratic, and cubic Lagrange polynomials, along with cubic Hermite and the cubic spline for (A) $k = 2\Delta x$, (B) $k = 4\Delta x$, and (C) $k = 8\Delta x$.

In Fig. 7.11 we have plotted the amplification factors by wavelength for the three cubic interpolation polynomials for $\alpha = 0.5$. As we have seen earlier, all three schemes damp out the features of wavelength π, but what is interesting is how each scheme approaches that wavelength. We can see that the most damping scheme is the cubic Lagrange, followed by the cubic Hermite, and then by the cubic spline approach. This indicates that as we introduce higher order constraints on the interpolation polynomial, the damping effects are reduced, but not eliminated.

However, just because we know that the semi-Lagrangian schemes presented here are unconditionally stable and can take larger time steps than the Eulerian schemes, we still have to

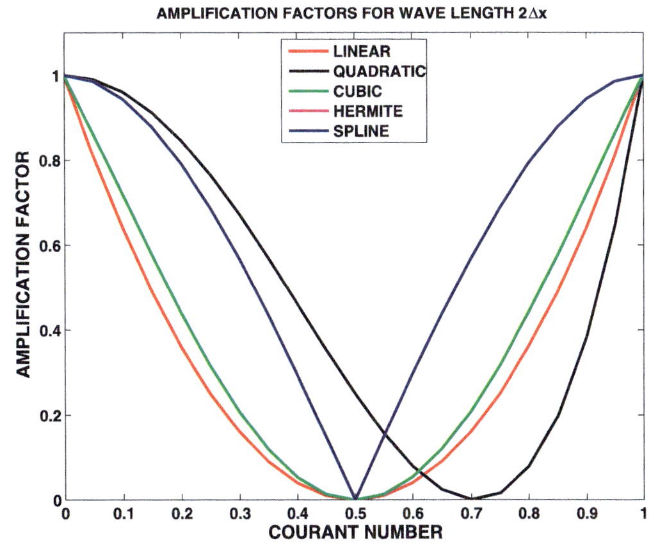

FIGURE 7.11 Plot of the amplification factors for the three cubic schemes against wavenumber.

address how accurate these schemes are. We therefore now move on to consistency analysis of semi-Lagrangian schemes.

7.5 Consistency analysis of semi-Lagrangian schemes

As we saw with the finite difference approximation to Eulerian formulation of the one-dimensional advection equation, we need to assess whether or not the approximations presented are consistent with the original partial differential equation, and if they are, how accurate are the semi-Lagrangian approximation schemes. We shall only consider the consistency analysis for the quadratic Lagrange interpolation polynomial here as the techniques are extendable to higher order Lagrange interpolation polynomials and to the other types of interpolation that we have mentioned in this chapter.

Recall the equation for the quadratic Lagrange interpolation semi-Lagrangian scheme

$$\psi_i^{n+1} = \frac{1}{2}\alpha\left(1+\alpha\right)\psi_{i-p-1}^n + \left(1-\alpha\right)\left(1+\alpha\right)\psi_{i-p}^n - \frac{1}{2}\alpha\left(1-\alpha\right)\psi_{i-p+1}^n, \tag{7.51}$$

where we have restricted α so that $-\frac{1}{2} \leq \alpha \leq \frac{1}{2}$. We have already shown that a quadratic scheme in this form is unconditionally stable for all choices of Δt and Δx, but we have not addressed the question of how accurate the scheme is.

To answer this question, we will need to show first that the scheme is consistent with the Lagrangian/Eulerian advection differential equation and then determine the order of the remainder term. As with the Eulerian based schemes, we consider a Taylor series expansion of the tracer field ψ about the arrival point but at time n, not $n+1$. Therefore, we have to expand all three terms in the quadratic interpolation formula with respect to x_i, which yields for each term

$$\psi_{i-p-1}^{n} \approx \left(\psi_i - (p+1)\,\Delta x \left[\frac{\partial \psi}{\partial x}\right]_i^n + \frac{(p+1)^2}{2}(\Delta x)^2 \left[\frac{\partial^2 \psi}{\partial x^2}\right]_i^n + \frac{(-(p+1))^3}{3!}(\Delta x)^3 \left[\frac{\partial^3 \psi}{\partial x^3}\right]_i^n\right), \tag{7.52a}$$

$$\psi_{i-p}^{n} \approx \left(\psi_i - (p)\,\Delta x \left[\frac{\partial \psi}{\partial x}\right]_i^n + \frac{(p)^2}{2}(\Delta x)^2 \left[\frac{\partial^2 \psi}{\partial x^2}\right]_i^n + \frac{(-(p))^3}{3!}(\Delta x)^3 \left[\frac{\partial^3 \psi}{\partial x^3}\right]_i^n\right), \tag{7.52b}$$

$$\psi_{i-p+1}^{n} \approx \left(\psi_i + (-p+1)\,\Delta x \left[\frac{\partial \psi}{\partial x}\right]_i^n + \frac{(-p+1)^2}{2}(\Delta x)^2 \left[\frac{\partial^2 \psi}{\partial x^2}\right]_i^n + \frac{((-p+1))^3}{3!}(\Delta x)^3 \left[\frac{\partial^3 \psi}{\partial x^3}\right]_i^n\right). \tag{7.52c}$$

Combining the Taylor series expansions in (7.52a)–(7.52c) with their associated factors in (7.51) results in a Taylor series expansion for the tracer at the departure point as

$$\psi(x, n\Delta t) = \psi_i^n - (p-\alpha)\,\Delta x \left[\frac{\partial \psi}{\partial x}\right]_i^n + (p+\alpha)^2 \frac{(\Delta x)^2}{2}\left[\frac{\partial^2 \psi}{\partial x^2}\right]_i^n + O\left(\Delta x^3\right), \tag{7.53}$$

where the leading third-order term in x is

$$O\left((\Delta x)^3\right) = \frac{(\Delta x)^3}{6}\left[\frac{\partial^3 \psi}{\partial x^3}\right]_i^n W_2^3,$$

with W_2^3 defined as

$$W_2^3(\alpha, p) \equiv \frac{1}{2}\alpha(1-\alpha)(-p-1)^3 + (1-\alpha)(1+\alpha)(-p)^3 - \frac{1}{2}\alpha(1-\alpha)(-p+1)^3; \tag{7.54}$$

for details see [117].

The next stage in the derivation of the consistency check is to note that with $x = \tilde{x} = i\Delta x - u\dfrac{\Delta t}{\Delta x}$, this then implies that $p + \alpha = u\dfrac{\Delta t}{\Delta x}$. We now consider the forward upwind approximation to the semi-Lagrangian form of advection as

$$\frac{\psi(i\Delta x, (n+1)\Delta t) - \psi(\tilde{x}, n\Delta t)}{\Delta t} = 0, \tag{7.55}$$

where substituting (7.53) evaluated at $p + \alpha = u\dfrac{\Delta t}{\Delta x}$ yields

$$\psi(\tilde{x}, n\Delta t) = \psi_i^n - u\Delta t \left[\frac{\partial \psi}{\partial x}\right]_i^n + u^2(\Delta t)^2 \left[\frac{\partial^2 \psi}{\partial x^2}\right]_i^n + O\left(\Delta t^3\right). \tag{7.56}$$

Next expanding the term $\psi\left(i\Delta x,(n+1)\Delta t\right)$ about $(i\Delta x, n\Delta t)$ in a Taylor series results in

$$\psi\left(x_i, t^{n+1}\right) \approx \psi_i^n + \Delta t\left[\frac{\partial \psi}{\partial t}\right]_i^n + \frac{\Delta t^2}{2}\left[\frac{\partial^2 \psi}{\partial t^2}\right]_i^n + O\left(\Delta t^3\right). \tag{7.57}$$

Substituting (7.56) and (7.57) into (7.55) results in

$$\begin{aligned}\frac{\psi\left(i\Delta x,(n+1)\Delta t\right)-\psi\left(\tilde{x}, n\Delta t\right)}{\Delta t} &= \left[\left(1+\frac{\Delta t}{2}\left(\frac{\partial}{\partial t}-u\frac{\partial}{\partial t}\right)\right)\left(\frac{\partial \psi}{\partial t}+u\frac{\partial \psi}{\partial x}\right)\right]_i^n \\ &+ O\left(\Delta t^2\right)+O\left(\frac{\Delta x^3}{\Delta t}\right)=0.\end{aligned} \tag{7.58}$$

While it may not appear obvious at first why Eq. (7.58) is written in this form, the reason is to illustrate that the first-order-in-time term contains the Eulerian form of the advection equation which, as we are considering the case of no forcing, is zero, and therefore the quadratic interpolation scheme is of order Δt^2 and $\frac{\Delta x^3}{\Delta t}$, where the leading third-order spatial term is given by

$$-\frac{\Delta x^3}{6\Delta t}\left[\frac{\partial^3 \psi}{\partial x^3}\right]_i^n W_2^3.$$

It is important that the behavior of this third-order term in space, which is divided by a first-order temporal term, be examined to ensure that the scheme does indeed converge to the differential equation as $\Delta x \to 0$ and $\Delta t \to 0$. There are three situations that can occur due to the spatial and temporal step sizes:

Case 1: The first situation that we consider is where the grid is refined such that the ratio $\frac{\Delta t}{\Delta x} \to 0$ as $\Delta t \to 0$ and $\Delta x \to 0$. Given this situation, we then have $p \to 0$ and, by association, the Courant number $\alpha \to u\frac{\Delta t}{\Delta x}$. In this situation the W_2^3 term would tend to $-u\frac{\Delta t}{\Delta x}$, and then the term that appears to be the ratio of the third-order space to first-order time would be reduced to $O\left(\Delta x^2\right)$.

Case 2: The second possible situation is where the grid is refined such that the ratio $\frac{\Delta t}{\Delta x}$ tends to a constant a, as Δt and Δx trend to zero. This then implies that $\Delta t = a\Delta x$, and as such $\frac{\Delta x}{\Delta t}W_2^3 \to aW_2^3\left(\alpha, \alpha - au\right)$, which is a constant, where again it appears that the term $O\left(\frac{\Delta x^3}{\Delta t}\right)$ is $O\left(\Delta x^2\right)$.

Case 3: The third, and final, possible situation with respect to the refining of the spacial and temporal grid is where $\frac{\Delta x}{\Delta t} \to 0$ as $\Delta t \to 0$ and $\Delta x \to 0$. Then we have that $|u|\frac{\Delta t}{\Delta x} \gg |\alpha|$ and $p \to u\frac{\Delta t}{\Delta x}$. In this case we have that $W_2^3 \to \left(u\frac{\Delta t}{\Delta x}\right)^3$ and the term which appears to be ratio of the third-order in space to first-order in time happens to be $O(\Delta t^2)$. Since the temporal step size is assumed to be much greater than the spatial step size, in this case we have that the quadratic interpolation results in $O(\Delta x^2)$ and $O(\Delta t^2)$.

Therefore, given the three scenarios described and analyzed above, we have shown that the quadratic Lagrange interpolation semi-Lagrangian approach is an $O(\Delta t^2)$ and $O(\Delta x^2)$ scheme. The analysis stated above comes from [120], where the motivation was to find a semi-Lagrangian scheme that had the desired rate of convergence, but also possessed a higher order accuracy than the equivalent Eulerian approach.

In [117] the authors state a general formula for the truncation error for the Lagrange based interpolation polynomials for the semi-Lagrangian advection case as

$$\frac{\Delta \psi}{\Delta t} = D_r \psi_p^n + \sum_{i=r+1}^{\infty} \frac{1}{i!} \left[\Delta t^{i-1} \left\{ \frac{\partial^i \psi}{\partial t^i} \right\}_p^n - \frac{\Delta^p}{\Delta t} \left\{ \frac{\partial i \psi}{\partial x^i} \right\}_p^n W_r^i(\alpha, p) \right], \tag{7.59}$$

where in the appendix of [117] there are expressions for D_r and $W_r(\alpha, p)$, namely

$$D_1 \psi = \frac{\partial \psi}{\partial t} + u \frac{\partial \psi}{\partial x}, \tag{7.60a}$$

$$D_2 \psi = D_1 \psi + \left(\frac{\partial^2 \psi}{\partial t^2} - u^2 \frac{\partial^2 \psi}{\partial x^2} \right) \frac{\Delta t}{2}, \tag{7.60b}$$

$$D_3 \psi = D_2 \psi + \left(\frac{\partial^3 \psi}{\partial t^3} + u^3 \frac{\partial^3 \psi}{\partial x^3} \right) \frac{(\Delta t)^2}{6}, \tag{7.60c}$$

where we can see that each term comprises the next order derivative version of the wave equation, and as such $\frac{D\psi}{Dt}$, which is equal to zero. Thus $D_r \psi = 0$, for $r = 1, 2, 3, \ldots$.

The expression for the remainder term W_r is not as elegant as those for the differential operators above, but for the linear, quadratic, and cubic Lagrange interpolation polynomials this term is

$$W_1(\alpha, p) = (1-\alpha)(-p)^I + \alpha(-p-1)^i, \tag{7.61a}$$

$$W_2(\alpha, p) = \frac{\alpha(1+\alpha)(-p-1)^i}{2} + (1-\alpha)(1+\alpha)(-p)^i - \frac{\alpha(1-\alpha)(-p+1)^i}{2}, \tag{7.61b}$$

$$\begin{aligned} W_3(\alpha, p) = & -\frac{\alpha\left(1-\alpha^2\right)(-p-2)^i}{6} + \frac{\alpha(1+\alpha)(2-\alpha)(-p-1)^i}{2} \\ & + \frac{\left(1-\alpha^2\right)(2-\alpha)(-p)^i}{2} - \frac{\alpha(1-\alpha)(2-\alpha)(-p+1)^i}{6}, \end{aligned} \tag{7.61c}$$

where we can see that the expressions above are associated with the grid points that are used to calculate the interpolation polynomial from.

Exercise 7.4. *Show that the cubic Lagrange interpolation based semi-Lagrange scheme is consistent with the advection equation and show that it is equivalent to* $O(\Delta x^4)$.

Exercise 7.5. *Determine the order of accuracy of the cubic Lagrange interpolation semi-Lagrangian scheme.*

Exercise 7.6. *Show that the central difference approximation for the derivative at the endpoints with the Hermite interpolation polynomial is consistent with the semi-Lagrangian advection differential equation.*

Exercise 7.7. *Determine the order of accuracy of the central difference based Hermite interpolation.*

Exercise 7.8. *Show that the cubic spline interpolation provides a consistent approximation to the advection equation.*

Exercise 7.9. *Determine the order of accuracy of the cubic spline approximation to the Lagrangian advection equation.*

7.6 Summary

In this chapter we have introduced the changes to the von Neumann stability analysis that we performed on the Eulerian based finite difference schemes to take into account that in the semi-Lagrangian formulation we are able to allow the trajectory to travel several grid points back in a single time step and have shown that there is no restriction on the size of the time and spatial step sizes for these methods, that is to say, they are **unconditionally stable**. We have shown that some of the schemes can cause damping effects for certain wavelengths and that the higher order polynomial can produce spurious oscillations. We have performed the stability analysis on three different types of interpolation polynomial, and have shown that it is possible to damp these oscillations but at the cost of introducing an implicit numerical interpolation which implies a matrix to invert at each time step. We have also provided the relative frequency or phase error analysis for the linear, quadratic, and cubic Lagrange interpolation polynomials and showed that there was no real reduction in phase error between the linear and the quadratic case, but there was an improvement when using a cubic approximation.

We have also introduced consistency analysis for the quadratic Lagrange interpolation polynomial and showed that this approach makes the scheme second order in time and space. We have also provided the expression for the truncation error for the Lagrange interpolation polynomial based semi-Lagrangian advection case from [117]. We recall at this point that we have only considered the case with zero forcing and constant velocity, where it may be the case that the actual problem to be solved for has both a linear and nonlinear forcing, as well as a possible linear or nonlinear advection velocity. We shall consider what happens to the semi-Lagrangian schemes and present different techniques that have been used to address these features. In the next chapter we consider semi-Lagrangian methods with nonconstant velocities.

CHAPTER

8

Advection with nonconstant velocities

Throughout all the previous chapters, for both the Eulerian and semi-Lagrangian formulations, we have been considering the case where the advection velocity has been constant with respect to both time and space. However, in the more advanced and realistic modeling of geophysical flows that have a pure advection component to them, it is highly likely that the advection velocity, u, is not a constant throughout the whole spatial domain, or for all time. This is important as we need to obtain a quite accurate estimate of the departure point to ensure that the interpolation of the values of the field to this point is the most accurate estimate for the field of interest at the arrival time.

We will see in later chapters that there are several real-world applications of a form of semi-Lagrangian advection where the field that is being advected is advected by itself, that is to say, the advection velocity is also the field that is being advected. This form of nonconstant velocity advection is referred to as **nonlinear** advection, and we shall introduce it and apply a semi-Lagrangian approximation to it later in this chapter. However, given this nonlinear behavior, we have to take into account what is referred to as nonlinear instability, which is where waves are introduced due to the nonlinear structure of the advection equation that may not be resolved by the numerical grid; this is also referred to as **aliasing**. We start this chapter by considering linear advection with a nonconstant advection velocity.

8.1 Semi-Lagrangian approaches for linear nonconstant advection velocity

In this section we shall present the summaries of the early attempts to deal with the case in a semi-Lagrangian formulation of the nonconstant advection velocity. We start by considering Eulerian formulation of the nonconstant velocity advection equation, which is given by

$$\frac{\partial \psi (x,t)}{\partial t} + u(x,t)\frac{\partial \psi}{\partial x} = 0, \tag{8.1}$$

where $u(x,t)$ is a given function of space and time. We should also note here that the field is still conserved following the parcel of fluid. This then implies that the semi-Lagrangian finite difference approach is still valid, but now the departure point is obtained through solving the

https://doi.org/10.1016/B978-0-12-817222-3.00012-8

equation

$$\frac{dx}{dt} = u(x,t). \tag{8.2}$$

It is possible to integrate (8.2) to obtain

$$x(t+\Delta t) - x_d(t) = \int_t^{t+\Delta t} u(x(\omega),\omega)\,d\omega, \tag{8.3}$$

where ω is a dummy time variable. When u is constant, we have the situation that the right-hand side of (8.3) becomes

$$x(t+\Delta t) - x_d(t) = (t+\Delta t)u - tu = (t+\Delta t - t)u = u\Delta t, \tag{8.4}$$

which, upon rearranging, yields $x_d(t) = x(t+\Delta t) - u\Delta t$, namely the expression that we were using earlier. However, for the situation when u is not constant, we have to apply some form of numerical integration technique to (8.3), or make an assumption about the behavior of the velocity at certain time levels along the trajectory. We shall now present a first-order approach that has been proposed to tackle this problem.

The expressions given in (8.3) is a highly implicit system that must be solved to obtain an estimate of the departure point, $x(t) = \tilde{x}$, assuming that the arrival point, $x(t+\Delta t)$, has a known location at $i\Delta x$. The following consideration comes from [120].

First consider the case where we assume the simplest possible approach, which is that the advection velocity $u(x(\omega),\omega)$ is a constant $u_i^{n+\frac{1}{2}}$. Thus from (8.3) we have that $\tilde{x_d} = x_i^{n+1} - \Delta t u_i^{n+\frac{1}{2}}$. Substituting the new expression for the departure point into the expression for the Courant number presented earlier yields $p + \alpha = u_i^{n+\frac{1}{2}}\frac{\Delta t}{\Delta x}$.

The next step is to recall the Taylor series expansion of the tracer at the arrival point's equivalent at time $t = t_n$, which is given by

$$\psi(x, n\Delta t) = \psi_i^n - (p+\alpha)\Delta x\frac{\partial \psi_i^n}{\partial x} + (p+\alpha)^2\frac{(\Delta x)^2}{2}\frac{\partial^2 \psi_i^n}{\partial x^2} + O\left((\Delta x)^3\right). \tag{8.5}$$

Substituting our new expression for $p + \alpha$ above into (8.5), along with assuming that we shall use a quadratic Lagrange interpolation formula for the estimation of the value of the tracer at the departure point, enables the value of the tracer at the departure point to be expanded in a Taylor series about $(i\Delta x, n\Delta t)$ [121]. Upon substituting all this information into the semi-Lagrangian forward upwind expression for the time derivative results in a scheme that is first order, and as such this estimate is not accurate enough according to [120].

8.1.1 Numerical integration

While we have not presented a diagram to show the following statement, when we are considering the constant advection velocity case, we have that the grid for the departure points is a regular grid that is offset from the main numerical model grid. When we allow for a nonconstant advection velocity, we have that the deformation of the grid of departure points is not uniform, and also dependent on the advection velocity.

As we have seen in the introduction to this chapter, the backward trajectory calculation is no longer trivial, and will involve complicated numerical integration. However, for us to be able to determine different properties of the possible approaches that we present in this section, we consider the truncation error for the general expression for the semi-Lagrangian equation which is obtained by considering the Taylor series expansion of the field ψ about the **estimated**, the meaning of which we shall explain soon, departure point, which has to be substituted into an expression of the form

$$\tau = \frac{1}{\Delta t}\left(\psi_i^{n+1} - \psi_d^n\right) + \frac{1}{\Delta t}\left(\psi_d^n - \sum_{k=r}^{s} f_i\left(\alpha_j\right)\psi_{i-p-k}^n\right), \tag{8.6}$$

where we have added and subtracted ψ_d^n in the truncation error definition given earlier to arrive at (8.6). This ψ_d^n is the **true departure point**, r represents the number points downstream of our estimate of the departure point, and s is the order of the interpolation polynomial that is being employed to find the value of ψ at the *departure point*, α is calculated as before, but now not the same for each grid point as u may not be the same at each grid point, and so the departure point will not be the same distance from the nearest point for each arrival point, and $f_r\left(\alpha_j\right)$ is the function of the interpolation weights for the chosen interpolation polynomial. The expression in (8.6) comes from [47].

According to [47], the expressions in (8.6) comprises two sets of error where the first term represents the error due to the calculation of the trajectory, while the second represents the error due to the interpolation to the departure point, which is the same as for the constant case. And we now explain what we mean by estimated departure point. As the advection velocity is no longer constant with respect to the spatial domain, if we follow the same back trajectory calculation as before, then we may not arrive at the true departure point, and this introduces an error into the advection calculations.

A first approach that we could take, as presented in [47], is to apply a forward Euler finite difference to the differential equation, rather than the integral equation in (8.2), applied between the estimate of the departure and arrival points, but with the advection velocity evaluated at the arrival point equivalent at time $n+1$, which results in

$$\widetilde{x}_i^n = x_i^{n+1} - u\left(x_i^{n+1}, t^n\right)\Delta t. \tag{8.7}$$

Next if we define the difference between the arrival point and the estimate of the departure point as $\nu \equiv x_i^{n+1} - \widetilde{x}_i^n$, then we know that this difference is equivalent to the numerical approximation $\nu = u\left(x_i^{n+1}, t^n\right)\Delta t$. If we now apply a Taylor series expansion to the right-hand side of (8.7), then we obtain

$$\nu = \Delta t\left(u\left(\widetilde{x}_i^n, t^n\right) + \nu\frac{\partial u}{\partial x}\left(\widetilde{x}_i^n, t^n\right) + O\left(\nu^2\right)\right) = u_d\Delta t + O\left(\left(\Delta t\right)^2\right), \tag{8.8}$$

where $u_d \equiv u\left(\widetilde{x}_i^n, t^n\right)$ is our estimate of the advection wind at our estimate of the departure point. We now expand ψ_j^{n+1} in a Taylor series about the estimate of the departure point at t^n, recalling that we are expanding in both x and t, which yields

$$\psi_i^{n+1} = \psi_d + \Delta t \frac{\partial \psi_d}{\partial t} + \nu \frac{\partial \psi_d}{\partial x} + \frac{(\Delta t)}{2}\frac{\partial^2 \psi_d}{\partial x^2} + \nu \Delta t \frac{\partial^2 \psi_d}{\partial x \partial t} + \frac{\nu^2}{2}\frac{\partial^2 \psi_d}{\partial x^2} + \cdots, \tag{8.9}$$

where $\psi_d \equiv \psi\left(\tilde{x}_i^n, t^n\right)$. Now recalling that $\nu = u_d \Delta t$, we can rearrange (8.9) to obtain

$$\psi_j^{n+1} = \psi_d + \Delta t \left(\frac{\partial}{\partial t} + u_d \frac{\partial}{\partial x}\right)\psi_d + \frac{(\Delta t)^2}{2}\left(\frac{\partial}{\partial t} + u_d \frac{\partial}{\partial x}\right)^2 \psi_d + \cdots, \tag{8.10}$$

which enables us to show that the estimate for the value at the arrival point has order Δt, and thus the contribution to the global error in (8.6) from the trajectory calculation is first order in time.

Given this disappointing setback, we consider an approach that was suggested in [161] where an iterative approach was used to solve the implicit relationship between the departure point and the advection velocity. The basic idea of the iterative approach is to try and obtain the best estimate of the advecting velocity at the center of the trajectory as possible, and then approximate (8.3) by

$$x_d = x_i^{n+1} - \Delta t u\left(x\left(t + \frac{\Delta t}{2}\right), t + \frac{\Delta t}{2}\right),$$

which enables us obtain a second order accuracy in time. The estimate of $x\left(t + \frac{\Delta t}{2}\right)$ is taken as half of the position of the arrival point plus the previous best estimate of the departure point position, which is given by

$$x^{(k)}\left(t + \frac{\Delta t}{2}\right) = \frac{1}{2}\left(x\left(t + \Delta t\right) + x^{(k)}\left(t\right)\right), \tag{8.11}$$

and thus we can approximate (8.3) to obtain the next estimate of $x^{(k)}$ as

$$x^{(k+1)}\left(t\right) = x\left(t + \Delta t\right) - \Delta t u\left(x^{(k)}\left(t + \frac{\Delta t}{2}\right), t + \frac{\Delta t}{2}\right). \tag{8.12}$$

This process is continued until we reach convergence.

Once the scheme has converged, and the best estimate of the departure point has been obtained, it is then possible to evaluate the interpolation polynomial for a best estimate of the tracer at that point. An interesting twist in the case of the iterative approach, suggested in [161] and then expanded upon in [120], is the situation where it is possible to obtain the same second-order approximation to the advection equation. This accuracy arises from considering the first guess for the departure point, which initially is

$$\tilde{x}^{(0)} = x^{(0)}\left(t\right) = i\Delta x - \Delta t u_i^{n+\frac{1}{2}},$$

and using (8.11) and (8.12) results in the first-order estimate of the departure point as

$$\tilde{x}^{(1)} = i\Delta x - \Delta t u\left(i\Delta x - \frac{\Delta t}{2}u_i^{n+\frac{1}{2}}, \left(n + \frac{1}{2}\right)\Delta t\right). \tag{8.13}$$

However, in the great work by McDonald throughout his papers from the 1980s, 1990s, and the early 2000s, he suggests to use a linear Lagrange interpolating to compute the first guess of $x^{(1)}$ in (8.13) as

$$u\left(i\Delta x - u_i^{n+\frac{1}{2}}\frac{\Delta t}{2}, \left(n+\frac{1}{2}\right)\Delta t\right) = \left(1-\hat{\gamma}\right)u_{i-m}^{n+\frac{1}{2}} + \hat{\gamma}u_{i-m-1}^{n+\frac{1}{2}}, \tag{8.14}$$

where $\hat{\gamma} = -m + \dfrac{\Delta t u_i^{n\frac{1}{2}}}{2\Delta x}$ and m is an integer that is chosen such that $0 < \hat{\gamma} < 1$ as a good estimate of the wind field at the departure point. See [120] for the proof that this approach yields a second-order scheme.

To be able to increase the accuracy in time of the backward trajectory, we need to include more time levels, but here we have the problem of where should the value of the advection wind be used as it is a function of time and space as mentioned earlier. The first technique to address this problem was suggested by André Robert in 1981 [161], and we shall summarize that approach in the next section, along with other approximations that have been developed and are referred to as **two and three time level schemes**.

8.2 Two and three time level schemes

As just mentioned, the first attempt to include a higher order accuracy for the trajectory calculations in time was suggested by Robert in 1981 [161], where he was interested in applying a semi-Lagrangian approach to the vorticity equation in two dimensions with a forcing term. We shall not consider that case here, but will take note of the fact that the equivalent one-dimensional problem considered here can be expressed in his notation as

$$\psi\left(x, t+\Delta t\right) = \psi\left(x-2a, t-\Delta t\right), \tag{8.15}$$

where

$$a = \Delta t u\left(x-a, t\right). \tag{8.16}$$

The important points to note here are as follows: (1) we are considering a second-order central difference to the differential equation $\dfrac{D\psi}{Dt}$, (2) the departure point is not dependent on the estimate of the velocity at the current time, but at a halfway point from the departure point at time $t = t - \Delta t \equiv t^{n-1}$, and (3) the relationship in (8.16) is implicit and, as noted in [161], will have to be solved iteratively.

A good explanation of what is going on here can be found in the summary paper of semi-Lagrangian approaches at the end of the 1980s by Staniforth and Côté in 1991 [176]. In [176] a schematic of the situation being presented here is given, and we have a recreation of that schematic in Fig. 8.1.

In Fig. 8.1 we have that the exact trajectory in the space–time (x, t) plane of the fluid particle that arrives at the grid point x_m at time $t^n + \Delta t$ is denoted by the red curve between the points A and C, and the approximate straight line trajectory is denoted by the black dashed line

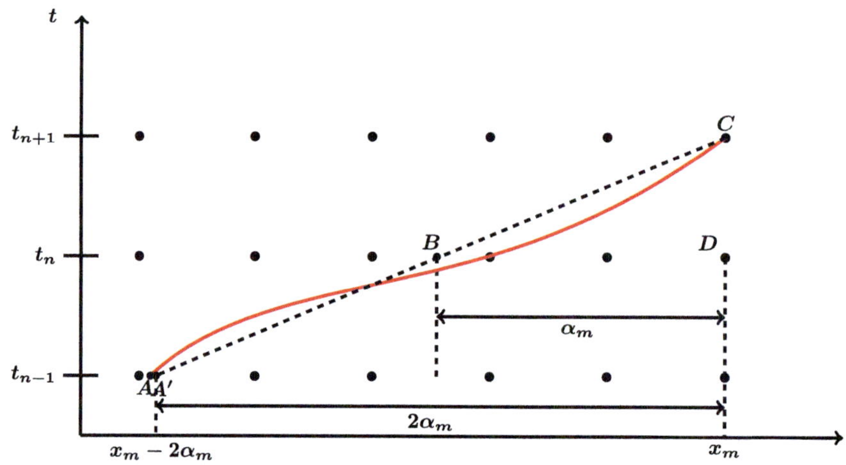

FIGURE 8.1 Recreation of the three time level advection schematic from Staniforth and Côté (1991).

between the points A' and C. It is now assumed that the values for $\psi(x,t)$ are known at the grid points, x_m, at times $t^n - \Delta t$ and at t^n, and so we are seeking the values at the same grid points at time $t^n + \Delta t$. It is stated in [176] that the essence of semi-Lagrangian advection is to approximately integrate $\frac{D\psi}{Dt} = 0$ along the approximated fluid trajectory $A'C$. Thus we have, as stated in [161],

$$\frac{\psi(x_m, t^n + \Delta t) - \psi(x_m - 2\alpha_m, t^n - \Delta t)}{2\Delta t} = 0, \tag{8.17}$$

where α_m is the distance BD in Fig. 8.1, which represents the distance that the particle has traveled in time Δt when following the approximate space time trajectory $A'C$. Thus, if we know α_m, then the value of ψ at the arrival point x_m at time $t^n + \Delta t$ is just its value at the upstream point $x_m - 2\alpha_m$ at time $t^n - \Delta t$. However, α_m has not yet been determined, and if we had determined its value, then we still would have the fact that we only know ψ at the grid points.

To determine the value of α_m, we note that the value of the advection velocity at the point B in Fig. 8.1 is the inverse of the slope of the straight line $A'C$, and is obtained through the approximation to

$$\frac{dx}{dt} = u(x,t)$$

as

$$\alpha_m = \Delta t u\left(x_m - \alpha_m, t^n\right). \tag{8.18}$$

As mentioned earlier, the relationship above has to be solved iteratively in the form

$$\alpha_m^{(k+1)} = \Delta t u\left(x_m - \alpha^{(k)}, t^n\right), \tag{8.19}$$

where we have some initial guess for $\alpha_m^{(0)}$, provided that u can be evaluated between grid points. To be available to evaluate u between grid points, we would have to employ some form of interpolation as we have to do with ψ, but it may not have to be the same order interpolation polynomials for the tracer and the advection velocity. We will add a caveat here: it has recently been shown to maybe not be the case for high resolution modeling of the atmosphere. The scheme that has been presented here is referred to as a **three-time-level** scheme, as we are using information at two previous time levels of the numerical model to obtain the value at t^{n+1}.

Another formulation that has been suggested, which is a three time level approach, is presented in Temperton and Staniforth (1987) [188], and is of the form

$$u\left(x,t+\frac{\Delta t}{2}\right)=\frac{1}{8}\left(15u\left(x,t\right)-10u\left(x,t-\Delta t\right)+3u\left(x,t-2\Delta t\right)\right), \tag{8.20}$$

but here the value of the wind is extrapolated to the halftime step. It is stated in [188] that the trajectory calculation that used the expression in (8.20) gave more accurate results than that using the two level scheme presented above. However, later work showed that this may not be the case when we have larger time steps.

We have to recall that are we are approximating the integral equation

$$\frac{\partial x}{\partial t}=u\left(x,t\right)\Longrightarrow x_i^{n+1}-x_d^n=\int_t^{t+\Delta t}u\left(x,t\right)dt, \tag{8.21}$$

and once we have solved the equation above in some form, we have to interpolate the fields at the surrounding grid points to the departure point. A very commonly used approximation in numerical weather prediction to solve the trajectory equation above is the midpoint rule, which is given by

$$x_i^{n+1}-x_d^n=\Delta t u\left(\frac{x_i^n+x_d^n}{2},t+\frac{\Delta t}{2}\right). \tag{8.22}$$

We have a replica of the schematic from Staniforth and Côté (1991) [176], to illustrate the difference between the three and two time level approximations in Fig. 8.2. We have changed the figure slightly in that we have changed the color of the points at the halftime level to indicate that these are calculated and are not known in advance, and as such these are **estimates** of the variables and fields at this *time level.*

This is the first time that we have introduced the midpoint formulation for the time component, as we are aware that we do not have values at $t=t_n+\frac{\Delta t}{2}$, and can again see that the expression in (8.22) is implicit for the departure point that we are seeking. However, for us to start to achieve this, we need the velocity field value at the trajectory *midpoint*, and at time $t=t+\frac{\Delta t}{2}$ first. A commonly used technique for this situation is a time extrapolation, and interpolation of the extrapolated velocity field at the estimated midpoint. Thus a second-order extrapolation formula can be derived as

$$u^{n+\frac{1}{2}}=u^n+\frac{\Delta t}{2}\frac{\partial u^n}{\partial t}+O\left(\left(\Delta t\right)^2\right)=u^n+\frac{\Delta t}{2}\frac{u^n-u^{n-1}}{\Delta t}+O\left(\left(\Delta t\right)^2\right)\approx\frac{3}{2}u^n-\frac{1}{2}u^{n-1}, \tag{8.23}$$

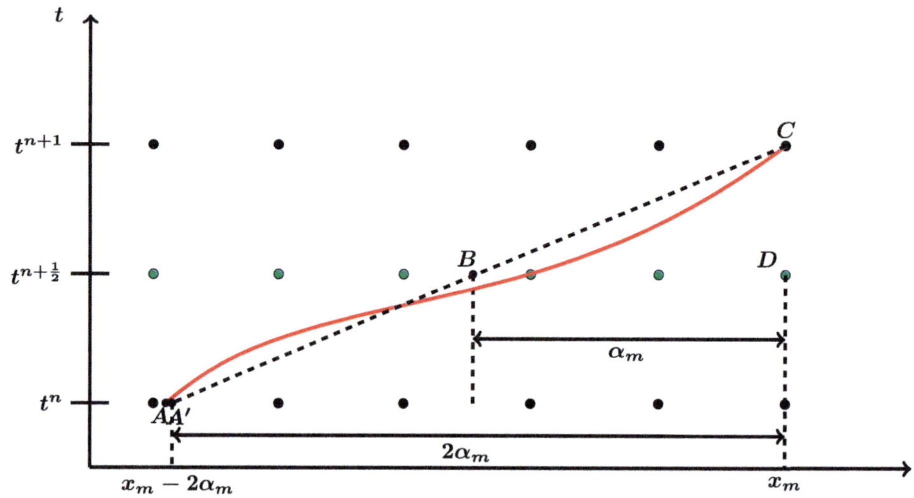

FIGURE 8.2 Replica of the figure from Staniforth and Côté (1991) illustrating the two-time-level approximation of the trajectory calculation.

where we can see that the right-hand side of (8.23) is already possible to be evaluated as we have the velocity field at the previous two time steps. Thus the procedure is to create the extrapolated velocity fields at the halfway time points and then apply a fixed point iteration algorithm to calculate the departure points as follows:

1. Initialize the first guess for the departure point as $x_d^{(0)} = x_i - \Delta t u^n$.
2. For $k = 1, \ldots, K$,
   (a) interpolate $u^{n+\frac{1}{2}}$ to the midpoint $x_m \equiv 0.5\left(x_i + x_d^{(k-1)}\right)$;
   (b) update the estimate for the departure point as $x_d^{(k)} = x_i - \Delta t u^{n+\frac{1}{2}}(x_m)$.

In McDonald (1999) [121], there is a series of *midpoint* schemes that are summarized from other approaches. To be able to compare these different schemes, we shall present the first two steps of the techniques from [121], which are given by

$$x_d^{(k+1)} = x_i^{n+1} - \Delta t u\left(\frac{1}{2}x_i^{n+1} + x_d^{(k)}, \left(n + \frac{1}{2}\right)\Delta t\right), \tag{8.24a}$$

$$x_d^{(1)} = x_i^{n+1} - \Delta t u_i^{n+\frac{1}{2}}, \tag{8.24b}$$

$$x_d^{(2)} = x_i^{n+1} - \Delta t u\left(x_i^{n+1} - u_i^{n+\frac{1}{2}}\frac{\Delta t}{2}, \left(n + \frac{1}{2}\right)\Delta t\right). \tag{8.24c}$$

Another approach that can be used to approximate the trajectory integration for the advecting wind is to consider a *trapezoidal rule* which is presented in both Bartello and Thomas (1996) [10] and in McDonald (1999) [121], where the departure point calculation is given by the implicit relationship

$$x_d^n = x_i^{n+1} - \frac{\Delta t}{2}\left(u\left(x_i^{n+1}\right) + u\left(x_d^n\right)\right). \tag{8.25}$$

We would have to solve (8.25) iteratively as for the midpoint approach, and these iterations would have to be applied to $u\left(x_d^n\right)$. We still have the problem of needing $u\left(x_i^{n+1}\right)$ to initiate the iteration. It is suggested in [10], and restated in [121], that the first step in this process would be to extrapolate the wind field to $t = t^{n+1}$ though the extrapolation formula given by

$$u_i^{n+1} = 2u_i^n - u_i^{n-1},$$

and then we have the equivalent iterations steps for this method given by

$$x_d^{(k+1)} = x_i^{n+1} - \frac{\Delta t}{2}\left(u_i^{n+1} + u\left(x_d^{(k)}, n\Delta t\right)\right), \tag{8.26a}$$

$$x_d^{(1)} = x_i^{n+1} - u_i^{n+1}\Delta t, \tag{8.26b}$$

$$x_d^{(2)} = x_i^{n+1} - \frac{\Delta t}{2}\left(u_i^{n+1} + u\left(x_i^{n+1} - u_i^{n+1}\Delta t, n\Delta t\right)\right). \tag{8.26c}$$

Another two integration approaches, referenced in [121], are referred to as approaches that are midpoint based of the general form

$$x_d^n = x_i^{n+1} - \Delta t u\left(x_d^{n+\frac{1}{2}}, t + \frac{\Delta t}{2}\right), \tag{8.27}$$

where they both use the same first estimate given by

$$x_d^{(1)} = x_i^{n+1} - u_i^n \Delta t, \tag{8.28}$$

but then apply different estimates for $u\left(x + \frac{\Delta t}{2}, t + \frac{\Delta t}{2}\right)$. The first of these approaches is from Temperton and Staniforth (1987) [188], and is given by

$$x_d^{(2)} = x_i^{n+1} - \frac{\Delta t}{2}\left(3u\left(x_i^{n+1} - u_i^n\Delta t, n\Delta t\right) - u\left(x_i^{n+1} - 2u_i^n\Delta t, (n-1)\Delta t\right)\right). \tag{8.29}$$

The other scheme that can be considered of a midpoint type approximation was suggested by Hortal in 1998 [81], and is defined by

$$x_d^{(2)} = x_i^{n+1} - \frac{\Delta t}{2}\left(u_i^n + 2u\left(x_i^{n+1} - u_i^n\Delta t, n\Delta t\right) - u\left(x_i^{n+1} - 2u_i^n\Delta t, (n-1)\Delta t\right)\right). \tag{8.30}$$

In [121] the two formulations in (8.29) and (8.30) are referred to as techniques, or methods, that are **extrapolating along the trajectory**. One important feature to note here is that both approaches are **three** time level schemes. An important difference between (8.30) and (8.29) is that (8.30) requires one less interpolation than (8.29).

The reason for considering these different approaches is that it can be shown through the truncation error calculations that all of these schemes have order of $O\left((\Delta t)^2\right)$. In McDonald (1999) it is stated that there are many different combination of arrival and departure point quantities that combine to yield $x_d^{(2)}$ to $O\left((\Delta t)^2\right)$ accuracy, and the author then goes on to state that any combination of the form

$$x_d^{(2)} = x_i^{n+1} - \Delta t \widehat{u}^{(1)}, \tag{8.31}$$

with

$$\begin{aligned} \widehat{u}_i^{(1)} &= au\left(x_i^{n+1}, n\Delta t\right) + cu\left(x_i^{n+1} - u_i^n \Delta t, n\Delta t\right) + eu\left(x_i^{n+1} - 2u_i^n \Delta t, n\Delta t\right) \\ &+ bu\left(x_i^{n+1}, (n-1)\Delta t\right) + du\left(x_i^{n+1} - u_i^n \Delta t, (n-1)\Delta t\right) \\ &+ fu\left(x_i^{n+1} - 2u_i^n \Delta t, (n-1)\Delta t\right), \end{aligned} \tag{8.32}$$

where if we have satisfied the following conditions

$$a = 1 + d + e + 2f, \tag{8.33a}$$

$$b = -\frac{1}{2} - d - f, \tag{8.33b}$$

$$c = \frac{1}{2} - d - 2e - 2f, \tag{8.33c}$$

then the associated scheme will be guaranteed to have second-order accuracy in time. As stated in [121], to show this result, we rewrite the semi-Lagrangian equation as

$$\psi\left(x_i^{n+1}, (n+1)\Delta t\right) - \psi\left(x_i^{n+1} - \Delta t \widehat{u}, n\Delta t\right) = 0, \tag{8.34}$$

and write a Taylor series of ψ about $\left(x_i^{n+1}, n\Delta t\right)$, giving a quadratic, or any higher-order, interpolation scheme

$$\left(\frac{\partial \psi}{\partial t}\right)_i^n + \widehat{u}\left(\frac{\partial \psi}{\partial x}\right)_i^n + \frac{\Delta t}{2}\left(\left(\frac{\partial^2 \psi}{\partial t^2}\right)_i^n - \widehat{u}^2\left(\frac{\partial^2 \psi}{\partial x^2}\right)_i^n\right) + O\left((\Delta t)^2\right) = 0. \tag{8.35}$$

Now if we have that

$$\widehat{u} = u_i^n + \frac{\Delta t}{2}\left(\frac{\partial u}{\partial t} - u\frac{\partial u}{\partial x}\right)_i^n, \tag{8.36}$$

then the scheme becomes second-order accurate in time, where this can be shown through substituting (8.36) into (8.35), which results in

$$\left[\left\{1 + \frac{\Delta t}{2}\left(\frac{\partial}{\partial t} - u\frac{\partial}{\partial x}\right)\right\}\left(\frac{\partial \psi}{\partial t} + u\frac{\partial \psi}{\partial x}\right)\right]_i^n + O\left((\Delta t)^2\right) = 0. \tag{8.37}$$

There has been much research into what order interpolation is best suited to achieve the value of the wind field at the midpoint; and what might be a surprise to the reader is that the linear interpolation is the most common used scheme here. It is stated in [176] that the ***order of the interpolation for step 1 is much less important, than the ones used to find the value of the field at the departure point.*** As mentioned above, this was true at the time, given the resolution of the numerical models.

In his initial work, McDonald in 1987 [119] states that one should use an interpolation polynomial of an order less than that being used for the field interpolation to the departure point. It is stated in [176] that in the work by Staniforth and Pudykiewicz in 1985 [177], Temperton and Staniforth in 1987 [188], and Bates et al. in 1990 [16], it is sufficient to use linear interpolation for the displacements, when used in conjunction with a cubic interpolation of the field ψ. Is also stated in [176] that it has been shown that there is no advantage in using more than two iterations for step 1. Again this statement was true at the time due to the features that could be resolved, given the resolution of the models. However, the work done in [119] showed theoretically that it is not necessary to use the same order interpolation for each iteration. It is more economical, but not less accurate, to perform the first integration with a linear interpolation, and then use a quadratic interpolation for the second iteration.

We now summarize the derivation from [119], showing how to obtain the accuracy of different choices of schemes that could be used in conjunction with the extrapolation of the velocity fields in the departure point calculations, and then the interpolations for the value of the field at the departure points.

8.2.1 Order of extrapolation versus order of interpolation

The derivation presented in [119] is in two dimensions, along with a forcing term, but we shall omit the extra dimension and the forcing term in this summary as they do not affect the results that we show.

In [119] McDonald presents six schemes that are denoted by (a), (b1), (b2), (c1), (c2), and (c3), where a is referring to a linear formulation, b is referring to quadratic formulation of some form, and c is referring to some form of cubic formulation.

Scheme (a)

We are going to use the standard linear Lagrange interpolation formulation to calculate

$$\psi\left(x_i^{n+1}-\widehat{u}\Delta t, n\Delta t\right)=(1-\widehat{\alpha})\,\psi_{i-p}^n+\widehat{\alpha}\psi_{i-p-1}^n, \tag{8.38}$$

where we have to update the value of $\widehat{\alpha}$ for each iteration, which is given by

$$\widehat{\alpha}=-p+\widehat{u}\frac{\Delta t}{\Delta x}. \tag{8.39}$$

The notation for ψ_{i-p}^n is the same as that used in the previous semi-Lagrangian chapters, and we will still have that $0\leq\widehat{\alpha}\leq 1$ holds. We now expand all of the quantities except $\widehat{u}$ about $\left(x_i^{n+1}, t^n\right)$, and use the first equation in the appendix in [119], which we have seen before in

Chapter 7 for the remainder terms for the linear, quadratic, and cubic Lagrange polynomial stability and accuracy analysis. Thus we have

$$\frac{\Delta_L \psi}{\Delta t} = \left(\frac{\partial \psi}{\partial t}\right)_i^n + \widehat{u}\left(\frac{\partial \psi}{\partial x}\right)_i^n + O(\Delta), \tag{8.40}$$

where McDonald has used the following notation:

$$\Delta_L \equiv \frac{\psi\left(x_i^{n+1}, t^{n+1}\right) - \psi\left(x_i^{n+1} - \widehat{u}\Delta t, t^n\right)}{\Delta t},$$

and $O(\Delta)$ represents the order of the accuracy of the scheme, where it is with respect to either Δt or Δx depending on how the ratio $\frac{\Delta t}{\Delta x}$ behaves in the limit as Δx and Δt approach zero, as we presented in Chapter 7, and as explained in McDonald (1984) [117]. Thus, as expected, this scheme is only first order, but it is stated in [119] that this statement is true if the velocity $\widehat{u}$ is estimated by using the grid point velocity u_i^n, or through iterating.

Scheme (b1)

We now consider two options using a quadratic interpolation to approximate the ψ field value at the departure point. Recalling the Lagrange quadratic interpolation formulation, we have

$$\psi\left(x_i^{n+1} - \widehat{u}\Delta t, n\Delta t\right) = \frac{1}{2}\widehat{\alpha}\left(1+\widehat{\alpha}\right)\psi_{i-p-1}^n + \left(1-\widehat{\alpha}\right)\left(1+\widehat{\alpha}\right)\psi_{i-p}^n - \frac{1}{2}\widehat{\alpha}\left(1-\widehat{\alpha}\right)\psi_{i-p+1}^n, \tag{8.41}$$

but here now we have $-0.5 \leq \widehat{\alpha} \leq 0.5$. We again expand all of the quantities except $\widehat{u}$ about the point $\left(x_i^{n+1}, t^n\right)$, and obtain the following expression for the semi-Lagrangian approximation:

$$\frac{\Delta_L \psi}{\Delta t} = \left(\frac{\partial \psi}{\partial t}\right)_i^n + \widehat{u}\left(\frac{\partial \psi}{\partial x}\right)_i^n + \frac{\Delta t}{2}\left[\left(\frac{\partial^2 \psi}{\partial t^2}\right)_i^n - \widehat{u}^2\left(\frac{\partial^2 \psi}{\partial x^2}\right)_i^n\right] + O\left(\Delta^2\right). \tag{8.42}$$

This scheme is an instance where $\widehat{u}$ is approximated by the grid point velocity u_i^n, and as such it will be $O(\Delta t)$ or $O\left((\Delta x)^2\right)$, whichever is larger.

Scheme (b2)

The second quadratic scheme does not change the interpolation formula, but now considers a centered in time and space approximation for $\widehat{u}$, which is given by the linear interpolation

$$\widehat{u}^1 = u\left(x_i^{n+1} - u_i^{n+\frac{1}{2}}\frac{\Delta t}{2}, t^{n+\frac{1}{2}}\right) = \left(1-\widehat{\gamma}\right)u_{i-m}^{n+\frac{1}{2}} + \widehat{\gamma}u_{i-m-1}^{n+\frac{1}{2}}, \tag{8.43}$$

where $\widehat{\gamma}=-m+\frac{\Delta t u_i^{n+\frac{1}{2}}}{2\Delta x}$ and m is an integer chosen such that $0<\widehat{\gamma}<1$. This then yields that the approximation is approximately

$$\widehat{u}^{(1)}=\left(u-\left(u\frac{\partial u}{\partial x}\right)\frac{\Delta t}{2}\right)_i^{n+\frac{1}{2}}+O\left(\Delta t\Delta\right). \tag{8.44}$$

If we use the standard extrapolation formula for $u_i^{n+\frac{1}{2}}$,

$$u_i^{n+\frac{1}{2}}=\frac{3}{2}u_i^n-\frac{1}{2}u_i^{n-1}, \tag{8.45}$$

then, after expanding u_i^{n-1} about (x_i,t^n), we have

$$\widehat{u}^{(1)}=\left(u+\left(\frac{\partial u}{\partial t}-u\frac{\partial u}{\partial x}\right)\frac{\Delta t}{2}\right)_i^n+O\left(\Delta t\Delta\right). \tag{8.46}$$

It can be shown that by using these approximations this scheme is now second order in time and space, and that this has happened through using one iteration of a linear interpolation to center in space, and then (8.45) enables the scheme to be centered in time, and as such gives an accuracy of $O\left(\Delta^2\right)$.

Scheme (c1)

As mentioned earlier, this is an instance of a cubic Lagrange interpolation scheme. We shall not present the interpolation formula again as it can be found in Chapter 6. We also refer the reader to [119] for the truncation error. As with scheme (a) and scheme (b1), we consider the case where we are using the grid point velocity u_i^n to approximate $\widehat{u}$, and so the scheme is only accurate to $O\left(\Delta t\right)+O\left(\left(\Delta x\right)^3\right)$.

Scheme (c2)

Now consider the case where we use one iteration of the linear interpolation to center $\widehat{u}$ in space, and then (8.45) to center in time, with the overall accuracy of the scheme being raised to $\left(\left(\Delta t\right)^2\right)$. An important point that is made in the discussion of this scheme in [119] is that from the truncation error equation it appears that it should be possible to construct an $\left(\Delta^3\right)$ accurate scheme, which will appear if $\widehat{u}$ can be calculated to $O\left(\Delta^3\right)$, and with that we now present the scheme that obtains this level of accuracy.

Scheme (c3)

To be able to center a numerical approximation in time, we shall require three time levels, and ironically the scheme that McDonald uses to achieve this is that from Temperton and Staniforth (1987),

$$u_i^{n+\frac{1}{2}}=\frac{1}{8}\left(15u_i^n-10u_i^{n-1}+3u_i^{n-2}\right), \tag{8.47}$$

but this scheme must be combined with a quadratic interpolation and iterated twice. The reader is again referred to [119] for the truncation error analysis, where it can be shown that

$$\frac{\Delta_L \psi}{\Delta t} = \left[\frac{\Delta_L \psi}{\Delta t}\right]_i^n + (\Delta t)^2 [T_2]_i^n +) \left(\Delta^3\right), \tag{8.48}$$

where $T_2 = t_1 + t_2$ and t_1 and t_2 are defined as follows:

$$t_1 = \frac{1}{6}\frac{\partial}{\partial t}\left(\frac{\partial}{\partial t} - 2u\frac{\partial}{\partial x}\right)\left(\frac{\partial \psi}{\partial t} + u\frac{\partial \psi}{\partial x}\right), \tag{8.49a}$$

$$t_2 = \frac{1}{6}u\frac{\partial}{\partial x}\left(\frac{\partial}{\partial t} - u\frac{\partial}{\partial x}\right)\left(\frac{\partial \psi}{\partial t} + u\frac{\partial \psi}{\partial x}\right) - \frac{1}{24}\frac{\partial \psi}{\partial x}\left(\frac{\partial}{\partial t} + u\frac{\partial u}{\partial x} - 3\frac{\partial u}{\partial x}\right)\left(\frac{\partial u}{\partial t} + u\frac{\partial u}{\partial x}\right). \tag{8.49b}$$

We should note here that there is another term appearing in the expression for (8.49b) in [119], but it concerns the forcing term which we do not deal with in this chapter.

Another important fact that McDonald states is that applying a third iteration would not increase the order of the accuracy unless it is accompanied with a cubic interpolation.

8.2.2 Construction of higher order schemes

In [119] McDonald poses the question of whether it is possible to obtain third-order accurate in time and space semi-Lagrangian approach to the advection equation, and not with two time levels; we had obtained the third order of spatial and temporal accuracy with a two-time-level scheme (c3). To address this question, McDonald presents a three-time-level approach defined by

$$\frac{\Delta_{2L}\psi}{2\Delta t} = \frac{\left(\psi\left(x_i, t^{n+1}\right) - \psi\left(x_i^{n+1} - 2\widetilde{u}\Delta t, t^{n-1}\right)\right)}{2\Delta t} = 0 \tag{8.50}$$

and

$$\widetilde{u}^{(k)} = u\left(x_i^{n+1} - \widetilde{u}^{(k-1)}\Delta t, t^n\right), \tag{8.51}$$

where the first guess for the iteration is given by $\widetilde{u}^{(0)}\left(x_i, t^n\right)$. By choosing this approach, we have that the scheme is centered in space. The next important points to note here are the following:

1. Field ψ must be interpolated to the departure point by means of some form of cubic interpolation polynomial or formula.
2. The first iteration for the estimate of $\widetilde{u}$ should use a linear interpolation to find the value at the departure point.
3. The second iteration for the estimate of $\widetilde{u}$ should use a quadratic interpolation to find the value at the departure point.

The truncation error derivation for the situation above can be shown to be equal to

$$\frac{\Delta_{2L}\psi}{2\Delta t} = \left[\left(1 - \Delta t u\frac{\partial}{\partial x}\right)\left(\frac{\partial \psi}{\partial t} + u\frac{\partial \psi}{\partial x}\right) + (\Delta t)^2 T_3\right]_i^n + O\left((\Delta t)^3\right), \tag{8.52}$$

where the T_3 term is defined as $T_3 = t_1 + 4t_2$, and t_1 and t_2 are as defined in (8.49a) and (8.49b), respectively. Thus the scheme presented above is second-order accurate in time. However, the errors associated with this scheme will be four times as large as those associated with (8.48), which, according to McDonald, can be seen by comparing the expressions for T_2 and T_3. We should also note here that this was first pointed out by Temperton and Staniforth in their 1987 paper [188] and could be a possible advantage of the two-time-level approaches over the three-time-level schemes.

Given the definitions from (8.49a) and (8.49b), it became obvious to McDonald that to be able to create a semi-Lagrangian scheme that was third-order accurate in space and time, one needs to take $\frac{4}{3}$(8.48)–$\frac{1}{3}$(8.52), which yields the following scheme:

$$\frac{7\psi\left(x_i, t^{n+1}\right) - 8\psi\left(x_i^{n+1} - \widehat{u}\Delta t, t^n\right) + \psi\left(x_i^{n+1} - 2\widetilde{u}\Delta t, t^{n-1}\right)}{6\Delta t} = 0. \tag{8.53}$$

When considering the scheme in (8.53), we need to ensure that the following steps are implemented:

1. A cubic interpolation polynomial or formula must be used to obtain the values of the field at the departure points.
2. A linear interpolation polynomial or formula for the two wind fields $\widehat{u}^{(1)}$ and $\widetilde{u}^{(1)}$ must be used.
3. A quadratic interpolation polynomial or formula for the two wind fields $\widehat{u}^{(2)}$ and $\widetilde{u}^{(2)}$ must be used.
4. The extrapolation scheme for $\widehat{u}$ at the halfway step given by (8.47) must be used.

All of the schemes presented in this chapter have so far been linear, that is to say, the advective velocity is not the field that is also being advected. However, there are many applications in modeling geophysical processes where this is not the case. So we now move on to consider the case where the advection is nonlinear.

8.3 Semi-Lagrangian approximations to nonlinear advection

As we have just mentioned, we now consider the case for the scalar situation, again without forcing terms, for the nonlinear formulation of advection. The nonlinear advection problem that we consider here is referred to as the inviscid Burgers equation, which in an Eulerian formulation is given by

$$\frac{\partial u}{\partial t} + u\frac{\partial u}{\partial x} = 0. \tag{8.54}$$

In a paper by Kuo and Williams in 1990 [95], the authors consider this equation on the infinite domain with the initial conditions

$$u(x, 0) = f(x) = \overline{u} - \tan^{-1}(x - x_0). \tag{8.55}$$

We know already that the general solution to the unforced advection equation is given by

$$u(x,t) = f(x - u(x,t)t).$$

Given the initial conditions stated in (8.55), it is then stated in [95] that the analytical solution for this problem is given by

$$u = \overline{u} - \tan^{-1}(x - ut - x_0). \tag{8.56}$$

The reason that Kuo and Williams are considering the Burgers equation in [95] is that at the time there were concerns about important phenomena in the atmosphere that are associated with sharp gradients. This is still a concern with many geophysical modeling and prediction situations today. One such example is frontogenesis which can be discontinuous, and that work by Williams and Kurth from 1976 has shown these discontinuities to be analogous to hydraulic jumps in a one-layer homogeneous fluid [204]. This phenomenon is referred to as a **scale collapse**, and that is why we are considering the Burgers equation here, as it can support shock formation, equivalent to this scale collapse.

Therefore, given the setup presented above, we need to determine at what time the scale collapse will occur. To achieved this, we need to differentiate the analytical solution in (8.56) with respect to x, which results in

$$\frac{\partial u}{\partial x} = -\frac{1 - t\dfrac{\partial u}{\partial x}}{1 + (x - x_0 - ut)^2}.$$

When $x = x_0 + \overline{u}t$, we have $u = \overline{u}$ and so obtain

$$\left(\frac{\partial u}{\partial x}\right)_{x=x_0+\overline{u}t} = -\left(1 - t\left(\frac{\partial u}{\partial x}\right)_{x=x_0+\overline{u}t}\right), \tag{8.57}$$

where upon factorizing,

$$\left(\frac{\partial u}{\partial x}\right)_{x=x_0+\overline{u}t} = \frac{1}{t-1} \rightarrow -\infty \quad \text{as} \quad t \rightarrow 1. \tag{8.58}$$

In [95] it is stated that physically the above problem provides a scale collapse formation with mean background advection at speed $\overline{u}$. The scale collapse, or shock formation, time is 1.0 with the position of the scale collapse at $x_0 + \overline{u}$.

In [95] the authors considered two Eulerian based schemes first. We shall not say much about them, but the first is a centered space formulation given by

$$\frac{\partial u_i}{\partial t} + u_i \frac{u_{i+1} - u_{i-1}}{2\Delta x} = 0, \tag{8.59}$$

whereas the second scheme is referred to as the energy-conserving second-order countered difference. The reason for this name is that the energy is given by $\frac{u^2}{2}$, and the Burgers equation can be written as

$$\frac{\partial u}{\partial t}+\frac{\partial}{\partial x}\left(\frac{u^2}{2}\right)=0. \tag{8.60}$$

This then results in the finite difference equation given by

$$\frac{\partial u_i}{\partial t}+\frac{\left(u_{i+1}+u_i+u_{i-1}\right)\left(u_{i+1}-u_{i-1}\right)}{6\Delta x}. \tag{8.61}$$

We now consider how to possibly apply the semi-Lagrangian formulation to the inviscid Burgers equation. We start by recasting (8.54) in terms of the total derivative as

$$\frac{D}{Dt}u\left(x\left(t\right)t\right)=0. \tag{8.62}$$

We now apply the three time level approximation introduced in the last section, which results in

$$\frac{u\left(x\left(t+\Delta t\right),t+\Delta t\right)-u\left(x\left(t-\Delta t\right),t-\Delta t\right)}{2\Delta t}=0. \tag{8.63}$$

Next we choose the location at the forecast time to correspond to a grid point, $x\left(t+\Delta t\right)=x_i^{n+1}$, and as before we let α_i represent the displacement during one time step, which leads to

$$u\left(x_i,t+\Delta t\right)=u\left(x_i-2\alpha_i,t-\Delta t\right). \tag{8.64}$$

As we have seen before, the α_is are determined by the approximate integration of

$$\frac{\partial x}{\partial t}=u \tag{8.65}$$

over the period of $[t-\Delta t,t+\Delta t]$. In [95] the authors apply the midpoint rule, which yields

$$\alpha_i=\Delta t u\left(x_i^{n+1}-\alpha_i,t\right). \tag{8.66}$$

The combination of (8.64) and (8.66) is the three-time-level semi-Lagrangian scheme that is second-order accurate in time. It is noted in [95] that the scheme just presented is quite similar to the bivariate one that was presented in [161].

The next important property to ascertain is the convergence and computational efficiency, where we have assumed that a fixed point iteration scheme has been used. We shall now briefly introduce some of the theory around fixed point iteration schemes and then go on to present two of the most common fixed point schemes, which are the secant and the different forms of the Newton methods.

Fixed point iteration methods

In this small subsection we will summarize the general theory associated with fixed point, or one point as they are referred to in [6], iteration methods, and present two of the most commonly used approaches. As we have seen above, we are considering the case where

$$x^{(m+1)} = g\left(x^{(m)}\right), \qquad m \geq 0, \tag{8.67}$$

with $x^{(0)}$ being the initial guess for the solution α. According to [6], each solution of $x = g(x)$ is a *fixed point* of g. We shall now present a couple of lemmas and a theorem to describe requirements and behaviors of different iterative schemes.

Lemma 8.1. *Let $g(x)$ be continuous on the interval $a \leq x \leq b$, and assume that $a \leq g(x) \leq b$ for every $a \leq x \leq b$, then $x = g(x)$ has at least one solution in $[a, b]$.*

Lemma 8.2. *Let $g(x)$ be continuous on $[a, b]$, and assume $g([a, b]) \subset [a, b]$. Furthermore, assume there is a constant $0 < \lambda < 1$ such that*

$$|g(x) - g(y)| \leq \lambda |x - y|, \qquad \forall x, y \in [a, b]. \tag{8.68}$$

Then $x = g(x)$ has a unique solution $\alpha \in [a, b]$. Also the iterates

$$x^{(m)} = g\left(x^{(m-1)}\right), \qquad m \geq 1, \tag{8.69}$$

will converge to α for any choice of $x^{(0)} \in [a, b]$.

Theorem 8.3. *Assume that $g(x)$ is continuously differentiable on $[a, b]$, $g([a, b]) \subset [a, b]$, and that*

$$\lambda = \max_{a \leq x \leq b} \left|g'(x)\right| < 1. \tag{8.70}$$

Then

(i) *$x = g(x)$ has a unique solution $\alpha \in [a, b]$.*
(ii) *For any choice of $x^{(0)} \in [a, b]$, with $x^{(m+1)} = g\left(x^{(m)}\right)$, $m \geq 0$,*

$$\lim_{n \to \infty} x^{(m)} = \alpha.$$

(iii)

$$\left|\alpha - x^{(0)}\right| \leq \lambda^m \left|\alpha - x^{(0)}\right| \leq \frac{\lambda^m}{1-\lambda}\left|x^{(1)} - x^{(0)}\right|, \tag{8.71}$$

$$\lim_{n \to \infty} \frac{\alpha - x^{(m+1)}}{\alpha - x^{(m)}} = g'(\alpha). \tag{8.72}$$

We refer the readers to Atkinson (1988) [6] for the proofs of the lemmas and theorem above. We now introduce a couple of commonly used fixed point iteration methods in semi-Lagrangian advection.

Newton–Raphson method

The first fixed point iteration method we present is the Newton–Raphson method which is defined as follows: For an initial estimate $x^{(0)}$ of a desired root α of $f(x)=0$, the Newton–Raphson iterative formula is

$$x^{(m+1)} = x^{(m)} - \frac{f\left(x^{(m)}\right)}{f'\left(x^{(m)}\right)}, \qquad m \geq 0. \tag{8.73}$$

The iteration method above is applied until some stopping criterion is met. A very common used stopping criterion is that the difference between two iterates is less than some prescribed small number.

Secant method

As we can see above, the Newton–Raphson method requires knowledge of the derivative of the function f, which may not always be possible. Another tangent-based iterative solver that is applied quite often in different geophysical applications is the secant method, where for two initial guesses for the root α, $x^{(0)}$ and $x^{(1)}$, this approach approximates the graph of the function, $y=f(x)$, by the secant line determined by $\left(x^{(0)}, f\left(x^{(0)}\right)\right)$ and $\left(x^{(1)}, f\left(x^{(1)}\right)\right)$ as follows:

$$\frac{f\left(x^{(1)}\right)-f\left(x^{(0)}\right)}{x^{(1)}-x^{(0)}} = \frac{f\left(x^{(1)}\right)-0}{x^{(1)}-x^{(2)}} \Longrightarrow x^{(2)} = x^{(1)} - f\left(x^{(1)}\right)\frac{x^{(1)}-x^{(0)}}{f\left(x^{(1)}\right)-f\left(x^{(0)}\right)}. \tag{8.74}$$

The expression above can be generalized to the $(m+1)$th iteration as

$$x^{(m+1)} = x^{(m)} - f\left(x^{(m)}\right)\frac{x^{(m)}-x^{(m-1)}}{f\left(x^{(m)}\right)-f\left(x^{(m-1)}\right)}. \tag{8.75}$$

Given the aside above, we now return to the semi-Lagrangian approximation to the non-linear Burgers equation. If we now consider that we have applied a fixed point iteration scheme k times, then we can write (8.66) as

$$\alpha_i^{(k+1)} = \Delta t u\left(x_i^{n+1} - \alpha_i^{(k)}, t\right). \tag{8.76}$$

The next step is to subtract (8.66) from (8.76), and apply the Lipschitz continuity condition, which yields

$$\left\|\alpha_i^{(k+1)} - a_i\right\| = \Delta t \left\|\frac{\partial u}{\partial x}\right\| \left\|\alpha_i^{(k)} - \alpha_i\right\|. \tag{8.77}$$

The next step in the analysis is to enact the contraction mapping principle which then indicates that from (8.77) we are able to infer that the iteration in (8.76) will converge provided

$$\Delta t \left\|\frac{\partial u}{\partial x}\right\| < 1. \tag{8.78}$$

An interesting property mentioned in [95] is that (8.77) indicates that the accuracy of the iteration is raised by $O(\Delta t)$ on each application of the iterative scheme. However, since the midpoint rule in (8.66) has third-order accuracy in time, no more than three iterations are needed to solve (8.66) [95].

8.4 Nonlinear instability I

In this section we shall be considering the continuous case, keeping in mind that we require the numerical approximation keep the properties of the continuous problem and also not introduce spurious waves. If we recall the inviscid version of Burgers equation,

$$\frac{\partial u}{\partial t} + u\frac{\partial u}{\partial x} = 0, \tag{8.79}$$

then we have to be aware that nonlinear problems create two important problems when modeling fluid dynamics. The first problem is that they can generate nonlinear instability, and second, new waves can be generated in nonlinear problems through nonlinear wave interaction. We know that for linear advection all of the wave move at the same speed, and that the shape of the wave is unchanged. However, for the nonlinear problem, as we saw earlier, the wave **advects itself** such that the local speed depends on the wave amplitude and the shape of the wave changes in time.

The process of the wave changing shape is referred to as nonlinear steepening, and as we mentioned earlier about the Burgers equation, the process can result in shock waves and overturning if the flow is inviscid. In this situation the characteristics coalesce into a group where we have the situation that there are multiple values of u that exist for a given x.

We have to be aware that the nonlinear problems can create new wave modes, and so the principle of superposition of waves does not apply here. As a simple illustration lets analytically consider the case where $u = \sin kx$, and apply the differential operator from the second part of the inviscid Burgers equation, which results in

$$u\frac{\partial u}{\partial x} = k\sin(kx)\cos(kx) = \frac{k\sin(2kx)}{2},$$

which implies that the system contains a new wave given by $-\sin(2kx)$ that has a wavenumber of $2k$, with a wavelength of $L = \frac{\pi}{k}$, which is half of the wavelength of the original $L = \frac{2\pi}{k}$. An important feature to note about the new wave here is that it can interact with itself and the original wave, and that this process can carry on to produce an entire spectrum of waves. This process is the source of **aliasing error**.

Aliasing

Before we progress we shall quickly explain what is meant by **aliasing error**.

Definition 8.4. Aliasing is an effect that causes different signals to become indistinguishable when sampled. That is to say, the two signals are aliases of each other. The aliasing error is defined as the distortion of an original continuous signal, or wave, when it is reconstructed from samples that are from different signal/waves than the original continuous signal/wave.

As an example of where aliasing could occur, we have plotted two sine waves in Fig. 8.3 that have different frequencies but pass through the same sample points. We can see that if we reconstructed the wave from the larger frequency, then we would not be correctly reconstructing the oscillation associated with the faster wave and hence in a numerical modeling situation we would be incorrectly reconstructing the smaller scale features.

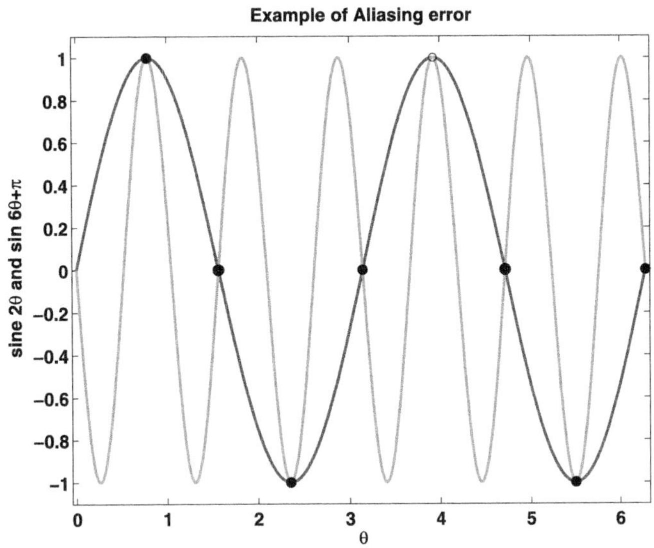

FIGURE 8.3 Plot of two different sine waves of different frequencies that both pass through our observation points, and this is aliasing.

Returning to nonlinear instability, we have to take note here that unlike in the linear case where linear instability occurs when the CFL condition is violated, quite often as a result of the time step Δt being too large compared to the spatial step Δx size, for the nonlinear situation the nonlinear instability occurs when waves that are shorter than $2\Delta x$ are generated by the numerical scheme and that these feed spurious energy into the wavelengths near and larger than $2\Delta x$. This generation of waves with wavelengths that are shorter than $2\Delta x$ is a consequence of aliasing.

If we return to our example of $u = \sin(kx)$, then we can see that, given a numerical spatial grid with step size Δx, the shortest wave that can be represented by the grid is of wavelength $2\Delta x$, whereas the largest wave number is $k_{\max} = \frac{2\pi}{2\Delta x} = \frac{\pi}{\Delta x}$. If we now reconsider the nonlinear advection term in the inviscid Burgers equation, then we have

$$u\frac{\partial u}{\partial x} = k\sin(kx)\cos(kx) = \frac{k\sin(2k\Delta x)}{2},$$

but if we substitute the largest wave number into the expression above, then we have a wavenumber of $2k_{\max}$ which corresponds to a wavelength $\frac{2\Delta x}{2} = \Delta x$, which is too short to be represented by the numerical grid. Thus the nonlinear interactions between waves can generate waves that cannot be resolved by the grid.

This now raises the question of what happens to these unresolvable waves. It should also be noted here that the waves that are generated by aliasing are always near $2\Delta x$, which indicates that energy starts to pile up in the form of short wave noise.

Therefore, we have to now consider how to control these spurious waves. One way to approach this problem is to consider that if a flow contains many modes, then it is useful to consider the distribution of energy, which is a measure of the amplitude of the modes, as a function of the wavenumber, which is

$$E = \frac{\sum u_k^2}{2}. \tag{8.80}$$

In a numerical approximation to the inviscid Burgers equation or some other form of nonlinear differential equation, aliasing occurs near $2\Delta x$, which implies that the energy is shifted to the small scales and the short waves grow with time, which itself implies nonlinear instability.

The first indication that these spurious waves could be a problem was discovered by Phillips in 1959 [143], where he showed that catastrophic growth of wave disturbances could be prevented in a two-dimensional geostrophic model by using a spectral filter that eliminates waves shorter than or equal to $4\Delta x$. The spectral approach decomposes the solution into Fourier modes, and recomposes them without the smaller waves, hence eliminating the shortest waves.

When considering the nonlinear advection problem, which is based upon the Burgers equation but with a nonzero forcing term, another technique to remove the spurious waves is through creating a change of variable to a conserving variable to reduce the risk of aliasing in the numerical models. We shall go into more details about this conserving change of variables in later chapters. Another way, similar to that mentioned in the work by Phillips, is to apply a time filter to remove spurious waves, as well as to damp computational modes created by a three level or higher time discretization.

8.5 Nonlinear instability II

The explanation that we present here is a summary from Haltiner and Williams' book from 1980 [71], where we start from what they refer to as an extremely simple, but surprisingly realistic, atmospheric prediction equation:

$$\frac{\partial \zeta}{\partial t} = -\mathbf{v} \cdot \nabla \zeta = -\nabla \cdot (\zeta \mathbf{v}), \qquad \mathbf{v} = \mathbf{k} \times \nabla \psi, \qquad \zeta = \nabla^2 \psi. \tag{8.81}$$

If we apply the Gauss divergence theorem to (8.81) over a region and there is no flux at the boundaries, or the region is closed, then the right-hand side of the equation vanishes. This then implies that we have the condition $\frac{\partial \overline{\zeta}}{\partial t} = 0$, where the bar represents the integral over the region here, later in other chapters this will be used in a somewhat similar fashion but there to represent the integration along a trajectory, and is indicating that the total, or mean, vorticity is conserved. If we were to multiply (8.81) by ζ, then the result can be shown to be

$$\frac{\partial \frac{\zeta^2}{2}}{\partial t} = -\frac{1}{2}\nabla \cdot \left(\zeta^2 \mathbf{v}\right). \tag{8.82}$$

Applying Gauss divergence theorem to (8.82) shows that the total, or mean, square vorticity, which is referred to as **enstrophy**, is also conserved. Finally, if we multiply (8.81) by ψ instead, then the resulting differential equation is

$$\psi \frac{\partial \nabla^2 \psi}{\partial t} = -\psi \nabla \cdot (\zeta \mathbf{v}). \tag{8.83}$$

The next step is to apply the following modifications:

$$\psi \frac{\nabla \cdot \nabla \psi}{\partial t} = \psi \nabla \cdot \nabla \frac{\partial \psi}{\partial t} = \nabla \cdot \left(\psi \nabla \frac{\partial \psi}{\partial t}\right) - \nabla \psi \cdot \nabla \frac{\partial \psi}{\partial t}, \tag{8.84a}$$

$$\psi \nabla \cdot (\zeta \mathbf{v}) = \nabla \cdot (\psi \eta \mathbf{v}) - \zeta \mathbf{v} \cdot \nabla \psi. \tag{8.84b}$$

Substituting (8.84a) and (8.84b) into (8.83) results in

$$\nabla \psi \cdot \nabla \frac{\partial \psi}{\partial t} = \nabla \cdot \left(\psi \nabla \frac{\partial \psi}{\partial t}\right) + \nabla \cdot (\psi \zeta \mathbf{v}) - \zeta \mathbf{v} \cdot \nabla \psi. \tag{8.85}$$

Now the term $-\zeta \mathbf{v} \cdot \nabla \psi$ vanishes as $\mathbf{v}$ is perpendicular to $\nabla \psi$; also $\nabla \psi \cdot \nabla \frac{\partial \psi}{\partial t} = \frac{1}{2}\frac{\nabla \psi \cdot \nabla \psi}{\partial t}$, where $\nabla \psi \cdot \nabla \psi$ is twice the **kinetic energy**. Applying the Gauss divergence theorem again results in the right-hand side of (8.85) vanishing, which yields

$$\overline{\frac{\partial \left(\nabla \psi \cdot \nabla \psi\right)}{\partial t}} = 0. \tag{8.86}$$

This equation is therefore stating that the **mean kinetic energy** is *conserved* over a closed region.

Summarizing what we have just presented; this model over a closed region conserves the mean kinetic energy, mean vorticity, and mean enstrophy. These properties hold for the differential system and therefore we would prefer for these properties to also be conserved by the discretization scheme.

While we have not moved on to the two dimensional application of the semi-Lagrangian approaches, the following derivation from [71] was undertaken in two dimensions, but we will be moving onto the multidimensional problem soon.

If we assume that the we are considering a rectangular domain of dimensions L_x and L_y, over which the streamfunction ψ is continuously represented by a double Fourier series of the form

$$\psi = \sum_{\substack{m=0 \\ n=1}}^{\infty} \left(a_{m,n} \cos \frac{2\pi n x}{L_x} + b_{m,n} \sin \frac{2\pi n x}{L_x} \right) \sin \frac{\pi n y}{L_y}, \tag{8.87}$$

where $a_{m,n}$ and $b_{m,n}$ are functions of time, then if this series is used to calculate the product $-\mathbf{v}\cdot\nabla\zeta$ in (8.81) together with the trigonometric identities to recover a series of the same form, then it can be shown that when a trigonometric term from the $\mathbf{v}$ series with wavenumbers m_1, n_1 is multiplied by a term from the $\nabla\zeta$ series with wavenumbers m_2, n_2 this results in the following four pairs of wavenumbers:

$$\begin{aligned} &m_1+m_2, n_1+n_2, \quad m_1-m_2, n_1+n_2, \\ &m_1+m_2, n_1-n_2, \quad m_1-m_2, n_1-n_2. \end{aligned} \tag{8.88}$$

A feature that is mentioned in [71] is that the term $\mathbf{v}\cdot\nabla\zeta$ causes the transfer of vorticity and energy between different wavenumbers. This **nonlinear interaction** therefore results in the exchange within the total spectrum and indeed may develop wave components that previously did not exist. Another important feature to note here is that the terms of the trigonometric series (8.87) are the orthogonal characteristics functions of the equation differential equation below for a rectangular domain,

$$\nabla^2\psi_{m,n} + \lambda_{m,n}^2\psi_{m,n} = 0, \tag{8.89}$$

where the eigenvalues of (8.89) are $\lambda_{m,n}^2 = \left(\frac{2\pi m}{L_x}\right)^2 + \left(\frac{\pi n}{L_y}\right)^2$. Due to orthogonality, the mean kinetic energy and the mean enstrophy, which are conserved over the closed region, can be shown to be

$$\begin{aligned} \overline{K} &= \tfrac{1}{2}\overline{\sum\left(\nabla\psi_{m,n}\right)^2} = \overline{\sum K_{m,n}} = \tfrac{1}{2}\overline{\sum \lambda_{m,n}^2\psi_{m,n}^2}, \\ \overline{\zeta^2} &= \overline{\sum\left(\nabla^2\psi_{m,n}\right)^2} = \overline{\sum \lambda_{m,n}^2 K_{m,n}} = \overline{\sum \lambda_{m,n}^4\psi_{m,n}^2}. \end{aligned} \tag{8.90}$$

To help illustrate these properties, Haltiner and Williams provide the following example: Consider three wavenumbers in one dimension such that $\lambda_1 > \lambda_2 > \lambda_3$, and

$$\begin{aligned} K_1 + K_2 + K_3 &= C_1, \\ \lambda_1^2 K_1 + \lambda_2^2 K_2 + \lambda_3^2 K_3 &= C_2, \end{aligned}$$

where C_1 and C_2 are constants. If we had the situation where there were only two wave components K_1 and K_2, then it would be possible to solve for K_1 and K_2 as we would have two equations in two unknowns. If we consider the three-component situation and eliminate K_1 from the first equation, and K_3 from the second, then we would have

$$\begin{aligned} &\left(\lambda_1^2-\lambda_2^2\right)K_2 + \left(\lambda_1^2-\lambda_3^2\right)K_3 = \lambda_1^2 C_1 - C_2, \\ &\left(\lambda_1^2-\lambda_3^2\right)K_1 + \left(\lambda_2^2-\lambda_3^2\right)K_2 = C_2 - \lambda_3^2 C_1. \end{aligned}$$

Because the numbers on the right-hand sides above are constant and the coefficients of the Ks are positive, it follows that if K_2 decreases then K_1 and K_3 must increase. Thus this is implying that energy is not transferred in one direction of the wavenumber spectrum. When more waves are present, the transfer processes are much more complex and, as stated in [71], it was shown in a paper by Fjørtoft in 1953 [57] that energy cannot flow consistently in one direction, for example, it cannot cascade towards higher wavenumbers, but the transfer is limited and the higher the wave number, the more it is limited. Given that this is a property of the continuous problem, we would then require that the numerical approximation also have this property.

8.5.1 Discrete approximations

If we now suppose that we have a grid that has equal spacing $\Delta x = \Delta y$ in two-dimensional Cartesian coordinates, (x, y), where the grid points are defined by x_i and y_j such that

$$x_i = i\Delta x, \quad i = 0, 1, \ldots, N, \quad N\Delta x = L_x,$$

$$y_j = j\Delta y, \quad j = 0, 1, \ldots, M, \quad M\Delta y = L_y,$$

and we assume that ψ vanishes at $y = 0$ and at $y = L_y$, with a periodic domain in x, then there are $N(M-1)$ arbitrary values of ψ that can be used to determine the coefficients of the finite Fourier series

$$\psi_{ij} = \sum_{n=0}^{\frac{N}{2}} \sum_{m=1}^{M-1} \left(a_{n,m} \cos \frac{2\pi ni\Delta x}{N\Delta x} + b_{n,m} \sin \frac{2\pi ni\Delta x}{N\Delta x} \right) \sin \frac{\pi mj\Delta y}{M\Delta y}. \tag{8.91}$$

A feature to note about the series in (8.91) is that at first glance it may appear that this series has $(2+N)(M-1)$ coefficients, but $\sin \frac{2\pi ni}{N}$ vanishes at $n = 0$ and $n = \frac{N}{2}$, hence $2(N-1)$ of the $b_{i,j}$ $b_{0,m}$ and $b_{\frac{N}{2},m}$ may be taken as zero, and so the correct number of coefficients equals the number of gridpoints.

Obviously in (8.91) we no longer have an infinite series, and only have a series that can be obtained from a finite set of grid points. The maximum wave number in the x-direction is $\frac{N}{2}$, which corresponds to the wavelength $2\Delta x$ representable on a finite grid. If nonlinear interactions were occurring between two wave numbers n_1 and n_2 and the sum $(n_1 + n_2)$ exceeds $\frac{N}{2}$, then the resulting wave cannot be resolved, and will be falsely represented in terms of a permissible wave number. This misrepresentation as we have seen is referred to as aliasing, and can be determined be letting $n_1 + n_2 = N - s$. Next if we consider the cosine term in (8.91) then this becomes

$$\cos \frac{2\pi (N-s)i}{N} = \cos 2\pi i \cos \frac{2\pi is}{N} + \sin 2\pi N \sin \frac{2\pi is}{N} = \cos \frac{2\pi is}{N}, \qquad \forall i \in \mathbb{N}^+.$$

Applying the argument to the sine term yields a similar result. Therefore, according to [71], if nonlinear interactions produce a wave with $(n_1 + n_2) > \frac{N}{2}$, then it will falsely appear as

$$s = N - (n_1 + n_2). \tag{8.92}$$

The important fact stated in [71] about these features is that feedback due to aliasing explains how the nonlinear instability can develop if energy is falsely generated and persistently channeled into the short resolvable wavelengths, $2\Delta x$ to $4\Delta x$, and is consistent with observations made by Phillips in 1959.

Therefore, when we are considering semi-Lagrangian approaches for nonlinear advection, we have to ensure that the interpolation and the extrapolation techniques do not create unresolvable waves to avoid aliasing.

Given this motivation to avoid spurious waves, we finish this chapter by considering the work by McDonald to avoid the introduction of these spurious waves when we are applying the semi-Lagrangian technique to a limited area problem, where these waves may have to exit the numerical domain without reflection, or be able to enter without being damped or blocked from doing so, but also to ensure that only the correct type of waves enter and leave the boundaries.

8.6 Boundary conditions for limit area models

In the experiments presented in Chapter 5 we applied the semi-Lagrangian techniques to a periodic domain where the scheme advected the bell curve and the step function out of a boundary and back in at the other end without introducing damping effects or creating extra spurious features. However, in numerical weather prediction, as well as ocean basin modeling, it may be that we are using a higher order version of the larger prediction model to produce more accurate forecast of smaller dynamical scale feature that does not need to be resolved globally. This then means that we have some form of linear, or nonlinear, advection that needs to know how to bring the information into the domain as well as leaving it, but where we may require points outside the domain of interest. In [122], McDonald introduces three different techniques to consider with the one-dimensional constant velocity semi-Lagrangian case to start with. These three approaches are **trajectory truncation, time interpolation, and a well-posed buffer zone**.

The setup for this work is based upon the semi-Lagrangian approximation

$$\frac{\partial \psi (x,t)}{\partial t} + \overline{u}\frac{\partial \psi (x,t)}{\partial x} = 0, \tag{8.93}$$

where $0 \leq x \leq L, t \geq 0$, and $\overline{u}$ is a constant. We have initial conditions given by $\psi (x,0) = f(x)$, but we also require boundary conditions. If $\overline{u} > 0$ then we have $\psi (0,t) = g(t)$, while if $\overline{u} < 0$ then we have $\psi (L,t) = h(t)$.

We now consider how to discretize (8.93) where if the solution enters the domain at $x = 0$ and exits at $x = L$, without distortions or reflections, then the discretization is well posed. We know from previous chapters that the solution to (8.93) is given by

$$\psi_i^{n+1} = \psi (i\Delta x - \overline{u}\Delta t, n\Delta t) = \psi_d^n. \tag{8.94}$$

If we apply the cubic Lagrange interpolation polynomial to obtain the value of ψ at the departure point, then we have

$$\begin{aligned}\psi_d^n &= -\frac{1}{6}\widehat{\alpha}\left(1-\widehat{\alpha}\right)\left(2-\widehat{\alpha}\right)\psi_{i-p+1}^n+\frac{1}{2}\left(1-\widehat{\alpha}\right)\left(1+\widehat{\alpha}\right)\left(2-\widehat{\alpha}\right)\psi_{i-p}^n \\ &+ \frac{1}{2}\widehat{\alpha}\left(1+\widehat{\alpha}\right)\left(2-\widehat{\alpha}\right)\psi_{i-p-1}^n-\frac{1}{6}\widehat{\alpha}\left(1-\widehat{\alpha}\right)\left(1+\widehat{\alpha}\right)\psi_{i-p-2}^n,\end{aligned} \tag{8.95}$$

were α, $\widehat{\alpha}$, and p are as defined before. If we have the situation where the departure point is outside the domain, then we cannot use (8.95), nor can we use it for the situation where the departure point lies in the grid box immediately next to the boundary. In these situation McDonald suggests using quadratic schemes for the following two situations:

$$\psi_d^n=-\frac{1}{2}\widehat{\alpha}\left(1-\widehat{\alpha}\right)\psi_{i-p+1}^n+\left(1+\widehat{\alpha}\right)\left(1-\widehat{\alpha}\right)\psi_{i-p}^n+\frac{1}{2}\widehat{\alpha}\left(1+\widehat{\alpha}\right)\psi_{i-p-1}^n,\quad 0\leq x_d^n\leq\Delta x, \tag{8.96a}$$

$$\psi_d^n=\frac{1}{2}\left(1-\widehat{\alpha}\right)\left(2-\widehat{\alpha}\right)\psi_{i-p}^n+\widehat{\alpha}\left(2-\widehat{\alpha}\right)\psi_{i-p-1}^n-\frac{1}{2}\widehat{\alpha}\left(1-\widehat{\alpha}\right)\psi_{i-p-2}^n,\qquad L-\Delta x\leq x_d^n\leq L. \tag{8.96b}$$

It is stated in [122] that for well-posedness we must **not** impose a value of ψ at the **outflow** boundary, and that we must extrapolate from the interior. We can see quite clearly from the expressions above that the semi-Lagrangian approach allows us to do this. If we have the situation where $|\alpha|<1$, then we can use (8.96b), and it yields an $O\left(\frac{(\Delta x)^3}{\Delta t}\right)$ accurate solution, while if $|\alpha|>1$ then we would use (8.95) which is $O\left(\frac{(\Delta x)^4}{\Delta t}\right)$ accurate [122].

If we now consider the inflow boundary and points in its immediate vicinity, then on the boundary itself we must impose a value of the field. If $|\alpha|>1$ then updating ψ at the grid points adjacent to the inflow boundary causes difficulties (see [122] for a full detail description of these difficulties). If $|\alpha|<1$ and $\overline{u}>0$, then McDonald imposes the value of the discrete version of the true solution that he is working with at the next time step. That is to say, for the initial conditions

$$\psi\left(x,0\right)=\widehat{\psi}\exp\left\{-\left(\frac{x-x_s}{0.1L}\right)^2\right\}, \tag{8.97}$$

where at the left boundary we have

$$\psi\left(0,t\right)=\widehat{\psi}\exp\left\{-\left(\frac{0-x_s-\overline{u}t}{0.1L}\right)^2\right\}, \tag{8.98}$$

for $t>0$, and $\Delta x=10$ km, $\overline{u}=20\ \text{ms}^{-1}$, $L=1000$ km, $\widehat{\psi}=10$. The point x_s is where the peak of the bell curve described above is located, and McDonald considers two tests:

Test 1: Choosing $x_s = \frac{L}{2}$ and a time such that at the end of the integration the bell curve will be centered at $\frac{3L}{2}$, having past out of the domain, to see that the scheme does not produce false reflections.

Test 2: Choosing $x_2 = -\frac{L}{2}$ and a time such that at the end of this integration the bell curve will be centered at $\frac{L}{2}$, to see if the curve can enter the domain through the boundary at $x = 0$.

Therefore, when $|\alpha| < 1$ and $\overline{u} > 0$, we impose ψ_0^{n+1} and compute ψ_1^{n+1} using (8.96a). In the interior we use (8.95), while at the other boundary we compute ψ_L^{n+1} using (8.96b).

McDonald shows in [122] that when $\alpha = 0.5$, recalling that this is the value of α that has the largest damping effect, the scheme just described passes both tests above very well, where there are no false reflections when the curve exits the domain, and the scheme is also able to successfully introduce the bell curve when it was initiated outside the domain.

We now turn our attention to the more challenging component of dealing with the boundary conditions, when $|\alpha| > 1$, which as we have seen earlier means that the trajectory has covered more than one grid space, and for the situation here this implies that when we are considering ψ_1^{n+1}, its departure point for $\overline{u} > 0$ is outside of the numerical domain. Given this setup, we now proceed to consider the three different approaches mentioned at the beginning of this section.

Trajectory truncation

If the departure point is beyond the boundary, then the first technique is to truncate the trajectory at the boundary. For example, if the departure point associated with ψ_2 were at $x = -\dfrac{\Delta x}{2}$, then the trajectory equation (8.94) would indicate that

$$\psi_2^{n+1} = \psi\left(-\frac{\Delta x}{2}, n\Delta t\right) \approx \psi\left(0, n\Delta t\right). \tag{8.99}$$

It is stated in [122] that this method is stable, but the solution is slightly out of phase with the true state, where the situation just described would make $\alpha = 2.5$, and so $\widehat{\alpha} = 0.5$, and we have seen earlier, this value is the most damped of all of the possibilities.

Time interpolation

Given the problems just encountered, we now seek a method to improve the accuracy of the trajectory truncation method. We should recall here that we know the value of the field at the inflow boundary for all time steps. We start with the solution of (8.93) which can be written as

$$\psi\left(x_i, t + \delta t\right) = \psi\left(x_i - \overline{u}\delta t, t\right). \tag{8.100}$$

Now choose $\overline{u}\delta t = \mu\Delta x$, where μ is a nonzero integer, along with $t + \delta t = (n+1)\Delta t$, which implies that if these are substituted into (8.100), then we obtain

$$\psi\left(x_i, (n+1)\Delta t\right) = \psi\left(x_{i-\mu}, \left(n + 1 - \frac{\mu}{\alpha}\right)\Delta t\right). \tag{8.101}$$

There is no loss of generality if we now assume that $\overline{u} > 0$. Therefore, if we let $i = \mu$, then this implies that (8.101) expresses the value of the field at $(n+1)\,\Delta t$ at the interior grid point μ in terms of the field on the boundary line $i = 0$, at time $\left(n + 1 - \frac{\mu}{\alpha}\right)$. The field is not known at this time, but it is possible to *interpolate in time* because, in principle, on the boundary line we know $\psi\,(x_0, m\Delta t)$ for all integers $m > 0$. For example, if we were to apply linear interpolation to the right-hand side of (8.101), then we would obtain

$$\psi_{mu}^{n+1} = \left(1 - \frac{\mu}{\alpha}\right)\psi_0^{n+1} + \frac{\mu}{\alpha}\psi_0^n. \tag{8.102}$$

It is possible to obtain second-order accuracy in time by using a quadratic Lagrange interpolation given by

$$\psi^{n+1-\frac{\mu}{\alpha}} = \frac{1}{2}\left(1 - \frac{\mu}{\alpha}\right)\psi_0^{n+1} + \frac{\mu}{\alpha}\left(2 - \frac{\mu}{\alpha}\right)\psi_0^n - \frac{1}{2}\frac{\mu}{\alpha}\left(1 - \frac{\mu}{\alpha}\right)\psi_0^{n-1}, \tag{8.103}$$

and McDonald was able to demonstrate that this scheme passed both of his two tests.

Well-posed buffer zone

As semi-Lagrangian methods require departure points upstream of the arrival point, being able to define a buffer zone around the domain of interest when applied to a limited area model would be ideal. This would mean that if the departure point were in this zone, then we could interpolate the same way as if it were in the interior.

In Eulerian modeling of a boundary-value problem, it is quite often the case that we introduce ghost points outside of the numerical domain which are then formulated in terms of the grid points associated with the full numerical domain. In this case we are using points of a nested model, where our domain of interest is nested in a larger one. Given that we are using a cubic interpolation polynomial and if we consider the inflow boundary line, then we only require at most two external points here, ψ_{-1} and ψ_{-2}.

McDonald states that a naive extrapolation from the interior to the inflow buffer zone will cause instability, which implies that we cannot apply (8.95) here. Therefore, since for well-posedness we must supply $\psi\,(0,t)$ for all t, we know $\frac{\partial\psi\,(0,t)}{\partial x}$, $\frac{\partial^2\psi\,(0,t)}{\partial x^2}$, and so on. Thus from (8.93) we know

$$\begin{aligned}
\frac{\partial\psi\,(0,t)}{\partial x} &= -\frac{1}{\overline{u}}\frac{\partial\,(0,t)}{\partial t},\\
\frac{\partial^2\psi\,(0,t)}{\partial x^2} &= \frac{1}{\overline{u}^2}\frac{\partial\,(0,t)}{\partial t^2}.
\end{aligned} \tag{8.104}$$

This expression enables us to estimate ψ_{-1} and ψ_{-2} using a Taylor series expansion and substituting from (8.104), which yields

$$\psi\,(-i\Delta x, t) = \psi\,(0,t) + i\frac{\Delta x}{\overline{u}}\frac{\partial\psi\,(0,t)}{\partial t} + \frac{(i\Delta x)^2}{2\overline{u}}\frac{\partial^2\psi\,(0,t)}{\partial t^2} + \cdots, \tag{8.105}$$

for $i = -1, -2$. We note that the same procedure could be applied at the other end of the domain if that was the inflow boundary.

8.7 Summary

In this chapter we have started to introduce some complexity into the semi-Lagrangian advection methods through first allowing the advection velocity to become both a function of time and space, and then for it to be the quantity that is being advected as well. We have shown that due to this velocity no longer being constant we cannot calculate the departure point exactly through the back trajectory, and that we now may have to solve an implicit equation to find the velocity at different time levels, but that this equation is a function of the departure point. Given that the departure point is not at a grid point, we have to interpolate the velocity to this point at a mid-level point if we are applying a two-time-level scheme, where we have to extrapolate certain variables to a halfway time step. We have also presented three-time-level schemes that require smaller time steps but do not require any extrapolation. We also dealt with nonlinear advection in the form of inviscid Burgers equation, as well as presented the theory around nonlinear instability to warn about spurious waves that can be introduced in the numerical approximations. Finally, we looked at the sensitivities and the need to ensure good boundary conditions, along with techniques to compute departure points outside the integration domain. The reason for showing this leads into the next chapter where we consider the forced equations, implying that the right-hand side of the advection equation is no longer zero.

CHAPTER

9

Nonzero forcings

Over the previous chapters we have considered the case where we are applying the semi-Lagrangian integration to the scalar, unforced advection equation in Lagrangian form, $\frac{D\psi}{Dt}=0$. We have considered the situation where the advecting velocity could be a constant, or a function of spatial coordinate and time, where for the latter we saw that we needed to approximate the wind field before we could find the departure point. There were several stages involved in this process, and these could affect both the accuracy and efficiency of the semi-Lagrangian scheme. We saw that we could consider either a two- or a three-time-level scheme to estimate the wind field, and that we needed to iterate a few times to find a good estimate of the departure point, and hence the associated advection velocity of wind through an implicit relationship between the two. We also saw that the order of the extrapolation polynomial to the halfway time levels for the two-time-level scheme did not dictate which order of interpolation polynomial should be used for the interpolation for the field value at the estimate of the departure point. The higher resolution models, as mentioned in the last chapter, we shall explain more in Chapter 11. It was possible to use a linear interpolation, followed by a quadratic interpolation, in a two iteration approach for the departure point and the velocity estimates, and then use a cubic Lagrange interpolation polynomial to find the value of the field at the departure point.

However, in many different geophysical situations where there is some form of either linear or nonlinear advection, it is quite often the case that there is a forcing involved as well. In this chapter we will extend the ideas just summarized above to the case when we have a forcing term, which means that the right-hand side of the advection differential equation is no longer zero. That is to say, the general linear advection differential equation in Eulerian form is given by

$$\frac{\partial\psi(x,t)}{\partial t}+u(x,t)\frac{\partial\psi(x,t)}{\partial x}=f(x,u,\psi,t),\tag{9.1}$$

where the Lagrangian formulation is

$$\frac{D\psi(x,t)}{Dt}=f(x,u,\psi,t),\tag{9.2}$$

with both (9.1) and (9.2) being valid over some spatial domain $[a,b]$, and for time $t\geq 0$. Note that we will also consider the nonlinear situation of

Semi-Lagrangian Advection Methods and Their Applications in Geoscience
https://doi.org/10.1016/B978-0-12-817222-3.00013-X

$$\frac{Du}{Dt} = f(x, u, t), \tag{9.3}$$

where u is also the advection velocity as well as advected quantity.

The situations above occur in many different applications of modeling of geophysical processes. However, work that was undertaken in the early development of the semi-Lagrangian approaches showed that it was quite important how these forcing terms are handled as they do not only affect the characteristic, and so the departure point, but also can excite spurious waves, as well as computational modes that need to be damped as they are not physical modes of the continuous problem. There are many different forcing situations, and we shall consider some of them in this chapter, but others will be considered later on in the two- and three-dimensional problems' chapters, as well as in the spherical problems' and applications' chapters.

The nonzero forcing term situation does occur in many different forms; just to name a few that we shall cover in the remainder of the book are the advection–diffusion, reaction–dispersion, advection–condensation, advection–reaction, reaction–advection-diffusion, and reaction–advection–diffusion–dispersion problems. Other situations are part of a larger set of equations such as the Navier–Stokes, shallow water, barotropic, and baroclinic equations, as well as the primitive equations.

We start by consider an approach to remove the forcing term through a change of variable approach based upon the characteristics of the advection equation itself.

9.1 Methods of characteristics approach

In this section we shall summarize the paper by Erbes from 1993 [53], where a method to introduce a transform of variables is applied to the linear and nonlinear one-dimensional shallow water equations, such that the forcing term is removed, which then implies that we can then apply the techniques from the last chapter associated with the nonconstant advection velocity with zero forcing.

We start with the linearized version of the one-dimensional shallow water equations that come from Holton (1992) [79], and are defined by the coupled differential equations in Eulerian form as

$$\frac{\partial u}{\partial t} + u_0\frac{\partial u}{\partial x} + g\frac{\partial h}{\partial x} = 0, \tag{9.4a}$$

$$\frac{\partial h}{\partial t} + u_0\frac{\partial h}{\partial x} + h_0\frac{\partial u}{\partial x} = 0, \tag{9.4b}$$

which in a Lagrangian formulation (9.4a) and (9.4b) become

$$\frac{Du}{Dt} = -g\frac{\partial h}{\partial x}, \tag{9.5a}$$

$$\frac{Dh}{Dt} = -h_0\frac{\partial u_0}{\partial x}, \tag{9.5b}$$

$$\frac{dx}{dt} = u_0, \tag{9.5c}$$

where u_0 is the basic-state wind velocity, h_0 is the basic-state height of the free upper surface. The perturbations about u_0 and h_0 are given by u and h, respectively, g is the acceleration due to gravity constant. We should note that the semi-Lagrangian advection is performed along the trajectories that are defined by the wind velocity u_0 in (9.5c).

The transform of variables comes about through writing (9.4a) and (9.4b) in matrix form as

$$\frac{\partial \mathbf{q}}{\partial t} + \mathbf{A}\frac{\partial \mathbf{q}}{\partial x} = 0, \tag{9.6}$$

where the vector $\mathbf{q}$ and the coefficient matrix $\mathbf{A}$ are given by

$$\mathbf{q} = \begin{pmatrix} u \\ h \end{pmatrix}, \qquad \mathbf{A} = \begin{pmatrix} u_0 & g \\ h_0 & u_0 \end{pmatrix}. \tag{9.7}$$

The next step is to apply an eigenvalue decomposition of the matrix $\mathbf{A}$, which results in the eigenvalue matrix $\mathbf{\Lambda}$, where the diagonal entries are the eigenvalues of $\mathbf{A}$, $\mathbf{R}$ is the eigenvector matrix, and $\mathbf{R}^{-1}$ is its inverse. It can be shown for (9.7) that the eigenvalues, eigenvector matrix, and its inverse, respectively, are

$$\mathbf{\Lambda} = \begin{pmatrix} u_0 + c_0 & 0 \\ 0 & u_0 - c_0 \end{pmatrix}, \quad \mathbf{R} = \begin{pmatrix} 1 & 1 \\ \frac{c_0}{g} & -\frac{c_0}{g} \end{pmatrix}, \quad \mathbf{R}^{-1} = \begin{pmatrix} 1 & \frac{g}{c_0} \\ 1 & -\frac{g}{c_0} \end{pmatrix}, \tag{9.8}$$

where $c_o \equiv \sqrt{gh_0}$ is the speed of the gravity waves. An important feature to note from (9.8) is that the two eigenvalues on the diagonal of $\mathbf{\Lambda}$ are both real and distinct.

Given that the system of differential equations can be diagonalized, if the eigenvalues are real and distinct then (9.6) is said to be a *strictly hyperbolic system*. Given a hyperbolic system, the variables $\mathbf{q}$ are referred to as the **primitive variables**, and they can be transformed into **characteristic variables** $\mathbf{p}$ through multiplying $\mathbf{q}$ with the left eigenvector matrix $\mathbf{R}^{-1}$, which results in

$$\mathbf{p} = \mathbf{R}^{-1}\mathbf{q}. \tag{9.9}$$

Inserting (9.7) and (9.8) into (9.9), we obtain for the linearized shallow water equations,

$$\mathbf{p} \equiv \begin{pmatrix} p_1 \\ p_2 \end{pmatrix} \equiv \begin{pmatrix} \frac{1}{2}\left(u + \frac{c^2}{c_0}\right) \\ \frac{1}{2}\left(u - \frac{c^2}{c_0}\right) \end{pmatrix}, \tag{9.10}$$

where $c \equiv \sqrt{gh}$ is the gravity-wave speed of the perturbation around the basic state.

The characteristic equations system can be found through multiplying (9.6) by $\mathbf{R}^{-1}$, and using the definition for the characteristic variables in (9.9), along with the property of diagonalization, $\mathbf{R}^{-1}\mathbf{A} = \mathbf{\Lambda}\mathbf{R}^{-1}$ results in

$$\frac{\partial\left(\mathbf{R}^{-1}\mathbf{q}\right)}{\partial t} + \mathbf{\Lambda}\frac{\partial\left(\mathbf{R}^{-1}\mathbf{q}\right)}{\partial x} = \frac{\partial \mathbf{p}}{\partial t} + \mathbf{\Lambda}\frac{\partial \mathbf{p}}{\partial x} = 0. \tag{9.11}$$

Through the properties of diagonalizable matrices we arrive at the system of characteristic equations that are independent, and so the Lagrangian formulation is given by

$$\frac{Dp_1}{Dt} = 0 \text{ along } \frac{dx}{dt} = \lambda_1, \tag{9.12}$$

$$\frac{Dp_2}{Dt} = 0 \text{ along } \frac{dx}{dt} = \lambda_2. \tag{9.13}$$

The important feature to notice here is that we have been able to **decouple** the original Lagrangian equations for the linearized shallow water model to be in terms of two independent scalar advection equations that are formulated in terms of the characteristic variables, p_1 and p_2, and the eigenvalues (characteristic velocities), λ_1 and λ_2. Given the definitions of the transforms, we are able to write (9.12) and (9.13) in terms of the velocity perturbation as

$$\frac{D}{Dt}\left(\frac{1}{2}\left(u + \frac{c^2}{c_0}\right)\right) = 0 \text{ along } \frac{dx}{dt} = u_0 + c_0, \tag{9.14}$$

$$\frac{D}{Dt}\left(\frac{1}{2}\left(u - \frac{c^2}{c_0}\right)\right) = 0 \text{ along } \frac{dx}{dt} = u_0 - c_0. \tag{9.15}$$

It is possible to recover the original values of the linearized shallow water equation model through

$$\mathbf{q} = \mathbf{R}\mathbf{p} \Longrightarrow \mathbf{q} = \begin{pmatrix} u \\ h \end{pmatrix} \equiv \begin{pmatrix} p_1 + p_2 \\ \frac{c_0}{g}\left(p_1 - p_2\right) \end{pmatrix}. \tag{9.16}$$

An important feature to note about the relationship between the original variables and the characteristic variables is that u and h are each dependent on both characteristic variables.

Returning to (9.12) and (9.13), these differential equations are in the form of independent scalar advection equations describing the advection of the variables p_1 and p_2 by the characteristic velocities λ_1 and λ_2. It is noted in [53] that if the characteristic velocities are of different signs, then the two characteristic variables will propagate in opposite directions. Another important feature to note about (9.14) and (9.15) is the fact that the *troublesome gravity-wave source* is not present here [53]. Therefore, it is possible to apply the two-time-level scheme of Bates and McDonald from their 1982 paper [14], to the characteristic advection equations in the form

$$\frac{p_1\left(x_i, t_n + \Delta t\right) - p_1\left(x_i - \alpha_{i,1}, t_n\right)}{\Delta t} = 0, \tag{9.17}$$

$$\frac{p_2\left(x_i, t_n + \Delta t\right) - p_2\left(x_i - \alpha_{i,2}, t_n\right)}{\Delta t} = 0, \tag{9.18}$$

where $\alpha_{i,1}$ and $\alpha_{i,2}$ are the distances that the characteristic variables propagate during the time interval Δt, given by

$$\begin{aligned} \alpha_{i,1} &= \Delta t \lambda_1, & (9.19)\\ \alpha_{i,2} &= \Delta t \lambda 2. & (9.20) \end{aligned}$$

As the characteristic velocities are constant in space for the linear case, it is possible to find the values of $\alpha_{i,1}$ and $\alpha_{i,2}$ exactly using techniques presented in earlier chapters, and thus the values at the departure point can be estimated through a quadratic, or higher, interpolation polynomial as before.

9.1.1 Nonlinear shallow water equations

We now move on to the nonlinearized, or rather the nonlinear, form of the shallow water equations, where in Eulerian form the governing equations are given by

$$\begin{aligned} \frac{\partial u}{\partial t} + u\frac{\partial u}{\partial x} &= -g\frac{\partial h}{\partial x}, & (9.21)\\ \frac{\partial h}{\partial t} + u\frac{\partial h}{\partial x} &= -h\frac{\partial u}{\partial x}, & (9.22) \end{aligned}$$

and the equivalent Lagrangian formulations are

$$\begin{aligned} \frac{Du}{Dt} &= -g\frac{\partial h}{\partial x}, & (9.23)\\ \frac{Dh}{Dt} &= -h\frac{\partial u}{\partial x}, & (9.24)\\ \frac{dx}{dt} &= u. & (9.25) \end{aligned}$$

As a quick note we should see that in the nonlinear shallow water equations model, (9.21) is a form of Burgers equation, but now with a forcing term on the right-hand side of the differential equation.

We are not able to apply the method that was described for the linearized version of these equations because the eigenvalues and eigenvectors are no longer constant. The technique that is described in [53] involves rewriting the *pressure* term as

$$\frac{\partial u}{\partial t} + u\frac{\partial u}{\partial x} + 2\sqrt{(gh)}\frac{\partial \sqrt{gh}}{\partial x} = 0. \quad (9.26)$$

The next step is to multiply (9.22) by $\sqrt{\frac{g}{h}}$, which yields

$$\sqrt{\frac{g}{h}}\frac{\partial h}{\partial x} + \sqrt{\frac{g}{h}}u\frac{\partial h}{\partial x} + \sqrt{\frac{g}{h}}h\frac{\partial u}{\partial x} = 2\frac{\partial \sqrt{gh}}{\partial t} + 2u\frac{\partial \sqrt{gh}}{\partial x} + \sqrt{gh}\frac{\partial u}{\partial x} = 0. \quad (9.27)$$

Recalling the definition for the gravity wave speed from earlier, $c=\sqrt{gh}$, one gets that

$$\frac{\partial u}{\partial t}+u\frac{\partial u}{\partial x} = -c\frac{\partial (2c)}{\partial x}, \tag{9.28}$$

$$\frac{\partial (2c)}{\partial t}+u\frac{\partial (2c)}{\partial x} = -c\frac{\partial u}{\partial x}. \tag{9.29}$$

If we now add and subtract (9.28) and (9.29), then we obtain

$$\frac{\partial}{\partial t}(u+2c)+(u+c)\frac{\partial}{\partial x}(u+2c) = 0, \tag{9.30}$$

$$\frac{\partial}{\partial t}(u-2c)+(u-c)\frac{\partial}{\partial x}(u-2c) = 0. \tag{9.31}$$

As we can see, (9.30) and (9.31) can now be expressed in the Lagrangian formulation as a system of unforced advection problems as

$$\frac{D}{Dt}(u+2c)=0 \ \text{ along } \ \frac{dx}{dt}=u+c, \tag{9.32}$$

$$\frac{D}{Dt}(u-2c)=0 \ \text{ along } \ \frac{dx}{dt}=u-c, \tag{9.33}$$

and so we can determine the characteristic variables and velocities as

$$p_1=u+2c, \qquad \lambda_1=u+c, \tag{9.34}$$

$$p_2=u-2c, \qquad \lambda_2=u-c. \tag{9.35}$$

It is possible to recover the u and h fields through the following relationship with the characteristic variables:

$$u=\frac{p_1+p_2}{2}, \qquad h=\left(\frac{p_1-p_2}{4\sqrt{g}}\right)^2. \tag{9.36}$$

A major difference between this approach and the linearized shallow water equations characteristic based approach is that the characteristic variables are given by $u\pm 2c$, where [53] refers to these as the Riemann invariants of the system; the characteristic velocities $u\pm c$ vary in space, and this spatial variation can give rise to wave breaking processes.

The semi-Lagrangian approaches described in the last chapter for the varying velocities can be applied to the nonlinear characteristic formulation that we have just derived, and an iterative scheme can be applied to the two departure points that we need. However, as we also saw in the last chapter, as the problem is based upon a nonlinear problem and since the velocity is a function space and time, there is a condition on the time step. It is stated in [53] that if the time step is too big, then the iterative process to find the two departure points will lose accuracy and the convergence of the iterative scheme will also deteriorate. Following the theory from [147] and [173], the following conditions have to be satisfied:

$$\Delta t\left\|\frac{\partial\lambda_1}{\partial x}\right\| \leq 1, \tag{9.37}$$

$$\Delta t \left\| \frac{\partial \lambda_2}{\partial x} \right\| \leq 1. \tag{9.38}$$

However, in many applications of the shallow water equations, along with the more advanced two- and three-dimensional systems, it is more often the case that we have to take into account the forcing term and how to discretize it, to ensure that the resulting semi-Lagrangian advection remains stable, so that we can use a large time step and do not introduce spurious waves. The more commonly used approach to address this is referred to as the Semi-Implicit Semi-Lagrangian method, which is quite often abbreviated to SISL. We start in the next section by introducing the original work to present the semi-implicit component before moving on to the semi-implicit semi-Lagrangian technique in the following section.

9.2 Semi-implicit integration

As mentioned in the last section, the goal of the semi-Lagrangian integration when applied to the forced linear, or nonlinear, advection equation is to avoid exciting nonphysical waves in the numerics, to remain stable for larger time steps than for an Eulerian approach, and to not allow computational modes to overwhelm the numerics either.

In a paper by Kwizak and Robert in 1972 [97], the authors start with the following statement:

> *Motion in compressible fluids can generate time oscillations extending over a very broad spectrum of frequencies. Numerical models concerned with a simulation of the essential features of the fluid motion need to carry only the part of the frequency spectrum that directly affects the large-scale circulations.*
>
> *The filtering approximation used in early atmospheric models eliminated both sound and gravity waves, but they produced equations that became difficult to integrate. Current models retain gravity modes, provides greater accuracy, and do not require any transformation of the basic equations for the numerical calculations.*
>
> *Unfortunately, the primitive equations models will not give stable solutions with long time steps. The short 10-min steps used in these models reduce the truncation error to a level two orders of magnitude lower than the error associated with the spatial resolution.*

We have to remember that Robert had not developed the semi-Lagrangian approach at that time yet, but in the following summary of the paper from 1972, we shall see some terms that look familiar to how we apply the semi-Lagrangian integration today.

The starting point is to note here is that oscillatory systems generally behave in accordance with relatively simple physical laws [97]. Trigonometric functions or complex exponential functions represent this type of motion adequately, and a first-order differential equation will describe the time variations accurately [97]. Thus the first-order differential equation that Kwizak and Robert consider is given by

$$\frac{\partial F}{\partial t} = -i\omega F, \tag{9.39}$$

where the constant ω represents the frequency of the oscillation. The next step is to break down this frequency into two components α and β such that $\omega \equiv \alpha + \beta$. In [97] the authors apply the following discretization to (9.39) with ω defined as above:

$$\frac{F(t+\Delta t) - F(t-\Delta t)}{2\Delta t} = i\alpha F(t) + i\beta\left[\frac{F(t+\Delta t) + F(t-\Delta t)}{2}\right]. \tag{9.40}$$

Two observations that are stated at this point in [97] are: (1) if the β term in (9.40) is absent then the difference equation has a centered step, (2) if α vanishes then we have a fully implicit step. The next statement is the part of this work that would eventually lead to the semi-implicit semi-Lagrangian approach: **"we may consider that β represents that part of the frequency ω that we will treat implicitly".**

Kwizak and Robert now introduce the notation X to be the *prediction operator* that transforms $F(t)$ to $F(t+\Delta t)$. This then results in (9.40) becoming the algebraic equation

$$(1 - i\beta\Delta t)X^2 - 2i\alpha\Delta t X - (1 + i\beta\Delta t) = 0, \tag{9.41}$$

where this equation has two solutions

$$X = \frac{i\alpha\Delta t \pm \sqrt{1 + \beta^2(\Delta t)^2 - \alpha^2(\Delta t)^2}}{1 - i\beta\Delta t}. \tag{9.42}$$

Looking at the square root in (9.42), we see that the property that the real component does not grow, and hence the numerical integration remains stable, is dependent on

$$\alpha^2(\Delta t)^2 \leq 1 + \beta^2(\Delta t)^2. \tag{9.43}$$

Given this derivation, Kwizak and Robert state that if we now consider a high-frequency perturbation where the restoring influence of gravity is accounted for by β, while the advective effect of a basic current u is included in α, then we have

$$k^2U^2(\Delta t)^2 \leq 1 + k^2gh_0(\Delta t)^2, \tag{9.44}$$

where h_0 represents the mean depth of the fluid, and k is the wavenumber of the perturbation. The condition above will be satisfied if

$$|U| \leq \sqrt{gh_0}, \tag{9.45}$$

and it is stated in [97] that this condition presents no problem in the atmosphere. The authors then go on to say that if we make an allowance for the rotation of the Earth in the model, or let the temperature vary with latitude, then these changes will alter the phase speed of the gravity waves. If these effects are incorporated into α, then the left-hand side of (9.43) will increase, but not sufficiently to violate this condition. They then state that **we conclude that the proposed treatment of the gravity waves will remain stable under nearly all conditions.**

If we now consider meteorological perturbations, then, since gravity has little effect on the phase speed of these waves [97], we can delete the corresponding term from (9.44). If we now apply a second-order centered finite difference in space, then (9.44) becomes

$$\frac{U^2 (\Delta t)^2}{(\Delta x)^2} \sin^2 k\Delta x \leq 1. \tag{9.46}$$

It is then stated in [97], and would motivate the semi-implicit semi-Lagrangian approaches in Roberts 1981 paper [161], and those following in the next section, that

> *The above analysis suggests that all terms in the meteorological equations should be evaluated explicitly with only a few exceptions. Gravity waves operate through pressure gradient terms and the divergence term. These quantities should be averaged in time as proposed in the β term in (9.40).*

9.3 Semi-implicit semi-Lagrangian (SISL)

An important feature that we have to note straightaway when the forcing term is present is that it has to be taken into account in the updating of the value of the tracer at the arrival point through the back trajectory. We have seen in the last section that we do not wish to excite nonphysical modes, but we also do not wish to damp correct physical modes, while controlling computational modes from the discretization. A general way to state the semi-implicit semi-Lagrangian approach is through

$$\frac{\psi\left(x_i, t^{n+1}\right) - \psi\left(\tilde{x}_i, t^n\right)}{\Delta t} = \theta f\left(x_i, t^{n+1}\right) + (1-\theta) f\left(\tilde{x}_i, t^n\right), \tag{9.47}$$

where if $\theta = 1$ then we have the equivalent to an implicit Euler method, while if we take $\theta = \frac{1}{2}$, then we have the trapezoidal rule approach; finally, if we set $\theta = 0$ then we have the explicit Euler method to integrate the forcing term.

However, we have to determine the value for the forcing term at the departure point at time t^n, and so we would use the same order of interpolation polynomial as that used for the Lagrange differential operator's numerical approximation for a constant advecting velocity.

Therefore, discretizing (9.47) with a simple forward upwind scheme for the time derivative results in the general update for ψ_i^{n+1} as

$$\psi_i^{n+1} = \tilde{\psi}_i^n + \Delta t f\left(\tilde{x}_i, t^n\right). \tag{9.48}$$

If we consider a semi-implicit, semi-Lagrange approach, then the associated discrete equation may be of the form

$$\frac{D\psi}{Dt} = \frac{1}{2} f\left(\tilde{x}_i, t^{n+1}\right) + \frac{1}{2} f\left(\tilde{x}, t^n\right).$$

The reason for considering the forcing terms as the sum of an implicit and an explicit component is motivated by the work by Kwizak and Robert in 1972 to separate the forcing term into a fast linear term and a slower nonlinear term, which is referred to as a *residual nonlinear term* in [17], such that we now can write the general forced Lagrangian differential equation as

$$\frac{D\psi}{Dt} = \mathsf{N} + \mathsf{L}. \tag{9.49}$$

It is assumed that the two terms on the right-hand side of (9.49) depend on the tracer $\psi(x,t)$, and it is assumed that we know the velocity u at each time step up to the current one. To solve (9.49), it is quite common for an explicit based discretization to be used for the nonlinear term, while implicit schemes are used for the linear term.

We now consider the scheme from [128], where although the formulation is for two-dimensional spherical coordinates, we shall go into numerical modeling on the sphere in Chapter 12, we will explain the theory for a one-dimensional Cartesian coordinate system here.

We start by assuming that a dynamical quantity is integrated in a semi-Lagrangian form, which means that Eq. (9.49), when discretized in time, has the solution

$$\psi^{n+1} - \widetilde{\psi}^{n} = \int_{t}^{t+\Delta t} \left(\mathsf{N}(x,t) + \mathsf{L}(x,t)\right) dt, \tag{9.50}$$

where $\widetilde{\psi}$ is the value of ψ at the departure point. However, unlike where we only search for the departure point at the current time step, it is stated that for a variable advecting wind, $u(x,t)$, the departure point is calculated as $x - \hat{u}\Delta t$, as we saw in the last chapter, where $\hat{u}$ is calculated as a space and time-centered estimate along a trajectory.

We should recall here that the integral in (9.50) is along a trajectory of a parcel of air. For the linear term, it is possible to use the following semi-Lagrangian estimate:

$$\int_{t}^{t+\Delta t} \mathsf{L}(x,t)\, dt = \left(\frac{\Delta t}{2}\right)\left(\mathsf{L}^{n+1} - \widetilde{\mathsf{L}}^{n}\right), \tag{9.51}$$

which is correct to $O\left((\Delta t)^2\right)$.

However, it is not really recommended to use the approach in (9.51) for the nonlinear term as it leads to a complicated system of equations which is difficult to solve [128]. It is stated in [128] that if the nonlinear terms are small enough, then it is possible to integrate through an explicit approach without causing instability. As an example, [128] introduces two different explicit schemes to approximate the integration of the nonlinear term. The first scheme is referred to as *centered Lagrangian explicit* (1) *scheme*, which comes from [188] and is defined as

$$\int_{t}^{t+\Delta t} \mathsf{N}(x,t)\, dt = \Delta t \widetilde{\mathsf{N}}_{\frac{1}{2}}^{n+\frac{1}{2}}, \tag{9.52}$$

where the nonlinear term is evaluated at the halfway point between the arrival point and departure points such that this would be the value for the nonlinear term at $t + \frac{1}{2}\Delta t$, i.e.,

$$\widetilde{\mathsf{N}}_{\frac{1}{2}}^{\frac{n+1}{2}} \equiv \mathsf{N}\left(x - \frac{\widehat{u}\Delta t}{2}, t^{n+\frac{1}{2}}\right).$$

The second scheme that has the same order of accuracy as the first was introduced in [128] and is referred to as the *centered Lagrangian explicit* (2) approximation, given by

$$\int_{t}^{t+\Delta t} \mathsf{N}(x,t)\,dt = \frac{\Delta t}{2}\left(\mathsf{N}^{n+\frac{1}{2}} + \widetilde{\mathsf{N}}^{n+\frac{1}{2}}\right). \tag{9.53}$$

The value of the fields at the time level $n + \frac{1}{2}$ are computed through the second order **Adams–Bashforth** scheme, which just to recap is given by

$$\psi^{n+\frac{1}{2}} = \frac{3}{2}\psi^{n} - \frac{1}{2}\psi^{n-1}. \tag{9.54}$$

Using the second-order Adams–Bashforth scheme means that the second centered scheme only requires one set of interpolation, i.e.,

$$\psi^{n+1} - \frac{\Delta t}{2}\mathsf{L}^{n+1} = \left[\psi^{n} + \frac{\Delta x}{2}\mathsf{L}^{n} + \frac{\Delta t}{2}\mathsf{N}^{n+\frac{1}{2}}\right]_{d} + \frac{\Delta t}{2}N^{n+\frac{1}{2}}, \tag{9.55}$$

whereas for the first centered scheme we have

$$\psi^{n+1} - \frac{\Delta t}{2}\mathsf{L}^{n+1} = \left[\psi^{n} + \frac{\Delta t}{2}\mathsf{L}^{n}\right]_{d} + \Delta t\mathsf{N}_{d/2}^{n+\frac{1}{2}}, \tag{9.56}$$

and therefore we require two sets of interpolations, the linear term to the departure point and then the nonlinear term to the halfway point on the trajectory from the departure point.

If we now consider how to determine the accuracy of the two approaches, we use equations (6), (A3), and (14) from McDonald 1987 [119], then, assuming a linear or higher order interpolation to estimate the departure point values of N and (9.56) to evaluate $\mathsf{N}^{n+\frac{1}{2}}$ to order $O\left((\Delta)^3\right)$ both $\Delta t\mathsf{N}_{\frac{d}{2}}^{n+\frac{1}{2}}$, we have

$$\frac{\Delta t}{2}\left[\mathsf{N}^{n+\frac{1}{2}} + \mathsf{N}_{d}^{n+\frac{1}{2}}\right] = \left[\Delta\mathsf{N} + \left(\frac{(\Delta t)^2}{2}\right)\left(\frac{\partial}{\partial t} - u\frac{\partial}{\partial x}\right)\mathsf{N}\right]_{i}^{n} + O\left((\Delta)^3\right). \tag{9.57}$$

Hence, if either (9.55) or (9.56) is used to discretize

$$\frac{D\psi(x,t)}{Dt} = \mathsf{L}(x,t) + \mathsf{N}(x,t), \tag{9.58}$$

with a quadratic or higher order interpolation to estimate the departure point values of ψ, L, and N, one iteration of a linear interpolation to center in space and (9.54) to center in time, then

$$\frac{D\psi}{Dt} - \mathsf{L} - \mathsf{N} = \left[\left[\left(1+\frac{\Delta t}{2}\right)\left(\frac{\partial}{\partial t} - u\frac{\partial}{\partial x}\right)\right]\left(\frac{\partial \psi}{\partial t} + u\frac{\partial \psi}{\partial x} - \mathsf{L} - \mathsf{N}\right)\right]_i^n + o\left(\left(\Delta\right)^2\right), \quad (9.59)$$

thus both schemed are equivalent to this level of accuracy.

It can easily be shown that the relative accuracy of the two centered approaches through using the equations in [128] and [126] are the same when a linear or a higher order interpolation polynomial is used to estimate the departure point values for the nonlinear term for the various trajectories.

9.4 Spatial averaging

The reason for the introduction of the semi-implicit semi-Lagrangian approach was to have a stable treatment of the fast moving gravity modes where it was shown early on in numerical weather prediction attempts that these spurious waves could cause unstable forecasts. However, it is stated in [186] when strong winds blow over steep mountains, instabilities may occur in part of the linear term of $\nabla\left(\phi + RT^* \ln p_s\right)$ and the nonlinear term $RT'\nabla \ln p_s$ of the horizontal pressure gradient term. We shall go into details about the two- and three-dimensional models over the next couple of chapters. An approach to overcome this instability is suggested in a series of papers about the same time in the early 1990s. Given the scalar advection equation

$$\frac{D\psi}{Dt} + G = R, \quad (9.60)$$

where G is the term responsible for the fast-moving waves, and the meteorological slower Rossby modes are represented by R, then the semi-implicit semi-Lagrangian approach consists in evaluating G as a time average along the trajectory which is given by

$$\overline{G}^t = \frac{G\left(x_i, t^{n+1}\right) + G\left(x_i - 2u\Delta t, t^{n-1}\right)}{2}, \quad (9.61)$$

and the nonlinear R term in an explicit manner, centered in time along trajectories, as follows:

$$R^n = R\left(x - u\Delta t, t^n\right) \quad (9.62)$$

evaluated at a geographical midpoint. Combining all of these terms results in the semi-Lagrangian approximation to (9.60) as

$$\frac{\psi\left(x_i, t^{n+1}\right) - \psi\left(x_i - 2u\Delta t, t^{n-1}\right)}{2\Delta t} + \overline{G}^t = R^n. \quad (9.63)$$

Regrouping the unknown terms at t^{n+1} on the left-hand side and the known terms at t^n and t^{n-1} on the right-hand side yields

$$(\psi + \Delta t G)\left(x_i, t^{n+1}\right) = (\psi - \Delta t G)\left(x_i - 2u\Delta t, t^{n-1}\right) + 2\Delta t R\left(x_i - u\Delta t, t^n\right). \tag{9.64}$$

The right-hand side expression interpolates the first term at the geographical departure point and the second term at the geographical midpoint of the trajectory.

In [186] they now define a spatial average operator as

$$\overline{R}^x \equiv \frac{R\left(x_i, t^n\right) + R\left(x_i - 2u\Delta t, t^n\right)}{2}, \tag{9.65}$$

where R is evaluated at the geophysical endpoints of the trajectory, but at the same time level t^n. The modified semi-implicit semi-Lagrangian formulation of (9.60) is obtained by replacing R^n with $\overline{R}^x$ in (9.63) such that

$$\frac{\psi\left(x_i, t^{n+1}\right) - \psi\left(x_i - 2u\Delta t, t^{n-1}\right)}{2\Delta t} + \overline{G}^n = \overline{R}^x. \tag{9.66}$$

It can be shown that this approximation introduces an extra term of the order of $O\left((\Delta x)^2\right)$ into the truncation error analysis. So if we now regroup the knowns and unknowns on either side of the equal sign, then we obtain

$$(\psi + \Delta t G)\left(x_i, t^{n+1}\right) = (\psi - \Delta t G)\left(x_i - 2u\Delta t, t^{n-1}\right) + \Delta t R\left(x_i - 2u\Delta t, t^n\right) + \Delta t R\left(x_i, t^n\right). \tag{9.67}$$

The modified right-hand side only requires one iteration for the first term at the geographical departure point of the trajectory. The second term is already known at the grid point x_i. If we now compare (9.64) with (9.67), we can see that the total number of interpolations required for the right-hand side calculations is reduced by half in (9.67).

9.5 Optimal accuracy associated with uncentering time averages

In a paper by Semazzi and Dekker in from 1994 [169], the authors performed an investigation into the sensitivity of the semi-implicit semi-Lagrangian approach for a linear model on (1) the order of interpolation applied to the departure point, (2) trajectory uncentering which was used to suppress computational gravitational noise, and (3) the optimal truncation conditions. The work by Tanguay et al. (1992), Bates et al. (1993), and McDonald and Haugen (1992) [186,15,128], respectively, showed that it was possible to use uncentered schemes to suppress the computational noise that can be written in terms of a general parameter ϵ to indicate the degree of uncentering, given by

$$\overline{L} \equiv \frac{1}{2}\left[(1+\epsilon) L\left(x_i, t^{n+1}\right) + (1-\epsilon) L\left(s - u\Delta t, t^n\right)\right],$$

$$\overline{N} \equiv \frac{1}{2}\left[(1+\epsilon)\widehat{N}\left(x_i, t^{n+1}\right) + (1-\epsilon)N\left(x_i - u\Delta t, t^n\right)\right], \tag{9.68}$$

where

$$\widehat{N}\left(x_i, t^{n-1}\right) \equiv 2N\left(x_i, t^n\right) - N\left(x_i, t^{n-1}\right). \tag{9.69}$$

Given this general form of uncentering, we now consider the following one-dimensional linear atmospheric model that can be used to investigate the Rossby wave solutions, which is given by

$$\left(f_0 + \beta y\right) v'(x,t) - \frac{\partial \phi'(x,y,t)}{\partial x} = 0, \tag{9.70}$$

$$\frac{Dv'(x,t)}{Dt} + \frac{\partial'(x,y,t)}{\partial y} = 0, \tag{9.71}$$

where the advective wind $\overline{u}$, which is the basic-state zonal wind, in the Lagrangian derivative above is a constant, ϕ' is a perturbation of the basic-state mass variable $\overline{\phi}(y)$, v' is a perturbation of the meridional wind who's constant basic-state meridional wind is equal to zero, β is the beta parameter at 45 N, f_0 is the Coriolis parameter at a reference latitude. In the stability analysis that is undertaken in [169], the authors assume a nonzero basic current, which is in geostrophic balance with the mean geopotential height $\overline{\phi}(y)$. It is also assumed that the basic state is in balance with a constant North–South slope in the bottom topography. Given the condition of nondivergence, this implies that gravitational oscillations have been suppressed [169].

Given the continuous model, we now apply the general semi-implicit semi-Lagrangian finite difference approximation to (9.70) and (9.71) as

$$(1+\epsilon)\left(\left(\frac{\partial\phi}{\partial x}\right)_{i,j}^{n+1} - f_j v_{i,j}^{n+1}\right) + (1+\epsilon)\left(\left(\frac{\partial\phi}{\partial x}\right)_d^n - f_d v_d^n\right) = 0, \tag{9.72}$$

$$v_{i,j}^{n+1} - v_d^n = -\frac{\Delta t}{2}\left((1+\epsilon)\left(\frac{\partial\phi}{\partial y}\right)_{i,j}^{n+1} + (1-\epsilon)\left(\frac{\partial\phi}{\partial y}\right)_d^n\right). \tag{9.73}$$

Linear, quadratic, and cubic interpolation polynomials are used for estimating the values at the departure points in (9.72) and (9.73). In the analysis presented in [169], the authors represent the departure point by the general formula

$$q_d^n = \sum_r C_r q_{i-p-r,j}^n, \tag{9.74}$$

where for the linear interpolation polynomial we have

$$\begin{aligned} r &= 0, \quad C_0 = 1 - \widehat{\alpha}, \\ r &= 1, \quad C_1 = \widehat{\alpha}, \end{aligned} \tag{9.75}$$

for the quadratic interpolation polynomial we have

$$\begin{aligned} r&=1, & C_1&=\frac{1}{2}\widehat{\alpha}\left(1-\widehat{\alpha}\right),\\ r&=0, & C_0&=\left(1-\widehat{\alpha}\right)\left(1+\widehat{\alpha}\right),\\ r&=-1, & C_{-1}&=-\frac{1}{2}\widehat{\alpha}\left(1-\widehat{\alpha}\right), \end{aligned} \tag{9.76}$$

and for the cubic case we have

$$\begin{aligned} r&=-1, & C_{-1}&=-\frac{1}{6}\widehat{\alpha}\left(1-\widehat{\alpha}\right)\left(2-\widehat{\alpha}\right),\\ r&=0, & C_0&=\frac{1}{2}\left(1-\widehat{\alpha}\right)\left(1+\widehat{\alpha}\right)\left(2-\widehat{\alpha}\right),\\ r&=1, & C_1&=\frac{1}{2}\widehat{\alpha}\left(1+\widehat{\alpha}\right)\left(2-\widehat{\alpha}\right),\\ r&=2, & C_2&=-\frac{1}{6}\widehat{\alpha}\left(1-\widehat{\alpha}\right)\left(1+\widehat{\alpha}\right). \end{aligned} \tag{9.77}$$

It should also be noted that $\widehat{\alpha}$ satisfies the inequality $0\leq\widehat{\alpha}\leq 1$ for the linear and cubic interpolation polynomials, and $-\frac{1}{2}\leq\widehat{\alpha}\leq\frac{1}{2}$ for the quadratic interpolation polynomial, as we saw for the Lagrange interpolation polynomials earlier. Now using (9.74), combined with a second-order accurate differencing for the spatial derivatives, results in the model equations (9.72) and (9.73) becoming

$$\begin{aligned} &(1+\epsilon)\left(\frac{\phi_{i+1,j}^{n+1}-\phi_{i-1,j}^{n+1}}{2\Delta x}-f_j v_{i,j}^{n+1}\right)\\ &\quad+(1-\epsilon)\left(\sum_r C_r\frac{\psi_{i-p-r+1,j}^{n}-\phi_{i-p-r-1,j}^{n}}{2\Delta x}-f_j\sum_r C_r v_{i-p-r,j}^{n}\right)=0, \end{aligned} \tag{9.78}$$

$$\begin{aligned} v_{i,j}^{n+1}-\sum_r C_r v_{i-p-r,j}^{n}=-\frac{\Delta t}{2}&\left((1+\epsilon)\left(\frac{\phi_{i,j+1}^{n+1}-\phi_{i,j-1}^{n+1}}{2\Delta y}\right)\right.\\ &\left.+(1-\epsilon)\left(\sum_r C_r\frac{\psi_{i-p-r,j+1}^{n}-\phi_{i-p-r,j-1}^{n}}{2\Delta y}\right)\right). \end{aligned} \tag{9.79}$$

To construct the one-dimensional Rossby wave solution, we have to assume that

$$(v,\phi)_{i,j}^{n}=\lambda^n\left(\widehat{v},\widehat{\phi}_j\right)e^{iki\Delta x}, \tag{9.80}$$

where to keep the indexing consistent throughout the book, i here is $\sqrt{-1}$. Substituting (9.80) into (9.78) results in

$$\begin{aligned} &(1+\epsilon)\lambda\left(\widehat{\phi}_j\left(\frac{\left(e^{ik\Delta x}-e^{-ik\Delta x}\right)}{2\Delta x}\right)-f_j\widehat{v}\right)\\ &\quad+(1+\epsilon)\sum_r C_r e^{-ik(p+r)\Delta x}\left(\widehat{\phi}_j\frac{\left(e^{ik\Delta x}-e^{-ik\Delta x}\right)}{2\Delta x}-f_j\widehat{v}\right)=0. \end{aligned} \tag{9.81}$$

Now making use, as always, of the trigonometric identity, $\sin k\Delta x \equiv \dfrac{e^{ik\Delta x} - e^{-ik\Delta x}}{2i}$, results in (9.81) becoming

$$(1+\epsilon)\lambda + (1-\epsilon)\sum_r C_r e^{-ik(p+r)\Delta x}\left(\widehat{\phi}_j - \left(\frac{\Delta x}{i\sin k\Delta x}\right) f_j \widehat{v}\right) = 0. \tag{9.82}$$

In general, the first term in (9.82) does not vanish, and the nontrivial solution is

$$\widehat{\phi}_j - \left(\frac{\Delta x}{i\sin k\Delta x}\right) f_j \widehat{v} = 0, \tag{9.83}$$

with the expected analytical limit when $\Delta x \to 0$ being

$$\widehat{\phi}_j - \left(\frac{f_j \widehat{v}}{ik}\right) = 0. \tag{9.84}$$

It is stated in [169] that the limiting condition may be obtained by assuming that the analytical solution satisfies

$$(v, \phi) = \left(\widehat{v}, \widehat{\phi}\right) e^{i(kx - \omega t)}. \tag{9.85}$$

Substituting (9.85) into (9.70) and (9.71) yields

$$\omega = \left(\overline{u}k - \frac{\beta}{k}\right), \tag{9.86}$$

and (9.84).

The next step is to focus on the equation of motion in the y direction, (9.79), where we again assume the solution to be of the form in (9.80), such that we arrive at

$$\begin{aligned}
&\left(\lambda - \sum_r C_r e^{-ik(p+r)\Delta x}\right)\widehat{v} \\
&= \frac{\Delta t}{2}\left((1+\epsilon)\lambda \frac{\left(\widehat{\phi}_{j+1} - \widehat{\phi}_{j-1}\right)}{2\Delta y} + (1-\epsilon)\sum_r C_r e^{-ik(p+r)\Delta x}\frac{\left(\widehat{\phi}_{j+1} - \widehat{\phi}_{j-1}\right)}{2\Delta y}\right).
\end{aligned} \tag{9.87}$$

It is stated in [169] that the left-hand side of (9.87) arises from the semi-Lagrangian contribution of the semi-implicit semi-Lagrangian scheme. Given this information, we now combine (9.83) with (9.87) to obtain

$$\begin{aligned}
\left(\lambda - \sum_r e^{-ik(p+r)\Delta x}\right)\Delta x \quad &= \quad (1+\epsilon)\lambda\left(\frac{i\Delta x\Delta t}{2\sin k\Delta x}\right)\widehat{v}\frac{\left(f_{j+1} - f_{j-1}\right)}{2\Delta y} \\
&+ \quad \left((1-\epsilon)\lambda\left(\frac{i\Delta x\Delta t}{2\sin k\Delta x}\right)\widehat{v}\frac{\left(f_{j+1} - f_{j-1}\right)}{2\Delta y}\right)\sum_r C_r e^{-ikp\Delta x} e^{-ikr\Delta x}.
\end{aligned} \tag{9.88}$$

We now define the parameter $B = \dfrac{\beta \Delta x \Delta t}{2} \sin k\Delta x$, where here $\beta = \dfrac{f_{j+1} - f_{j-1}}{2\Delta y}$, which enables us to write (9.88) as

$$\lambda\left(1 - \mathrm{i}\left(1+\epsilon\right)B\right) = \sum_r C_r\left(1 + \mathrm{i}\left(1-\epsilon\right)B\right)e^{\mathrm{i}kp\Delta x}e^{-\mathrm{i}kr\Delta x}. \tag{9.89}$$

Recalling the trigonometric identity $e^{\pm \mathrm{i}\psi} \equiv \cos\psi \pm \mathrm{i}\sin\psi$, it is possible to write (9.89) as

$$\lambda = \frac{\mu_1\cos\theta + \mu_2\sin\theta + \mathrm{i}\left(\mu_2\cos\theta - u_1\sin\theta\right)}{1 + B_+^2}, \tag{9.90}$$

where

$$\begin{aligned}
\theta &\equiv kp\Delta x, \\
B_{\pm} &\equiv \left(1 \pm \epsilon\right)B, \\
\mu_1 &\equiv \sum_r C_r\left(\cos kr\Delta x + 2B\sin kr\Delta x - \left(1-\epsilon\right)\left(1+\epsilon\right)B^2\cos kr\Delta x\right), \\
\mu_2 &\equiv \sum_r C_r\left(-\sin kr\Delta x + 2B\cos kr\Delta x\left(1-\epsilon\right)\left(1+\epsilon\right)B^2\sin kr\Delta x\right).
\end{aligned} \tag{9.91}$$

The corresponding amplification factor is

$$\left|\lambda\right|_{SISL}^2 = \frac{1+B_-^2}{1+B_+^2}\left(\left(\sum_r C_r\cos\varphi_r\right)^2 + \left(\sum_r C_r\sin\varphi_r\right)^2\right), \tag{9.92}$$

where $\varphi_r = rk\Delta x$, and we should note that $B_- + B_+ = 2B$.

When $\widehat{\alpha} = p = \epsilon = 0$ in (9.92), we see that $\left|\lambda\right|_{SISL} = 1$ and so neutral stability is guaranteed. Setting $\epsilon \neq 0$, linear computational stability is assured if $\epsilon \geq 0$. In this case $\left|\lambda\right|_{SISL} < 1$, thus the Rossby wave solutions are being damped in time.

In [169] the authors state that, to complete the solution for (9.72) and (9.73), temporal dependency of the form $\lambda = \left|\lambda\right|_{SISL}e^{\mathrm{i}\omega_{SISL}\Delta t}$ is assumed. Thus the SISL wave frequency ω_{SISL} is given by

$$\omega_{SISL} = -\frac{1}{\Delta t}\tan^{-1}\left(\frac{\mu_2\cos\theta - \mu_1\sin\theta}{\mu_1\cos\theta + \mu_2\sin\theta}\right). \tag{9.93}$$

Applying (9.92) and (9.93) in (9.80) yields

$$v_{i,j}^n = \widehat{v}\left(\left|\lambda\right|_{SISL}\right)^n\exp\left\{\mathrm{i}\left(n\tan^{-1}\left(\omega_{SISL}\right) + ki\Delta x\right)\right\} \tag{9.94}$$

and

$$\phi_{i,j}^n = \left(\frac{\Delta x}{\mathrm{i}\sin k\Delta x}\right)f_j\widehat{v}\left(\left|\lambda\right|_{SISL}\right)^n\exp\left\{\mathrm{i}\left(n\tan^{-1}\left(\omega_{SISL}\right) + ki\Delta x\right)\right\}, \tag{9.95}$$

where $\widehat{v}$ is a free parameter of $O\left(10\,\mathrm{ms}^{-1}\right)$. From (9.85) and (9.86) the analogous analytical solutions are

$$v_{i,j}^{n} = \widehat{v}\exp\left\{\mathrm{i}\left(n\Delta t\left(\frac{\beta}{k}-u_0k\right)+ki\Delta x\right)\right\}, \tag{9.96}$$

$$\phi_{i,j}^{n} = \left(\frac{f_j\widehat{v}}{\mathrm{i}k}\right)\exp\left\{\mathrm{i}\left(n\Delta t\left(\frac{\beta}{k}-u_0k\right)+ki\Delta x\right)\right\}. \tag{9.97}$$

In [169] the authors go on to perform a series of experiments with the three interpolations schemes described earlier, and show that for the linear model we have just derived, the Rossby wave solution for the linear and quadratic interpolations for the estimate of the field at the departure point showed that the truncation error associated with the semi-Lagrangian component of the semi-implicit semi-Lagrangian scheme dominates the error characteristics. It is also shown in [169] that the rate of error growth is large and may significantly exceed the acceptable levels in operational numerical weather prediction.

However, for the case where Semazzi and Dekker are using the cubic Lagrange interpolation, this scheme drastically reduced the overall truncation errors, and the accuracy appeared to be within acceptable limits, which is still true today. They also showed that the truncation errors in this case are almost entirely due to the semi-implicit part of the semi-implicit semi-Lagrangian scheme.

Semazzi and Dekker do have a warning about the uncentering of the calculations of the semi-implicit components. They state that in the case of planetary-scale waves, uncentering tends to impose a more stringent upper limit on the choice of the time step compared to the fully centered scheme, and that unlike the Eulerian formulation, this constraint appears due to the consideration of accuracy, and not that of limitation of stability. In a two-dimensional case, which we will present later in the book, the uncentering approach had been shown to be effective in suppressing undesirable gravitational computational noise, but in [169] the authors point out that the uncentering scheme must be applied with caution as it may also degrade the low-frequency component of the flow that we wish to preserve and predict.

9.6 Semi-Lagrangian trajectories and discrete modes

In a paper by Cordero et al. [38], the authors are concerned with analyzing the stability effects of the trajectory computation, where the authors do state that in the paper by Bates et al. [13], there was a Rossby-wave instability that was detected in their global shallow water equations model which was reformulated in terms of potential vorticity. However, when Bates et al. reformulated the model in a stable way, by including the effects of the trajectory calculation, their formulation indicated that the instability was due to advection of the Coriolis term, due to the extrapolation used in the trajectory calculations.

Given this initial indication of an instability in the trajectory calculation, Cordero extended the work from Staniforth et al. (2003) [178], where the authors had demonstrated the importance that trajectory computations can have for model accuracy. We do not summarize the work from [178] here as there have been more recent developments at the Environment and

Climate Change Center in Canada to address this sensitivity (see Chapter 11). In [38] the authors indicate that they are going to analyze the impact various trajectory schemes have on the temporally and spatially discrete normal modes of a standard two-time-level semi-implicit semi-Lagrangian discretization of a prototypical vertical-column model. In this work they study stability via the growth rate of the normal modes, and the impact of the discretization on the vertical normal mode structure.

The model that we are going to present here from [38] is slightly different to what we have seen so far in that it is a vertical column model, which means that the governing equations in continuous form are

$$\frac{Dw}{Dt} = -g - c_p\theta\frac{\partial \Pi}{\partial z}, \tag{9.98a}$$

$$\frac{D\theta}{Dt} = 0, \tag{9.98b}$$

$$\frac{D\rho}{Dt} = -\rho\frac{\partial w}{\partial z}, \tag{9.98c}$$

$$\frac{p_0}{\kappa c_p}\Pi^{\frac{1-\kappa}{\kappa}} = \rho\theta, \tag{9.98d}$$

subject to the bottom and top boundary conditions

$$w|_{z=0} = w|_{z=z_T} = 0, \tag{9.99}$$

where the Lagrangian total derivative here is given by

$$\frac{D}{Dt} \equiv \frac{\partial}{\partial t} + w\frac{\partial}{\partial z},$$

w is vertical velocity, $\Pi \equiv \left(\frac{p}{p_0}\right)^{\kappa}$ is the Exner pressure, p is pressure, $p_0 = p\,(z=0)$, $\kappa \equiv \frac{R}{c_p}$, R is the gas constant, c_p is the specific heat of dry air at a constant pressure, θ is the potential temperature, g is gravity, ρ is density, z is the height above the surface, and z_T is the height of the top boundary.

In [38] the authors introduce a centered semi-implicit scheme to not add any extra damping. The discretization obtained is through introducing an isothermal, hydrostatically balanced, semi-implicit reference profile, that is denoted by ref and represents the variables θ, Π, and ρ as the sum of a reference state component and a finite amplitude perturbation such that $\theta = \theta_{ref} + \theta'$, $\Pi = \Pi_{ref} + \Pi'$, and $\rho = \rho_{ref} + \rho'$, but unlike in a linearization the perturbation product terms are still included. This then results in

$$\begin{aligned}\frac{w'^{n+1} - w'^{n}_{d}}{\Delta t} &= -\frac{1}{2}c_p\left(\left(\theta_{ref}\left(\delta_z\Pi'\right)^{n+1} + \left(\theta_{ref}\delta_z\Pi'\right)\Big|_d^n\right)\right.\\ &+ \left(\theta'^{n+1}\delta_z\Pi_{ref} + \left(\theta'\delta_z\Pi_{ref}\right)\Big|_d^n\right)\\ &+ \left.\left(\left(\theta'\delta_z\Pi'\right)^{n+1} + \left(\theta'\delta_z\Pi'\right)\Big|_d^n\right)\right),\end{aligned} \tag{9.100}$$

$$\frac{\theta'^{n+1} - \theta'^{n}_{d}}{\Delta t} = -\frac{1}{2}\left(w'^{n+1}\delta_{2z}\theta_{ref} + \left(w'\delta_{2z}\theta_{ref}\right)\Big|_{d}^{n}\right), \tag{9.101}$$

$$\begin{aligned}\frac{\rho'^{n+1} - \rho'^{n}_{d}}{\Delta t} &= -\frac{1}{2}\left(w'^{n+1}\left(\delta_{z}\rho_{ref}\right)^{n+1} + \left(w'\delta_{z}\rho_{ref}\right)\Big|_{d}^{n}\right)\\ &- \frac{1}{2}\left(\left(\rho_{ref} + \rho'\right)^{n+1}\left(\delta_{z}w'\right)^{n+1} + \left(\rho_{ref} + \rho'\right)\Big|_{d}^{n}\left(\delta_{z}w'\right)\Big|_{d}^{n}\right),\end{aligned} \tag{9.102}$$

$$\frac{p_0}{\kappa c_p}\left(\Pi_{ref} + \Pi'^{n+1}\right)^{\frac{1-\kappa}{\kappa}} = \left(\rho_{ref} + \rho'\right)\left(\overline{\theta_{ref}}^{z} + \overline{\theta'^{n+1}}^{z}\right). \tag{9.103}$$

The departure point in this discretization is in the vertical direction and as such $z_d \equiv z - \alpha$, where α is the vertical displacement, defined by

$$\alpha = \Delta t w\left(z - \frac{\alpha}{2}, t^{n+\frac{1}{2}}\right). \tag{9.104}$$

The variables are stored on a Charney–Phillips grid which consists of $N+1$, including the top and bottom boundaries, integer model levels where θ and w are held, and N half-integer levels where Π and ρ are held. Since the variables are staggered in the vertical direction, vertical averages, denoted by $\overline{(\cdot)}^{z}$, are introduced in order to evaluate the variables at the levels they are not stored. This average, and the two vertical derivatives δ_z and δ_{2z}, are defined as follows:

$$\overline{G}^{z}\Big|_{z_k} \equiv \frac{1}{2}\left(G\left(z_{k+\frac{1}{2}}\right) + G\left(z_{k-\frac{1}{2}}\right)\right), \tag{9.105a}$$

$$\delta_z G|_{z_k} \equiv \frac{G\left(z_{k+\frac{1}{2}}\right) - G\left(z_{k-\frac{1}{2}}\right)}{\Delta z}, \tag{9.105b}$$

$$\delta_{2z} G|_{z_k} \equiv \frac{G\left(z_{k+1}\right) - G\left(z_{k-1}\right)}{\Delta 2z}, \tag{9.105c}$$

where G is any quantity, and Δz is the uniform vertical increment between adjacent integer, or half-integer, model levels.

Linearization of the discrete equations

In [38] the authors introduce a linearization of (9.100)–(9.103) about an isothermal hydrostatically balanced basic state, which is generally different from the semi-implicit reference state. This is achieved through rewriting the finite-amplitude perturbations as the sum of a finite perturbation component, which represents the difference between its basic and reference state values, and an infinitesimal perturbation, denoted by a double prime in [38], such that for any dependent variable G we have $G' = \Delta G_{ref} + G''$ where $\Delta G_{ref} \equiv G_s - G_{ref}$, so that, since $G = G_{ref} + G''$, it follows that $G = G_s + G''$. Nonlinear terms in the infinitesimal perturbations are neglected in these equations.

The linearized discrete equations are at the appropriate model levels, where all the terms that do not have a subscript are the double prime quantities. The **vertical momentum equa-**

tion at an integer model level $k = 0, 1, \dots, N$ is given by

$$\begin{aligned}
w^{n+1}\Big|_{z=0} &= 0, \\
\frac{w^{n+1} - w^d}{\Delta t} &= -\frac{1}{2}c_p\left(\left((\theta_s \delta z \Pi)^{n+1} + (\theta_s \delta_z \Pi)|_d^n\right) + (\theta \delta_z \Pi_s)^{n+1} + (\theta \delta_z \Pi)|_d^n\right), \\
&\qquad k = 1, \dots, N-1, \\
w^{n+1}\Big|_{z=z_T} &= 0.
\end{aligned} \tag{9.106}$$

The thermodynamics equation is at an integer model level $k = 0, 1, \dots, N$ is given by

$$\begin{aligned}
\theta^{n+1}\Big|_{z=0} &= \theta_d^n\Big|_{z=0}, \\
\frac{\theta^{n+1} - \theta_d^n}{\Delta t} &= -\frac{1}{2}\left(\left(w\delta_{2z}\theta_{ref}\right)^{n+1} + \left(w\delta_{2c}\theta_{ref}\right)|_d^n\right) - w|_M^{n+\frac{1}{2}}\delta_{2z}\left(\Delta\theta_{ref}\right), \quad k = 1, \dots, N-1, \\
\theta^{n+1}\Big|_{z_T} &= \theta_d^n\Big|_{z_T}.
\end{aligned} \tag{9.107}$$

The continuity equation at a half-integer model level $k = \frac{1}{2}, \frac{3}{2}, \dots, N - \frac{1}{2}$ is given by

$$\begin{aligned}
\frac{\rho^{n+1} - \rho_d^n]}{\Delta t}\Bigg|_{z_{\frac{1}{2}}} &= -\frac{1}{2}\left(w|_M^{n+\frac{1}{2}}\delta_z\Delta\rho_{ref}\Big|_{z_1} + \frac{1}{2}\left(\left(w\delta_z\rho_{ref}\right)^{n+1} + \left(w\delta_z\rho_{ref}\right)_d^n\right)\Big|_{z_1}\right) \\
&\quad - \frac{1}{2\Delta z}\left(\left(\rho_s|_{z_{\frac{1}{2}}} w\right)^{n+1} + \left(\rho_s|_{z_{\frac{1}{2}}} w\right)_d^n\right)\Bigg|_{z_1}, \\
\frac{\rho^{n+1} - \rho_d^n}{\Delta t}\Bigg|_{z_k} &= = -\overline{w|_M^{n+\frac{1}{2}}\delta_z\Delta\rho_{ref}}^z\Bigg|_{z_k} - \frac{1}{2}\overline{\left(\left(w\delta_z\rho_{ref}\right)^{n+1} + \left(w\delta_z\rho_{ref}\right)_d^n\right)}^z\Bigg|_{z_k} \\
&\quad - \frac{1}{2}\left((\rho_s\delta_z w)^{n+1} + (\rho_s\delta_z w)_d^n\right)\Big|_{z_k}, \quad k = \frac{3}{2}, \dots, N - \frac{3}{2}, \\
\frac{\rho^{n+1} - \rho_d^n]}{\Delta t}\Bigg|_{z_{N-\frac{1}{2}}} &= -\frac{1}{2}\left(w|_M^{n+\frac{1}{2}}\delta_z\Delta\rho_{ref}\Big|_{z_{N-1}} + \frac{1}{2}\left(\left(w\delta_z\rho_{ref}\right)^{n+1} + \left(w\delta_z\rho_{ref}\right)_d^n\right)\Big|_{z_{N-1}}\right) \\
&\quad - \frac{1}{2\Delta z}\left(\left(\rho_s|_{z_{\frac{1}{2}}} w\right)^{n+1} + \left(\rho_s|_{z_{\frac{1}{2}}} w\right)_d^n\right)\Bigg|_{z_{N-1}}.
\end{aligned} \tag{9.108}$$

The equation of state at a half integer model level $k = \frac{1}{2}, \frac{3}{2}, \dots, N - \frac{1}{2}$ is given by

$$\left(\frac{1-\kappa}{\kappa}\right)\frac{\Pi^{n+1}}{\Pi_s} = \frac{\rho^{n+1}}{\rho_s} + \frac{\overline{\theta^{n+1}}^z}{\overline{\theta_s}^z}. \tag{9.109}$$

The term $w|_M^{n+\frac{1}{2}}$ above denotes the estimate of the vertical velocity at the **midpoint** of the trajectory. This term arises from the linearization of the semi-Lagrangian derivative as follows:

$$\frac{G'^{n+1} - G_d'^{n}}{\Delta t} = \frac{G''^{n+1} - G''^{n}}{\Delta t} + \frac{\Delta G_{ref}(z) - \Delta G_{ref}(z-\alpha)}{\Delta t}. \tag{9.110}$$

The left-hand side of (9.110) constitutes the left-hand side of the linearized equations above. The second term is linearized as

$$\begin{aligned} \frac{\Delta G_{ref}(z) - \Delta G_{ref}(z-\alpha)}{\Delta t} &= \frac{\Delta G_{ref}(z) - \Delta G_{ref}\left(z - \Delta t w'\left(z - \frac{\alpha}{2}, t^{n+\frac{1}{2}}\right)\right)}{\Delta t}, \\ &= w'\left(z, t^{n+\frac{1}{2}}\right)\delta z \Delta G_{ref} + H.O.T, \\ &\approx w|_M^{n+\frac{1}{2}} \delta_z \Delta G_{ref}. \end{aligned} \tag{9.111}$$

It is this quantity above that is the subject of the study in [38], where the authors considered one- and two-term extrapolation trajectory computation schemes given by

$$w|_M^{n+\frac{1}{2}} = w^n, \tag{9.112}$$

$$w|_M^{n+\frac{1}{2}} = \frac{3}{2}w^n - \frac{1}{2}w^{n-1}. \tag{9.113}$$

Given this derivation, it is now possible to write the linearized equations above in matrix form as

$$\mathbf{A}\mathbf{x}^{n+1} = \mathbf{B}\mathbf{x}^n + \mathbf{C}\mathbf{x}^{n-1}, \tag{9.114}$$

where **A**, **B**, and **C** are square matrices of order $N_x = 4N + 2$ and

$$\mathbf{x} = \begin{bmatrix} \Pi \\ \theta \\ w \\ \rho \end{bmatrix} \tag{9.115}$$

is an array of dimension N_x composed of subarrays Π and ρ, both of order $N-1$, and θ and w are of order of N, where **A**, **B**, and **C** are defined by

$$\mathbf{A} = \begin{pmatrix} \mathbf{I} & \mathbf{A}_{\Pi,\theta} & \mathbf{0} & \mathbf{A}_{\Pi,\rho} \\ \mathbf{0} & \mathbf{I} & \mathbf{A}_{\theta,w} & \mathbf{0} \\ \mathbf{A}_{w,\Pi} & \mathbf{A}_{w,\theta} & \mathbf{I} & \mathbf{0} \\ \mathbf{0} & \mathbf{0} & \mathbf{A}_{\rho,w} & \mathbf{I} \end{pmatrix}, \tag{9.116}$$

$$\mathbf{B} = \begin{pmatrix} \mathbf{0} & \mathbf{0} & \mathbf{0} & \mathbf{0} \\ \mathbf{0} & \mathbf{I} & \mathbf{B}_{\theta,w} & \mathbf{0} \\ \mathbf{B}_{w,\Pi} & \mathbf{B}_{w,\theta} & \mathbf{I}_0 & \mathbf{0} \\ \mathbf{0} & \mathbf{0} & \mathbf{B}_{\rho,w} & \mathbf{I} \end{pmatrix}, \tag{9.117}$$

$$\mathbf{C} = \begin{pmatrix} 0 & 0 & 0 & 0 \\ 0 & 0 & c_{\theta,w} & 0 \\ 0 & 0 & 0 & 0 \\ 0 & 0 & \mathbf{C}_{\rho,w} & 0 \end{pmatrix}, \tag{9.118}$$

where $\mathbf{I}_0$ is the identity matrix for that order, but with the first and last rows being all zeros. We now reformulate (9.114) as an eigenproblem such that

$$\mathbf{M}\mathbf{y}^{n+1} = \mathbf{N}\mathbf{y}^{n}, \tag{9.119}$$

is formally a two-time-level matrix problem, through the variable

$$\mathbf{y}^{n} = \begin{pmatrix} \mathbf{x}^{n} \\ \mathbf{x}^{n-1} \end{pmatrix}, \tag{9.120}$$

which is an array of dimension $2N_x$, composed of the values of the variables stored at the appropriate model levels at the current and previous time steps, where

$$\mathbf{M} = \begin{pmatrix} \mathbf{A} & \mathbf{0} \\ \mathbf{0} & \mathbf{I} \end{pmatrix}, \quad \mathbf{N} = \begin{pmatrix} \mathbf{B} & \mathbf{C} \\ \mathbf{I} & \mathbf{0} \end{pmatrix}. \tag{9.121}$$

The normal modes of the generalized system satisfy

$$\mathbf{y}^{n+1} = \lambda \mathbf{y}^{n}, \tag{9.122}$$

and so substituting this into the matrix equation above results in a generalized eigenvalue problem

$$\mathbf{N}\mathbf{y} = \lambda \mathbf{M}\mathbf{y}. \tag{9.123}$$

Given this setup, it is identified in [38] that there are external and internal normal modes of the 1D system (9.98a)–(9.98d) that are degenerate due to the assumption of no horizontal variation.

In [38] the authors run a set of experiments with two different setups of the model, where the first is a more idealized configuration with a limited number of degrees of freedom associated with the number of vertical levels, and a second configuration that was nearer to the operational model at the UK Met Office at the time. The conclusions from these experiments are as follows:

1. Both the one- and two-term extrapolation schemes are unstable for at least one internal model.
2. The growth rate of the instability is generally greater for the two-term extrapolation than for the one-term extrapolation.
3. The growth rate also increases as $\left|T_s - T_{ref}\right|$ increases for the idealized configuration.
4. In the operational configuration, this is only true when comparing modes for which the difference $T_s - T_{ref}$ has the same sign.
5. The two-term extrapolation scheme introduces computational modes due to using information at two time levels. However, it is shown that these are damped in [38] and not the source of the instability, but that it is coming form the extrapolation itself.

6. There was significant distortion of the vertical structure of the normal modes.
7. At larger time steps, additional spurious zeros appear that migrate in time. This produces a change to both the spatial and temporal aspects of the vertical structure of the normal modes, as a result of the phase of the modes becoming a function of height due to the extrapolation procedure.

Finally, as a remark to other models that employ a semi-Lagrangian technique Cordero et al. say

> *Whilst here only demonstrated for acoustic modes, it is likely that the underpinning mechanism for instability is a general one and would apply to SL discretizations of more general equations. In particular, the gravity modes of the full Euler equations, and also of the hydrostatic primitive equations, are likely to be susceptible to the instability when extrapolation is used for trajectories.*
>
> *For both the hydrostatic primitive equations and the non-hydrostatic Euler equations, the term* $w\frac{\partial\theta}{\partial z}$ *is an important contributor to horizontal propagation of gravitational oscillations — and similarly for the analogous vertical advection of potential temperature in terrain-following pressure-based coordinate systems.*

9.7 Time-splitting

There are many different ways to deal with the nonlinear and linear forcing terms. In a paper from 1998 by Caya, Laprise, and Zwack [31], the authors looked at the consequences of using the splitting method for approximating the physical forcing in the advection. The time splitting method was introduced into a semi-Lagrangian formulation by Bates in 1984 [12], where he considered the shallow water equations on the sphere in two dimensions. Bates separated out the advection from the momentum equations component of the shallow water equations and split the equation in the advective and adjustment phases where for the two-dimensional system this would be

$$\begin{aligned}
\frac{Du}{Dt} &= 0, && (9.124)\\
\frac{Dv}{Dt} &= 0, && (9.125)\\
\frac{D\phi}{Dt} &= 0, && (9.126)\\
\frac{\partial u}{\partial t} &= -\frac{\partial\phi}{\partial x}, && (9.127)\\
\frac{\partial v}{\partial t} &= -\frac{\partial\phi}{\partial y}, && (9.128)\\
\frac{\partial\phi}{\partial t} &= -\left(\overline{\phi}+\phi'\right)\left(\frac{\partial u}{\partial x}-\frac{\partial v}{\partial y}\right), && (9.129)
\end{aligned}$$

where $\overline{\phi}$ is the mean geopotential and ϕ' is the deviation therefrom.

Bates then took the advective equations (9.124)–(9.126) and integrated these using semi-Lagrangian techniques similar to those developed in earlier chapters, where the advective time step was denoted by Δt_{adv}. The adjustment equations (9.127)–(9.129) were integrated over two steps of $\frac{\Delta t_{adv}}{2}$. In the first, the x-derivatives were integrated in linear terms and treated implicitly, while y-derivatives were treated explicitly. In the second, the y-derivatives were treated implicitly, while the x-derivatives were treated explicitly. All nonlinear terms were treated explicitly. This scheme is referred to as the **Semi-Lagrangian Alternating Direction Implicit (SLADI)**.

As we have been presenting so far in this chapter, the semi-implicit method is an efficient method to treat linear terms that are responsible for high frequency oscillations in the meteorological equations, and, when used with the semi-Lagrangian approach, it has been shown to allow for longer time steps as compared to Eulerian based discretizations. There is also the case that there are physical forcing terms in the governing equations, where there are two different approaches examined in [31]:

- The physical forcing term can be combined with the nonlinear terms in the semi-implicit formulation or
- An inviscid interim time step can be performed without forcing, applying forcing correction afterwards.

If we consider the canonical equation

$$\frac{D\Psi}{Dt} = L(\Psi) + R(\Psi) + P(\Psi), \tag{9.130}$$

where $L(\Psi)$ is a linear function of Ψ, $R(\Psi)$ is a nonlinear function of Ψ, and $P(\Psi)$ is a physical parameterization forcing applied to the system.

The semi-Lagrangian scheme used in [31] is the three-time-level scheme where a traditional method to apply $P(\Psi)$ is through a single step such that

$$\frac{\Psi(t+\Delta t) - \Psi(t-\Delta t)}{2\Delta t} = L\left(\frac{\Psi(t+\Delta t) + \Psi(t-\Delta t)}{2}\right) + R(\Psi(t)) + P(\Psi(t-\Delta t)), \tag{9.131}$$

whereas the splitting method operates in two steps:

Step 1:

$$\frac{\Psi^*(t+\Delta t) - \Psi(t-\Delta t)}{2\Delta t} = L\left(\frac{\Psi(t+\Delta t) + \Psi(t-\Delta t)}{2}\right) + R(\Psi(t)); \tag{9.132}$$

Step 2:

$$\Psi(t+\Delta t) = \Psi^*(t+\Delta t) + 2\Delta t P\left(\Psi^*(t+\Delta t)\right), \tag{9.133}$$

where $\Psi^*(t+\Delta t)$ is an interim result of the inviscid time step.

Given the two setups just presented, we now consider their numerical analysis to see if they are consistent. We start by considering the simplified model

$$\frac{DF(t)}{Dt} + \beta F(t) = G, \tag{9.134}$$

where β and G are constant. The dependent variable $F(t)$ for this equation is a function of time only. The term G is a forcing corresponding to adiabatic heating. The term βF may represent a damping mechanism, where β is a positive real number, or an oscillation if β is purely imaginary. The analytical steady state for this set up can be shown to be

$$F_a = \frac{G}{\beta}. \tag{9.135}$$

If we now apply a semi-implicit semi-Lagrangian integration of (9.134), then we obtain

$$\frac{F(t+\Delta t) - F(t-\Delta t)}{2\Delta t} + \beta\left(\frac{F(t+\Delta t) + F(t-\Delta t)}{2}\right) = G, \tag{9.136}$$

and the resulting recursive formula is

$$F(t+\Delta t) = \frac{F(t-\Delta t)(1-\beta\Delta t) + 2\Delta t G}{(1+\beta\Delta t)}. \tag{9.137}$$

To find the steady-state solution, we set $F(t+\Delta t) = F(t-\Delta t) = F_{SIT}$, a constant, where the subscript SIT stands for semi-implicit and traditional, thus

$$F_{SIT} = \frac{G}{\beta}, \tag{9.138}$$

which is identical to the analytical solution.

Now if we consider using the SM approach then we have

Step 1:

$$\frac{F^*(t+\Delta t) - F(t-\Delta t)}{2\Delta t} + \beta\left(\frac{F^*(t+\Delta t) + F(t-\Delta t)}{2}\right) = 0, \tag{9.139}$$

Step 2:

$$F(t+\Delta t) = F^*(t+\Delta t) + 2\Delta t G \tag{9.140}$$

eliminating F^*, we obtain the recursive formula

$$F(t+\Delta t) = \frac{F(t-\Delta t)(1-\beta\Delta t)}{(1+\beta\Delta t)} + 2\Delta t G. \tag{9.141}$$

The steady state solution resulting from this method is

$$F_{SISM} = \frac{G(1+\beta\Delta t)}{\beta}. \tag{9.142}$$

An important feature to note about (9.142) is that it is not equal to the analytical steady state, and is dependent on Δt, but it will tend to the true state as $\beta \Delta t \to 0$. However, for longer time steps the further F_{SISM} departs from the exact solution, the more the magnitude of F_{SISM} exceeds that of the analytical solution by the factor $|1 + \beta \Delta t|$. If β is considered to mimic gravity waves' oscillations triggered by thermal forcing, and taking $|\beta| = N \simeq 10^2 \text{ s}^{-1}$ as the Brunt–Väisälä frequency (9.142), then this implies a spurious amplification by a factor 10 of the response when using the split method with a 15-min time step. This effect is what was the motivation for the work in [31], as it is consistent with the effect that the authors were observing in the Canadian Regional Climate Model. Thus the summary presented here is a caveat about using the time splitting method with larger time steps.

9.8 Boundary conditions for the advection-adjustment equation

Recall from the last chapter that McDonald showed in [122] that it was still possible to use the semi-Lagrangian approach with large values for $|\alpha|$ stably through using a time-integration interpolation, or a well-posed buffer zone. It was also shown that while the time truncation approach was stable, it was not, however, as accurate, or consistent, as the other two approaches just mentioned. That being said, the situation considered in the last chapter was for an unforced problem, and given that we are considering the forced problems in this chapter, we now take in to account rotation, as well as adjustment, in large scale geophysical modeling. Therefore, to test if the three methods from the last chapter could be extended to more complex equations, McDonald in [122] introduces the three variable in one-dimension advection-adjustment equations

$$\frac{\partial \psi(x,t)}{\partial t} + \overline{u}\frac{\partial \psi(x,t)}{\partial x} + \overline{\psi}\frac{\partial u(x,t)}{\partial x} = 0, \tag{9.143}$$

$$\frac{\partial u(x,t)}{\partial t} + \overline{u}\frac{\partial u(x,t)}{\partial x} - \overline{f}v(x,t) = 0, \tag{9.144}$$

$$\frac{\partial v(x,t)}{\partial t} + \overline{u}\frac{\partial v(x,t)}{\partial x} + \overline{f}u(x,t) = 0, \tag{9.145}$$

where the advection velocity $\overline{u}$, the Coriolis parameter $\overline{f}$, and the mean geopotential height $\overline{\psi}$ are constants.

It can be shown that by rearranging (9.143) and (9.144) we can write these equations as

$$\left(\frac{\partial}{\partial t} + (\overline{u} + \overline{c})\frac{\partial}{\partial x}\right)\left[\psi + \overline{c}u\right] = \overline{c}\overline{f}v, \tag{9.146a}$$

$$\left(\frac{\partial}{\partial t} + (\overline{u} - \overline{c})\frac{\partial}{\partial x}\right)\left[\psi - \overline{c}u\right] = -\overline{c}\overline{f}v, \tag{9.146b}$$

where $\overline{c} = \sqrt{\overline{\psi}}$. Therefore, the characteristic variables of these equations, and their associated characteristics, are

$$
\begin{aligned}
&v, && (\overline{u}),\\
&p=\psi+\overline{c}u, && (\overline{u}+\overline{c}),\\
&q=\psi-\overline{c}u, && (\overline{u}-\overline{c}).
\end{aligned}
\tag{9.147}
$$

It is stated in [122] that for each inward-pointing characteristics a field must be imposed on the boundary. Therefore, if we substitute $\psi=\overline{\psi}\exp\{i(kx-\omega t)\}$ into (9.143)–(9.145), then this yields three solutions; the first is referred to as a **slow** solution with $\omega=k\overline{u}$, and the other two are **fast** solutions with $\omega=k(\overline{u}\pm c_k)$, where

$$
c_k=\sqrt{\overline{\psi}+\frac{\overline{f}^2}{k^2}}. \tag{9.148}
$$

As this is a book on semi-Lagrangian methods, we now apply a two-time-level technique and an implicit discretization on (9.143)–(9.145) using what is referred to as a C-grid; we shall explain this grid in Chapter 10 for two dimensions, but in the one-dimensional case here it means that the wind field u and the geopotential field ϕ are stored half a grid size apart from each other, see Fig. 9.1 for the set up for this case. Thus we obtain the following discrete

$$
\begin{array}{ccccccccccccc}
 & & \mathbf{0} & & & & & & & & & \boldsymbol{L} & \\
\phi_{-1} & u_{-\frac{1}{2}} & \phi_0 & u_{\frac{1}{2}} & \phi_1 & u_{\frac{3}{2}} & \cdots & \phi_{iL-2} & u_{iL-\frac{3}{2}} & \phi_{iL-1} & u_{iL-\frac{1}{2}} & \phi_{iL} & u_{iL+\frac{1}{2}}\\
v_{-1} & & v_0 & & v_1 & & & v_{iL-2} & & v_{iL-1} & & v_{iL} &
\end{array}
$$

FIGURE 9.1 Illustration of the C-grid in a one-dimensional domain.

implicit equations:

$$
\phi_i^{n+1}+\frac{\overline{\phi}\Delta t}{2\Delta x}\left(u_{i+\frac{1}{2}}^{n+1}-u_{i-\frac{1}{2}}^{n+1}\right)=R_i^{\phi}=\left(X^{\phi}\right)_{id}^{n}, \tag{9.149}
$$

$$
u_{i+\frac{1}{2}}^{n+1}+\frac{\Delta t}{2\Delta x}\left(\phi_{i+1}^{n+1}-\phi_i^{n+1}\right)-\frac{\overline{f}\Delta t}{4}\left(v_{i+1}^{n+1}+v_i^{n+1}\right)=R_{i+\frac{1}{2}}^{n}=\left(X^{u}\right)_{i+\frac{1}{2}d}^{n}, \tag{9.150}
$$

$$
v_i^{n+1}+\frac{\overline{f}\Delta t}{4}\left(u_{i+\frac{1}{2}}^{n+1}+u_{i-\frac{1}{2}}^{n+1}\right)=R_i^{v}=\left(X^{v}\right)_{id}^{n}, \tag{9.151}
$$

where

$$
\left(X^{\phi}\right)_i^n\equiv\phi_i^n-\frac{\overline{\phi}\Delta t}{2\Delta x}\left(u_{i+\frac{1}{2}}^{n}-u_{i-\frac{1}{2}}^{n}\right), \tag{9.152a}
$$

$$
\left(X^{u}\right)_i^n\equiv u_{i+\frac{1}{2}}^{n}-\frac{\Delta t}{2\Delta x}\left(\phi_{i+1}^{n}-\phi_i^{n}\right)+\frac{\overline{f}\Delta t}{4}\left(v_{i+1}^{n}+v_i^{n}\right), \tag{9.152b}
$$

$$
\left(X^{v}\right)_i^n\equiv v_i^n-\frac{\overline{f}\Delta t}{4}\left(u_{i+\frac{1}{2}}^{n}+u_{i-\frac{1}{2}}^{n}\right), \tag{9.152c}
$$

where X_i^n represents that we interpolate X to the departure point associated with grid point i. Once R^{ϕ}, R^u, and R^v are known, an approach for solving the system of equations is to rearrange (9.149)–(9.151) to obtain a tridiagonal system for u^{n+1}. After this is solved, it is possible to obtain the values of ϕ^{n+1} and v^{n+1} from (9.149) and (9.151), respectively. However, McDonald does make the comment here that if this approach were applied to the two-dimensional problem, then it would generate the doubly averaged terms $\overline{f\overline{u}}$ and $\overline{f\overline{v}}$, each of which has a nine-point template, which could cause problems for stability near the boundaries.

As an alternative, McDonald presents an iterative approach which is shown below:

$$\left[\phi_i + \frac{\overline{\phi}\Delta t}{2\Delta x}\left(u_{i+\frac{1}{2}} - u_{i-\frac{1}{2}}\right)\right]^{(k+1)} = R_i^{\phi}, \tag{9.153a}$$

$$\left[u_{i+\frac{1}{2}} + \frac{\overline{\phi}\Delta t}{2\Delta x}\left(\phi_{i+1} - \phi_i\right)\right]^{(k+1)} = R^u_{i+\frac{1}{2}} + \frac{\overline{f}\Delta t}{4}\left(v_{i+1} + v_i\right)^{(k)} \equiv \widetilde{R}^u_{i+\frac{1}{2}}, \tag{9.153b}$$

$$v_i^{(k+1)} = R_i^v - \frac{\overline{f}\Delta t}{4}\left(u_{i+\frac{1}{2}} + u_{i-\frac{1}{2}}\right)^{(k)}. \tag{9.153c}$$

Having computed R^{ϕ}, R^u, and R^v, substituting (9.153b) into (9.153a) results in

$$\left[\phi_i - \overline{\phi}\left(\frac{\Delta t}{2\Delta x}\right)^2\left(\phi_{i+1} - 2\phi_i + \phi_{i-1}\right)\right]^{(k+1)}$$
$$= R_i^{\phi} - \frac{\overline{\phi}\Delta t}{2\Delta x}\left(R^u_{i+\frac{1}{2}} - R^u_{i-\frac{1}{2}}\right) - \frac{\overline{\phi}\overline{f}\left(\Delta t\right)^2}{8\Delta x}\left(v_{i+1} - v_i\right)^{(k)}. \tag{9.154}$$

Once ϕ has been solved for with appropriate boundary conditions, $u_{i+\frac{1}{2}}^{(k+1)}$ can be computed from (9.153b), and $v_i^{(k+1)}$ from (9.153c).

It is stated in [122] that it is not possible to use (9.151) for (9.145) on the boundary line, and so it must be replaced with an equation that uses an accurate, stable extrapolation from the interior of the required fields.

Trajectory truncation

First we impose boundary conditions on ϕ_0, and ϕ_{i_l} and v at the inflow. These fit naturally in (9.154), but we also need to know v on the outflow boundary line in order to solve (9.154). There are also some problems with v at the inflow boundary.

We consider the problem of computing v at the outflow. We assume that an approximation to (9.145) is obeyed at the outflow boundary point, given that we must extrapolate unknown fields form the interior, McDonald considers the simplest approximation:

$$v_0^{n+1} + \frac{\overline{f}\Delta t}{4}\left(3u_{\frac{1}{2}}^{n+1} - u_{\frac{3}{2}}^{n+1}\right) = R_0^v = \left(X^v\right)_{0d}^n, \qquad \overline{u} < 0, \tag{9.155a}$$

$$v_{i_L}^{n+1} + \frac{\overline{f}\Delta t}{4}\left(3u_{i_L-\frac{1}{2}}^{n+1} - u_{i_l-\frac{3}{2}}^{n+1}\right) = R_{i_L}^v = \left(X^v\right)_{i_Ld}^n, \qquad \overline{u} > 0. \tag{9.155b}$$

TABLE 9.1 Table of boundary conditions for different departure points.

x_d	$x_d < \frac{\Delta x}{2}$	$x_d > L - \frac{\Delta x}{2}$	$x_d < \Delta x$	$x_d > L - \Delta x$	$x_d < 0$	$x_d > L$
BC	$R^u = (X^u)^n_{\frac{1}{2}}$	$R^u = (X^u)^n_{i_L - \frac{1}{2}}$	$R^\phi = \left(X^\phi\right)^n_1$	$R^\phi = \left(X^\phi\right)^n_{i_L - 1}$	$R^v = (X^v)^n_0$	$R^v = (X^v)^n_{i_L}$

When the departure point is within two grid points of the boundary, McDonald indicates that we require $(X^v)^n_0$ or $(X^v)^n_{i_L}$ for the computation of R^v, where for consistency McDonald defines these quantities as

$$\left(X^v\right)^n_0 = v^n_0 - \frac{\overline{f}\Delta t}{4}\left(3u^n_{\frac{1}{2}} - u^n_{\frac{3}{2}}\right), \tag{9.156a}$$

$$\left(X^v\right)^n_{i_L} = v^n_{i_L} - \frac{\overline{f}\Delta t}{4}\left(3u^n_{i_L-\frac{1}{2}} - u^n_{i_L-\frac{3}{2}}\right). \tag{9.156b}$$

In the work presented in [123], which McDonald refers to in [122], the integration is inaccurate unless it is also applied at the inflow equivalent right-hand sides. McDonald identifies what needs to happen at either boundary depending on where the departure point is. We summarize this in Table 9.1.

Time interpolation

We have to see if this approach is still feasible when adjustment and rotation are added to the advection equation. We start by following the same arguments for this approach from the last chapter, which is to discretize (9.143) with a semi-Lagrangian approach, yielding

$$\frac{\phi\left(x_i, t+\delta t\right) - \phi\left(x_i - \overline{u}\delta t, t\right)}{\delta t} + \frac{\overline{\phi}}{2}\left(\frac{\partial u}{\partial x}\left(x_i, t+\delta t\right) + \frac{\partial u}{\partial x}\left(x_i - \overline{u}\delta t, t\right)\right) = 0, \tag{9.157}$$

which, following the same arguments as before, becomes

$$\left[\phi + \frac{\mu}{\alpha}\frac{\overline{\phi}\Delta t}{2}\frac{\partial u}{\partial x}\right]^{n+1}_{b+\mu} = \left[\phi - \frac{\mu}{\alpha}\frac{\overline{\phi}\Delta t}{2}\frac{\partial u}{\partial x}\right]^{n+1-\frac{\mu}{\alpha}}_{b} \equiv \left[\widehat{X}^\phi\left(\frac{\mu}{\alpha}\right)\right]^{n+1-\frac{\mu}{\alpha}}_b, \tag{9.158}$$

where b represents boundary line and μ is the integer number of points the ϕ point of interest is from the boundary. Following the same argument as above, we obtain the discretizations of (9.144) and (9.145) as

$$\left[u + \frac{v}{\alpha}\frac{\Delta t}{2}\left(\frac{\partial \phi}{\partial x} - \overline{f}v\right)\right] = \left[u - \frac{v}{\alpha}\frac{\Delta t}{2}\left(\frac{\partial \phi}{\partial x} - \overline{f}v\right)\right]^{n+1-\frac{v}{\alpha}}_b \equiv \left[\widehat{X}^u\left(\frac{v}{\alpha}\right)\right]^{n+1-\frac{v}{\alpha}}_b, \tag{9.159}$$

$$\left[v + \frac{\mu}{\alpha}\frac{\overline{f}\Delta t}{2}u\right]^{n+1}_{b+\mu} = \left[v - \frac{\mu}{\alpha}\frac{\overline{f}\Delta t}{2}u\right]^{n+1-\frac{\mu}{\alpha}}_b \equiv \left[\widehat{X}^v\left(\frac{\mu}{\alpha}\right)\right]^{n+1-\frac{\mu}{\alpha}}_b, \tag{9.160}$$

respectively. We should note here that v is a half-odd-integer from the boundary.

If we now assume that $\overline{c} > \overline{u}$, then well-posedness implies, according to [122], that two fields on the inflow boundary and one on the outflow must be imposed. McDonald assumes that ϕ and v are imposed on the inflow boundary and ϕ is also imposed on the outflow boundary. However, it is clear that we do not know most of the fields that are required to calculate $\widehat{X}^{\phi}$, $\widehat{X}^{u}$, and $\widehat{X}^{v}$ on the boundary, even at t^n. We also know that the terms u, $\frac{\partial u}{\partial x}$, and $\frac{\partial \phi}{\partial x}$ are not known on the boundary and thus must be extrapolated, where the simplest extrapolation at $x = 0$ is

$$u_0 = \frac{3}{2}u_{\frac{1}{2}} - \frac{1}{2}u_{\frac{3}{2}}, \qquad \left(\frac{\partial u}{\partial x}\right)_0 = \frac{1}{\Delta x}\left(u_{\frac{3}{2}} - u_{\frac{1}{2}}\right), \qquad \left(\frac{\partial \phi}{\partial x}\right)_0 = \frac{1}{\Delta x}\left(\phi_1 - \phi_0\right). \tag{9.161}$$

The same form of extrapolation and finite differences can be applied at the L boundary but form a different direction. In [122] McDonald ponders three different ways to implement this form of approach stably, but also computationally efficiently to go to vector computing. We refer the reader to [122] for that discussion. What we shall summarize here is the iterative solution that McDonald presents as the path forward. Therefore, the iteration equations are given by

$$\left[\phi + \overline{\phi}\frac{\mu}{2}\frac{\Delta t}{\alpha}\frac{\partial u}{\partial x}\right]^{(k+1)}_{b+\mu} = C_+\left(\frac{\mu}{\alpha}\right)\left[\widehat{X}^{\phi}\left(\frac{\mu}{\alpha}\right)\right]^{(k)}_b + C_0\left(\frac{\mu}{\alpha}\right)\left[\widehat{X}^{\phi}\left(\frac{\mu}{\alpha}\right)\right]^{n}_b + C_-\left(\frac{\mu}{\alpha}\right)\left[\widehat{X}^{\phi}\left(\frac{\mu}{\alpha}\right)\right]^{n-1}_b, \tag{9.162a}$$

$$\begin{aligned}\left[u + \frac{v}{2}\frac{\Delta t}{\alpha}\frac{\partial \phi}{\partial x}\right]^{(k+1)}_{b+v} &= \frac{\overline{f} v \Delta t}{2\alpha} v^{(k)}_{b+v} + C_+\left(\frac{v}{\alpha}\right)\left[\widehat{X}^{u}\left(\frac{v}{\alpha}\right)\right]^{(k)}_b + C_0\left(\frac{v}{\alpha}\right)\left[\widehat{X}^{u}\left(\frac{v}{\alpha}\right)\right]^{n}_b \\ &\quad + C_-\left(\frac{v}{\alpha}\right)\left[\widehat{X}^{u}\left(\frac{v}{\alpha}\right)\right]^{n-1}_b,\end{aligned} \tag{9.162b}$$

$$\begin{aligned}v^{(k+1)}_{b+\mu} &= -\frac{\overline{f}\mu\Delta t}{2\alpha}u^{(k)}_{b+\mu} + C_+\left(\frac{\mu}{\alpha}\right)\left[\widehat{X}^{v}\left(\frac{\mu}{\alpha}\right)\right]^{(k)}_b + C_0\left(\frac{\mu}{\alpha}\right)\left[\widehat{X}^{v}\left(\frac{\mu}{\alpha}\right)\right]^{n}_b \\ &\quad + C_-\left(\frac{\mu}{\alpha}\right)\left[\widehat{X}^{v}\left(\frac{\mu}{\alpha}\right)\right]^{n-1}_b,\end{aligned} \tag{9.162c}$$

where

$$C_+\left(\frac{\mu}{\alpha}\right) \equiv \frac{1}{2}\left(1 - \frac{\mu}{\alpha}\right)\left(2 - \frac{\mu}{\alpha}\right), \quad C_0\left(\frac{\mu}{\alpha}\right) = \frac{\mu}{\alpha}\left(2 - \frac{\mu}{\alpha}\right), \quad C_-\left(\frac{\mu}{\alpha}\right) = -\frac{1}{2}\frac{\mu}{\alpha}\left(1 - \frac{\mu}{\alpha}\right).$$

It is stated in [122] that once the main equations above are discretized, they form a tridiagonal system, with variable coefficients, along with a few extra terms that have be evaluated during each iteration.

Well-posed buffer zone

We start by assuming that $\overline{u} > 0$, and thus require a buffer zone for $x < 0$. We therefore extrapolate q_0 from the interior using (9.146b). For $\overline{u} < 0$, we require a buffer zone for the region $x > L$, therefore, here we extrapolate p_{i_L} from the interior using (9.146a). It is stated in

[122] that to maintain stability we solve (9.146b) and (9.146a) through finding the value of the field at the departure points $x_{dq} = x_0 - (\overline{u} - \overline{c})\,\Delta t$ and $x_{dp} = x_{i_L} - (\overline{u} + \overline{c})\,\Delta t$, respectively, as

$$q_0^{n+1} = \left(q - \frac{\Delta t}{2}\overline{f}\overline{c}v\right)_{dq}^{n} - \frac{\Delta t}{2}\overline{f}\overline{c}v_0^{n+1}, \tag{9.163a}$$

$$p_{i_L}^{n+1} = \left(p + \frac{\Delta t}{2}\overline{f}\overline{c}v\right)_{dq}^{n} + \frac{\Delta t}{2}\overline{f}\overline{c}v_{i_L}^{n+1}. \tag{9.163b}$$

If, as before, we assume that we are imposing ϕ at both boundaries and v at the inflow boundary, then for $\overline{u} > 0$ we impose v_0^{n+1}, and thus we know q^{n+1} at x_0 from (9.163a), and so it is possible to approximate $\left(\frac{\partial q}{\partial t}\right)^n$ and $\left(\frac{\partial^2 q}{\partial t^2}\right)^n$ there. It is also possible to estimate $\left(\frac{\partial p}{\partial t}\right)^n$ and $\left(\frac{\partial^2 p}{\partial t^2}\right)^n$ there through $p^{n+1} = 2\psi^{n+1} - q^{n+1}$. If we have $\overline{u} < 0$, then $v_{i_L}^{n+1}$ is imposed, and then it is possible to approximate $\left(\frac{\partial p}{\partial t}\right)^n$ and $\left(\frac{\partial^2 p}{\partial t^2}\right)^n$ there through (9.163b). It is also possible to estimate $\left(\frac{\partial q}{\partial t}\right)^n$ and $\left(\frac{\partial^2 q}{\partial t^2}\right)^n$ there using $q^{n+1} = 2\phi^{n+1} - p^{n+1}$. Therefore, it is possible to use (9.163a) and (9.163b) to compute $\frac{\partial p}{\partial x}$ and $\frac{\partial q}{\partial x}$ as

$$\frac{\partial p}{\partial x} = \frac{1}{\overline{u} + \overline{c}}\left(-\frac{\partial p}{\partial t} + \overline{f}\overline{c}v\right), \tag{9.164a}$$

$$\frac{\partial q}{\partial x} = \frac{1}{\overline{u} - \overline{c}}\left(-\frac{\partial q}{\partial t} - \overline{f}\overline{c}v\right), \tag{9.164b}$$

$$\frac{\partial v}{\partial x} = \frac{1}{\overline{u}}\left(-\frac{\partial v}{\partial t} - \frac{\overline{f}}{2\overline{c}}(p - q)\right). \tag{9.164c}$$

From (9.164a)–(9.164c) it is possible to deduce $\frac{\partial \phi}{\partial x}$ and $\frac{\partial u}{\partial x}$ using the definitions of p and q in (9.147). Differentiating (9.164a)–(9.164c) with respect to x makes it possible to compute the second-order term. Thus we use the Taylor series expansion to compute ϕ, u, and v in the buffer zone accurate to second order in time.

The concepts presented here have been developed further for the semi-implicit semi-Lagrangian case in [195], and the reader is referred to that paper for the details.

9.9 Summary

In this chapter we have extended the concepts for the nonconstant velocity theory for the semi-Lagrangian approaches to allow for nonzero forcing terms that can be linear, nonlinear, or a combination of both. We started by considering a change of variable to the characteristics of the underlying hyperbolic differential equation to rewrite the equations in terms of a nonzero forcing. We showed this for both the linearized and nonlinear version of the one-dimensional shallow water equations model. Next we introduced the work by Kwizak and Robert in 1972 [97] that introduced the concept of the semi-implicit integration to control the exciting of spurious gravity fast waves that are introduced through considering an explicit form of the integration of the forcing term associated with this part of the frequency spectrum. Next we considered the advantages of spatial averaging to control the stability and accuracy of the semi-implicit approach, we also considered the accuracy of the using the uncentering approach of the integration of the trajectory. We also saw that the trajectory calculation in the semi-implicit semi-Lagrangian approach can cause an instability in the integration. We then considered the optimal accuracy for these uncentering approaches. Next we introduced the concept of time splitting where the shallow water equations model was separated out into advection and adjustment equations and integrated separately, and then showed that for large time steps this may not be very efficient. We completed this chapter by considering the boundary condition approaches that were introduce in the last chapter, but now with the adjustment forcing term for rotation, where we showed that it was possible to extend the three approaches to the forcing situation.

CHAPTER

10

Semi-Lagrangian methods for two-dimensional problems

In the previous chapters on semi-Lagrangian methods, we have been considering the one-dimensional problem for the constant and nonconstant velocity, nonlinear, and with forcing, cases. As a natural progression from this, we move on to summarize different techniques, and some different models, that contain a form of advection, of the total derivative, in two dimensions. We now consider the case where the departure point is no longer guaranteed to be on a line between two grid points at time t^n or t^{n-1}, but rather this point is inside of an area, element, or a cell, depending on whether we are using a finite difference, finite element, or a finite volume approach, respectively.

10.1 Bivariate interpolation methods

The Eulerian form of the general linear advection equation in Cartesian coordinates that we are considering in this chapter is given by

$$\frac{\partial \psi}{\partial t} + u(x, y, t)\frac{\partial \psi}{\partial x} + v(x, y, t)\frac{\partial \psi}{\partial y} = f(x, y, u, v, \psi), \tag{10.1}$$

which can be expressed in vectorial form as

$$\frac{\partial \psi}{\partial t} + (\mathbf{u}\cdot\nabla)\psi = f(x, y, u, v, \psi, t), \tag{10.2}$$

where $\mathbf{u} \equiv \begin{pmatrix} u & v \end{pmatrix}^T$ and the Lagrangian form equivalent is given by

$$\frac{D\psi}{dt} = f(x, y, u, v, \psi, t), \tag{10.3}$$

but now we have to numerically integrate

$$\frac{\partial x}{\partial t} = u, \tag{10.4a}$$
$$\frac{\partial y}{\partial t} = v, \tag{10.4b}$$

Semi-Lagrangian Advection Methods and Their Applications in Geoscience
https://doi.org/10.1016/B978-0-12-817222-3.00014-1

as we have advection velocities in both the x and y directions and they are not assumed to always be the same.

The first application of the two-dimensional semi-Lagrangian approach was developed by André Robert, whom we will see was a pioneer in the development of semi-Lagrangian methods. As mentioned earlier, Robert implemented the semi-Lagrangian method into the two-dimensional barotropic vorticity model where nonconstant velocities and nonzero forcings are involved. However, we start with the approach that was introduced by Bates and McDonald for the constant velocity and nonforced case in [14].

In [14] the authors set up the situation with equal distance in x and y, with gird spacing Δx and Δy, and where the tracer arrives at point (x_i, y_j) at time t^{n+1}. We have a simple illustration of this situation in Fig. 10.1 where we can see that the departure point is inside an area that is bounded by (x_{i-p}, y_{j-q}), (x_{i-p-1}, y_{j-q}), (x_{i-p}, y_{j-q-1}), and (x_{i-p-1}, y_{j-q-1}) at time t^n, where p represents the number of points upwind of the arrival point in the x direction, and q represents the number of points upwind of the arrival point in the y direction.

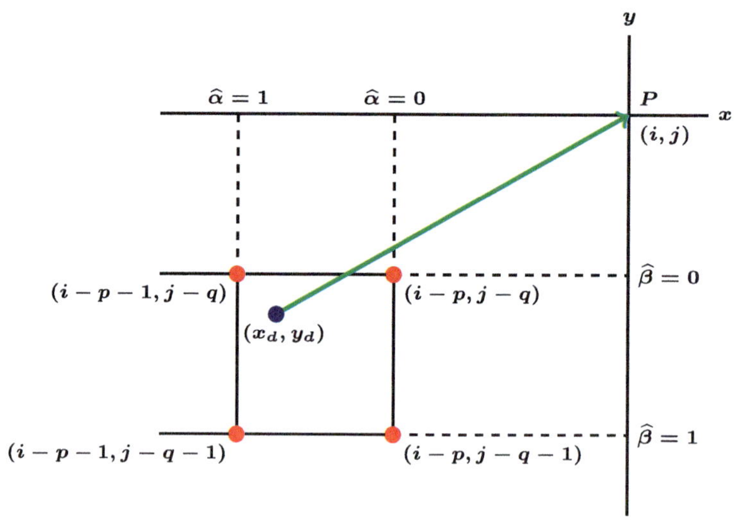

FIGURE 10.1 Schematic of the area surrounding the departure point in a two-dimensional constant velocities case.

Now we integrate (10.4a) with respect to x, but also (10.4b) with respect to y. As the velocities are constant and not functions of the other coordinate, these equations simplify to

$$x_d^n = x_i^{n+1} - \Delta x u, \tag{10.5a}$$
$$y_d^n = y_j^{n+1} - \Delta y v. \tag{10.5b}$$

The question now is how do we find the values of the tracer at the departure point? In [14] the formulae for the bilinear and biquadratic Lagrange interpolation polynomials are presented, and we shall summarize these approaches, along with their stability analyses, over the remainder of this section. We shall also introduce the bicubic Lagrange interpolation. The biquartic Lagrange interpolation analysis is left as an exercise.

10.1.1 Bilinear Lagrange interpolation

In the one-dimensional case we required the Courant number to comprise the two components, $\alpha = \frac{u\Delta t}{\Delta x}$ and $\widehat{\alpha} = \alpha - p$. This is still true for the x component here, but now we define the equivalent for the y direction as $\beta = \frac{v\Delta t}{\Delta x}$ and $\widehat{\beta} = \beta - q$. Thus the bilinear Lagrange interpolation polynomial is given by

$$\begin{aligned} \psi_{i,j}^{n+1} &= (1-\widehat{\alpha})\left(1-\widehat{\beta}\right)\psi_{i-p,j-q}^{n} + \widehat{\alpha}\left(1-\widehat{\beta}\right)\psi_{i-p-1,j-q}^{n} \\ &+ \widehat{\beta}(1-\widehat{\alpha})\psi_{i-p,j-q-1}^{n} + \widehat{\alpha}\widehat{\beta}\psi_{i-p-1,j-q-1}^{n}. \end{aligned} \tag{10.6}$$

We can see the points used for this interpolation in Fig. 10.1. As with the one-dimensional case, we have to consider if the scheme is stable for any choices of Δt, Δx, but now we also have to consider Δy. This is achieved through considering a von Neumann stability analysis with test solution

$$\psi_{i,j}^{n} = \psi^{0}\lambda^{n}\exp\left\{i\left(kx+ly\right)\right\}. \tag{10.7}$$

If we now substitute (10.7) into (10.6), then we obtain

$$\lambda = \lambda_1\lambda_2,$$

where

$$\lambda_1 = \left(1-\widehat{\alpha}\left(1-\exp\left\{-ik\Delta x\right\}\right)\right)\exp\left\{-ipk\Delta x\right\}, \tag{10.8a}$$

$$\lambda_2 = \left(1-\widehat{\beta}\left(1-\exp\left\{-il\Delta x\right\}\right)\right)\exp\left\{-iqk\Delta y\right\}, \tag{10.8b}$$

implying that the amplification factor is the product of two factors of the same form as for the one-dimensional case. It is then stated in [14] that it is theoretically possible to apply a splitting method, even though no splitting method has been applied to the terms in the equation. Thus a sufficient condition for stability is

$$0 \leq \widehat{\alpha} \leq 1 \quad \text{and} \quad 0 \leq \widehat{\beta} \leq 1, \tag{10.9}$$

which holds if the departure point $\left(x_d^n, y_d^n\right)$ lies within the interpolation box as in Fig. 10.1. As with the one-dimensional case, this condition is guaranteed by the choice of interpolation points. Thus the scheme is unconditionally stable. It is mentioned in [14] that this scheme does suffer from heavy damping for the shortest resolvable waves as we saw for the scalar advection semi-Lagrangian case earlier.

Exercise 10.1. *Derive the amplifications factors in* (10.8a) *and* (10.8b) *and verify that they form the product shown above, along with the condition for stability, and verify that it is the inequalities above.*

10.1.2 Biquadratic Lagrange interpolation

The situation here now becomes a bit more complicated since, to form a biquadratic approximation, we require three points in each direction, which results in an interpolation area that involves **nine** grid points. An illustration of the interpolation boxes is given in Fig. 10.2.

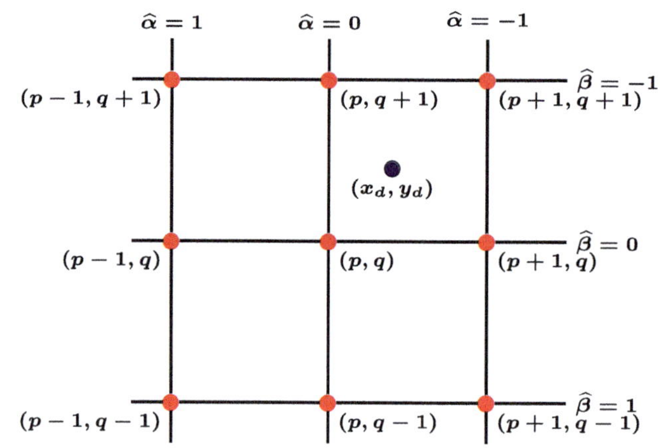

FIGURE 10.2 Schematic of the nine-point interpolation area for the bivariate Lagrange interpolation polynomial.

The grid point, which is nearest the departure point, is chosen as the center of the nine points used in the interpolation. The general expression for a biquadratic Lagrange interpolation polynomial is given by

$$\psi^n(x, y) \sum_{\substack{i=-1 \\ j=-1}}^{\substack{j=1 \\ i=1}} W_{i,j}\psi^n_{i,j}, \tag{10.10}$$

where the (i, j) range over the nine interpolation points, and

$$W_{i,j} = \prod_{\substack{k=-1 \\ k \neq i}}^{k=1} \frac{x - x_k}{x_i - x_k} \prod_{\substack{l=-1 \\ l \neq j}}^{l=1} \frac{(y - y_l)}{(y_j - y_l)}.$$

Bates and McDonald then make the comment that this formula reproduces the exact solution at the grid points, and, when applied to our situation, yields the rather lengthy expression:

$$\begin{aligned}
\psi^{n+1}_{i,j} &= \frac{1}{4}\widehat{\alpha}(1+\widehat{\alpha})\widehat{\beta}\left(1+\widehat{\beta}\right)\psi^n_{i-p-1,j-q-1} + \frac{1}{2}\left(1-\widehat{\alpha}^2\right)\widehat{\beta}\left(1+\widehat{\beta}\right)\psi^n_{i-p,j-q-1} \\
&- \frac{1}{4}\widehat{\alpha}(1-\widehat{\alpha})\widehat{\beta}\left(1+\widehat{\beta}\right)\psi^n_{i-p+1,j-q-1} + \frac{1}{2}\widehat{\alpha}(1+\widehat{\alpha})\left(1-\widehat{\beta}^2\right)\psi^n_{i-p-1,j-q}
\end{aligned}$$

$$
\begin{aligned}
&+ \left(1-\widehat{\alpha}^2\right)\left(1-\widehat{\beta}^2\right)\psi^n_{i-p,j-q} - \frac{1}{2}\widehat{\alpha}\left(1-\widehat{\alpha}\right)\left(1-\widehat{\beta}^2\right)\psi^n_{i-p+1,j-q} \\
&- \frac{1}{4}\widehat{\alpha}\left(1+\widehat{\alpha}\right)\widehat{\beta}\left(1-\widehat{\beta}\right)\psi^n_{i-p-1,j-q+1} - \frac{1}{2}\left(1-\widehat{\alpha}^2\right)\widehat{\beta}\left(1-\widehat{\beta}\right)\psi^n_{i-p,j-q-1} \\
&+ \frac{1}{4}\widehat{\alpha}\left(1-\widehat{\alpha}\right)\widehat{\beta}\left(1-\widehat{\beta}\right)\psi^n_{i-p+1,j-q+1}. \qquad (10.11)
\end{aligned}
$$

For the stability analysis, we again assume a solution of the form in (10.7). It can be shown that the amplification factor is again a product of the amplification factors in each direction, which implies that the amplification factors in the x and y directions are

$$\lambda_1 = \left(1-\widehat{\alpha}^2\left(1-\cos k\Delta x\right) - i\widehat{\alpha}\sin k\Delta x\right)\exp\{-ipk\Delta x\}, \qquad (10.12a)$$

$$\lambda_2 = \left(1-\widehat{\beta}^2\left(1-\cos l\Delta y\right) - i\widehat{\beta}\sin l\Delta y\right)\exp\{-iql\Delta y\}, \qquad (10.12b)$$

respectively. We can see that the amplification factors in (10.12a) and (10.12b) are the same as that of the one-dimensional quadratic Lagrange interpolation, and so we can again theoretically use the simplicity of a splitting methods. This then implies that a sufficient condition for stability is

$$-1 \leq \widehat{\alpha} \leq 1 \qquad \text{and} \qquad -1 \leq \widehat{\beta} \leq 1. \qquad (10.13)$$

However, given the way that the quadratic interpolation is set up, the initial choice of interpolation points is such that

$$-\frac{1}{2} \leq \widehat{\alpha} \leq \frac{1}{2} \qquad \text{and} \qquad -\frac{1}{2} \leq \widehat{\beta} \leq \frac{1}{2}, \qquad (10.14)$$

and thus this scheme is unconditionally stable. With respect to damping effects with this scheme, as with the scalar case, it is stated in [14] that complete extinction never occurs and the damping is in all cases much less that given by the bilinear interpolation, consistent with the results from the one-dimensional case.

10.1.3 Bicubic Lagrange interpolation

The bicubic, biquartic, and general bivariate Lagrange interpolation polynomials appear in [117], where the general form is given by

$$\psi\left(x_d, y_d, t^n\right) = \sum_u \sum_v W_{uv}\psi^n_{uv},$$

where now we have W_{uv} given by

$$W_{rs} = \prod_{u\neq r}\frac{\left(x_d - x_u\right)}{\left(x_r - x_u\right)}\prod_{v\neq s}\frac{\left(y_d - y_v\right)}{\left(y_s - y_v\right)}. \qquad (10.15)$$

We saw for the biquadratic case that we require nine points to be able to fit the associated Lagrange interpolation polynomial, which implies that there are going to be quite a few more points for the bicubic case. As we saw for the one-dimensional case, we require four points to fit a cubic polynomial, therefore we are going to need the square of that, i.e., 16 points!

We shall be using an upwind polynomial, so we shall be staring at the points two grid points before the arrival point in both the x and y direction. Therefore, the points that we are using are

$$u : i-2, i-1, i, i+1, \qquad v : j-2, j-1, j, j+1. \tag{10.16}$$

We have created a schematic of the interpolation area for this approach in Fig. 10.3, showing the box where the departure point is located with respect to the total interpolation area.

As just mentioned, the bicubic Lagrange interpolation polynomial contains 16 terms and do not present that equation here, but rather recall that as with the bilinear and biquadratic Lagrange interpolation polynomials, the coefficients are the products of the equivalent one-dimensional terms, evaluated at $\widehat{\beta}$ and $\widehat{\alpha}$. We have created Tables 10.1 and 10.2 where the coefficients are the related terms at that grid point in the interpolation area in Fig. 10.3 in the i and j directions, so to form the coefficients of the interpolation polynomial one has to multiply the coefficient from the index in each table.

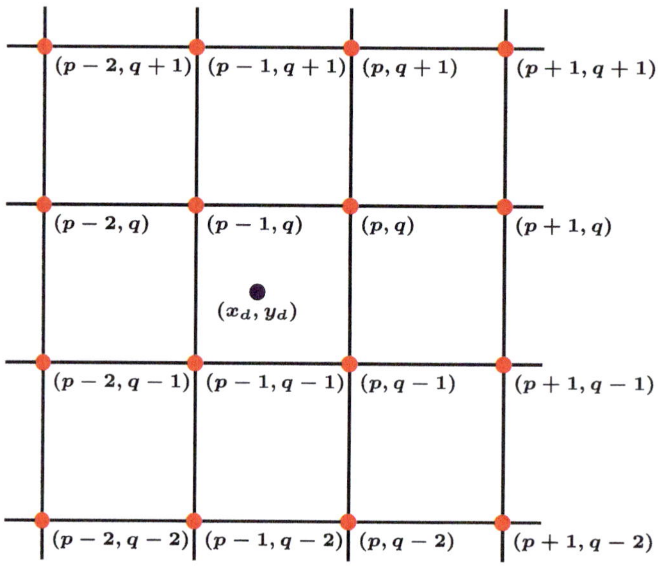

FIGURE 10.3 Schematic of the 16 grid points that are used for bicubic Lagrange interpolation polynomial, as well as the departure point.

TABLE 10.1 Table of the x direction component coefficients of the bicubic Lagrange interpolation polynomial.

x index	$p-2$	$p-1$	p	$p+1$
Coefficient	$-\dfrac{\widehat{\alpha}\left(1-\widehat{\alpha}^2\right)}{6}$	$\dfrac{\widehat{\alpha}(1+\widehat{\alpha})(2-\widehat{\alpha})}{2}$	$\dfrac{\left(1-\widehat{\alpha}^2\right)(2-\widehat{\alpha})}{2}$	$-\dfrac{\widehat{\alpha}(1-\widehat{\alpha})(2-\widehat{\alpha})}{6}$

TABLE 10.2 Table of the y direction component coefficients of the bicubic Lagrange interpolation polynomial.

y **index**	$q-2$	$q-1$	q	$q+1$
Coefficient	$-\frac{\widehat{\beta}\left(1-\widehat{\beta}^2\right)}{6}$	$\frac{\widehat{\beta}(1+\widehat{\beta})(2-\widehat{\beta})}{2}$	$\frac{\left(1-\widehat{\beta}^2\right)(2-\widehat{\beta})}{2}$	$-\frac{\widehat{\beta}(1-\widehat{\alpha})(2-\widehat{\beta})}{6}$

Therefore, to find the coefficient of the term multiplying $\psi_{i-p,j-q}$, we see that this is

$$\frac{\left(1-\widehat{\alpha}^2\right)(2-\widehat{\alpha})}{2}\frac{\left(1-\widehat{\beta}^2\right)(2-\widehat{\beta})}{2}.$$

A stability analysis performed in [117], and as shown for the bilinear and the biquadratic Lagrange interpolation polynomial approximations, becomes a product of the one-dimensional equivalents, and the same spatial conditions for the bilinear case apply here, and thus this scheme is unconditionally stable. It is stated in [117] that through the phase error analysis the bicubic approach has the best phase representation.

10.1.4 Biquartic Lagrange interpolation

In this small subsection we only present the figure for the interpolation area for this approach and set as exercises the derivation of the coefficients of the biquartic Lagrange interpolation, its amplification factors, and the associated spatial analysis. As we saw for the three previous interpolation polynomials, when we increase the order of the polynomial, the number of points needed to determine that polynomial increases rapidly, compared to the one-dimensional case. As we require five points to fit a quartic polynomial, we will require the square of that for the biquartic Lagrange interpolation polynomial, therefore we require 25 points! As with the other polynomials, we have a schematic to illustrate where the departure point is located and the interpolation area that is required to obtain the value of $\psi\left(x_d, y_d, t^n\right)$; see Fig. 10.4. Here we are using $p-2, p-1, p, p+1, p+2$ in the x-direction, and $q-2, q-1, q, q+1, q+2$ in the y direction.

Exercise 10.2. *Derive the coefficients for the x and y components of the Lagrange biquartic interpolation polynomial. Determine the amplification factors λ_1 and λ_2 and show that the scheme is unconditionally stable if $-\frac{1}{2}\leq\widehat{\alpha}\leq\frac{1}{2}$ and $-\frac{1}{2}\leq\widehat{\beta}\leq\frac{1}{2}$.*

In the interpolation formulas above, we have shown the grid points to be equally distributed in both the x and y direction, this is to say, the fields of interest are stored at the same location. We saw the one-dimensional c grid at the end of the last chapter, where the velocity fields were stored half a grid space away from the geopotential heights. The grid used in this section is referred to as the A-grid, or the **unstaggered** grid. In the next section we move on to present a summary of a series of staggered grids that were developed by Arakawa and Lamb in 1977 [3].

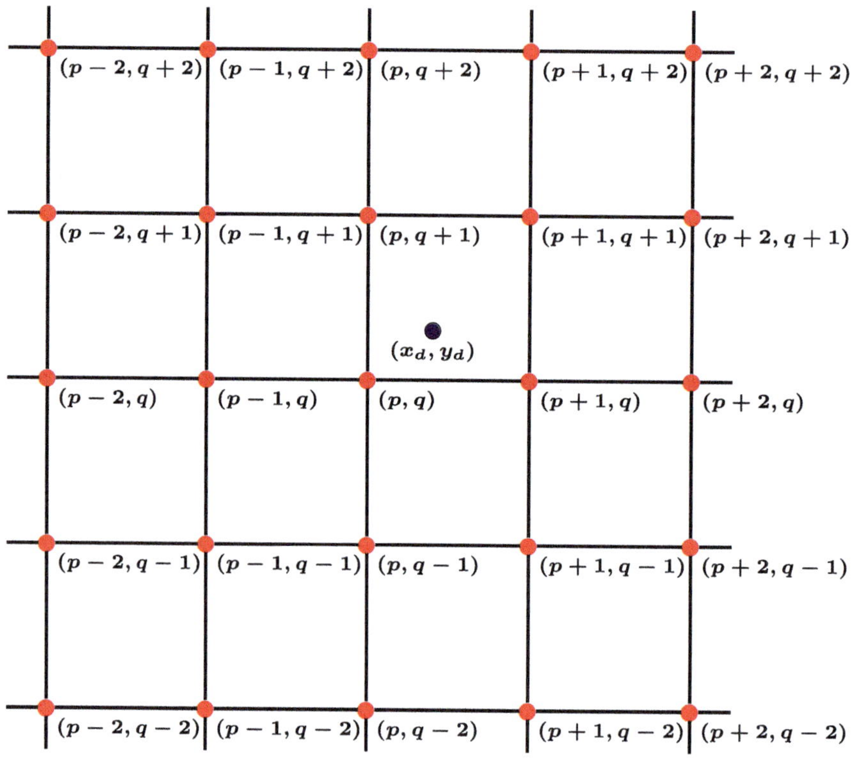

FIGURE 10.4 Schematic of the 25 points required to generate the biquartic Lagrange polynomial.

10.2 Grid configurations

There are five different grids that are regularly used in numerical modeling for the arrangement of the prognostic and diagnostic variables. As we mentioned at the end of the last section, these different grid configurations were presented by Arakawa and Lamb in 1977 [3] for horizontal staggering and are referred to as the A, B, C, D, and E staggered grids. The reason why these grids are referred to as staggered is due to the fact that the points with the height fields h and the wind components u and v are stored at different locations. We shall present all of the five **Arakawa** grids below.

10.2.1 Arakawa A grid

The first of the Arakawa grids is not surprisingly referred to as the A grid, but it is not a staggered grid, which is the grid from the last section. We have plotted the Arakawa A grid in Fig. 10.5.

The A grid is referred to as an *unstaggered* grid. An advantage of this grid is that it enables the Coriolis terms of the momentum equations to be easily evaluated since the wind compo-

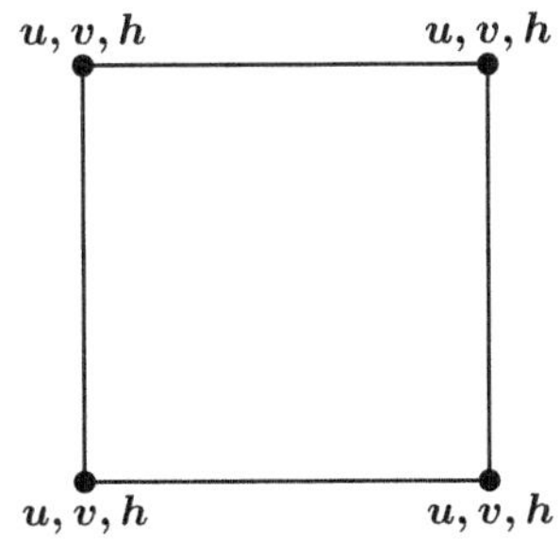

FIGURE 10.5 Diagram of the Arakawa unstaggered A grid.

nents are defined at the same points. While the *A* grid may look like a good configuration for the grid points, it can be shown that this configuration can support two solutions that do not communicate with each other and can be considerably different. As a result of this, the grid introduces noise at the smallest scales. As a result this grid is not used in many applications nowadays. However, there was a lot of work undertaken by Lance Leslie and James Purser in the 1990s using the unstaggered grid and then using the associated interpolation polynomial to overcome the problem above; see [151,152,109,153,110].

10.2.2 Arakawa B grid

The next grid configuration that we consider is the Arakawa B grid. In this grid configuration, we have that the height fields h are stored at the corner of the grid cells, while the horizontal wind fields (u, v) are stored at the center of the cells. We have plotted an illustration of this grid in Fig. 10.6.

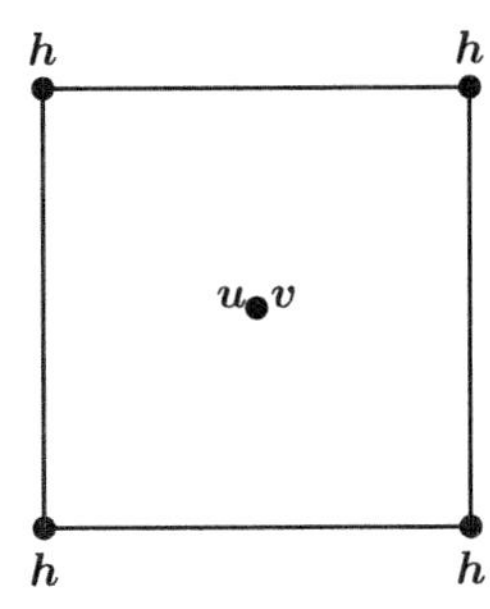

FIGURE 10.6 Diagram of the Arakawa staggered B grid.

Given this configuration for the B grid, because we have the wind fields at the same points, it makes the evaluation of the Coriolis terms easier. However, the evaluation of the pressure gradient terms requires averaging as is the case with the A grid.

10.2.3 Arakawa C grid

The C grid is the first of two grids that do not have the wind fields collocated at the same grid point. For the C grid, we have the height fields located at the center of the grid cell, while the u component of the wind fields is stored at the center of the vertices to the left and right of the center point, the v component of the wind fields is stored at the center vertices above and below the center point. If we denote the center point as (i, j), where i represents the x direction and j represents the y direction, then the u fields are located at $\left(i \pm \frac{\Delta x}{2}, j\right)$ and the v fields are stored at $\left(i, j \pm \frac{\Delta y}{2}\right)$. See Fig. 10.7 for the schematic of this grid. An advantage of the C grid is that it enables the gradients of the prognostic variables to only be over Δx compared to $2\Delta x$ for the A grid.

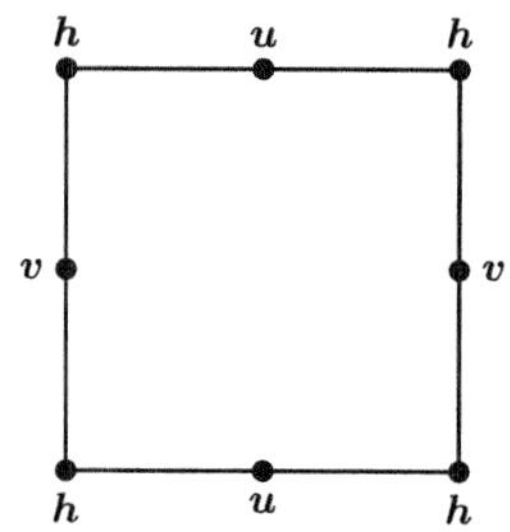

FIGURE 10.7 Diagram of the Arakawa staggered C grid.

10.2.4 Arakawa D grid

The Arakawa D grid is staggered the same as the C grid, but now the wind fields u and v are swapped. Therefore, we have that u fields are at $\left(i, j \pm \frac{\Delta y}{2}\right)$ and the v fields are stored at $\left(i \pm \frac{\Delta x}{2}, j\right)$. In [91] it is stated that there is no particular merit to this grid configuration, however, it is quite often used in finite volume flux form approaches. We have presented an illustration of the D grid in Fig. 10.8.

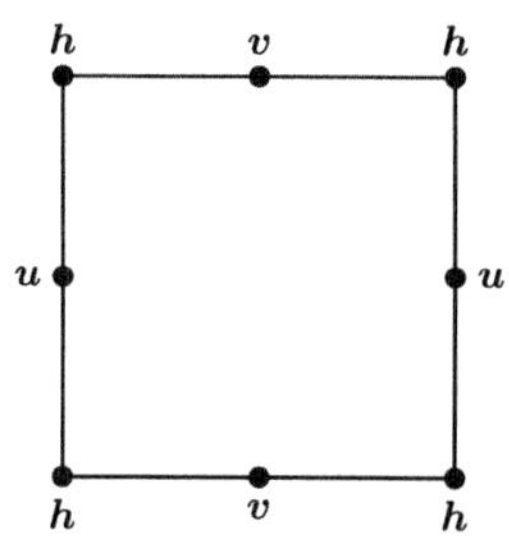

FIGURE 10.8 Diagram of the Arakawa staggered D grid.

10.2.5 Arakawa E grid

The last of the Arakawa grids is referred to as the E grid, and is defined the same as the B grid but with the points where the wind fields are stored rotated by 45°. The E grid can also be considered equivalent to superimposed, but shifted C grids. We have plotted the schematic for the E grid in Fig. 10.9.

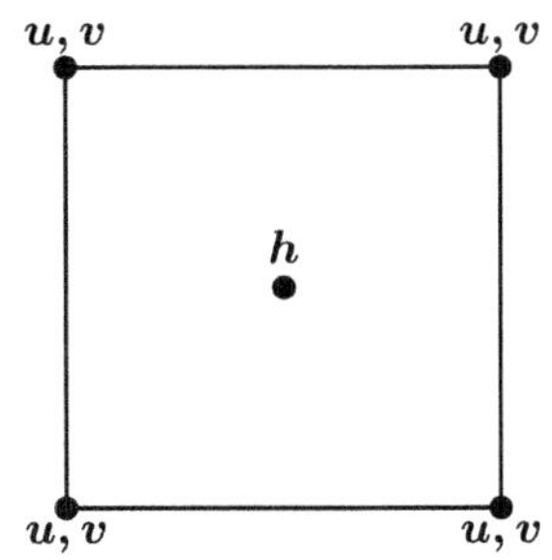

FIGURE 10.9 Diagram of the Arakawa staggered E grid.

10.2.6 Vertical staggered grids

When considering models that contain derivatives, as well as variables in the vertical direction, it becomes important to define the vertical coordinate system so that the numerical approximation is consistent with the dynamical situation. The most commonly used vertical coordinate systems are the z, pressure levels, p, normalized pressure, σ, as presented in Phillips in 1957 [142], potential temperature, θ [52], along with combination of hybrid coordinate systems.

The normalized pressure coordinate system is referred to as the σ coordinates, where $\sigma \equiv \frac{p}{p_s}$, p_s is the pressure at the surface, so that $\sigma = 1$ at the surface, and 0 at $p = 0$.

The grids that we have presented in the last subsection were associated with the staggering of the grid points in the horizontal direction. Most of the world's numerical weather prediction models of different scales also have a staggered grid in the vertical direction. Two of the most regularly used vertical staggered grids are: (1) Lorenz grid from [116], where the vertical velocity is defined at the boundary of layers, while the prognostic variables are stored at the center of the layers. This grid configuration allows for simple quadratic conservation, and the boundary conditions of no fluxes across the boundaries at the top and bottom are fulfilled. However, it was pointed out in [4] that this configuration allows the development of a spurious computational mode; (2) The other possible staggered vertical grid which has become used more frequently for numerical weather predictions models is the configuration presented by Charney and Phillips in 1953 [33]. In this formulation we have the height fields at the σ levels while the wind and temperature fields are stored at the halfway points between the σ levels. This formulation is consistent with the hydrostatic equation. We have presented a schematic of these two vertical staggering approaches in Fig. 10.10.

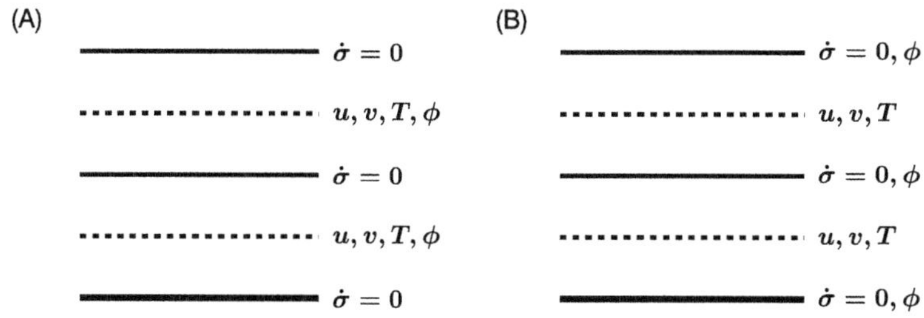

FIGURE 10.10 (A) Lorenz staggered vertical grid, (B) Charney–Phiilips staggered vertical grid.

10.3 Semi-implicit semi-Lagrangian finite differences in two dimensions

The first instance of the semi-Lagrangian approach with a nonconstant velocity and a nonzero forcing appears in Robert (1981) [161], and is extended in Robert (1982) [162]. The model that we shall consider here is from [161], which is the barotropic vorticity equation

$$\frac{\partial Q}{\partial t} + u\frac{\partial Q}{\partial x} + v\frac{\partial Q}{\partial y} = R, \tag{10.17}$$

where Q represents the absolute vorticity, and R represents the remaining terms in the vorticity equation. Robert introduced the three-time-level scheme for the vorticity as

$$Q(y, y, t + \Delta t) = Q(x - 2\alpha, y - 2\beta, t - \Delta t) + 2\Delta t R(x - \alpha, y - \beta, t), \tag{10.18}$$

$$a = \Delta t u(u - \alpha, y - \beta, t), \tag{10.19}$$

$$b = \Delta t v(x - \alpha, y - \beta, t), \tag{10.20}$$

where we can see from the equations above that the departure point is estimated at the mid-time-point along with the forcing term. We need to note that α and β have to be determined from a pair of equations and that they also appear on the right-hand side of these equations. Approximate values of α and β are used to evaluate the right-hand side of (10.19) and (10.20). This process generates a set of new values which are assumed to be more accurate. The calculation is repeated until sufficiently accurate values are obtained.

Robert does not explain much of the actual semi-Lagrangian implementation beyond what we have above for this model, other than to say that upstream values of u, v, Q, and R are obtained from interpolation from values of these variables specified over a regular grid, where he employs a bicubic interpolation polynomial, but with R set to zero. In [161] Robert goes on to consider the approach just described with the absolute vorticity based version of the shallow water equations on the sphere.

As mentioned before, Staniforth and Côté in 1991 [176] provided a very good summary of the state of semi-Lagrangian methods in the early 1990s, and one part of that summary is for the advection problem with forcing in two-dimensions, where the general form of this problem is given by

$$\frac{DF}{Dt} + G(\mathbf{x}, t) = R(\mathbf{x}, t), \tag{10.21}$$

where

$$\frac{DF}{Dt} \equiv \frac{\partial F}{\partial t} + \mathbf{v}(\mathbf{x}, t) \cdot \nabla F, \tag{10.22a}$$

$$\frac{\partial \mathbf{x}}{\partial t} = \mathbf{u}(\mathbf{x}, t), \tag{10.22b}$$

$$\mathbf{x} \equiv \begin{pmatrix} x & y \end{pmatrix}^T.$$

The semi-Lagrangian approach presented in [176] is discretized by the three-time-level scheme as

$$\frac{F^{n+1} - F^{n-1}}{2\Delta t} + \frac{G^{n+1} + G^{n-1}}{2} = R^n, \tag{10.23a}$$

$$\boldsymbol{\alpha} = \Delta t \mathbf{u}(\mathbf{x} - \boldsymbol{\alpha}, t), \tag{10.23b}$$

where the midpoint of the trajectory is at $(\mathbf{x} - \boldsymbol{\alpha}, t)$, and the departure point is at $(\mathbf{x} - 2\boldsymbol{\alpha}, t - \Delta t)$. The approximation above is a centered approximation, which is $O\left((\Delta t)^2\right)$, and applied to (10.21) and (10.22b), and G is evaluated as a time average of its values at the end points of the trajectory, while R is evaluated at the midpoint of the trajectory. Finally, $\boldsymbol{\alpha}$ is the vector displacement given by $\boldsymbol{\alpha} \equiv \begin{pmatrix} \alpha \\ \beta \end{pmatrix}$.

To implement this scheme, we assume that G is unknown, and R should be known as it is being evaluated at the current time, which then results in the following algorithm:

- Solve (10.23b) iteratively for the vector displacement $\boldsymbol{\alpha}$ for all mesh points $\mathbf{x}$, using some initial guess. It is suggested in [176] that the value of the field at the previous time step and an interpolation formula should be used.
- Evaluate $F - \Delta G$ at the **upstream** points $\mathbf{x} - 2\boldsymbol{\alpha}$ at time $t - \Delta t$ using an interpolation formula at the departure point. Evaluate $2\Delta t R$ at the midpoints $\mathbf{x} - \boldsymbol{\alpha}$ of the trajectories at time t using an interpolation formula.
- Evaluate F at the arrival points $\mathbf{x}$ at time $t + \Delta t$ using

$$\begin{aligned} F(\mathbf{x}, t + \Delta t) &= (F - \Delta t G)|_{(\mathbf{x} - 2\boldsymbol{\alpha}, t - \Delta t)} + 2\Delta t \, R|_{(\mathbf{x} - \boldsymbol{\alpha}, t)} - \Delta t \, G|_{(\mathbf{x}, t + \Delta t)} \\ &= (F - \Delta t G)^{t - \Delta t} + 2\Delta t R^n - \Delta t G^{t + \Delta t}. \end{aligned} \tag{10.24}$$

It is possible that G is not known at time $t + \Delta t$, which leads to a coupling to other equations.

10.3.1 Advection–diffusion in two dimensions

Advection–diffusion plays a very important part in many different geophysical modeling applications. The summary that we present here is from an early application of the semi-Lagrangian approach to this formulation. The paper was written by Pudykiewicz and Staniforth in 1984 [148]. The differential equation that we are considering here is

$$\frac{\partial \phi}{\partial t}(\mathbf{x}, t) + \mathbf{u}(\mathbf{x}, t) \cdot \nabla \phi(\mathbf{x}, t) = \nu \nabla^2 \phi(\mathbf{x}, t), \tag{10.25}$$

where ϕ is a scalar field and ν is the diffusion coefficient. In [148] the authors introduce the *semi-Lagrangian approach following Robert (1981)–(1982)* [161,162], where (10.25) can be written as a finite difference approximation

$$\frac{\psi(\mathbf{x}, t+\Delta t) - \phi(\mathbf{x}-\boldsymbol{\alpha}, t)}{\Delta t} = \frac{1}{2}\left[\left(\nu\nabla^2\phi\right)\Big|_{(\mathbf{x},t+\Delta t)} + \left(\nu\Delta^2\phi\right)\Big|_{(\mathbf{x}-\boldsymbol{\alpha},t)}\right], \tag{10.26}$$

with

$$\boldsymbol{\alpha}(\mathbf{x}) = \Delta t\mathbf{u}\left(\mathbf{x} - \frac{\boldsymbol{\alpha}}{2}, t + \frac{\Delta t}{2}\right). \tag{10.27}$$

We can see from (10.26) and (10.27) that Pudykiewicz and Staniforth are applying a temporal average for the diffusion term, but now the difference is that this scheme is a two-time-level scheme, similar to that introduced in [117]. Thus the associated algorithm for this approach is given by

1. Solve (10.27) iteratively for the displacement $\boldsymbol{\alpha}(\mathbf{x})$ using

$$\boldsymbol{\alpha}^{(m+1)}(\mathbf{x}) = \Delta t\mathbf{u}\left(\mathbf{x} - \frac{\boldsymbol{\alpha}^{(m)}}{2}, t + \frac{\Delta t}{2}\right), \tag{10.28}$$

   where m is the number of iterations and where an interpolation is used to evaluate $\mathbf{u}$ between grid points.
2. Evaluate $\left(\nu\nabla^2\phi\right)\big|_{(\mathbf{x},t+\Delta t)}$ at mesh points using some form of discretization of choice, and then use an interpolation formula to evaluate $\left(\phi + \frac{\nu\Delta t\nabla^2\phi}{2}\right)\Big|_{(\mathbf{x}-\boldsymbol{\alpha},t)}$.
3. Solve the **Helmholtz equation**

$$\left|\left(\phi - \frac{\nu\Delta t\nabla^2\phi}{2}\right)\right|_{(\mathbf{x},t+\Delta t)} = \left|\left(\phi + \frac{\nu\Delta t\nabla^2\phi}{2}\right)\right|_{(\mathbf{x}-\boldsymbol{\alpha},t)}. \tag{10.29}$$

In [148] the authors state that the convergence of the iterative procedure is assured by the fixed-point theorem, provided that $\mathbf{u}$ and its first derivatives are continuous, Δt is sufficiently small, and finally, the first guess is **sufficiently close** to the solution. They also say that it is possible to show that it is easy to derive a sufficient condition for the convergence of the form

$$\Delta t \max\left(|u_x|, |u_y|, |v_x|, |v_y|\right) < 1, \tag{10.30}$$

which we have seen from the last chapter, and this inequality is regularly cited in many papers on semi-implicit semi-Lagrangian approaches.

An interesting result from [148] comes from the 2D stability analysis that they perform on the scheme above, using a bicubic spline for their interpolation scheme. We shall derive this condition now where the starting point is to consider the stability for the one-dimensional case with uniform velocity, U. We have

$$\phi(x, t+\Delta) - \phi(x-\alpha, t) = \frac{\nu}{2}\left(\frac{d^2\phi}{dx^2}\bigg|_{(x,t+\Delta t)} + \frac{d^2\phi}{dx^2}\bigg|_{(x-\alpha,t)}\right), \tag{10.31}$$

with $\alpha = U\Delta t$ and $C = \frac{U\Delta t}{\Delta x} = \widehat{C} + \beta$, where in the notation from [148] $\widehat{C}$ is the integer part of C, and $0 \leq \beta \leq 1$, and now assume that we have a solution of the form

$$\phi(x,t) = B\exp\{-ikx + i\omega t\}. \tag{10.32}$$

We can show that

$$\phi\left(x_i, t^{n+1}\right) = \left(\frac{1 - \frac{\nu\Delta t k^2 R}{2}}{1 + \frac{\nu\Delta t k^2 R}{2}}\right)\lambda\left(K, \widehat{C}, \beta\right)\phi\left(x_i, t^n\right), \tag{10.33}$$

where

$$\begin{aligned}\lambda\left(K, \widehat{C}, \beta\right) &= \exp\{i\omega\Delta t\}\exp\left\{i\widehat{C}K\right\}\left(1 + iG\beta - \left(3\left(1 - e^{iK}\right) + iG\left(2 + e^{iK}\right)\right)\beta^2\right.\\ &+ \left.\left(iG\left(1 + e^{iK}\right) + 2\left(1 + e^{iK}\right)\right)\beta^3\right),\end{aligned} \tag{10.34}$$

with $K = k\Delta x$, $G = \frac{3\sin K}{2 + \cos K}$, and R being the response of the discretization of the operator $\frac{d^2}{dx^2}$ depending upon which approach is used here. It is then stated that for most (if not all) discretizations, R is real and varies between 0 and 1, where it takes the value 1 for a spectral expansion (we will introduce spectral methods in Chapter 12), $\frac{2(1 - \cos K)}{K^2}$ for a centered second-order finite difference, and $\frac{6(1 - \cos K)}{(2 + \cos K)K^2}$ for the linear finite element discretization. Thus for stability we require

$$\left|\frac{\left(1 - \frac{\nu\Delta t k^2 R}{2}\right)}{\left(1 + \frac{\nu\Delta t k^2 R}{2}\right)}\lambda\left(K, \widehat{C}, \beta\right)\right| \leq 1, \tag{10.35}$$

and this approach is unconditionally stable independent of Courant number, since the ratio term in (10.35) is less than or equal to 1 for $R \geq 0$ and $|\lambda| \leq 1$.

When we set ν equal to zero, we have the pure advection case, and in a series of plots in [148] the authors see virtually no damping of the large scales; some damping of the small scales occurs, while maximum damping occurs midway between the mesh points, which is consist with what we have seen earlier. For wavelengths of six grid point lengths or greater, the damping is equivalent to 1% of the time step.

The next step is to include the diffusion which introduces a damping factor $\frac{\left(1 - \frac{\nu\Delta t k^2 R}{2}\right)}{\left(1 + \frac{\nu\Delta t k^2 R}{2}\right)}$, and this damping increases as either ν or Δt, or both, increase. This damping factor is a third-order in time approximation to the true damping factor of $\exp\left\{-\nu\Delta t k^2\right\}$, and so the numerical scheme here damps a little less than the exact solution, according to [148].

Moving on to the two-dimensional stability analysis, we will come to a very important finding that is still used today in some operational numerical weather prediction centers and in research institutions. We start by defining a trial solution of the form

$$\phi(x,y,t) = B\exp\{-ikx - ily + \omega t\}, \tag{10.36}$$

similar to earlier for the one-dimensional case, where it can be shown that

$$\phi\left(x_i, y_j, t^{n+1}\right) = \Lambda\phi\left(x_i, y_j, t^n\right), \tag{10.37}$$

$$\Lambda = \frac{\left(1 - \dfrac{\nu\Delta t\left(R_x k^2 + R_y l^2\right)}{2}\right)}{\left(1 + \dfrac{\nu\Delta t\left(R_x k^2 + R_y l^2\right)}{2}\right)}\lambda\left(K_x\widehat{C}_x, \beta_x\right)\lambda\left(K_y, \widehat{C}_y, \beta_y\right). \tag{10.38}$$

An important feature to notice here is that the amplification factor is independent of the spatial mesh length and time step, provided R_x and R_y are positive. We can see that (10.38) is the product of two terms: the first component is associated with the diffusion in both x and y directions, and the second term is associated with the advection in both spatial directions. Thus in [148] the authors state that the effects of advection and diffusion are separable, and so it is possible to perform the discretizations in (10.26) and (10.27) as the following two-step algorithm, which may be more convenient in practical three-dimensions:

1. Advection step:

$$\frac{\widetilde{\phi}\left(\mathbf{x}, t^{n+1}\right) - \phi\left(\mathbf{x} - \boldsymbol{\alpha}, t^n\right)}{\Delta t} = 0, \tag{10.39}$$

   where $\boldsymbol{\alpha}$ is given by (10.27).
2. Diffusion step:

$$\frac{\phi\left(x, t^{n+1}\right) - \widetilde{\phi}\left(\mathbf{x}, t^{n+1}\right)}{\Delta t} = \frac{\nu}{2}\left[\left(\nabla^2\phi\right)\Big|_{(\mathbf{x}, t+\Delta t)} + \left(\nabla^2\phi\right)\Big|_{(\mathbf{x}-\boldsymbol{\alpha}, t)}\right]. \tag{10.40}$$

   However, this approach may not be a very accurate approximation for too large a time step.

This statement is consistent with the work that we presented in the last chapter on time-splitting and a warning about the unsuitability of this approach with large time steps.

10.4 Nonlinear shallow water equations

It is very common in most courses on numerical modeling of geophysical flows, especially for the atmosphere, ocean and hydrology, to consider the two-dimensional nonlinear, but also maybe the linearized, version of the shallow water equations. From a semi-Lagrangian point of view, this model contains what is referred to as the nonlinear momentum equations for u and v, along with the continuity equation for the height field, sometimes recast as the geopotential. An advantage of the shallow water equations is that they are useful for testing the viability of new numerical schemes as they share many of the properties of the full three-dimensional system, but without many of their complexities. The two-dimensional shallow water equations in Cartesian coordinates are given by

$$\frac{Du}{Dt} = -\phi_x + fv, \tag{10.41}$$

$$\frac{Dv}{Dt} = -\phi_y - fu, \tag{10.42}$$

$$\frac{D\phi}{Dt} = -\phi\delta, \tag{10.43}$$

where u and v are the velocity components of the wind, $\delta = u_x + v_y$ is the divergence of the velocity fields $\mathbf{u}$, f is the Coriolis parameter, and ϕ is the geopotential height, which is assumed to be a positive function.

There are many formulations of the shallow water equations. One approach is to define vorticity, η, as

$$\eta = v_x - u_y \equiv \mathbf{k} \cdot \nabla \times \mathbf{u}. \tag{10.44}$$

If we now subtract the y-derivative of (10.42) from the x-derivative of (10.41), then we obtain

$$\frac{D}{Dt}(\eta + f) = (\eta + f)\,\delta. \tag{10.45}$$

Solving for the divergence in (10.43), and then substituting this expression into (10.45), results in

$$\frac{D}{Dt}\left(\frac{\eta + f}{\phi}\right) = 0. \tag{10.46}$$

An important feature of (10.46) is that it asserts that the physical quantity $\frac{\eta + f}{\phi}$, which is referred to as the potential vorticity, is **conserved** in time along the Lagrangian trajectory.

The next step in this formulation is to add the x-derivative of (10.41) to the y-derivative of (10.42) which yields

$$\frac{D\delta}{Dt} = -\nabla^2\phi - \nabla \cdot \left(f\mathbf{k} \times \mathbf{u}\right) - N, \tag{10.47}$$

where $\mathbf{k} = \begin{pmatrix} 0 & 0 & 1 \end{pmatrix}^T$ is the z-direction unit vector, $\mathbf{u} = \begin{pmatrix} u & v & 0 \end{pmatrix}^T$, and $N = u_x^2 + v_y^2 + 2u_x v_y$. The last step here is to now reformulate (10.43), (10.46), and (10.47) in terms of the geopotential, ϕ, the streamfunction ψ, and the velocity potential, χ, where we have the following relationships:

$$\mathbf{u} = \mathbf{k} \times \nabla\psi + \nabla\chi, \tag{10.48a}$$

$$\eta = \nabla^2 \psi, \tag{10.48b}$$

$$\delta = \nabla^2 \chi. \tag{10.48c}$$

Therefore, the shallow water equations can be written as

$$\frac{D}{Dt}\left(\frac{\nabla^2 \psi + f}{\phi}\right) = 0, \tag{10.49a}$$

$$\frac{D\nabla^2 \chi}{Dt} = -\nabla^2 \phi - \nabla \cdot \left(f\mathbf{k} \times \nabla \chi - f \nabla \psi\right) - N, \tag{10.49b}$$

$$\frac{D\phi}{Dt} = -\phi \nabla^2 \chi, \tag{10.49c}$$

where now $N = \left(\psi_{xy} - \chi_{xx}\right)^2 + \left(\psi_{xy} - \chi_{yy}\right)^2 - 2\left(\psi_{yy} - \psi_{xy}\right)\left(\psi_{xx} + \chi_{xy}\right)$.

A method to solve these equations is to consider a cylindrical domain, where we have a periodicity boundary condition for the x direction with length d, and y is in the range $[0, L]$. A good approach is to assume that $\chi = \psi = 0$ and $\phi = \phi_0$ at the y boundary, where ϕ_0 is a given constant. Another approximation is to use what is referred to as a β-plane, which enables the Coriolis parameter to be approximated by

$$f = f_0 + \beta y, \tag{10.50}$$

where f_0 and β are constants.

If we now apply a two-time-level semi-Lagrangian approach to (10.49a)–(10.49c), we need to find the departure point at the previous time level, $\mathbf{x}_d\,(t - \Delta t)$, where we require $\boldsymbol{\alpha}_d$. Thus, if we assume that the velocity field is constant between $t^n - \Delta t$ and t^n, then $\boldsymbol{\alpha}_d$ satisfies

$$\boldsymbol{\alpha}_d = \Delta t \mathbf{u}\left(\mathbf{x}_d - \frac{\boldsymbol{\alpha}}{2}, t^n - \frac{\Delta t}{2}\right). \tag{10.51}$$

To find the velocity at time $t - \dfrac{\Delta t}{2}$, we employ the extrapolation from the two previous time levels at

$$\mathbf{u}\left(\mathbf{x}, t^n - \frac{\Delta t}{2}\right) = \frac{3}{2}\mathbf{u}\left(\mathbf{x}, t^n - \Delta t\right) - \frac{1}{2}\mathbf{u}\left(\mathbf{x}, t - 2\Delta t\right). \tag{10.52}$$

Once $\boldsymbol{\alpha}_d$ is known, the departure point values of the variables in the equations above are given by

$$\phi_d\left(t^n - \Delta t\right) = \phi\left(\mathbf{x}_d - \boldsymbol{\alpha}_d, t^n - \Delta t\right). \tag{10.53}$$

We should note here that these values must be interpolated from known values at the grid points, and it is quite often the case that a form of cubic interpolation is best for this. Once we have the value at the departure point, we apply a upwind finite difference to the Lagrangian derivatives as we have seen earlier, whereas the non-derivative based quantities are represented by the average

$$\phi = \frac{1}{2}\left(\phi\left(t^n\right) + \phi_d\left(t^n - \Delta t\right)\right). \tag{10.54}$$

We should note here that in the early development of the semi-implicit semi-Lagrangian approaches the off-centering had not been thought of yet and that is why it is not here in this formulation.

Using this discretization, the shallow water equations can be manipulated into the form

$$\nabla^2\psi^{n+1} + f^{n+1} - f_1\phi^{n+1} = 0, \tag{10.55}$$

$$\nabla^2\chi^{n+1} + \tau\left(\nabla^2\phi^{n+1} + \beta\chi_x^{n+1} - \beta\psi_y^{n+1} - f\nabla^2\psi^{n+1}\right) = f_2, \tag{10.56}$$

$$\phi^{n+1}\left(1 + \tau\nabla^2\chi^{n+1}\right) = f_3, \tag{10.57}$$

where

$$f_1 \equiv \frac{\nabla^2\psi_d + f_d}{\phi_d}, \tag{10.58}$$

$$f_2 \equiv \nabla\cdot\left(\mathbf{u}_d - \tau\left(f\mathbf{k}\times\mathbf{u}_d + \nabla\phi_d\right)\right), \tag{10.59}$$

$$f_3 \equiv \phi_d\left(1 - \tau\nabla^2\chi_d\right), \tag{10.60}$$

and $\tau \equiv \frac{\Delta t}{2}$. Equations (10.55)–(10.57) are sometimes referred to as the *static equations.*

Therefore, given this setup, it is possible to integrate these forms of the shallow water equations in two parts. First, we compute the departure point quantities needed to define f_1, f_2, and f_3, which is achieved through using the information from the previous two time levels. The velocity field at any time level may be obtained from χ and ψ using

$$u = -\psi_y + \chi_x, \qquad v = \psi_x + \chi_y.$$

Once the departure points quantities are known, then second step is to solve the static equations above, which can be done through a multigrid method, which we shall not go into details about here, as we will see other methods in the spherical model chapter to solve this model, but have this approach here to inform the reader of the different approaches that are available. We now move on to consider a different version of the semi-Lagrangian approach with finite elements.

10.5 Finite element based semi-Lagrangian method

The summary that we present in this section is from [108], where the authors introduce a finite element based shallow water ocean model. The starting point is the inviscid shallow water equations in Cartesian coordinates, which are given by

$$\frac{D\mathbf{u}}{Dt} + f\mathbf{k}\times\mathbf{u} + g\nabla\eta = 0, \tag{10.61a}$$

$$\frac{D\ln(H+\eta)}{Dt} + \nabla\cdot\mathbf{u} = 0, \tag{10.61b}$$

where $\eta(x, y, t)$ is the surface elevation which represents the height of the fluid surface above the reference level $z = 0$ and the rigid bottom is defined by $z = -H(x, y)$, where $H(x, y)$ is the equilibrium fluid depth between the bottom and the reference height. We also have the lateral boundary conditions

$$\mathbf{u} \cdot \mathbf{n} = 0, \tag{10.62}$$

where $\mathbf{n}$ is the outward pointing normal vector at the boundary. It is assumed that the total fluid depth, $H + \eta$, is strictly positive. The model also has initial conditions for the velocity $\mathbf{u}(x, y, t = 0)$ and surface elevation $\eta(x, y, t = 0)$.

If we now consider applying a semi-Lagrangian discretization for the temporal dimension to (10.61a) and (10.61b), then we would obtain

$$\frac{\mathbf{u}^{n+1} - \mathbf{u}_d}{\Delta t} + \frac{1}{2}\left[\left(f\mathbf{k} \times \mathbf{u}\right)^{n+1} + \left(f\mathbf{k} \times \mathbf{u}\right)_d\right] + \frac{g}{2}\left[(\nabla\eta)^{n+1} + (\nabla\eta)_d\right] = 0, \tag{10.63}$$

$$\frac{\ln(H + \eta)^{n+1} - \left[\ln(H + \eta)\right]_d}{\Delta t} + \frac{1}{2}\left[(\nabla \cdot \mathbf{u})^{n+1} + (\nabla \cdot \mathbf{u})_d\right] = 0, \tag{10.64}$$

where the subscript d represents the departure points $(\mathbf{x}_m - \boldsymbol{\alpha}_m, t^n)$. If we now introduce remainder terms $\mathbf{R}^{\mathbf{u}}$ and R^{η}, then we obtain

$$\mathbf{u}^{n+1} \frac{\Delta t}{2} f\mathbf{k} \times \mathbf{u}^{n+1} + \frac{g\Delta t}{2}\nabla\eta^{n+1} = \left(\mathbf{u} - \frac{\Delta t}{2} f\mathbf{k} \times \mathbf{u} - \frac{g\Delta t}{2}\nabla\eta\right)_d \equiv \left(\mathbf{R}^u\right)_d, \tag{10.65}$$

$$\ln(H + \eta)^{n+1} + \frac{\Delta t}{2}(\nabla \cdot \mathbf{u})^{n+1} = \left[\ln(H + \eta) - \frac{\Delta t}{2}\nabla \cdot \mathbf{u}\right]_d \equiv \left(R^{\eta}\right)_d. \tag{10.66}$$

The right-hand sides of (10.65) and (10.66) are computed following a two step process:

1. Solve for the displacement $\boldsymbol{\alpha}_m$ by approximating the integration of the displacement equation

$$\frac{D\mathbf{x}}{Dt} = \mathbf{u}(\mathbf{x}, t) \tag{10.67}$$

that defines the trajectories, where $\mathbf{x} = (x, y)$ is the position vector. Applying the two-time-level scheme we require the extrapolation of the velocity field at all nodes at time $\left(t^n + \frac{1}{2}\right)$, which is achieved through the familiar extrapolation formula

$$\tilde{\mathbf{u}}\left(\mathbf{x}_m, t^{n+\frac{1}{2}}\right) = \frac{3}{2}\mathbf{u}\left(\mathbf{x}_m, t^n\right) - \frac{1}{2}\mathbf{u}\left(\mathbf{x}_m, t^{n-1}\right). \tag{10.68}$$

To integrate (10.67), a first-order estimate $\boldsymbol{\alpha}^{(0)} = \Delta t\mathbf{u}(\mathbf{x}_m, t^n)$ is combined with a number of iterations, of a second order midpoint Runge–Kutta **corrector**

$$\boldsymbol{\alpha}_m^{(k+1)} = \Delta t\tilde{\mathbf{u}}\left(\mathbf{x}_m - \frac{\boldsymbol{\alpha}_m^{(k)}}{2}, t^{n+\frac{1}{2}}\right). \tag{10.69}$$

2. Once the values of $\boldsymbol{\alpha}_m$ are known, an interpolator must be used to evaluate the right-hand sides of (10.65)–(10.66) at the upstream point departure point $\mathbf{x}_m - \boldsymbol{\alpha}_m$. The choice of the interpolator for the upstream evaluation of advected quantities has a critical impact of the accuracy of the method, this is because the accuracy of the advection terms, and thus the slow Rossby modes, depends strongly on the choice of interpolation procedure.

10.5.1 Weak Galerkin finite-element discretization

Let Ω be the model domain with boundary Γ. Functions spaces are now defined, where $H^1(\Omega)$ is the space of square integrable functions from the space $L^2(\Omega)$ whose first derivatives belong to $L^2(\Omega)$. Let $\mathbf{u}$ be in a subspace V of $H^1(\Omega) \times H^1(\Omega)$ such that $\mathbf{u} \cdot \mathbf{n} = 0$ on Γ for all $\mathbf{u}$ belonging to V, and let η be a sufficiently regular scalar function. The finite element weak formulation of the temporarily discretized Eqs. (10.65)–(10.66) requires test functions $\boldsymbol{\varphi}$ and ψ to belong to the same function space as $\mathbf{u}$ and η, respectively, such that

$$\int_\Omega \mathbf{u} \cdot \boldsymbol{\varphi} d\Omega + \frac{\Delta}{2} \int_\Omega f (\mathbf{k} \times \mu) \cdot \boldsymbol{\varphi} d\Omega + \frac{g\Delta t}{2} \int_\Omega \nabla \eta \cdot \boldsymbol{\varphi} d\Omega = \int_\Omega (\mathbf{R}^\mathbf{u})_d \cdot \boldsymbol{\varphi} d\Omega, \quad (10.70)$$

$$\int_\Omega \ln (H + \eta) \psi d\Omega + \frac{\Delta t}{2} \int_\Omega \nabla \cdot \mathbf{u} \psi d\Omega = \int_\Omega (R^\eta)_d d\Omega, \quad (10.71)$$

where the area element is $d\Omega = dxdy$.

The Galerkin method approximates the solution of (10.70)–(10.71) in a finite dimensional subspace. The discrete approximation represents each variable as a linear sum of appropriate basis functions over a given triangular element. To avoid computing η derivative, the third term in (10.70) and the corresponding one appearing on the right-hand side of (10.70) are integrated by parts using Green's theorem. In this way only $\boldsymbol{\varphi}$ derivatives are required, and no ψ derivative. The former are well defined since $\boldsymbol{\varphi}$ belongs to V, which is piecewise differentiable. The integration by parts leads to the *weak* form:

$$\int_\Omega \nabla \eta \cdot \boldsymbol{\varphi} d\Omega = - \int_\Omega \nabla \cdot \boldsymbol{\varphi} d\Omega + \int_\Gamma \boldsymbol{\varphi} \cdot \mathbf{n} \eta d\Gamma, \quad (10.72)$$

where $d\Gamma$ denotes the counterclockwise integration around the boundary Γ. The second term on the right-hand side of (10.72) vanishes since $\boldsymbol{\varphi}$ belongs to V, and so $\boldsymbol{\varphi} \cdot \mathbf{n} = 0$, therefore (10.70) is written as

$$\int_\Omega \mathbf{u} \cdot \boldsymbol{\varphi} d\Omega + \frac{\Delta t}{2} \int_\Omega f (\mathbf{k} \times \mathbf{u}) \cdot \boldsymbol{\varphi} d\Omega - \frac{g\Delta t}{2} \int_\Omega \eta \nabla \cdot \boldsymbol{\varphi} d\Omega = \int_\Omega (\mathbf{R}^\mathbf{u})_d \cdot \boldsymbol{\varphi} d\Omega, \quad (10.73)$$

where

$$\int_\Omega \mathbf{R}^\mathbf{u} \cdot \boldsymbol{\varphi} d\Omega = \int_\Omega \mathbf{u} \cdot \boldsymbol{\varphi} d\Omega - \frac{\Delta t}{2} \int_\Omega f (\mathbf{k} \times \mathbf{u}) \cdot \boldsymbol{\varphi} d\Omega + \frac{g\Delta t}{2} \int_\Omega \eta \nabla \cdot \boldsymbol{\varphi} d\Omega. \quad (10.74)$$

In [108] the authors introduce a finite-element triangulation $\mathcal{T}_h$ of the polygonal domain Ω where h is a representative mesh length parameter measuring resolution. Each triangle K in

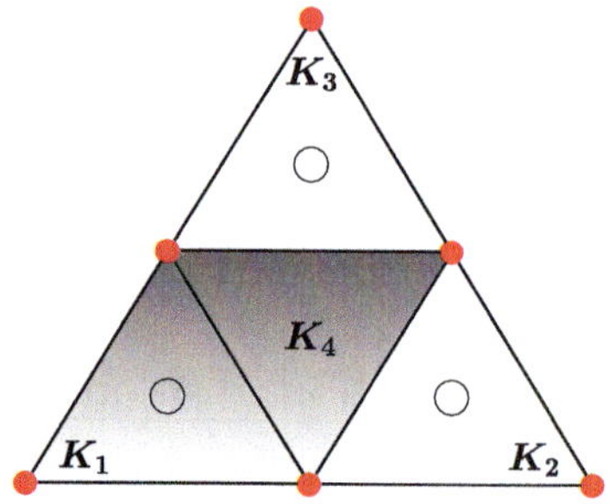

FIGURE 10.11 A triangle K of the triangulation $\mathcal{T}_h$ is subdivided into four triangles.

the triangulation $\mathcal{T}_h$ is now divided into four subtriangles $K_i, i = 1, \ldots, 4$ by joining midpoints of all sides, denoted $\mathcal{T}_{\frac{h}{2}}$. We have a diagram of the subtriangles in Fig. 10.11.

We now introduce the discrete velocity, $\mathbf{u}_h$, which is represented over a triangle K_i by a linear interpolating function of the form $\varphi(x, y) = a + bx + cy$. The vectors $\mathbf{u}_h$ and $\mathbf{R}^{\mathbf{u}}$ are expanded over triangle K_i using the linear basis function φ_s so that

$$\mathbf{u}_h = \sum_{s=1}^{3} \mathbf{u}_s \varphi_s, \tag{10.75}$$

$$\mathbf{R}^{\mathbf{u}} = \sum_{s=1}^{3} \mathbf{R}_s^{\mathbf{u}} \varphi_s, \tag{10.76}$$

where the sum is over the three vertices of triangles K_i with $\varphi_s = 1$ at vertex s and zero at the remaining two vertices. A similar linear approximation is better to be used here for the Coriolis term, rather than the quadratic expansion that arises from the product of $\mathbf{u}_h$ with f. This approach simplifies the algebra while having a negotiable impact on accuracy [108], since f varies smoothly, and so

$$f\mathbf{u}_h = \sum_{s=1}^{3} f_s \mathbf{us} \varphi_s, \tag{10.77}$$

over a triangle K_i of the triangulation $\mathcal{T}_{\frac{h}{2}}$.

The discrete solution η_k is sought in a finite-dimensional subspace Q^k of $L^2(\Omega)$ consisting of piecewise-constant polynomials over the triangle of the triangulation $\mathcal{T}_{\frac{h}{2}}$. For each triangle K of $\mathcal{T}_h$, three constant basis functions ψ_i, $i = 1, 2, 3$ are defined, corresponding to three degrees of freedom, such that

$$\psi_i = \chi_{K_i} + \frac{1}{3}\chi_{K_4}, \tag{10.78}$$

where χ_{K_i} for $i = 1, \ldots, 4$ is a basis function over triangle K_i. The latter function is defined as 1 on K_i and zero elsewhere. The basis functions ψ_i for $i = 1, \ldots, 3$ vanish outside a given triangle K of the triangulation $\mathcal{T}_h$. We should note here that this choice of basis functions has the property that $\sum_{i=1}^{3} \psi_i(\mathbf{x}) = 1$ over triangle K, ensuring that a constant field is properly represented [108]. The variables η_k, R^η, and H are expanded over triangle K in terms of basis functions ψ_i, resulting in

$$\eta_h = \sum_{i=1}^{3} \eta_{K_i} \psi_i, \tag{10.79}$$

$$R^\eta = \sum_{i=1}^{3} R^\eta_{K_i} \psi_i, \tag{10.80}$$

$$H = \sum_{i=1}^{3} H_{K_i} \psi_i. \tag{10.81}$$

We now turn our attention to evaluating the expressions $\mathbf{R}^u$ and R^η on the right-hand sides of (10.73) and (10.71), which need to be first evaluated at the previous time step at mesh nodes. By using (10.65), (10.66), and (10.72), and applying the weak Galerkin procedure to orthogonalize the error to the basis, $\mathbf{R}^u$ and R^η are sought in V^h and Q^h, respectively, such that

$$\int_\Omega \mathbf{R}^\mathbf{u} \cdot \boldsymbol{\varphi}_p d\Omega = \int_\Omega \mathbf{u}_h \cdot \boldsymbol{\varphi}_p d\Omega - \frac{\Delta t}{2} \int_\Omega f\left(\mathbf{k} \times \mathbf{u}_h\right) \cdot \boldsymbol{\varphi}_p d\Omega + \frac{g\Delta t}{2} \int_\Omega \eta_h \nabla \cdot \boldsymbol{\varphi}_p d\Omega, \tag{10.82}$$

$$\int_\Omega R^\eta \psi_j d\Omega = \int_\Omega \ln\left(H + \eta_h\right) \psi_j d\Omega - \frac{\Delta t}{2} \int_\Omega \nabla \cdot \mathbf{u}_h \psi_j d\Omega, \tag{10.83}$$

for all basis functions $\boldsymbol{\varphi}_p$ and ψ_j belonging to V^h and Q^h, respectively.

It is now stated in [108] that the dependent variables in (10.82) are expanded in terms of the basis functions using (10.75)–(10.77) and (10.79). This then results in a linear system for $\mathbf{R}^\mathbf{u}$ for all vertices of $\mathcal{T}_{\frac{h}{2}}$ since $\mathbf{u}_h$ and η_h are known at the previous times step.

If we now consider (10.83) then Le Roux et al. state that the mass matrix appearing on the left-hand side is block diagonal with all 3×3 blocks being identical. The dependent variables in (10.83) are expanded in terms of the basis function using (10.75) and (10.79)–(10.81). For each triangle K of $\mathcal{T}_h$ shown in Fig. 10.11, this then leads to the three equations

$$R^\eta_{K_i} + \frac{1}{3} R^\eta_{K_4} = \ln\left(H_{K_i} + \eta_{K_i}\right) + \frac{1}{3} \ln\left(H_{K_4} + \eta_{K_4}\right) - \frac{\Delta t}{2}\left(\left(\nabla \cdot \mathbf{u}_h\right)_{K_i} + \frac{1}{3}\left(\nabla \cdot \mathbf{u}_h\right)_{K_4}\right), \tag{10.84}$$

for $i = 1, 2, 3$. Since $R^\eta_{K_4} = \frac{1}{3} \sum_{i=1}^{3} R^\eta_{K_i}$, (10.84) can be rewritten as

$$\frac{1}{9}\begin{pmatrix} 10 & 1 & 1 \\ 1 & 10 & 1 \\ 1 & 1 & 10 \end{pmatrix}\begin{pmatrix} R^\eta_{K_1} \\ R^\eta_{K_2} \\ R^\eta_{K_3} \end{pmatrix} = \begin{pmatrix} \Phi_{K_1} \\ \Phi_{K_2} \\ \Phi_{K_3} \end{pmatrix}, \tag{10.85}$$

where

$$\Phi_{K_i} = \ln\left(\left(H_{K_i} + \eta_{K_i}\right)\left(H_{K_4} + \eta_{K_4}\right)^{\frac{1}{3}}\right) - \frac{\Delta t}{2}\left((\nabla \cdot \mathbf{u}_h)_{K_i} + \frac{1}{3}(\nabla \cdot \mathbf{u}_h)_{K_4}\right), \tag{10.86}$$

for $i = 1, 2, 3$. Here $\eta_{K_4} = \frac{1}{3}\sum_{i=1}^{3}\eta_{K_i}$ and $H_{K_4} = \frac{1}{3}\sum_{i=1}^{3} H_{K_i}$. The local divergence $(\nabla \cdot \mathbf{u}_h)_{K_i}$, $i = 1, \ldots, 4$ is constant on each triangle K_i of $\mathcal{T}_{\frac{h}{2}}$, since $\mathbf{u}_h$ is represented as a piecewise linear function. Finally, R^η is obtained over triangle K of $\mathcal{T}_h$ by explicitly solving (10.85) as

$$\begin{pmatrix} R^\eta_{K_1} \\ R^\eta_{K_2} \\ R^\eta_{K_3} \end{pmatrix} = \frac{1}{12}\begin{pmatrix} 11 & -1 & -1 \\ -1 & 11 & -1 \\ -1 & -1 & 11 \end{pmatrix}\begin{pmatrix} \Phi_{K_1} \\ \Phi_{K_2} \\ \Phi_{K_3} \end{pmatrix}. \tag{10.87}$$

The fields $\mathbf{R}^{\mathbf{u}}$ and R^η are then interpolated at the upstream positions $(\mathbf{x}_m - \alpha_m)$ through a kriging interpolator to obtain $\left(\mathbf{R}^{\mathbf{u}}\right)_d$ and $(R^\eta)_d$. The reader is referred to [107] for details about this form of interpolation. Finally, we have to consider how to solve (10.71) and (10.73), to obtain estimates of the velocity and surface elevation at the new time step.

We recall that (10.71) and (10.73) form a coupled system, and to solve these Le Roux et al. again apply the weak Galerkin procedure. We are seeking discrete solutions $\mathbf{u}_h$ and η_h in V^h and Q^h, respectively, such that

$$\int_\Omega \mathbf{u}_h \cdot \boldsymbol{\varphi}_p d\Omega + \frac{\Delta t}{2}\int_\Omega f\left(\mathbf{k} \times \mathbf{u}_h\right) \cdot \boldsymbol{\varphi}_p d\Omega - \frac{g\Delta t}{2}\int_\Omega \eta_h \nabla \cdot \boldsymbol{\varphi}_p d\Omega = \int_\Omega \left(\mathbf{R}^{\mathbf{u}}\right)_d \cdot \boldsymbol{\varphi}_p d\Omega, \tag{10.88}$$

$$\int_\Omega \ln\left(H + \eta_h\right)\psi_j d\Omega + \frac{\Delta t}{2}\int_\Omega \nabla \cdot \mathbf{u}_h \psi_j d\Omega = \int_\Omega \left(R^\eta\right)_d \psi_j d\Omega, \tag{10.89}$$

for all basis functions $\boldsymbol{\varphi}_p$ and ψ_j belonging to V^h and Q^h, respectively. The fields $\left(\mathbf{R}^{\mathbf{u}}\right)_d$ and $(R^\eta)_d$ are expanded in terms of the basis functions $\boldsymbol{\varphi}_s$ and ψ_i as

$$\left(\mathbf{R}^{\mathbf{u}}\right)_d = \sum_{s=1}^{3}\left(\left(\mathbf{R}^{\mathbf{u}}\right)_d\right)_s \varphi_s, \tag{10.90a}$$

$$\left(R^\eta\right)_d = \sum_{i=1}^{3}\left(\left(R^\eta\right)_d\right)_{K_i} \psi_i. \tag{10.90b}$$

Le Roux et al. point out that (10.89) is weakly nonlinear due to the logarithmic term on the left-hand side of this equation, and they refer to sets of different solvers that could be used to solve such a nonlinear algebraic equation. In [108] the authors use a Newton-based procedure that leads to a search for a sequence $\eta_h^{(k)}$, where for a given initial guess $\eta^{(0)_h}$, we have

$$\begin{aligned} \int_\Omega \frac{\eta_h^{(k)}}{H + \eta_h^{(k-1)}}\psi_j d\Omega + \frac{\Delta t}{2}\int_\Omega \nabla \cdot \mathbf{u}_h \psi_j d\Omega &= \int_\Omega \left(R^\eta\right)_d \psi_j d\Omega + \int_\Omega \frac{\eta_h^{(k-1)}}{H + \eta_h^{(k-1)}}\psi_j d\Omega \\ &\quad - \int_\Omega \ln\left(H + \eta_h^{(k-1)}\right)\psi_j d\Omega, \end{aligned} \tag{10.91}$$

for each iterative step $k = 1, 2, 3, \ldots$ The mass matrix on the left-hand side of (10.91) is still block diagonal as in (10.83), but the 3×3 blocks are no longer identical. At each iteration, the surface elevation η in (10.91) is represented on each triangle K of $\mathcal{T}_h$ in terms of local divergence of $\mathbf{u}$. This then leads, for each triangle K of $\mathcal{T}_h$, to rewriting (10.91) as

$$\frac{\eta_{K_i}^{(k)}}{H_{K_i} + \eta_{K_i}^{(k-1)}} + \frac{1}{3} \frac{\eta_{K_4}^{(k)}}{H_{K_4} + \eta_{K_4}^{(k-1)}} = \Psi_{K_i}, \tag{10.92}$$

where

$$\begin{aligned} \Psi_{K_i} &= \left[(R^\eta)_d\right]_{K_i} + \frac{1}{3}\left[(R^\eta)_d\right]_{K_4} + \frac{\eta_{K_i}^{(k-1)}}{H_{K_i} + \eta_{K_i}^{(k-1)}} + \frac{1}{3} \frac{\eta_{K_4}^{(k-1)}}{H_{K_4} + \eta_{K_4}^{(k-1)}} \\ &- \ln\left(\left(H_{K_i} + \eta_{K_i}^{(k-1)}\right)\left(H_{K_4} + \eta_{K_4}^{(k-1)}\right)^{\frac{1}{3}}\right) - \frac{\Delta t}{2}\left((\nabla \cdot \mathbf{u}_h)_{K_i} + \frac{1}{3}(\nabla \cdot \mathbf{u}_h)_{K_4}\right), \end{aligned} \tag{10.93}$$

for $i = 1, 2, 3$. Since $\eta_{K_4}^{(k-1)} = \frac{1}{3}\sum_{i=1}^{3} \eta_{K_i}^{(k-1)}$ and $H_{K_4} = \frac{1}{3}\sum_{i=1}^{3} H_{K_i}$, (10.92) can be rewritten in vector form as

$$\begin{pmatrix} \frac{1}{a_1} - \frac{4}{3a_4} & -\frac{4}{3a_4} & -\frac{4}{3a_4} \\ -\frac{4}{3a_4} & \frac{1}{a_2} - \frac{4}{3a_4} & -\frac{4}{3a_4} \\ -\frac{4}{3a_4} & -\frac{4}{3a_4} & \frac{1}{a_3} - \frac{4}{3a_4} \end{pmatrix} \begin{pmatrix} \eta_{K_1}^{(k)} \\ \eta_{K_2}^{(k)} \\ \eta_{K_3}^{(k)} \end{pmatrix} = \begin{pmatrix} \Psi_{K_1} \\ \Psi_{K_2} \\ \Psi_{K_3} \end{pmatrix}, \tag{10.94}$$

where $a_i = H_{K_i} + \eta_{K_i}^{(k-1)}$, $i = 1, 2, 3$, $a_4 = -4\sum_{i=1}^{3} a_i$. The surface elevation can be explicitly obtained over triangle K of $\mathcal{T}_h$ as

$$\begin{pmatrix} \eta_{K_1}^{(k)} \\ \eta_{K_2}^{(k)} \\ \eta_{K_3}^{(k)} \end{pmatrix} = \frac{1}{a_4} \begin{pmatrix} a_1(a_1 + a_4) & a_1 a_2 & a_1 a_3 \\ a_1 a_2 & a_2(a_2 + a_4) & a_2 a_3 \\ a_1 a_3 & a_2 a_3 & a_3(a_3 + a_4) \end{pmatrix} \begin{pmatrix} \Psi_{K_1} \\ \Psi_{K_2} \\ \Psi_{K_3} \end{pmatrix}. \tag{10.95}$$

The values of $\eta_{K_i}^{(k)}$ for each triangle K_i, $i = 1, 2, 3$ are substituted in the third term on the left-hand side of (10.88). This procedure leads to a linear system for the velocity components only, and thus to a significant gain in computational efficiency and storage.

An important feature of this approach is that we can write (10.88) in a matrix form as

$$(\mathbf{M} + \mathbf{C} + \mathbf{G})\,\mathbf{u} = \mathbf{B} \tag{10.96}$$

where $\mathbf{M}$ and $\mathbf{C}$ are the mass and Coriolis operators, respectively, and $\mathbf{B}$ denotes the right-hand side. The matrices $\mathbf{M}$ and $\mathbf{C}$ are computed once at the beginning since they are time independent. Here $\mathbf{G}$ is the gradient operator obtained after substitution of η from (10.95) in the third term on the left-hand side of (10.88), and it involves the computation of products of $\nabla \cdot \boldsymbol{\varphi}_s$ and $\nabla \cdot \boldsymbol{\varphi}_p$; the latter are in turn constant on each subtriangle of $\mathcal{T}_{\frac{h}{2}}$.

Given all of the derivation presented above, Le Roux et al. run a series of experiments in [108] and in their conclusions do state that they have been able to successfully implement the finite-element semi-implicit semi-Lagrangian approach in the unforced shallow water equations on an unstructured mesh, combined with the kriging interpolator to accurately represent the slow Rossby modes.

10.6 Semi-Lagrangian integration in flux form

In this section we consider how to implement a semi-Lagrangian approach with an atmospheric chemistry and transport model. We really do not want the numerical model to be too diffusive and/or dispersive, and in a paper by Lin and Rood in 1996 [113], the authors develop an algorithm that is built upon the generalization of one-dimensional finite-volume schemes.

When we are dealing with tracer advection, it is stated in [113] that a numerical scheme should meet the following physical constraints:

1. Conserve mass without posteriori restoration.
2. Compute fluxes based on the subgrid distribution in the upwind direction.
3. Generate no new maxima or minima.
4. Preserve tracer correlations.
5. Be computationally efficient in spherical geometry.

Another important feature that has to be considered when dealing with transport advection is that there must be consistency between the tracer continuity equation and the underlying equation of continuity of the air, so we start by considering the advective form of the conservation law for a conservative scalar as

$$\frac{\partial q}{\partial t} + \mathbf{u} \cdot \nabla q = 0, \tag{10.97}$$

where q represents a *mixing ratio-like* quantity, and $\mathbf{u}$ is the horizontal velocity field again here. The conservation law in **flux form** is

$$\frac{\partial \pi q}{\partial t} + \nabla \cdot (\pi q \mathbf{u}) = 0, \tag{10.98}$$

where π could be the density of a nonhydrostatic system, the surface pressure of a σ-coordinate hydrostatic system, or the depth of the fluid of a shallow water system [113]. The conservation law for π in flux form can be obtained by setting $q = 1$ in (10.98),

$$\frac{\partial \pi}{\partial t} + \nabla \cdot (\pi \mathbf{u}) = 0. \tag{10.99}$$

It is then stated in [113] that the quantities to be conserved are π and πq, and not q. As we have mentioned at the end of the last section, numerical discretizations of (10.97) will not in general conserve πq, but conservation of πq is almost automatic if the flux-form conservation law (10.98) is discretized.

In [113] the authors mention that operator splitting is often used to extend 1D schemes to higher dimensions, where splitting is the process of replacing a complex problem by a series of simpler subproblems and that this often introduces errors, in the worse case scenario leading to a failure in multidimensional applications. In applications where the flow is nonhydrostatic and highly compressible, it is stated in [113] that it has not been found necessary to eliminate splitting error to have a successful simulation. However, applications in which the flow is hydrostatic and nearly incompressible, the splitting error must be explicitly considered.

To extend 1D schemes to multiple dimensions, and to maintain the required attributes as we mentioned above, [113] postulates three rules for algorithm development:

1. ***Conservation*** – The total constituent mass, that is, the global integral of πq, must be exactly conserved in the discretized form.
2. ***Consistency*** – For a spatially uniform q field, the discretized form of (10.98) must degenerate to (i) the discretized form of (10.99) for a general flow, or (ii) the discretized incompressibility condition $\nabla \cdot \mathbf{u} = 0$ when the flow is incompressible. Fulfillment of these conditions is *sufficient* to guarantee that a spatially uniform q field will remain uniform regardless of the compressibility, or divergence, of the flow.
3. ***Stability*** – The CFL linear stability condition should be the same for a directional split scheme, that is,

$$\max\left(C^x, C^y\right) \leq 1, \tag{10.100}$$

instead of the more stringent condition

$$C^x + C^y \leq 1, \tag{10.101}$$

where $C^x = u\dfrac{\Delta t}{\Delta x}$ and $C^y = v\dfrac{\Delta t}{\Delta y}$ are the Courant numbers in the East–West and North–South directions, respectively. It is mentioned that the stability condition (10.100) must also be met for large time steps as well.

To develop the multidimensional algorithm, Liu and Rood define F and G as the $1D$ operators representing increments or updates for one time step Δt to the constituent density filed $Q = \pi q$ due to the flux form transport from the East–West and North–South directions, respectively. They now introduce the difference operator as

$$\delta_\sigma S \equiv S\left(\sigma + \frac{\Delta\sigma}{2}\right) - S\left(\sigma - \frac{\Delta\sigma}{2}\right), \tag{10.102}$$

such that F and G can be written as

$$F\left(\overline{u}^t, \Delta t, \Delta x; Q^n\right) = -\delta_x\left[\mathcal{X}\left(\overline{u}^t, \Delta t, \Delta x; Q^n\right)\right], \tag{10.103a}$$

$$G\left(\overline{v}^t, \Delta t, \Delta y; Q^n\right) = -\delta_y\left[\mathcal{Y}\left(\overline{v}^t, \Delta t, \Delta y; Q^n\right)\right], \tag{10.103b}$$

where $\mathcal{X}$ and $\mathcal{Y}$ are the time-averaged fluxes of Q in the East–West and North–South direction, respectively, defined by

$$\mathcal{X}\left(\overline{u}^t, \Delta t, \Delta x; Q^n\right) = \frac{1}{\Delta x}\int_t^{t+\Delta t} uQdt - HOT, \tag{10.104a}$$

$$\mathcal{Y}\left(\overline{v}^t, \Delta t, \Delta y; Q^n\right) = \frac{1}{\Delta y}\int_t^{t+\Delta t} vQdt - HOT, \tag{10.104b}$$

where $\left(\overline{u}^t, \overline{v}^t\right)$ are the time averaged winds. In [113] they state that the very essence of 1D finite volume schemes is to use a properly defined time-averaged, or time-centered, wind and the cell-averaged fields at time level n, Q^n to approximate the time-averaged fluxes, the integrals in (10.104a) and (10.104b) across the boundaries of the grid cell to predict the cell-averaged Q field at time level $n+1$, Q^{n+1}. It should be noted here that in this formulation the winds and Q field are staggered as in the C grid from before. Lin and Rood do then make an assumption that the flow is nondivergent, and π is a constant, which means that Q and q are interchangeable, and as such F and G must satisfy the following condition when $Q=\alpha$:

$$F(\alpha) + G(\alpha) = -\alpha\Delta t\left(\frac{\delta_x\overline{u}^t}{\Delta x} + \frac{\delta_y\overline{v}^t}{\Delta y}\right) = 0. \tag{10.105}$$

It should be clear that (10.105) is a 2D finite difference representation of the divergence of the velocity field being zero. Finally, F and G are required to satisfy the following relations:

$$H(Q+\alpha) = H(Q) + H(\alpha), \tag{10.106a}$$

$$H(\alpha Q) = \alpha H(Q), \tag{10.106b}$$

where H represents a generic 1D flux-form increment operator.

It should be noted here that, according to Lin and Rood, (10.106a) and (10.106b) are the minimum conditions that must be met in order to preserve linear constituent correlations. Next the authors introduce sequential splitting to advance the Q field one time step; this is achieved by applying the F operator first in the x direction and then this result is operated on by the y-direction operator G such that:

$$Q^x = Q^n + F\left(Q^n\right), \tag{10.107a}$$

$$Q^{yx} = Q^x + G\left(Q^x\right). \tag{10.107b}$$

Upon substituting (10.107a) into (10.107b) results in

$$Q^{yx} = Q^n + F\left(Q^n\right) + G\left(Q^n\right) + GF\left(Q^n\right). \tag{10.108}$$

Therefore, Q^{yx} is the predicted value at time t^{n+1}, where we have assumed that

$$G\left(Q^n + F\left(Q^n\right)\right) = G\left(Q^n\right) + GF\left(Q^n\right). \tag{10.109}$$

The second and third terms on the right-hand side of (10.108) are the convergence of the orthogonal fluxes in the x and y directions, respectively. It is then stated in [113] that the last term in (10.108) is critical to the stability of the scheme. This term is referred to as the *splitting term* in [113], and it should vanish under the assumption that $\nabla \cdot \mathbf{u} = 0$ and $Q = \alpha$, thus

$$\alpha = \alpha + \left(F(\alpha) + G(\alpha)\right) + GF(\alpha). \tag{10.110}$$

Using (10.105), the two terms in the parentheses above vanish, and the error term $GF(\alpha)$ can be written as

$$GF(\alpha) = G\left(\frac{-\alpha \Delta t \delta_x \overline{u}^t}{\Delta x}\right) \approx \frac{\alpha (\Delta t)^2 \delta_y \left(\overline{v}^t \delta_x \overline{u}^t\right)}{\Delta x \Delta y}. \tag{10.111}$$

It is possible to reverse the order of the operators to obtain

$$Q^{xy} = Q^n + F\left(Q^n\right) + G\left(Q^n\right) + FG\left(Q^n\right), \tag{10.112}$$

where the splitting error term can be written as

$$FG(\alpha) = F\left(\frac{-\alpha \Delta t \delta_y \overline{v}^t}{\Delta y}\right) \approx \frac{\alpha (\Delta t)^2 \delta_x \left(\overline{u}^t \delta_y \overline{v}^t\right)}{\Delta x \Delta y}, \tag{10.113}$$

which is operator-wise antisymmetric to (10.111). An important feature to note here is that (10.108) and (10.112) are both only first-order accurate, due to the existence of the directional bias and the splitting error. It is stated in [113] that it is possible to achieve second-order accuracy through a symmetric sequence of the operator splitting method, that it, if we averaged the two antisymmetric schemes, then we would obtain

$$\overline{Q} = \frac{1}{2}\left(Q^{xy} + Q^{yx}\right) = Q^n + F\left(Q^n\right) + G\left(Q^n\right) + \frac{1}{2}\left(GF\left(Q^n\right) + FG\left(Q^n\right)\right). \tag{10.114}$$

In [113] the authors present a discussion of whether or not this scheme satisfies the three conditions from earlier, where they now introduce the *advective* form operators f and g to obtain

$$Q^{n+1} = Q^n + F\left(Q^n + \frac{1}{2} g\left(v_{av}, \Delta t; Q^n\right)\right) + G\left(Q^n + \frac{1}{2} f\left(u_{av}, \Delta t; Q^n\right)\right) \tag{10.115}$$

where

$$u_{av} \equiv \overline{\overline{u}^t}^x, \quad v_{av} \equiv \overline{\overline{v}^t}^y \tag{10.116}$$

are the winds at the cell center used by the inner advection operators. The upper overbar represents the following operator:

$$\overline{\overline{S}^t}^\sigma = \frac{1}{2}\left(\overline{S}^t\left(\sigma + \frac{\Delta\sigma}{2}\right) + \overline{S}^t\left(\sigma - \frac{\Delta\sigma}{2}\right)\right). \tag{10.117}$$

The inner operators f and g are approximations to $-\Delta t u_{av}\dfrac{\partial Q}{\partial x}$ and $-\Delta t v_{av}\dfrac{\partial Q}{\partial y}$, respectively.

For the more general case when π is not a constant, (10.115) can be written in terms of the mixing ratio q as

$$q^{n+1} = \frac{1}{\pi^{n+1}}\left(\pi^n q^n + F\left(q^n + \frac{1}{2}g\left(q^n\right)\right) + G\left(q^n + \frac{1}{2}f\left(q^n\right)\right)\right). \tag{10.118}$$

This derivation above is for the Eulerian formulation of the problem, but we shall extend some of these ideas to be more in the form of semi-Lagrangian format. It is stated in [113] that their version of semi-Lagrangian advection is slightly different to the forms above and is referred to as **flux-form semi-Lagrangian**, or FFSL.

10.6.1 Flux-form semi-Lagrangian (FFSL) advection

We start by considering the 1D advection problem with constant velocity as shown in Fig. 10.12, where we are assuming that the tracer distribution is known continuously and the 1D space has been divided into uniform grid **cells** of width Δx. The velocity moves the field from left to right and the advective process should simply translate the distribution while preserving the shape. The amount of material that is in the grid box, which is bounded by $x_{i-\frac{1}{2}}$ and $x_{i+\frac{1}{2}}$, is updated by the difference of flux into and out of the box. For the constant velocity case, the tracer distribution is translated by the distance $u\Delta t$ as we saw earlier, and the spatial translation for both grid points is the same.

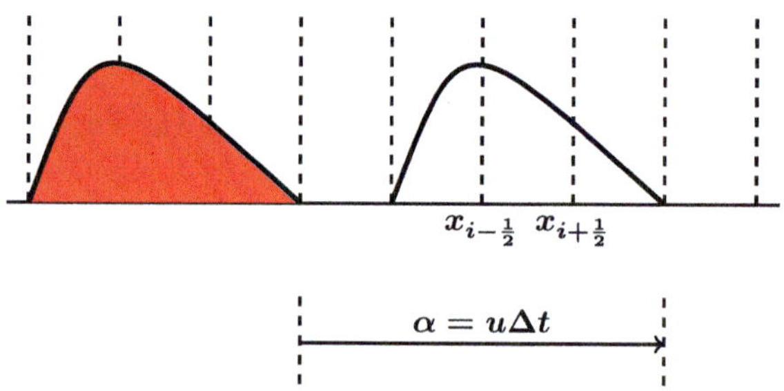

FIGURE 10.12 Illustration of the shape preserving of the distribution from t^n to t^{n+1}.

If we now consider the case where we no longer require the velocity field to be constant, so we now consider Fig. 10.13 where, according to [113], on the abscissa the spatial dimension x is divided into equal intervals of length Δx. On the ordinate we have time at steps n, $n+\frac{1}{2}$, and $n+1$. The algorithm the authors are developing in [113] requires the calculation the average density distribution at time $n+1$ between grid points $x_{i-\frac{1}{2}}$ and $x_{i+\frac{1}{2}}$ based only on the distribution at time n.

In Fig. 10.13 the lines $A'A$ and $B'B$ are the trajectories that arrive at our points $i+\frac{1}{2}$ and $i-\frac{1}{2}$ at $n+1$, respectively. We know that the distance between the arrival points is Δx but because we are not assuming that u is not constant, the distance between A' and B' is not guaranteed to be Δx, and so we denote this distance at time n by $\Delta x'$.

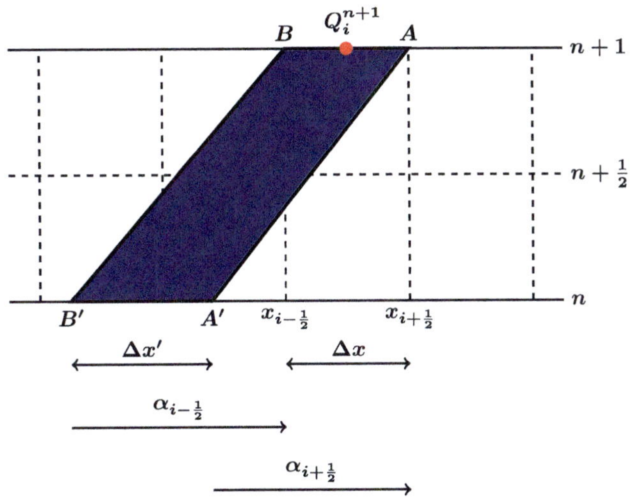

FIGURE 10.13 The mass conservation law in the discretized Lagrangian space.

The spatial displacement from point A' to A is denoted as $D_{i+\frac{1}{2}}$ in [113], which is equal to the *average velocity* along the trajectory, $u_{i+\frac{1}{2}}^{n+\frac{1}{2}}$, multiplied by the time increment Δt. There is a similar velocity for the trajectory $B'B$. Therefore, according to Lin and Rood, the problem in Fig. 10.13 focuses on the question; What spatial increment upstream at time n contains the mass that will be in spatial cell of Δx size at time $n+1$? In the absence of sources and sinks, the mass is conserved between trajectories, which is the shaded area in Fig. 10.13. It is then stated in [113] that if this question can be answered, then mass will be conserved at time $n+1$.

Now for an arbitrarily long time step in the x direction, the Courant number at $i-\frac{1}{2}$ can be written as

$$C_{i-\frac{1}{2}}^{x} = K_{i-\frac{1}{2}} + c_{i-\frac{1}{2}}, \tag{10.119}$$

where $K_{i-\frac{1}{2}}$ is the integer part of the Courant number, and $c_{i-\frac{1}{2}}$ is the fractional Courant number as we have seen before. In [113] the authors state that the advection can then be broken down into an integer and fractional flux. Using the fractional Courant number, the fractional flux can be computed by most Eulerian upstream flux form scheme that utilizes cell-averaged values to construct subgrid distributions.

If we now denote the cell-averaged density between $x_{i-\frac{1}{2}}$ and $x_{i+\frac{1}{2}}$ at future time $n+1$ as Q_i^{n+1}, then the 1D transport problem is

$$Q_i^{n+1} = Q_i^n + F_i = Q_i^n - (\delta_x \mathcal{X})_i, \tag{10.120}$$

where F is the *semi-Lagrangian* extension of the flux-form operator from (10.103a), and $\mathcal{X}$ is the *time-averaged* flux in (10.104a) at the left edge of the cell, decomposed into a integer flux and a fractional flux. The integer fluxes are computed exactly as

$$\begin{cases} \sum_{k=1}^{K_{i-\frac{1}{2}}} Q_{i-k}^n, & K_{i-\frac{1}{2}} \geq 1, \\ 0, & K_{i-\frac{1}{2}} = 0, \\ \sum_{k=1}^{-K_{i-\frac{1}{2}}} Q_{i-1+k}, & K_{i-\frac{1}{2}} \leq -1. \end{cases}$$

It is then stated in [113] that they remain compliant with (10.105), so that condition II from earlier is met, and the averaged velocity along the trajectory used to compute the displacement $D_{i-\frac{1}{2}} = u_{i-\frac{1}{2}} \Delta t$ of the cell interfaces is approximated by the velocity given at $x = x_{i-\frac{1}{2}}$ at time $n + \frac{1}{2}$.

In [113] the authors state that for stability of the scheme proposed, they enforce the condition that the trajectories of the left and right interfaces of a cell do not cross each other, which is the same as saying that the cell does not collapse during one time step, which mathematically is equivalent to

$$\frac{\Delta t \delta_x u^{n+\frac{1}{2}}}{\Delta x} \leq 1. \tag{10.121}$$

Given all of the derivation above, the trajectory calculation for the inner operator f uses the cell-averaged velocity defined in (10.117) as

$$u_{av}^{n+\frac{1}{2}} = \frac{1}{2}\left(u_{i-\frac{1}{2}}^{n+\frac{1}{2}} + u_{i+\frac{1}{2}}^{n+\frac{1}{2}} \right).$$

The 1D projection of the *departure point* x_d^n is computed from

$$x_d^n = x_i^{n+1} - \Delta t u_{av}^{n+\frac{1}{2}}.$$

The values from two nearest grid cell to x_d^n at time t^n are then used in a linear interpolation procedure to obtain the advective update.

Extension to 2D

Following directly from (10.115) and (10.120), the two-dimensional version of FFSL for an incompressible flow is given by

$$Q^{n+1} = q^n - \delta_x \left(\mathcal{X} \left(Q^n + \frac{1}{2} g\left(Q^n\right) \right) \right) - \delta_y \left(\mathcal{Y} \left(Q^n + \frac{1}{2} f\left(Q^n\right) \right) \right). \tag{10.122}$$

The construction of a 2D FFSL scheme using (10.122), where focusing on the second term in this equation on the right-hand side to update $Q^n_{i,j}$ to $Q^{n+1}_{i,j}$, the contribution from the first order advective-form operator g at (i, j) can be written as

$$g_{i,j} = \left(Q^n_{i,J} - Q^n_{i,j}\right) + \left|c^y_{i,j}\right| \left(Q^n_{i,J*} - Q^n_{i,J}\right), \tag{10.123}$$

where

$$C^y_{i,j} = \frac{\Delta t}{2\Delta y}\left(v_{i,j-\frac{1}{2}} + v_{i,j+\frac{1}{2}}\right)$$

is the Courant number at (i, j) in the y-direction, $c^y_{i,j} = C^y_{i,j} - \mathrm{INT}\left(C^y_{i,j}\right)$ is the fractional Courant number at (i, j) in the y direction, $J = j - \mathrm{INT}\left(C^y_{i,j}\right)$, and $J* = J - \mathrm{SIGN}\left(1, C^y_{i,j}\right)$.

To complete the construction of the second term on the right-hand side of (10.122), Lin and Rood apply a second-order van Leer scheme as the outer flux-form operator. We now define $Q^g_{i,j}$ as the input tracer density field to $\mathcal{X}$, that is,

$$Q^g_{i,j} = Q^n_{i,j} + \frac{1}{2}g_{i,j}\left(Q^n\right) = \frac{1}{2}\left(\left(Q^n_{i,J} + Q^n_{i,j}\right) + \left|c^y_{i,j}\right|\left(Q^n_{i,J*} - Q_{i,J}\right)\right), \tag{10.124}$$

which implies that the flux $\mathcal{X}$ across the left interface of the cell (i, j) is then computed as the sum of the integer flux $\widehat{\mathcal{X}}$ and the fractional flux $\widetilde{\mathcal{X}}$ such that

$$\mathcal{X}_{i-\frac{1}{2},j} = \widehat{\mathcal{X}}_{i-\frac{1}{2},j} + \widetilde{\mathcal{X}}_{i-\frac{1}{2},j}, \tag{10.125}$$

where the fractional flux is given by

$$\widetilde{\mathcal{X}}_{i-\frac{1}{2},j} = c^x_{i-\frac{1}{2},j}\left(Q^g_{I,j} + \frac{\Delta Q^g_{I,j}}{2}\left(\mathrm{SIGN}\left(1, c^x_{i-\frac{1}{2},j}\right) - c^x_{i-\frac{1}{2},j}\right)\right), \tag{10.126}$$

where $C^x_{i-\frac{1}{2},j} = \frac{\Delta t u_{i-\frac{1}{2},j}}{\Delta x}$ is the Courant number at $\left(i - \frac{1}{2}, j\right)$, $I = \mathrm{INT}\left(i - C^x_{i-\frac{1}{2},j}\right)$, $M_{i-\frac{1}{2},j} = \mathrm{INT}\left(C^x_{i-\frac{1}{2},j}\right)$, $\Delta Q^g_{I,j} = \frac{1}{2}\left(Q^g_{I+1,j} - Q^g_{I-1,j}\right)$, and $c^x_{i-\frac{1}{2},j} = C^x_{i-\frac{1}{2},j} - M_{i-\frac{1}{2},j}$ is the fractional Courant number at $\left(i - \frac{1}{2}, j\right)$.

While we will not show it here, Lin and Rood do perform a stability analysis of the scheme in their appendix, and show that the scheme is unconditionally stable if the flow is nondeformational, but if this is not the case then (10.121) should be the limit on stability. Lin and Rood do perform a series of numerical experiments with this scheme, and this scheme was recommended to be the advection scheme for the NASA chemical transport model at the time.

10.7 Semi-Lagrangian integration with finite volumes

The method that we present here comes from a paper by Peter Lauritzen [99], which opens with the statement:

> *Finite-volume methods for numerical solutions to conservation laws, in particular the continuity equation stating the* ***conservation of mass****, have received increased attention in the meteorological literature during the last two decades.*

However, Lauritzen does state that these approaches have been used for atmospheric constituents as standard in transport modules in atmospheric models as they do not have **spurious** numerical sources or sinks. While there is a whole chapter on approaches for mass conservation, as well as shape preserving, in Chapter 13 we present the technique from [99] as it is seen as a cell-integrated advection scheme.

10.7.1 One-dimensional cell-integrated semi-Lagrangian (CISL) schemes

We start by considering the one-dimensional cell-integrated continuity equation for air as a whole and an atmospheric constituent, which is given by

$$\frac{D}{Dt}\int_{\delta x}\psi\left(x\right)dx=0, \tag{10.127}$$

where $\psi=\rho$ and $\psi=\rho Q$, respectively, with ρ being the density of air, and Q is the mixing ratio for the constituent in question, δx is a length interval moving with the flow. In [99] the author defines the operator I to be

$$I_A^x\left[\psi\left(A\right)\right]\equiv\frac{1}{A}\int_A\psi\left(x\right)dx=\overline{\psi}, \tag{10.128}$$

where A is the length interval over which $\psi\left(x\right)$ is integrated, the superscript X refers to the coordinate direction in which the integral is taken, and $\overline{\psi}$ is the average value of $\psi\left(x\right)$ over A. The upstream CICL discretization of (10.127) can be written as

$$\overline{\psi}^{n+1}\Delta x=I_{\delta x_d^n}^x\left[\psi\left(x\right)\right]\delta x_d^n, \tag{10.129}$$

where $\overline{\psi}^{n+1}$ is the average of $\psi^{n+1}\left(x\right)$ over Δx, and $\delta_{x_d^n}$ is the length interval whose boundaries end up at the boundaries of Δx after being transported by the wind over one time step.

In the discretization here, $\psi\left(x\right)^n$ is not known and must be constructed from the known grid cell average $\overline{\psi}^n$. In [99] this process as is referred to as **subgrid cell reconstruction**. To ensure that the scheme conserves mass, the subgrid cell reconstruction must satisfy

$$I_{\Delta x}^x\left[\psi^n\left(x\right)\right]=\overline{\psi}^n, \tag{10.130}$$

and the cell walls of any particular **departure cell** must be the left and right cell walls of the two neighboring cells, respectively. In [99] the author then makes the statement that, contrary to cell-integrated schemes, an advective-form discretization is based on gridpoint values rather than cell averages. Hence the velocity components can be interpreted as unique to a particular control volume rather than individual cell faces [140]. Hence each cell moves as a solid body, and as a result neighboring cells in divergent flows may overlap and/or have cracks between them, causing a violation of mass conservation.

If we now assume an equidistant grid with grid spacing Δx, and let the ith Eulerian grid cell extend from $x_i = i\Delta x$ to $x_{i+1} = (i+1)\Delta x$, then this Eulerian cell is referred to as the **arrival cell**. The corresponding upstream Lagrangian **departure cell walls** are located at $x_{i,d}$ and $x_{i+1,d}$, respectively. The width of the departure cell is $\delta^n_{i,d} = x^n_{i+1,d} - x^n_{i,d}$. The departure points as we have seen for the semi-Lagrangian approaches are computed using iterative methods, whereas for a Eulerian approach we may use the Euler's method of

$$x^n_{i,d} = x_i - u^n(x_i)\Delta t. \tag{10.131}$$

The three different schemes that [99] considered, referred to as the first-order **Piecewise Constant Method (PCM)**, the second-order **Piecewise Linear Method (PLM)**, and the third-order **Piecewise Parabolic Method (PPM)** for subgrid-scale reconstruction in cell i, are given by

$$\psi_i(\xi) = \overline{\psi}_i, \tag{10.132}$$

$$\psi_i(\xi) = \overline{\psi}_i + \frac{1}{2}\left(\overline{\psi}_{i+1} - \overline{\psi}_{i-1}\right)\left(\xi - \frac{1}{2}\right), \tag{10.133}$$

$$\psi_i(\xi) = \psi^L_i + \xi\left(\Delta\psi + \widetilde{\psi}_1(1-\xi)\right), \tag{10.134}$$

respectively, where ξ is a normalized coordinate in the ith cell such that $\xi \in [0,1]$, where

$$\xi = \frac{x - x_i}{2}, \quad x \in \left[x_i, x_{i+1}\right], \tag{10.135}$$

and where for the PPM the slope and curvature of the polynomial are given by

$$\Delta\psi_i = \psi^R_i - \psi^L_i, \tag{10.136}$$

$$\widetilde{\psi}_i = 6\psi\overline{\psi}_i - 3\left(\psi^L_i + \psi^R_i\right), \tag{10.137}$$

respectively, where $\psi^L_i = \psi_i(0)$ and $\psi^R_i = \psi_i(1)$ are the values of $\psi_i(\xi)$ at the left and right cell walls. It should be noted here that performing the subgrid scale reconstruction from cell averages in all cells defined a global piecewise-continuous function.

10.7.2 Von Neumann stability analysis

In [99] Lauritzen performs a stability analysis following the same notation as before, such that if we assume a constant wind $u = u_0 < 0$, then this implies that the exact departure cell is simply the translation of the arrival cell $\Delta t u_0$ units upstream, implying that the departure point corresponding to the arrival point x_i is located in x_{i-p-1} and x_{i-p}. We have recreated the figure from [99] in Fig. 10.14.

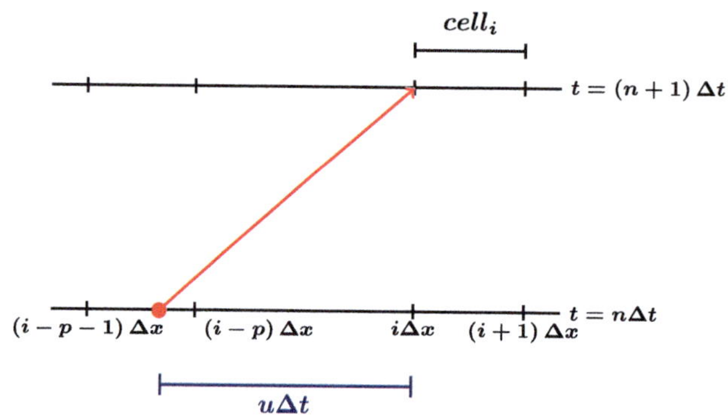

FIGURE 10.14 Reconstruction of the space time representation of the departure and arrival positions of the ith left cell wall.

As the field is nondivergent, (10.129) becomes

$$\overline{\psi}_i^{n+1} = \int_{1-\widehat{\alpha}}^{1} \psi_{i-p-1}^{n}(\xi)\,d\xi + \int_{0}^{1-\widehat{\alpha}} \psi_{i-p}^{n}(\xi)\,d\xi, \tag{10.138}$$

where $\psi_i(\xi)$ is the subgrid-scale reconstruction polynomial in the cell bounded by x_i and x_{i+1}. When using the PCM, PLM, or PPM for this calculation, (10.138) becomes

$$\begin{aligned}
\left(\overline{\psi}_i\right)_{PCM}^{n+1} &= \widehat{\alpha}\overline{\psi}_{i-p-1} + (1-\widehat{\alpha})\,\overline{\psi}_{i-p}, && (10.139)\\
\left(\overline{\psi}_i\right)_{PLM}^{n+1} &= \frac{1}{4}\left(-\widehat{\alpha}\,(1-\widehat{\alpha})\,\overline{\psi}_{i-p-2} + \alpha\,(5-\widehat{\alpha})\,\overline{\psi}_{i-p-1} + (4+\widehat{\alpha})\,(1-\widehat{\alpha})\,\overline{\psi}_{i-p}\right.\\
&- \left.\widehat{\alpha}\,(1-\widehat{\alpha})\,\overline{\psi}_{i-p+1}\right), && (10.140)\\
\left(\overline{\psi}_i\right)_{PPM}^{n+1} &= \frac{1}{12}\left(\widehat{\alpha}^2\,(1-\widehat{\alpha})\,\overline{\psi}_{i-p-3} - \widehat{\alpha}\,(1+7\widehat{\alpha})\,(1-\widehat{\alpha})\,\overline{\psi}_{i-p-2} - 4\widehat{\alpha}\left(4\widehat{\alpha}^2 - 5\widehat{\alpha} - 2\right)\overline{\psi}_{i-p-1}\right.\\
&- 4\,(1-\widehat{\alpha})\left(4\widehat{\alpha}^2 - 3\widehat{\alpha} - 3\right)\overline{\psi}_{i-p} + \widehat{\alpha}\,(1-\widehat{\alpha})\,(7\widehat{\alpha} - 8)\,\overline{\psi}_{i-p+1}\\
&+ \left.\widehat{\alpha}\,(1-\widehat{\alpha})^2\,\overline{\psi}_{i-p+2}\right), && (10.141)
\end{aligned}$$

respectively. Note that (10.140) and (10.141) are not formally equivalent to a traditional semi-Lagrangian scheme using cubic and quintic Lagrange interpolation, respectively, although they require the same stencil.

Lauritzen provides the von Neumann stability analysis for the schemes above, which results in the following amplification factors and numerical frequencies:

$$\begin{aligned}
|\Gamma|_{PCM}^2 &= 1 - 2\widehat{\alpha}\,(1-\widehat{\alpha})\,C, && (10.142)\\
|\Gamma|_{PLM}^2 &= 1 - \frac{1}{2}C^2\widehat{\alpha}\,(1-\widehat{\alpha})\,(2 - \widehat{\alpha}\,(1-\widehat{\alpha})\,(2-C)), && (10.143)
\end{aligned}$$

$$\begin{aligned}|\Gamma|^2_{PPM} &= 1-\frac{1}{9}\widehat{\alpha}\left(1-\widehat{\alpha}\right)^2 C^2\left(3-2\left(\widehat{\alpha}^2-\widehat{\alpha}-5\right)C-\left(4\widehat{\alpha}^2-4\widehat{\alpha}-1\right)C^2\right.\\&+\left.2\widehat{\alpha}\left(1-\widehat{\alpha}\right)C^3\right),\end{aligned} \tag{10.144}$$

and

$$\omega_{PCM}\Delta t = pk\Delta x+\tan^{-1}\left(\frac{\widehat{\alpha}\sin k\Delta x}{1-\widehat{\alpha}C}\right), \tag{10.145}$$

$$\omega_{PLM}\Delta t = pk\Delta x+\tan^{-1}\left(\frac{\widehat{\alpha}\sin\left(k\Delta x\right)\left(1+\frac{C}{2}\left(1-\widehat{\alpha}\right)\right)}{1-\widehat{\alpha}C\left(\widehat{\alpha}+\frac{C}{2}\left(1-\widehat{\alpha}\right)\right)}\right), \tag{10.146}$$

$$\omega_{PPM}\Delta t = pk\Delta x+\tan^{-1}\left(\frac{\widehat{\alpha}\sin\left(k\Delta x\right)\left(1+\frac{C}{3}\left(1-\widehat{\alpha}^2\right)+\frac{C^2\widehat{\alpha}}{3}\left(1-\widehat{\alpha}\right)\right)}{1-C\widehat{\alpha}^2\left(1+\frac{C}{3}\left(1+C\right)\left(1-\widehat{\alpha}\right)\right)}\right), \tag{10.147}$$

where $C = 1-\cos k\Delta x$ as before.

Exercise 10.3. *Verify the amplification factors in* (10.142)–(10.144) *and the numerical frequencies in* (10.145)–(10.147).

It is stated in [99] that the routines above have the following properties: The cell-integrated scheme based on PPM is less damping than the traditional semi-Lagrangian scheme and cell-integrated scheme based on PLM, and the damping is much more scale selective; that is, the $2\Delta x$ wave that is often associated with noise and/or ripples is heavily damped while the longer wavelengths are much less diffused. Regarding dispersion properties, the cell integrated scheme based on PPM and the traditional semi-Lagrangian scheme are very similar. The relative phase speed error for the cell-integrated scheme using PLM is smaller and different in structure compared to the other schemes.

10.7.3 Two-dimensional finite volume schemes

It is stated in [99] that two-dimensional finite-volume schemes can be divided into two categories; first are the CICL schemes in which the integral over the departure area is approximated explicitly, and second are the flux-form Eulerian schemes. Lauritzen states that the scheme from [113], which we presented in the last section, is a member or this latter family.

For the two-dimensional subgrid cell reconstruction, the following notation is used: let Δx and Δy be the equidistant grid spacing in the x- and y-coordinate directions, with the (i, j)th cell vertices located at $(x, y) = (x_i, y_j), (x_i, y_{j+1}), (x_{i+1}, y_j)$, and (x_{i+1}, y_{j+1}), respectively. As with the one-dimensional case, the normalized spatial position variables inside the (i, j)th cell are defined by

$$(\xi, \eta) = \left(\frac{x-x_i}{\Delta x}, \frac{y-y_j}{\Delta y}\right), \quad (x, y) \in \left[x_i, x_{i+1}\right] \times \left[y_j, y_{j+1}\right], \tag{10.148}$$

such that $\xi, \eta \in [0, 1]$. A quasi-biparabolic subgrid cell reconstruction scheme that was presented in [135] is stated in [99] to be based upon on the directional fitting of two parabolas based upon PPM. The slope and curvature of the parabola in the x- and y-coordinate directions are denoted $\Delta^{\xi}\psi_{ij}$ and $\widetilde{\psi}_{ij}^{\xi}$, and $\Delta^{\eta}\psi_{ij}$ and $\widetilde{\psi}_{ij}^{\eta}$, respectively. Lauritzen now introduces the standard compass point notations, which then enables the quasi-biparabolic subgrid cell reconstruction to be written as

$$\psi_{ij}(\xi,\eta) = \overline{\psi}_{ij}^{n} + \psi_{ij}^{W} + \xi\left(\Delta^{xi}\psi_{ij} + \widetilde{\psi}_{ij}^{\xi}(1-\eta)\right) + \psi_{ij}^{S} + \eta\left(\Delta^{\eta}\psi_{ij} + \widetilde{\psi}_{ij}^{\eta}(1-\eta)\right). \tag{10.149}$$

In Fig. 10.15 we have a schematic of the locations of the variables that are used in the quasi-parabolic subgrid cell reconstitution.

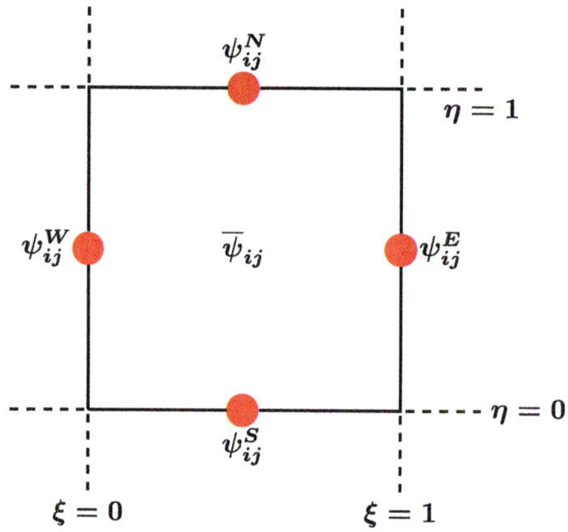

FIGURE 10.15 Graphical illustration of the variables that are used in the quasi-parabolic subgrid cell reconstruction.

Given this setup, we leave as exercises to derive the associated interpolation polynomials for the advection of the area to the arrival cell. To help illustrate the way the integration advects the cell with constant velocities, we have a replica of the grid setup from [99] in Fig. 10.16.

Exercise 10.4. *Given the polynomial in* (10.149)*, integrate over the area to find the expression for* $\overline{\psi}_{ij}^{n+1}$*, and derive the associated amplification factor and numerical frequency for* $\overline{\psi}_{ij}^{n+1}$ *with the quasi-biparabolic cell reconstruction.*

The reader is referred to [99] for the more detailed analysis of the stability of the schemes presented here, where there are several plots of the stability regions, as well as a comparison with the stability of the bicubic Lagrange interpolation in a traditional semi-Lagrangian formulation. There is also a more detailed explanation of the flux form from earlier where Lauritzen defines the two-dimensional flux form as

$$\begin{aligned} \overline{\psi}_{ij}^{n+1} &= (1-\widehat{\alpha})\left(1-\widehat{\beta}\right)\overline{\psi}_{i-p,j-q}^{n} + \widehat{\alpha}\left(1-\widehat{\beta}\right)\overline{\psi}_{i-p-1,j-q} \\ &+ \widehat{\beta}(1-\widehat{\alpha})\overline{\psi}_{i-p,j-q-1} + \widehat{\alpha}\widehat{\beta\overline{\psi}}_{i-p-1,j-q-1}, \end{aligned} \tag{10.150}$$

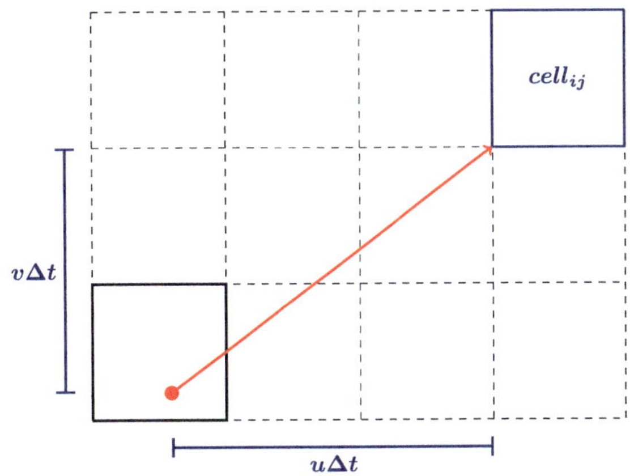

FIGURE 10.16 Same type of setup as for the 1D case earlier, where the cell bounded by the box in the top right-hand corner is the arrival cell. The departure point corresponding to the southwest corner is plotted in the departure cell area.

which we can see is the same as the bilinear Lagrange interpolation formula from earlier.

We now return to finite difference methods to consider a problem that was detected with semi-Lagrangian advection for rotational and deforming flows and a solution that was developed to overcome this problem.

10.8 Semi-Lagrangian advection in flows with rotation and deformation

In this section we provide a summary of a paper by Brassington and Sanderson from 1999 [22], where they present a warning about a first-order error that is introduced when semi-Lagrangian schemes advect scalars with high-order accuracy when calculating **field accelerations** in a steady rotational flow or in a steady deformational flow. This error, according to [22], is attributed to the neglect of the correlation between the advective change of the velocity field and the velocity field during the advection. In general, the field acceleration varies along a Lagrangian path. The authors suggest considering several points along the flow path to extend the calculation of field acceleration to higher order accuracy and to suppress this first-order error.

To show the derivation from [22], we consider a fluid undergoing a steady uniformly rotational motion in a clockwise sense with angular velocity ω. The authors of [22] state that, without loss of generality, the field acceleration can be calculated at the apex of a trajectory of radius R where the velocity is $(u, 0)$. This is shown in the copy of Fig. 1 from [22] in Fig. 10.17. At this apex point, $(x, y) = (0, R)$, the exact field acceleration is known to be

$$\left(\frac{Du}{Dt}, \frac{Dv}{Dt}\right) = \left(0, -\frac{u^2}{R}\right). \tag{10.151}$$

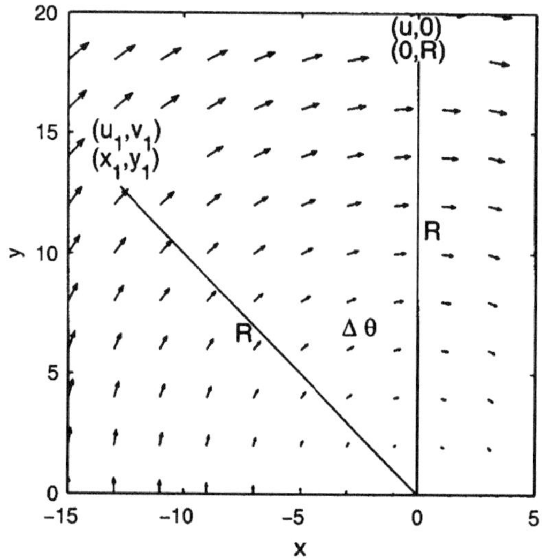

FIGURE 10.17 Copy of Fig. 1 from [22] illustrating rotating flow with the departure point vector (x_1, y_1) and velocity (u_1, v_1). Figure comes from Brassington, G. and Sanderson, B. (1999), *Semi-Lagrangian and COSMIC advection in flows with rotation or deformation*, Atmosphere-Ocean, **37**, 369–388. ©La Societe Canadienne de Meterologie et d'Oceanographie, reprinted by permission of Taylor and Francis Ltd, www.tandfonline.com on behalf of La Societe Canadienne de Meterologie et d'Oceanographie.

In a Δt time interval, the fluid sweeps through an angle $\Delta\theta = \frac{u}{R}\Delta t = \omega\Delta t$. The particle that arrives at the point $(0, R)$ at time $t + \Delta t$ leaving a departure point (x_1, y_1) at time t. The departure point can be calculated as shown in Fig. 10.17.

If we now consider the semi-Lagrangian integration applied to this situation, then, given the approximations of the departure point, we estimate the velocity (u_1, v_1) at the departure point (x_1, y_1) using a high-order interpolation. It is possible to calculate the departure point analytically as

$$u_1 = u\cos\Delta\theta, \qquad v_1 = u\sin\Delta\theta. \tag{10.152}$$

The field acceleration is numerically calculated as

$$\frac{Du}{Dt} = \frac{u - u_1}{\Delta t} = \frac{u\,(1 - \cos\Delta\theta)}{\Delta t}, \tag{10.153}$$

$$\frac{Dv}{Dt} = \frac{v - v_1}{\Delta t} = -\frac{u\sin\Delta\theta}{\Delta t}. \tag{10.154}$$

If we now apply a Taylor series expansion to $\cos\Delta\theta$ and $\sin\Delta\theta$, then (10.153) and (10.154) become

$$\frac{Du}{Dt} = \frac{u^3\Delta t}{2R^2} + HOT, \tag{10.155}$$

$$\frac{Dv}{Dt} = -\frac{u^2}{R} + \frac{u^4 \Delta t^2}{6R^3} + HOT, \tag{10.156}$$

where HOT stands for higher order terms, and where $\Delta\theta$ has been rewritten as $\frac{u\Delta t}{R}$. It is then stated in [22] that there is a first-order truncation error in the tangential acceleration and a second-order in Δt error in the radial acceleration, and that this error is present regardless of how accurately the departure point and field interpolations are determined.

As the flow is steady, this error is not an error in the local time derivative; rather it is an error that results from the spatial operators in the field acceleration. It is stated in [22] that this error cannot be simply removed by embedding the calculation of field accelerations within a higher-order time stepping scheme. Following the same arguments above, it is possible to show that the deformation flow also has a first-order Δt advection errors.

As stated in [22], a first-order Δt truncation error associated with steady rotational and deformational flow places severe accuracy restrictions on the time step, which diminishes the advantage of the semi-Lagrangian scheme. Brassington and Sanderson in [22] consider several points along the flow path traversed in Δt to achieve higher order accuracy with respect to time.

10.8.1 Higher-order time accuracy

A second-order estimate of the field accelerations could be obtained as follows according to [22]:

1. Compute the departure point (x_2, y_2) as being the position of a particle at time t that arrives at $(0, R)$ at time $t + \Delta t$.
2. Next compute departure point (x_1, y_1) as being the position of a particle at time $t + \frac{\Delta t}{2}$ that arrives at $(0, R)$ at time $t + \Delta t$.
3. Calculate the velocity fields (u_2, v_2) and (u_1, v_1) using the usual interpolations polynomial at the departure points (x_2, y_2) and (x_1, y_1), respectively.

We have a copy of Fig. 3 from [22] to illustrate the positions of the departure points along the arc in the rotational flows, Fig. 10.18.

Given this setup, Brassington and Sanderson introduce a verified second-order accurate estimate of the acceleration as

$$\frac{Du}{Dt} = \frac{3u - 4u_1 + u_2}{\Delta t}, \tag{10.157}$$

$$\frac{Dv}{Dt} = \frac{3v - 4v_1 + v_2}{\Delta t}. \tag{10.158}$$

For time varying flow fields, both of these backed–displaced trajectories would have to be computed at each time step. In the case of rotational flow, the acceleration at the apex $(x, y) = (0, R)$ now becomes

$$\frac{Du}{Dt} = \frac{3u^5 (\Delta t)^3}{96R^4} + HOT,$$

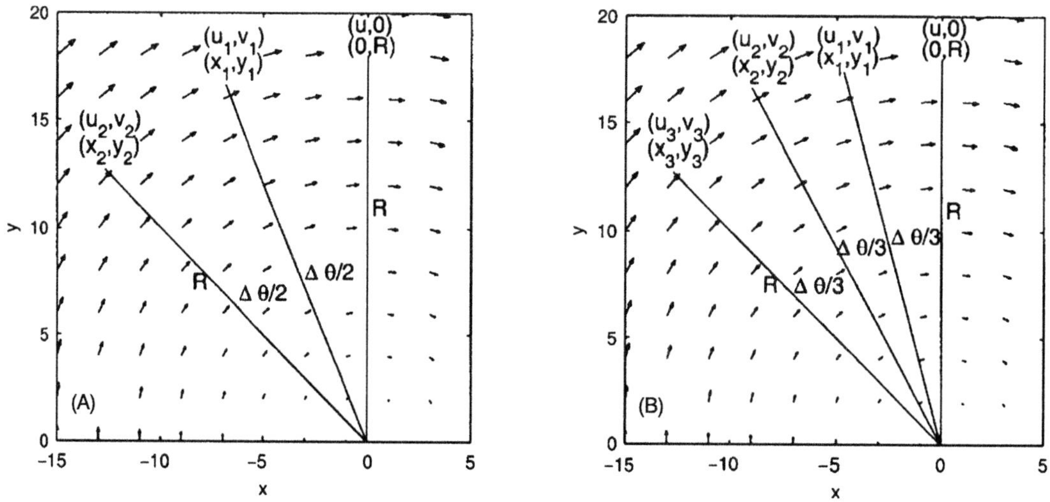

FIGURE 10.18 Copy of Fig. 3 from [22] illustrating the departure points along the arc. Figure comes from Brassington, G. and Sanderson, B. (1999), *Semi-Lagrangian and COSMIC advection in flows with rotation or deformation,* Atmosphere-Ocean, **37**, 369–388. ©La Societe Canadienne de Meterologie et d'Oceanographie, reprinted by permission of Taylor and Francis Ltd, www.tandfonline.com on behalf of La Societe Canadienne de Meterologie et d'Oceanographie.

$$\frac{Dv}{Dt} = -\frac{u^2}{R} - \frac{u^4(\Delta t)^2}{12R^3} + HOT,$$

which is more accurate than that obtained from (10.155) and (10.156). An important feature to note here is that for the first-order estimate it was the tangential acceleration that dominated the error, while for the second-order approximation, it is the radial acceleration that dominates.

A third-order estimate can be achieved through considering departure points (x_3, y_3), (x_2, y_2), and (x_1, y_1) at times t, $t + \frac{\Delta t}{3}$, and $t + \frac{2\Delta t}{3}$ for the particle arriving at (x, y) at time $t + \Delta t$. Again, given the interpolations for the velocities at the these departure points, (u_3, v_3), (u_2, v_2), and (u_1, v_1), respectively, the third-order accurate discrete approximation to the field accelerations is then

$$\frac{Du}{Dt} = \frac{11u - 18u_1 + 9u_2 - 2u_3}{2\Delta t}, \tag{10.159}$$

$$\frac{Dv}{Dt} = \frac{11v - 18v_1 + 9v_2 - 2v_3}{2\Delta t}. \tag{10.160}$$

The acceleration at the apex now becomes

$$\frac{Du}{Dt} = -\frac{2u^5(\Delta t)^3}{216R^4} + \frac{50u^7(\Delta t)^5}{58320R^6} + HOT,$$

$$\frac{Dv}{Dt} = -\frac{u^2}{R} - \frac{108u^6(\Delta t)^4}{29160R^5} + HOT,$$

which is much more accurate than that in (10.155) and (10.156). The dominant error here is the tangential component of the acceleration, and it is stated in [22] that this is the case when the number of interpolation points is odd. The dominant error is a radial component of acceleration when the order is even.

Brassington and Sanderson present a fourth-order scheme that uses departure points (x_4, y_4), (x_3, y_3), (x_2, y_2), and (x_1, y_1) at departure times $t + \frac{3\Delta t}{4}$, $t + \frac{\Delta t}{2}$, $t + \frac{\Delta t}{4}$, and t, with associated velocity fields such that

$$\frac{Du}{Dt} = \frac{25u - 48u_1 + 36u_2 - 16u_3 + 3u_4}{3\Delta t} = -\frac{45u^7(\Delta t)^5}{138240R^6} + HOT, \tag{10.161}$$

$$\frac{Dv}{Dt} = \frac{25v - 48v_1 + 36v_2 - 16v_3 + 3u_4}{3\Delta t} = -\frac{u^2}{R} + \frac{3u^6(\Delta t)^4}{3840R^5} + HOT. \tag{10.162}$$

It is pointed out in [22] that the fourth-order calculation requires four times the computational cost of the usual first-order calculation, when the same size time step is used. However, the additional accuracy is worth the cost as the fourth-order approximation has reduced the error by 2×10^7 for the same time step. A first-order scheme, for which the error is proportional to Δt requires that the time step be divided by 2×10^7 in order to achieve the same error reduction.

In [22] the authors perform a series of tests with the different schemes for a rotational flow as well as with the baratropic model, and the reader is referred to this paper to see the performance of this approach.

10.9 Eliminating the interpolation

One of the major sticking points with the semi-Lagrangian approaches is the interpolations that are involved when the departure points, or the mid-points, are not at a grid point. In 1986 Harold Ritchie came up with an approach to remove the interpolation. In this section we shall present this theory. We start with the one-dimensional advection equation

$$\frac{\partial \psi}{\partial t} + u_0 \frac{\partial \psi}{\partial x} = R, \tag{10.163}$$

where u_0 is a constant advecting wind in the x-direction. As we know, the Lagrangian form of (10.163) is given by

$$\frac{D}{Dt}\left(\psi\left(x\left(t\right), t\right)\right) = R\left(x\left(t\right), t\right). \tag{10.164}$$

The usual approach is to discretize (10.164) is to apply a central difference to the left-hand side and calculate the displacement α such that

$$\frac{\psi\left(x_i, t^{n+1}\right) - \psi\left(x_i - 2\alpha, t^{n-1}\right)}{2\Delta t} = R\left(x_i - \alpha, t^n\right), \tag{10.165}$$

and as we have seen many times that now the upstream locations $x_i - 2\alpha$ and $x_i - \alpha$ will lie between grid points, and thus require an interpolation to evaluate $\psi\left(x_i - 2\alpha, t^{n-1}\right)$ and $R\left(x_i - \alpha t^n\right)$. Ritchie in [157] instead suggests that the advecting wind be broken into two parts by adding and subtracting $\frac{p\Delta x}{2\Delta t}$, resulting in

$$\frac{\partial \psi}{\partial t} + \frac{p\Delta x}{2\Delta t}\frac{\partial \psi}{\partial x} = u_0' \frac{\partial \psi}{\partial x} + R\left(x, t\right), \tag{10.166}$$

where $u_0' = u_0 - \frac{p\Delta x}{2\Delta t}$. The left-hand side now represents the total derivative of ψ following a point that moves with speed $\frac{p\Delta x}{2\Delta t}$, so that over two time steps the displacement will be p grid lengths away from the arrival point. The **residual** advection associated with u_0' appears as an extra term on the right-hand side, and is therefore accounted for in the same way that the total advection would be in an Eulerian treatment; p is chosen to give the displacement to the grid point nearest $x_i - 2\alpha$, which is to say that p is the integer nearest to $\frac{2\Delta u_0}{\Delta x}$. After applying the centered semi-Lagrangian approach, (10.166) becomes

$$\frac{\psi\left(x_i, t^{n+1}\right) - \psi\left(x_i - p\Delta x, t^{n-1}\right)}{2\Delta t} = \left(R - u_0' \frac{\partial \psi}{\partial x}\right)\left(x_i - \frac{p\Delta x}{2}, t^n\right). \tag{10.167}$$

Since $x_i - p\Delta x$ is a grid point, no interpolation is required to evaluate $\psi\left(x_i - p\Delta x, t^{n-1}\right)$. The right-hand side of (10.167) will be evaluated either at grid points for p even, or midway between them for p odd.

We now consider the stability analysis of this approach but for the two-dimensional unforced problem, where now we have the advection equation

$$\frac{\partial \psi}{\partial t} + u_0 \frac{\partial \psi}{\partial x} + v_0 \frac{\partial \psi}{\partial y} = 0, \tag{10.168}$$

with u_0 and v_0 being the constant advection winds in the x and y directions, respectively. We now apply the noninterpolating technique to both the u_0 and v_0 winds to yield

$$\frac{\partial \psi}{\partial t} + \frac{p\Delta x}{2\Delta t}\frac{\partial \psi}{\partial x} + \frac{q\Delta y}{2\Delta t}\frac{\partial \psi}{\partial y} = -\left(u_0' \frac{\partial \psi}{\partial x} + v_0' \frac{\partial \psi}{\partial y}\right), \tag{10.169}$$

where

$$u_0' = u_0 - \frac{p\Delta x}{2\Delta t}, \quad v_0' = v_0 - \frac{q\Delta y}{2\Delta t}.$$

The left-hand side of (10.169) represents the total derivative of ψ following a point that moves with velocity

$$\left(\frac{p\Delta x}{2\Delta t}, \frac{q\Delta x}{2\Delta t}\right),$$

so that the displacement over two time steps will be p in the x-direction and q in the y-direction. In two time steps the actual displacement of ψ via (10.168) will be $(2\Delta t u_0, 2\Delta t v_0)$; here p and q are chosen as the integers nearest to $\frac{2\Delta t u_0}{\Delta x}$ and $\frac{2\Delta t v_0}{\Delta y}$, respectively. Thus the centered Lagrangian approximation of (10.168) is

$$\frac{\psi\left(x_i, y_i t^{n+1}\right) - \psi\left(x_i - p\Delta x, y_j - q\Delta y, t^{n-1}\right)}{2\Delta t}$$
$$= -\left(u_0'\frac{\partial\psi}{\partial x} + v_0'\frac{\partial\psi}{\partial y}\right)\left(x_i - \frac{p\Delta x}{2}, y_j - \frac{q\Delta y}{2}, t^n\right). \tag{10.170}$$

As p and q are integers, no interpolation is required to evaluate $\psi\left(x_i - p\Delta x, y_j - q\Delta y, t^{n-1}\right)$.

To examine the stability of this scheme, we introduce the undamped solution of the form

$$F = Ce^{i(kx+ly+\omega t)}. \tag{10.171}$$

As an aside, we should note here that

$$\frac{u_0'\Delta t}{\Delta x} = \frac{1}{2}\left(\frac{2\Delta t u_0}{\Delta x} - p\right),$$

which must be less that a quarter in absolute value, because of the choice we have made for p. The equivalent statement for v_0' is also true, thus we have that

$$\left|\frac{u_0'\Delta t}{\Delta x}\right| \leq \frac{1}{4}, \quad \left|\frac{v_0'\Delta t}{\Delta y}\right| \leq \frac{1}{4}. \tag{10.172}$$

There are now four possibilities that we have to consider, that are dependent on the parities of p and q:

(i) If p and q are both even, the residual advection terms are evaluated at grid points and the derivative can be approximated by centered space discretizations as

$$-\left(u_0'\frac{\psi^n_{i+1-\frac{p}{2},j-\frac{q}{2}} - \psi^n_{i-1-\frac{p}{2},j-\frac{q}{2}}}{2\Delta x} + v_0'\frac{\psi^n_{i-\frac{p}{2},j+1-\frac{q}{2}} - \psi^n_{i-\frac{p}{2},j-1-\frac{q}{2}}}{2\Delta y}\right).$$

Substituting into (10.170) and looking for solutions of the form in (10.171) leads to

$$\sin\left(\frac{kp\Delta x}{2} + \frac{lq\Delta y}{2} + \omega\Delta t\right) = -\frac{u_0'\Delta t}{\Delta x}\sin k\Delta x - \frac{v_0'\Delta t}{\Delta y}\sin l\Delta y. \tag{10.173}$$

The scheme is guaranteed to be stable if the **residual Courant number**

$$\left|\frac{u_0'\Delta t}{\Delta x}\right| + \left|\frac{v_0'\Delta t}{\Delta y}\right|$$

is less than 1, which is satisfied due to (10.172).

(ii) If p is even and q is odd, the point $\left(x_i - \frac{p\Delta x}{2}, y_j - \frac{q\Delta y}{2}\right)$ is midway along a line joining adjacent grid points in the y-direction. Therefore, we apply the following second-order finite differences to the x and y derivatives as

$$\frac{1}{2}\left(\frac{\psi^n_{i+1-\frac{p}{2},j-\frac{q-1}{2}} - \psi^n_{i-1-\frac{p}{2},j-\frac{q-1}{2}}}{2\Delta x} + \frac{\psi^n_{i+1-\frac{p}{2},j-\frac{q+1}{2}} - \psi^n_{i-1-\frac{p}{2},j-\frac{q+1}{2}}}{2\Delta x}\right),$$

and

$$\frac{\psi^n_{i-\frac{p}{2},j-\frac{q-1}{2}} - \psi^n_{i-\frac{p}{2},j-\frac{q+1}{2}}}{\Delta y},$$

respectively.

The stability analysis yields

$$\begin{aligned}
\sin\left(\frac{kp\Delta x}{2} + \frac{lq\Delta y}{2} + \omega\Delta t\right) &= -\frac{u_0'\Delta t}{\Delta x}\sin k\Delta x \cos\frac{l\Delta y}{2} - 2\frac{v_0'\Delta t}{\Delta y}\sin\frac{l\Delta y}{2} \\
&= \sqrt{\left(\frac{u_0'\Delta t}{\Delta x}\sin l\Delta x\right)^2 + \left(\frac{2v_0'\Delta t}{\Delta y}\right)^2} \\
&\times \sin\left(\frac{l\Delta y}{2} + \tan^{-1}\left(\frac{u_0'\Delta y}{2v_0'\Delta x}\sin k\Delta x\right)\right).
\end{aligned}$$

Stability is guaranteed if the residual Courant number here, namely

$$\sqrt{\left(\frac{u_0'\Delta t}{\Delta x}\right)^2 + \left(\frac{2v_0'\Delta t}{\Delta y}\right)^2},$$

is less than 1.

(iii) If p is odd and q is even, the point $\left(x_i - \frac{p\Delta x}{2}, y_j - \frac{q\Delta y}{2}\right)$ is midway along the line joining two adjacent points in the x-direction. Reversing the treatments of the x and y-derivatives of what we have shown above yields the residual Courant number here to be

$$\sqrt{\left(\frac{2u_0'\Delta t}{\Delta x}\right)^2 + \left(\frac{v_0'\Delta t}{\Delta y}\right)^2}.$$

(iv) The final situation is where p and q are both odd, this gives the point $\left(x_i - \frac{p\Delta x}{2},\right.$ $\left. y_j - \frac{q\Delta y}{2}\right)$ at the corner of a grid rectangle. The two second-order approximations to the x and y derivatives are given by

$$\frac{1}{2}\left(\frac{\psi^n_{i+\frac{p-1}{2},j-\frac{q-1}{2}} - \psi^n_{i-\frac{p+1}{2},j-\frac{q-1}{2}}}{\Delta x} + \frac{\psi^n_{i-\frac{p-1}{2},j-\frac{q+1}{2}} - \psi^n_{i-\frac{p+1}{2},j-\frac{q+1}{2}}}{\Delta x}\right)$$

and

$$\frac{1}{2}\left(\frac{\psi^n_{i+\frac{p-1}{2},j-\frac{q-1}{2}} - \psi^n_{i-\frac{p-1}{2},j-\frac{q+1}{2}}}{\Delta y} + \frac{\psi^n_{i-\frac{p+1}{2},j-\frac{q-1}{2}} - \psi^n_{i-\frac{p+1}{2},j-\frac{q+1}{2}}}{\Delta y}\right),$$

respectively. Stability analysis shows that residual Courant number must satisfy

$$\max\left\{2\left|\frac{u_0'\Delta t}{\Delta x}\right|, 2\left|\frac{v_0'\Delta t}{\Delta y}\right|\right\} < 1.$$

Given the four cases above, it is stated in [157] that this now completes the proof that the scheme is unconditionally stable and nondamping for two-dimensional advection by a constant wind field.

We now move on to consider the noninterpolation formulation with the shallow water equations model.

10.9.1 Application in the shallow water equations model

In [157] Ritchie extends the testing of the noninterpolating semi-Lagrangian method to the shallow water equations model of the form

$$\frac{DU}{Dt} - fV + \overline{\frac{\partial \phi^n}{\partial X}} + K\frac{\partial S}{\partial X} = 0, \tag{10.174}$$

$$\frac{DV}{Dt} + fU + \overline{\frac{\partial \phi^n}{\partial Y}} + K\frac{\partial S}{\partial Y} = 0, \tag{10.175}$$

$$\frac{D\phi}{Dt} + \phi_0\overline{\delta^n} + (\phi - \phi_0)\,\delta = 0, \tag{10.176}$$

where there is a map scale factor m associated with a conformal projection from the Cartesian (x, y) coordinate system to the (X, Y) coordinate system of the model, $(U, V) = \left(\frac{u}{m}, \frac{v}{m}\right)$ represents the model wind vector with its components along the (X, Y) coordinates of the model, the Lagrangian derivative is given by

$$\frac{D}{Dt} \equiv \frac{\partial}{\partial t} + m^2\left(U\frac{\partial}{\partial X} + V\frac{\partial}{\partial Y}\right), \tag{10.177}$$

ϕ is the geopotential height of the free upper surface of the fluid, f is the Coriolis parameter, $S = m^2$ is the square of the map scale factor, and

$$K = \frac{1}{2}\left(U^2 + V^2\right), \qquad \delta = S\left(\frac{\partial U}{\partial X} + \frac{\partial V}{\partial Y}\right),$$

ϕ_0 is a constant that is roughly equal to the mean of ϕ, and as before the overline represents a time average to ensure the semi-implicit treatment of the inertia gravity waves, K is the kinetic energy, and δ is the divergence.

In the noninterpolating semi-Lagrangian model, the Lagrangian term here is handled as

$$\begin{aligned} &\frac{\psi\left(X_i, y_j, t^{n+1}\right) - \psi\left(X_i - p\Delta X, Y_j - q\Delta Y, t^{n-1}\right)}{2\Delta t} \\ &\quad = -\left(U'\frac{\partial \psi}{\partial X} + V'\frac{\partial \psi}{\partial Y}\right)\left(X_i - \frac{p\Delta X}{2}, Y_j - \frac{q\Delta Y}{2}, t^n\right), \end{aligned} \tag{10.178}$$

where

$$U' = \left(m^2 U\right)\left(X_i - \frac{p\Delta X}{2}, Y_j - \frac{q\Delta Y}{2}, t^n\right) - \frac{p\Delta X}{2\Delta t}, \tag{10.179}$$

$$V' = \left(m^2 V\right)\left(X_i - \frac{p\Delta X}{2}, Y_j - \frac{q\Delta Y}{2}, t^n\right) - \frac{q\Delta Y}{2\Delta t}. \tag{10.180}$$

The time-averaged term in the noninterpolating formulation is calculated by

$$\overline{\psi^n} \equiv \frac{1}{2}\left(\psi\left(X_i, Y_j, t^{n+1}\right) + \psi\left(X_i - p\Delta X, Y_j - q\Delta Y, t^{n-1}\right)\right). \tag{10.181}$$

This then implies that the shallow water equations model can be written as

$$\begin{aligned} \left(U + \Delta t\frac{\partial \phi}{\partial X}\right)\left(X_i, Y_j, t^{n+1}\right) &= \left(U - \Delta t\frac{\partial \phi}{\partial X}\right)\left(X_i - p\Delta X, Y_j - q\Delta Y, t^{n-1}\right) \\ &+ 2\Delta t\left(fV - K\frac{\partial S}{\partial X} - \left(U'\frac{\partial U}{\partial X} + V'\frac{\partial U}{\partial Y}\right)\right) \\ &\times \left(X_i - \frac{p\Delta X}{2}, Y_j - \frac{q\Delta Y}{2}, t^n\right), \end{aligned} \tag{10.182}$$

$$\begin{aligned} \left(V + \Delta t\frac{\partial \phi}{\partial Y}\right)\left(X_i, Y_j, t^{n+1}\right) &= \left(V - \Delta t\frac{\partial \phi}{\partial Y}\right)\left(X_i - p\Delta X, Y_j - q\Delta Y, t^{n-1}\right) \\ &+ 2\Delta t\left(-fU - K\frac{\partial S}{\partial Y} - \left(U'\frac{\partial V}{\partial X} + V'\frac{\partial V}{\partial Y}\right)\right) \\ &\times \left(X_i - \frac{p\Delta X}{2}, Y_j - \frac{q\Delta Y}{2}, t^n\right), \end{aligned} \tag{10.183}$$

$$
\begin{aligned}
(\phi+\Delta\phi_0\delta)\left(X_i,Y_j,t^{n+1}\right) &= (\phi-\Delta t\phi_0\delta)\left(X_i-p\Delta X,Y_j-q\Delta Y,t^{n-1}\right)\\
&+\ 2\Delta t\left(-(\phi-\phi_0)\,\delta-\left(U'\frac{\partial\phi}{\partial X}+V'\frac{\partial\phi}{\partial Y}\right)\right)\\
&\times\ \left(X_i-\frac{p\Delta X}{2},Y_j-\frac{q\Delta Y}{2},t^n\right).
\end{aligned}
\tag{10.184}
$$

As pointed out earlier, depending upon the parities of p and q, the point $\left(X_i-\frac{p\Delta X}{2},\right.$ $\left.Y_j-\frac{q\Delta Y}{2}\right)$ may be a grid point, lie midway along a line joining adjacent grid points in either the X or Y directions, or lie at the corner of a grid rectangle. When the location is between grid points, a second-order approximation to the required fields values is found using an approximate average of the values at neighboring grid points.

Therefore, given U, V, and ϕ on grid points at times t^{n-1} and t^n, the following two equations:

$$
p=NINT\left(\frac{2\Delta t}{\Delta X}\left(m^2U\right)\left(X_i-\frac{p\Delta X}{2},Y_j-\frac{q\Delta Y}{2},t\right)\right),
\tag{10.185}
$$

$$
q=NINT\left(\frac{2\Delta t}{\Delta Y}\left(m^2V\right)\left(X_i-\frac{p\Delta X}{2},Y_j-\frac{q\Delta Y}{2},t\right)\right),
\tag{10.186}
$$

are solved for the integer displacement, which then enables us to evaluate the right-hand sides of (10.182)–(10.184). If we now let Q_1, Q_2, and Q_3 denote these right-hand sides, respectively, then the system of equations can be written as

$$
\left(U+\Delta t\frac{\partial\phi}{\partial X}\right)\left(X_i,Y_j,t^{n+1}\right) = Q_1,
\tag{10.187}
$$

$$
\left(V+\Delta t\frac{\partial\phi}{\partial Y}\right)\left(X_i,Y_j,t^{n+1}\right) = Q_2,
\tag{10.188}
$$

$$
(\phi+\Delta t\phi_0\delta)\left(X_i,Y_j,t^{n+1}\right) = Q_3.
\tag{10.189}
$$

Eliminating δ leads to the following form of the Helmholtz equation, which we shall soon see is the process that is followed quite frequently in semi-Lagrangian applications with the shallow water equations, the primitive equations, and the Euler equations models, which is solved for $\phi\left(X_i,Y_j,t^{n+1}\right)$ through some iterative scheme,

$$
\frac{\partial^2\phi}{\partial X^2}+\frac{\partial^2\phi}{\partial Y^2}-\frac{1}{\Delta t^2\phi_0S}\phi=\frac{1}{\Delta t}\left(\frac{\partial Q_1}{\partial X}+\frac{\partial Q_2}{\partial Y}\right)-\frac{Q_3}{\Delta t^2\phi_0S}.
\tag{10.190}
$$

The time step is then completed by determining $U\left(X_i,Y_i,t^{n+1}\right)$ and $V\left(X_i,Y_j,t^{n+1}\right)$ via (10.187) and (10.188).

10.10 Semi-Lagrangian approach with ocean circulation models

So far all of the theory presented, along with the implementation, has been in atmospheric models. In 1995 Das and Weaver [93] published a paper introducing a semi-Lagrangian approach in to a y–z slice two-dimensional, or meridional-plane, ocean circulation model. Their motivation for implementing the semi-Lagrangian advection approach was motivated by two primary considerations: first, was the need to increase the size of the time step in the numerical model, especially for long climate runs of the model, and second, to help increase the resolution of the models. Another strong reason for this approach was that in coastal regions where strong upwelling and downwelling occur gives rise to strong advection-dominated flow.

The numerical model that is used in [93] approximates the two-dimensional thermohaline circulation in an ocean of constant depth H and zonal width D. As mentioned above, the model is in a y–z Cartesian coordinate system, where the y axis is positive northward and axis z is positive upwards. The momentum dynamics of the model are governed by a linear relationship

$$A^*\overline{v_{zz}} = \frac{1}{\rho_0}\overline{P}_y, \tag{10.191}$$

where v is the meridional velocity component, $\overline{P}$ is the zonally averaged pressure, and A^* is referred to as the Fickian viscosity coefficient [93]. It is also stated in [93] that it is possible to define a meridional streamfunction, ψ, through zonally averaging of the continuity equation

$$u_x + u_y + w_z = 0, \tag{10.192}$$

and we need to note here that $u = 0$ at both eastern and western boundaries, such that

$$\psi_y = \overline{w}, \quad \psi_z = -\overline{v}. \tag{10.193}$$

To eliminate the pressure term in (10.191), a zonally hydrostatic relation,

$$\frac{\partial \overline{P}}{\partial z} = -\overline{\rho}g, \tag{10.194}$$

is applied, where a linear equation of state for the zonally averaged density is given by

$$\overline{\rho} = \rho_0\left(1 - \alpha_h\overline{T} - \beta_s\overline{S}\right), \tag{10.195}$$

and $\overline{T}$ and $\overline{S}$ are the zonally averaged temperature and salinity, respectively.

Given the setup above, it is possible to obtain a diagnostic equation for the streamfunction of the form

$$\frac{\partial^4\psi}{\partial z^4} = \frac{g}{A^*}\left(-\alpha_h\frac{\partial T}{\partial y} + \beta_s\frac{\partial S}{\partial y}\right), \tag{10.196}$$

where α_h and β_s are coefficients of heat and salt expansions, respectively.

The prognostic equations for the **conservation** of heat and salt are given by

$$\frac{\partial}{\partial t}\begin{pmatrix} T \\ S \end{pmatrix} + J\left[\psi, \begin{pmatrix} T \\ s \end{pmatrix}\right] = k\frac{\partial^2}{\partial z^2}\begin{pmatrix} T \\ S \end{pmatrix}, \tag{10.197}$$

k is the vertical eddy diffusivity, and J is the Jacobian which is defined as

$$J(\psi, \mu) = -\frac{\partial \psi}{\partial z}\frac{\partial \mu}{\partial y} + \frac{\partial \psi}{\partial y}\frac{\partial \mu}{\partial z}.$$

In [93] the authors state a series of boundary conditions for this model, where at the ocean surface $\psi = 0$ and $\psi_{zz} = 0$, which is equivalent to saying $w = 0$ at $z = 0$, $v_z = 0$ and $z = 0$, which physically means that there is no surface stress. At the bottom of the domain, where $z = -H$, they enforce a no-slip condition, which is $\psi = 0$, $\psi_z = 0$. At the northern and southern walls, the boundary condition is $\psi = 0$ at $y = \pm L$.

At the surface, Dirichlet boundary conditions for the temperature and salinity are employed such that

$$T = T_*\left(1 + \cos\frac{\pi y}{L}\right), \quad S = S_*\left(1 + \cos\frac{\pi y}{L}\right), \tag{10.198}$$

where $2T_*$ and $2S_*$ are the equator to pole temperature and salinity contrasts, respectively. A final condition that is enforced on the model is that there are no fluxes through the lateral walls, which is to say $T_z = S_z = 0$ at $z = -H$, and $T_y = S_y = 0$ at $y = \pm L$.

10.10.1 Implementing the semi-Lagrangian approach

In [93] the authors present the three-time-level scheme for the advection and diffusion of the scalars $T(\mathbf{x}, t)$ and $S(\mathbf{x}, t)$ in a velocity field $\mathbf{u}(\mathbf{x}, t)$, where $\mathbf{x}$ is the position vector and t is time. The authors then generalize and let $f(\mathbf{x}, t)$ represent either T or S, such that we have

$$\begin{aligned} \frac{f\left(\mathbf{x}, t^{n+1}\right) - f\left(\mathbf{x} - 2\mathbf{x}, t^{n-1}\right)}{2\Delta t} &= \frac{1}{2}k\left[\left(\frac{\partial^2}{\partial z^2}f\right)_{x, t^{n+1}} + \left(\frac{\partial^2}{\partial z^2}f\right)_{\mathbf{x}-2\boldsymbol{\alpha}, t^{n-1}}\right], && (10.199) \\ \alpha(\mathbf{x}) &= \Delta t u\left(x - \alpha, t^n\right). && (10.200) \end{aligned}$$

The algorithmic description for the approach adopted in [93] is as follows:

1. Solve (10.200) iteratively for the displacement $\alpha(\mathbf{x})$ through using

$$\alpha^{(m+1)}\mathbf{x} = \Delta t\mathbf{u}\left(\mathbf{x} - \alpha^{(m)}, t^n\right), \tag{10.201}$$

   and an interpolation formula to evaluate $\mathbf{u}$ between mesh points.

2. Evaluate $\left(k\frac{\partial^2}{\partial z^2}f\right)_{\mathbf{x}, t^{n-1}}$ at the mesh point through some form of discretization, and then use the interpolation formula to evaluate

$$\left(f + \frac{1}{2}k\Delta t\frac{\partial^2}{\partial z^2}f\right)_{\mathbf{x}-2\boldsymbol{\alpha}, t^{n-1}}.$$

3. Solve the Helmholtz equation

$$\left(f+\frac{1}{2}k\Delta t\frac{\partial^2}{\partial z^2}f\right)_{\mathbf{x},t^{n+1}}=\left(f+\frac{1}{2}k\Delta t\frac{\partial^2}{\partial z^2}f\right)_{\mathbf{x}-2\alpha,t^{n-1}} \tag{10.202}$$

at mesh points.
4. Repeat steps 1–3 to obtain $f(\mathbf{x},t+2\Delta t)$.

An important feature of the temperature and salinity is mentioned in [93] that both of these fields are **positive definite**, which means that they cannot go below zero, and so we require the interpolation formula not to introduce over- or undershoots. To ensure that the positive definiteness property holds, the authors of [93] apply the quasi-monotone scheme introduced in Bermejo and Staniforth in 1992 [20], where this scheme at each grid point x_j the algorithm is changed to:

1. Compute the departure point and identify the grid elements that surround it.
2. Compute the higher-order interpolation.
3. Compute the local maximum and minimum of the four nearest grid points surrounding the departure point.
4. Clip the spurious maxima and minima of the higher order interpolation to the local maximum and minimum values evaluated in step 3.

As with all of the implementations of the semi-Lagrangian methods, the interpolation scheme plays a vital part in many aspects, from the dispersions, dissipation, and damping aspects to the phase error, but what we have not considered so far is the conservation of mass. It is stated in [93] that they opt for the interpolation scheme that conserves mass for divergence-free flows. The scheme that the authors opt for is that from Bermejo 1990 [19], which uses cubic splines.

In [93] they employ the three-time-level extrapolation scheme

$$\mathbf{u}\left(\mathbf{x},t^{n+\frac{1}{2}}\right)=\frac{1}{8}\left(15\mathbf{u}\left(\mathbf{x},t^{n}\right)-10\mathbf{u}\left(\mathbf{x},t^{n-1}\right)+3\mathbf{u}\left(\mathbf{x},t^{n-2}\right)\right),$$

so that a two-time-level scheme for the time derivative could be used. As an aside, we should mention that in this study Das and Weaver also consider the noninterpolating approach that we described in the last chapter, and if the reader is interested in their result with this approach as well, we recommend reading [93].

10.11 Transparent boundary conditions

As we have seen over the last couple of chapters, we have the need to able to defined boundary conditions on the inflow and outflow boundaries to ensure that the problem is well-posed [122]. McDonald went on to develop more theory for the two-dimensional limited area models in both Cartesian and spherical coordinates. Here we shall present the work from McDonald (2002) and (2003), where asks the question of what is the best way to treat the lateral boundaries. The answer is that these should be **transparent**. First, all waves approaching them from the interior of the limited area should exit without reflection; and second, in a nested environment, where the boundary data are being supplied by a *host* model, all meteorologically important waves impinging on the boundaries from the exterior of the limited area should enter without their amplitude or phase being changed, and without exciting spurious high-frequency waves of noise.

McDonald states that full transparency is probably unattainable in most practical integrations, but then states that, since most boundary strategies contain difficulties that cannot be easily identified when they are considered in the framework of realistic models, it is a good idea to first test them on simple problems with known solutions. As such he considers the linearized shallow water equations on an f-plane which are described as follows:

$$\frac{\partial u}{\partial t}+u_0\frac{\partial u}{\partial x}+v_0\frac{\partial u}{\partial y}+\frac{\partial \Phi}{\partial x}-fv=0, \tag{10.203a}$$

$$\frac{\partial v}{\partial t}+u_0\frac{\partial v}{\partial x}+v_0\frac{\partial v}{\partial y}+\frac{\partial \Phi}{\partial y}+fu=0, \tag{10.203b}$$

$$\frac{\partial \Phi}{\partial t}+u_0\frac{\partial \Phi}{\partial x}+v_0\frac{\partial \Phi}{\partial y}+\Phi_0\left(\frac{\partial u}{\partial x}+\frac{\partial v}{\partial y}\right)=0, \tag{10.203c}$$

where u and v are the x and y components of the winds, respectively, $\Phi=gz$ is the geopotential height, $g=9.81\,\mathrm{ms}^{-2}$. The Coriolis parameter, f, the advecting velocities, u_0 and v_0, and the mean geopotential height, Φ_0, are all constants.

Substituting $(u,y,\Phi)=\left(\widehat{u},\widehat{v},\widehat{\Phi}\right)\exp\{i\left(kx+ly+\omega t\right)\}$ results in the equations

$$i\left(-\omega+ku_0+lv_0\right)\widehat{u}-f\widehat{v}+ik\widehat{\Phi}=0, \tag{10.204a}$$

$$f\widehat{u}+i\left(-\omega+ku_0+lv_0\right)\widehat{v}+il\widehat{\Phi}=0, \tag{10.204b}$$

$$ik\Phi_0\widehat{u}+il\Phi_0\widehat{v}+i\left(-\omega+ku_0+lv_0\right)\widehat{\Phi}=0, \tag{10.204c}$$

which have solutions, provided the following dispersion relation is valid:

$$\left(-\omega+ku_0+lv_0\right)\left(\left(-\omega+ku_0+lv_0\right)^2-f^2-\left(k^2+l^2\right)\Phi_0\right)=0. \tag{10.205}$$

According to McDonald, there are three types of waves associated with this setup; advection waves with frequency

$$\omega_a=ku_0+lv_0, \tag{10.206}$$

and two adjustment waves with frequencies

$$\omega_{\pm}=ku_0+lv_0\pm\sqrt{f^2+K^2\Phi_0},\qquad K^2\equiv k^2+l^2, \tag{10.207}$$

and the solutions to (10.204a)–(10.204c) are

$$\widehat{\Phi}\;=\;\widehat{\Phi}_{+}+\widehat{\Phi}_{-}+\widehat{\Phi}_{a}, \tag{10.208}$$

$$\widehat{u} = \frac{ilf}{K^2\Phi_0}\left(\widehat{\Phi}_+ + \widehat{\Phi}_-\right) + \frac{k\sqrt{f^2+K^2\Phi_0}}{K^2\Phi_0}\left(\widehat{\Phi}_+ - \widehat{\Phi}_-\right) - \frac{il}{f}\widehat{\Phi}_a, \quad (10.209)$$

$$\widehat{v} = -\frac{ikf}{K^2\Phi_0}\left(\widehat{\Phi}_a + \widehat{\Phi}_-\right) + \frac{i\sqrt{f^2+K^2\Phi}}{K^2\Phi_0}\left(\widehat{\Phi}_a - \widehat{\Phi}_+\right) + \frac{ik}{f}\widehat{\Phi}_a. \quad (10.210)$$

Here we have

$$\begin{aligned} \Phi_+ &= \widehat{\Phi}_+ \exp\{i\,(kx+ly-\omega_+ t)\}, \\ \Phi_- &= \widehat{\Phi}_- \exp\{i\,(kx+ly-\omega_- t)\}, \\ \Phi_a &= \widehat{\Phi}_a \exp\{i\,(kx+ly-\omega_a t)\}. \end{aligned} \quad (10.211)$$

It is stated in [124] that the following combination of the fields is a pure advection wave:

$$\frac{k}{K}\widehat{v} - \frac{l}{K}\widehat{u} + \frac{if}{K\sqrt{\Phi_0}}\frac{\widehat{\Phi}}{\sqrt{\Phi_0}} = \frac{iK\sqrt{\Phi_0}}{f}\left(1+\frac{f^2}{K^2\Phi_0}\right)\frac{\widehat{\Phi}}{\sqrt{\Phi_0}}, \quad (10.212)$$

whereas the following combinations of fields project out the two adjustment waves:

$$\frac{\widehat{\Phi}}{\sqrt{\Phi_0}} + \sqrt{1+\frac{f^2}{K^2\Phi_0}}\left(\frac{k}{K}\widehat{u} + \frac{l}{K}\widehat{v}\right) - \frac{if}{K\sqrt{\Phi_0}}\left(\frac{l}{K}\widehat{u} - \frac{k}{K}\widehat{v}\right) = 2\left(1+\frac{f^2}{K^2\Phi_0}\right)\frac{\widehat{\Phi}_+}{\sqrt{\Phi_0}}, \quad (10.213)$$

$$\frac{\widehat{\Phi}}{\sqrt{\Phi_0}} - \sqrt{1+\frac{f^2}{K^2\Phi_0}}\left(\frac{k}{K}\widehat{u} + \frac{l}{K}\widehat{v}\right) - \frac{if}{K\sqrt{\Phi_0}}\left(\frac{l}{K}\widehat{u} - \frac{k}{K}\widehat{v}\right) = 2\left(1+\frac{f^2}{K^2\Phi_0}\right)\frac{\widehat{\Phi}_-}{\sqrt{\Phi_0}}. \quad (10.214)$$

The group velocities of the advective waves are

$$\left(C_a^g\right)_x = u_0, \qquad \left(C_a^g\right)_y = v_0, \quad (10.215)$$

and the group velocities of the adjustment waves are

$$\left(C_\pm^g\right)_x = u_0 + \pm\sqrt{\Phi_0}\left(1+\frac{l^2}{k^2}+\frac{f^2}{k^2\Phi_0}\right)^{-\frac{1}{2}}, \quad (10.216)$$

$$\left(C_\pm^g\right)_y = v_0 + \pm\sqrt{\Phi_0}\left(1+\frac{k^2}{l^2}+\frac{f^2}{l^2\Phi_0}\right)^{-\frac{1}{2}}. \quad (10.217)$$

Given the setup just described, we shall now summarize the derivation of the transparent boundary conditions from [124]. We start by defining the domain as $0 \le x \le L_x$ and $0 \le y \le L_y$. It is assumed that $\sqrt{\Phi_0} > \sqrt{u_0^2+v_0^2}$, where it is stated in [124] that we have to impose two fields at inflow and one field at outflow.

Next, if we consider the coefficient $\frac{f}{K\sqrt{\Phi_0}}$, and take $f = 10^{-4}\ \mathrm{s}^{-1}$, then we have to determine what would be a typical range of the values of the coefficient in the context here of a limited area meteorological model. For the external mode which corresponds to $\sqrt{\Phi_0}\ 300\ \mathrm{ms}^{-1}$, this coefficient is ≤ 0.05, assuming the longest wavelength to be 1000 km. The internal modes can have much smaller values of $\sqrt{\Phi_0}$. When the wavelength is 1000 km, and $\sqrt{\Phi_0} = 16\ \mathrm{ms}^{-1}$, then $\frac{f}{K\sqrt{\Phi_0}} \sim 1$. Therefore, if we were modeling over a 1000 km $\times$ 1000 km area, then $\frac{f}{K\sqrt{\Phi_0}} \ll 1$ for the fast moving short wavelength modes that we particularly do not want to be reflected from the boundaries.

It is stated in [124] that at the x boundaries the group velocity for the advection waves, (10.215), shows that the energy of the advection wave propagates in the same direction as the advection velocity. From (10.216) it is stated that the energy of $\omega_\pm$ waves always propagates in the $\pm x$ direction as long as

$$\sqrt{\Phi_0} > |u_0|\sqrt{\left(1 + \frac{l^2}{k^2} + \frac{f^2}{k^2\Phi_0}\right)}, \tag{10.218}$$

where this will hold for a large collection of the waves in a meteorological environment, provided the ratio $\frac{l^2}{k^2}$ does not become large, and this ratio is small for waves that are almost perpendicular to these boundaries.

10.11.1 Waves at the boundaries

We shall start by assuming that $u_0 > 0$, which implies that the L_x boundary is an outflow boundary. If we choose to extrapolate u and v, then how is the third field imposed? At this boundary we have ω_a and ω_+ wave packets impinging from the interior, and so we require the boundary to facilitate this transmission as best as possible, to avoid reflection, and to **not excite** any ω_- wave. The approach McDonald shows is to drop the terms of order $\frac{l^2}{k^2}$, $\frac{f^2}{k^2\Phi_0}$, and $\frac{fl}{k^2\sqrt{\Phi_0}}$, which using (10.214) implies that

$$\frac{\widehat{\Phi}_a}{\sqrt{\Phi_0}} - \widehat{u}_a - \frac{l}{k}\widehat{v}_a + \frac{if}{k\sqrt{\Phi_0}}\widehat{v}_a = 0, \tag{10.219}$$

$$\frac{\widehat{\Phi}_+}{\sqrt{\Phi_0}} - \widehat{u}_+ - \frac{l}{k}\widehat{v}_+ + \frac{if}{k\sqrt{\Phi_0}}\widehat{v}_+ = 0. \tag{10.220}$$

Therefore, if we impose $ik\widehat{\Phi} - \sqrt{\Phi_0}(ik\widehat{u} + il\widehat{v}) - f\widehat{v} = 0$ on this boundary, then we will be describing the outgoing ω_a and ω_+ waves correctly **to this level of approximation**. From (10.214) we see that to this level of approximation we require $\widehat{\Phi}_- = 0$ on this boundary, which is what is being sought in a meteorological application where such a wave is unwanted *noise*.

The next stage is to replace ik and il in the condition above for the advective and the positive adjustment waves by the differential operators $\frac{\partial}{\partial x}$ and $\frac{\partial}{\partial y}$, respectively, which gives

$$\frac{\partial \Phi}{\partial x} - fv - \sqrt{\Phi_0}\left(\frac{\partial u}{\partial x} + \frac{\partial v}{\partial y}\right).$$

However, through using (10.203a) and (10.203c), the expression above can be rewritten as

$$\left(\frac{\partial}{\partial t} + u_0\frac{\partial}{\partial x} + v_0\frac{\partial}{\partial y}\right)\left(\Phi - \sqrt{\Phi_0 u}\right) = 0, \tag{10.221}$$

which we can see is in a form that can have a semi-Lagrangian discretization applied to it.

If we now consider the inflow boundary, then here we must impose two fields and extrapolate the third, and McDonald states that we wish to do this in such a way that the host model's advective waves enter the area accurately without stimulating any ω_+ waves, while simultaneously allowing the ω_- waves impinging from the interior to exit without reflection. Now McDonald makes a strong assumption of the fields coming into the limited area model, namely that model fields are in **geostrophic balance**, which implies the following conditions on to the host fields, where superscript h stands for the host fields:

$$ik\widehat{\Phi}^h - f\widehat{v}^h = 0, \qquad il\widehat{\Phi}^h + f\widehat{u}^h = 0, \qquad ik\widehat{u}^h + il\widehat{v}^h = 0,$$

with the first two expressions above representing the definitions of the geostrophic winds in spectral space, and the third equation being the requirement that the geostrophic winds are nondivergent, again in spectral space. As such (10.212) informs us that the advective wave is dominated by $\widehat{v}^h$ when $\frac{l^2}{k^2} \ll 1$ and $\frac{f^2}{l^2\Phi_0} \ll 1$. Therefore, we shall take $\widehat{v}^h$ as one of the imposed fields. Now if we consider the equation

$$\frac{\widehat{\Phi}}{\sqrt{\Phi_0}} + \widehat{u} + \frac{l}{k}\widehat{v} + \frac{if}{k\sqrt{\Phi_0}}\widehat{v} = 0, \tag{10.222}$$

then from (10.213) we can see that the ω_a and ω_- waves obey this equation to the order of accuracy we are considering here. At the same time it forces $\widehat{\Phi}_+$ to be zero. Thus we must impose a second field on this boundary such that this equation holds. McDonald states that the simplest way to achieve this is to impose $\left(\psi^h + \sqrt{\Phi_0}u^h\right)$ such that Φ^h and u^h obey the geostrophic wind definitions and the nondivergence conditions

$$\begin{aligned}\frac{\partial \Phi^h}{\partial x} - fv^h &= 0,\\ \frac{\partial u^h}{\partial x} + \frac{\partial v^h}{\partial y} &= 0.\end{aligned}$$

Finally, the third field u is extrapolated from the interior to complete the boundary conditions at inflow.

Following similar arguments as above, we can arrive at similar boundary conditions for $u_0 < 0$, but also for the two y boundaries depending on whether $v_0 < 0$ or $v_0 > 0$.

10.11.2 Semi-Lagrangian discretization of transparent boundaries

The numerical grid that McDonald uses in [124] is the C grid as described earlier, where we have $x_i = i\Delta x$, $y_j = j\Delta y$, and $t^n = n\Delta t$, $x_N = L_x$, $y_M = L_y$. Applying a semi-implicit, semi-Lagrangian discretization of (10.203a)–(10.203c) on the Arakawa C grid that is second-order in time yields

$$\left[u_{i+\frac{1}{2},j} + \frac{\Delta t}{2\Delta x}\left(\Phi_{i+1,j} - \Phi_{i,j}\right) - \frac{\Delta t}{2} f\overline{v}_{i+\frac{1}{2},j}\right]^{n+1} = \left(Y_u\right)^n_{i+\frac{1}{2},j,d} \equiv \left(R_u\right)_{i+\frac{1}{2},j}, \tag{10.223}$$

$$\left[v_{i,j+\frac{1}{2}} + \frac{\Delta t}{2\Delta y}\left(\Phi_{i,j+1} - \Phi_{i,j}\right) + \frac{\Delta t}{2} f\overline{u}_{i,j+\frac{1}{2}}\right]^{n+1} = \left(Y_v\right)^n_{i,j+\frac{1}{2},d} \equiv \left(R_v\right)_{i,j+\frac{1}{2}}, \tag{10.224}$$

$$\left[\Phi_{i,j} + \frac{\Phi_0 \Delta t}{2}\delta_{i,j}\right]^{n+1} = \left(Y_\Phi\right)^n_{i,j,d} \equiv \left(R_\Phi\right)_{i,j}, \tag{10.225}$$

where

$$\overline{v}_{i+\frac{1}{2},j} = \frac{1}{4}\left(v_{i+1,j+\frac{1}{2}} + v_{i,j+\frac{1}{2}} + v_{i+1,j-\frac{1}{2}} + v_{i,j-\frac{1}{2}}\right), \tag{10.226}$$

$$\overline{u}_{i,j+\frac{1}{2}} = \frac{1}{4}\left(u_{i+\frac{1}{2},j+1} + u_{i+\frac{1}{2},j} + u_{i-\frac{1}{2},j+1} + u_{i-\frac{1}{2},j}\right), \tag{10.227}$$

$$\delta_{i,j} = \frac{1}{\Delta x}\left(u_{i+\frac{1}{2},j} - u_{i-\frac{1}{2}}\right) + \frac{1}{\Delta y}\left(v_{i,j+\frac{1}{2}} - v_{i,j-\frac{1}{2}}\right), \tag{10.228}$$

and

$$\left(Y_u\right)_{i+\frac{1}{2},j} = u_{i+\frac{1}{2},j} - \frac{\Delta t}{2\Delta x}\left(\Phi_{i+1,j} - \Phi_{i,j}\right) + \frac{\Delta t}{2} f\overline{v}_{i+\frac{1}{2},j}, \tag{10.229}$$

$$\left(Y_v\right)_{i,j+\frac{1}{2}} = v_{i,j+\frac{1}{2}} - \frac{\Delta t}{2\Delta y}\left(\Phi_{i,j+1} - \Phi_{i,j}\right) - \frac{\Delta t}{2} f\overline{u}_{i,j+\frac{1}{2}}, \tag{10.230}$$

$$\left(Y_\Phi\right)_{i,j} = \Phi_{i,j} - \frac{\Phi \Delta t}{2}\delta_{i,j}. \tag{10.231}$$

We now define $\alpha = u_0 \dfrac{\Delta t}{\Delta x}$ and $\beta = v_0 \dfrac{\Delta t}{\Delta y}$ such that $Y^n_{i,j,d} \equiv Y\left(i - \alpha, j - \beta, n\Delta t\right)$ is the field at the departure point, and McDonald uses a bicubic Lagrange interpolation polynomial to estimate Y here.

What we must note here is the indices that (10.223)–(10.225) are valid for, which are $i = 0$, $N-1$, $j = 1, M-1$ for (10.223), $i = 1, N-1$, $j = 0, M-1$ for (10.224), and $i = 1, N-1$, $j = 1$, $M-1$ for (10.225). We should also note that we may have to estimate Y_u, Y_v and Y_Φ outside these ranges of indices.

The next step introduced by McDonald is to consider applying an iterative approach to solve (10.223)–(10.225), so the equations are rearranged into the following iterative forms:

$$\left[u_{i+\frac{1}{2},j} + \frac{\Delta t}{2\Delta x}\left(\Phi_{i+1,j} - \Phi_{i,j}\right)\right]^{(k+1)} = \left(R_u\right)_{i+\frac{1}{2},j} + \frac{\Delta t}{2} f\overline{v}^{(k)}_{i+\frac{1}{2},j} \equiv \left(C_u\right)^{(k)}_{i+\frac{1}{2},j}, \tag{10.232}$$

$$\left[v_{i,j+\frac{1}{2}} + \frac{\Delta t}{2\Delta y}\left(\Phi_{i,j+1} - \Phi_{i,j}\right)\right]^{(k+1)} = (R_v)_{i,j+\frac{1}{2}} - \frac{\Delta t}{2} f\overline{u}^{(k)}_{i,j+\frac{1}{2}} \equiv (C_v)^{(k)}_{i,j+\frac{1}{2}}, \tag{10.233}$$

$$\left[\Phi_{i,j} + \frac{\Phi_0 \Delta t}{2}\delta_{i,j}\right]^{(k+1)} = (R_\Phi)_{i,j}. \tag{10.234}$$

Before proceeding on how to implement the iterative solver, we consider the characteristics on the boundaries from the appendix in [124].

10.11.3 Characteristic boundaries

This section describes the procedure for solving the equations when the fields corresponding to the ingoing characteristics are imposed. To make this section applicable to all of the various options, McDonald considers the generic u equation:

$$\left[u_{i+\frac{1}{2},j} + a^u_{i+\frac{1}{2},j}\left(\Phi_{i+1,j} - \Phi_{i,j}\right)\right]^{(k+1)} = (A_u)^{(k)}_{i+\frac{1}{2},j}. \tag{10.235}$$

Let p be imposed at $x = \frac{\Delta x}{2}$ such that

$$p^{n+1}_{\frac{1}{2},j} = \frac{1}{2}\left(\Phi_{0,j} + \Phi_{1,j}\right)^{n+1} + \sqrt{\Phi_0} u^{n+1}_{\frac{1}{2},j}, \tag{10.236}$$

which is used to eliminate $\Phi^{(k+1)}_{0,j}$ from (10.235) to yield

$$\left(u_{\frac{1}{2},j} + 2\tau_{\frac{1}{2},j} a^u_{\frac{1}{2},j}\Phi_{1,j}\right)^{(k+1)} = \tau_{\frac{1}{2},j}\left[(A_u)^{(k)}_{\frac{1}{2},j} + 2a^u_{\frac{1}{2},j} p^{n+1}_{\frac{1}{2},j}\right], \tag{10.237}$$

where

$$\tau_{\frac{1}{2},j} = \left(1 + 2\sqrt{\Phi_0} a^u_{i+\frac{1}{2},j}\right)^{-1}. \tag{10.238}$$

We can impose q at $x = L_x - \dfrac{\Delta x}{2}$ as

$$q^{n+1}_{i-\frac{1}{2},j} = \frac{1}{2}\left(\Phi_{N-1,j} + \Phi_{N,j}\right)^{n+1} - \sqrt{\Phi_0} u^{n+1}_{N-\frac{1}{2},j}, \tag{10.239}$$

which when substituted into (10.235) results in

$$\left(u_{N-\frac{1}{2},j} - 2\tau_{N-\frac{1}{2},j} a^u_{N-\frac{1}{2},j}\Phi_{N-1,j}\right)^{(k+1)} = \tau_{N-\frac{1}{2},j}\left[(A_u)^{(k)}_{N-\frac{1}{2},j} - 2a^u_{N-\frac{1}{2},j} q^{n+1}_{N-\frac{1}{2},j}\right]. \tag{10.240}$$

Employing the same procedure to impose fields corresponding to the ingoing characteristic at the other two boundaries, we consider the generic v equation

$$\left[v_{i,j+\frac{1}{2}} + a^v_{i,j+\frac{1}{2}}\left(\Phi_{i,j+1} - \Phi_{i,j}\right)\right]^{(k+1)} = (A_v)^{(k)}_{i,j+\frac{1}{2}}. \tag{10.241}$$

At $y = \frac{\Delta y}{2}$ and $y = L_y - \frac{\Delta y}{2}$ we have

$$s^{n+1}_{i,\frac{1}{2}} = \frac{1}{2}\left(\Phi_{i,0} + \Phi_{i,1}\right)^{n+1} + \sqrt{\Phi_0} v^{n+1}_{i,\frac{1}{2}}, \tag{10.242}$$

$$r^{n+1}_{i,M-\frac{1}{2}} = \frac{1}{2}\left(\Phi_{i,M-1} + \Phi_{i,M}\right)^{n+1} - \sqrt{\Phi_0} v^{n+1}_{i,M-\frac{1}{2}}, \tag{10.243}$$

respectively. When these expressions are substituted into (10.241), they yield

$$\left(v_{i,\frac{1}{2}} + 2\sigma_{i,\frac{1}{2}} a^v_{i,\frac{1}{2}} \Phi_{i,1}\right)^{(k+1)} = \sigma_{i,\frac{1}{2}}\left[(A_v)^{(k)}_{i,\frac{1}{2}} + 2a^v_{i,\frac{1}{2}} s^{n+1}_{i,\frac{1}{2}}\right], \tag{10.244}$$

$$\left(v_{i,M-\frac{1}{2}} + 2\sigma_{i,\frac{1}{2}} a^v_{i,M-\frac{1}{2}} \Phi_{i,M-1}\right)^{(k+1)} = \sigma_{i,M-\frac{1}{2}}\left[(A_v)^{(k)}_{i,M-\frac{1}{2}} + 2a^v_{i,M-\frac{1}{2}} r^{n+1}_{i,M-\frac{1}{2}}\right], \tag{10.245}$$

where

$$\sigma_{i,j+\frac{1}{2}} = \left(1 + 2\sqrt{\Phi_0} a^v_{i,j+\frac{1}{2}}\right)^{-1}. \tag{10.246}$$

Given the expressions above, we now consider how to solve the semi-Lagrangian equations from before.

10.11.4 Implementing semi-Lagrangian approach

When we are imposing fields that correspond to the ingoing characteristic, then (10.237), (10.240), (10.244), and (10.245) must be used in the vicinity of the boundaries. It is stated in [124] that this introduces a positional dependency to the coefficient multiplying $\Phi^{(k+1)}$, $u^{(k+1)}$, and $v^{(k+1)}$ in (10.232)–(10.234). Thus the equations that need to be solved can formally be written as

$$(B_u)_{i+\frac{1}{2},j} = u_{i+\frac{1}{2},j} + b^{u^+}_{i+\frac{1}{2},j} \Phi_{i+1,j} + b^{u_0}_{i+\frac{1}{2},j} \Phi_{i,j}, \tag{10.247}$$

$$(B_v)_{i,j+\frac{1}{2}}, = v_{i,j+\frac{1}{2}} + b^{v^+}_{i,j+\frac{1}{2}} \Phi_{i,j+1} + b^{v_0}_{i,j+\frac{1}{2}} \Phi_{i,j}, \tag{10.248}$$

$$(B_\phi)_{i,j} = \Phi_{i,j} + b^{\Phi_u}_{i,j}\left(u_{i+\frac{1}{2},j} - u_{i-\frac{1}{2},j}\right) + b^{\Phi_v}_{i,j}\left(v_{i,j+\frac{1}{2}} - v_{i,j-\frac{1}{2}}\right). \tag{10.249}$$

Substituting u from (10.247) and for v from (10.248) into (10.249) results in

$$\begin{aligned}
\Phi_{i,j} &- b^{\Phi_u}_{i,j}\left(b^{u^+}_{i+\frac{1}{2},j}\Phi_{i+1,j} + b^{u_0}_{i+\frac{1}{2},j}\Phi_{i,j} - b^{u^+}_{i-\frac{1}{2},j}\Phi_{i,j} - b^{u_0}_{i-\frac{1}{2},j}\Phi_{i-1,j}\right) \\
&- b^{\Phi_v}_{i,j}\left(b^{v^+}_{i,j+\frac{1}{2}}\Phi_{i,j+1} + b^{v_0}_{i,j+\frac{1}{2}}\Phi_{i,j} - b^{v^+}_{i,j-\frac{1}{2}}\Phi_{i,j} - b^{v_0}_{i,j-\frac{1}{2}}\Phi_{i,j-1}\right) \\
&= (B_\Phi)_{i,j} - b^{\Phi_u}_{i,j}\left[(B_u)_{i+\frac{1}{2},j} - (B_u)_{i-\frac{1}{2},j}\right] - b^{\phi_v}_{i,j}\left[(B_v)_{i,j+\frac{1}{2}} - (B_v)_{i,j-\frac{1}{2}}\right].
\end{aligned} \tag{10.250}$$

McDonald states in [124] that he solves (10.250) through using successive overrelaxation solver for Φ, and recovers u and v from (10.247) and (10.248). For more details about the matrix equation solver method of successive overrelaxation, see [6].

10.11.5 Discretization on the boundaries

The experiments that McDonald presents in [124] impose the tangential velocity, v_τ, at inflow, and he assumes that $\sqrt{\Phi_0} > |\mathbf{v}_0|$. This implies that one more field at inflow and one at outflow still need to be defined. McDonald mentions three different options:

1. Imposing Φ at all boundary points;
2. Imposing the field corresponding to the ingoing characteristic, $\Phi - \sqrt{\Phi_0}v_n$, when v_n is the normal velocity;
3. Imposing $\Phi - \sqrt{\Phi_0}v_n$ at the inflow, and $\dfrac{D\left(\Phi - \sqrt{\Phi_0}v_n\right)}{Dt} = 0$ at outflow.

The normal velocities at both the inflow and outflow emerge naturally when we solve the equations because of the C grid staggering. Thus we still have to develop an expression for the tangential velocities at the outflow. McDonald states that these must be extrapolated from the interior.

The tangential velocity is updated at the outflow at $x = 0$ and $x = L_x$ by assuming that (10.203b) is valid; we can use (10.233) to compute v there. At $y = 0$ and $y = L_x$, we assume that (10.203a) is valid and use (10.233) to compute u there.

It could be that Y_u and Y_v are required on the boundary lines. Different trajectories are used depending on whether Φ or $\Phi - \sqrt{\Phi_0}v_n$ are imposed on the boundaries. If the Coriolis terms are required on the boundary, then we must use the following extrapolations. At $y = 0$ and $y = L_y$, we have

$$\overline{v}_{i+\frac{1}{2},0} = \frac{1}{4}\left(3v_{i+1,\frac{1}{2}} - v_{i+1,\frac{3}{2}} + 3v_{i,\frac{1}{2}} - v_{i,\frac{3}{2}}\right), \tag{10.251}$$

$$\overline{v}_{i+\frac{1}{2},M} = \frac{1}{4}\left(3v_{i+1,M-\frac{1}{2}} - v_{i+1,M-\frac{3}{2}} + 3v_{i,M-\frac{1}{2}} - v_{i,M-\frac{3}{2}}\right), \tag{10.252}$$

and at $x = 0$ and $x = L_x$,

$$\overline{u}_{0,j+\frac{1}{2}} = \frac{1}{4}\left(3u_{\frac{1}{2},j+1} - u_{\frac{3}{2},j+1} + 3u_{\frac{1}{2},j} - u_{\frac{3}{2},j}\right), \tag{10.253}$$

$$\overline{u}_{N,j+\frac{1}{2}} = \frac{1}{4}\left(3u_{N-\frac{1}{2},j+1} - u_{N-\frac{3}{2},j+1} + 3u_{N-\frac{1}{2},j} - u_{N-\frac{3}{2},j}\right). \tag{10.254}$$

As we mentioned earlier, it is possible that the departure point could be beyond the boundary. In [124] McDonald makes the assumption to restrict the time steps to be of a modest length, and as such he employs the time interpolation from [122]. Therefore, there are 12 situations that we have to consider, three at each of the four boundaries, and they are as follows:

- If $x_d < \frac{\Delta x}{2}$ then $(Y_u)^n_{i+\frac{1}{2},j,d} = Y_u\left(\frac{1}{2}, j - \beta_{i+\frac{1}{2},j}, n\Delta t\right)$.
- If $x_d < 0$ then $(Y_v)^n_{i,j+\frac{1}{2},d} = Y_v\left(0, j + \frac{1}{2} - \beta_{i,j+\frac{1}{2}}, n\Delta t\right)$.
- If $x_d < \Delta x$ then $(Y_\Phi)^n_{i,j,d} = Y_\Phi\left(1, j - \beta_{i,j}, n\Delta t\right)$.
- If $x_d > \left(N - \frac{1}{2}\right)\Delta x$ then $(Y_u)^n_{i+\frac{1}{2},j,d} = Y_u\left(N - \frac{1}{2}, j - \beta_{i+\frac{1}{2},j}, n\Delta t\right)$.
- If $x_d > N\Delta x$ then $(Y_v)^n_{i,j+\frac{1}{2},d} = Y_v\left(N, j + \frac{1}{2} - \beta_{i,j+\frac{1}{2}}, n\Delta t\right)$.
- If $x_d > (N-1)\Delta x$ then $(Y_\Phi)^n_{i,j,d} = Y_\Phi\left(N-1, j - \beta_{i,j}, n\Delta t\right)$.
- If $y_d < 0$ then $(Y_u)^n_{i+\frac{1}{2},j,d} = Y_u\left(i + \frac{1}{2} = \alpha_{i+\frac{1}{2},j}, 0, n\Delta t\right)$.
- If $y_d < \frac{\Delta y}{2}$ then $(Y_v)^n_{i,j+\frac{1}{2},d} = Y_v\left(i - \alpha_{i,j+\frac{1}{2}}, \frac{1}{2}, n\Delta t\right)$.
- If $y_d < \Delta y$ then $(Y_\Phi)^n_{i,j,d} = Y_\Phi\left(i - \alpha_{i,j}, 1, n\Delta t\right)$.
- If $y_d > M\Delta y$ then $(Y_u)^n_{i+\frac{1}{2},j,d} = Y_u\left(i + \frac{1}{2} - \alpha_{i+\frac{1}{2},j}, M, n\Delta t\right)$.
- If $y_d > \left(M - \frac{1}{2}\right)\Delta y$ then $(Y_v)^n_{i,j+\frac{1}{2},d} = Y_v\left(i - \alpha_{i,j+\frac{1}{2}}, M - \frac{1}{2}, n\Delta t\right)$.
- If $y_d > (M-1)\Delta y$ then $(Y_\Phi)^n_{i,j,d} = Y_\Phi\left(N - \alpha_{i,j}, M-1, n\Delta t\right)$.

1. Imposing Φ on the boundary fits naturally into (10.232)–(10.234), where the only extrapolation required are those of (10.225)–(10.232).
2. To impose fields corresponding to the ingoing characteristics, proceed as above, except at the lines half a grid step size in from the boundary. There use the equations from the last section, where $a^u_{i+\frac{1}{2},M} = \frac{\Delta t}{2\Delta x}$ and $a^v_{i,j+\frac{1}{2}} = \frac{\Delta t}{2\Delta y}$.
3. Using $\frac{D(\Phi - \Phi_0 v_n)}{Dt} = 0$ at the outflow is a minor variation on step 2 in a semi-Lagrangian discretization.

The last part that we have to consider here are the **corners** of the limited area model. If we consider the case $x = y = 0$, then to compute $(Y_v)_{0,\frac{1}{2}}$ and $(Y_u)_{\frac{1}{2},0}$, we require $\Phi(0,0)$, which is an unknown for option 2 or 3. To overcome this, McDonald states that in the experiments that are shown in [124] he used $(Y_v)_{0,\frac{1}{2}} = v_{0,\frac{1}{2}}$ and $(Y_u)_{\frac{1}{2},0} = u_{\frac{1}{2},0}$, and the equivalent fields at the other three corners.

For the balanced quantities in Y_u and Y_v whenever Φ is required on the boundary, McDonald suggests the following for options 2 and 3:

$$
\begin{aligned}
\left(\frac{\partial \Phi}{\partial x} - fv\right)_b &= \left(\frac{\partial \Phi}{\partial x} - fv\right)_{b+1}, \\
\left(\frac{\partial \Phi}{\partial y} + fu\right)_b &= \left(\frac{\partial \Phi}{\partial y} + fu\right)_{b+1},
\end{aligned}
$$

where $b+1$ means one grid point removed perpendicular to the boundary towards the interior.

To assess the performance of these schemes McDonald runs a series of numerical experiments with an adjustment whose asymptotic solution is known, where the second instance is with an advective case that has an analytical solution, where he is considering the bell curve under many different situations. The reader is referred to [124] to see the performance of the different boundary schemes that we have described, as well sensitivity experiments with the time step.

However, in 2003 McDonald published a paper with a different approach for the boundary conditions [125]. If we recall the linearized shallow water equations that we started this section with, but now with a periodic domain, then the advection waves are given by

$$\widehat{u}_a = -\frac{il}{f}\widehat{\Phi}_a, \qquad \widehat{v}_a = \frac{ik}{f}\widehat{\Phi}_a, \tag{10.255}$$

with group velocity in the x-direction as

$$\left(C_a^g\right)_x = u_0, \tag{10.256}$$

and two adjustment waves for which

$$\widehat{u}_\pm = \frac{\widehat{\Phi}}{\sqrt{\Phi_0}}\left(\pm\left(1-\frac{\epsilon_l^2}{2}+\frac{\epsilon_f^2}{2}\right)+i\epsilon_l\epsilon_f+O\left(\epsilon^4\right)\right), \tag{10.257}$$

$$\widehat{v}_\pm = \frac{\widehat{\Phi}}{\sqrt{\Phi_0}}\left(\pm\epsilon_l - i\epsilon_f + O\left(\epsilon^3\right)\right), \tag{10.258}$$

whose group velocities in the x-direction are

$$\left(C_\pm^g\right) = u_0 \pm \frac{\sqrt{\Phi_0}}{\sqrt{\left(1+\epsilon_l^2+\epsilon_f^2\right)}}, \tag{10.259}$$

where

$$\epsilon_l \equiv \frac{l}{k}, \qquad \epsilon_f \equiv \frac{\overline{f}}{k\sqrt{\Phi_0}}. \tag{10.260}$$

Given this setup, McDonald presents three alternatives for (10.203c) to try and control the ω_- wave at the L_x boundary.

The first alternative that McDonald suggests is

$$\Phi - u\sqrt{\Phi_0} = 0, \tag{10.261}$$

where this should be a good approximation for two reasons: (1) the removal of the time derivative means that the number of waves will be reduced from three to two; (2) ω_+ still satisfies (10.261) to lowest order as can be seen from (10.257) [125]. Upon substituting the wave solution as before, we obtain a solution that consists of two waves, an ω_2 wave which looks like an ω_+ wave if we neglect the $O(\epsilon^2)$ and higher terms. However, McDonald is not happy with this setup for the accuracy of the third wave.

We saw earlier that more accuracy can be achieved through using

$$\left(\frac{\partial \Phi}{\partial x} - \overline{f} v\right) - \sqrt{\Phi_0}\left(\frac{\partial u}{\partial x} + \frac{\partial v}{\partial y}\right) = 0. \tag{10.262}$$

Now the solution consists of two waves, one of which looks like an ω_+ wave, and a second wave with amplitude and group velocity x component agreeing with ω_a to all orders.

It is possible to continue this process according to McDonald, where if we applied a perturbation computation then combining (10.203a) and (10.203b) with

$$\sqrt{\Phi_0}\frac{\partial}{\partial x}\left[\frac{\partial \Phi}{\partial x} - \overline{f} v - \sqrt{\Phi_0}\left(\frac{\partial u}{\partial x} + \frac{\partial v}{\partial y}\right)\right] + \sqrt{\Phi_0}\frac{\partial}{\partial y}\left(\frac{\partial \Phi}{\partial y} + \overline{f} u\right) - \frac{\Phi_0}{2}\frac{\partial^2 u}{\partial y^2} + \frac{\overline{f^2} u}{2} = 0 \tag{10.263}$$

yields two waves whose group velocity x-components agree with $\left(C_a^g\right)_x$ and $\left(C_+^g\right)_x$ up to $O(\epsilon^3)$.

The final form that McDonald considers in [125] is the case that we presented in the last section, where we replace (10.203c) with

$$\frac{D\left(\Phi - u\sqrt{\Phi_0}\right)}{Dt} = 0,$$

and this approach yields three waves, two of whose solutions are the same as those found through using (10.203a) and a third whose solutions are the same as those of the advective wave.

Given all this derivation in Cartesian coordinates, McDonald tests these conditions in the nonlinear shallow water equations model in spherical coordinates with a full varying Coriolis parameter.

10.12 Test cases for two-dimensional semi-Lagrangian methods

There are many papers that are associated with different developments of the semi-Lagrangian approach in multiple dimensions, and they use different test cases to qualify the performance of their new theory.

Gaussian bell

In [124] a bell shape at the center of the domain is used, which is defined by

$$\Phi(x, y, 0) = \Phi_0 + \breve{\Phi} \exp\left\{-\left[\frac{\left(x - \frac{L_x}{2}\right)}{\frac{L_x}{10}}\right]^2\right\} \exp\left\{-\left[\frac{\left(y - \frac{L_y}{2}\right)}{\frac{L_y}{10}}\right]^2\right\}, \tag{10.264}$$

where the winds for the adjustment experiment are $u(x, y, 0) = 0$ and $v(x, y, 0) = 0$, as well as u_0 and v_0, but also one also assigns minuscule positive values to define the inflow and outflow boundaries. In the adjustment process the adjustment waves radiate away radially from the center of the area. The values of the parameters used here are: $\Phi_0 = (5000\ \text{m})g$, $\breve{\Phi} = (500\ \text{m})\,g$, $f = 0.729 \times 10^{-4}\ \text{s}^{-1}$, $\Delta x = \Delta y = 10$ km, with $L_x = L_y = 10000$ km with $\Delta t = 15$ minutes. We have provided a plot of this set of initial conditions in Fig. 10.19.

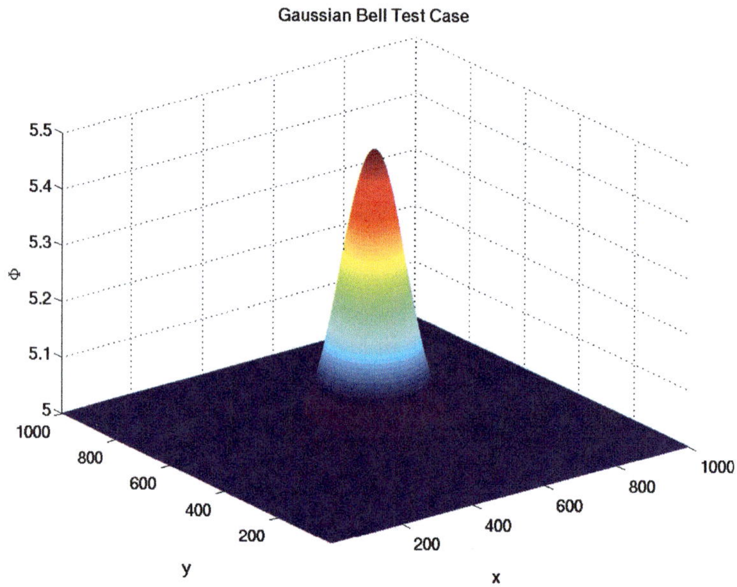

FIGURE 10.19 Mesh plot of the Gaussian bell initial conditions.

Solid body rotation

In the solid body rotation experiments, the density field rotates with constant angular velocity ω about a point (x_c, y_c). One test case for this situation is the **slotted cylinder**, which has an analytical solution

$$\psi' = \begin{cases} \psi_0, & |\xi| \geq s_w \quad \text{and} \quad r \leq \sigma, \\ \psi_0, & \zeta \geq s_l - \sigma \quad \text{and} \quad r \leq \sigma, \\ 0, & \text{otherwise}, \end{cases} \tag{10.265}$$

with ψ_0 as a constant, σ the cylinder radius, and s_l and s_w being the length and width of the slot, respectively. The relative position coordinates ξ and ζ are defined with respect to the rotating center of the cylinder:

$$\xi = x - x_c + \gamma \cos\omega t, \qquad \zeta = y - y_c + \gamma \sin\omega t,$$

where γ is the distance from the enter of the flow (x_c, y_c) to the center of the cylinder $(x_c - \gamma \cos \omega t, y_c - \gamma \sin \omega t)$. The distance r is defined as $r = \sqrt{\xi^2 + \zeta^2}$. These initial conditions are plotted in Fig. 10.20.

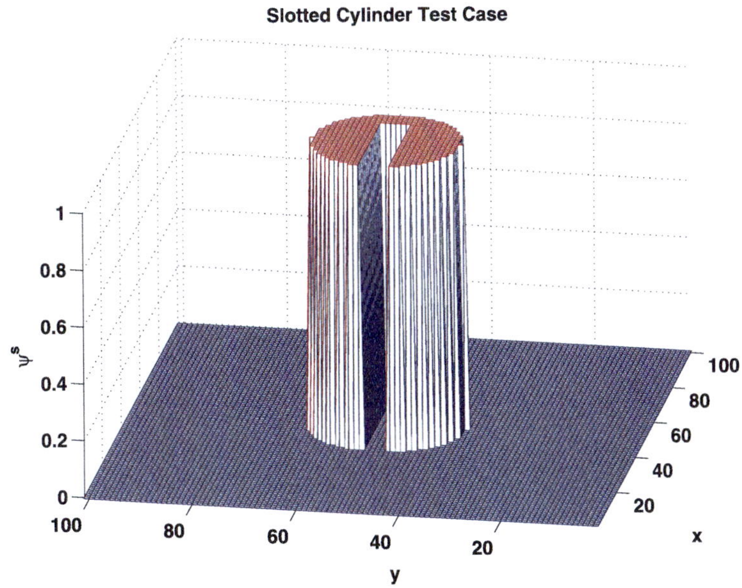

FIGURE 10.20 Mesh plot of the slotted cylinder initial conditions.

Another set of initial conditions are those for the **cosine hill** which is similar to the slotted cylinder, but now the analytical solution here is defined by

$$\psi'(x, y, t) = \begin{cases} 0.5\psi_0 \left(1 + \cos \dfrac{\pi r}{\sigma}\right), & r \leq \sigma, \\ 0, & \text{otherwise.} \end{cases} \tag{10.266}$$

C^0 cone

In [193] the authors present a flux-form version of semi-Lagrangian integration with a spectral element formulation, but one of the test cases that they considered is referred to as a C^0 cone, which means that unlike the smooth Gaussian bell shape earlier, this structure has a point at its center that should not be flattened by the advection scheme, so this test case is good for testing damping properties of a scheme, but also the mass conserving nature of the scheme in the absence of diffusion or dispersion. If we center the cone at coordinates $\left(x_0^c, y_0^c\right)$, and the cone has a radius r_0^c, then mathematically this test case is given by

$$\psi^c = \begin{cases} 1 - \dfrac{r^c(\mathbf{x})}{r_0^c}, & r^c(\mathbf{x}) < r_0^c, \\ 0, & \text{otherwise,} \end{cases} \tag{10.267}$$

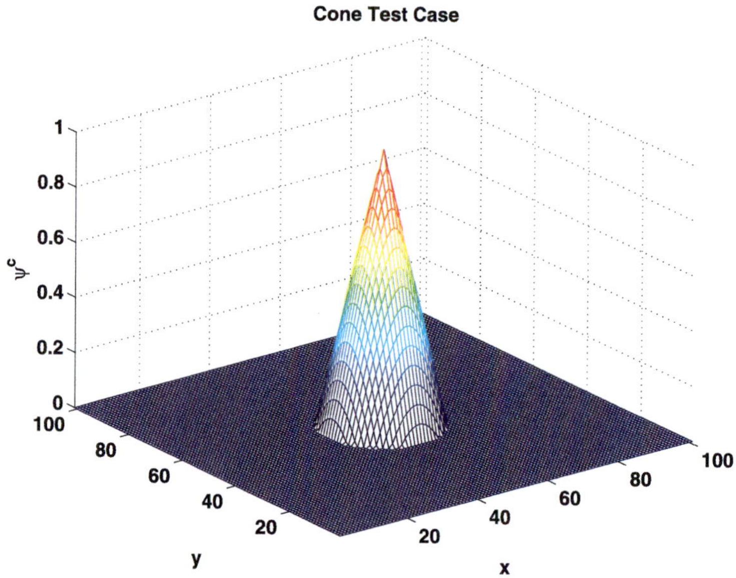

FIGURE 10.21 Mesh plot of the cone initial conditions.

where $r^c(\mathbf{x}) \equiv \sqrt{(x - x_0^c)^2 + (y - y_0^c)^2}$ denotes the radius from the center of the cone. This test case comes from [193]. We have provided a plot of this set of initial conditions in Fig. 10.21.

Idealized cyclogenesis

This flow is a steady circular vortex with a tangential velocity $v_\tau = v_0 \dfrac{\tanh r}{\cosh^2 r}$, where r is the radial distance from the center of the domain and v_0 is a value chosen such that the maximum value of v_τ never exceeds unity. The analytical solution here is

$$\psi'(x, y, t) = -\tanh\left(\frac{y - y_c}{\delta}\cos\omega t - \frac{x - x_c}{\delta}\sin\omega t\right), \tag{10.268}$$

where δ is a constant. These test cases are from [87].

Two deformational flows

In [193] the authors also present two types of flow to test the performance of different versions of semi-Lagrangian algorithms. The first flow is the **nondivergent** velocity field defined by

$$u(x, y, t) = \sin^2(\pi x)\sin(2\pi y)\cos\left(\frac{\pi t}{T}\right), \qquad v(x, u, t) = -\sin^2(\pi y)\sin(2\pi x)\cos\left(\frac{\pi t}{T}\right),$$

where T is set to 5 time units in [193].

The second flow presented in [193] is the **divergent flow**, where this test case has been formulated to contain a divergent flow component overlaid on top of a steady flow in the x direction, defined as

$$u\left(x', y', t\right) = -\sin^2\left(\frac{x'}{2}\right)\sin y' \cos^2 y' \cos\left(\frac{\pi t}{T}\right) + \frac{1}{T}, \qquad v\left(x', y', t\right) = \frac{1}{2}x' \cos^3 y' \cos\left(\frac{\pi t}{T}\right),$$

$$x' = 2\pi\left(x - \frac{1}{T}\right), \qquad y' = \pi\left(y - \frac{1}{2}\right),$$

where $T = 1$ is the period of the flow.

To help illustrate the performance of the semi-Lagrangian approach with different order of interpolation polynomials and also show how some Eulerian schemes perform, we present some results from the **Eady model** which is a baroclinic instability model. See [59,61] for a derivation of this model.

10.13 Semi-Lagrangian methods with the 2D quasi-geostrophic potential vorticity (Eady model)

In this section we introduce a two-dimensional model, where the model is a $x - z$ strip that is an approximation to the atmosphere circling the Earth. The two-dimensional Eady model [50] is the simplest theoretical model that is used to study baroclinic instability in the x–z plane. It is a linear Quasi-Geostrophic (QG) model that can be used to describe the vertical coupling between waves at the tropopause and the surface, for a meteorological setup. The Eady model equations support two types of what are referred to as **normal mode** solutions [55]: neutral modes that correspond to boundary waves with short wavelengths and unstable long waves that grow, or decay, exponentially.

The 2D Eady model is a simple linear model for QG flows derived from considering simplifications to the **primitive equations**. The primitive equations describe horizontal and vertical momentum, mass conservation, and the thermodynamic equations, which for a nonspherical coordinate system are given by

$$\frac{D\mathbf{v}}{Dt} + f\mathbf{k} \times \mathbf{v} + \frac{1}{\rho}\nabla p = 0, \tag{10.269a}$$

$$\frac{Dw}{Dt} + \frac{1}{\rho}\frac{\partial p}{\partial z} = -g, \tag{10.269b}$$

$$\frac{D\rho}{Dt} + \nabla.\mathbf{v} = 0, \tag{10.269c}$$

$$\frac{D\theta}{Dt} = 0, \tag{10.269d}$$

where f is the Coriolis force defined as $2\Omega \sin\phi$, Ω is the Earth's rotation rate, and ϕ is the angle of latitude where $\phi \in \left[-\frac{\pi}{2}, \frac{\pi}{2}\right]$, p is pressure, ρ is the density, and θ is the temperature.

The basis of the Eady model is to assume that there is a large static state in the atmosphere that has no real effect on the way weather systems move and develop. What is of interest then is what are the effects of perturbations away from this static state. Three of the main variables that are of interest, temperature, pressure, and density, can be rewritten in the form of a basic state that is only dependent on height, z, and a perturbation dependent on the horizontal plane coordinate, height and time. It is the effect that these perturbations have on Eqs. (10.269a)–(10.269d) that eventually leads to the Eady model. Thus we have

$$\theta(x, y, z, t) = \overline{\theta(z)} + \theta'(x, y, z, t), \tag{10.270a}$$

$$p(x, y, z, t) = \overline{p(z)} + p'(x, y, z, t), \tag{10.270b}$$

$$\rho((x, y, z, t)) = \overline{\rho(z)} + \rho'(x, y, z, t). \tag{10.270c}$$

The Eady model is used to investigate how perturbations evolve with time from thermal wind balance. The perturbations that are of interest are believed to generate mid-latitude cyclones and anticyclones, in the x–z plane. The purpose of the Eady model is to describe the behavior of the QGPV in the interior of the domain. Through the manipulation of the vorticity and thermodynamics equations, we eventually arrive at the full Eady model that links the perturbations of QGPV and the buoyancy as

$$q' = \nabla_h^2 \psi + \frac{\partial}{\partial z}\left(\frac{f_0^2}{N^2}\frac{\partial \psi}{\partial z}\right), \tag{10.271a}$$

$$D_g b' = -wN^2. \tag{10.271b}$$

What is important here are the two different sets of partial differential equations that are numerically approximated with the forward upwind and centered-time centered-space schemes, as well as with the linear, quadratic, and cubic Lagrange interpolation polynomials in the expressions above with two sets of initial conditions; the first is where the PV is zero in the interior and we are simply modeling the advection of the buoyancy on the z axis, but we shall see that the two boundaries communicate with each other. The second test case is for a dipole of PV in the interior and where these positive and negative balls are advected and conserved, but there is a vertical shear which implies that one ball goes away, one stays still, and one ball goes in the opposite direction to the first ball.

To be able to solve the second-order partial differential equation, we require boundary conditions on both the x and z boundaries. The setup we have for the model approximates a vertical strip in the atmosphere that encompasses the whole length of a mid-latitude longitudinal ring. It is assumed that the streamfunction, meridional winds, and the QGPV are periodic in the x-direction, while on the z-boundaries it is assumed that the vertical velocity is zero; that is to say, there is no motion out of the atmosphere or through the ground. On these boundaries the quasi-geostrophic thermodynamic equation is applied. The equation is also linearized to give

$$\left(\frac{\partial}{\partial t} + \overline{U}\frac{\partial}{\partial z}\right)\frac{\partial \psi}{\partial x} = A\frac{\partial \psi}{\partial x}. \tag{10.272}$$

Recalling that $\frac{\partial \psi}{\partial z} \sim b'$ implies that this equation can be written in terms of the buoyancy and the meridional wind in the form

$$\left(\frac{\partial}{\partial t} + \bar{U}\frac{\partial}{\partial x}\right) b' = A\frac{\partial \psi}{\partial x}. \tag{10.273}$$

The equations that we are considering here for the model can be transformed onto $\hat{z} \in [0, 1]$ and $\hat{x} \in [0, 1]$. Therefore, the equations that define the Eady model in two dimensions, and the set of equations that we shall consider different numerical approximations to in the first set of experiments, are given by

$$\frac{\partial^2 \psi}{\partial \hat{x}^2} + \frac{\partial^2 \psi}{\partial \hat{z}^2} = 0 \qquad \text{in} \quad \Omega, \tag{10.274}$$

$$\left(\frac{\partial}{\partial \hat{t}} + \hat{z}\frac{\partial}{\partial \hat{x}}\right)\frac{\partial \psi}{\partial \hat{z}} = \frac{\partial \psi}{\partial \hat{x}} \qquad \text{on} \quad \partial\Omega_1, \tag{10.275}$$

$$\psi\left(0, \hat{z}\right) = \psi\left(1, \hat{z}\right) \qquad \text{on} \quad \partial\Omega_2, \tag{10.276}$$

where $\partial\Omega_1$ and $\partial\Omega_2$ are the z and x boundaries, respectively, and Ω is the interior unit square here.

10.13.1 Buoyancy advection on the boundaries: $b'_0 = 0$, $b'_1 = \alpha \sin(K\Delta x)$

Due to the periodic condition on the buoyancy and the streamfunction in the x direction, in this test problem we choose a sine wave of wavelength $K = 2n\pi$ and amplitude α, which is set to 20 for this test case.

The problem for this test case is defined as

$$\begin{aligned}
&\frac{\partial^2 \psi}{\partial z^2} + \frac{\partial \psi^2}{\partial x^2} = 0 && \text{in} \quad \Omega = (0,1) \times (0,1), \\
&\frac{\partial \psi}{\partial z} = \alpha \sin(K\Delta x) && \text{on} \quad z = 1, \\
&\frac{\partial \psi}{\partial z} = 0 && \text{on} \quad z = 0.
\end{aligned} \tag{10.277}$$

From the equations presented in (10.277), it is possible to find the initial expression for the streamfunction. We consider a solution of the form $\psi = \left(Ae^{Kz} + Be^{-Kz}\right)\sin(kx) + C$, where C is the constant of integration, at $t = 0$. From the extra condition to overcome the ill-posedness, $\psi(0,0) = 0$, it follows then that $C = 0$. It can easily be shown that this expression for the initial streamfunction satisfies the interior differential equation in (10.277).

From the initial conditions for the buoyancy on the two z boundaries relating them to the z derivative of the streamfunction, $b' = \dfrac{\partial \psi}{\partial z}$, the two constants A and B can be found through rearranging these initial conditions as

$$\frac{\partial \psi}{\partial z}(x, 0) = 0 \implies A - B = 0, \tag{10.278}$$

$$\frac{\partial \psi}{\partial z}(x,1) = \alpha \sin(Kx) \quad \Longrightarrow \quad K\left(Ae^{K} - Be^{-K}\right) = \alpha. \tag{10.279}$$

From (10.278) we obtain the condition that

$$A = B. \tag{10.280}$$

Substituting (10.280) into (10.279) results in

$$KA\left(e^{K} - e^{-K}\right) = \alpha \Longrightarrow \quad A = \frac{\alpha}{2K\sinh(K)}. \tag{10.281}$$

Hence the final form for the initial streamfunction is

$$\psi(x,z,0) = \frac{\alpha}{K\sinh(K)}\cosh(Kz)\sin(Kx). \tag{10.282}$$

In the numerical experiments that we present results form, we have 64 grid points in the x direction, with 8 vertical levels in the z direction. This then gives a mesh containing 512 grid points. To ensure stability of the two Eulerian schemes, we use a time step of $\Delta t = \frac{\Delta x}{10}$. The model is run for 640 time steps.

In Figs. 10.22–10.24 we have plotted the solutions for the three numerical approximation to the buoyancy advection on the $z = 0$ and $z = 1$ boundaries, in the order that they have been introduced. We have plotted the buoyancy at five different times between the initial time and at $T = 640$ time steps, which is equivalent to 1 day, to illustrate how the wave is moving but

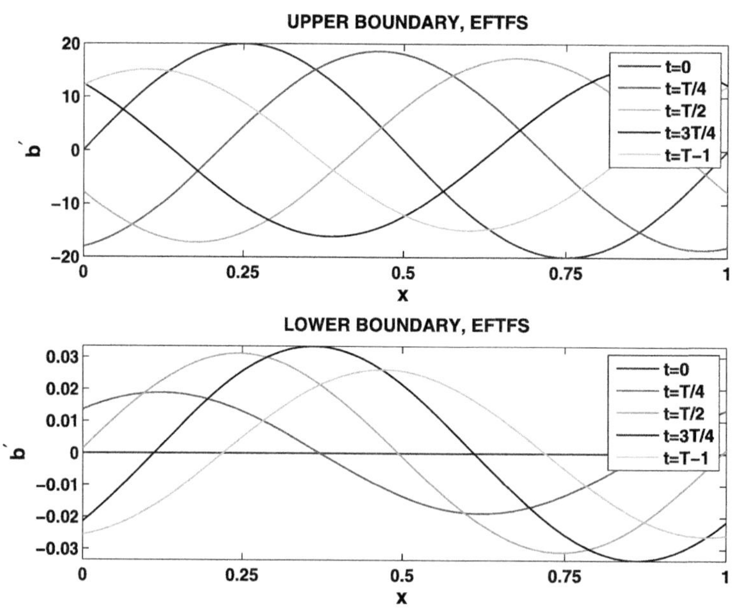

FIGURE 10.22 Plot of the buoyancy advection by the forward upwind scheme.

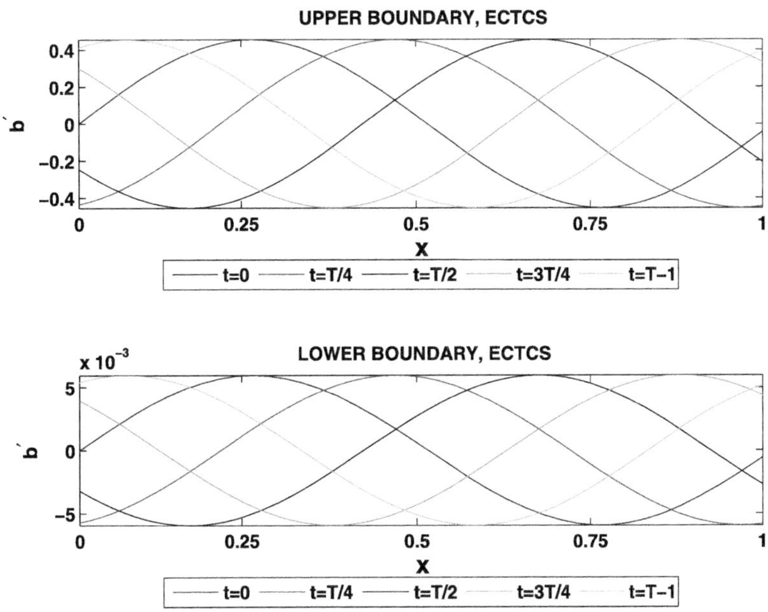

FIGURE 10.23 Plot of the buoyancy advection by the centered-time centered-space scheme.

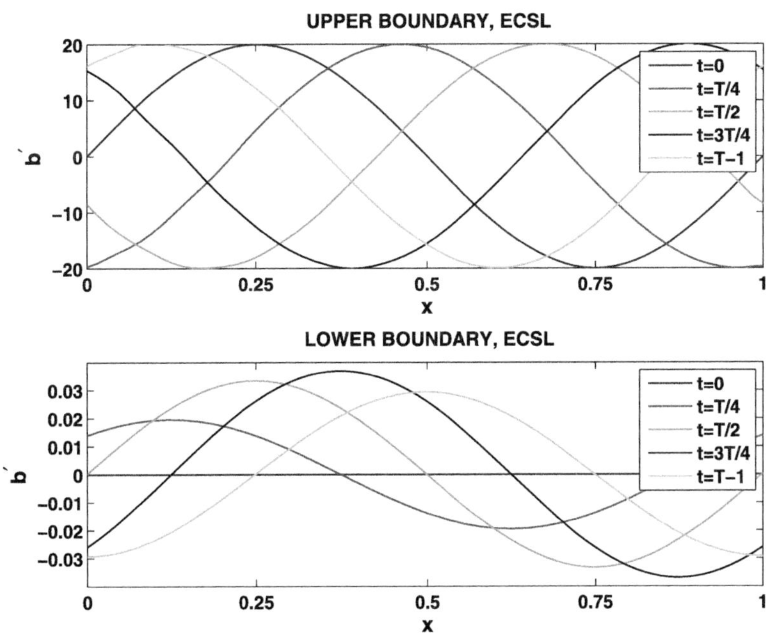

FIGURE 10.24 Plot of the buoyancy advection by the cubic semi-Lagrange scheme.

also to highlight any effects the numerical schemes are having on the shape of the buoyancy. The five times are: the initial time, $T/4 \equiv 160$, $T/2 \equiv 320$, $3T/4 \equiv 480$, and $T-1 \equiv 639$ time steps.

In Fig. 10.22 we have the results from the EFU scheme, where we can see that the buoyancy wave on the upper boundary is being advected, however, it can clearly be seen that the buoyancy is beginning to be damped by the numerical scheme. The damping becomes obvious when Fig. 10.22 is compared to the solution for the ECTCS results in Fig. 10.23. An important feature that can be inferred from the plots of the upper and lower boundary buoyancy is that the numerical model in the interior is communicating between the two boundaries. Recall that the initial condition for the lower boundary was $b' = 0$.

When comparing the solutions on the lower boundary between the EFU and the ECTCS there are subtle differences in both the amplitude and location of the buoyancy wave. Therefore, the first-order scheme appears to be suffering from an artificial damping affect, which is consistent with the analysis of the amplification factor for this scheme. Finally, in Fig. 10.24 we have plotted the solution from using the CESL scheme for the advection of the buoyancy. It is clear that the solution from the CESL is not suffering from the damping effect that the EFU scheme does. The results from the CESL and the ECTCS are almost identical.

10.13.2 QGPV $\neq 0$

We now move on to the more advance numerical model where the initial perturbations here are based upon those from [7]. The initial perturbation is for the meridional wind $v' = \dfrac{\partial \psi'}{\partial x}$. This perturbation is defined in the form $v' = VF_x(x)G(z)$. The functions $F_x(x)$ and $G(z)$ are defined as

$$F_x(x) = \begin{cases} -\sin^2(Kx), & -\dfrac{L}{2} \leq x < -\dfrac{L}{4}, \\ \sin(Kx), & -\dfrac{L}{4} \leq x < \dfrac{L}{4}, \\ \sin^2(Kx), & \dfrac{L}{4} \leq x < \dfrac{L}{2}, \\ 0, & \text{elsewhere}, \end{cases} \tag{10.283}$$

and the function $G(z)$ is given by

$$G(z) = \begin{cases} \cos^2(m(z - z_{\max})), & -\dfrac{D}{2} \leq z - z_{\max} < \dfrac{D}{2}, \\ 0, & \text{elsewhere}, \end{cases} \tag{10.284}$$

where L denotes the length of the perturbation in the horizontal direction. For the experiments, L was taken to be 4000 km where now the domain is no longer the unit square but a realistic approximation in scale to the x–z plane in the atmosphere. The horizontal length scale is taken to be 16000 km and the vertical length scale is 10 km. It is as a result of this scaling that the Coriolis parameter f_0 and the static stability parameter N have to be included;

$z_{\max}$ represents where the center of the disturbance is in the z direction, and D is the vertical length scale of the disturbance, here taken to be 4 km; V is constant and for this example is set to 10.

We now use the property that the meridional wind is the x-derivative of the streamfunction, which then enables us to find a mathematical expression for the initial streamfunction perturbation through integrating $F(x)$ above. Thus the associated initial streamfunction perturbation, ψ', is defined by

$$\psi(x,z)=\begin{cases}\left(-\frac{L}{4}-\frac{x}{2}+\frac{\sin(2Kx)}{4K}\right)\cos^2(m(z-z_{\max})), & -\frac{L}{2}\leq x<-\frac{L}{4},\\ & -\frac{D}{2}\leq z-z_{\max}<\frac{D}{2},\\ \left(-\frac{L}{8}-\frac{\cos(Kx)}{K}\right)\cos^2(m(z-z_{\max})) & -\frac{L}{4}\leq x<\frac{L}{4},\\ & -\frac{D}{2}\leq z-z_{\max}<\frac{D}{2},\\ 0, & \text{elsewhere.}\end{cases} \tag{10.285}$$

Finally, given the expressions for the initial perturbations to the meridional winds and the streamfunction, it is possible to derive the expression for the QGPV through taking the second derivative of $F(x)$ with respect to x, and the second z derivative of $G(z)$. This then results in

$$F_{xx}(x) = \begin{cases}-2K\sin(Kx)\cos(Kx), & -\frac{L}{2}\leq x<-\frac{L}{4},\\ K\cos(Kx), & -\frac{L}{4}\leq x<\frac{L}{4},\\ 2K\sin(Kx)\cos(Kx), & \frac{L}{4}\leq x<\frac{L}{2},\\ 0, & \text{elsewhere,}\end{cases} \tag{10.286}$$

$$G_{zz}(z) = \begin{cases}-2m^2\left(\sin^2(m(z-z_{\max}))+\cos^2(m(z-z_{\max}))\right), & -\frac{D}{2}\leq z-z_{\max}<\frac{D}{2},\\ 0, & \text{elsewhere.}\end{cases} \tag{10.287}$$

Therefore combining (10.287) and (10.286) with the $F(x)$ component of (10.285), the initial perturbation to the QGPV can be defined by

$$q'=\left(F_{xx}(x)G(z)+F(x)G_{zz}(z)\right)V.$$

The initial perturbations to the streamfunction, meridional wind, and the QGPV defined above are presented in Fig. 10.25. For the streamfunction we can see that the perturbation is located in the center of the domain. The perturbation for the meridional wind are on either

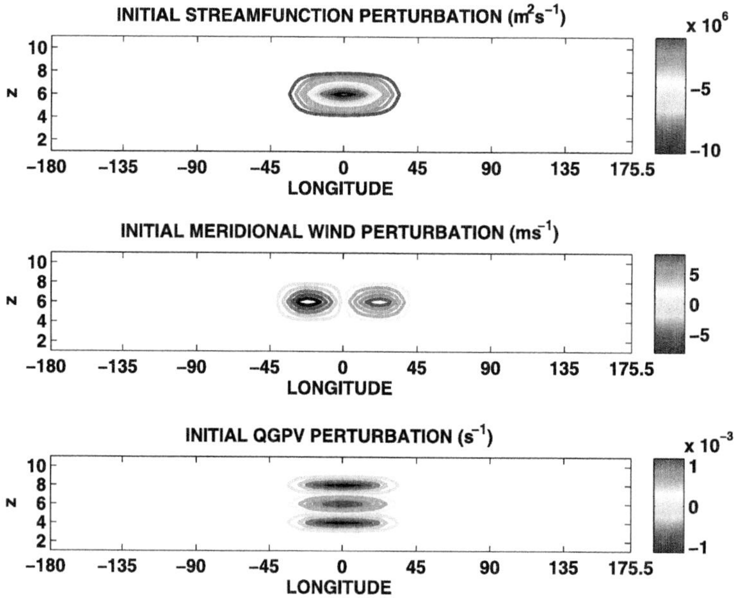

FIGURE 10.25 Plot of initial perturbations to (a) the meridional winds, v', (b) the streamfunction, ψ', and (c) the QGPV, q'.

side of the center of the domain and of opposite sign. Finally, the QGPV perturbation is a dipole of negative, positive, and then negative PV.

The continuous equations that we are numerically approximating in this section are (10.271a) in the interior of the domain and (10.273) for the buoyancy advection on the z boundaries. For the parameters required for (10.271a) we take $\overline{U}$ to be in the form of $\overline{U} = \Lambda z - u_0$, where $\Lambda = 4 \times 10^{-4}\ \mathrm{s}^{-1}$ and $u_0 = 20\ \mathrm{ms}^{-1}$. These choices for the parameters in $\overline{U}$ result in a linear shear with respect to the z direction that starts at $-20\ \mathrm{ms}^{-1}$ on the lower z boundary and increases linearly to $20\ \mathrm{ms}^{-1}$ on the upper z boundary; A is taken to be equal to Λf_0 where $f_0 = 1 \times 10^{-4}$. Finally, we set the static stability parameter N to be $N = 1 \times 10^{-2}$.

The PV in the interior is to be advected by the ECTCS and ECSL numerical approximations, but we will show results from the linear and quadratic Lagrange interpolation to show their damping effects on the QGPV later as well.

In the vertical direction we use a grid spacing of 1 km, resulting in 11 levels, and in the horizontal direction we use a grid spacing of 200 km which results in 80 grid points. We use $\Delta t = 0.05$ days, which is equivalent to 4320 s. In the definitions for the meridional wind perturbations, we set $L = 2000$ km, which then makes $k = \frac{2\pi}{L}$, and we set $D = 4$ km, which then makes $m = \frac{\pi}{D}$ and $z_{\max} = 5$ km. A more detailed meteorological justification for this setup can be found in [7]. We are interested in the numerical modeling component here and the effects different numerical schemes have on the accuracy of the shape preserving of the advection of the QGPV increment. A final comment about the setup of this problem is that we are using the constraint approach, where we set $\psi_{1,1} = 0$ for the duration of the numerical experiment.

We are considering two sets of experiments for the model setup just described. The first set of experiments involves comparing the advection of the PV for a long integration of 45 days, which is equivalent to 800 time steps. One of the reasons for the long integrations is to show that the numerical schemes can wrap around through the periodic boundary conditions. For this long integration run, we compare the performance of the linear, quadratic, and cubic Lagrange interpolation polynomial based advection, as well as the CTCS scheme.

The results from the long integration for the semi-Lagrangian scheme with the linear Lagrange polynomial are presented in Fig. 10.26. We can clearly see the damping effects that the linear scheme has, which were identified earlier in the stability analysis. For the model setup considered here the Courant numbers for each level are presented in Table 10.3.

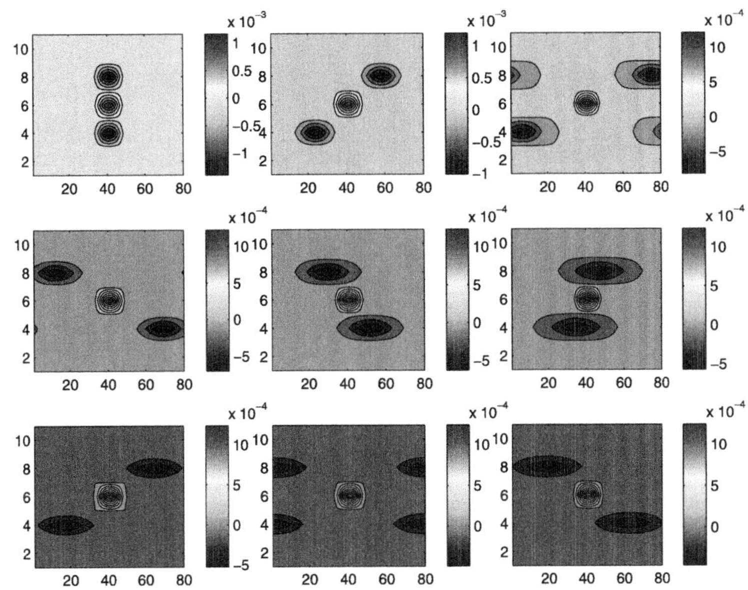

FIGURE 10.26 Plot of the long time integration for the linear semi-Lagrangian scheme.

TABLE 10.3 Table of the Courant numbers for the different vertical levels associated with the different advecting wind speeds.

Level	0	1	2	3	4	5	6	7	8	9	10
α	0.42	0.35	0.26	0.17	0.09	0	0.09	0.17	0.26	0.35	0.42

We have swapped the direction of the interpolation schemes for levels where we have a negative advecting velocity to ensure that the Lagrange interpolation polynomials-based schemes are always upwind of the departure point.

The damping affects on the two negative PV balls is quite severe, such that the balls have lost over an order of magnitude. There is no diffusion in this model setup so the schemes

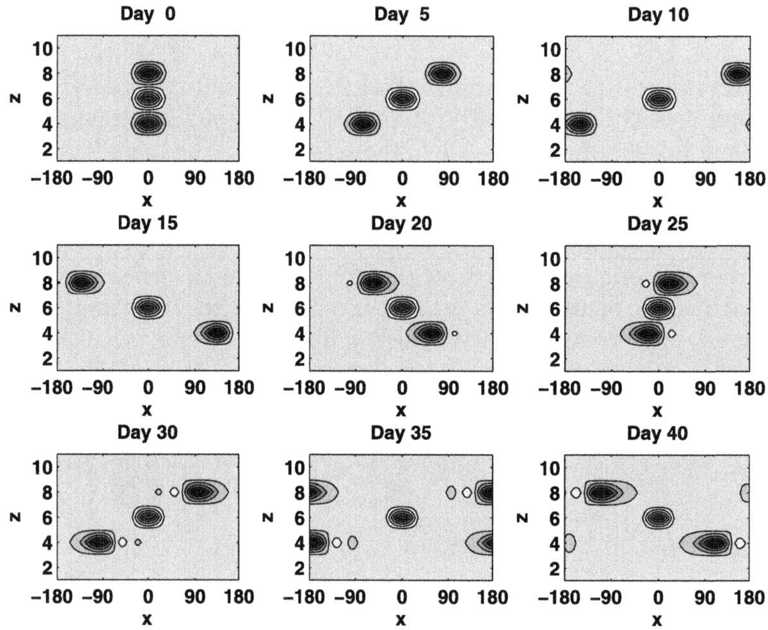

FIGURE 10.27 Plot of the long-time integration for the quadratic semi-Lagrangian scheme.

should be as close to shape preserving as possible. These results indicate that the linear interpolation approach should not be used for these scales involved in the Eady model.

We now consider the results from the quadratic Lagrange interpolation, where at the same 5-day intervals as the linear case are presented in Fig. 10.27. As indicated in the stability analysis for this scheme, there is still some damping occurring but not as severe as for the linear case, however, we can see that some spurious detachments from the main PV negative balls do occur later on in the integration.

In Fig. 10.28 we present the results for the same setup as for the linear and quadratic semi-Lagrangian schemes, but now for the ECTCS. We can see that the negative PV balls are being stretched in the direction which they are being advected, which again should not be occurring.

We can see in the plots for the ECTCS scheme that at day 20 we are starting to see a spurious area of positive PV cutting off downwind of the negative PV balls. Recall that q' is the right-hand side term of the Poisson equation that we have to solve in the interior and on the boundaries to obtain the streamfunction, and hence by association the buoyancy perturbations as well. Ten days later there appears to be a small spurious ball of negative PV that is separated from behind the main ball, which is not physical.

Finally, moving onto the cubic Lagrange interpolation semi-Lagrangian scheme, who's results are presented in Fig. 10.29, we can see that the distortion of the negative PV ball that is observed with the ECTCS scheme, and to some extent with the quadratic semi-Lagrangian scheme, is not as pronounced with the cubic Lagrange interpolation. This is indicating that

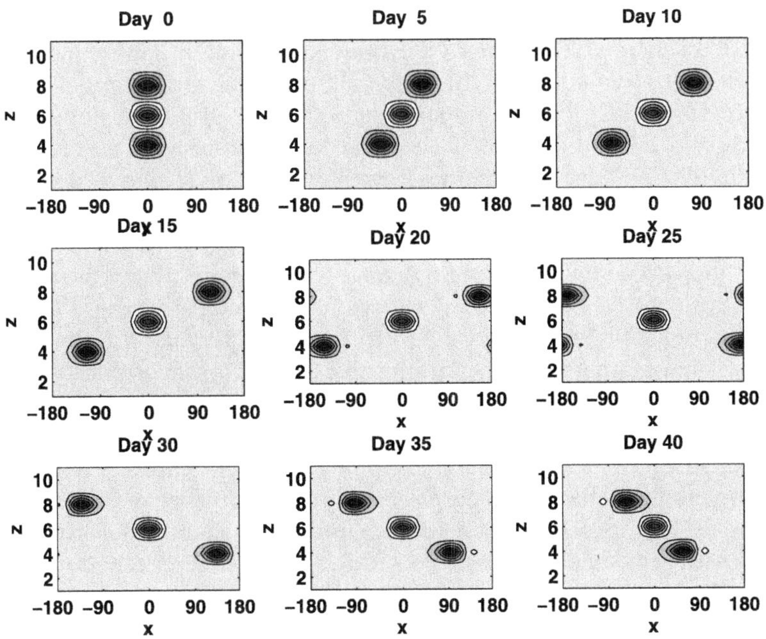

FIGURE 10.28 Plot of the long-time integration for the explicit centered-time centered-space scheme.

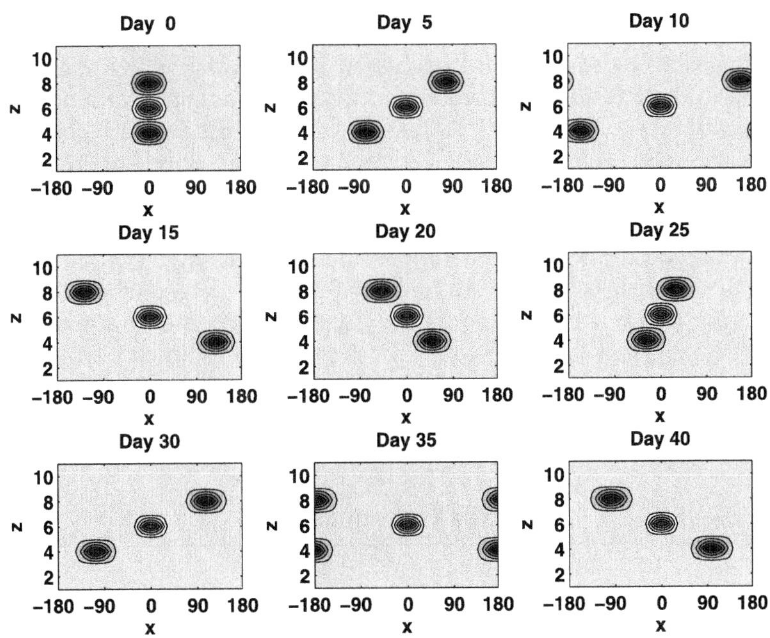

FIGURE 10.29 Plot of the long time integration for the cubic semi-Lagrangian scheme.

the cubic interpolation-based semi-Lagrangian approach is more shape-preserving than if we use the ECTCS scheme. While there are more computations involved with the semi-Lagrangian approaches, we still have the advantage that semi-Lagrange schemes can use any time step for stability, but there are still restrictions on the time step for the accuracy. Therefore, to reduce computation time, we can take larger time steps but then perform fewer evaluations of the scheme. Just as a reminder, we have used the same time step for all of the numerical schemes. Halving the time step size did not eliminate the spurious areas of QGPV in the model.

However, it should be noted that while it might appear that the cubic semi-Lagrangian approach may appear to be shape-preserving we have plotted the value of the PV for the $z = 4$ level which cuts straight through the lower negative PV ball for all four schemes for the first 5 days, 24 hours apart, of integration to illustrate the severity of the overshoots in Fig. 10.30 for the cubic semi-Lagrangian approach, and for the ECTCs.

From Fig. 10.30 we can see that the overshoots from the ECTCS scheme are growing with time, which is consistent with the stretching we saw of the PV earlier. We see that there are some overshoots with the cubic semi-Lagrangian approach but not as severe. We should also note here that the ECTCS scheme is initiated with the FTCS scheme which we saw earlier can introduce wiggles into the scheme. These plots indicate that neither scheme is preserving the PV or mass completely, which we have seen is a problem for some applications.

The final sets of plots that we present in this chapter are of the streamfunction at 24, 48, 72, and 96 hours for the ECTCS and the cubic semi-Lagrangian scheme. These results are presented in Fig. 10.31. We can see from the plots that there are quite significant differences between the two solutions, recalling that the QGPV is the right-hand side of the Poisson equation that we have to invert to obtain new values for the streamfunction. We can see that there are differences in the slanting of the streamfunctions, as well as the magnitude. The process going on in the interior is referred to as **PV shielding**, and we recommend reading [7] for a good description of the effects of the shielding. As the ECTCS scheme is stretching the PV, it is extending the area that the PV covers and as such prevents the streamfunction from growing into this area. This feature is again an example of model error as this extra shielding is an effect of the numerical scheme and not a physical process.

We can therefore see that there was an improvement with using the cubic semi-Lagrangian approach over the explicit centered-time centered-space scheme, when initialized by the EFTCS scheme, and that if we had used a Lax–Wendroff scheme to initiate then the overshoots may have not been as bad. The codes for these schemes are available, and the reader can play with the time steps to see what happens to the performances of the scheme over long time.

10.14 Summary

In this chapter we have introduced extensions of the one-dimensional versions of semi-Lagrangian integration, where we have presented the bivariate versions of the Lagrange interpolation formulas, bilinear, biquadratic, and bicubic. We have also shown that it is possible to apply semi-Lagrangian approaches with finite element discretization, with an ocean

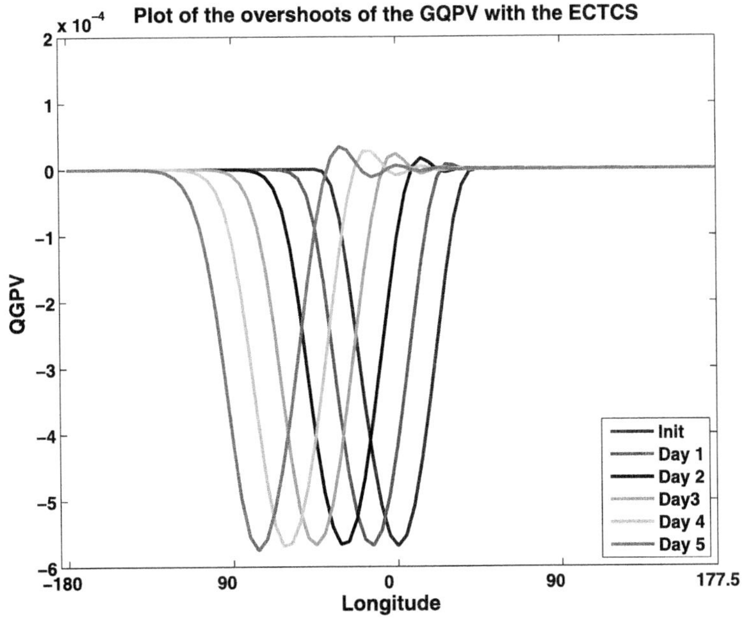

(A)

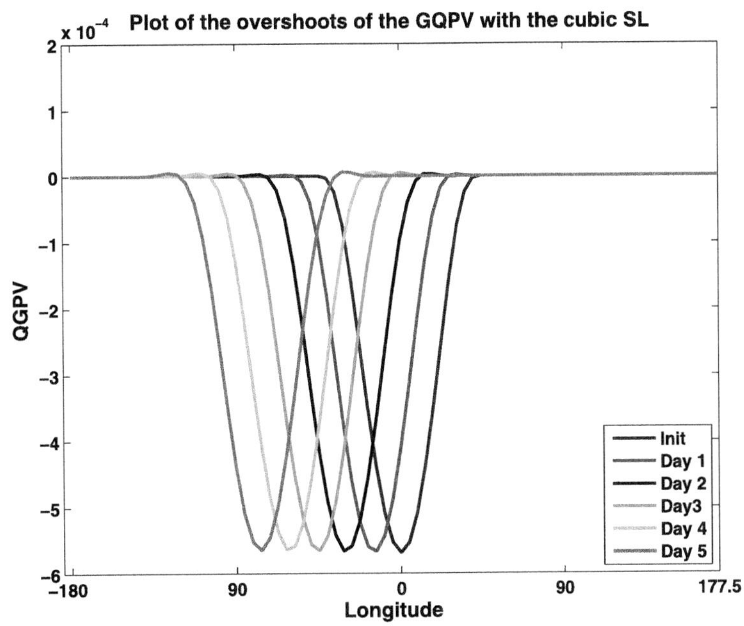

(B)

FIGURE 10.30 Plot of the overshoots for the different schemes for the first 5 days of integration 24 hours apart for the ECTCS and the cubic semi-Lagrangian methods.

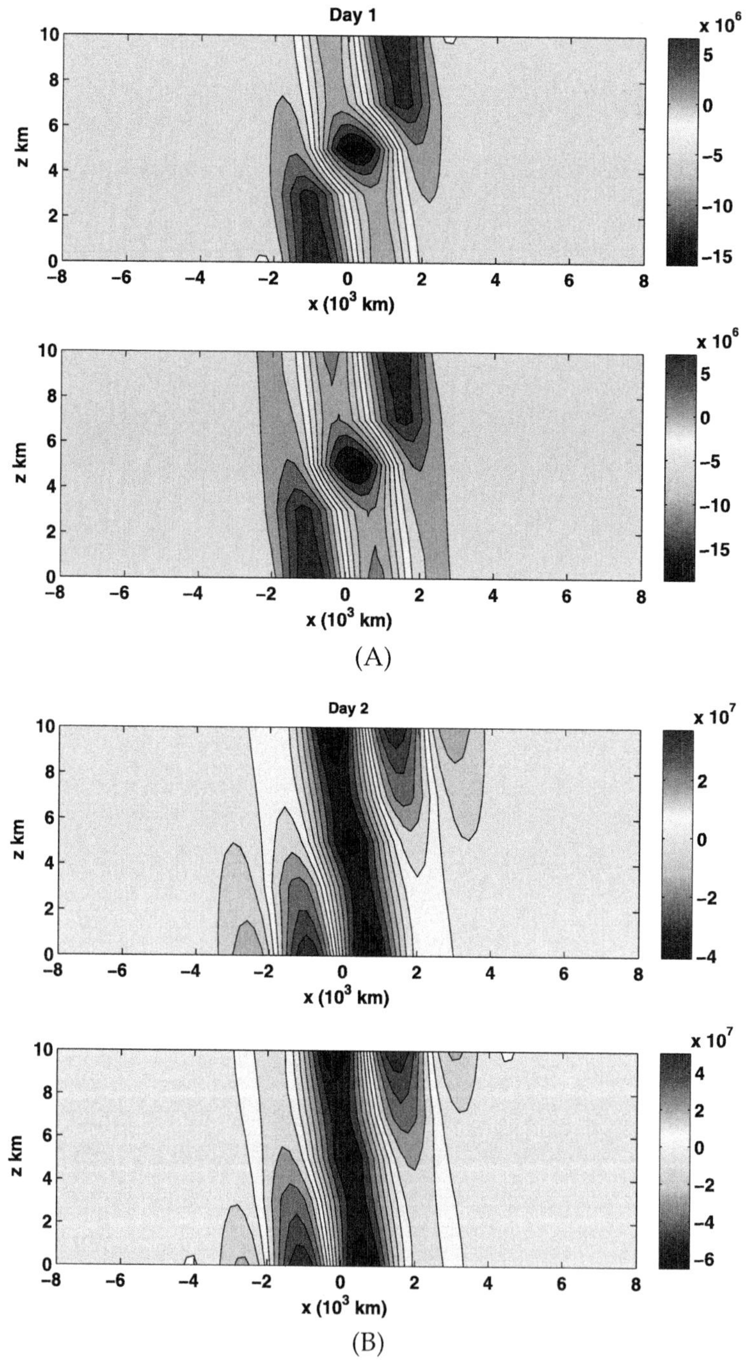

FIGURE 10.31 Plot of the evolution of the streamfunction perturbation for the explicit centered time centered space and the cubic Lagrange interpolation semi-Lagrange schemes at 24, 48, 72, and 96 h.

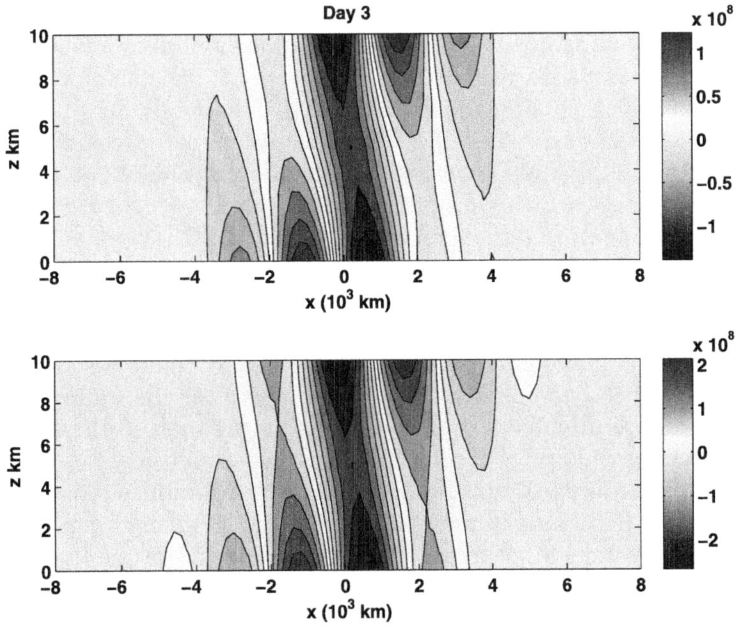

(C)

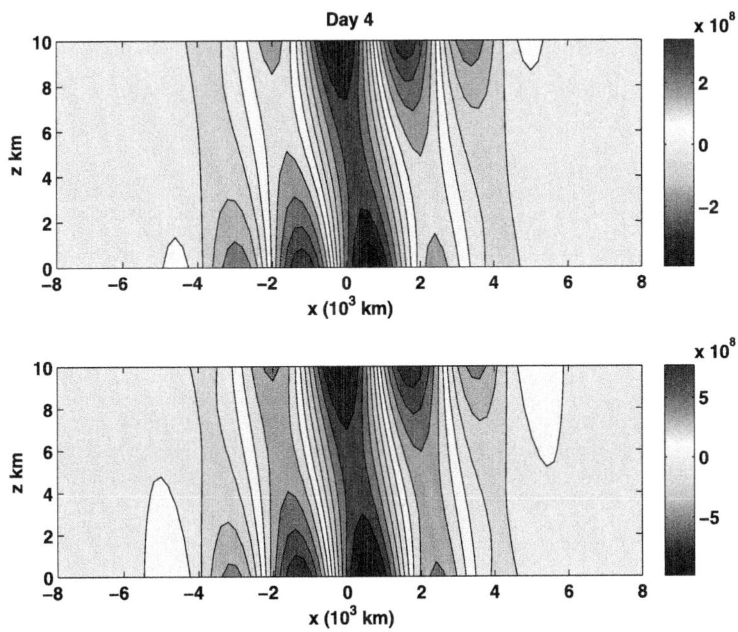

(D)

FIGURE 10.31 *(continued).*

model. We have also introduced the two-dimensional nonlinear shallow water equations model where we have momentum equations containing nonlinear advection of the wind fields u and v by themselves, as well as the forced version. The shallow water equations models are used extensively in atmospheric and oceanic modeling and data assimilation to test the effects of new techniques, where it is often the case that if the technique does not work in this model, then it is highly unlikely to work in the more sophisticated three-dimensional primitive equations models, which we shall introduce in the next chapter. We have also introduced the flux-form of semi-Lagrangian integration, as well as considered rotational and deforming flows, where we saw that we need to include more points along the trajectory to remove a first-order error that is introduced through considering one point to determine the velocity. The next important feature that we had to consider was the impact of different boundary conditions in a limited area model, where it could be the case that the departure points could be outside the domain, but also that points near the inflow boundary could require points outside the domain to complete the interpolation to these points. We then presented some formulaes and plots of some standard two-dimensional Cartesian initial conditions for test problems to work with semi-Lagrangian techniques. Finally, we presented some findings with the Eady model and showed that the cubic Lagrange semi-Lagrangian approach performed much better than the Eulerian approaches or the linear and quadratic Lagrange interpolation polynomial-based schemes.

CHAPTER

11

Semi-Lagrangian methods for three-dimensional problems

In the previous chapters we have been concerned with the one- or two-dimensional version of some form of advection, but in many different geophysical applications of advection it could quite possibly be that the particle is not moving just in a horizontal plane, but could instead be moving in the vertical direction as well. Therefore, we may have to consider finding a departure point that is in a volume of some sort. Therefore, this departure point could very well be nowhere near an x or y point, or be at a model vertical level, z, and so we have to find this point through some interpolation in **three dimensions**.

In the next section we introduce a couple of interpolations in three dimensions to highlight the difficulties in this process. We then move on to consider different sets of partial differential equations that have been used at many operational numerical weather prediction centers, flows in the ocean and fluids in general. We shall look at recent work at the ECMWF and ECCC centers where we reassess that only two iterations are required to find the departure point, and that we can have different order interpolation methods for the velocity and prognostic field. We shall also consider the Semi-Implicit Eulerian Lagrangian Finite Element (SELFE) ocean model. We start this chapter with a look at interpolation polynomials.

11.1 Trivariate interpolation methods

As with the 1D and 2D cases, we start by considering the general form of the Eulerian formulation of the linear advection equation in Cartesian coordinates, which is given by

$$\frac{\partial \psi}{\partial t}+u\left(x,y,z,t\right)\frac{\partial \psi}{\partial x}+v\left(x,y,z,t\right)\frac{\partial \psi}{\partial y}+w\left(x,y,z,t\right)\frac{\partial \psi}{\partial z}=f\left(x,y,z,u,v,w,\psi,t\right), \quad (11.1)$$

where the advection equation can again be expressed in vectorial form as

$$\frac{\partial \psi}{\partial t}+\left(\mathbf{u}\cdot\nabla\right)\psi=f\left(x,y,z,u,v,w,\psi,t\right), \quad (11.2)$$

Semi-Lagrangian Advection Methods and Their Applications in Geoscience
https://doi.org/10.1016/B978-0-12-817222-3.00015-3

where $\mathbf{u} \equiv \begin{pmatrix} u & v & w \end{pmatrix}^T$, with w being the vertical advecting wind, the Lagrangian form of which is

$$\frac{D\psi}{Dt} = f\left(x, y, u, v, w, \psi, t\right), \tag{11.3}$$

but now we also have to numerically integrate the three following displacement equations:

$$\frac{\partial x}{\partial t} = u, \tag{11.4a}$$
$$\frac{\partial y}{\partial t} = v, \tag{11.4b}$$
$$\frac{\partial z}{\partial t} = w. \tag{11.4c}$$

Note that the three wind fields may not be equal, they are just constant.

As mentioned above, the departure point may not coincide with a grid point in any of the three dimensions and as we have to interpolate to this point in 3D. As with the two-dimensional case, or for the one-dimensional case, we have to integrate (11.4a)–(11.4c) to find the departure point upstream, to then find the value of ψ there. This then implies for a constant velocity in all three directions that we have

$$x_d^n = x_i^{n+1} - \Delta x u, \tag{11.5a}$$
$$y_d^n = y_j^{n+1} - \Delta y v, \tag{11.5b}$$
$$z_d^n = z_k^{n+1} - \Delta z w. \tag{11.5c}$$

Recall from the last chapter that, to be able to implement a bilinear interpolation, we required 4 points; to interpolate in three dimensions, we will require a cube where we have values at 8 points at the corners of the cube. We have an illustration of the cube needed to form the trilinear Lagrange interpolation polynomial in Fig. 11.1.

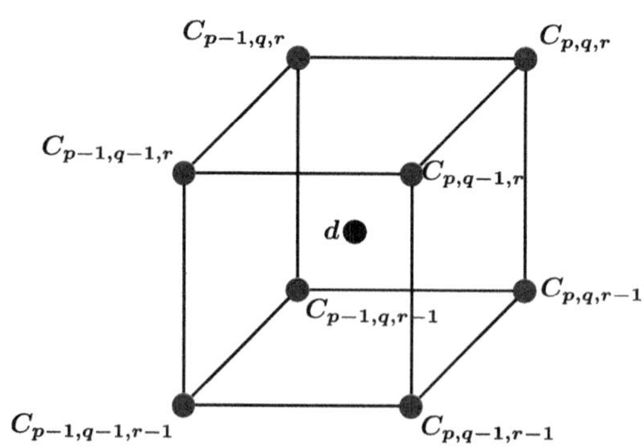

FIGURE 11.1 Illustration of the cube and the eight corner points to interpolate to the departure point.

As we keep mentioning, it is highly likely that the departure point will not be at a grid point, and so we will have a Courant number in all three directions, where now

$$\alpha = \frac{u\Delta t}{\Delta x}, \qquad \widehat{\alpha} = \alpha - p, \tag{11.6}$$

$$\beta = \frac{v\Delta t}{\Delta y}, \qquad \widehat{\beta} = \beta - q, \tag{11.7}$$

$$\gamma = \frac{w\Delta t}{\Delta z}, \qquad \widehat{\gamma} = \gamma - r, \tag{11.8}$$

where γ is the z direction's displacement and r is the nearest integer in the z direction to the departure point.

One way of applying the trilinear interpolation is to first interpolate along the x direction, which could be imagined as pushing the face of the cube defined by C_{1jk} to the opposing face defined by C_{0jk}, which gives

$$c_{00} = (1-\widehat{\alpha})\,C_{p,q-1,r-1} + \widehat{\alpha}C_{p-1,q-1,r-1}.$$

This approach is also applied along the other three x directions on the two z levels such that

$$\begin{aligned} c_{01} &= (1-\widehat{\alpha})\,C_{p,q-1,r} + \widehat{\alpha}C_{p-1,q-1,r}, \\ c_{10} &= (1-\widehat{\alpha})\,C_{p,q,r-1} + \widehat{\alpha}C_{p-1,q,r-1}, \\ c_{11} &= (1-\widehat{\alpha})\,C_{p,q,r} + \widehat{\alpha}C_{p-1,q,r}. \end{aligned}$$

Thus we have four points to perform two linear interpolations in the y directions on each z level such that

$$c_0 = \left(1-\widehat{\beta}\right)c_{10} + \widehat{\beta}c_{00}, \qquad c_1 = \left(1-\widehat{\beta}\right)c_{11} + \widehat{\beta}c_{01}.$$

Finally, we apply a linear interpolation between the two points from the y interpolation for the points on each z level such that

$$d = (1-\widehat{\gamma})\,c_1 + \widehat{\gamma}c_0.$$

We should note that it does not matter in which order we perform the interpolations, we could have gone a long the y axis first and then along the x to finish at z. Another feature to note here is that the trilinear interpolation is identical to two bilinear interpolations combined with a linear interpolation. We have an illustration of where the new points end up to find d in Fig. 11.2.

In the expressions above, the Cs are referring to the values of ψ at those corner locations, while cs are referring to the interpolated values so far of ψ at those locations in the cube.

Exercise 11.1. *Expand the terms for the coefficients above to find the one equation for the trilinear Lagrange interpolation polynomial. Given the expression above, using the von Neumann stability approach, find the stability condition for the trilinear interpolation.*

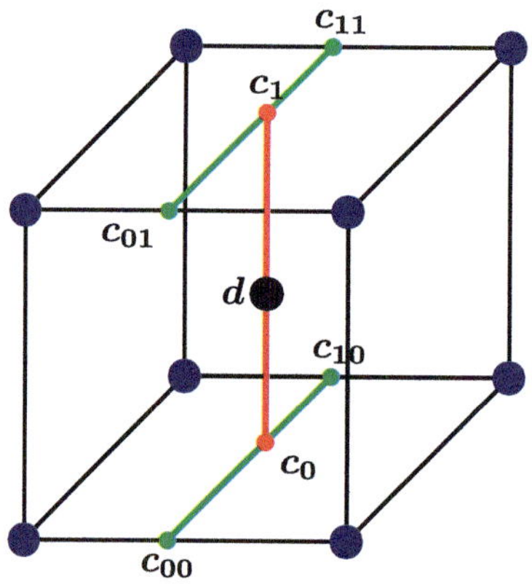

FIGURE 11.2 Depiction of the three-dimensional interpolation.

However, we have to recall that the linear interpolation can have a damping effect for smaller wave numbers, especially if the departure point is halfway between grid points. In applications in numerical weather and climate prediction, it is quite often the case that some form of tricubic interpolation is applied, either with the Lagrange interpolation polynomial or some form of tricubic splines. The general expression for the tricubic interpolation polynomial is given by

$$\psi\left(x, y, z\right) = \sum_{i=0}^{3}\sum_{j=0}^{3}\sum_{k=0}^{3} a_{i,j,k} x^i y^j z^k, \tag{11.9}$$

where we require 64 coefficients $a_{i,j,k}$, which will be functions of the fractional displacements $\widehat{\alpha}$, $\widehat{\beta}$, and $\widehat{\gamma}$.

We have a diagram of the equidistant grid of a cube that would be needed to form the tricubic interpolation polynomial in Fig. 11.3, where each intersecting line would be a grid point that would be used to form the interpolation polynomial.

If we let $s\left(i, j, k\right)$ be the field at the grid point $\left(i, j, k\right)$, and if we have the same upstream approach as for the cubic and bicubic interpolation polynomials where $i, j, k \in \{-2, -1, 0, 1\}$, then the procedure to fit the polynomial would be similar to that of the trilinear approach, but now interpolating in the following way:

1. For each vertical level fit a cubic interpolation polynomial such that $t\left(i, j, \widehat{\gamma}\right) = \text{CUBE}_z\left(s\left(i, j, -2\right), s\left(i, j, -1\right), s\left(i, j, 0\right), s\left(i, j, 1\right)\right)$.
2. Given these interpolated points for the z direction, fit a cubic interpolation polynomial in the y direction to these points such that $p\left(i, \widehat{\beta}, \widehat{\gamma}\right) = \text{CUBE}_y\left(t\left(i, -2, \widehat{\gamma}\right), t\left(i, -1, \widehat{\gamma}\right), t\left(i, 0, \widehat{\gamma}\right), t\left(i, 1, \widehat{\gamma}\right)\right)$.
3. Finally, given the interpolated points in the y and z directions, fit a cubic interpolation polynomial in the x direction such that $\psi\left(\widehat{\alpha}, \widehat{\beta}, \widehat{\gamma}\right) = \text{CUBE}_x\left(p\left(-2, \widehat{\beta}, \widehat{\gamma}\right), p\left(-1, \widehat{\beta}, \widehat{\gamma}\right), p\left(0, \widehat{\beta}, \widehat{\gamma}\right), p\left(1, \widehat{\beta}, \widehat{\gamma}\right)\right)$.

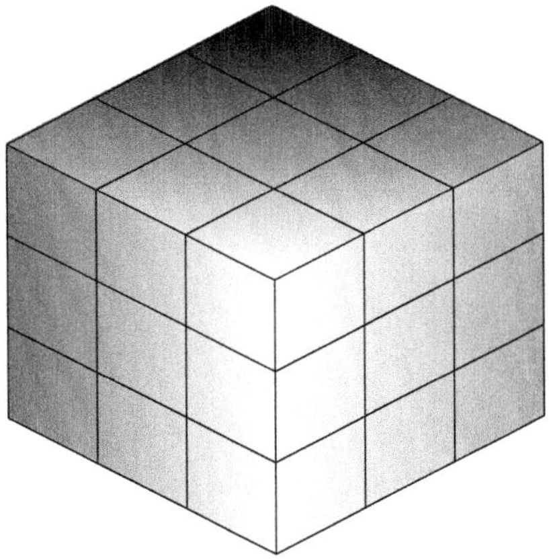

FIGURE 11.3 Diagram of the cubes that would be needed to form the 64 grid points to fit a tricubic interpolation polynomial, where each intersection point would be a grid point.

This procedure would require 16 interpolations in the vertical direction, followed by four in the meridional, followed by 1 in the zonal direction, which implies 21 interpolations, or calls of the function CUBE.

Given this setup for the constant velocity case, we now move on to consider more realistic models for numerical weather and ocean prediction. Over the previous chapters on semi-Lagrangian development we have considered the shallow water equations in Cartesian coordinates for both the one- and two-dimensional case. However, as the name suggests, we do not have shallow water equations in three dimensions. Therefore, we now look at the primitive equations in some of their forms.

11.2 Semi-Lagrangian advection in the primitive equations

In this chapter we shall provide an overview of the semi-implicit semi-Lagrangian integration that was implemented into the Canadian Regional Finite-Element Forecast model [185]. The model comprises the hydrostatic primitive equations on a polar-stereographic projection using the sigma, σ, normalized pressure, vertical coordinate. The equations comprise the horizontal momentum, thermodynamic continuity, vorticity, and moisture equations, which are given by

$$\frac{DU}{Dt} + \overline{\frac{\partial P}{\partial X}}^t = r_1^n, \tag{11.10}$$

$$\frac{Dv}{Dt} + \overline{\frac{\partial P}{\partial Y}}^t = r_2^n, \tag{11.11}$$

$$\frac{DT'}{Dt} - \frac{RT^*}{c_p}\left(\frac{D\ln p_s}{Dt} + \left(\frac{\overline{\dot{\sigma}}^t}{\sigma}\right)\right) = r_3^n, \tag{11.12}$$

$$\frac{D \ln p_s}{Dt} + \overline{\delta + I}^t = 0, \quad (11.13)$$

$$\frac{D\xi}{Dt} = r_4^n, \quad (11.14)$$

$$\frac{Dq}{Dt} = 0, \quad (11.15)$$

where

$$r_1 = fV - \left(\frac{U^2+V^2}{2}\right)\frac{\partial S}{\partial X} - RT'\frac{\partial \ln p_s}{\partial X}, \quad (11.16)$$

$$r_2 = -fU - \left(\frac{U^2+V^2}{2}\right)\frac{\partial S}{\partial Y} - RT'\frac{\partial \ln p_s}{\partial Y}, \quad (11.17)$$

$$r_3 = -\frac{RT'}{c_p}\left((\delta + I) - \frac{\dot{\sigma}}{\sigma}\right), \quad (11.18)$$

$$r_4 = S\left(-\frac{\partial}{\partial X}\left(RT'\frac{\partial \ln p_s}{\partial Y}\right) + \frac{\partial}{\partial Y}\left(RT'\frac{\partial \ln p_s}{\partial X}\right)\right) - \xi\left(S\left(\frac{\partial U}{\partial X} + \frac{\partial V}{\partial Y}\right)\right) + S\left(\frac{\partial U}{\partial \sigma}\frac{\partial \dot{\sigma}}{\partial Y} - \frac{\partial V}{\partial \sigma}\frac{\partial \dot{\sigma}}{\partial X}\right). \quad (11.19)$$

The symbols above have been summarized in Table 11.1.

TABLE 11.1 Summary of the terms in the primitive equations.

Symbol	Name
$\sigma = \frac{p}{p_s}$	vertical coordinates
p	pressure
p_s	surface pressure
$\dot{\sigma} \equiv \frac{d\sigma}{dt}$	
$I \equiv \frac{\partial \dot{\sigma}}{\partial \sigma}$	
$P \equiv \phi + RT^* \ln p_s$	
ϕ	geopotential
T	virtual temperature $\equiv T^* + T' = 300\text{ K} + T'$
q	specific humidity
$S = m^2$	
$m = \dfrac{1+\sin\frac{\pi}{3}}{1+\sin\theta}$	map scale factor
$(U, V) = \left(\frac{u}{m}, \frac{v}{m}\right)$	wind images
(u, v)	wind components along the (X, Y) axes of the polar stereographic projection
$\delta = S(U_X + V_Y)$	horizontal divergence
$\xi = S(U_X - V_y) + f$	vertical component of absolute vorticity
R	gas constant
c_p	specific heat of air

The temporal operators are defined for the field F as

$$\frac{DF}{Dt} \equiv \frac{F^{n+1} - F^{n-1}}{2\Delta t} + A\left(F^{n}\right), \tag{11.20a}$$

$$\overline{F}^{t} \equiv \frac{F^{n+1} + F^{n-1}}{2}, \tag{11.20b}$$

where

$$A\left(F^{n}\right) \equiv \left(S\left(U\frac{\partial F}{\partial X} + V\frac{\partial F}{\partial Y}\right) + \dot{\sigma}\frac{\partial F}{\partial \sigma}\right)(X, Y, \sigma, t),$$

is the three-dimensional advection of F, and

$$F^{n+1} \equiv F\left(X, Y, \sigma, t^{n+1}\right), \tag{11.21a}$$

$$F^{n-1} \equiv F\left(X, Y, \sigma, t^{n-1}\right), \tag{11.21b}$$

$$F^{n} \equiv F\left(X, Y, \sigma, t^{n}\right). \tag{11.21c}$$

The space discretization is achieved through a finite element integration, which consists of expanding the dependent variables of the problem in terms of piecewise-defined polynomial base where the details can be found in Staniforth (1987) [175]. We now consider the formulation of the solution to the equations above using the Eulerian notation, where we will use these to complete the semi-Lagrangian integration afterwards:

$$\frac{\delta^{n+1}}{S} + \Delta t\left(\frac{\partial^2 P^{n+1}}{\partial X^2} + \frac{\partial^2 P^{n+1}}{\partial Y^2}\right) = R_1, \tag{11.22}$$

$$T'^{n+1} - \left(\frac{RT^*}{c_p}\left(\ln p_s + \Delta t\frac{\dot{\sigma}}{\sigma}\right)\right)^{n+1} = R_2, \tag{11.23}$$

$$(\ln p_s)^{n+1} + \Delta t\,(\delta + I)^{n+1} = R_3, \tag{11.24}$$

$$\xi^{n+1} = R_4, \tag{11.25}$$

$$q^{n+1} = R_5, \tag{11.26}$$

where the right-hand sides of the equations above are given by

$$\begin{aligned} R_1 \equiv\; & \frac{\partial}{\partial X}\left(\left(U - \Delta t\frac{\partial P}{\partial X}\right)^{n-1} + 2\Delta t\,(r_1 - A(U))^{n}\right) \\ & + \frac{\partial}{\partial Y}\left(\left(V - \Delta t\frac{\partial P}{\partial Y}\right)^{n-1} + 2\Delta t\,(r_2 - A(V))^{n}\right), \end{aligned} \tag{11.27}$$

$$R_2 \equiv \left(T' - \frac{RT^*}{c_p}\left(\ln p_s - \Delta t\frac{\dot{\sigma}}{\sigma}\right)\right)^{n-1} + 2\Delta t\left(r_3 - A\left(T'\right) + \frac{RT^*}{c_p}A\left(\ln p_s\right)\right)^{n}, \tag{11.28}$$

$$R_3 \equiv (\ln p_s - \Delta t\,(\delta + I))^{n-1} - 2\Delta t\, A\,(\ln p_s)^{n}, \tag{11.29}$$

$$R_4 \equiv \xi^{n-1} + 2\Delta t \left(r_4 - A(\xi)\right)^n, \tag{11.30}$$

$$R_5 \equiv q^{n-1} - 2\Delta t A(q)^n. \tag{11.31}$$

In [185] the authors state that the hydrostatic equation $\frac{\partial \phi}{\partial \sigma} = -\frac{RT}{\sigma}$ is vertically discretized as

$$P_{k+1}^{n+1} - P_k^{n+1} = -R\left(\frac{T_{k+1}^{n+1} + T_k^{n+1}}{2}\right)\ln\left(\frac{\sigma_{k+1}}{\sigma_k}\right), \tag{11.32}$$

where (11.32) is valid for $1 \leq k \leq N-1$; at the lowest vertical level, we have $P_N^{n+1}\phi_s + RT^* \ln p_s^{n+1}$, ϕ_s is the geopotential height of the topography, N is the number of vertical levels, and the levels are numbered from the top $\sigma = \sigma_1$ to the bottom $\sigma = \sigma_N = 1$.

At this point Tanguay et al. introduce a new variable W which is related to the divergence as $\delta = -\frac{\partial W}{\partial \sigma}$, where the vertical discretization is

$$\left(W_{k+1}^{n+1} - W_k^{n+1}\right) = -\frac{\delta_{k+1}^{n+1} + \delta_k^{n+1}}{2}\left(\sigma_{k+1} - \sigma_k\right), \quad 1 \leq k \leq N-1, \tag{11.33}$$

and $W_1^{n+1} = \sigma_1 W_N^{n+1}$.

The vertical discretization of the variable I is similar to that above and is given by

$$\left(\dot{\sigma}_{k+1}^{n+1} - \dot{\sigma}_k^{n+1}\right) = \frac{I_{k+1}^{n+1} + I_k^{n+1}}{2}\left(\sigma_{k+1} - \sigma_k\right), \quad 1 \leq k \leq N-1, \tag{11.34}$$

and $\dot{\sigma}_1^{n+1} = \dot{\sigma}_N^{n+1} = 0$.

It is stated in [185] that the system formed by (11.22)–(11.34) can be shown to be equivalent to resolving a three-dimensional Helmholtz equation for the variable W with backsubstitution to obtain the variables of the model $\left(\delta, T', \dot{\sigma}, I, \ln p_s, \xi, q, U, V\right)$. The wind images U and V are obtained from the streamfunction and potential velocity χ. Two elliptic equations,

$$\nabla_H^2 \psi = \xi + f \quad \text{and} \quad \nabla_H^2 \chi = \delta,$$

are solved to obtain the fields ψ and χ at time t^{n+1}, where ∇_H^2 is the horizontal Laplacian. The boundary conditions are $\psi = 0$ and $\frac{\partial \chi}{\partial n} = 0$, where $\frac{\partial}{\partial n}$ is the normal derivative at the boundary, and these conditions correspond to a wall at the lateral boundaries.

We now consider the physical processes where, after the adiabatic step, forcing terms are added to the momentum, temperature, and humidity equations to represent **subgrid-scale** effects and parameterized processes. The horizontal diffusion as stated in [185] is rather weak and is discretized as

$$\frac{F^H - F^{n+1}}{2\Delta t} = \nu_H \nabla_H^2 F^{n+1},$$

where ν_H is a constant, and F^{n+1} denotes the momentum, temperature, or humidity variables after the adiabatic step described above. The vertical diffusion is treated implicitly as

$$\frac{F^V - F^H}{2\Delta t} = \frac{\partial}{\partial \sigma}\left(\nu_V \frac{\partial F^V}{\partial \sigma}\right),$$

where ν_V is a function of space and time.

11.2.1 3D semi-Lagrangian integration

For the semi-Lagrangian integration of the primitive equation presented earlier, the temporal operators (11.20a)–(11.21c) are redefined to be

$$\frac{DF}{Dt} = \frac{F^{n+1} - F^{n-1}}{2\Delta t}, \tag{11.35a}$$

$$\overline{F}^t = \frac{F^{n+1} + F^{n-1}}{2}, \tag{11.35b}$$

where now we have the field F at the different departure points at the three time levels as

$$F^{n+1} = F\left(X, Y, \sigma, t^{n+1}\right), \tag{11.36a}$$

$$F^{n} = F\left(X - \alpha, Y - \beta, \sigma - \gamma, t^{n}\right), \tag{11.36b}$$

$$F^{n-1} = F\left(X - 2\alpha, Y - 2\beta, \sigma - 2\gamma, t^{n-1}\right). \tag{11.36c}$$

We can see that the expressions in (11.36a)–(11.36c) are functions of the **three-dimensional displacements,** (α, β, γ), where here the advection velocity vector, given by

$$\mathbf{U} = \begin{pmatrix} U & V & \dot{\sigma} \end{pmatrix}^T,$$

is **not constant** and we now have to iteratively solve for the three displacements through the following iterative procedure:

$$\alpha^{(l+1)} = \Delta t\,(SU)\left(X - \alpha^{(l)}, Y - \beta^{(l)}, \sigma - \gamma^{(l)}, t^{n}\right), \tag{11.37a}$$

$$\beta^{(l+1)} = \Delta t\,(SV)\left(X - \alpha^{(l)}, Y - \beta^{(l)}, \sigma - \gamma^{(l)}, t^{n}\right), \tag{11.37b}$$

$$\gamma^{(l+1)} = \Delta t\dot{\sigma}\left(X - \alpha^{(l)}, Y - \beta^{(l)}, \sigma - \gamma^{(l)}, t^{n}\right), \tag{11.37c}$$

where l is the number of iterations, and the first guess for the displacement computed at the preceding timestep. It is stated in [185], and as we have seen in other papers, that in general two iterations are sufficient. The next step is to use a three-dimensional cubic spline interpolation to evaluate the wind at the midpoint $\left(X - \alpha^{(l)}, Y - \beta^{(l)}, \sigma - \gamma^{(l)}, t^{n}\right)$ of the trajectory.

This is then followed by evaluating the R_i terms in (11.27)–(11.31) ignoring the advection terms. The advection is implicitly included by evaluating the n and $n-1$ terms through

(11.36b) and (11.36c). It is stated in [185] that the implementation of this form of three-dimensional semi-implicit semi-Lagrangian integration allowed for the Canadian model to run with larger time steps and remain stable, but also that it allowed for a large increase in the horizontal resolution from 150 to 100 km.

11.3 3D flux form semi-Lagrangian

As we mentioned in the last chapter, it is possible to introduce one- and two-dimensional flux-form semi-Lagrangian integration for a chemical transport model, which is currently in use at NASA in the Goddard Earth Observing System General Circulation Model and Data Assimilation System (GOES-GCM) and (GEOS-DAS). In [113] the authors state that in global general circulation, or chemistry transport, model, the large scale motions are **quasihorizontal** and nearly incompressible.

In [113] the authors state that the vertical velocity, as diagnosed from the hydrostatic continuity equation, is more than an order of magnitude smaller than the horizontal velocity. The vertical transport can, therefore, be **decoupled** from the horizontal transport and computed **independently**, where this is implying that the cross-derivative terms associated with the vertical advection are ignored.

Now let H denote the 1D operator for the vertical transport, similar to the F and G operators defined in the last chapter. Thus the 3D transport flux-form semi-Lagrangian integration can be written as

$$q^{n+1} = \frac{1}{\pi^{n+1}}\left(\pi^n q^n + F\left[q^n + \frac{1}{2}g\left(q^n\right)\right] + G\left[q^n + \frac{1}{2}f\left(q^n\right)\right] + H\left(q^n\right)\right). \tag{11.38}$$

In computing the mass convergence terms, F and G, the time-centered, or time-averaged, winds $\left(\overline{u}^t, \overline{v}^t\right)$ and surface pressure $\overline{\pi}^t$ defined at time $n+\frac{1}{2}$ are to be used. We can verify that condition II from earlier is satisfied here by setting $q=1$ in (11.38) and integrating from the top to the bottom of the σ domain to obtain

$$\begin{aligned} \pi^{n+1} &= \pi^n \int_{\sigma_{\text{top}}}^{\sigma_{\text{bottom}}} \left(F\left(1\right) + G\left(1\right)\right) d\sigma \\ &\approx \pi^n \Delta t \sum \left(\frac{\delta_x \overline{\overline{\pi}^t}^x \overline{u}^t}{\Delta x} + \frac{\delta_y \overline{\overline{\pi}^t}^y \overline{v}^t}{\Delta y}\right)\Delta\sigma_k, \end{aligned} \tag{11.39}$$

where the summation is over the whole σ coordinate domain. It is then stated in [113] that (11.39) is simply a second-order center-difference scheme for updating the surface pressure on a C grid.

The derivation above in summary form will be expanded upon in the next chapter, as this theory is what will go on to become what is referred to as the **finite volume three**, or **FV3**, model which will be the next generation of models for the numerical weather prediction of NOAA.

11.4 Three-dimensional fully elastic Euler equations with semi-Lagrangian

In 1999 there was a paper by Caya and LaPrise [32], explaining the next generation of the Canadian Regional Climate Model CRCM, where now they are using the fully elastic three-dimensional Euler equations to describe the atmospheric flow on the rotating Earth. As with the primitive equations earlier, the model projects onto the local Cartesian coordinates, and with a different vertical coordinate.

The equations here are now given by:

$$\frac{DU}{Dt} = fV - K\frac{\partial S}{\partial X} - RT\frac{\partial q}{\partial X} + F_X, \tag{11.40}$$

$$\frac{DV}{Dt} = -fU - K\frac{\partial S}{\partial Y} - RT\frac{\partial q}{\partial Y} + F_Y, \tag{11.41}$$

$$\frac{Dw}{Dt} = -g - RT\frac{\partial q}{\partial z} + F_z, \tag{11.42}$$

$$\frac{Dq}{Dt} = \left(-S\left(\frac{\partial U}{\partial X} + \frac{\partial V}{\partial Y}\right) - \frac{\partial w}{\partial z} + \frac{L}{T}\right)(1-\kappa)^{-1}, \tag{11.43}$$

$$\frac{DT}{Dt} = \kappa T\frac{Dq}{Dt} + L, \tag{11.44}$$

$$\frac{DM}{Dt} = E, \tag{11.45}$$

where

U, V, w, T, and M are the velocity components along the X, Y, and z coordinates, temperature, and specific humidity, respectively;

$q \equiv \ln\frac{p}{p_0}$ where p is pressure and p_0 is a constant;

$K \equiv \frac{U^2 + V^2}{2}$ is the specific pseudokinetic energy;

L and E represent heat and moisture sources or sinks;

F_X, F_Y, and F_z correspond to sources or sinks of momentum for their respective wind components;

all the other symbols are as defined earlier. The total derivative in conformal projection is interpreted as

$$\frac{D}{Dt} \equiv \frac{\partial}{\partial t} + S\left(U\frac{\partial}{\partial X} + V\frac{\partial}{\partial Y}\right) + w\frac{\partial}{\partial z}, \tag{11.46}$$

where the (X, Y) components of the polar stereographic coordinates are defined as follows:

$$\begin{aligned} dX &= -m\,(dx\sin\lambda + dy\cos\lambda), \\ dY &= m\,(dx\cos\lambda - dy\sin\lambda). \end{aligned} \tag{11.47}$$

The image wind components (U, V) are defined in terms of the true horizontal wind components (u, v) expressed in a local (x, y, z) Cartesian reference system as

$$
\begin{aligned}
U &= -m^{-1}\left(u \sin\lambda + v\cos\lambda\right), \\
V &= m^{-1}\left(u\cos\lambda - v\sin\lambda\right).
\end{aligned}
\tag{11.48}
$$

Vertical coordinate

In [32] the authors state that (11.40)–(11.45) are modified by the introduction of a terrain-following vertical coordinates employed to facilitate the implementation of the topography in the model. The vertical coordinate used in this model is the Gal–Chen coordinate defined as

$$
Z(X, Y, z) = \left(\frac{z - h_0(X, Y)}{H - h_0(X, Y)}\right) H, \tag{11.49}
$$

where Z is the scaled height coordinate, $h_0(X, Y)$ is the topographic height, and H is the top of the model atmosphere. This coordinate is defined such that the surface $Z = 0$ corresponds to the ground $z = h_0(X, Y)$, and the surface $Z = H$ is at a constant height $z = H$. The advantage of this coordinate is that the kinematic boundary condition at the surface, expressed by the following relation in z-coordinate form:

$$
w(X, Y, z = h_0, t) = \mathbf{U}(X, Y, z = h_0, t) \cdot \nabla h_0, \tag{11.50}
$$

takes the form of a homogenous boundary condition in the Z coordinate:

$$
W(X, Y, Z = 0, t) = \left(\frac{DZ}{DT}\right)_{(X,Y,Z=0,t)} = 0. \tag{11.51}
$$

The equations are recast using the new vertical coordinate system as

$$
\begin{aligned}
\frac{DU}{Dt} &= fV - K\frac{\partial S}{\partial X} - RT\left(\frac{\partial q}{\partial X} + \frac{G_1}{G_0}\frac{\partial q}{\partial Z}\right) + F_X, && (11.52)\\
\frac{DV}{Dt} &= -fU - K\frac{\partial S}{\partial Y} - RT\left(\frac{\partial q}{\partial Y} + \frac{G_2}{G_0}\frac{\partial q}{\partial Z}\right) + F_Y, && (11.53)\\
\frac{Dw}{Dt} &= -g - \frac{RT}{G_0}\frac{\partial q}{\partial z} + F_z, && (11.54)\\
(1-\kappa)\frac{Dq}{Dt} &= S\left(F_1 U + F_2 V\right) - S\left(\frac{\partial U}{\partial X} + \frac{\partial V}{\partial Y}\right) - \frac{1}{G_0}\left(\frac{\partial G_0 W}{\partial Z}\right) + \frac{L}{T}, && (11.55)\\
\frac{DT}{Dt} &= \kappa T\frac{Dq}{Dt} + L, && (11.56)\\
\frac{DM}{Dt} &= E, && (11.57)\\
W &= \frac{S\left(G_1 U + G_2 V\right) + w}{G_0}, && (11.58)
\end{aligned}
$$

where now we have

$$
\begin{aligned}
F_1 &= \frac{1}{G_0 H}\left(\frac{\partial h_0}{\partial X}\right), &(11.59)\\
F_2 &= \frac{1}{G_0 H}\left(\frac{\partial h_0}{\partial Y}\right), &(11.60)\\
G_0 &= \frac{H-h_0}{H}, &(11.61)\\
G_1 &= -\frac{(H-Z)}{H}\left(\frac{\partial h_0}{\partial X}\right), &(11.62)\\
G_2 &= -\frac{(H-Z)}{H}\left(\frac{\partial h_0}{\partial Y}\right). &(11.63)
\end{aligned}
$$

The total derivative must be now be interpreted as follows:

$$
\frac{D}{Dt} \equiv \left(\frac{\partial}{\partial t}\right)_Z + S\left(U\left(\frac{\partial}{\partial X}\right)_Z + V\left(\frac{\partial}{\partial Y}\right)_Z\right) + W\frac{\partial}{\partial Z}. \tag{11.64}
$$

It is stated in [32] that the generalized vertical velocity in the new coordinate system is *interpreted* as

$$
\begin{aligned}
W &\equiv \frac{DZ}{Dt}\\
&= \left(\frac{\partial Z}{\partial t}\right)_z + S\left(U\left(\frac{\partial Z}{\partial X}\right)_z + V\left(\frac{\partial Z}{\partial Y}\right)_z\right) + w\frac{\partial Z}{\partial z}. &(11.65)
\end{aligned}
$$

The next part of [32] is concerned with the implementation of the semi-implicit approach to control the spurious gravity modes, where they split the temperature and q field, which we have to recall is **not** the moisture field here, into two parts consisting of a basic state (T^*, q^*) and the deviation away from this state (T', q'). The basic state mentioned in [32] is an isothermal atmosphere in hydrostatic equilibrium. We do not present the semi-implicit derivation, but move onto the semi-Lagrangian implementation in three-dimensions.

The semi-Lagrangian integration procedure in this case consists of evaluating the total derivative as

$$
\frac{D\psi}{Dt} = \frac{\psi\left(X, Y, Z, t^{n+1}\right) - \psi\left(X - 2\alpha, Y - 2\beta, Z - 2\gamma, t^{n-1}\right)}{2\Delta t}. \tag{11.66}
$$

The displacements (α, β, γ) are calculated by solving iteratively again the following implicit coupled system:

$$
\alpha(X, Y, Z, t) = \Delta t S U(X - \alpha, Y - \beta, Z - \gamma, t), \tag{11.67a}
$$

$$
\beta(X, Y, Z, t) = \Delta t S V(X - \alpha, Y - \beta, Z - \gamma, t), \tag{11.67b}
$$

$$
\gamma(X, Y, Z, t) = \Delta t W(X - \alpha, Y - \beta, Z - \gamma, t), \tag{11.67c}
$$

Once we have the displacement values to calculate the three-dimensional departure point, we have to apply some form of interpolation polynomial to the field ψ to this point. One technique that ECMWF has employed here is to assume that we have a set of equidistant points around the departure point, and then only to apply the cubic interpolation using the rows/columns in the immediate vicinity of the departure point, and finally, use a linear interpolation for the points further away [160]. We have an illustration of this approach in Fig. 11.4, where the departure point is the yellow sphere.

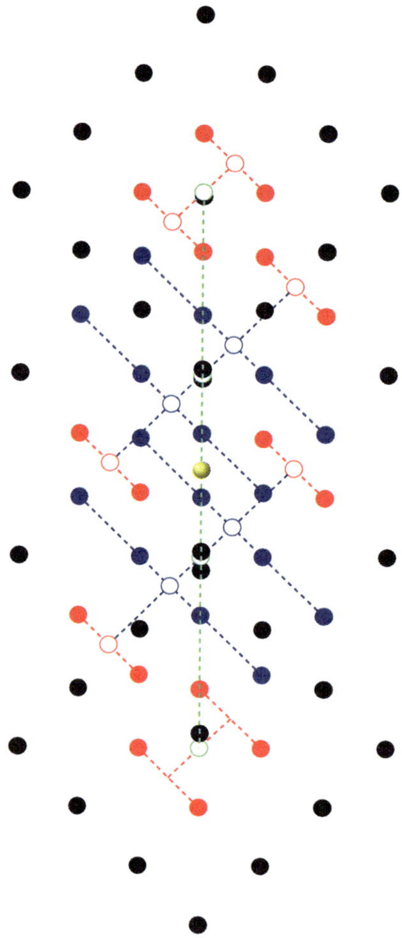

FIGURE 11.4 Schematic of the points that are used for the ECMWF quasimonotone, quasicubic three-dimensional interpolation for the field to the grid point, where red is linear interpolation, blue is cubic, and green is the cubic interpolation in the vertical, where the departure point is the yellow sphere. The black points are not used for this interpolation, but would be for the full tricubic interpolation.

The approach depicted in Fig. 11.4 is referred to as a **quasimonotone, quasicubic** Lagrange interpolation. As we saw earlier, there are 64 points involved if we were to implement a true tricubic Lagrange interpolation of the field to the departure point. To save computation costs,

ECMWF only uses a cubic Lagrange interpolation for the nearest points in all three directions, which creates a three-dimensional plus sign for the points used. We can see in Fig. 11.4 that on the top and the bottom z layers we only require points here, and apply 3 linear interpolations; 2 in the x direction and 1 in the y direction, taking our linear interpolation count to 6.

For the two layers on either side of the departure point, we apply two cubic interpolations in the x direction on the nearest points, and then two linear interpolations for the points on either edge of the plane in x direction here. In the y-direction, we apply a cubic interpolation through the 4 points we have in the x-direction on both layers to obtain the final two points we need to apply a cubic interpolation in the z direction. This means that we have three cubic interpolations on each of these two levels, taking that count to 6, and we have four more linear interpolations, taking the count to 10. When we apply the final interpolation, which is a cubic Lagrange in the vertical, we have applied 10 linear interpolations, and 7 cubic, which is much less than the 21 we require for the full tricubic interpolation. In Fig. 11.4 the black points are the extra points that would be used to fit the full tricubic Lagrange interpolation polynomial. Thus the total number of interpolation points is halved to 32. In Fig. 11.5 we have a schematic of the interpolation points, as well as the interpolated points to form the interpolation point for the z direction interpolation, for one of the two levels near the departure point.

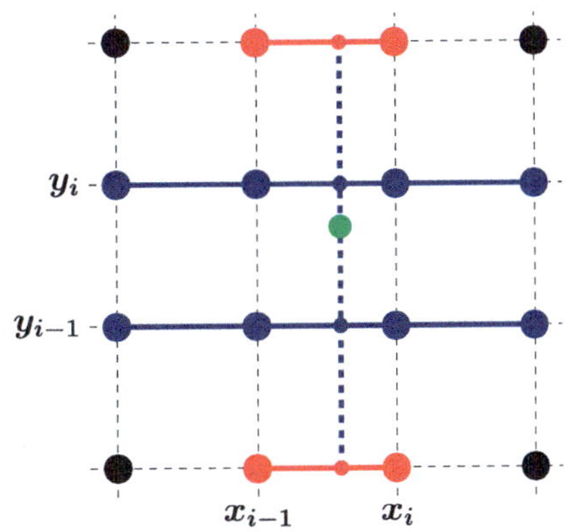

FIGURE 11.5 Schematic of the points that are used for the ECMWF quasimonotone, quasicubic three-dimensional interpolation on a z level near the departure point, where the red lines indicate linear interpolation, and the blue lines represent cubic interpolations, where the smaller circles are the points for the cubic interpolation in the y direction, and the green circle is the interpolation for this level to be used in the final z direction cubic interpolation.

Returning to the next step of the implementation of a three-dimensional semi-Lagrangian integration into the Euler equations [32], the authors introduce the definition of the three-dimensional averaging in time for the linear components for the average at forecast time $\left(\psi_{X,Y,Z,t^{n+1}}\right)$ and their upstream values at past time $\left(\psi_{X-2\alpha,Y-2\beta,Z-2\gamma,t^{n-1}}\right)$. They introduce a *partly uncentered time average* of the following form:

$$\overline{\psi}^t \equiv \frac{(1-\epsilon)\,\psi\left(X,Y,Z,t^{n+1}\right)+(1-\epsilon)\,\psi\left(X-2\alpha,Y-2\beta,Z-2\gamma,t^{n-1}\right)}{2}, \tag{11.68}$$

where, as we have seen before, ϵ represents the degree by which the average is uncentered. It is then stated that this approach is used to reduce the problems of instability of topographic waves associated with rapid flows when using large time steps, as we mentioned earlier.

The last approximation that is introduced in the derivation in [32] is associated with the right-hand sides, R_i, or rather the nonlinear components of the semi-implicit semi-Lagrangian implementation where these terms are evaluated based upon current time values $\psi(t^n)$ but as uncentered averages of their values at the end points of the Lagrangian trajectory, which comes from [186] as follows:

$$\overline{R_i}^{traj} \equiv \frac{(1+\epsilon)\, R_i\,(X, Y, Z, t) + (1-\epsilon)\, R_i\,(X - 2\alpha, Y - 2\beta, Z - 2\gamma, t)}{2}. \tag{11.69}$$

We will not present the implementation of all of the approach described above; the reader is referred to [32] for the exact expressions to be coded. One comment we will make about the implementation is that Caya and LaPrise do the elimination of variables to obtain one Helmholtz equation for the pressure increment, but this time it is in three dimensions.

We shall finish this section with an explanation of the physical processes that are taken into account in the formulation presented in this section; mainly terms F_X, F_Y, F_Z, L, and E. Starting with the F terms, these are introduced to represent the effect of the **subgrid-scale** processes affecting the momentum variables. These processes are the vertical turbulence fluxes of momentum, the gravity wave drag, and the horizontal diffusion. Vertical turbulent fluxes of horizontal momentum are evaluated as a stability-dependent diffusion using an eddy diffusivity formulation in the free atmosphere with a drag coefficient at the surface. The effect of **unresolved** orographic gravity waves is parameterized as an additional drag force on the flow.

Once an adiabatic time step of the dynamics has been completed, followed by the correction to incorporate physical forcings, the horizontal diffusion is said to have been applied in the third place in the time splitting procedure in this model. The horizontal diffusion is applied in the general form

$$\psi_H = \psi + 2\Delta t k \nabla^{2m} \psi, \tag{11.70}$$

where k and m are free to be adjusted, and the subscript H represents the variable after diffusion.

The L term in the equations represents combined adiabatic heating, resulting from the effects of radiation, both solar and terrestrial, the release and absorption of latent heat, and the turbulent and convective vertical heat fluxes. The surface albedo is defined for two spectral bands and is a function of surface type, humidity, snow cover, and snow wage. Cloud effects are also taken into account in this term, which are determined diagnostically. Latent heat released by condensation, or freezing, of water vapor in a grid box is also included in this term.

The E term deals with the vertical heat fluxes of moisture in a similar way to that of temperature. The water in the atmosphere is controlled in one part by precipitation, which is a sink, and by evaporation, which is a source, from the surface. The convective flux of water vapor associated with the convective adjustment between two layers is the minimum required to saturate the upper layer or to remove supersaturation from the bottom layer.

As the semi-Lagrangian scheme implemented in this model is the **three-time-level** scheme, through a central difference, the finite difference scheme introduces a computational mode which is not associated with the physical dynamical system, and as such this mode needs to be controlled in some form to prevent spurious fast oscillations that are generated by the scheme. The way this mode is controlled is through what is referred to as a **time filter**. We shall briefly present a very well known filter that was updated recently due to an instability it was causing.

11.4.1 Time filters

As we saw earlier, the three-time-level scheme used in the Canadian regional climate model generates a computational, or parasitic mode, which is not one of the physical modes of the continuous system. The three-level-time level scheme is equivalent to using the leap-frog scheme and, according to [201], there are more accurate schemes that could be used, for example, the Runge–Kutta or Adams–Bashforth schemes.

In [201] Williams states that the most widely used mechanism to control this spurious mode is to apply a time filter during the time-stepping procedure. The original design of the filter came from the same Robert that initiated the semi-Lagrangian approach, as well as the semi-implicit treatment of the fast waves, in 1966 [163] where he designed a filter for the leap-frog scheme. In 1972 Asselin showed that the time filter selectively suppresses the computational mode but leaves the physical mode relatively undamped at low frequencies [5]. This filter was referred to for quite a while as the **Robert–Asselin filter**. In [32] the authors state that the version they are using to control this spurious mode is given by

$$\psi_t^n = \psi_N^n + \chi\left(\psi_N^{n+1} - 2\psi_N^n + \psi_t^{n-1}\right), \tag{11.71}$$

where the subscript t is referring to a variable that has been time-filtered, while subscript N means that the filed has not been time-filtered. The variable χ is referred to as the filtering coefficient. We should note that the numerical approximation above is similar to a central difference to the second-order time derivative of ψ.

In [201] Williams states that the Robert–Asselin filter is used in many different applications from the atmosphere to the ocean and climate, as well as the modeling of fluids in rotating annulus in a laboratory. However, Williams points out two problems with this filter:

> In addition to suppressing the computational mode, the scheme also weakly suppresses the physical mode. Therefore, physical quantities, for example, energy, that are conserved by the time-continuous equations are not necessarily conserved by the time-discretized equations when the filter is activated. The damping and nonconservation may be benign for sufficiently short integrations, but possibly not for longer ones.
> Second, the Robert–Asselin filter severely degrades the leap-frog scheme's numerical accuracy, measured as the rate at which the error tends to zero as the time step is progressively refined.

Williams provides an illustration of what the filter is trying to do: if we have three values of a dependent variable x at three successive and equally spaced times, t^{n+1}, t^n, and t^{n-1}, then from a geometrical perspective, the Robert–Asselin filter operates by moving the inner point, which has coordinates $[t^n, x^n]$, a fraction χ towards the midpoint $\left[t^n, \frac{x^{n-1}+x^{n+1}}{2}\right]$ of the two outer points. It is stated in [201] that the value for χ is typically between 0.01 and 0.2.

Williams now introduces two relevant properties of the three points: the first is a property of their mean,

$$\mu^n = \frac{x^{n+1}+x^n+x^{n-1}}{3}, \tag{11.72}$$

while the second is the curvature which is given approximately by

$$C^n \sim x^{n-1} - 2x^n + x^{n+1}. \tag{11.73}$$

By displacing x^n through the amount d, the Robert–Asselin filter reduces the magnitude of the curvature of the three points. When used with the leap-frog scheme, this feature filter strongly suppresses the computational mode. However, the application does not conserve the mean of the three-time-level mean which has been shown when used with the leap-frog scheme to severely degrade the numerical accuracy.

The modification that Williams suggests in [201] is designed to attempt to consider the three-time-level mean through acting on the outer point, as well as the inner point. That is to say, for any α satisfying $0 \leq \alpha \leq 1$, the modified filter, which will eventually be called the Robert–Asselin–Williams (RAW) filter [203], displaces x^n by the amount αd, where d is calculated through (11.71) but with $\chi/2$ instead of χ. The other part of the filter is to move the point x^{n+1} by the amount $(\alpha-1)d$. It is stated in [201] that all members of this filter family reduce the magnitude of the curvature of the three points, with α controlling the relative contribution of the reduction from the displacement of the inner and outer points.

Given this extra step that we have to be aware of when dealing with three-time-level schemes, we now move on to consider a problem that was recently discovered by ECMWF associated with the dependence of the accuracy of the departure point calculation and the associated forecast.

11.5 Sensitivity to departure point calculations

As we have seen through the book from when we introduced the nonzero forcing terms, the semi-implicit semi-Lagrangian technique is quite efficient as it allows unconditional stability without the loss of accuracy, except for small dispersion errors, for larger time steps. This technique has been in operational use at the ECMWF for nearly 30 years now. As we know, nonconstant velocity and nonconstant forcing require an iterative approach to find the displacement to calculate the departure point, where it was quite often the case that we only required two or three iterations to provide sufficient accuracy for the coordinate of the departure point. In [45] Diamantakis and Magnusson indicate that two iterations may not be sufficient to produce a good forecast. When the guidelines of two to three iterations were published in the 1980s and 1990s, the resolution of the models were quite coarse compared to today's high resolution global models. In [45] Diamantakis and Magnusson discover that it was possible to decrease the extratropical geopotential errors, but just as important they show that in a simulated hurricane they were able to improve the vertical structure of the hurricane, which is of huge importance. In [45] they develop a stopping criterion for the departure point iterations based upon estimated convergence rates. We now present this new approach below.

Let ψ be a scalar field such that $\psi = \psi(\mathbf{r}, t)$ in Cartesian coordinate system and such that

$$\frac{D\psi(\mathbf{r}, t)}{Dt} = 0, \tag{11.74}$$

where we have the total derivative defined by

$$\frac{D}{Dt} \equiv \frac{\partial}{\partial t} + \mathbf{u} \cdot \nabla,$$

with

$$\mathbf{u} = \begin{pmatrix} u & v & w \end{pmatrix}^T \quad \text{and} \quad \mathbf{r} = \begin{pmatrix} x & y & z \end{pmatrix}^T.$$

As we have seen many times before, the semi-Lagrangian solution to (11.74) is

$$\psi\left(\mathbf{r}_a, t^{n+1}\right) = \psi(\mathbf{r}_d, t). \tag{11.75}$$

An important feature to note about (11.75) is that it is dependent on our estimate of the departure point $\mathbf{r}_d$. To compute the departure point, the following trajectory equations need to be solved backwards:

$$\frac{D\mathbf{r}}{Dt} = \mathbf{u}(\mathbf{r}, t). \tag{11.76}$$

Integrating (11.76) in $[t, t + \Delta t]$ through discretizing the integral using the midpoint rule, we can obtain

$$\mathbf{r}_d = \mathbf{r}_a - \Delta t \mathbf{u}(\mathbf{r}_m, t_m), \tag{11.77}$$

where $\mathbf{r}_m = \frac{\mathbf{r}_a + \mathbf{r}_d}{2}$ and $t_m = t + \frac{\Delta t}{2}$. We can see that (11.77) is an implicit equation for the departure point position $\mathbf{r}_d$ and can be solved for iteratively. The wind field at the mid-time is usually estimated through a time extrapolation of some form and the derived field is linearly interpolated at $\mathbf{r}_m$. As we mentioned earlier, for the coarse resolution of the global atmospheric models in the 1980s and 1990s, it was shown that two or three iterations were sufficient for convergence. We saw that in [148] the authors provide a sufficient condition for the convergence of the semi-Lagrangian departure point iteration in horizonal advection which is given by

$$\Delta t \max\left\{|u_x|, \left|u_y\right|, |v_x|, \left|v_y\right|\right\} < 1, \tag{11.78}$$

stating that this condition would be satisfied for Δt less than three hours, given an upper bound for the wind shear is $10^{-4}\ \mathrm{s}^{-1}$.

As we are dealing with the three-dimensional problem now, we have to consider how this condition extends to include the vertical direction. In [173] the authors provide the corresponding condition for this case as

$$\mathscr{L} = \Delta t \left\| \frac{\partial \tilde{\mathbf{u}}}{\partial \mathbf{r}} \right\| < 1, \tag{11.79}$$

where $\tilde{\mathbf{u}}$ is an approximation of the wind field at the trajectory midpoint, and $||\cdot||$ is a matrix norm. The positive number $\mathscr{L}$ is referred to in [45] as essentially being a Lipschitz constant for the fixed point iteration method to solve (11.77), and is then referred to as the **Lipschitz number**.

The method for calculating the departure point at the ECMWF is based upon what is referred to as the stable extrapolation two-time-level scheme SETTLS [82]; we shall go into more details about this scheme in the next chapter. The three-dimensional version of SETTLS, which is second-order accurate, is given by

$$\mathbf{r}_d = \mathbf{r}_a - \frac{\Delta t}{2}\left(\mathbf{u}\left(br_a, t^n\right) + 2\mathbf{u}\left(\mathbf{r}_d, t^n\right) - \mathbf{u}\left(\mathbf{r}_d, t^{n-1}\right)\right), \tag{11.80}$$

where the wind field values at $\mathbf{r}_d$ are obtained by interpolation. The term

$$\frac{1}{2}\left(\mathbf{u}\left(\mathbf{r}_a, t^n\right) + 2\mathbf{u}\left(\mathbf{r}_d, t^n\right) - \mathbf{u}\left(\mathbf{r}_d, t^{n-1}\right)\right)$$

provides a second-order approximation to the velocity field at the midpoint $(\mathbf{r}_m, t^m)$ of the semi-Lagrangian trajectory. As we can see, this scheme defines an implicit equation that is solved through iterating

$$\mathbf{r}_d^{(\nu)} = \mathbf{r}_a - \frac{\Delta t}{2}\left(\mathbf{u}\left(\mathbf{r}_a, t^n\right) + \left(2\mathbf{u}\left(\mathbf{r}, t^n\right) - \mathbf{u}\left(\mathbf{r}, t^{n-1}\right)\right)\Big|_{\mathbf{r}=\mathbf{r}_d^{(\nu-1)}}\right), \quad \nu = 2, 3, \ldots, \nu_{\max}, \tag{11.81}$$

where the term evaluated at $\mathbf{r}_d^{(\nu-1)}$ is interpolated at the latest available departure point estimate. A starting value for this scheme is provided using an Euler scheme backwards [45], which is to say that

$$\mathbf{r}_d^{(1)} = \mathbf{r}_a - \Delta t \mathbf{u}\left(\mathbf{r}_a, t\right), \tag{11.82}$$

and it is then stated in [45] that two more additional iterations are used, this makes $\mu_{\max} = 3$, and is referred to as the **number of departure point iterations**.

The analysis for the accuracy and convergence of the departure point iteration comes from [173], and is centered around the midpoint time integration method. It is stated in [45] that the same analysis as from [173] is valid here due to the ECMWF Integrated Forecast System (IFS) model, given that SETTLS is an extrapolation to obtain the midpoint velocities and it can be considered to be equivalent to the midpoint scheme with the limits of second-order accuracy. As a result of this, we have

$$\mathbf{r}_d^{(\nu)} + \left(\left(\Delta t\right)^3, \Delta t A\right) = \mathbf{r}_i - \Delta t \tilde{u}\left(\mathbf{r}_d^{(\nu-1)}, t\right), \tag{11.83}$$

where term A denotes the advective scheme truncation error, so effectively the interpolation error will hold. Through following the arguments from [173], where we subtract (11.79) from the continuous equations that are implied by the algorithm, taking the norm and assuming $\widetilde{\mathbf{u}}$ is differentiable, we obtain in [45] the inequality

$$\left\|\mathbf{r}_d - \mathbf{r}_d^{(\nu)}\right\| \leq \mathscr{L}\left\|\mathbf{r}_d - \mathbf{r}_d^{(\nu-1)}\right\| + \left\|O\left((\Delta t)^3, \Delta t A\right)\right\|, \tag{11.84}$$

where $\mathscr{L}$ is given by (11.79). If we now apply recursively the previous inequality, we obtain

$$\left\|\mathbf{r}_d - \mathbf{r}_d^{(\nu)}\right\| \leq \mathscr{L}^{\nu-1}\left\|\mathbf{r}_d - \mathbf{r}_d^{(1)}\right\|, \tag{11.85}$$

where we have ignored the higher-order truncation error term.

In [45] Diamantakis and Magnusson state that (11.85) is a complementary condition to (11.79), where they say that it appears that $\mathscr{L}$ is an upper bound for the rate of convergence of the departure point iteration. Diamantakis and Magnusson go on to show through plotting the Lipschitz number for winter wind and typhoon cases that for the first situation the Lipschitz number is quite small, and so the number of iterations required could be as low as 2 still. However, when Diamantakis and Magnusson plot the Lipschitz number for the typhoon case, they see a lot larger values, suggesting slow or even no convergence, and so more iterations are required to obtain a better estimate of the departure point.

Diamantakis and Magnusson go on to develop a method that is dynamically based to select the number of iterations required for the departure point calculation, and say that departure points in different areas on the globe convergence at different rates, and there is a small percentage that may not converge at all. They show that this happens in cases of extreme winds and very high wind shear, as these are associated with large Lipschitz numbers, which determine the theoretical upper bounds for the rate of convergence of the implicit system which is being solved. Therefore, this implies that it would be beneficial to have a control mechanism in the model to stop the departure point iterations.

To address this concern, Diamantakis and Magnusson introduce a *dynamic* iteration strategy where for the (i,k)th gridpoint estimate of the convergence rate at iteration ν:

$$cr_{ik} = \frac{\left\|x_{d_{ik}^{(\nu)}} - x_{d_{ik}^{(\nu-1)}}\right\|}{\left\|x_{d_{ik}^{(\nu-1)}} - x_{d_{ik}^{(\nu-2)}}\right\|}, \quad \nu = 3, 4, \ldots, \tag{11.86}$$

where $x_{d_{ik}^{(\nu)}}$ refers to the horizontal or vertical part of the departure point estimate at νth iteration. Here, the norm $\|\cdot\|$ is an absolute/relative tolerance distance:

$$\left\|x_{d_{ik}^{(\nu)}} - x_{d_{ik}^{(\nu-1)}}\right\| = \frac{\left|x_{d_{ik}^{(\nu)}} - x_{d_{ik}^{(\nu-1)}}\right|}{\text{atol} + \text{rtol} \times \left|x_{d_{ik}^{(\nu-1)}}\right|}. \tag{11.87}$$

The parameters "atol" and "rtol" are two predefined numerical tolerances, the absolute and relative, respectively. These parameters express the level of error that is considered sufficient to accept that the departure point iteration has converged to the solution.

We finish this section with the summary from [45] on the dynamical iteration strategy from (11.86):

1. A minimum of 2 iterations is done with a maximum up to $\nu_{\max}$ where the latter can be, for example, 5 or 10.
2. A convergence test is applied both in the horizontal and vertical direction.
3. Iterations stop updating the solution when both tests have been satisfied.
   (a) When the convergence rate of iterations begins to deteriorate, that is, when cr_{ik} exceeds 0.5 at grid point (i, k), either in the horizontal or vertical direction, updates of the departure point in this direction freeze, but they continue in the other direction.
   (b) Once decided that at a specific grid point the procedure should terminate at the νth iteration (either because of convergence or divergence of the iterative procedure), iterations continue to the maximum allowed number $\nu_{\max}$ but $\nu+1, \ldots, \nu_{\max}$ will reproduce identical departure point estimates with ν. These "idle" iterations are done for the purposes of avoiding load-balancing problems.

However, it is not just the departure point calculation that can affect the three-dimensional semi-Lagrangian integration work at Environment and Climate Change Canada (ECCC), now they have the current version, which is in spherical coordinates and global.

11.6 Consistency of semi-Lagrangian trajectory calculations

The summary that we present here is the from the work of Husain and Girard from 2017 [84]. As we have seen, the Canadians were quite instrumental in the development of semi-Lagrangian approaches, starting with the work of Robert, and then onto Tanguay.

The model that is in use now for the numerical weather prediction at ECCC is referred to as the Global Environmental Multiscale (GEM) model, where the vertical coordinate is log-hydrostatic pressure defined by

$$\zeta = \ln \pi - Bs, \tag{11.88}$$

where π is the hydrostatic pressure, B is a metric term, $s = \ln \pi_s - \zeta_s$, with π_s being the hydrostatic pressure at the surface, and $\zeta_s = \ln P_{\text{ref}}$, where the reference pressure $p_{\text{ref}} = 10^3$ hPa. The term B is defined as

$$B \equiv \left(\frac{\zeta - \zeta_T}{\zeta_s - \zeta_T} \right)^{\chi}, \tag{11.89}$$

where $\zeta_T = \ln p_T$, p_T refers to the pressure at the model top, and χ is a variable exponent providing control over the stretching of the coordinate with height over orography, defined as

$$\chi = \chi_{\max} - (\chi_{\max} - \chi_{\min}) \left(\frac{\zeta - \zeta_T}{\zeta_s - \zeta_T} \right).$$

The main equations are similar to those presented earlier in this chapter, but now with some more parameterizations, as well as extra forcing terms, but the suite of differential equations consists of Newton's second law for momentum, the second law of thermodynamics, and mass continuity, along with the diagnostic equation of the ideal gas law. After changing the vertical coordinate system to that described above and applying the perturbation around a stationary and isothermal basic state, we obtain the continuous equations. We are not going to present them here (the reader is referred to [84] for them); what we are concerned with here is the semi-Lagrangian implementation that is done.

The first noticeable change from the two models earlier is that the coordinate system is in terms of latitude and longitude, θ and λ, where now the vertical coordinate is ζ, but the horizontal and vertical directions are separated out such that these coordinates are expressed as $\mathbf{x} \equiv (\mathbf{r}, \zeta) = (\lambda, \theta, \zeta)$. Given these coordinates, we now focus on the trajectory calculations and the changes that have been undertaken at ECCC.

11.6.1 Trajectory calculations

In the previous version of the CRM we saw that the authors solved simultaneously for the three displacement through integrating the one kinematic displacement equations

$$\frac{D\mathbf{x}}{Dt} = \mathbf{u}\left(X, Y, Z, t\right).$$

In the new GEM, they are solving a **pair** of kinematic displacement equations

$$\frac{D\mathbf{x}}{Dt} = \mathbf{u}_H\left(\mathbf{r}, \zeta, t\right), \tag{11.90a}$$

$$\frac{D\zeta}{Dt} = \dot{\zeta}\left(\mathbf{r}, \zeta, t\right). \tag{11.90b}$$

The two equations can be discretized in the usual format

$$\mathbf{r}_a - \mathbf{r}_d = \int_t^{t+\Delta t} \mathbf{u}_H\left(\mathbf{r}, \zeta, t\right) dt, \tag{11.91a}$$

$$\zeta_a - \zeta_d = \int_t^{t+\Delta t} \dot{\zeta}\left(\mathbf{r}, \zeta, t\right) dt, \tag{11.91b}$$

with the integrals on the right-hand side being approximated numerically, where using the midpoint rule gives

$$\mathbf{r}_a - \mathbf{r}_d = \Delta t \mathbf{u}_H\left(\mathbf{r}^m, \zeta^m, t + \frac{\Delta t}{2}\right) \cong \Delta t \frac{\mathbf{u}_H\left(\mathbf{r}^m, \zeta^m, t+\Delta t\right) + \mathbf{u}_H\left(\mathbf{r}^m, \zeta^m, t\right)}{2}, \tag{11.92a}$$

$$\zeta_a - \zeta_d = \Delta t \dot{\zeta}\left(\mathbf{r}^m, \zeta^m, t + \frac{\Delta t}{2}\right) \cong \Delta t \frac{\dot{\zeta}\left(\mathbf{r}^m, \zeta^m, t+\Delta t\right) + \dot{\zeta}\left(\mathbf{r}^m, \zeta^m, t\right)}{2}, \tag{11.92b}$$

where m is referring to the midpoint again. A note to make here is that in the discretization for the other equations in the model, Husain and Girard have used the trapezoidal rule, but also to make it consistent with the off-centering that is still used in the new model, thus

$$\frac{\mathbf{r}_a - \mathbf{r}_d}{\Delta t} = \left(\frac{1+b_u}{2}\mathbf{u}_H^a + \frac{1-b_u}{2}\mathbf{u}_H^d \right), \tag{11.93a}$$

$$\frac{\zeta_a - \zeta_d}{\Delta t} = \left(\frac{1+b_\zeta}{2}\dot{\zeta}^a + \frac{1-b_\zeta}{2}\dot{\zeta}^d \right), \tag{11.93b}$$

where in [84] they determine that the values for b_u and b_ζ through numerical experiments should be the same as those used for the off-centering of horizontal momentum and vertical velocity equation, respectively. It is shown in [84] that when these values are not the same, or there is no off-centering in the trajectory equations, as opposed to the dynamical equations, then numerical inconsistencies are introduced.

Also as part of the study in [84] they show that numerical inconsistencies can be introduced through not using the same order of polynomial for the winds' and fields' interpolation. Again the linear interpolation is a feature of the coarser resolutions, and with higher resolution models we need to resolve these features more accurately.

We now move on to consider a finite element approach for the three-dimensional semi-Lagrangian integration, where the application is in an ocean model, which is referred to as the **Semi-implicit Eulerian Lagrangian Finite Elements (SELFE)** model.

11.7 Semi-implicit Eulerian Lagrangian finite elements (SELFE)

The work that we present here comes from a paper by Zhang and Baptista in 2008 [218], where the authors state that numerical modeling of ocean circulation at scales ranging from estuaries to ocean basins quite often involves codes that solve some form of 3D Navier–Stokes equations, complemented with conservation equations for water volume, salt, and heat. The motivation for the development of the Semi-implicit Eulerian Lagrangian Finite Element, or SELFE, model was to be able to model circulation *across* scales, for example, the modeling of river–estuary–plume–shelf systems. The starting point in [218] is to consider the 3D shallow water equations with hydrostatic and Boussinesq approximation. The prognostic variables that SELFE model solves for are the free-surface elevation, 3D velocity, 3D salinity, and 3D temperature of the water. In Cartesian coordinates the model equations in continuous form are given by:

$$\nabla \cdot \mathbf{u} + \frac{\partial w}{\partial z} = 0, \tag{11.94}$$

$$\frac{\partial \eta}{\partial t} + \nabla \cdot \int_{-h}^{\eta} \mathbf{u} dz = 0, \tag{11.95}$$

$$\frac{D\mathbf{u}}{Dt} = \mathbf{f} - g\nabla\eta + \frac{\partial}{\partial z}\left(\nu \frac{\partial \mathbf{u}}{\partial z} \right), \tag{11.96}$$

$$\mathbf{f} = -f\mathbf{k} \times \mathbf{u} + \alpha g \Delta \widehat{\psi} - \frac{1}{\rho_0}\nabla p_A - \frac{g}{\rho}\int_z^{\eta} \nabla \rho d\eta + \nabla \cdot (\mu \nabla \mathbf{u}), \tag{11.97}$$

$$\frac{DS}{Dt} = \frac{\partial}{\partial z}\left(\kappa\frac{\partial S}{\partial z}\right) + F_s, \tag{11.98}$$

$$\frac{DT}{Dt} = \frac{\partial}{\partial z}\left(\kappa\frac{\partial T}{\partial z}\right) + \frac{\dot{Q}}{\rho_0 C_p} + F_h, \tag{11.99}$$

where $\eta(x, y, t)$ is the free surface in meters, $h(x, y)$ is the bathymetric depth in meters, $\psi(\theta, \lambda)$ is the Earth-tidal potential in meters, α is the effective Earth-elasticity factor, $\rho(\mathbf{x}, t)$ is water density, where ρ_0 is set to 1025 $\mathrm{kg\,m^{-3}}$, $p_A(x, y, t)$ is the atmospheric pressure at the free surface in $\mathrm{Nm^{-2}}$, ν is the vertical eddy viscosity in $\mathrm{m^2\,s^{-1}}$, μ is the horizontal eddy viscosity, κ is the vertical eddy diffusivity for salt and heat, F_s and F_h are the horizontal diffusions for transport equations which are not considered in SELFE, $\dot{Q}$ is the rate of absorption of solar radiation in $\mathrm{WM^{-2}}$, and finally, C_p is the specific heat of water in $\mathrm{J\,kg^{-1}\,K^{-1}}$.

As we can see from the system of partial differential equations above, SELFE is considering two different sets of advection type problems, the advection of momentum for the horizontal wind and the three-dimensional transport of salinity and temperature. SELFE solves the system of partial differential equations above using finite elements and finite volumes. In [218] the authors state that no mode splitting is used in SELFE, thus eliminating the errors associated with the splitting term between internal and external modes. Semi-implicit schemes are applied to **all** equations, the continuity equation (11.95) and the momentum equation (11.97) are solved simultaneously, which means bypassing the most severe stability restriction from the CFL requirements.

A key step in SELFE is to decouple the continuity and momentum equations just mentioned via the bottom boundary layer. The reason to mentioned SELFE at all is that is uses an Eulerian–Lagrangian method, which Zhang and Baptista refer to as ELM, to treat the advection in the momentum equations. We should recall here that we mentioned very early on that semi-Lagrangian is also know as quasi-Lagrangian, but also as Eulerian–Lagrangian, using this second name here obviously allowed for a memorable acronym for the model. As a final note here, the transport equations for salinity and temperature are treated with either the ELM approach, or a finite volume upwind scheme FVUM, and, as we have seen earlier, the finite volume approaches can be mass conserving which is not always the case for certain semi-Lagrangian approaches.

11.7.1 Numerical formulation of SELFE

In SELFE, unstructured triangular grids are used in the horizontal direction, while hybrid vertical coordinates – partly terrain-following S coordinates and partly Z coordinates – are used in the vertical direction. The origin of the z-axis is at the undisturbed Mean Sea Level (MSL). The terrain-following S layers are placed on top of a series of Z layers with the demarcation line between S and Z layers located at level k^z $(z = -h_s)$. That is to say, the vertical grid is allowed to follow the terrain up to a maximum depth of h_s. The free surface is at level N_z throughout the domain (for all wet points), but the bottom level indices, k^b, may vary in space due to the staircase representation of the bottom in Z layers. It is then mentioned in [218] that we should note that $k^b \leq k^z$ and the equality occurs when the local depth $h \leq h_s$. A *pure* S representation is a special case with $k^b = k^z = 1$ and h_s being greater than the maximum depth in the domain. We have a copy of the structure of the vertical grid system from [218] in Fig. 11.6.

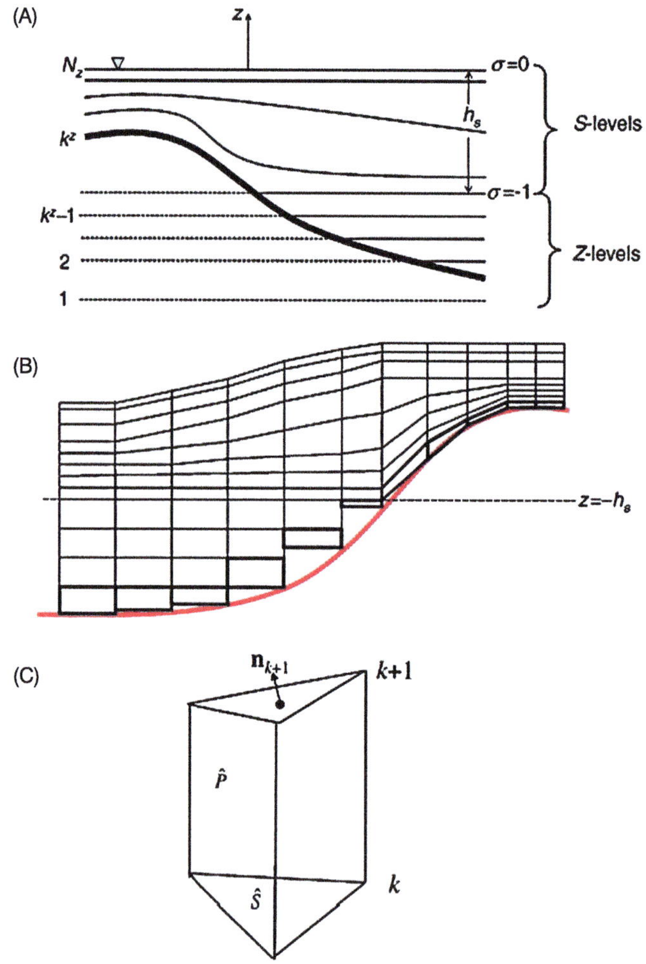

FIGURE 11.6 Vertical grid and hybrid coordinate system: (A) Schematic view. S-levels are always on top of Z-levels. (B) Vertical transected view of the discretized domain. Bottom cells are highlighted. (C) Basic computational unit of a triangular prism, with uneven top and bottom surfaces. Reprinted from Ocean-Modelling, Volume 21, Zhang, Y. and Baptista, A. M. (2008): SELFE: A semi-implicit semi-Lagrangian finite element model for cross-scale ocean circulation, 71–96, ©(2008) with permission from Elsevier.

In [218], Zhang and Baptista state that in the horizontal direction, unstructured triangular grids are used, and the connectivity of the grid is defined as follows: the three sides of an element i are enumerated as $js(i,l)$ for $l = 1, 2, 3$. The surrounding elements of a particular node i are enumerated as $ine(i,l)$ for $l = 1, \ldots, nne(i)$, where $nne(i)$ is the total number of elements in the *ball* of the node.

Given the descriptions of the discretizations above, the most common 3D computational units of SELFE are triangular prisms, where this is shown in Fig. 11.6 taken from [218]. In the original Z space, the prisms may not have level bottom and top surfaces. A staggering scheme is used to define variables. The surface elevations are defined at the nodes. The horizontal velocities are defined at the side centers and whole levels. The vertical velocities are defined at the element centers and whole levels as they are solved with a finite-volume method. The linear shape functions are used for elevations and velocities. An important remark that is made here in [218] is that for velocities, shape functions are only used for interpolation at the feet of characteristic lines, where here they are referring to the departure point interpolation. The locations where salinities and temperatures are defined depend on the method used to solve the transport equations; they are defined at the prism centers if the FVUM is used, and at both nodes and side centers, at whole levels, if the ELM is used.

The part of the description of the SELFE model in [218] is referred to as the *barotropic module*, where Zhang and Baptista state that SELFE solves the barotropic equations (11.94)–(11.97) first as the transport and turbulence closure equations, see [218] for the full derivation, lagged one time step behind, which is to say that the baroclinic pressure gradients terms in the momentum equation is treated explicitly in this model. Due to the hydrostatic approximation, the vertical velocity w is solved from (11.94) after the horizontal velocity is found. To solve the coupled equations (11.95) and (11.97), they are first discretized along with the vertical boundary conditions

$$\nu\frac{\partial \mathbf{u}}{\partial z}=\tau_w, \quad z=\eta, \qquad \nu\frac{\partial \mathbf{u}}{\partial z}=\tau_b, \quad z=-h,$$

where τ_w is the stress factor, and the bottom layer is often not well resolved in ocean models, the no-slip condition at the sea/river bottom $\mathbf{u}=w=0$ is replaced by a balance between the internal Reynolds stress and the bottom frictional stress, semi-implicity in time as

$$\frac{\eta^{n+1}-\eta^n}{\Delta t}+\theta\nabla\cdot\int_{-h}^{\eta}\mathbf{u}^{n+1}dz+(1-\theta)\nabla\cdot\int_{-h}^{\eta}\mathbf{u}^{n}dz=0, \tag{11.100}$$

$$\frac{\mathbf{u}^{n+1}-\mathbf{u}_d}{\Delta t}=\mathbf{f}^n-g\theta\nabla\eta^{n+1}-g(1-\theta)\nabla\eta^n+\frac{\partial}{\partial z}\left(\nu^n\frac{\partial \mathbf{u}^{n+1}}{\partial z}\right), \tag{11.101}$$

$$\begin{cases} \nu^n\dfrac{\partial \mathbf{u}^{n+1}}{\partial z}=\tau_w^{n+1}, & z=\eta^n, \\ \nu^n\dfrac{\partial \mathbf{u}^{n+1}}{\partial z}=\chi^n\mathbf{u}_b^{n+1}, & z=-h, \end{cases} \tag{11.102}$$

where θ is referred to the implicitness factor such that $0\leq\theta\leq 1$ $\mathbf{u}_d$ is the departure point for the horizontal wind, $\chi^n=C_D\left|\mathbf{u}_b^n\right|$, and C_D is the drag coefficient such that

$$C_D=\left(\frac{1}{\kappa_0}\ln\frac{\delta_b}{z_0}\right)^{-2},$$

κ_0 is the von Karman's constant equal to 0.4, δ_b is the thickness of the bottom computational cell.

The finite element approach that is used in [218] is the Galerkin method which results in residual statement in weak form of (11.100) as

$$\int_{\Omega}\frac{\eta^{n+1}-\eta^{n}}{\Delta t}d\Omega \;+\; \theta\left[-\int_{\Omega}\nabla\phi_i\mathbf{U}^{n+1}d\Omega+\int\Gamma_v\phi_i\widehat{U}_n^{n+1}d\Gamma_v+\int_{\overline{\Gamma}_v}\phi_i U_n^{n+1}d\overline{\Gamma}_v\right]$$
$$+\;(1-\theta)\left[-\int_{\Omega}\nabla\phi_i\cdot\mathbf{U}^{n}d\Omega+\int_{\Gamma}\phi_i U_n^{n}d\Gamma\right]=0, \tag{11.103}$$

where (11.103) is valid for $i=1,2,\ldots,N_p$, N_p being the total number of nodes, $\Gamma\equiv\Gamma_v+\overline{\Gamma}_v$ is the boundary of the entire domain Ω, with Γ_v corresponding to the boundary segments where natural boundary conditions are specified, $\mathbf{U}=\int_{-h}^{\eta}\mathbf{u}dz$ is the depth-integrated velocity, U_n is its normal component along the boundary, and $\widehat{U}_n$ is the boundary condition. SELFE uses linear shape function, which implies that ϕ_i are conventional *hat functions*.

While it may not be obvious that this is a semi-Lagrangian problem, when we integrate the momentum equation (11.101) along the vertical direction, we obtain

$$\mathbf{U}^{n+1} = \mathbf{G}-g\theta H^n\Delta t\nabla\eta^{n+1}-\chi^n\Delta t\mathbf{u}_b^{n+1}, \tag{11.104}$$

$$\mathbf{G}^n = \mathbf{U}_d+\Delta t\left(\mathbf{F}^n+\tau_w^{n+1}-g(1-\theta)H^n\nabla\eta^n\right), \tag{11.105}$$

$$H^n = h+\eta^n, \qquad \mathbf{F}^n=\int_{-h}^{\eta^n}\mathbf{f}dz, \qquad \mathbf{U}_d=\int_{-h}^{\eta^n}\mathbf{u}_d dz, \tag{11.106}$$

where now we have the departure points involved and this is a form of finite element based three-dimensional semi-Lagrangian scheme. In [218] the authors next explain the elimination of the unknown $\mathbf{u}_b^{n+1}$ though invoking the discretized momentum equation applied to the top and bottom cell, along with other manipulations for details of which the reader is referred to [218].

After the elevations are found, SELFE solves the momentum equation (11.97), along each vertical column at side centers. In [218] Zhang and Baptista implement a semi-implicit Galerkin finite-element method with the pressure gradient and the vertical viscosity terms being treated implicitly, and the other terms treated explicitly, as we have kind of seen with the other model developments. This then leads to

$$\int_{-h}^{\eta}\gamma_k\left[\mathbf{u}-\Delta t\frac{\partial}{\partial z}\left(\nu\frac{\partial\mathbf{u}}{\partial z}\right)\right]_{j,k}^{n+1}dz=\int_{-h}^{\eta}\gamma_k\left(\mathbf{u}_d+\Delta t\left(\mathbf{f}_{j,k}^n-g\theta\nabla\eta_j^{n+1}-g(1-\theta)\nabla\eta_j^n\right)\right)dz, \tag{11.107}$$

where $\gamma_k(z)$ is the hat function in the vertical dimension. In [218] the authors state that the two terms that are treated implicitly would have imposed the most severe stability constraints. The explicit treatment of the baroclinic pressure gradient and the horizontal viscosity terms, however, does impose mild stability constraints.

After the velocities at all sides are found, the velocity at a node, which is needed in ELM, is evaluated by a weighted average of all surrounding sides in its ball, aided by proper interpolation in the vertical direction. The procedure to average the velocities (or alternatively calculating the velocity at a node based on a least-squares fit from all surrounding sides) introduces numerical diffusion of the same order as the ELM [218], where Zhang and Baptista state that this is because the velocities at nodes are not used anywhere else in the model except in ELM tracking and interpolation. As an alternative to the averaging procedure, the velocity at a node is computed within each element from the three sides using the linear shape function and is kept discontinuous between elements. This approach leads to parasitic oscillations, but a Shapiro filter [170] can be used to suppress the noise, with minimum distortion of physical features. Our preliminary results indicate that the filter approach induces less numerical diffusion.

11.7.2 Semi-Lagrangian advection in SELFE

In [218] the authors state that SELFE treats the advection in the momentum and transport equations with ELM. In this method, a fluid particle is tracked from its position at step $n+1$ backwards in time and in 3D space to find its originating position at step n, while they never present the displacement equation in [218], followed by an interpolation at the foot of the characteristic line to evaluate the variable of interest. In SELFE, backtracking, which is what we have been referring to as the displacement or trajectory calculation, is done using either an Euler scheme or a more accurate, but expensive, fifth-order embedded Runge–Kutta scheme which can be found in Oliveria and Baptista (1998) [138].

Zhang and Baptista then go on to highlight that the order of interpolation in ELM determines whether the leading-order truncation error is diffusion- or dispersion-dominant [137]. For velocity, the linear interpolation is what they use, which is based upon the interpolation using the velocity information at the three nodes of the element that contains the foot of the characteristic line, i.e., the departure point. In one dimension, this ELM introduces a diffusion-like leading order truncation error in the following form [9,30,218]:

$$\begin{aligned} \varepsilon_1 &= \frac{v'_d}{2\Delta t}\left(x_{i+1}-x_d\right)\left(x_d-x_i\right) \\ &= \frac{(\Delta x)^2}{2\Delta t} v'_d\left[C_u\right]\left(1-\left[C_u\right]\right)\left|\varepsilon_1\right| \leq \frac{(\Delta x)^2}{8\Delta t}\left|v''_d\right|, \end{aligned} \tag{11.108}$$

where v is the analytical solution for velocity, $\left[x_i, x_{i+1}\right]$ is the interval encompassing the *foot* as stated in [218], and $[C_u]$ is the fractional part of the Courant number $C_u = v\frac{\Delta t}{\Delta x}$, so we are back into familiar territory here, but we have been denoting $[C_u]$ as $\widehat{\alpha}$. However, since the velocities at the two nodes of the interval are not exact but have errors from the averaging procedure mentioned earlier, an additional truncation error occurs, the leading-order term of which is

$$\varepsilon_2 = \frac{(\Delta x)^2}{8\Delta t} v''_d. \tag{11.109}$$

However, the two errors ε_1 and ε_2 are of the same order of magnitude so there is no deterioration of the accuracy.

An important feature to note here is that the numerical diffusion is controlled by the Courant number, where Zhang and Baptista point out that if the numerical diffusion becomes excessive due to a **small Courant number**, then a larger time step, or smaller grid size, needs to be used!

Zhang and Baptista go on to describe the implementation of the horizontal viscosity discretization in a finite element form, as well as for the baroclinic pressure gradient, which can use either the ELM approach or the FVUM approach to ensure the mass conservation in the temperature and salinity transport. They then go onto to run a series of experiments to test the validity of this approach in a quarter annulus with a linear bottom slope and a depth varying by over 15 meters at the outer boundary to the inner boundary. The reader is referred to [218] to see the results of this approach.

11.8 Summary

In this section we have extended the theory from the one- and two-dimensional cases to three dimensions. We have introduced the trilinear and tricubic interpolation methods, as well as an operational work around the 64 point needed to fit a full tricubic Lagrange interpolation polynomial, through considering the fact that points further away from the departure points should not be given the same weight as those nearer to it. We have looked at different versions of the Canadian regional, but now multiscale global model which uses uncentering, spatial averaging semi-implicit semi-Lagrangian integration. We have seen that time filters are needed to remove the noise that comes from the computational mode. We have revisited the assumptions and saw that we only require two iterations to solve for the departure point, as well as only a linear interpolation for the wind field interpolation to the departure point. We saw that as the resolution of the global numerical weather prediction models has increased there are now regions of certain flows that require more iterations for the departure point value, and that the stopping criterion should be dynamically dependent, while also showing that the uncentering of the velocity interpolation as well as the semi-implicit term needs to be the same as the order of the interpolation. We have also briefly introduced the 3D version of the flux-form semi-Lagrangian integration, but we shall go into more details in the spherical model chapter. Finally, we finished this chapter by introducing SELFE, which is a finite element semi-Lagrangian, or Eulerian–Lagrangian, technique in the ocean and hydrology communities, to show that it can work in three dimensions with this discretization for the spatial derivatives.

However, we do not live on a cube, nor are we surrounded by a cube atmosphere, and so we now move on to the spherical coordinates to present versions of many models that have been introduced throughout the book so far in this more consistent coordinate system for the spherical Earth.

CHAPTER

12

Semi-Lagrangian methods on a sphere

In the first few sections of this chapter we shall introduce different aspects of the general numerical setup for modeling the semi-Lagrangian equation in its many different forms in numerical modeling of geophysical flows. We start by considering how the differential vector calculus transfers to spherical coordinates compared to Cartesian coordinates, before presenting a series of projections that are used in many different applications of the semi-Lagrangian approach. We shall also introduce spectral modeling before presenting the advection equation in spherical coordinates, and then present the shallow water equations model and the primitive equations in two and three dimensions, along with advection–diffusion. We will also introduce the flux form/finite volume approaches. We start now with presenting the changes that occur in the vector operators.

12.1 Vector operators in spherical coordinates

When transforming from Cartesian to spherical coordinates, the direction in which the unit vectors are pointing changes and this has to be taken into account in the vector operators. We therefore now consider how the definition for the unit vectors are different in spherical coordinates compared to Cartesian.

12.1.1 Spherical unit vectors

In spherical coordinates there is a local approximation through a tangential plane relative to the spherical surface, where the unit vectors in spherical coordinates are defined. In Fig. 12.1 there is an illustration of the tangential plane to the sphere, where θ is the angle of colatitude such that $\theta \in \left[-\frac{\pi}{2}, \frac{\pi}{2}\right]$, λ is the angle of longitude such that $\lambda \in [0, 2\pi)$, and r is the radial distance satisfying $r \in [0, R]$, where R is a set distance from the center of the sphere.

It is from this tangential plane approximation that the following nine derivatives for the unit vectors in a 3D framework arise

$$\frac{\partial \mathbf{i}}{\partial r} = 0, \qquad \frac{\partial \mathbf{i}}{\partial \theta} = 0, \qquad \frac{\partial \mathbf{i}}{\partial \lambda} = -\cos\theta \mathbf{k} + \sin\theta \mathbf{j}, \tag{12.1}$$

https://doi.org/10.1016/B978-0-12-817222-3.00016-5

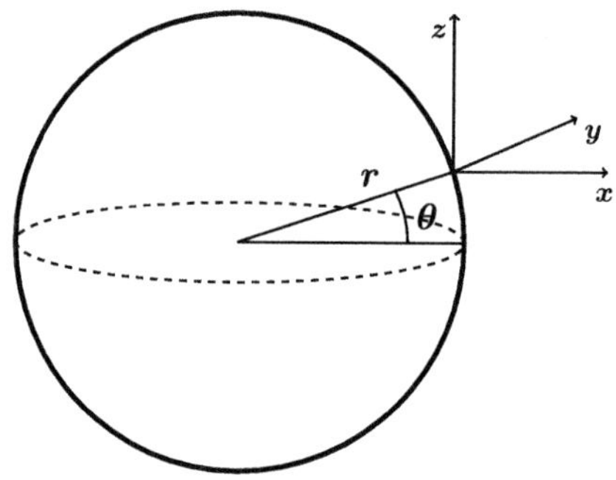

FIGURE 12.1 Schematic of the spherical coordinate system.

$$\frac{\partial \mathbf{j}}{\partial r}=0, \qquad \frac{\partial \mathbf{j}}{\partial \theta}=-\mathbf{k}, \qquad \frac{\partial \mathbf{j}}{\partial \lambda}=-\sin\theta\mathbf{i}, \tag{12.2}$$

$$\frac{\partial \mathbf{k}}{\partial r}=0, \qquad \frac{\partial \mathbf{k}}{\partial \theta}=\mathbf{j}, \qquad \frac{\partial \mathbf{k}}{\partial \lambda}=\cos\theta\mathbf{i}, \tag{12.3}$$

where the full derivation of these expression can be found in [11]. If one considers a 2D framework then this implies that there is a constant radial distance and hence there is no change along the radial, **k**, direction. This then makes all the terms in the derivatives that contain **k** equal to zero, leaving

$$\frac{\partial \mathbf{i}}{\partial \lambda}=\sin\theta\mathbf{j}, \qquad \frac{\partial \mathbf{j}}{\partial \lambda}=-\sin\theta\mathbf{i}. \tag{12.4}$$

Given the definitions of what are referred to as the metric terms in many of the papers associated with the development of semi-Lagrangian models, we now look at how these affect the vector calculus definitions.

12.1.2 Spherical vector derivative operators

We now consider how to alter the Cartesian definition for the gradient operator ∇, the divergence operator $\nabla\cdot$, the curl operator $\nabla\times$, the Laplacian operator ∇^2, and finally, the Jacobian operator, in spherical coordinates. We start by defining the expression for the derivatives of a vector **G**. For the Cartesian coordinates we have $\boldsymbol{G}=\left(G_x, G_y, G_z\right)^T$ while for spherical coordinates we have $\boldsymbol{G}=(G_\lambda, G_\theta, G_r)^T$.

The definition of the gradient operator of a scalar function F in Cartesian coordinates is given by

$$\nabla F=\mathbf{i}\frac{\partial F}{\partial x}+\mathbf{j}\frac{\partial F}{\partial y}+\mathbf{k}\frac{\partial F}{\partial z},$$

which in spherical coordinates becomes

$$\nabla F=\frac{\mathbf{i}}{r\cos\theta}\frac{\partial F}{\partial \lambda}+\frac{\mathbf{j}}{r}\frac{\partial F}{\partial \theta}+\mathbf{k}\frac{\partial F}{\partial r}.$$

The next operator is the divergence, which for the vector field $\mathbf{G}$ is defined in Cartesian coordinates as

$$\nabla \cdot \mathbf{G} = \frac{\partial G_x}{\partial x} + \frac{\partial G_y}{\partial y} + \frac{\partial G_z}{\partial z},$$

and in spherical coordinates by

$$\nabla \cdot \mathbf{G} = \frac{1}{r^2 \cos\theta} \left\{ \frac{\partial r G_\lambda}{\partial \lambda} + \frac{\partial r \cos\theta G_\phi}{\partial \theta} + \frac{\partial r^2 \cos\theta G_r}{\partial r} \right\}.$$

The curl operator in Cartesian coordinates is defined as

$$\nabla \times \mathbf{G} = \left(\frac{\partial G_z}{\partial y} - \frac{\partial G_y}{\partial z} \right) \mathbf{i} - \left(\frac{\partial G_z}{\partial x} - \frac{\partial G_x}{\partial z} \right) \mathbf{j} + \left(\frac{\partial G_y}{\partial x} - \frac{\partial G_x}{\partial y} \right) \mathbf{k},$$

whereas in spherical coordinates it is defined as

$$\nabla \times \mathbf{G} = \frac{\mathbf{i}}{r} \left(\frac{\partial G_r}{\partial \theta} - \frac{\partial r G_\theta}{\partial r} \right) + \frac{\mathbf{j}}{r\cos\theta} \left(\cos\theta \frac{\partial r G_\lambda}{\partial r} - \frac{\partial G_r}{\partial \lambda} \right) + \frac{\mathbf{k}}{r\cos\theta} \left(\frac{\partial G_\theta}{\partial \lambda} - \frac{\partial \cos\theta G_\lambda}{\partial \theta} \right).$$

The Laplacian of the scalar field F in Cartesian coordinates is defined as

$$\nabla^2 F = \frac{\partial^2 F}{\partial x^2} + \frac{\partial^2 F}{\partial y^2} + \frac{\partial^2 F}{\partial z^2},$$

which in spherical coordinates becomes

$$\nabla^2 F = \frac{1}{r^2 \cos^2\theta} \frac{\partial^2 F}{\partial \lambda^2} + \frac{1}{r^2} \frac{\partial^2 F}{\partial \theta^2} + \frac{\partial^2 F}{\partial r^2} - \frac{\tan\theta}{r^2} \frac{\partial F}{\partial \theta} + \frac{2}{r} \frac{\partial F}{\partial r}.$$

Finally, the spherical version of the Jacobian operator is given by

$$\frac{\partial (G_\lambda, G_\theta)}{\partial (\lambda, \theta)} = \frac{1}{a\cos\theta} \frac{\partial G_\lambda}{\partial \lambda} \left(\frac{1}{a} \frac{\partial G_\theta}{\partial \theta} \right) - \frac{1}{a} \frac{\partial G_\theta}{\partial \lambda} \left(\frac{1}{a\cos\theta} \frac{\partial G_\lambda}{\partial \lambda} \right).$$

A full explanation for all the operators can be found in [174]. Given these new expressions for the differential operators, we move on to present techniques for starting to develop numerical approximations to them.

12.2 Grid development for a sphere

So far when we have been considering different discretizations for the semi-Lagrangian approach they have either been in the horizontal plane, a vertical slice, or a box, in x, y, and z coordinates, but we have to consider that we live on a near-spherical planet, and so for certain applications of geophysical modeling we have to include the curvature of the Earth. In this section we explain different factors that can affect implementing semi-Lagrangian schemes on the sphere where we consider different projection of the sphere.

12.2.1 Map projections

There are many ways to represent data points on the sphere through different forms of canonical projections. In this section we shall briefly present some of the more commonly used projections for either numerical modeling of the sphere, or for the storing of observational data. There exist several desirable properties for any map projection, some of these are as follows:

- Preservation of angles – This implies that the angles on the flat surface and the sphere should be equivalent.
- Preservation of area – Areas on the flat surface and the sphere should be identical.
- Preservation of shape – Shapes on the flat surface should match those on the sphere.
- Correct direction – Cardinal directions on the flat surface and the sphere should be equal.
- Shortest distance between two lines should be a great circle.

Unfortunately, there does not exist an individual map projection that can satisfy all of the five properties above. Map projections that preserve angles are referred to as **conformal** map projections. The most commonly used map projections are the polar stereographic, Mercator, and the Lambert conic projections.

Stereographic projection

The stereographic projection is a mapping that projects the sphere onto a plane. This projection is defined on the whole sphere, except at the projection point. If we first consider the unit sphere in Cartesian coordinates, which is given by $x^2 + y^2 + z^2 = 1$, then we denote $N = (0, 0, 1)$, which are the coordinates for a north pole, and then $\mathcal{M}$ denotes the remainder of the sphere. If we now assume a plane $z = 0$ that runs through the center of the sphere, then the sphere's equator is the intersection of the sphere with the plane.

For any point P on $\mathcal{M}$ there is a unique line through N and P, and this line intersects the plane $z = 0$ at one point P'. Thus the stereographic projection of P is set to be the point P' in the plane.

If Cartesian coordinates (x, y, z) on the sphere and then (X, Y) on the plane of a point are given, then we can define the projection and its inverse as

$$\begin{aligned} (X, Y) &= \left(\frac{x}{1-z}, \frac{y}{1-z} \right), \\ (x, y, z) &= \left(\frac{2X}{1+X^2+Y^2}, \frac{2Y}{1+X^2+Y^2}, \frac{-1+X^2+Y^2}{1+X^2+Y^2} \right). \end{aligned}$$

If we have the situation where we have spherical coordinates on the sphere denoted by (θ, λ) where $\theta \in [0, \pi]$, and $\lambda \in [0, 2\pi]$, and polar coordinates (R, Θ) on the plane, then the projection in these coordinates and its inverse, assuming $R = 0$ when $\theta = \pi$, are given by

$$(R, \Theta) = \left(\frac{\sin\theta}{1-\cos\theta}, \lambda \right) = \left(\cot\frac{\theta}{2}, \lambda \right), \qquad (\theta, \lambda) = \left(2\arctan\left(\frac{1}{R} \right), \lambda \right).$$

We have presented some examples of the use of this projection in displaying model output from different geophysical situations in Fig. 12.2.

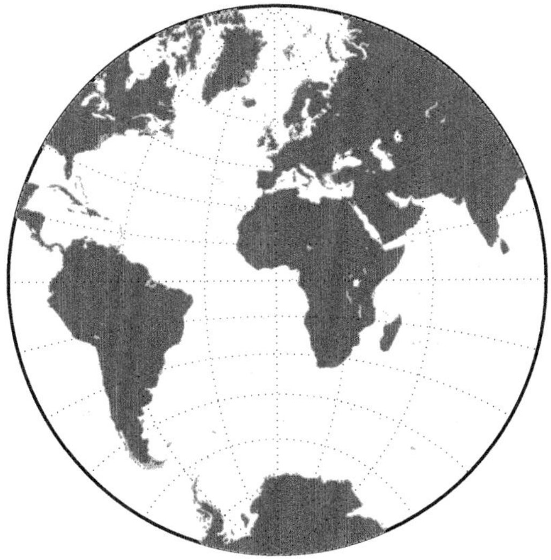

FIGURE 12.2 Example of a stereographic projection.

Mercator projection

The Mercator projection is a cylindrical map projection presented by the Flemish geographer and cartographer Gerardus Mercator in 1569. This map projection is practical for nautical applications due to its ability to represent lines of constant course, known as rhumb lines, as straight segments that conserve the angles with the meridians. Although the linear scale is equal in all directions around any point, thus preserving the angles and the shapes of small objects, the Mercator projection distorts the size of objects as the latitude increases from the equator to the poles, where the scale becomes infinite. A classic example of the distortion that this projection causes is that it projects Greenland and Antarctica to appear much larger than they actually are relative to land masses nearer the equator, such as Central Africa. Another distortion this projection causes is that it makes Greenland appear larger than Australia, while Australia is approximately three-and-a-half times larger than Greenland. We have plotted some examples of the Mercator projection to show the distortions in shapes that it causes in Fig. 12.3.

The following equations place the x-axis of the projection on the equator and the y-axis at longitude λ_0, such that

$$x = \lambda - \lambda_0, \qquad y = \ln\left(\tan\theta + \sec\theta\right).$$

The formulas for the inverse of this projection are given by

$$\theta = 2\tan^{-1}\left(e^{y}\right) - \frac{1}{2}\pi, \qquad \lambda = x + \lambda_0.$$

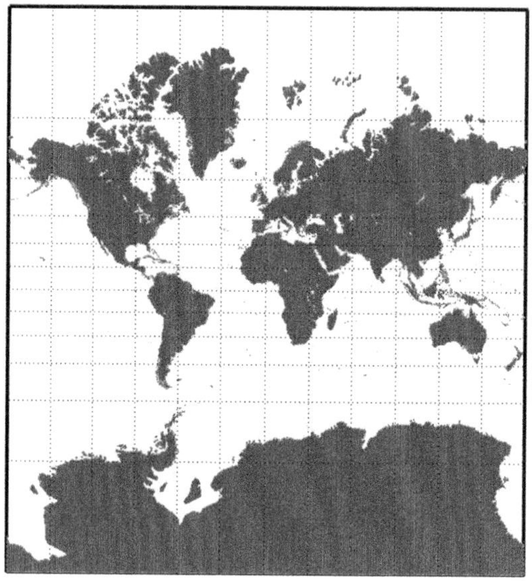

FIGURE 12.3 Examples of a Mercator projection.

Lambert conic projection

A Lambert conformal conic projection is a conic map projection used for aeronautical charts, portions of the State Plane Coordinate System, and many national and regional mapping systems. The projection places a cone over the sphere of the Earth and projects the surface conformably onto the cone. The cone is unrolled, and the parallel that was touching the sphere is assigned unit scale. That parallel is called the reference parallel or standard parallel.

By scaling the resulting map, two parallels can be assigned unit scale, with scale decreasing between the two parallels and increasing outside them. As a result of scaling, deviation from unit scale can be minimized within a region of interest that lies largely between the two standard parallels.

Mathematically the projection and its inverse are defined as follows: Let λ be the longitude, λ_0 the reference longitude, θ the latitude, where ϕ_0 is the reference latitude, and θ_1 and θ_2 are the standard parallels. The transformation of spherical coordinates to the plane via the Lambert conformal conic projection is given by

$$x = \rho \sin\left(n\left(\lambda - \lambda_0\right)\right), \qquad y = \rho_0 - \rho\cos\left(n\left(\lambda - \lambda_0\right)\right),$$

where

$$F = \frac{\cos\theta_1 \tan^n\left(\frac{1}{4\pi} + \frac{1}{2\theta_1}\right)}{n},$$

$$
\begin{aligned}
n &= \frac{\ln(\cos\theta_1 \sec\theta_2)}{\ln\left(\tan\left(\frac{1}{4\pi}+\frac{1}{2\theta_2}\right)\cot\left(\frac{1}{4\pi}+\frac{1}{2\theta_1}\right)\right)},\\
\rho &= F\cot^n\left(\frac{1}{4\pi}+\frac{1}{2\theta}\right),\\
\rho_0 &= F\cot^n\left(\frac{1}{4\pi}+\frac{1}{2\theta_0}\right).
\end{aligned}
$$

The inverse formulas are

$$\theta = 2\tan^{-1}\left(\left(\frac{F}{\rho}\right)^{\frac{1}{n}}\right)-\frac{1}{2\pi}, \qquad \lambda = \lambda_0 + \frac{\phi}{n},$$

where

$$\rho = \operatorname{sign}(n)\sqrt{x^2+(\rho_0-y)^2}, \qquad \phi = \tan^{-1}\left(\frac{x}{\rho_0-y}\right),$$

while F, ρ_0, and n remain as defined above. An example of this projection can be seen in Fig. 12.4.

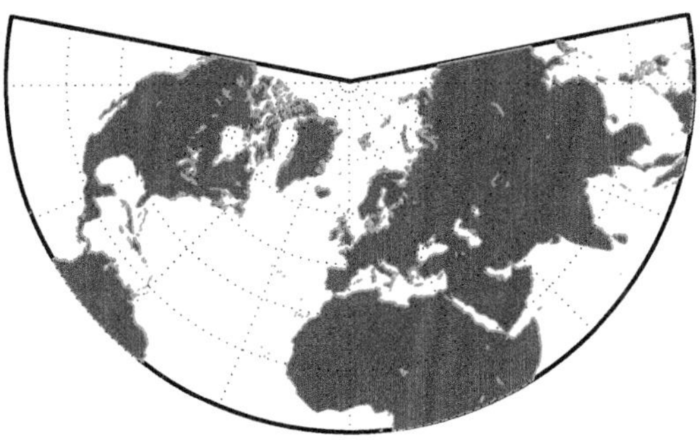

FIGURE 12.4 Example of a Lambert projection.

Sinusoidal projection

The sinusoidal projection is a pseudocylindrical equal-area map projections, it is some times referred to as the Mercator equal area projection. The projection is defined by:

$$x = (\lambda-\lambda_0)\cos\theta, \qquad y = \theta,$$

where λ_0 is the central median. With this projection we have that scale is constant along the central median, and the East–West scale is constant throughout the map. This then implies that the length of each parallel on the map is proportional to the cosine of the latitude. We have a plot of this projection in Fig. 12.5.

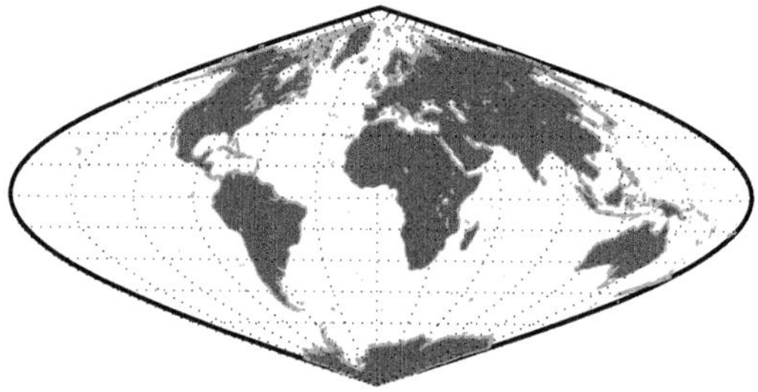

FIGURE 12.5 Example of a sinusoidal projection.

12.3 Grid-point representations of the sphere

The idea of producing a weather forecast by means of numerical approximations using physically based model was first proposed by Lewis Fry Richardson in 1922 [156], where he based his theory through Bjerknes's work which had identified that numerical weather prediction is an initial value problem [21].

Richardson, recognizing that a weather forecast could be seen as an initial value problem, stated that

> *If the values of certain environmental variables are known, then the physical equations can be used to calculate their value at some later time in the future.*

To achieve this goal, Richardson proposed to divide the Earth's surface into a **grid**, where each grid cell was the base of a vertical column in the atmosphere. Each vertical column was then divided into several layers, that resulted in a three-dimensional grid of atmospheric boxes. The horizontal resolution that Richardson used was 3° in the longitudinal direction, and 200 km in the latitudinal direction, which then resulted in 12, 000 columns to cover the globe. Given the location of each of these volumes, the vertical layers were at approximately 2, 4.2, 7.2, and 11 km above the Earth's surface which is approximately equal to 800, 600, 400, and 200 hPa. The values of the prognostic variables were stored at the center of each box. Unfortunately, the initial forecast that Richardson was aiming for failed due to gravity waves that created noise in the observed data set, that then lead to errors in the initial conditions to initialize the numerical model – a problem that can still plague numerical modeling today, but not as severe. We have presented a copy of the schematic of the original Richardson grid over northern Europe in Fig. 12.6.

As a result of Richardson formulation of grid point approximations to the sphere, there has been much development of different grids that comprise different shapes. To be able to accommodate the spherical shape of the Earth, and to be able to represent the equations that govern the motion of the atmosphere and other geophysical systems, more accurately and more efficiently, given the computational power available, there has been development of grids that are based upon different shapes, where these shapes have been rectangular, triangular, and hexagonal, to name but a few. Such spherical grid approximations are referred to as *structured grids*.

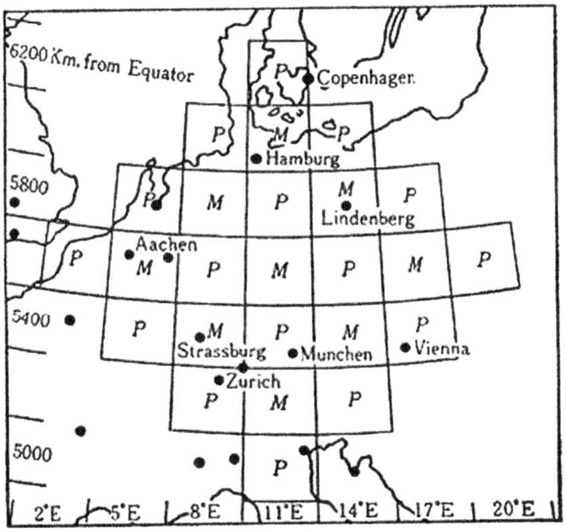

FIGURE 12.6 Copy of the initial Richardson grid reprinted from the Journal of Wind Engineering and Industrial Aerodynamics, Kimura, R. (2002), Numerical weather prediction, **90**, 1403–1414. ©(2002) with permission from Elsevier.

12.3.1 Rectangular/square grids

If we consider the rectangular/square grid approximations, then this grid, or mesh, has equal number of points on each longitudinal line, which means that as we approach the poles, the distance between the grid points is decreasing, implying that the resolution of the numerical approximation is increasing, but this introduces what is referred to as the *pole/polar problem* where the meridians converge to a point.

An alternative approach to the equal point grid formation was proposed by Kurihara in 1965 [96], where he sought a formulation to keep the grid spacing uniform, and still inside the latitude–longitude framework. The idea in [96] was to have a grid where the number of points along a latitudinal circle varied with the latitude. As a result of placing fewer points at the higher latitudes, the grid covered the sphere more homogeneously. The formulation of the Kurihara grid starts with a single point at the pole and this latitude circle, which is a single point anyway, is given the label $j = 1$. The next latitude circle chosen is given the label $j = 2$, but now there are $4(j-1)$ grid points along the line of longitude of this circle. This carries on till we reach the equator where this circle is labeled $j = N + 1$, which represents the last of the **evenly** spaced points along the longitudinal meridian. The Southern Hemisphere is a mirror configuration of the Northern Hemisphere configuration just described. A copy of the illustration from Kurihara's grid configuration for an octant of the sphere is presented in Fig. 12.7.

The Kurihara grid configuration just described leads to spuriously high pressure to develop at the poles, but it has been shown that it is possible to remove this problem through using more accurate numerical schemes [150].

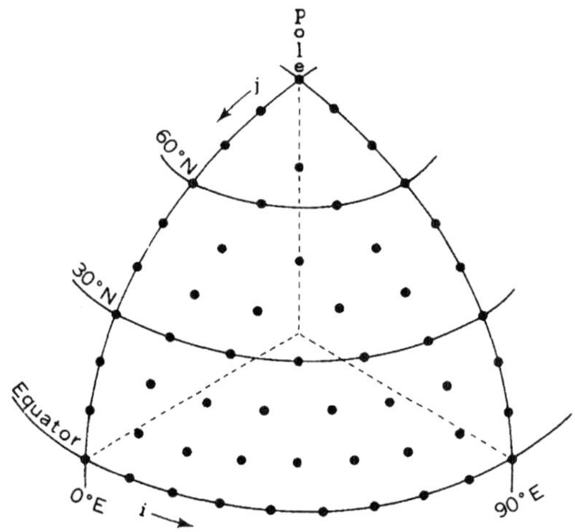

FIGURE 12.7 Copy of Fig. 1 from [96] of the schematic of the proposed Kurihara grid. Kurihara, Y., 1965: Numerical integration of the primitive equations on primitive grids, *Monthly Weather Review*, **93**, 399–415. ©American Meteorological Society. Used with permission.

12.3.2 Triangular grids

It should be noted that triangular grids are not used as commonly as rectangular grids in numerical models. One such form of a quasiuniform grid whose base element is a triangle is referred to as the **spherical geodesic grid**. The idea of using triangles from an icosahedron was proposed in the 1960s by both Williamson and Sadourny, see [205] and [166], respectively.

The spherical geodesic grid is constructed from an icosahedron which has 20 faces and 12 vertices. A simple scheme for constructing the geodesic grid is to divide the edges of the icosahedral triangular faces into smaller triangles, where the vertices of these triangles are the grid points. This process is run recursively until the desired resolution is met. Each point on the face or edge of one of the faces of the icosahedron is surrounded by six triangles making each grid point the center of a hexagon. The triangular faces of the icosahedrons are arranged in pairs to form five rhombuses around both poles; here the actual pole points are pentagonal points where the rhombuses meet. An example of the different resolutions are presented in Fig. 12.8.

12.3.3 Hexagonal grid

We now present a summary of the hexagonal grid that is presented in Thuburn (1997) [190]. The grid that is developed in [190] is similar to the twisted icosahedral grid from [73,74] but without the twisting. The sphere is covered by pentagonal and hexagonal grid boxes; the number of hexagons depends on the resolution, while there are always exactly 12 pentagons. Thuburn states that there is a hierarchy of possible grids that are of different resolutions, where each member of the hierarchy has approximately four times as many grid boxes as the previous member.

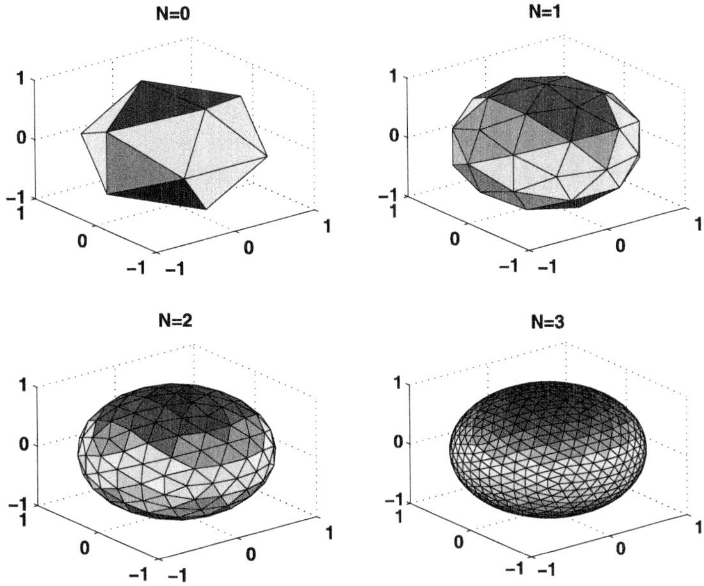

FIGURE 12.8 Plot of the buildup of the resolution of a triangular mesh.

The starting point for the construction of the grid is the original regular icosahedron that is presented in Fig. 12.8 in the top left plot. Given a grid of triangles, a new finer grid of triangles is generated by placing new vertices at the midpoints of the existing edges and then projecting these new vertices onto the surface of the sphere (Fig. 12.8 (top right)). This process is repeated, see Fig. 12.8, to obtain higher resolution grids. However, before we progress, we introduce the definition of the **Voronoi grid**.

Definition 12.1. Let X be a metric space with a distance function d. Let K be a set of indices and let P_k for $k \in K$ be an ordered collection of nonempty subsets in the space X. Then Voronoi grid, or Voronoi cell, or Voronoi region, V_k, associated with the subset P_k is the set of all points in X whose distance to P_k is not greater than their distance to the other subsets P_j, where $j \neq k$.

Returning to the description from [190], the hexagonal grid is a Voronoi grid associated with the triangular grid described above, where each vertex of the triangular grid corresponds to a face, that is to say, a grid box, of the Voronoi grid; each face of the triangular grid corresponds to a vertex of the Voronoi grid; each edge of the triangular grid corresponds to an edge of the Voronoi grid. The edges of the Voronoi grid are the perpendicular bisectors of the edges of the triangular grid.

Note that the hexagonal grid boxes vary slightly in their exact shape and size. The most distorted hexagons are those immediately adjacent to the pentagons. However, the pentagons are perfectly regular.

A more mathematical description of the hexagonal grid can be found in [73] which we summarize here: Given a set of N grid points $\{\mathbf{P}_1, \mathbf{P}_2, \ldots, \mathbf{P}_N\}$ on the unit sphere S, the definition of the Voronoi cell k associated with $\mathbf{P}_k$ is

$$\text{cell}_k = \left\{ \left| \mathbf{p} - \mathbf{P}_k \right| \leq \left| \mathbf{p} - \mathbf{P}_l \right|,\ \forall \mathbf{p} \in S \text{ and } l \neq k \right\}, \tag{12.5}$$

where $|x_1 - x_2|$ is the distance function mentioned in the definition of the Voronoi grid, where here this distance is a measure along the surface of the sphere. If we to consider two neighboring grid points $\mathbf{P}_0$ and $\mathbf{P}_1$, then it is stated in [73] that the cell wall that is shared by the two points is the perpendicular bisector of the circular arc between the two points. We should also note that a cell wall is a segment of a great circle. The end points of a wall are the Voronoi corners, the points were three cells intersect. The position of a corner is computed using three grid points. If we now consider the three points $\mathbf{P}_0$, $\mathbf{P}_1$, and $\mathbf{P}_2$ in three-dimensional Cartesian coordinates, $\mathbf{P}_0 = (x_0, y_0, z_0)$, $\mathbf{P}_1 = (x_1, y_1, z_1)$, and $\mathbf{P}_2 = (x_2, y_2, z_2)$, then the Voronoi corner, denoted by $\mathbf{C}$ in [73], is the point on the sphere that is equidistant from the three grid points, and defined as

$$\mathbf{C} \equiv \frac{(\mathbf{P}_2 - \mathbf{P}_0) \times (\mathbf{P}_1 - \mathbf{P}_0)}{|(\mathbf{P}_2 - \mathbf{P}_0) \times (\mathbf{P}_1 - \mathbf{P}_0)|}. \tag{12.6}$$

In [73] it is noted that the points are ordered in a clockwise fashion to ensure that $\mathbf{C}$ lies in the same hemisphere as the three points. A copy of the illustration of the cells and the full grid on the sphere, Figs. 3 and 4 in [73], are provided in Fig. 12.9.

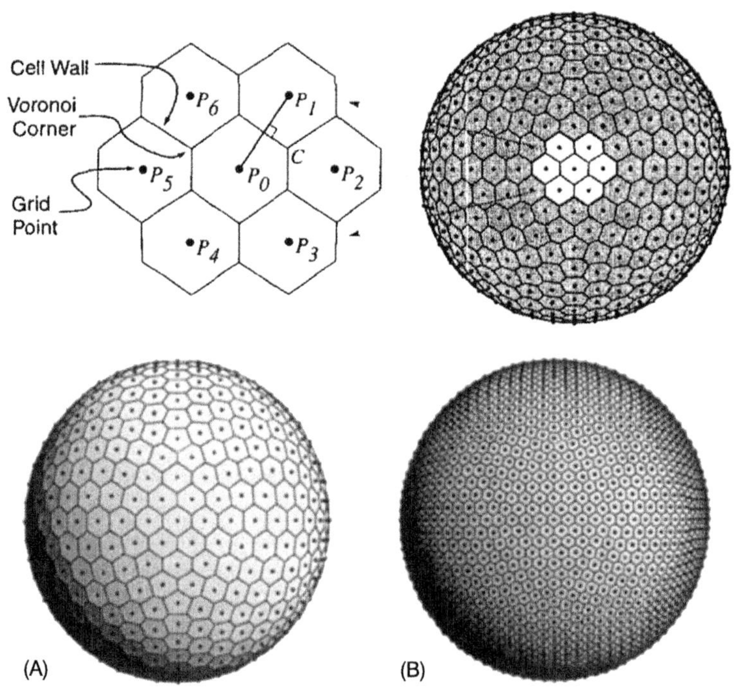

FIGURE 12.9 Copy of Figs. 3 and 4 from [73] of the properties of the Voronoi grid and the twisted icosahedron grid. Heikes, R. and Randall, D. A., 1995: Numerical Integration of the shallow water equations on a twisted icosahedral grid: Part I: Basic design and results of test, *Monthly Weather Review*, **123**, 1862–1880. ©American Meteorological Society. Used with permission.

12.3.4 Cubed sphere

The **cubed sphere** is based upon techniques similar to those of the geodesic grids just presented but now the initial shape is a cube. The derivation of the cubed sphere follows the same approach as that of the derivation of the triangular and hexagonal grids where we subdivide each face into smaller squares and then project onto the unit sphere. To show that the progression of this subdividing arrives at a shape appearing approximately like a sphere, we have plotted the grids for $N = 2, 4, 8, 16, 32$, and 64 in Fig. 12.10, where N is the number

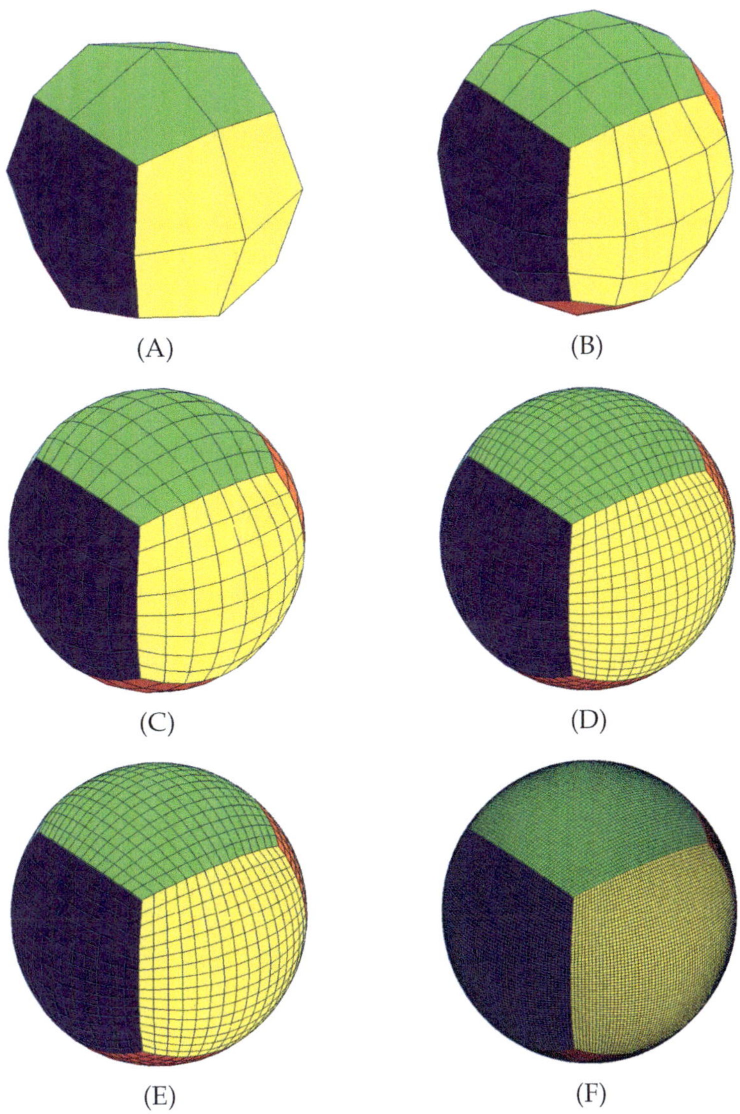

FIGURE 12.10 Plots of the buildup of the resolution of a cubed sphere grid from [89].

of squares on each face of the cube in each direction. The cubed sphere was presented by Sadourny in 1972 as a way of conserving finite differences approximations to the primitive equations [165]. It has also recently been used to model a mantle convection [89].

12.3.5 Yin–Yang

A quite recently derived approximation to the sphere was presented in Kageyama and Satoh (2004) [89] and Kagayama (2005) [90], where the motivation to developed this type of grid was not from a meteorological point of view, but from geodynamics and mantle convection simulations.

The type of grid that the Yin–Yang grid is referred to is an **overset grid**. The reason for the name is due to the grid appearing similar to the Chinese philosophy symbol of complementarity. The Yin–Yang grid is composed of two identical and complementary component grids.

The two component grids that we just mentioned are geometrically identical. The two component grids are referred to as the Yin grid, or n grid, and the Yang grid, or e grid. These two grids are combined to cover a spherical surface with partial overlap on their borders. The Yin–Yang grid can be seen to be similar in appearance to the two patches that are stitched together to form a baseball. Each component grid comprises a part of a latitude–longitude grid. The two grids overlap slightly and so an interpolation is used to patch them together as two polar stereographic grids. We have presented copies of the Yin–Yang grid from [90] in Fig. 12.11.

12.4 Spectral modeling

When we are using finite differences, these schemes only provide information of our solution at the individual grid points, and so no information about the solution between the points is provided. An alternative approach is to expand the dependent variables in terms of a finite series of smooth orthogonal functions. As a result of assuming this form for the dependent variables, the problem is then reduced to solving a set of ordinary differential equations that determine the behavior in time of the expansion coefficients.

If we start with a linear one-dimensional evolutionary differential equation given by

$$\frac{\partial \psi}{\partial t} = \mathrm{L}(\psi), \tag{12.7}$$

where L is a linear differential operator, then the spectral approach is to assume that we can expand the function ψ in terms of a set of orthogonal functions, which we shall denote by $e_k(x)$, where $k = k_1, \ldots, k_n$, as

$$\psi(t, x) = \sum_k \psi_k(t)\, e_k(x), \tag{12.8}$$

where the terms ψ_k are the expansion coefficients that we wish to determine. For the expression in (12.8) there are two possible techniques that could be used to determine the expansion

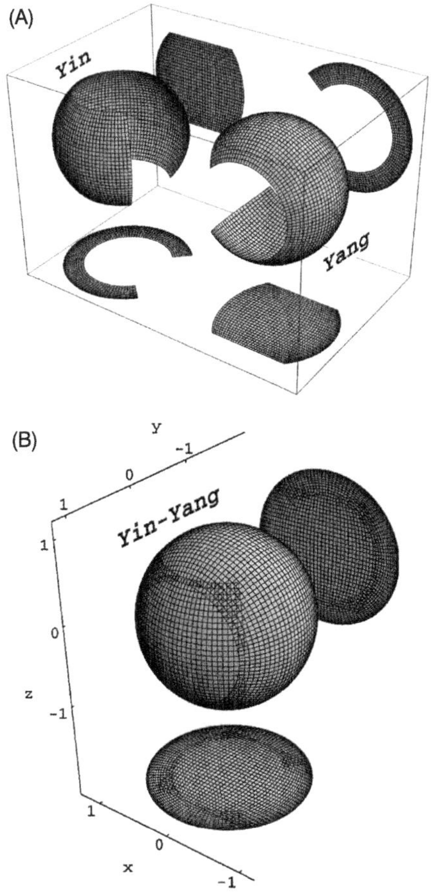

FIGURE 12.11 Copy of the Yin–Yang grid, Fig. 1 from Kageyama, A. and Sato, T., 2004: Yin–Yang grid: an overset grid in spherical geometry. *Geochem. Geophys. Geosyst.*, **4**, issue 9, 1–15. ©(2004) American Geophysical Union.

coefficients. The first is based upon minimizing a *residual* in a least-squares formulation. Here we shall present the general expression where the functions e_k are linearly independent functions. The first step is to substitute (12.8) into (12.7) and then define the residual as

$$\mathcal{R} = \sum_{k_1}^{k_2} \frac{d\psi_k}{dt} e_k - \sum_k \psi_k \mathrm{L}\left(e_k\right). \tag{12.9}$$

The next step is to choose the time derivative $\frac{d\psi_k}{dt}$ through minimizing the residual. As we mentioned earlier, this minimization can be achieved through a least-squares approach,

$$\mathcal{I} = \int \mathcal{R}^2 dx, \tag{12.10}$$

with respect to the time derivative. Therefore, we have

$$\begin{aligned}\int \mathcal{R}^2 dx &= \int \left(\sum_k \frac{d\psi_k}{dt} e_k - \sum_k \psi_k \mathrm{L}(e_k)\right)\left(\sum_l \frac{d\psi_l}{dt} e_l - \sum_l \psi_l \mathrm{L}(e_l)\right) dx \\ &= \int \left(\frac{d\psi_k}{dt}\right)\left(\frac{d\psi_l}{dt}\right) e_k e_l - \frac{d\psi_k}{dt}\psi_l \mathrm{L}(e_l) - \frac{d\psi_l}{dt}\psi_k \mathrm{L}(e_k) + \psi_k \psi_l \mathrm{L}(e_k)\mathrm{L}(e_l)\, dx.\end{aligned} \tag{12.11}$$

Now we differentiate (12.11) with respect to $\frac{d\psi_k}{dt}$ and set the derivative equal to zero, then this results in

$$\sum_l \frac{\partial \psi_l}{dt} \int e_k e_l dx = \sum_l \int e_k H(e_l)\, dx, \quad k = k_1, \ldots, k_2. \tag{12.12}$$

The second approach is referred to as the **Galerkin method** where we are seeking an approximation so that the **residual is orthogonal** to the space from which $\psi(x,t)$ comes. We start by setting

$$\int \mathcal{R}\psi_i dx = 0, \quad i = 1, 2, \ldots, N+1, \tag{12.13}$$

where ψ_i can be any set of linearly independent test functions. If the expansion functions are used as test functions, then we obtain (12.12). Since the expansion functions are known, Eq. (12.8) can be used to provide the expansion coefficients ψ_k, given the gridpoint values ψ_l. The integral in (12.12) can be calculated exactly for all possible values of k and l. This then implies that (12.12) reduces to a set of coupled ordinary differential equations that can be solved for the $\frac{d\psi_k}{dt}$ given the values of ψ_k. Thus the complete solution is

$$\frac{\partial \psi}{\partial t} = \sum_k \frac{d\psi_k}{dt} e_k. \tag{12.14}$$

Returning to the spectral approach where e_k are orthogonal, we then have the property that the expansion functions are orthonormal and so they satisfy the conditions

$$\int_0^L \overline{e_l} e_k dx = \begin{cases} 1, & l = k, \\ 0, & l \neq k, \end{cases} \tag{12.15}$$

where $\overline{e_l}$ is the complex conjugate of e_l. Applying this condition above, we obtain

$$\frac{d\psi_k}{dt} = \sum_l \psi_l \int_0^L \overline{e_k} \mathrm{L}(e_l)\, dx, \quad \forall k. \tag{12.16}$$

That is to say, we now have a set of ordinary differential equations for the rate of change in time of the expansion coefficient.

We now consider how the choice of expansion functions can greatly simply the problem of interest.

a) If the expansion functions are eigenfunctions of L, then we have that $\mathrm{L}(e_l) = \lambda_l e_l$, where the λ_l are the eigenvalues. This then makes (12.16) to be $\dfrac{d\psi_l}{dt} = \lambda_l \psi_l$.

b) If the original ordinary differential equation is of the form

$$\mathrm{M}\left(\frac{\partial \psi}{\partial t}\right) = \mathrm{L}(\psi),$$

where M is a linear operator, then the problem is simplified through using the eigenfunctions of M with eigenvalues λ_m as expansion functions, so that we obtain

$$\lambda_m \frac{d\psi}{dt} = \sum_l \psi_l \int_0^L \bar{e}_m \mathrm{L}(e_l)\, dx. \tag{12.17}$$

12.4.1 One-dimensional linear advection equation example

We can write the advection equation in terms of the longitude λ and angular velocity $\gamma = \dfrac{2\pi u_0}{L}$, where u_0 is the advecting wind speed, such that the partial differential equation is

$$\frac{\partial \varphi}{\partial t} + \gamma \frac{\partial \varphi}{\partial \lambda} = 0, \tag{12.18}$$

which has the periodicity boundary condition $\varphi(\lambda, t) = \varphi(\lambda + 2\pi K, t)$, for some integer K, and the initial condition of $\varphi(\lambda, t) = f(\lambda)$.

Given the partial differential equation in (12.18), if we are going to apply a series expansion approximation for the solution to (12.18), then we have to choose a suitable set of expansion functions. The most common choice is to apply a finite Fourier series approximation in the form of

$$\varphi(\lambda, t) = \sum_{m=-M}^{M} \varphi_m(t)\, e^{im\lambda}, \tag{12.19}$$

because the expansion functions are the eigenfunctions of the spatial differential operator. The number M is the maximum wavenumber and the φ_ms are the complex expansion coefficients. We have the property that $\varphi_{-m}(t) = \overline{\varphi_m(t)}$, and so we only have to solve for the φ_m for $0 \leq m \leq M$.

If we substitute (12.19) into (12.18), then equating coefficients of the expansion factors yields

$$\frac{d\varphi_m}{dt} + im\gamma\varphi_m = 0, \quad 0 \leq m \leq M, \tag{12.20}$$

which then results in $2M+1$ equations for the φ_ms. It is possible to integrate (12.20) exactly, which results in

$$\varphi_m = \varphi_m(0)\, e^{im\gamma t}, \quad 0 \leq m \leq M. \tag{12.21}$$

We now have to consider the initial conditions for φ. If we approximate this function with a truncated Fourier series with respect to just the spatial coordinate by

$$f(\lambda) = \sum_{m=-M}^{M} a_m e^{im\lambda},$$

then we obtain the complete solution in the form

$$\varphi(\lambda, t) = \sum_{m=-M}^{M} a_m e^{im(\lambda-\gamma t)}. \tag{12.22}$$

An important property of the solution in (12.22) is that it does not suffer from the dispersion problem that plagues some finite difference methods due to the spatial derivative being calculated analytically and not approximated, which is the case for the finite difference schemes. However, there are certain geophysical problems that involve nonlinear differential equations, and so we now look at the nonlinear advection equation to ascertain what changes, if any, are needed when considering such problems.

If we now multiply (12.19) by each of the basis functions and apply the orthogonality property of Fourier basis, then we obtain at the initial time

$$\varphi(0) = A_m \int_0^{2\pi} \varphi(\lambda, 0)\, e^{-im\lambda} d\lambda, \tag{12.23}$$

where A_m are the normalization factors. We should note that we can apply (12.19) to obtain the space distribution of the solution at any future time. This process is an application of the inverse Fourier transform.

It is quite often the case that the initial conditions are given at a set of $N+1$ grid points that have a spacing of Δx. This then means that the truncated Fourier series could be interpreted as representing an interpolating function that exactly fits the values of φ at the $N+1$ grid points. Therefore, it is possible to compute (12.23) as a discrete sum as

$$\varphi(0) = A'_m \sum_{i=1}^{K} \varphi(\lambda_i)\, e^{-im\lambda_i}, \tag{12.24}$$

which we have already seen is the discrete Fourier transform. The corresponding inverse Fourier transform for this setup is defined as

$$\varphi(\lambda_i, 0) = \sum_{m=-M}^{M} \varphi_m(0)\, e^{im\lambda_i}. \tag{12.25}$$

Both expressions in (12.24) and (12.25) are calculated through the fast Fourier transforms. It can be shown that if we start from the set of $\varphi_m(0)$, move to the set of $\varphi(\lambda_i, 0)$ for $i = 1, 2, \ldots, K$, and return to $\varphi_m(0)$, then we recover the original values as long as $K \geq 2M + 1$, with the associated grid points equally spaced. The distribution of these $2M + 1$ points is referred to as the **linear grid**.

However, it can be shown that when we have the product of two functions, it is possible to avoid aliasing errors as long as we have minimum of $K \geq 3M + 1$ grid points. The distribution of these points is referred to as the **quadratic grid**.

If we now consider the time component, then, given the derived initial conditions for the spectral coefficients, we have to integrate the ordinary differential equation for the expansion coefficients to some future time. This goal is usually achieved through some form of time-stepping finite difference procedure.

12.4.2 Nonlinear advection (Burgers equation)

As we saw earlier, the nonlinear advection equation can be written in the form

$$\frac{\partial \varphi}{\partial t} = -\varphi \frac{\partial \varphi}{\partial \lambda}. \tag{12.26}$$

We will again assume a truncated Fourier series as our estimate of a solution to (12.26) as defined in (12.19), then the right-hand side of (12.26) becomes

$$F = \sum_{m=-2M}^{2M} F_m e^{im\lambda} \text{ where } F_m = i \sum_{\widehat{m}=m-M}^{M} (m - \widehat{m}) \varphi_{\widehat{m}} \varphi, \text{ for } m \geq 0. \tag{12.27}$$

Substituting (12.19) into the left-hand side of (12.26) results in

$$\frac{\partial \varphi}{\partial t} = \sum_{m=-M}^{M} \frac{d\varphi_m}{dt} e^{im\lambda}. \tag{12.28}$$

A feature that we have to note here is that the two series presented above as approximations to (12.26) are summations that are truncated at two different wavenumbers, and so there will always be a residual associated with this approach. Given that there is a residual, denoting it by $\mathcal{R}$ again, we will then select the $\frac{d\varphi_m}{dt}$ terms such that they satisfy the condition

$$\int_0^{2\pi} \mathcal{R} e^{im\lambda} d\lambda = 0, \ \forall m,$$

where this condition is the result of applying a Galerkin method to the residuals. It can be shown that the required time derivatives are

$$\frac{d\varphi_m}{dt} = F_m, \text{ for } -M \leq m \leq M.$$

This then means that the Fourier components F_m with wavenumbers that are larger than M are neglected. The implication of this is that there is no aliasing of the smaller scales components outside of the original truncation, which implies that there is no nonlinear instability of the type we introduced earlier.

We now address a technique that is used to calculate the nonlinear term $\varphi\frac{\partial\varphi}{\partial x}$ with spectral methods. The technique that we shall consider here is referred to as the **transform method**. As the name suggests, it involves the use of transforms, especially what are referred to the **discrete Fourier transform** and then the **fast Fourier transforms** which we shall briefly introduce here.

12.4.3 Discrete Fourier transform

The discrete Fourier transform is an equivalent of the continuous Fourier transform for signals that are known at only N points, and are separated by Δx in space. We let $f(x)$ be the continuous function, which is the source of the data, and let N samples be denoted as $f(x_0), f(x_1), \ldots, f(k), \ldots f(x_{N-1})$, along with recalling that the Fourier transform of the original function, $f(x)$, is

$$F(t) = \int_{-\infty}^{\infty} f(x)\, e^{-2\pi ixt}\,dx, \tag{12.29}$$

and that the grid points contain a value of the function $f(k)$ representing an area. Since the integrand only exists at the grid point, we have that

$$\begin{aligned} F(t) &= \int_0^{(N-1)\Delta x} f(x)\, e^{-2\pi ixt}\,dx \\ &= f(x_0)\, e^{-i0} + f(x_1)\, e^{-2\pi i\Delta x} + f(x_2)\, e^{-2\pi i2\Delta x} + \cdots \\ &+ f(k)\, e^{-2\pi ik\Delta x} + \cdots + f(x_{n-1})\, e^{-2\pi i(N-1)\Delta x} \\ &= \sum_{k=0}^{N-1} f(x_k)\, e^{-2\pi ik\Delta x}. \end{aligned} \tag{12.30}$$

Note that the inverse discrete Fourier transform can easily be shown to be

$$f(x) = \frac{1}{N}\sum_{n=0}^{N-1} F(x_n)\, e^{2\pi in\Delta x}. \tag{12.31}$$

12.4.4 Fast Fourier transforms

As the name of this subsection suggests, there have been developments over the last 60 years to find ways to be able to perform the multiplications required for the discrete Fourier transform, and its inverse, so that the number of operations is drastically reduced. This then enables the movement between the grid point and frequency spaces be quick and efficient, and that it then allows for a more practical implementation of the spectral model in the modern day supercomputer infrastructure.

The fast Fourier transform reduces the number of computation needed for N points from $2N^2$ to $2N \log_2 N$, where $\log_2$ is the base 2 logarithm. Fast Fourier Transforms were first officially discussed in a paper by Cooley and Tukey in 1965 [37]. However, there is evidence that Gauss derived a similar formulation in 1805.

The definition of a fast Fourier transform is that it computes the discrete Fourier transform and produces exactly the same result as evaluating the discrete Fourier transform definition directly, but as the name suggests, these algorithms are faster than the direct evaluation. The algorithm for the fast Fourier transform from [37] takes the discrete Fourier transform and first rearranges the input data into bit-reversed order, then it builds the output transform. The basic idea of the Cooley and Tukey's fast Fourier transform algorithm is to break up the transform of length N into two transforms of length $\frac{N}{2}$ using the identity

$$\begin{aligned}\sum_{n=0}^{N-1} a_n e^{\frac{-2\pi ink}{N}} &= \sum_{n=0}^{\frac{N}{2}-1} a_{2n} e^{\frac{-2\pi i(2n)k}{N}} + \sum_{n=0}^{\frac{N}{2}-1} a_{2n+1} e^{\frac{-2\pi i(2n+1)k}{N}} \\ &= \sum_{n=0}^{\frac{N}{2}-1} a_n^{\text{even}} e^{\frac{-2\pi i(2n)k}{N}} + e^{-\frac{2\pi ik}{N}} \sum_{n=0}^{\frac{N}{2}-1} a_n^{\text{odd}} e^{\frac{-2\pi i(2n)k}{N}}. \end{aligned} \tag{12.32}$$

The reason for introducing the discrete and fast Fourier transforms is because of the application in spectral modeling. We start this subsection considering the nonlinear advection equation and so we now explain why we require these transforms.

The advantage of the fast Fourier transform is that it is easy to move from the spectral representation (spectral space) to a grid-point representation, often referred to as the physical space. The essence of the transform method is to calculate derivatives in the spectral space, but to transform to physical space using the fast Fourier transform whenever a product is required. Once all of the products have been computed at grid points, the spectral coefficients of this product field are calculated.

Therefore, given φ_m, we wish to compute the spectral coefficients of the nonlinear term, $-\varphi\frac{\partial\varphi}{\partial\lambda}$, which is F_m in (12.27). To achieve this goal, we have the following three step process to follow:

1. Calculate the functions φ and $\frac{\partial\varphi}{\partial\lambda}$ at the grid points λ_i by using the spectral coefficients

$$\varphi(\lambda_i) = \sum_m \varphi_m e^{im\lambda_i}, \qquad \frac{\partial\varphi}{\partial\lambda_i} = \sum_m im\varphi_m e^{im\lambda_i}.$$

2. Calculate the advection term at each grid point in the physical space as

$$F(\lambda_i) = -\varphi(\lambda_i)\frac{\partial\varphi}{\partial\lambda_i}.$$

3. Finally, return this product to the spectral space through calculating the Fourier coefficients

$$F_m = \frac{1}{2\pi} \sum_i F(\lambda_i) e^{-im\lambda_i}.$$

We should note here that the procedure described above has to be employed to calculate the spectral coefficient of the nonlinear advection term at each time step. Another feature we have to make a note of here is that as the product of two functions is computed in the grid-point space and not in the spectral space, thus we obtain aliasing unless the number of grid points corresponds to the quadratic grid.

We have shown the basis for spectral method here, but we now move on to present the theory that will enable us to extend the spectral methods to spherical coordinates.

12.4.5 Sturm–Liouville theory

Another important set of theories that we need to introduce to be able to understand the implementation of the spectral method are those that lead to the Sturm–Liouville theory. The **Sturm–Liouville equation** is a real second-order linear differential equation of the form

$$\frac{d}{dx}\left(p(x)\frac{dy}{dx}\right) = q(x) = -\lambda w(x) y, \tag{12.33}$$

where the functions $p(x)$, $q(x)$, and $w(x)$ are specified. In the simplest case all the coefficients are continuous on the finite closed interval $[a, b]$, and the function $p(x)$ has a continuous first derivative; finally, we have boundary conditions on the solution $y(x)$ at $x = a$ and $x = b$. The value of λ is not specified in the equation and thus finding the values for λ, so that there exist nontrivial solutions of (12.33), is part of the set of problems referred to as **Sturm–Liouville** problems.

A Sturm–Liouville problem is said to be regular if $p(x)$, $w(x) > 0$ and $p(x)$, $p'(x)$, $q(x)$, and $w(x)$ are continuous functions over the finite interval $[a, b]$, and has separated boundary conditions of the form

$$\alpha_1 y(a) + \alpha_2 y'(a) = 0 \qquad \left(\alpha_1^2 + \alpha_2^2 > 0\right), \tag{12.34a}$$

$$\beta_1 y(b) + \beta_2 y'(b) = 0 \qquad \left(\beta_1^2 + \beta_2^2 > 0\right). \tag{12.34b}$$

An important property that we now consider is the definition of a **regular singular point**.

Definition 12.2. Consider the second-order ordinary differential equation of the form

$$y'' + P(x) y' + Q(x) y = 0. \tag{12.35}$$

If $P(x)$ and $Q(x)$ remain finite at $x = x_0$, then x_0 is called an **ordinary point**. If either $P(x)$ or $Q(x)$ diverges as $x \to x_0$, then x_0 is called a **singular point**. If either $P(x)$ or $Q(x)$ diverges as $x \to x_0$, but $(x - x_0) P(x)$ and $(x - x_0)^2 Q(x)$ remain finite as $x \to x_0$, then the point $x = x_0$ is called a **regular singular point.**

Given the assumption that the Sturm–Lioville problem is regular, as just defined, we have the following properties of this problem:

- The eigenvalues $\lambda_1, \lambda_2, \ldots$ of the regular Sturm–Lioville problem defined by (12.33), (12.34a), and (12.34b) are real, and can be ordered such that

$$\lambda_1 < \lambda_2 < \cdots < \lambda_n < \cdots \to \infty.$$

- For each eigenvalue, λ_n, there is a corresponding unique eigenfunction, $y_n(x)$, which has exactly $n-1$ zeros in the interval (a, b). The eigenfunction $y_n(x)$ is referred to as the nth fundamental solution satisfying the regular Sturm–Lioville problem given by (12.33), (12.34a), and (12.34b).
- The normalized eigenfunctions form an orthonormal basis

$$\int_a^b y_n(x)\, y_m(x)\, w(x)\, dx = \delta_{mn} = \begin{cases} 1, & m = n, \\ 0, & m \neq n. \end{cases} \tag{12.36}$$

We now move to consider another form of differential equation that plays an important part in spherical spectral modeling.

12.4.6 Legendre differential equation

If we consider the basic spherical coordinate system, then we start with a point P that is specified by the distance r from the origin, the angle θ between the position vector and the z axis, and the angle ϕ from the x axis to the projection of the position vector onto the x–y plane. We have presented an illustration of this coordinate system in Fig. 12.12.

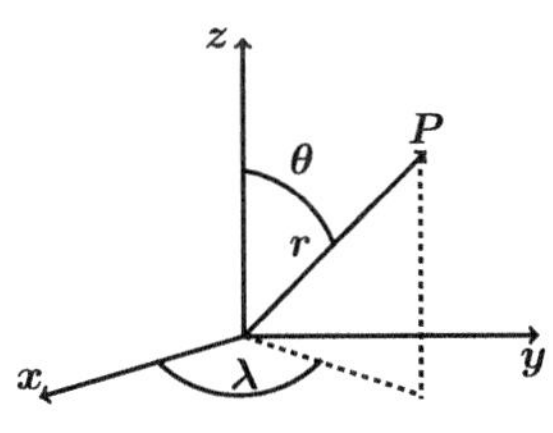

FIGURE 12.12 Schematic of the spherical coordinate system.

The Laplacian equation for a function $F(r, \theta, \phi)$ is given by

$$\frac{1}{r^2}\frac{\partial}{\partial r}\left(r^2\frac{\partial F}{\partial r}\right) + \frac{1}{r^2\sin\theta}\frac{\partial}{\partial \theta}\left(\sin\theta\frac{\partial F}{\partial \theta}\right) + \frac{1}{r^2\sin^2\theta}\frac{\partial^2 F}{\partial \lambda^2} = 0. \tag{12.37}$$

If we only consider solutions that are dependent on r and θ, then (12.37) simplifies to

$$\frac{1}{r^2}\frac{\partial}{\partial r}\left(r^2\frac{\partial F}{\partial r}\right) + \frac{1}{r^2\sin\theta}\frac{\partial}{\partial \theta}\left(\sin\theta\frac{\partial F}{\partial \theta}\right) = 0. \tag{12.38}$$

We now assume that it is possible to apply separation of variables technique to solve (12.38), which implies that we are seeking a solution of the form $Y(r,\theta) = R(r)\Theta(\theta)$. Assuming a solution of this form results in the following two uncoupled ordinary differential equations:

$$r^2\frac{d^2R}{dr^2} + 2r\frac{dR}{dr} - \lambda R = 0, \tag{12.39a}$$

$$\frac{d}{d\theta}\left(\sin\theta\frac{d\Theta}{d\theta}\right) + \lambda\sin\theta\Theta = 0, \tag{12.39b}$$

where λ is referred to as the **separation constant**. The ordinary differential equation for r is equidimensional and thus has solutions that are easily found and are powers of r. The ordinary differential equation for θ is **Legendre's equation** which can be solved through the following means:

First we introduce the change of variable $\tau = \cos\theta$, and substitute this information into (12.39b). Recognizing that $\frac{d\tau}{d\theta} = -\sin\theta$, we apply the chain rule $\frac{d\Theta}{d\tau} = \frac{d\Theta}{d\tau}\frac{d\tau}{d\theta}$, which results in

$$\begin{aligned}
\frac{d}{d\tau}\left(\frac{d\tau}{d\theta}\right)\left(\sin\theta\frac{d\tau}{d\theta}\frac{d\Theta}{d\tau}\right) + \lambda\sin\theta\Theta &= 0,\\
\frac{d}{d\tau}(-\sin\theta)\left((\sin\theta)(-\sin\theta)\frac{d\Theta}{d\tau}\right) + \lambda\sin\theta\Theta &= 0,\\
\frac{d}{d\tau}\left(\sin^2\theta\frac{d\Theta}{d\tau}\right) + \lambda\Theta &= 0,\\
\frac{d}{d\tau}\left(\left(1-\cos^2\theta\right)\theta\frac{d\Theta}{d\tau}\right) + \lambda\Theta &= 0,\\
\frac{d}{d\tau}\left(\left(1-\tau^2\right)\theta\frac{d\Theta}{d\tau}\right) + \lambda\Theta &= 0,
\end{aligned} \tag{12.40}$$

where (12.40) has regular singular points at the endpoints $\tau = \pm 1$. The ordinary differential equation in (12.40), combined with the boundary condition just mentioned, makes this problem into a singular Sturm–Lioville equation, which we just presented in the last subsection. Therefore, there is a set of λ for which there are well behaved solutions corresponding to the eigenvalues of this problem.

We now turn our attention to a method for solving (12.40), where we consider a power series solution for Θ about $\tau = 0$, which results in

$$\Theta(\tau) = \sum_{n=0}^{\infty} a_n\tau^n. \tag{12.41}$$

Substituting (12.41) into (12.40), using the fact that the derivatives of a power series are defined as

$$\Theta'(\tau) = \sum_{n=1}^{\infty} a_n n\tau^{n-1} = \sum_{n=0}^{\infty} a_{n+1}(n+1)\tau^n, \tag{12.42a}$$

$$\Theta''(\tau) = \sum_{n=2}^{\infty} a_n n(n-1)\tau^{n-2} = \sum_{n=0}^{\infty} a_{n+2}(n+1)(n+2)\tau^n, \tag{12.42b}$$

and substituting (12.42a) and (12.42b) into (12.40) enables us to derive a recursive relationship between a_{n+2} and a_n as follows:

$$\begin{aligned} \Theta'' - \tau^2\Theta'' - 2\tau\Theta' + \lambda\Theta &= 0, \\ \sum_{n=2}^{\infty} a_n n(n-1)\tau^{n-2} - \sum_{n=2}^{\infty} a_n n(n-1)\tau^n - 2\sum_{n=1}^{\infty} a_n\tau^n + \lambda\sum_{n=0}^{\infty} a_n\tau^n &= 0, \\ (n+1)(n+2)a_{n+2} - (n(n-1) + 2n - \lambda)a_n &= 0, \\ \Rightarrow a_{n+2} &= \frac{n(n+1) - \lambda}{(n+1)(n+2)} a_n. \end{aligned} \tag{12.43}$$

For very large values of n, we may approximate the recursive relationship in (12.43) by ignoring λ compared to $n(n+1)$. We see that the relationship simplifies to $(n+2)a_{n+2} \approx na_n$. From this it is possible to show that any such solution will be logarithmically singular at one of the endpoints. The only way to avoid this is for the recurrence relationship in (12.43) to terminate. We see that for this to happen we require the numerator to be equal to zero. This occurs when $\lambda = k(k+1)$ for some value of k, which is a nonnegative integer.

Given the termination condition, we now have a terminating solution of the form of polynomial of degree k. Such solutions are referred to as the **Legendre Polynomials**.

12.4.7 Legendre polynomials

The set of polynomials that we introduce here play an important part in defining the solutions to the spherical harmonic differential equation which leads to an important part in the application of spectral methods for modeling ordinary and partial differential equations in spherical coordinates.

We start with the Legendre functions that are the solutions to the Legendre differential equations defined by

$$\frac{d}{dx}\left(\left(1-x^2\right)\frac{d}{dx}P_n(x)\right) + n(n+1)P_n(x) \equiv \left(1-x^2\right)\frac{d^2P_n}{dx^2} - 2x\frac{dP_n(x)}{dx} + n(n+1) = 0, \tag{12.44}$$

which we have shown is a Sturm–Lioville problem. We have determined the constants $n(n+1)$ as the necessary conditions to enable the solution to (12.44) to have a regular solution in the interval $[-1, 1]$. We have seen that the solutions to (12.44) are polynomials of order n. An important property of the solutions to (12.44) is that they satisfy the orthogonality relationship

$$\int_{-1}^{1} P_n(x)P_m(x)\,dx = \begin{cases} 1, & n = m, \\ 0, & n \neq m. \end{cases} \tag{12.45}$$

A second condition that the Legendre polynomials satisfy is

$$\int_{-1}^{1} P_n^2(x)\,dx = \frac{2}{2n+1}.$$

The Legendre polynomials have two different methods for generating the associated polynomials. The first method is through a generating function that is given by

$$G(x,t) = \sum_{n=0}^{\infty} t^n P_n(x) = \frac{1}{\sqrt{1-2xt+t^2}}. \tag{12.46}$$

The second approach for generating the polynomials is through **Rodrigue's formula**, which is given by

$$P_n(x) = \frac{1}{2^n n!}\frac{d^n}{dx^n}\left(x^2-1\right)^n. \tag{12.47}$$

Given the two expressions for generating the Legendre polynomials in (12.46) and (12.47), it is possible to derive recursive relationships between the polynomials and their derivatives

$$\begin{aligned} P'_{n+1} - P'_{n-1} &= (2n+1)\,P_n, \\ (n+1)\,P_{n+1} &= (2n+1)\,x P_n + n P_{n-1} = 0. \end{aligned}$$

To help illustrate the mathematical expressions for the Legendre polynomials, we have presented the first 6 polynomials in Table 12.1. There we can see that if n is even then the associated Legendre polynomial is only a function of the even powers of x, while if n is an odd number then the polynomials are only functions of the odd powers of x. The first six Legendre polynomials are presented in Fig. 12.13.

TABLE 12.1 Expressions for the first six Legendre polynomials.

Order	Polynomial
$P_0(x)$	1
$P_1(x)$	x
$P_2(x)$	$\frac{1}{2}\left(3x^2-1\right)$
$P_3(x)$	$\frac{1}{2}\left(5x^3-3x\right)$
$P_4(x)$	$\frac{1}{8}\left(35x^4-30x^2+3\right)$
$P_5(x)$	$\frac{1}{8}\left(63x^5-70x^3+15x\right)$

Exercise 12.3. *Derive the coefficients for the first six Legendre polynomials in Table 12.1.*

An interesting feature of the plots of the Legendre polynomials in Fig. 12.13 is that each polynomial has the same number of zeros as the order of the polynomial. That is to say,

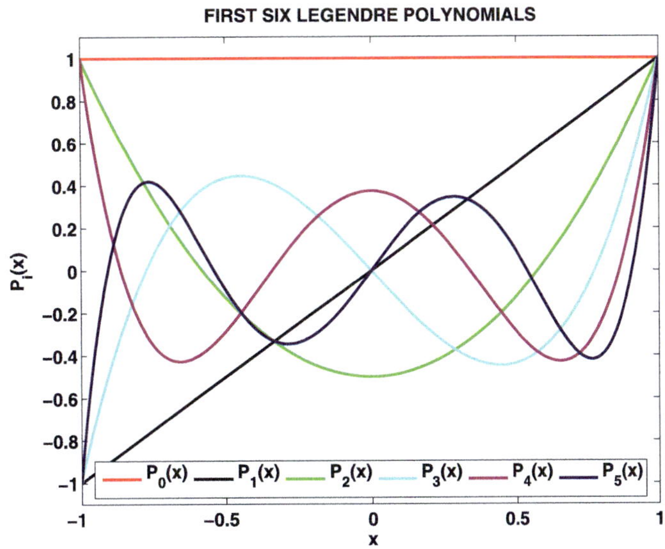

FIGURE 12.13 Plot of the first 6 Legendre polynomials.

P_0 never takes the value of zero, for P_1 we can see that this polynomial has one zero, while for P_2 there two values for $x \in [-1, 1]$ where $P_2(x) = 0$, and also we see that this it true for other three polynomials that we have presented in Fig. 12.13.

12.4.8 Spherical harmonics

The starting point for deriving the associated basis functions for spherical harmonics is to consider Laplace's equation in three dimensions on the sphere. Laplace's equation is equivalent to imposing the condition that the divergence of the gradient of a scalar field, f, is zero. Therefore, Laplace's equation in spherical coordinates is given by

$$\nabla^2 f = \frac{1}{r^2}\frac{\partial}{\partial r}\left(r^2\frac{\partial f}{\partial r}\right) + \frac{1}{r^2\sin\theta}\frac{\partial}{\partial\theta}\left(\sin\theta\frac{\partial f}{\partial\theta}\right) + \frac{1}{r^2\sin^2\theta}\frac{\partial^2 f}{\partial\lambda^2} = 0, \tag{12.48}$$

where r is the radius in the outward direction, θ is the colatitude such that $\theta \in [0, \pi]$, and λ is the longitude satisfying $\lambda \in [0, 2\pi)$.

We start to solve (12.48) by assuming a separation of variables in the form of

$$f(r, \theta, \lambda) = R(r)\,Y(\theta, \lambda), \tag{12.49}$$

and, through this separation of variables technique for the Laplace equation in (12.48), we obtain two differential equations in the form

$$\frac{1}{R}\frac{d}{dr}\left(r^2\frac{dR}{dr}\right) = \alpha, \tag{12.50a}$$

$$\frac{1}{Y}\frac{1}{\sin\theta}\frac{\partial}{\partial\theta}\left(\sin\theta\frac{\partial Y}{\partial\theta}\right)+\frac{1}{Y}\frac{1}{Y}\frac{1}{\sin^2\theta}\frac{\partial^2 Y}{\partial\lambda^2}=-\alpha, \tag{12.50b}$$

where we have used the property from the Laplace equation that the sum of these two equations must equal zero and as such they are equal to a constant but of opposite sign, and where we have made the assumptions that (1) f is not identically equal to zero, and then, by association, (2) neither R nor Y are equal to zero.

We now go a step further and assume that we can apply the separation of variable technique to the function Y to separate it into the product to two functions such that $Y(\theta,\lambda)=\Theta(\theta)\Lambda(\lambda)$, which then means that we can rewrite (12.50b) in terms of two differential equations given by

$$\frac{1}{\Lambda}\frac{d^2\Lambda}{d\lambda}=-m^2, \tag{12.51a}$$

$$\alpha\sin^2\theta+\frac{\sin\theta}{\Theta}\frac{d}{d\theta}\left(\sin\theta\frac{d\Theta}{d\theta}\right)=m^2, \tag{12.51b}$$

for some number m which is a complex constant, but because Λ must be a periodic function whose period evenly divides 2π, m is necessarily an integer, and Λ is a linear combination of the complex exponentials $e^{\pm im\lambda}$, that is to say, the solution to the ordinary differential equation in (12.51a) is $e^{\pm im\lambda}$.

For the ordinary differential equation in (12.51b), we have already shown that we have a regular condition for Θ at the poles at $\theta=0$ and $\theta=\pi$. Therefore, enforcing the regular condition in the solution Θ makes (12.51b) a Strum–Liouville problem that we have already shown forces the parameter α to be equal to $\alpha=l(l+1)$ for some nonnegative integer l such that $l\geq|m|$. Also if we use the change of variable, $\tau=\cos\theta$, then we now have the Legendre equation, whose solution is a multiple of the associated Legendre polynomial $P_l^m(\cos\theta)$. Given the expression for α, substituting it into the ordinary differential equations for r yields the solution $R(r)=Ar^l+Br^{-l-1}$. However, requiring the solution to be regular throughout $\mathbb{R}^3$ forces $B=0$.

Recall that we assumed that the solution has the special form $Y(\theta,\lambda)=\Theta(\theta)\Lambda(\lambda)$, and so for a given value of l, there are $2l+1$ independent solutions of this form, one for each integer m with $-l\leq m\leq l$. These solutions are referred to as **angular solutions** and are the products of trigonometric functions and the associated Legendre polynomials in the form of

$$Y_l^m(\theta,\lambda)=Ne^{\pm im\lambda}P_l^m(\cos\theta), \tag{12.52}$$

which satisfies

$$r^2\nabla^2Y_l^m(\theta,\lambda)=-(l+1)Y_l^m(\theta,\lambda). \tag{12.53}$$

Here Y_l^m is called a **spherical harmonic function** of degree l and order m, P_l^m is an associated Legendre polynomial, while N is a normalization constant. For a fixed integer l, every solution $Y(\theta,\lambda)$ of the eigenvalue problem

$$r^2\nabla^2Y=-l(l+1)Y, \tag{12.54}$$

is a linear combination of Y_l^m.

12.4.9 Legendre transforms

We start by considering a convex function, where by convex we mean that a real-valued function defined on an interval is convex if the line segment between any two points on the graph of the function lies above or on the graph. Given this requirement for $f(x)$, we define a new function, $f^*(p)$, such that

$$f^*(p) \equiv \max_x (px - f(x)), \tag{12.55}$$

which is referred to as the **Legendre transform** of $f(x)$. If the function f is also differentiable, then we can calculate the maximum through

$$0 = \frac{d}{dx}(px - f(x)) = p - \frac{df(x)}{dx},$$

whose solution depends on p, and as such this solution is denoted as $x(p)$, thus we have

$$\left.\frac{d(x)}{dx}\right|_{x=x(p)} = p, \tag{12.56}$$

where upon substituting (12.56) into (12.55) yields

$$f^*(p) = px(p) - f(x(p)). \tag{12.57}$$

If we now consider the Legendre transform of $f^*(p)$, i.e., of (12.57), then we have

$$\left(f^*\right)^*(y) = \max_p \left(yp - f^*(p)\right). \tag{12.58}$$

Now if we assume that the Legendre transform is differentiable with respect to p, then we obtain

$$\left.\frac{df^*(p)}{dp}\right|_{p=p(y)} = y. \tag{12.59}$$

However, if we substitute all of the known expressions from above into (12.59), then we arrive at

$$\frac{dy^*(p)}{dp} = \frac{d}{dp}(px(p) - f(x(p))) = x(p) + p\frac{df(x)}{dp} - \left.\frac{df(x)}{dx}\right|_{x=x(p)} \frac{dx(p)}{dp} = x(p), \tag{12.60}$$

where we have used (12.56) for the cancelation in (12.60). Given the result in (12.60), we now have that

$$f^{**}(y) = yp(y) - f^*(p(y)) = yp(y) - p(y)xp(y) + f(x(p(y))) = f(y). \tag{12.61}$$

The importance of (12.61) is that it tells us that the inverse of the Legendre transform is the Legendre transform itself, that is to say, the transform is its own inverse.

In numerical modeling we shall be dealing with multivariate problems and so we require the transform to be extendable to this situation, and it is. The Legendre transform for the multivariate case is given by

$$f^* (p_1, p_2, \ldots, p_n) = \sum_{n=1}^{m} x_n p_n - f (x_1, x_2, \ldots, x_n), \text{ where } p_n = \frac{\partial f}{\partial x_n}. \tag{12.62}$$

Given all the theory that we have presented so far, we now move on to introduce its applications in the spectral methods for spherical coordinates.

12.4.10 Spectral methods on the sphere

The use of spectral methods on the sphere was first suggested in Silberman (1954) [171]. As we have seen for the theory for spectral methods in Cartesian coordinates, we are able to represent a function F that is defined on the sphere as

$$F (\lambda, \theta) = \sum_{m=-M}^{M} \sum_{n=|m|}^{K} F_n^m Y_n^m (\lambda, \theta), \tag{12.63}$$

where m is referred to as the zonal wavenumber, n is the total wavenumber, and $n - |m|$ represents the effective meridional wavenumber. Given the expression in (12.63), it is possible to consider different truncations. Two of the most well known and used truncations are the **triangular** and **rhomboidal** truncations. They are defined as follows:

- If $K = M$, then the truncation is referred to as **triangular**, and for some applications of this truncation when $m = 100$, the model is denoted by $T100$.
- If $K = M + |m|$, then the truncation is referred to as being a **rhomboidal** truncation.

To help illustrate why the two different truncations have these labels, we have presented the first five rows of both grids in Fig. 12.14.

Given that we are considering spectral approximation on the sphere, it seems natural to use the spherical harmonic as the expansion functions. This then leads to

$$Y_n^m (\lambda \theta) = e^{im\lambda} P_n^m (\sin \theta), \tag{12.64}$$

where P_n^m are the associated Legendre polynomials with the property that

$$P_{-m}^n = P_m^n.$$

The spherical harmonics also have the property that

$$\nabla^2 Y_n^m = -\frac{n (n+1)}{a^2} Y_n^m, \tag{12.65}$$

where a is the Earth's radius.

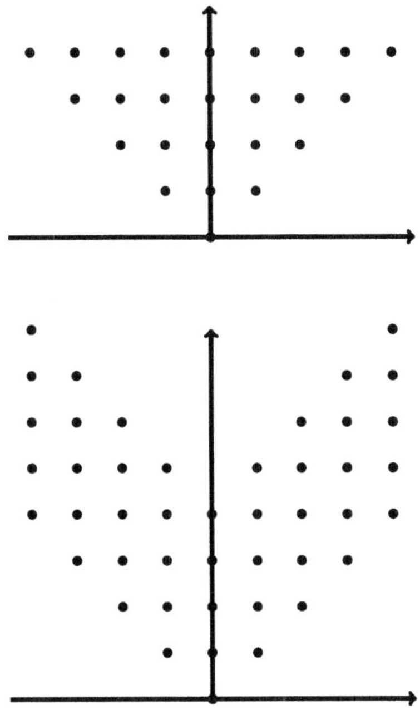

FIGURE 12.14 Triangular and rhomboidal truncations for $M = 4$.

We now recall the feature that the Legendre polynomials are orthogonal and therefore they satisfy

$$\frac{1}{2}\int_{-1}^{1} P_n^m(y)\, P_s^m(y)\, dy = \delta_{ns}. \tag{12.66}$$

The space derivative of the spherical harmonic can be computed analytically as

$$\frac{\partial}{\partial \lambda} Y_n^m = im Y_n^m, \tag{12.67}$$

and by the property of the Legendre polynomial we have

$$\left(1 - y^2\right)\frac{\partial}{\partial y} Y_n^m = -n\varepsilon_{n+1}^m Y_{n+1}^m + (n+1)\,\varepsilon_n^m Y_{n-1}^m, \text{ where } \varepsilon_n^m \equiv \sqrt{\frac{n^2 - m^2}{4(n-1)}}, \tag{12.68}$$

for $m \geq 0$. Another important property is that for $m > 0$ we have that $Y_n^{-m} = \overline{Y_n^m}$.

Given the properties just described, it is possible to calculate the space derivatives exactly, which leaves a set of ordinary differential equations for the time rate of change of the spherical harmonics coefficients, F_n^m.

However, as we saw with the Cartesian spectral theory, it is highly likely that we shall have nonlinear terms that we need to approximate, where two spherical harmonics interact to produce a third spherical harmonic. Unless a truncation is quite small, the calculations involved for this nonlinear interaction is very time-consuming. As we saw in the Cartesian section, we can overcome this problem through introducing transformations. The three steps involved in this process are:

(a) Starting in the spectral space, the spectral coefficients are used to calculate the dependent variables on a latitude–longitude grid through an inverse spectral transform. If the regularly spaced longitudinal grid has at least $2M+1$ grid points and a specific latitude grid is chosen, which we shall go into more details soon, then the transformations can be preformed exactly.
(b) The nonlinear dynamics and physical process terms are calculated in real space.
(c) The nonlinear terms are transformed back to the spectral domain, which is referred to as the direct spectral transform.

To be able to perform the spectral transform, we introduce the Fourier coefficients as

$$F_n^m(y,t)=\frac{1}{2\pi}\int_0^{2\pi}F(\lambda,y,t)\,e^{-im\lambda}d\lambda=\sum_{n=|m|}^{N}F_n^m(t)\,P_n^m(y). \tag{12.69}$$

Next we multiply (12.63) by each of the spherical harmonics and use the orthogonality properties of the Fourier basis functions and the Legendre polynomials to obtain

$$F_n^m(t)=\frac{1}{4\pi}\int_0^1\int_0^{2\pi}F_n^m(\lambda,y,t)\,P_n^m(y)\,e^{-im\lambda}d\lambda dy. \tag{12.70}$$

The transform defined in (12.70) is referred to as the **direct spectral transform**.

To be able to calculate this transform, we first evaluate the integral with respect to λ. This is a Fourier transform and as such, if the data is at grid point in the longitudinal direction, then this can be achieved through a fast Fourier transform of the discrete Fourier transform. The transform in the longitudinal direction will be exact if the number of points in this direction is at least $2M+1$ and they are also equally spaced.

For the evaluation of the integral in the latitude direction, we use the Fourier coefficients by means of a **Gaussian quadrature** formula, and it can be shown that this integral will be exact if the data points are taken at the points where the Legendre polynomials are zero. This transform in the θ direction is the **Legendre transform**. We saw that for each Legendre polynomial the number of its zeros is the same as its order. The points in the longitudinal direction that correspond to these zeros of the Legendre polynomial are referred to as the **Gaussian latitudes** and the number of the Gaussian latitude points for the solution to be exact must be $N_G\geq\frac{(2M+1)}{2}$. If we were to multiply two functions and wish the approximation to be free of aliasing errors, then it can be shown that we would require $N_G\geq\frac{3M+1}{2}$.

As we saw in the plot of the first six Legendre polynomials, the zeros of the polynomials were **not** equally spaced. The grid associated with the requirements just mentioned is referred to as the **Gaussian grid**. The grid associated with the single function when $N_G \geq \frac{2M+1}{2}$ is the **Gaussian linear grid**, while the grid associated with the product of two functions, provided $N_G \geq \frac{3M+1}{2}$, is referred to as the **Gaussian quadratic grid**. An important feature of the Legendre polynomials which transfers to the grid points of the Gaussian grid is that, due to the regularity condition of the solution to the Sturm–Lioville equation, there are no points at the poles.

We now provide a short algorithmic description of how to implement the spectral method on the sphere.

Step 1: Select a specific spectral truncation. Given this truncation, we identify the minimum number of grid points required in the longitudinal and latitudinal directions to avoid aliasing due to the quadratic nonlinearities.
Step 2: Identify the highest degree Legendre polynomial required with the chosen spectral truncation, and then find the roots of this polynomial which will be the Gaussian latitudes.
Step 3: Evaluate the horizontal derivatives in the spectral domain.
Step 4: Transform from the spectral space to the Gaussian grid point space through the use of the inverse discrete Fourier transform, by applying the inverse fast Fourier transform techniques, and also the inverse Legendre transform.
Step 5: Evaluate the nonlinear terms in the approximation to the differential equation along with model physics on the Gaussian grid.
Step 6: Apply the Fast Fourier transform and the Legendre transform to the grid point values to obtain the spectral coefficients.
Step 7: Apply a time-stepping scheme to advance the model and then repeat from step 3 at the next time level.

We now move on to consider how to implement the semi-Lagrangian method on the sphere.

12.5 Semi-Lagrangian and alternating direction implicit (SLADI) scheme

In this section we shall present an alternative scheme that was proposed in [12] for a limited area shallow water equations model on the sphere to the semi-implicit approach. To start we have to consider how the systems of partial differential equations translate onto the sphere. We shall start with the shallow water equations which, according to [12], are defined in spherical coordinates as

$$\frac{Du}{Dt} = -\frac{1}{a\cos\theta}\frac{\partial\phi}{\partial\lambda} + \left(f + \frac{u\tan\theta}{a}\right)v, \tag{12.71}$$

$$\frac{DV}{Dt} = -\frac{1}{a}\frac{\partial\phi}{\theta} - \left(f + \frac{u\tan\theta}{a}\right)u, \tag{12.72}$$

$$\frac{D\phi}{Dt} = -\phi\left(\frac{1}{a\cos\theta}\frac{\partial u}{\partial\lambda} + \frac{1}{a\cos\theta}\frac{\partial}{\partial\theta}(v\cos\theta)\right). \tag{12.73}$$

As we mentioned in Chapter 10, in this model Bates employs the time-splitting approach where to obtain the numerical solution the equations are split into advective and adjustment phases as

$$\frac{Du}{Dt} = 0, \tag{12.74a}$$

$$\frac{Dv}{Dt} = 0, \tag{12.74b}$$

$$\frac{D\phi}{Dt} = 0, \tag{12.74c}$$

$$\frac{\partial u}{\partial t} = -\frac{1}{a\cos\theta}\frac{\partial \phi}{\partial \lambda} + \left(f + \frac{u\tan\theta}{a}\right)v, \tag{12.74d}$$

$$\frac{\partial v}{\partial t} = -\frac{1}{a}\frac{\partial \phi}{\theta} - \left(f + \frac{u\tan\theta}{a}\right)u, \tag{12.74e}$$

$$\frac{\partial \phi}{\partial t} = -\left(\overline{\phi} + \phi'\right)\left(\frac{1}{a\cos\theta}\frac{\partial u}{\partial \lambda} + \frac{1}{a\cos\theta}\frac{\partial}{\partial \theta}(v\cos\theta)\right), \tag{12.74f}$$

where $\overline{\phi}$ is the mean geopotential and ϕ' is the deviation therefrom.

It is stated in [12] that the advection equations (12.74a) and (12.74c) are integrated using the multiply-upstream, semi-Lagrangian schemes with biquadratic interpolation as in [14] with an advective time step of Δt_{adv}. The adjustment equations (12.74d)–(12.74f) are integrated in two time steps of $\frac{\Delta t_{\text{adj}}}{2}$. As we mentioned earlier, in the first step the λ-derivatives in linear terms are treated implicitly, while the θ-derivatives are treated explicitly, whereas in the second step the θ-derivatives in linear terms are treated implicitly while λ-derivatives are treated explicitly. All nonlinear terms are treated in an effectively explicit manner. We assume that the two adjustment steps are executed in immediate succession:

Step I:

$$v_{i,j}^{n+\frac{1}{2}} = v_{i,j}^{n} - \mu_2\left(\phi_{i,j+1}^{n} - \phi_{i,j-1}^{n}\right) - F_{i,j}^{n}u_{i,j}^{n}, \tag{12.75}$$

$$u_{i,j}^{n+\frac{1}{2}} = u_{i,j}^{n} - \mu_1(j)\left(\phi_{i+1,j}^{n} - \phi_{i-1,j}^{n}\right) + F_{i,j}^{n}v_{i,j}^{n+\frac{1}{2}}, \tag{12.76}$$

$$\begin{aligned} \phi_{i,j}^{n+\frac{1}{2}} &= \phi_{i',j}^{n} - \overline{\phi}\left(\mu_1(j)\left(u_{i'+1,j}^{n+\frac{1}{2}} - u_{i'-1,j}^{n+\frac{1}{2}}\right) + \mu_3(j)\left(\widehat{v}_{i',j+1}^{n} - \widehat{v}_{i',j-1}^{n}\right)\right) \\ &- (\phi')_{i',j}^{n}\left(\left(\mu_1(j)\left(u_{i'+1,j}^{n} - u_{i'-1,j}^{n}\right)\right) + \mu_3(J)\left(\widehat{v}_{i',j+1}^{n} - \widehat{v}_{i',j-1}^{n}\right)\right); \end{aligned} \tag{12.77}$$

Step II:

$$u_{i,j}^{n+1} = u_{i,j}^{n+\frac{1}{2}} - \mu_1(j)\left(\phi_{i+1,j}^{n+\frac{1}{2}} - \phi_{i-1,j}^{n+\frac{1}{2}}\right) + F_{i,j}^{n+\frac{1}{2}}v_{i,j}^{n+\frac{1}{2}}, \tag{12.78}$$

$$v_{i,j}^{n+1} = v_{i,j}^{n+\frac{1}{2}} - \mu_2\left(\phi_{i,j+1}^{n+1} - \phi_{i,j-1}^{n+1}\right) - F_{i,j}^{n+\frac{1}{2}}u_{i,j}^{n+1}, \tag{12.79}$$

$$
\begin{aligned}
\phi_{i,j}^{n+1} &= \phi_{i,j'}^{n+\frac{1}{2}} - \overline{\phi}\left(\mu_1\left(j'\right)\left(u_{i+1,j'}^{n+\frac{1}{2}} - u_{i-1,j'}^{n+\frac{1}{2}}\right) + \mu_3\left(j'\right)\left(\widehat{v}_{i,j'+1}^{n+1} - \widehat{v}_{i,j'-1}^{n+1}\right)\right) \\
&- \left(\phi'\right)_{i,j'}^{n+\frac{1}{2}}\left(\left(\mu_1\left(J'\right)\left(u_{i+1,j'}^{n+\frac{1}{2}} - u_{i-1,j'}^{n+\frac{1}{2}}\right)\right) + \mu_3\left(j'\right)\left(\widehat{v}_{i,j'+1}^{n+\frac{1}{2}} - \widehat{v}_{i,j'-1}^{n+\frac{1}{2}}\right)\right),
\end{aligned}
\tag{12.80}
$$

where

$$
\begin{aligned}
\mu_1(j) &= \frac{\Delta t}{4a}\cos\theta\cos\theta_j\Delta\lambda, \\
\mu_2 &= \frac{\Delta t_{\text{adj}}}{4a\Delta\theta}, \\
\mu_3(j) &= \frac{\mu_2}{\cos\theta_j}, \\
\widehat{v} &= v\cos\theta, \\
F_{i,j}^{n} &= \frac{\Delta t_{\text{adj}}}{2}\left(f + \frac{u\tan\theta}{a}\right)_{i,j}^{n}.
\end{aligned}
$$

The indices (i, j) denote vector points, while (i', j) and (i, j') denote scalar points in the semi-staggered E grid.

The two sets of Eqs. (12.75)–(12.77) and (12.78)–(12.80) are consistent approximations to the differential equations (12.74d)–(12.74f) [12], the individual equations begin first-order accurate in time and second-order accuracy in space. In Step I, the v-equation, being explicit, is integrated first and the u and ϕ equations are then integrated together as a coupled implicit pair, which results in as a coupled implicit pair, giving rise to a linear tridiagonal system. Correspondingly in Step II, the u equation is integrated first and the v and ϕ equations are integrated together. It is this implicit step that gives rise to the name, or acronym, of this system as the Semi-Lagrangian Alternating Direct Implicit, or SLADI, as we mentioned earlier. The alternating comes from switching the terms in the adjustments that are being considered implicitly when integrating in each of the two time steps.

The domain of integration is shown in a copy of the domain from [12], where the values of the variables on the inner and outer boundary lines are held fixed during each stage of the interior integration. We have a copy of the domain from [12] in Fig. 12.15. The values of the boundary points are then updated in the procedure that is described in Appendix A in [12].

As this is a limited area model, we have to consider the boundary conditions, and [12] provided the following details about this:

1. **Geopotential**
    1. At outer boundary points, ϕ is held fixed.
    2. At inner boundary points, ϕ is updated as a four-point spatial average of the values at the four surrounding scalar points.
2. 1. At outer boundary points, there are two possibilities:
        (a) At inflow points, (u, v) are held fixed.

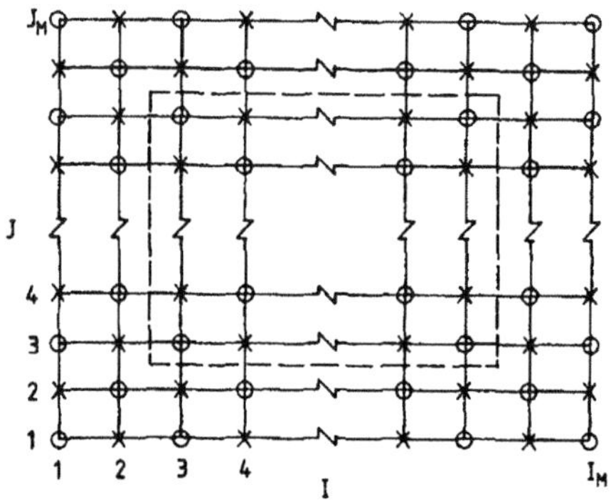

FIGURE 12.15 Distribution of points on the integration area (E grid with circles denoting scalar points and the crosses denoting vector points). The equations of motion are applied only at points within the dashed lines. Bates, J. R. 1984: An efficient semi-Lagrangian and alternating direction implicit method for integrating the shallow water equations, **112**, 2033–2047. ©American Meteorological Society. Used with permission.

(b) At outflow points, the components of velocity perpendicular to the boundary are held fixed, thus an outflow point always remains an outflow point, while the components of velocity parallel to the boundary are updated by a spatial extrapolation from the values at the two nearest vector points along the normal.

2. At inner boundary points, both velocity components are updated by taking a four-point spatial average of the corresponding values at the four surrounding vector points.

In [12] the authors go on to perform a stability analysis, as well as a consistency analysis, showing that this approach is quite accurate, although as mentioned earlier we have to be careful with the time splitting methods with a too large time step. However, the conclusions in [12] state that the scheme is computationally simpler than the semi-implicit scheme for grid point models, involving only the solution of linear tridiagonal systems of equations as compared to the Helmholtz equation that we have seen in the two and three dimension chapter. Through a linearization analysis, the E grid shows that it is unconditionally stable for simple Rossby waves and has a very lenient stability criterion for gravity–inertia waves.

We now move on to consider how to implement the shallow water equations in two dimensions and the primitive equations in 3D in spherical coordinates.

12.6 Global semi-Lagrangian modeling of the shallow water equations

In the last section we presented a limited area spherical model of the shallow water equations with a semi-Lagrangian approach for the advection that was separated from the adjustment equations. We also had boundary conditions but these could cause problems at the pole. We now move on to consider how to implement different sets of equations that involve advection in spherical coordinates, but what we have to be prepared for are the following two problems:

1. The departure point corresponding to a near polar grid point cannot be accurately located by going back along the coordinate line using the known velocity component at a single point, because of the strong curvature of the coordinate lines near the polar singularity.
2. The momentum equation in component form cannot be accurately integrated along a near polar trajectory corresponding to a long time step, because of the rapid spatial variation of all the terms in the equations, especially the curvature along such a trajectory [127].

One approach to overcome the two problems mentioned above was introduced by Ritchie in 1988 [159], and we shall summarize that approach below before moving onto the gridpoint approach introduced by MacDonald and Bates in 1989.

12.6.1 Ritchie (1988) approach

The starting point for the summary of the Ritchie (1988) paper [159] is the paper that he published in 1987 titled "*Semi-Lagrangian advection on a Gaussian Grid*" [158], which, as we have seen earlier, is implying some form of spectral modeling again.

The starting point is to consider the two-dimensional advection equation in Lagrangian vector form,

$$\frac{D}{Dt}\psi\left(\mathbf{r}\left(t\right),t\right)=0, \tag{12.81}$$

where

$$\frac{D}{Dt}\left(\cdot\right)=\frac{\partial}{\partial t}\left(\cdot\right)+\mathbf{u}_H\cdot\nabla_H\left(\cdot\right),$$

with $\mathbf{r}(t)$ being the position vector as a function of time t, and $\mathbf{v}_H$ being the horizontal wind vector advecting the field ψ. In longitude–latitude (λ,θ) coordinates, (12.81) becomes

$$\frac{\partial\psi}{\partial t}=-\frac{1}{\cos^2\theta}\left[U\frac{\partial\psi}{\partial\lambda}+V\cos\theta\frac{\partial\psi}{\partial\theta}\right], \tag{12.82}$$

where the zonal wind component u and meridional wind component v are converted to model wind images $(U,V)=\left(\frac{u\cos\theta}{a},\frac{v\cos\theta}{a}\right)$, and a is the radius of the Earth. Ritchie goes on to introduce a mapping similar to that we described from [68] but for the Eulerian formulation. Here we summarize the part about the implementation of the semi-Lagrangian approach.

The first step in the semi-Lagrangian approach is to apply a centered approximation for the Lagrangian derivative as

$$\frac{\psi\left(\mathbf{g},t+\Delta t\right)-\psi\left(\mathbf{r}\left(t-\Delta t\right),t-\Delta t\right)}{2\Delta t}, \tag{12.83}$$

where the position at forecast time $t+\Delta t$ is chosen to be a Gaussian gridpoint, and the forecast value at this point is found by picking off the values of ψ at the upstream location $\mathbf{r}(t-\Delta t)$ at an old time $t-\Delta t$. The origin of the position vector is taken to be the center of the sphere $|\mathbf{r}|=a$.

To be able to find the upstream location $\mathbf{r}(t-\Delta t)$ at time $t=t-\Delta t$, we first find the upstream position $\mathbf{r}(t)$ at time t via a Taylor series expansion as

$$\mathbf{r}(t)=\mathbf{g}-\Delta t\dot{\mathbf{r}}(t)+O\left((\Delta t)^2\right), \tag{12.84}$$

where $\dot{\mathbf{r}}$ represents the velocity vector at $(\mathbf{t})$, therefore (12.84) is an implicit equation. Since $\dot{\mathbf{r}}(t)$ is parallel to the surface of the sphere, starting from the Gaussian gridpoint $\mathbf{g}$ which is on the sphere, (12.84) would lead to a position that is slightly off the sphere. Ritchie introduces a correction factor b to avoid this problem, which is designed to guarantee that $|\mathbf{r}(t)|=a$ by modifying (12.84) to yield

$$\mathbf{r}(t)=b\left(\mathbf{g}-\Delta t\dot{\mathbf{r}}(t)\right), \tag{12.85}$$

where

$$b=\sqrt{1+\frac{(\Delta t)^2\dot{\mathbf{r}}(t)^2}{a^2}-\frac{2\Delta t\dot{\mathbf{r}}(t)\cdot\mathbf{g}}{a^2}}. \tag{12.86}$$

From (12.84) we have that $\mathbf{g}=\mathbf{r}(t)+\Delta t\dot{\mathbf{r}}(t)+O\left((\Delta t)^2\right)$, and substituting this into (12.86) yields

$$b=1+\frac{(\Delta t)^2}{2}\frac{\dot{\mathbf{r}}(t)^2}{a^2}+O\left((\Delta t)^3\right), \tag{12.87}$$

where Ritchie has used the property

$$\dot{\mathbf{r}}(t)\cdot\mathbf{r}(t)=\frac{1}{2}\frac{\partial}{\partial t}\left(\left|r(t)^2\right|\right)=0. \tag{12.88}$$

Therefore, the introduction of the b term only introduces an $O\left((\Delta t)^2\right)$ change, and (12.85) and (12.84) agree to this order.

Once we have solved for $\mathbf{r}(t)$, if we consider position vectors along a great circle trajectory, then $\mathbf{r}(t-\Delta t)$ can be calculated as

$$\mathbf{r}(t-\Delta t)=2\left(\frac{\mathbf{r}(t)\cdot\mathbf{g}}{a^2}\right)\mathbf{r}(t)-\mathbf{g}. \tag{12.89}$$

To be able to solve (12.85) and (12.89), they are expressed in a Cartesian coordinate system (x,y,z) whose origin is the center of the sphere. The semi-Lagrangian approach, as we have seen, effectively approximates the trajectories by straight lines which is a good approximation in Cartesian coordinates. While it may seem that a better choice might be to work in the spherical coordinates, Ritchie points out that if we were to calculate $(\Delta\lambda,\Delta\theta)$, the angular displacement over one time step would be

$$\Delta\lambda=\Delta t\frac{D\lambda}{Dt}(t)+O\left((\Delta t)^2\right),\qquad \Delta\theta=\Delta t\frac{D\theta}{Dt}(t)+O\left((\Delta t)^2\right), \tag{12.90}$$

with

$$\frac{D\lambda}{Dt}(t)=\frac{u(\lambda(t),\theta(t),t)}{a\cos\theta(t)},\qquad \frac{D\theta}{Dt}(t)=\frac{v(\lambda(t),\theta(t),t)}{a}. \tag{12.91}$$

However, here Ritchie introduces a caveat about assuming this form, where he points out that the expression above would correspond to approximating the trajectory by a straight line in (λ, θ) space, which would be a poor approximation, especially near the poles. This is illustrated in Fig. 12.16 which is a diagram of the Gaussian grid near the North pole. If we now suppose that $\mathbf{r}(t)$ corresponds to $(\lambda(t), \theta(t))$ and is represented by point A in Fig. 12.16, and if $v(\lambda(t), \theta(t), t) = 0$, then, to a good approximation, the trajectory is a straight line tangent to the latitude circle at point A. However, from (12.90) and (12.91), if a semi-Lagrangian scheme was applied directly in (λ, θ) coordinates, then this trajectory would be approximated by an arc of the latitude circle through the point A. This could lead to a significant error in calculating the trajectories, especially near the poles where the curvature of the latitude circles is large.

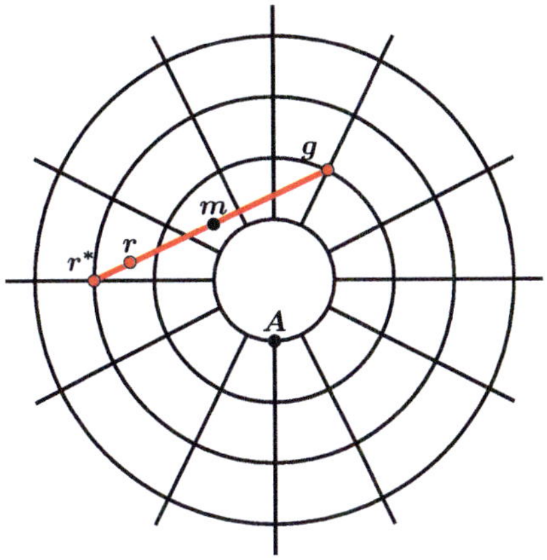

FIGURE 12.16 Replica of schematic from Ritchie (1987) which is illustrating the Gaussian grid near the North pole.

The next step is to determine how to perform the first iteration of (12.85), which is written in its general iterative form as

$$\mathbf{r}^{(m+1)} = b^{(m)}\left(\mathbf{g} - \Delta t \dot{\mathbf{r}}^{(m)}\right). \tag{12.92}$$

However, now the value $\mathbf{r}^{(m)}$ corresponds to $\left(\lambda^{(m)}, \theta^{(m)}\right)$, and thus the normalized velocities are

$$\widetilde{u}^{(m)} = \frac{u\left(\lambda^{(m)}, \theta^{(m)}, t\right)}{a}, \qquad \widetilde{v}^{(m)} = \frac{v\left(\lambda^{(m)}, \theta^{(m)}, t\right)}{a}.$$

Next we introduce what are referred to as the normalized Cartesian coordinates, $(X, Y, Z) \equiv \left(\frac{x}{a}, \frac{y}{a}, \frac{z}{a}\right)$, which have the following normalized Cartesian velocity components:

$$\begin{aligned}
\dot{X}^{(m)} &= -\widetilde{u}^{(m)} \sin\lambda^{(m)} - \widetilde{v}^{(m)} \cos\lambda^{(m)} \sin\theta^{(m)}, \\
\dot{Y}^{(m)} &= \widetilde{u}^{(m)} \cos\lambda^{(m)} - \widetilde{v}^{(m)} \sin\lambda^{(m)} \sin\theta^{(m)}, \\
\dot{Z}^{(m)} &= \widetilde{v}^{(m)} \cos\theta^{(m)}.
\end{aligned}$$

Now if we assume that the longitude and latitude of the Gaussian gridpoint **g** are (λ_i, θ_j), then the corresponding Cartesian coordinates are

$$\begin{aligned} X_G &= \cos\lambda_i \cos\theta_j, \\ Y_G &= \sin\lambda_i \cos\theta_j, \\ Z_G &= \sin\theta_j. \end{aligned}$$

Substituting all this information into (12.86) yields

$$b^{(m)}\sqrt{1+(\Delta t)^2\left(\dot{X}^{(m)^2}+\dot{Y}^{(m)^2}+\dot{Z}^{(m)^2}\right)-2\Delta t\left(\dot{X}^{(m)}X_G+\dot{Y}^{(m)}Y_G+\dot{Z}^{(m)}Z_G\right)}, \tag{12.93}$$

and (12.92) becomes

$$\begin{aligned} X^{(m+1)} &= b^{(m)}\left(X_G - \Delta t\dot{X}^{(m)}\right), \\ Y^{(m+1)} &= b^{(m)}\left(Y_G - \Delta t\dot{Y}^{(m)}\right), \\ Z^{(m+1)} &= b^{(m)}\left(Z_G - \Delta t\dot{Z}^{(m)}\right). \end{aligned}$$

The iteration is complete by calculating

$$\lambda^{(m+1)} = \tan^{-1}\frac{Y^{(m+1)}}{X^{(m+1)}} \quad \text{and} \quad \theta^{(m+1)} = \sin^{-1} Z^{(m+1)}. \tag{12.94}$$

It is stated in [158] that the first guess is usually taken as $\mathbf{r}_0 = \mathbf{g}$, which corresponds to a zero displacement, where in [158] the author makes the comments that for the first time step it could take around 5 iterations to find the upstream locations, whereas during subsequent steps two or three iterations are usually sufficient to find the uprated values of $\mathbf{r}(t)$, starting with the values found on the previous time steps as a first guess. We should also note here that the normalized (X, Y, Z) Cartesian coordinates are also used in solving (12.89) for $(\lambda(t-\Delta t), \theta(t-\Delta t))$ corresponding to $\mathbf{r}(t-\Delta t)$.

As we have seen in the different forms of Cartesian coordinate based semi-Lagrangian methods, it is quite often the case that the position $\left(\lambda^{(m)}, \theta^{(m)}\right)$ will be between the Gaussian gridpoints (λ_i, θ_j), so that interpolation is required in evaluating $\widetilde{u}^{(m)}$ and $\widetilde{v}^{(m)}$. We should then see that the upstream position $\mathbf{r}(t-\Delta t)$ will also not normally be a Gaussian gridpoint, as illustrated by point r in Fig. 12.16. This then implies that we require an interpolation polynomial for evaluating $\psi(\mathbf{r}(t-\Delta t), t-\Delta t)$.

In [158] the interpolation just mentioned was accomplished through using a bicubic scheme on the Gaussian grid to find the value of field H at position (λ, θ). To find the grid box in which these spherical coordinates lie, in the following three expressions and approximations from spectral space are utilized:

$$\lambda_i = 2\pi\frac{i-1}{K_1}, \quad 1 \le K_1, \tag{12.95}$$

$$\theta_j \approx \frac{2K_2+2-4j}{4K_2+2}\pi, \quad 1 \le j \le K_2, \tag{12.96}$$

$$\theta_j - \theta_{j-1} \approx \frac{2\pi}{2K_2+1} = \Delta\theta, \tag{12.97}$$

where $K_1 \ge 3N+1$, N is the maximum East–West wavenumber associated with a triangular truncation, ensuring that quadratic aliasing is avoided, and K_2 is the number of roots of the Legendre polynomial $P_{K_2}(\mu) = \sin\theta_j$. Again to avoid quadratic aliasing, $K_2 \ge \frac{3N+1}{2}$.

If we now impose periodicity in the longitudinal direction, then to interpolate for points (λ, θ) that are near the poles, it is necessary to extend the meridians to allow $\theta > \frac{\pi}{2}$ near the North pole, and $\theta < -\frac{\pi}{2}$ at the South pole. Thus for scalar fields ψ, U, and V, the extension rule for completing the set of gridpoint values required for interpolation is

$$H\left(\lambda, \frac{\pi}{2}+\epsilon\right) = H\left(\lambda \pm \pi, \frac{\pi}{2}-\epsilon\right) \tag{12.98}$$

near the North pole, and

$$H\left(\lambda, \frac{\pi}{2}+\epsilon\right) = H\left(\lambda \pm \pi, \frac{\pi}{2}-\epsilon\right) \tag{12.99}$$

near the South pole.

We now move on to consider how the theory presented above can be extended into a semi-Lagrangian representation in the spectral space, but as mentioned earlier, there is a problem at the pole that has to be dealt with, which we shall show here now. The derivations below are from [159]. The starting point is the vorticity-divergence version of the shallow water equations model in spherical coordinates as before, but for the semi-Lagrangian application it is stated in [159] that it is better to work with the **momentum** (U, V). Thus the vector form of the momentum equation is given by

$$\frac{\partial \mathbf{v}_H}{\partial t} + (\mathbf{v}_H \cdot \nabla_H) + f\mathbf{k} \times \mathbf{v}_H + \nabla_H \Phi = 0, \tag{12.100}$$

$$\Longrightarrow \frac{\partial U}{\partial t} + \frac{1}{a^2}\frac{\partial \overline{\Phi'}^t}{\partial \lambda} = -\left(A(U,V,U) - fV\right), \tag{12.101}$$

$$\frac{\partial V}{\partial t} + \frac{\cos\theta}{a^2}\frac{\partial \overline{\Phi'}^t}{\partial \theta} = -\left(A(U,V,V) + fU + 2\sin\theta E\right), \tag{12.102}$$

where

$$A(U,V,F) = \frac{1}{\cos^2\theta}\left(U\frac{\partial F}{\partial \lambda} + V\cos\theta\frac{\partial F}{\partial \theta}\right),$$

$$E = \frac{U^2+V^2}{2\cos^2\theta}.$$

Finally, the continuity equation is expressed as

$$\frac{\partial \Phi}{\partial t} + \Phi^* \overline{\delta}^t = -\left(A\left(U, V, \Phi\right) + \Phi\delta\right), \tag{12.103}$$

where the overline operator with the superscript t represent a time average as we showed before.

It is stated in [159] that when the author applied a semi-Lagrangian approach to (12.102), a problem arose near the poles, $\theta = \pm\frac{\pi}{2}$, due to an instability with the metric term $2\sin\theta E$. Therefore, Ritchie proposes using a vector discretization of the vector form of the momentum equation (12.100). Thus we start by expressing the wind vector using two parts

$$\mathbf{u}_H = \mathbf{u}_H^* + \mathbf{u}_H',$$

in the horizontal advection operator $\mathbf{v}_H \cdot \nabla_H$, which results in

$$\frac{\partial \mathbf{u}_H}{\partial t} + \left(\mathbf{v}_H^* \cdot \nabla\right)\mathbf{v}_H + \overline{\nabla_H \Phi}^t = -\left(\left(\mathbf{u}_H' \cdot \nabla\right)\mathbf{u}_H + f\mathbf{k} \times \mathbf{u}_H\right), \tag{12.104}$$

and the continuity equation becomes

$$\frac{\partial \Phi}{\partial t} + \mathbf{u}_H^* \cdot \nabla_H \Phi + \Phi^* \overline{\delta}^t = -\left(\mathbf{u}_H' \cdot \nabla_H \Phi + \Phi\Delta\right), \tag{12.105}$$

where the operator

$$\frac{\partial}{\partial t} + \mathbf{u}_H^* \cdot \nabla_H$$

provides the total derivative following a point that moves with velocity $\mathbf{v}_H^*$. Thus $\mathbf{v}_H^*$ represents the portion of the advecting velocity that is treated with a semi-Lagrangian approach, while $\mathbf{v}_H' \cdot \nabla_H$ represents the residual advection operator associated with the residual velocity $\mathbf{v}_H'$ and treated in an Eulerian fashion.

The evaluation of the semi-Lagrangian approach is done at the Gaussian gridpoint $\mathbf{g}$ at forecast time $t + \Delta t$, and the Lagrangian time derivatives are approximated by a centered time-difference as

$$\frac{\partial}{\partial t} + \mathbf{v}_H^* \cdot \nabla_H \approx \frac{(\cdot)^{n+1} - (\cdot)^{n-1}}{2\Delta t},$$

while the time average is approximated by the average along the trajectory

$$\overline{(\cdot)} \approx \frac{1}{2}\left((\cdot)^{n+1} + (\cdot)^{n-1}\right).$$

Thus (12.105) becomes

$$\begin{aligned}\left(\Phi + \Delta t \Phi^* \delta\right)^{n+1} &= \left(\Phi - \Delta t \phi^* \delta\right)^{n-1} \\ &- 2\Delta t\left(\left(A\left(U, V, \Phi\right) + \Phi\delta\right)^n - \frac{u^*}{a}\left(\frac{1}{\cos\theta}\frac{\partial \Phi}{\partial \lambda}\right)^n - \frac{v^*}{a}\left(\frac{\partial \Phi}{\partial \theta}\right)^n\right)\end{aligned} \tag{12.106}$$

and

$$\mathbf{u}_H^* = u^*\widehat{\lambda} + v^*\widehat{\theta}$$

where $\widehat{\lambda}$ and $\widehat{\theta}$ are unit vectors in spherical coordinates.

Since $\dfrac{\partial \mathbf{u}_H}{\partial t} + \left(\mathbf{u}_H^* \cdot \nabla\right)\mathbf{u}_h$ is the horizontal part for the total derivative of $\mathbf{u}_H$ following a point that moves with velocity $\mathbf{u}_H^*$, it is the derivative as evaluated in a tangential Cartesian coordinate system [159]. In order to find the horizontal vectors on a trajectory, we let $\mathbf{c}$ be a unit vector along $\mathbf{u}_H^*$ and let $\mathbf{n}$ be a horizontal unit vector normal to the left of $\mathbf{c}$. We have a replica of the figure from Ritchie (1988) of the tangential plane in Fig. 12.17, where α is the angular displacement, which is positive in the direction of $\mathbf{c}$, along a great circle trajectory between two positions $\mathbf{R}$ and $\mathbf{R}_1$; $\mathbf{R}$ is chosen to be the position at time t^n, and we use the corresponding $\widehat{\lambda}^0$ and $\widehat{\theta}^0$ as the unit vectors of the tangential Cartesian plane.

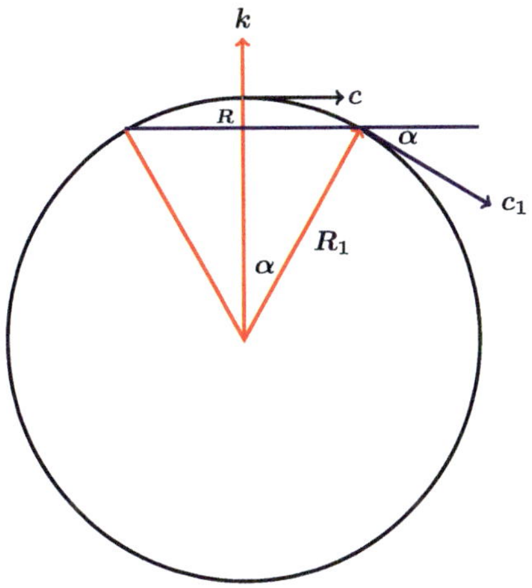

FIGURE 12.17 Replica of the illustration of the tangential plane from Ritchie (1988).

Therefore, for a general horizontal vector, $\mathbf{h}_1 = h_\lambda\widehat{\lambda}_1 + h_\theta\widehat{\theta}_1$ at position $\mathbf{R}_1$, which transforming to the tangential Cartesian place consists of finding the vector in the $\left(\widehat{\lambda}^0, \widehat{\theta}^0\right)$ plane having the same magnitude as $\mathbf{h}_1$, and whose direction is the projection of $\mathbf{h}_1$ in this plane. This is achieved by first expressing $\widehat{\lambda}_1$ and $\widehat{\theta}_1$ in terms of $\mathbf{c}_1$ and $\mathbf{n}_1 = \mathbf{n}$, then rotating $\mathbf{c}_1$ into $\mathbf{c}$ to produce vectors $\left(\widehat{\lambda}_R\right)_1$ and $\left(\widehat{\theta}_R\right)_1$ in the tangential Cartesian plane. Thus $h_\lambda\left(\widehat{\lambda}_R\right)_1 + h_\theta\left(\widehat{\theta}_R\right)_1$ has the same magnitude as $\mathbf{h}_1$ and its direction is the projection of $\mathbf{h}_1$ in the $\left(\widehat{\lambda}^0, \widehat{\theta}^0\right)$ plane. Finally, $\mathbf{c}$ and $\mathbf{n}$ are expressed in terms of $\widehat{\lambda}^0$ and $\widehat{\theta}^0$ in order to find the required component of $\mathbf{h}_1$ in the tangential Cartesian plane.

After all of this discussion, we can now express the semi-Lagrangian semi-implicit time discretization of the horizontal momentum equation (12.103) in tangential Cartesian coordinates as

$$\left(\left(\frac{u}{a}+\frac{\Delta t}{a^2\cos\theta}\frac{\partial\Phi}{\partial\lambda}\right)\widehat{\lambda}_R+\left(\frac{v}{A}+\frac{\Delta t}{a^2}\frac{\partial\Phi}{\partial\theta}\right)\widehat{\theta}_R\right)^{n+1}=\left(p_\lambda\widehat{\lambda}_R+p_\theta\widehat{\theta}_R\right)^{n-1}-2\Delta t\left(r_\lambda\widehat{\lambda}+r_\theta\widehat{\theta}\right)^n, \tag{12.107}$$

with

$$p_\lambda = \frac{P_1}{\cos\theta}, \tag{12.108}$$

$$p_\theta = \frac{P_2}{\cos\theta}, \tag{12.109}$$

where

$$P_1=U-\frac{\Delta t}{a^2}\frac{\partial\Phi}{\partial\lambda},\qquad P_2=V-\frac{\Delta t}{a^2}\cos\theta\frac{\partial\Phi}{\partial\theta},$$

along with

$$(r_\lambda)^0 = \left(\frac{A(U,V,U)}{\cos\theta}-f\frac{v}{a}\right)^n-\frac{u^*}{a}\left(\frac{1}{\cos^2\theta}\frac{\partial U}{\partial\lambda}-\frac{v}{a}\tan\theta\right)^n-\frac{v^*}{a}\left(\frac{1}{\cos\theta}\frac{\partial U}{\partial\theta}+\frac{u}{a}\tan\theta\right)^n, \tag{12.110}$$

$$\begin{aligned}(r_\theta)^0 &= \left(\frac{A(U,V,V)}{\cos\theta}+\frac{u^2+v^2}{a^2}\tan\theta+f\frac{u}{a}\right)^n-\frac{U^*}{a}\left(\frac{1}{\cos^2\theta}\frac{\partial V}{\partial\lambda}+\frac{U}{a}\tan\theta\right)^n\\ &- \frac{v^*}{a}\left(\frac{1}{\cos\theta}\frac{\partial V}{\partial\theta}+\frac{v}{a}\tan\theta\right)^n.\end{aligned} \tag{12.111}$$

For a full explanation of the calculations to the tangential Cartesian plane, the reader is referred to [159], but beware of the notation which is very confusing and this is why it is not presented here. As we can see this is quite a complicated method to overcome the problem at the pole for the semi-Lagrangian methods. Another approach was suggested by Côté and Staniforth in 1988 [40].

12.6.2 Côte and Staniforth (1988) approach

The starting point in [40] is again the shallow water equations in vectorial form on a rotating sphere of radius a, but this time we have

$$\frac{D\mathbf{V}}{Dt} = \mathbf{F}+\mu\mathbf{r}, \tag{12.112}$$

$$\frac{D\ln\phi}{Dt}+\nabla\cdot\mathbf{V} = 0, \tag{12.113}$$

where $\mathbf{r}$ is the normalized position vector whose components in Cartesian frame of reference, fixed at the center of the sphere, are

$$x=\cos\lambda\cos\theta,\quad y=\sin\lambda\cos\theta,\quad z=\sin\theta, \tag{12.114}$$

where λ and θ are the longitude and latitude, respectively, and where

$$\mathbf{V} = a\frac{d}{dt}\mathbf{r}, \tag{12.115}$$

$$\mathbf{F} = -2\Omega\sin\theta\mathbf{r}\times\mathbf{V}-\nabla\phi, \tag{12.116}$$

μ is a Lagrange multiplier determined by requiring that the fluid elements remain on the sphere, which is to say that for all time

$$\mathbf{r}\cdot\mathbf{r}=1. \tag{12.117}$$

The Lagrange multiplier that we just mentioned will be determined after time discretizing has been applied; for the reasoning see [39]. The constraint in (12.117) implies that **V** is tangent to the sphere **r**, and therefore

$$\mathbf{r}\cdot\mathbf{V}=0, \tag{12.118}$$

which implies that the three-dimensional wind vector has only two independent components which here will be written in terms of the wind images U and V. Thus

$$\mathbf{V}\equiv\frac{1}{\cos\theta}\left(U\widehat{e_\lambda}+V\widehat{e_\theta}\right), \tag{12.119}$$

where $\widehat{e_\lambda}$ and $\widehat{e_\theta}$ are eastward and northward pointing unit vectors, respectively. As seen a couple of times now, we introduce the Helmholtz decomposition of **V** in terms of the streamfunction ψ and the velocity potential χ, so that we have

$$\mathbf{V} = \mathbf{r}\times\nabla\psi+\nabla\chi, \tag{12.120}$$

$$U = \frac{1}{a}\left(\frac{\partial\chi}{\partial\lambda}-\cos\theta\frac{\partial\psi}{\partial\theta}\right), \tag{12.121}$$

$$V = \frac{1}{a}\left(\frac{\partial\psi}{\partial\lambda}+\cos\theta\frac{\partial\chi}{\partial\theta}\right), \tag{12.122}$$

$$\zeta = \mathbf{r}\cdot(\nabla\times\mathbf{V})=\nabla^2\psi, \tag{12.123}$$

$$\delta = \nabla\cdot\mathbf{V}=\nabla^2\chi, \tag{12.124}$$

where due to the constraint in (12.117) the differential operators above are in the horizontal direction only.

If we now consider the semi-Lagrangian discretization, and if we know the trajectory of the fluid element that is at position **r** at forecast time t^n, then (12.112) and (12.113) may be integrated following the motion over a time interval Δt, through

$$\mathbf{V}-\frac{1}{2}\Delta t\mathbf{F} = \mathbf{V}^{n-1}+\frac{1}{2}\Delta t\mathbf{F}^{n-1}+\frac{1}{2}\mu\Delta t\left(\mathbf{r}^n+\mathbf{r}^{n-1}\right)=\mathbf{R}, \tag{12.125}$$

$$\ln\phi+\frac{1}{2}\Delta t\delta = \ln\phi^{n-1}-\frac{1}{2}\Delta t\delta^{n-1}. \tag{12.126}$$

At the upstream point $t=t^n-\Delta t\equiv t^{n-1}$ all of the values of the vectors, fields, and scalars that are required at the position at $\mathbf{r}^{n-1}$ are found through a bicubic interpolation as in [159].

The Lagrange multiplier is determined by projecting the discretized equation (12.125) onto $\mathbf{r}$. Since for constrained motion $\mathbf{V}$ and $\mathbf{F}$ are both orthogonal to $\mathbf{r}$, such is also $\mathbf{R}$, and

$$\mu = -\frac{2}{\Delta t}\frac{\mathbf{r}\cdot\left(\mathbf{V}^{n-1}+\frac{1}{2}\Delta t\mathbf{F}^{n-1}\right)}{\left(1+\mathbf{r}^{n}\cdot\mathbf{r}^{n-1}\right)}. \tag{12.127}$$

The next part of the process of considering a semi-Lagrangian approach is the trajectory calculations, where in [40] the authors state that a sufficiently accurate estimate of the trajectory can be obtain by integrating (12.112) and (12.115) subject to (12.117) over the interval $\left[t^{n}-\Delta t, t^{n}\right]$ while neglecting $\mathbf{F}$. The solutions are great circles whose parameters are determined by the field at $\left(t^{n}-\frac{1}{2}\Delta t\right)$ for a centered scheme. The midpoint $\mathbf{r}^{n}$ of the trajectory is obtained through iteratively solving

$$\mathbf{r}^{n} = \cos\Theta\mathbf{r} + \sin\Theta\mathbf{r}\times\frac{\left(\mathbf{r}^{n+1}\times\mathbf{V}^{n}\right)}{\left|\mathbf{r}^{n+1}\times\mathbf{V}^{n}\right|}, \tag{12.128}$$

where

$$\mathbf{V}^{n} = \mathbf{V}\left(\mathbf{r}^{n}, t^{n+1}-\frac{\Delta t}{2}\right), \tag{12.129}$$

$$\Theta = \frac{\left|\mathbf{V}^{n}\right|}{a}\frac{\Delta t}{2}. \tag{12.130}$$

In [40] the authors provide a geometrical interpretation of what is above, where they say that we search for a point $\mathbf{r}^{n}$ situated a distance $\frac{1}{2}\left|\mathbf{V}^{n}\right|\Delta t$ upstream on the sphere, such that $\mathbf{r}^{n+1}$, $\mathbf{r}^{n}$, and $\mathbf{V}^{n}$ are coplanar, which then makes the point $\mathbf{r}^{n-1}$ to be

$$\mathbf{r}^{n-1} = 2\left(\mathbf{r}^{n+1}\cdot\mathbf{r}^{n}\right)\mathbf{r}^{n} - \mathbf{r}^{n+1}. \tag{12.131}$$

In Fig. 12.18 we have produced a replica of the coplanar illustration from [40] to help visualize this situation.

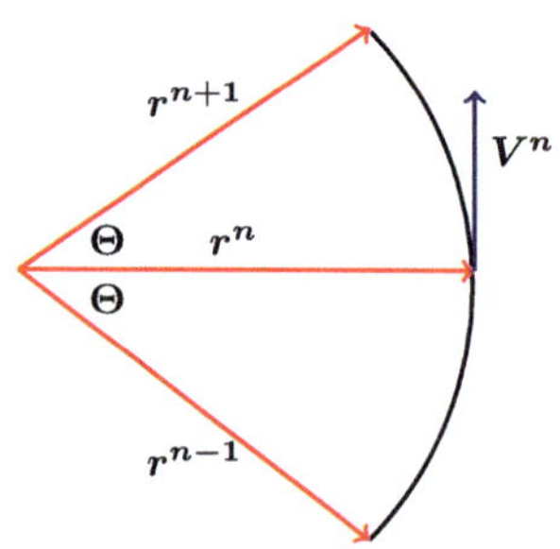

FIGURE 12.18 Replica of the schematic of the diagram for the geometry of the trajectory calculations from Côté and Staniforth (1988) [40].

The next step is to use a time extrapolation, as we have seen before, to find values of $\mathbf{v}^n$. This approach allows a saving of a factor of 2 over the usual semi-implicit semi-Lagrangian algorithms where it is required to forecast the wind at time $t^{n+1} - \frac{1}{2}\Delta t$. In [40] the authors introduce a general expression for the time extrapolation for the wind field as

$$\begin{aligned}\mathbf{V}\left(\mathbf{r}^{n+1}, t - \frac{1}{2}\Delta t\right) &= c_1 V\left(\mathbf{r}^{n+1}, t - \Delta t\right) + c_2\mathbf{V}\left(\mathbf{r}^{n+1}, t - 2\Delta t\right) \\ &+ (1 - c_1 - c_2)\mathbf{V}\left(\mathbf{r}^{n+1}, t - 3\Delta t\right), \end{aligned} \tag{12.132}$$

where for a constant, linear, and quadratic extrapolation, the values of the coefficients c_i are $(1, 0)$, $\left(\frac{3}{2}, \frac{1}{2}\right)$, and $\left(\frac{15}{8}, -\frac{10}{8}\right)$, respectively. An important feature to note here, which is pointed out in [40], is that this is not a traditional two-level scheme as claimed in the title of this paper, and the authors point this out with the following comment:

> *This step involves the wind at the previous time steps and the scheme is no longer formally a two-time-level scheme. However, once the trajectories have been determined, the momentum and continuity equations are discretized at time differences and averages over just two time levels t^n and $t^n - \Delta t$, and we subsequently refer to the scheme as being two-level. In effect we replace the full-cost odd time steps of the three-time-level scheme by an almost-free time extrapolation of the winds.*

However, to put these equations into a more suitable form to apply spectral approaches, we introduce a reference geopotential ϕ^* and define ϕ' as the departure of ϕ from ϕ^*. Next we apply the curl $\cdot\nabla\times$ operator and the divergence $\nabla\cdot$ to (12.125), and multiply (12.126) by ϕ^* which results in

$$\eta + \alpha\sin\theta\delta + \alpha\frac{V}{a} = \mathbf{r}\cdot(\nabla\times\mathbf{R}), \tag{12.133}$$

$$\delta - \alpha\sin\theta\eta + \alpha\frac{U}{a} + \beta\nabla^2\phi' = \nabla\cdot\mathbf{R}, \tag{12.134}$$

$$\phi' + \gamma\delta = \left(\phi' - \phi^*\ln\left(1 + \frac{\phi'}{\phi^*}\right)\right) + \left(\phi^*\ln\left(1 + \frac{\phi'}{\phi^*}\right) - \gamma\delta\right)^{n-1}, \tag{12.135}$$

where

$$\alpha = \Omega\Delta t, \quad \beta = \frac{\Delta t}{2}, \quad \gamma = \frac{1}{2}\phi^*\Delta t. \tag{12.136}$$

In the appendix of [40] is a derivation to show that through a generalization of the Helmholtz solver to include the Coriolis terms it is possible to solve (12.133) and (12.134) through a form of spectral model that uses truncated *Hough* functions. The second feature to notice here is that (12.135) is nonlinear in ϕ' and this can be overcome through an iterative process with an initial guess for the right-hand side $(\phi' - \gamma\delta)^{n-1} - \Delta t\,(\phi'\delta)^n$, where ϕ'^n and δ^n are obtained through a time extrapolation of mesh values of ϕ' and $\mathbf{V}$ followed by an interpolation to the midpoint, $\mathbf{r}^n$, of the trajectory.

12.6.3 McDonald and Bates (1989) approach

The starting point in McDonald and Bates (1989) is the statement of the continuous problem that we are going to approximate, which in the horizonal formulation is given by

$$\frac{D_H \Psi}{Dt} = \xi, \tag{12.137}$$

where the operator $\frac{D_H}{Dt}$ is the Lagrangian operator in spherical coordinates in the horizontal direction and will be approximated in the form

$$\Psi^{n+1} - \Psi_d^n = \int_{n\Delta t}^{(n+1)\Delta t} \xi dt. \tag{12.138}$$

We note that the right-hand side of (12.138) represents an integral along the trajectory over the time interval Δt. The simplest method of locating the departure point is to go back from the arrival point along the coordinate line, which in spherical coordinates gives the formulas

$$\lambda_d = \lambda - \left(\frac{\widehat{u}}{a\cos\widehat{\theta}}\right)\Delta t, \tag{12.139}$$

$$\theta_d = \theta - \left(\frac{\widehat{v}}{a}\right)\Delta t, \tag{12.140}$$

where

$$\widehat{\mathbf{v}}_{ij}^{(m)} = \begin{pmatrix} \widehat{u}_{ij}^{(m)} \\ \widehat{v}_{ij}^{(m)} \end{pmatrix} = \begin{bmatrix} u\left(i\Delta\lambda - \dfrac{\widehat{u}_{ij}^{(m-1)}\Delta t}{2a\cos\theta}\right) \\ v\left(j\Delta\theta - \dfrac{\widehat{v}_{ij}^{(m-1)}}{2a}\right) \end{bmatrix}, \tag{12.141}$$

the velocity is evaluated at $t = \left(n + \frac{1}{2}\Delta t\right)$, and

$$\widehat{\mathbf{v}}_{ij}^{(0)} = v\left(i\Delta\lambda, j\Delta\theta, \left(n + \frac{1}{2}\right)\Delta t\right).$$

To obtain the estimates of the velocity fields at the time level $\left(n + \frac{1}{2}\right)$ in [126], the authors apply a two-time-level approximation that uses the extrapolation formula which we saw for the Cartesian cases but now it is in spherical coordinates and given by

$$\mathbf{v}\left(i\Delta\lambda, j\Delta\theta, \left(n + \frac{1}{2}\right)\Delta t\right) = \frac{3}{2}\mathbf{v}\left(i\Delta\lambda, j\Delta\theta, n\Delta t\right) - \frac{1}{2}\mathbf{v}\left(i\Delta\lambda, j\Delta\theta, (n-1)\Delta t\right). \tag{12.142}$$

As with the Cartesian case, we have to iterate and interpolate to find the velocity at the midpoint. In [126] there is a discussion about the costs and accuracy of using a two-level scheme compared to the three-level version, and which order of interpolation to use; see [126] for the exact details.

Returning again to [127], we have a warning from the authors stating that when we are in the neighborhood of a polar singularity, Eqs. (12.139) and (12.140) will no longer provide accurate estimates of the position of the departure points. Another problem is if ξ varies rapidly along the trajectory (as do the components of the velocity, gradient operator, and curvature near the pole), the right-hand side of (12.138) cannot be accurately estimated using any simple numerical formula such as $\frac{\left(\xi^{n+1}+\xi_d^n\right)\Delta t}{2}$. Thus McDonald and Bates suggest considering two different integrations approximations, one for areas near the poles, and one for regions away from the poles.

As this is a shallow water equations model section, we shall consider the integrations of the shallow water equations with divergence damping included. Thus the equations that we are considering are given by

$$\frac{D_H u}{Dt} = -\frac{1}{a\cos\theta}\frac{\partial}{\partial\lambda}(\phi - c\delta) + fv + C_u, \tag{12.143}$$

$$\frac{D_H v}{Dt} = -\frac{1}{a}\frac{\partial}{\partial\theta}(\phi - c\delta) - fu + C_v, \tag{12.144}$$

$$\frac{D_H \phi}{Dt} = -\phi\delta, \tag{12.145}$$

where δ is the divergence given by

$$\delta = \frac{1}{a\cos\theta}\left(\frac{\partial u}{\partial\lambda} + \frac{\partial}{\partial\theta}(v\cos\theta)\right).$$

Here c is the coefficient of divergence damping, and C_u and C_v are the curvature terms given by

$$C_u = \frac{uv\tan\theta}{a}, \qquad C_v = -\frac{u^2\tan\theta}{a}.$$

An important property to note here is that the curvature terms can easily exceed the Coriolis terms in magnitude near the poles.

As mentioned earlier, we are going to consider two different regions to apply the semi-implicit semi-Lagrangian technique. First we shall consider the region away from the poles which is denoted by a latitude of θ_M. Note this is the same for both hemispheres. Therefore for $|\theta| \leq \theta_M$, we shall apply a two-step approach for two different $\frac{\Delta t}{2}$ steps, where for the first halftime step we shall consider Coriolis terms implicitly, while the pressure and divergence damping terms will be treated explicitly. In the second halfstep the converse is true. Thus for the first step we have

Step 1:

$$\frac{\left(u^{n+\frac{1}{2}}-u_d^n\right)}{\frac{\Delta t}{2}} = \left[-\frac{1}{a\cos\theta}\frac{\partial}{\partial\lambda}(\phi-c\delta)+2C_u\right]_d^n + f_d v^{n+\frac{1}{2}}, \tag{12.146}$$

$$\frac{\left(v^{n+\frac{1}{2}}-v_d^n\right)}{\frac{\Delta t}{2}} = \left[-\frac{1}{a}\frac{\partial}{\partial\theta}(\phi-c\delta)+2C_v\right]_d^n - f_d u^{n+\frac{1}{2}}. \tag{12.147}$$

Step 2:

$$\frac{\left(u^{n+1}-u^{n+\frac{1}{2}}\right)}{\frac{\Delta t}{2}} = -\left[\frac{1}{a\cos\theta}\frac{\partial}{\partial\lambda}(\phi-c\delta)+2C_u\right]^{n+1} + f v^{n+\frac{1}{2}}, \tag{12.148}$$

$$\frac{\left(v^{n+1}-v^{n+\frac{1}{2}}\right)}{\frac{\Delta t}{2}} = -\left[\frac{1}{a}\frac{\partial}{\partial\theta}(\phi-c\delta)+2C_v\right]^{n+1} - f u^{n+\frac{1}{2}}. \tag{12.149}$$

The continuity equation given by (12.145) is integrated in a single time step of Δt as

$$\frac{\left(\phi^{n+1}-\phi_d^n\right)}{\Delta t} = -\frac{\overline{\phi}}{2}\left(\delta^{n+1}-\delta_d^n\right) - \left(\phi'\delta\right)_d^n. \tag{12.150}$$

We should note that $\overline{\phi}$ is a constant such that $\phi=\overline{\phi}+\phi'$, and $\left|\phi'\right| \ll \overline{\phi}$.

Given this setup, we now solve (12.146) and (12.147) for $(u,v)^{n+\frac{1}{2}}$ and eliminate these quantities from (12.148) and (12.149). If we now bring all of the $(n+1)$-level quantities to the left in the resulting equations and treat (12.150) similarly, then we obtain

$$\left[u+\frac{\Delta t}{2}\frac{1}{a\cos\theta}\frac{\partial}{\partial\lambda}(\phi-c\delta)\right]^{n+1} = B_u, \tag{12.151}$$

$$\left[v+\frac{\Delta t}{2}\frac{1}{a}\frac{\partial}{\partial\theta}(\phi-c\delta)\right]^{n+1} = B_v, \tag{12.152}$$

$$\left[\phi+\frac{\Delta t}{2}\overline{\phi}\delta\right]^{n+1} = B_\phi, \tag{12.153}$$

where

$$\begin{aligned} B_u &= \left(W_u\right)_d^n + F\left(W_v\right)_d^n, \\ B_v &= \left(W_v\right)_d^n - F\left(W_u\right)_d^n, \\ B_\phi &= \left[\phi-\left(\frac{\Delta t}{2}\overline{\phi}+\Delta t\phi'\right)\delta\right]_d^n, \end{aligned}$$

with

$$
\begin{aligned}
W_u &= \frac{Y_u + F\overline{Y}_v}{1+F^2},\\
W_v &= \frac{Y_v - F\overline{Y}_u}{1+F^2},\\
Y_u &= u - \frac{\Delta t}{2}\frac{1}{a\cos\theta}\frac{\partial}{\partial\lambda}(\phi - c\delta) + \Delta t C_u,\\
Y_v &= v - \frac{\Delta t}{2}\frac{1}{a}\frac{\partial}{\partial\theta}(\phi - c\delta) + \Delta t C_v,\\
F &= = \frac{f\Delta t}{2}.
\end{aligned}
$$

We should note that we still need to apply spatial discretizations in the derivation above. In [127] the authors are using a C grid on a spherical grid. We have a copy of the C grid at the poles figure from [127] which is presented in Fig. 12.19.

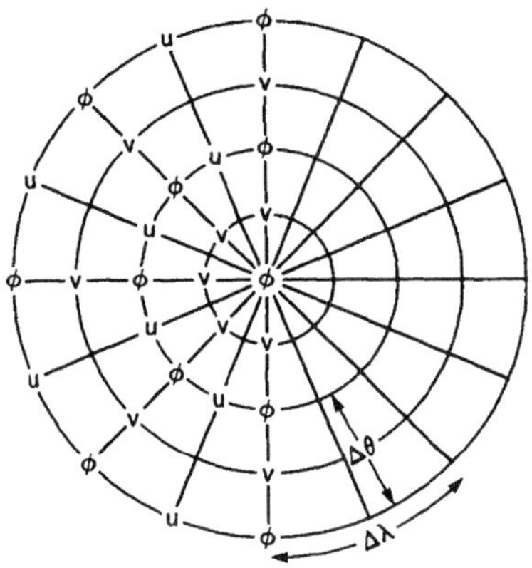

FIGURE 12.19 Copy of Fig. 1 from [127] of the polar C grid illustrating the distribution of the variables on the C grid near the poles. McDonald, A. and Bates, J. R., 1989: Semi-Lagrangian integration of a gridpoint shallow water model on the sphere, **117**, 130–137. ©American Meteorological Society. Used with permission.

We should note here that the quantities (B_u, B_v, B_ϕ) are expressed at (u, v, ϕ) points, respectively. The d subscripts that occur in the definitions of these quantities denote quantities that are evaluated at the departure points of trajectories that arrive at u, v, or ϕ point. The **overbars** in the definitions of W_u and W_v denote 16 point-averages evaluated at (u, v) points, respectively. From (12.151) and (12.152) we find

$$\delta^{n+1} = \nabla\cdot\mathbf{B} - \frac{\Delta t}{2}\nabla^2(\phi - c\delta)^{n+1}, \tag{12.154}$$

where $\mathbf{B} = \begin{pmatrix} B_u & B_v \end{pmatrix}^T$. Eliminating δ^{n+1} between (12.153) and (12.154) results in the Helmholtz equation

$$\nabla^2\phi^{n+1} - \mu^2\phi^{n+1} = H, \tag{12.155}$$

where

$$\mu^2 = \left(\left(\frac{\Delta t}{2}\right)^2 \overline{\phi} + c\frac{\Delta t}{2}\right)^{-1}, \qquad H = \mu^2\left(\left(\frac{\Delta t}{2}\overline{\phi}\right)\nabla\cdot\mathbf{B} + \left(c\frac{\Delta t}{2}\nabla^2 - 1\right)B_\phi\right).$$

Once ϕ^{n+1} is known, it is then possible to determine $(\delta, u, v)^{n+1}$ from (12.153), (12.151), and (12.152).

We now consider the neighborhoods of a pole where we continue to integrate the shallow water equations using the grid points of the original (λ, θ) C-grid. However, to be able to locate the departure points, and when integrating the momentum equations in a system in which their terms are smoothly varying, McDonald and Bates introduced for each grid point near-polar gridpoint A, where $\lambda = \lambda_i$ and $\theta = \theta_j$, an auxiliary spherical coordinate system (λ', θ') was also introduced such that in the new system the point A lies at $\lambda' = 0$ and $\theta' = 0$, and the coordinate unit vector $(\mathbf{i}, \mathbf{j})$ at point A in both coordinate systems coincide. We have a copy of the figure illustrating this coordinate system from [127] in Fig. 12.20.

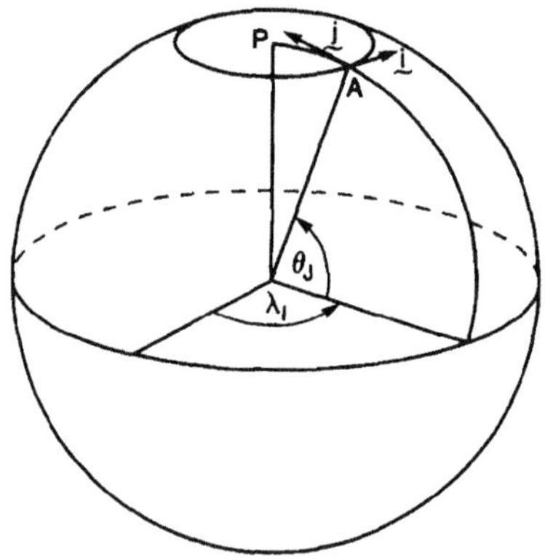

FIGURE 12.20 Copy of Fig. 2 from [127] highlighting the auxiliary coordinate system at point A which is obtained by rotating the geographical spherical system through angles λ_i and θ_j. McDonald, A. and Bates, J. R., 1989: Semi-Lagrangian integration of a gridpoint shallow water model on the sphere, **117**, 130–137. ©American Meteorological Society. Used with permission.

Therefore, we have that the coordinates of an arbitrary point in the auxiliary and the original coordinate systems are related by the following formulas:

$$\cos\theta'\cos\lambda' = \cos\theta\cos(\lambda - \lambda_i)\cos\theta_j + \sin\theta\sin\theta_j, \tag{12.156}$$

$$\cos\theta'\sin\lambda' = \cos\theta\sin(\lambda - \lambda'), \tag{12.157}$$

$$\sin\theta' = \sin\theta\cos\theta_j - \cos\theta\cos(\lambda - \lambda')\sin\theta_j, \tag{12.158}$$

where the inverse formulas are given by

$$\lambda = \lambda_i + \tan^{-1}\left(\frac{\cos\theta'\sin\lambda'}{\cos\theta'\cos\lambda'\cos\theta_j - \sin\theta'\sin\theta_j}\right), \tag{12.159}$$

$$\theta = \sin^{-1}\left(\cos\theta'\cos\lambda'\sin\theta_j + \sin\theta'\cos\theta_j\right). \tag{12.160}$$

The derivation of (12.156)–(12.160) can be found in Appendix A of [127]. It is also shown in the same appendix how to derive the following expression for the velocity components:

$$\begin{pmatrix} u' \\ v' \end{pmatrix} = \begin{pmatrix} \mathcal{G} & -\mathcal{S} \\ \mathcal{S} & \mathcal{G} \end{pmatrix}\begin{pmatrix} u \\ v \end{pmatrix}, \tag{12.161}$$

where

$$\mathcal{G} = \frac{\cos\theta\cos\theta_j + \sin\theta\sin\theta_j\cos(\lambda-\lambda_i)}{\cos\theta'}, \qquad \mathcal{S} = \frac{\sin\theta_j\sin(\lambda-\lambda_i)}{\cos\theta'},$$

and an important feature to note here is that at the point A, $\mathcal{G}=1$ and $\mathcal{S}=0$. As an aside in [127], it is stated that it is possible to show that the component of $\nabla\phi$ transforms in the same manner as the velocity components.

Given this transform, we now have that the curvature lines in the auxiliary system are quite small in the neighborhood of point A, and so the departure point corresponding to point A can be accurately located using the formulas

$$\lambda'_d = -\left(\frac{\widehat{u'}}{a\cos\widehat{\theta'}}\right)\Delta t, \tag{12.162}$$

$$\theta'_d = -\left(\frac{\widehat{v}}{a}\right)\Delta t. \tag{12.163}$$

The procedure to implement this system follows a quite similar setup to that described earlier, but we have to deal with the point at the pole where ϕ is evaluated. In [127] the authors consider the North pole first, where they assume that if the vector wind is constant in the immediate neighborhood of the pole, then its components can be written at any point within this neighborhood as

$$u_p = V_p\sin(\lambda-\lambda_0), \tag{12.164}$$

$$v_p = V_p\cos(\lambda-\lambda_0), \tag{12.165}$$

where V_p is the magnitude of the wind vector and λ_0 is the longitude from which it blows towards the pole. They state that (12.165) can be identified with the first Fourier component of $v^{n+\frac{1}{2}}\left(\lambda_i,\left(\frac{\pi}{2}-\frac{\Delta\theta}{2}\right)\right)$, the time-centered estimate of v at the nearest v-point to the pole. Hence it is stated that

$$V_p = \sqrt{a^2+b^2}, \qquad \lambda_0 = \tan^{-1}\frac{b}{a},$$

where

$$a = \frac{2}{N}\sum_{i=1}^{N} v^{n+\frac{1}{2}}\left(\lambda_i,\frac{\pi}{2}-\frac{\Delta\theta}{2}\right)\cos\lambda_i, \qquad b = \frac{2}{N}\sum_{i=1}^{N} v^{n+\frac{1}{2}}\left(\lambda_i,\frac{\pi}{2}-\frac{\Delta\theta}{2}\right)\sin\lambda_i,$$

with N being the number of grid points on a latitude circle.

The departure points for the North pole corresponding to the time interval Δt is then given by

$$\lambda_d = \tan^{-1}\frac{b}{a}, \qquad \theta_d = \frac{\pi}{2} - \frac{V_p \Delta t}{a}.$$

It is stated in [127] that a similar consideration to the South pole shows that the departure point is given by

$$\lambda_d = \pi + \tan^{-1}\frac{b}{a}, \qquad \theta_d = -\frac{\pi}{2} + \frac{V_p \Delta t}{a},$$

where (a, b, V_p) are evaluated as above using the time-centered estimates of v at the nearest v-point to the South pole.

Returning to the semi-Lagrangian problem slightly away from the poles, we have that it is possible to integrate the momentum equations in the auxiliary coordinates system using the two-step method from earlier. It is possible to make a simplification that the curvature terms can be ignored in the transformed system since they are quite small as shown by

$$\left|\frac{C'_u}{f v'}\right| = \left|\frac{C'_v}{f u'}\right| = \left(\frac{u'}{2\Omega a}\right)\left(\frac{\tan\theta}{\sin\theta}\right) \ll 1, \quad \left|\theta'\right| \ll 1.$$

Using (12.161) to transform the components of the velocity field and $\nabla\phi$ at the arrival and departure points back to the original system, the equations corresponding to (12.151) and (12.152) become

$$\left[u + \frac{\Delta t}{2}\frac{1}{a\cos\theta}\frac{\partial}{\partial\lambda}(\phi - c\delta)\right]^{n+1} = A_u, \tag{12.166}$$

$$\left[v + \frac{\Delta t}{2}\frac{1}{a}\frac{\partial}{\partial\theta}(\phi - c\delta)\right]^{n+1} = A_v, \tag{12.167}$$

where

$$\begin{aligned} A_u &= (\mathcal{G}_d + F\mathcal{S}_d)(W_u)_d^n - (\mathcal{S}_d - F\mathcal{G}_d)(W_v)_d^n, \\ A_v &= (\mathcal{G}_d + F\mathcal{S}_d)(W_v)_d^n + (\mathcal{S}_d - F\mathcal{G}_d)(W_u)_d^n, \end{aligned}$$

with the curvature terms in W_u and W_v being omitted in this region.

Finally, for the continuity equation (12.153) stays the same except for the forcing function which becomes

$$H = \mu^2\left[\left(\frac{\Delta t}{2}\bar{\phi}\right)\nabla\cdot\mathbf{A} + \left(c\frac{\Delta t}{2}\nabla^2 - 1\right)B_\phi\right], \tag{12.168}$$

where $\mathbf{A} = \begin{pmatrix} A_u \\ A_v \end{pmatrix}$. An important statement that is made in [127] just after this equation is that through the auxiliary coordinate system it is now possible to extend the Helmholtz equation to the whole sphere, while overcoming the problems associated with the polar singularity.

12.7 Spectral modeling of the shallow water equations

In this section we introduce the shallow water equations first in spectral form, to show how these methods are used in a semi-Lagrangian framework. We start through summarizing the derivation from Guo and Drake (2005) paper [68], where they start by making a reference to a smooth grid transformation in spectral atmospheric models that were introduced in a paper by Schmidt in 1977 [168]. An important conformal mapping defined in [168] was a conformal mapping of the sphere onto itself, with the property that the spherical Laplacian and Jacobian in the mapped coordinate system are simply modified by a mapping factor.

Thus if we consider the shallow water equations in advective form in spherical geometry, then we have

$$\begin{aligned}\frac{D\mathbf{v}}{Dt} &= -f\mathbf{k}\times-\nabla\phi, &(12.169)\\ \frac{D\phi^*}{Dt} &= -\phi^*\nabla\cdot\mathbf{v}, &(12.170)\end{aligned}$$

where $\mathbf{v}=\mathbf{i}u+\mathbf{j}v$ is the horizontal velocity, and is orthogonal to the unit vector $\mathbf{k}$ in the radial direction. The free surface geopotential is denoted by $\phi=\phi^*+\phi_s=g\left(h^*+h_s\right)$ and $\phi^*=\overline{\phi}+\phi'$, where g is the gravitational acceleration, and h^* is the height of the free surface above the bottom height h_s. The bottom surface height specifies orography as a time-invariant function and $\overline{\phi}$ is the time-invariant spatial mean. The Coriolis term, $f=2\Omega\sin\theta$, incorporates the rotation effects at latitude θ of the sphere with angular velocity Ω. An important feature to notice here is that this model does not require any boundary conditions, but does require an initial field for both $\mathbf{v}$ and ϕ.

To avoid problems with the vector representation of the velocities at the poles and to maintain compatibility with the scalar spectral transform method, the equations are reformulated in terms of vorticity $\xi=\mathbf{k}\cdot\nabla\times\mathbf{v}$ and divergence $\delta=\nabla\cdot\mathbf{v}$. To achieve this goal, we require the following two identities for spherical geometry:

$$\begin{aligned}(\mathbf{v}\cdot\nabla)\mathbf{v} &= \nabla\left(\frac{\mathbf{v}\cdot\mathbf{v}}{2}\right)+\xi\mathbf{k}\times\mathbf{v},\\ \nabla\times(\nabla\times\mathbf{v}) &= \nabla(\nabla\cdot\mathbf{v})-\frac{1}{\cos\theta}\nabla^2(\mathbf{v}\cos\theta)-\frac{2\sin\theta}{a\cos\theta}\mathbf{F},\end{aligned}$$

where $\mathbf{F}=\begin{pmatrix}-\xi & \delta\end{pmatrix}^T$.

By applying $\mathbf{k}\cdot\nabla\times$ and $\nabla\cdot$ to (12.169), the advective form of the vorticity and divergence equations are

$$\begin{aligned}\frac{D\eta}{Dt} &= -\eta\delta, &(12.171)\\ \frac{D\delta}{Dt}+\nabla^2\phi^* &= \Psi, &(12.172)\end{aligned}$$

where

$$\Psi = \eta(\eta-f)-\mathbf{k}\cdot\mathbf{v}\times\nabla f+\mathbf{v}\cdot\widetilde{\mathbf{F}}-\nabla^2\left(\phi_s+\frac{\mathbf{v}\cdot\mathbf{v}}{2}\right), \quad (12.173)$$

$$\begin{aligned} \widetilde{\mathbf{F}} &= \frac{1}{\cos\theta}\nabla^2(\mathbf{v}\cos\theta) + \frac{2\sin\theta}{a\cos\theta}\mathbf{F}, &(12.174)\\ \eta &= \xi + f. &(12.175) \end{aligned}$$

To recover the wind velocity from ξ and δ, the Helmholtz theorem separates the horizonal velocity vector $\mathbf{v}$ into a scaler stream function ψ and a scalar velocity potential χ as

$$\mathbf{v} = \mathbf{k} \times \nabla\psi + \nabla\chi. \tag{12.176}$$

The relationship between ξ and δ to ψ and χ is given by

$$\nabla^2\psi = \xi, \qquad \nabla^2\chi = \delta.$$

12.7.1 Conformal transformation

Let L be a mapping of the sphere to itself defined by $(\lambda', \mu') = L(\lambda, \mu)$, $\mu = \sin\theta$, and λ and θ are the spherical coordinates, longitude and latitude, as before, where (λ', μ') are the transformed coordinates, and any function with a prime for the remainder of this section is associated with the transformed coordinates.

We now let F' be the mapping factor and is the determinant of the Jacobian matrix of mapping L, given by

$$F'(\lambda', \mu') = \begin{vmatrix} \frac{\partial\lambda'}{\partial\lambda} & \frac{\partial\mu'}{\partial\lambda} \\ \frac{\partial\lambda'}{\partial\mu} & \frac{\partial\mu'}{\partial\mu} \end{vmatrix} = \begin{vmatrix} \frac{\partial\lambda}{\partial\lambda'} & \frac{\partial\mu}{\partial\lambda'} \\ \frac{\partial\lambda}{\partial\mu'} & \frac{\partial\mu}{\partial\mu'} \end{vmatrix}^{-1}.$$

An important property that we have to note here is that if the mapping is conformal then we have the property

$$\nabla^2 A = F'(\lambda', \mu')\nabla'^2 A', \tag{12.177}$$

where ∇'^2 is the standard Laplacian operator with respect to the transformed coordinates (λ', μ') and A is a scalar function.

The mapping that is considered here consists of a rotation and a stretching. Therefore, let R be a rotation transformation and $P(\lambda_0, \theta_0)$ be a given point on the unit sphere, where after applying R, one of the new coordinate axes will go through this point. The equations of the rotation that relate the two coordinate systems are given by

$$\begin{aligned} \cos\theta'\sin(\lambda'-\lambda_0) &= \cos\theta\sin(\lambda-\lambda_0),\\ \cos\theta'\cos(\lambda'-\lambda_0) &= \cos\theta\cos(\lambda-\lambda_0)\sin\theta_0 + \sin\theta\cos\theta_0, \qquad (12.178)\\ \sin\theta' &= -\cos\theta\cos(\lambda-\lambda_0)\cos\theta_0 + \sin\theta\sin\theta_0. \end{aligned}$$

The inverse transform of (12.178) can be shown to be

$$\cos\theta\sin(\lambda-\lambda_0) = \cos\theta'\sin(\lambda'-\lambda_0),$$

$$\begin{aligned} \cos\theta\cos(\lambda-\lambda_0) &= \cos\theta'\cos(\lambda'-\lambda_0)\sin\theta_0-\sin\theta'\cos\theta_0, \\ \sin\theta &= \cos\theta'\cos(\lambda'-\lambda_0)\cos\theta_0+\sin\theta'\sin\theta_0. \end{aligned} \tag{12.179}$$

Finally, the Jacobian of the transformation involves the following derivatives:

$$\begin{aligned} \frac{\partial\lambda}{\partial\lambda'} &= \frac{\cos^2\theta'\sin\theta_0-\cos\theta'\sin\theta'\cos(\lambda'-\lambda_0)\cos\theta_0}{\cos^2\theta}, \\ \frac{\partial\lambda}{\partial\mu'} &= \frac{\sin(\lambda'-\lambda_0)\cos\theta_0}{\cos\theta'\cos^2\theta}, \\ \frac{\partial\mu}{\partial\lambda'} &= -\cos\theta'\sin(\lambda'-\lambda_0)\cos\theta_0, \\ \frac{\partial\mu}{\partial\mu'} &= -\frac{\sin\theta'\cos(\lambda'-\lambda_0)\cos\theta_0+\cos\theta'\sin\theta_0}{\cos\theta'}. \end{aligned} \tag{12.180}$$

We have to note here that if $\theta=\pm\frac{\pi}{2}$ then $\frac{\partial\lambda}{\partial\mu'}$ and $\frac{\partial\mu}{\partial\mu'}$ are not defined. By changing the stretching factor such that there are no grid points located at the poles, this problem can be avoided without adding any restrictions on the focusing area and stretching factor.

Stretching

If we now denote the stretching transformation by S, which is defined by

$$\begin{aligned} \lambda' &= \lambda, \\ \mu' &= \frac{(c^2-1)+\mu(c^2+1)}{(c^2+1)+\mu(c^2-1)}, \end{aligned} \tag{12.181}$$

where c is the stretching factor, then depending on whether the value of c is greater than or less than 1, we stretch the mesh towards the North or South pole, respectively.

The inverse transformation is obtained for c^{-1} and is given by

$$\begin{aligned} \lambda &= \lambda', \\ \mu &= -\frac{(c^2-1)+\mu'(c^2+1)}{(c^2+1)-\mu'(c^2-1)}. \end{aligned} \tag{12.182}$$

The partial derivatives in the Jacobian are given by

$$\frac{\partial\lambda}{\partial\lambda'}=1,\quad \frac{\partial\lambda}{\partial\mu'}=0,\quad \frac{\partial\mu}{\partial\lambda'}=0,\quad \frac{\partial\mu}{\partial\mu'}=\frac{4C^2}{\left(c^2+1-\mu'(c^2-1)\right)^2}, \tag{12.183}$$

and the associated mapping factor of S is then

$$F'_S(\lambda',\mu')=\frac{\left(c^2+1-\mu'(c^2-1)\right)^2}{4c^2}. \tag{12.184}$$

Thus the mapping that puts the finest grid on a specific region is the product of the mapping R and S which results in a mapping factor $F'(\lambda', \mu')$ of L as

$$F'(\lambda', \mu') = F_R'(\lambda', \mu') F_s'(\lambda', \mu') = \left(\frac{c^2+1}{2c} - \frac{c^2-1}{2c}\right)^2. \tag{12.185}$$

12.7.2 Semi-implicit semi-Lagrangian scheme in spectral space

The semi-implicit semi-Lagrangian scheme for the shallow water model in vorticity-divergence is expressed in the form

$$\eta_a^{n+1} = N_\eta, \tag{12.186}$$
$$\delta_a^{n+1} + \Delta t (\nabla\phi)_a^{n+1} = N_\delta, \tag{12.187}$$
$$\phi_a^{n+1} + \Delta t \left(\overline{\phi}\delta\right)_a^{n+1} = N_\phi \tag{12.188}$$

where $N = (N_\eta, N_\delta, N_\phi)^T$ and $N = N_a + N_d$, the arrival and departure components of N, are defined by

$$N_a = \begin{pmatrix} -\Delta t (\eta\delta)_a^n \\ \Delta t (\Psi)_a^n \\ -\Delta t [\phi\delta)_a^n \end{pmatrix}, \tag{12.189}$$

$$N_d = \begin{pmatrix} \eta_d^{n-1} - \Delta (\eta\delta)_d^n \\ \delta_d^{n-1} + \Delta t (\Psi)_d^n - \Delta t \left(\nabla^2\phi\right)_d^{n-1} \end{pmatrix}, \tag{12.190}$$

where we can see that this is a three-time-level scheme.

We now move on to consider the spectral formulation of the semi-implicit part of the integration. The spectral transform, which is based upon the spherical harmonic transform, is based upon representing a scalar field as a linear combination of spherical harmonics as we saw earlier, and is given by

$$\xi(\lambda, \mu) = \sum_{m=-M}^{M} \sum_{n=|M|}^{N(m)} \xi^m P_n^m(\mu) e^{im\lambda}, \tag{12.191}$$

where

$$\xi_n^m = \int_{-1}^{1} \frac{1}{2\pi} \int_0^{2\pi} \xi(\lambda, \mu) d^{im\lambda} d\lambda P_n^m(\mu) d\mu, \tag{12.192}$$

with P_n^m being the normalized associated Legendre functions.

All of the spectral transforms are computed on the Gaussian grid, which is the grid that is mapped to the variable resolution, physical space grid, and is necessary to exactly calculate the spectral coefficients using Gaussian quadrature as we saw earlier. Once the spectral coefficients are known, the transformation from the spectral to the uniform grid space is accomplished using (12.191) and is referred to as **harmonic synthesis** [68]. The transform from the physical to spectral space uses a discrete version of the continuous transform, which we saw earlier, given by

$$\xi^m(\mu) = \frac{1}{2\pi}\int_0^{2\pi} \xi(\lambda,\mu)\, e^{-im\lambda}\, d\lambda\, P_n^m(\mu)\, d\mu = \frac{1}{Q}\sum_{i=1}^{P} \xi(\lambda_i,\mu)\, e^{-im\lambda_i} \tag{12.193}$$

and

$$\xi_n^m = \int_{-1}^{1} \xi^m(\mu)\, P_n^m(\mu)\, d\mu = \sum_{j=1}^{R} \xi^m(\mu_j)\, P_n^m(\mu_j)\, w_j, \tag{12.194}$$

where

$$\lambda_i = \frac{2\pi i}{Q}, \qquad w_j = \frac{2\left(1-\mu_j^2\right)}{\left(R P_{R-1}(\mu_j)\right)^2}, \qquad \sum_{j=1}^{R} w_j = 2, \tag{12.195}$$

and μ_j denotes the Gaussian latitudes, Q is the number of grid points in the East–West direction, R is the number of Gaussian latitudes from pole to pole, w_j is the Gaussian weight at latitude μ_j, and finally, the Gaussian latitudes μ_j are determined from the roots of the Legendre polynomial $P_R(\mu)$.

By now introducing a conformal mapping L and applying the property of conformal mapping that was given in (12.177), we can obtain the transformed equations from (12.186)–(12.188) as

$$\eta_a^{n+1} = N_\eta, \tag{12.196}$$

$$\delta_a^{n+1} + \Delta t\left[\nabla^2 \phi F\right]_a^{n+1} = N_\delta, \tag{12.197}$$

$$\phi_a^{n+1} + \Delta t\left[\overline{\phi\delta}\right]_a^{n+1} = N_\phi, \tag{12.198}$$

where

$$\begin{aligned} N_\eta &\equiv \eta_d^{n-1} - \Delta t\left(\left[\eta\delta\right]_a^n + \left[\eta\delta\right]_d^n\right), \\ N_\delta &\equiv \delta_d^{n-1} + \Delta t\left(\Psi_a^n + \Psi_d^n - \left[\nabla^2\phi\right]_d^{n-1}\right), \\ N_\phi &\equiv \phi_d^{n-1} - \Delta t\left(\left[\phi\delta\right]_a^n + \left[\phi\delta\right]_a^n + \left[\overline{\phi\delta}\right]_d^{n-1}\right). \end{aligned} \tag{12.199}$$

In Guo and Drake (2005), the authors apply a series of identities and integration by parts to the Laplacian operator and they arrive at three equations for the spectral form of the governing equations; we refer the reader to their paper [68] for more details. An important feature that they highlight is that the solution of the two-by-two system of equations is a semi-implicit treatment of the fast wave terms.

12.7.3 Treatment of Coriolis terms

In Temperton (1997) [187], the author states that the first approach is to treat the Coriolis terms in the shallow water equations in spherical coordinates in a semi-implicit form, through averaging them in time and space along the trajectory. The equations take the form

$$U^{n+1} - f\Delta t V^{n+1} + \frac{\Delta t}{a}\frac{\partial \phi^{n+1}}{\partial \lambda} = R_1, \tag{12.200}$$

$$V^{n+1} + f\Delta t U^{n+1} + \frac{\Delta t}{a}\cos\theta \frac{\partial \phi^{n+1}}{\partial \theta} = R_2, \tag{12.201}$$

$$\phi^{n+1} + \Phi \Delta t \delta^{n+1} = S_3, \tag{12.202}$$

where the left-hand sides of (12.200)–(12.202) are unknown, while the right-hand sides contain all the known terms, U and V are defined as $U = u\cos\theta$, and $V = v\cos\theta$, and Φ is the mean geopotential.

The next step in [187] is to transform (12.200)–(12.202) to the spectral space and apply the curl and divergence operators of the equations for the wind components, yielding, for each zonal wavenumber l and for $l \le k \le K$, where K is the maximum total wavenumber,

$$\left(1 - i\alpha_k^l\right)\zeta_k^{n+1} + 2\Omega\Delta t\left(\frac{(k+1)}{k}\varepsilon_k^l \delta_{k-1}^n + n - 1 + \frac{k}{(k+1)}\varepsilon_{k+1}^l \delta_{k+1}^{n+1}\right) = (S_1)_k, \tag{12.203}$$

$$\left(1 - i\alpha_k^l\right)\delta_k^{n+1} - 2\Omega\Delta t\left(\frac{(k+1)}{k}\varepsilon_k^l \eta_{k-1}^{n+1} + \frac{k}{(k+1)}\varepsilon_{k+1}^l \eta_{k+1}^{n+1}\right) - \frac{\Delta t k(k+1)}{a^2}\phi_k = (S_2)_k, \tag{12.204}$$

$$\phi_k^{n+1} + \Phi\Delta t \delta_k^{n+1} = (S_3)_k, \tag{12.205}$$

where

$$\alpha_k^l = \frac{2\Omega l \Delta t}{k(k+1)}, \qquad \varepsilon_k^l = \sqrt{\left(\frac{k^2 - l^2}{4k^2 - 1}\right)},$$

$$S_1 = \operatorname{curl}(R_1, R_2), \qquad S_2 = \operatorname{div}(R_1, R_2).$$

Two important properties to notice here are that, first, the vorticity and divergence equations are coupled together, and, second, although an independent set of equations is obtained for each zonal wavenumber m, the vorticity components at total wavenumber k are coupled to the divergence components at total wavenumbers $(k-1)$ and $(k+1)$, and vice versa.

To solve the equations, all the spectral components ($l \le k \le K$) for each variable are assembled into a vector for each wavenumber l. Thus we obtain

$$\mathbf{J}\boldsymbol{\zeta}^{n+1} + \mathbf{F}\boldsymbol{\delta}^{n+1} = \mathbf{S}_1, \tag{12.206}$$

$$\mathbf{J}\boldsymbol{\delta}^{n+1} - \mathbf{F}\boldsymbol{\zeta}^{n+1} - \mathbf{L}\boldsymbol{\phi}^{n+1} = \mathbf{S}_2, \tag{12.207}$$

$$\boldsymbol{\phi}^{n+1} + \boldsymbol{\Phi}\Delta t \boldsymbol{\delta}^{n+1} = \mathbf{S}_3, \tag{12.208}$$

where

$$\mathbf{J} = \operatorname{diag}\left(1 - i\alpha_l^k\right),$$

$$
\begin{aligned}
\mathbf{F} &= 2\Omega\Delta t \begin{pmatrix} 0 & f_l^{n+1} & & & & & \\ f_{l+1}^{n-1} & 0 & f_{l+1}^{n+1} & & & & \\ & f_{l+2}^{n-1} & 0 & f_{l+2}^{n+1} & & & \\ & & \ddots & \ddots & \ddots & & \\ & & & f_{K-1}^{n-1} & 0 & f_{K-1}^{n+1} \\ & & & & f_K^{n-1} & 0 \end{pmatrix}, \\
f_k^{n+1} &= \frac{k}{(k+1)}\varepsilon_{k+1}^{l}, \\
f_k^{n-1} &= \frac{(k+1)}{k}\varepsilon_k^{l}, \\
\mathbf{L} &= \frac{\Delta t}{a^2}\operatorname{diag}\left(k\left(k+1\right)\right).
\end{aligned}
$$

Eqs. (12.206)–(12.208) can then be combined to give

$$
\left(\left(\mathbf{J}+\mathbf{F}\mathbf{J}^{-1}\mathbf{F}+\mathbf{\Phi}\Delta t\mathbf{L}\right)\delta\right)^{n+1} = \mathbf{S}_2+\mathbf{F}\mathbf{J}^{-1}\mathbf{S}_1+\mathbf{L}\mathbf{S}_3. \tag{12.209}
$$

The matrix on the left-hand side of (12.209) is pentadiagonal with zeros on the sub- and superdiagonals, thus (12.209) decouples into two tridiagonal systems, one for the even components and one for the old components. The tridiagonal matrices have complex entries, but they are diagonally dominant and therefore the usual solution algorithm is stable. While this setup is for the three-time-level semi-Lagrangian scheme, Temperton indicates that the two-level semi-implicit treatment of the Coriolis terms would also be viable, however, if we have rotated and stretched the coordinates: the rotation would destroy the horizontal separability of the equations to be solved in the spectral space.

The second approach that Temperton mentioned in [187] is based upon considering the treatment of the Coriolis terms in shallow water equations where in semi-Lagrangian schemes on the sphere he uses the momentum equations in vector form to avoid an instability due to the metric term [159]. This means that we are considering the momentum equations as

$$
\frac{D\mathbf{v}}{Dt} + 2\Omega\times\mathbf{v} + \nabla\phi = 0. \tag{12.210}
$$

When we apply a semi-Lagrangian discretization of (12.210), we are advecting the vector quantity $\mathbf{v} = \begin{pmatrix} u & v \end{pmatrix}^T$ on the sphere, where we have to remember that the orientation of the local coordinate system changes as we move along the trajectory. Therefore, since $\mathbf{v} = \frac{D\mathbf{r}}{Dt}$ where $\mathbf{r}$ is the radial position as before, this implies that (12.210) can be written as

$$
\frac{D}{Dt}\left(\mathbf{v} + 2\Omega\times\mathbf{r}\right) + \nabla\phi = 0. \tag{12.211}
$$

In a semi-Lagrangian discretization of (12.211), the Coriolis terms are absorbed into the advection. The change does not affect the trajectory itself, but only the quantity being advected.

Since the term $2\Omega \times \mathbf{r}$ is known everywhere, which in component form is $\begin{pmatrix} 2\Omega a\cos\theta \\ 0 \end{pmatrix}$, the extra term is added to $\mathbf{v}$ at the departure point of the trajectory and subtracted again at the arrival gridpoint. The equations to be solved for the variable at the new time-level $(t+\Delta t)$ have exactly the same form as in the case of a simple explicit scheme for the Coriolis terms. The reader is referred to Temperton (1997) to look at the results of using the two approaches above in an earlier version of the Integrated Forecasting System (IFS) at ECMWF, where Temperton's conclusions indicate there are advantages to both approaches, but if the model has a rotated coordinate system, then the semi-implicit approach is more difficult to solve compared to including the Coriolis term in the advected term.

12.8 Semi-implicit semi-Lagrangian scheme on the sphere

In this section we shall summarize the derivation of the semi-implicit semi-Lagrangian approach from the McDonald (1986) paper, where he is implementing a dry adiabatic atmosphere model on the sphere with the following differential equations:

$$\frac{D_H u}{Dt} + \dot{\sigma}\frac{\partial u}{\partial \sigma} + \frac{1}{a\cos\theta}\frac{\partial \Phi}{\partial \lambda} + \frac{RT}{a\cos\theta}\frac{\ln p_s}{\partial \lambda} - \left(f + \frac{u\tan\theta}{a}\right)v = 0, \tag{12.212}$$

$$\frac{D_H v}{Dt} + \dot{\sigma}\frac{\partial v}{\partial \sigma} + \frac{1}{a}\frac{\partial \Phi}{\partial \theta} + \frac{RT}{a}\frac{\ln p_s}{\partial \theta} + \left(f + \frac{u\tan\theta}{a}\right)u = 0, \tag{12.213}$$

$$\frac{D_H T}{Dt} + \dot{\sigma}\frac{\partial T}{\partial \sigma} - \frac{RT}{c_p}\left(\frac{D\ln p_s}{Dt} + \frac{\dot{\sigma}}{\sigma}\right) = 0, \tag{12.214}$$

$$\frac{D_H \ln p_s}{Dt} + \delta + \frac{\partial \dot{\sigma}}{\partial \sigma} = 0, \tag{12.215}$$

$$\frac{\partial \Phi}{\partial \ln \sigma} = -RT, \tag{12.216}$$

where

$$\frac{D_H}{Dt} = \frac{\partial}{\partial t} + \frac{u}{a\cos\theta}\frac{\partial}{\partial \lambda} + \frac{v}{a}\frac{\partial}{\partial \theta}, \tag{12.217}$$

$$\delta = \frac{1}{a\cos\theta}\left(\frac{\partial u}{\partial \lambda} + \frac{\partial v\cos\theta}{\partial \theta}\right). \tag{12.218}$$

We have summarized the terms used in (12.212)–(12.218) in Table 12.2 to simplify things.

Given the table above, we now rewrite the equations so as to separate out the large linear terms from the rest by substituting for the temperature and geopotential as follows:

$$T = T_0 + T', \tag{12.219}$$

$$\Phi = G - RT_0 \ln p_s, \tag{12.220}$$

TABLE 12.2 List of variables and constants.

a	Radius of the Earth (6.371×10^6 m)
c_p	Specific heat ($1004.6\,\mathrm{J\,kg^{-1}K^{-1}}$)
g	Acceleration due to gravity ($9.80665\,\mathrm{ms^{-1}}$)
Ω	Angular speed of the Earth ($7.292 \times 10^{-5}\,\mathrm{s^{-1}}$)
R	Gas constant ($287.04\,\mathrm{J\,kg^{-1}K^{-1}}$)
δ	Divergence ($\mathrm{s^{-1}}$)
$f = 2\Omega\sin\theta$	Coriolis parameter ($\mathrm{s^{-1}}$)
p	Pressure (Pa)
p_s	Surface pressure (Pa)
Φ	Geopotential ($g \times$ height) ($\mathrm{m^2\,s^{-2}}$)
Φ_s	Surface geopotential ($\mathrm{m^2\,s^{-2}}$)
$\sigma = \dfrac{p}{p_s}$	Vertical coordinate
$\dot{\sigma} = \dfrac{d\sigma}{dt}$	σ-vertical velocity
σ_T	Value of σ at the top of the atmosphere
T	Temperature (K)

where T_0 is a function of σ only and is chosen such that $T \gg T'$. This then enables us to write (12.212)–(12.214) as

$$\frac{D_H u}{Dt} = -\frac{1}{a\cos\theta}\frac{\partial G}{\partial \lambda} + fv + N_u, \tag{12.221}$$

$$\frac{D_H v}{Dt} = -\frac{1}{a}\frac{\partial G}{\partial \theta} - fu + N_v, \tag{12.222}$$

$$\frac{D_H T}{Dt} = \frac{RT_0}{c_p}\left(-\delta - \frac{\partial \dot{\sigma}}{\partial \sigma} + \frac{\dot{\sigma}}{\sigma}\right) - \dot{\sigma}\frac{\partial T_0}{\partial \sigma} + N_T, \tag{12.223}$$

where

$$N_u = -\frac{RT'}{a\cos\theta}\frac{\ln p_s}{\lambda} - \dot{\sigma}\frac{\partial u}{\partial \theta} + \frac{uv\tan\theta}{a}, \tag{12.224}$$

$$N_v = -\frac{RT'}{a}\frac{\partial \ln p_s}{\partial \theta} - \dot{\sigma}\frac{\partial v}{\partial \sigma} - \frac{u^2\tan\theta}{a}, \tag{12.225}$$

$$N_T = -\dot{\sigma}\frac{\partial T'}{\partial \sigma} + \frac{RT'}{c_p}\left(-\delta - \frac{\partial \dot{\sigma}}{\sigma} + \frac{\dot{\sigma}}{\sigma}\right). \tag{12.226}$$

Given the model just described, where we can see several linear and nonlinear forcing terms in the horizontal advection terms, we now start with the time integration in a semi-Lagrangian manner for a general field, which implies that

$$\frac{D_H \psi(\lambda, \theta, \sigma, t)}{Dt} = \xi(\lambda, \theta, \sigma, t), \tag{12.227}$$

such that, when (12.227) is discretized in time, we obtain

$$\psi^{n+1} - \psi_d^n = \int_t^{t+\Delta t} dt \xi \left(\lambda, \theta, \sigma, t\right), \tag{12.228}$$

but now we are in 3D, where $\psi^n = \psi\left(\lambda, \theta, \sigma, n\Delta t\right)$ and the value of ψ at the departure point is given by

$$\psi_d^n = \psi\left(\lambda - \frac{u\Delta t\cos\theta}{a}, \theta - \frac{v\Delta t}{a}, \sigma, n\Delta t\right). \tag{12.229}$$

The integral in (12.228) is along the trajectory of a parcel of air. In this version of a semi-implicit approach on the sphere McDonald integrated (12.215) and (12.223), with an implicit approximation $\dfrac{\left(\xi^{n+1} + \xi_d^n\right)\Delta t}{2}$ for the linear terms and $\xi_d^n \Delta t$ for the nonlinear terms given by

$$\begin{aligned} T^{n+1} &= T_d^n + \frac{\Delta t}{2}\left(\left(L_T\right)^{n+1} + \left(L_T\right)_d^n\right) + \Delta t\left(N_T\right)_d^n, && (12.230)\\ \left(\ln p_s\right)^{n+1} &= \left(\ln p_s\right)_d^n + \frac{\Delta t}{2}\left(\left(L_p\right)^{n+1} + \left(L_p\right)_d^n\right), && (12.231) \end{aligned}$$

where the linear terms L_T and L_p are defined as

$$\begin{aligned} L_T &= \frac{RT_0}{c_p}\left(-\delta - \frac{\partial\dot{\sigma}}{\partial\sigma} + \frac{\dot{\sigma}}{\sigma}\right) - \dot{\sigma}\frac{\partial T_0}{\partial\sigma}, && (12.232)\\ L_p &= -\delta - \frac{\partial\dot{\sigma}}{\partial\sigma}. && (12.233) \end{aligned}$$

When integrating (12.221) and (12.222), McDonald suggests a different approach to that just presented, so that the final set of equations we arrive at are both unconditionally stable, but also that they could be readily solved. The first term that is dealt with is the Coriolis-implicit/gravity-explicit step:

$$\begin{aligned} u^{n+\frac{1}{2}} &= \left(u - \frac{\Delta t}{2a\cos\theta}\frac{\partial G}{\partial\lambda} + \frac{\Delta t}{2}N_u\right)_d^n + \frac{\Delta t}{2}fv^{n+\frac{1}{2}}, && (12.234)\\ v^{n+\frac{1}{2}} &= \left(v - \frac{\Delta t}{2a}\frac{\partial G}{\partial\theta} + \frac{\Delta t}{2}N_v\right)_d^n - \frac{\Delta t}{2}fu^{n+\frac{1}{2}}, && (12.235) \end{aligned}$$

where (12.234) and (12.235) can immediately be solved to yield $u^{n+\frac{1}{2}}$ and $v^{n+\frac{1}{2}}$.

The next step is to consider the Coriolis-explicit/gravity-implicit step:

$$\begin{aligned} u^{n+1} &= u^{n+\frac{1}{2}} - \frac{\Delta t}{2a\cos\theta}\frac{\partial G^{n+1}}{\partial\lambda} + \frac{\Delta t}{2}\left(N_u\right)_d^n + \frac{\Delta}{2}fv^{n+\frac{1}{2}}, && (12.236)\\ v^{n+1} &= v^{n+\frac{1}{2}} - \frac{\Delta t}{2a}\frac{\partial G}{\partial\theta} + \frac{\Delta t}{2}\left(N_v\right)_d^n - \frac{\Delta t}{2}fu^{n+\frac{1}{2}}, && (12.237) \end{aligned}$$

Given all of the changes that have been introduced, Eqs. (12.236), (12.237), (12.230), and (12.231) are now in the desired form, according to [118], and so we can solve them in two stages:

Step (i) In each equation just mentioned, the terms that are valid at time-level n are grouped together and evaluated to yield, what are referred to as B_u, B_v, A_T, and B_p, and are defined as

$$B_u = \frac{X_u\left(1-F^2\right)+\Delta t f X_v}{\left(1+F^2\right)} + N_u\frac{\Delta t}{2}, \tag{12.238}$$

$$B_v = \frac{X_v\left(1-F^2\right)-\Delta t f X_u}{\left(1+F^2\right)} + N_v\frac{\Delta t}{2}, \tag{12.239}$$

where $F = \dfrac{f\Delta t}{2}$ and

$$\begin{aligned}
X_u &= u - \frac{\Delta t}{2a\cos\theta}\frac{\partial G}{\partial\lambda} + \frac{\Delta t}{2}N_u,\\
X_v &= v - \frac{\Delta t}{2a}\frac{\partial G}{\partial\theta} + \frac{\Delta t}{2}N_v,\\
A_T &= T + \frac{L_T\Delta t}{2} + N_T\Delta t,\\
B_p &= \ln p_s + \frac{L_p\Delta t}{2}.
\end{aligned}$$

The values of the four parameters at the departure point, $(B_u)_d^n$, $(B_v)_d^n$, $(A_T)_d^n$, and $(B_p)_d^n$, are computed by a suitable polynomial interpolation from the surrounding grid points. This is the **semi-Lagrangian step.**

Step (ii) According to McDonald, the problem has now been reduced to that of solving the following equations, subject to the boundary conditions that $\dot{\sigma} = 0$ at the bottom and top of the atmosphere, thus we have

$$\left(u + \frac{\Delta t}{2a\cos\theta}\frac{\partial G}{\partial\lambda}\right)^{n+1} = (B_u)_d^n, \tag{12.240}$$

$$\left(v + \frac{\Delta t}{2a}\frac{\partial G}{\partial\theta}\right)^{n+1} = (B_v)_d^n, \tag{12.241}$$

$$\left(T - \frac{RT_0}{c_p}\ln p_s - \frac{\Delta t}{2}\left(\frac{RT_0}{c_p}\frac{\dot{\sigma}}{\sigma} - \dot{\sigma}\frac{\partial T_0}{\partial\sigma}\right)\right)^{n+1} = (B_T)_d^n, \tag{12.242}$$

$$\left(\ln p_s + \frac{\Delta t}{2}\left(D + \frac{\partial\dot{\sigma}}{\partial\sigma}\right)\right)^{n+1} = \left(B_p\right)_d^n, \tag{12.243}$$

$$(B_T)_d^n = (A_T)_d^n - \frac{RT_0}{c_p}\left(B_p\right)_d^n. \tag{12.244}$$

This step is referred as the **semi-implicit step.**

We have to recall that we are dealing with a three-dimensional problem here and so we have to introduce a vertical discretization scheme now. An important feature to note here is that the vertical grid being used here is similar to the Charney and Phillips grid from earlier, where u, v, T, and G are evaluated at what are referred to as the *principle levels*, while Φ and $\dot{\sigma}$ are evaluated at the *intermediate levels*. When $\sigma = 1$, $\dot{\sigma}$ vanishes identically. At the top of the atmosphere, $\sigma = \sigma_{\text{top}}$, $\dot{\sigma} = 0$ throughout the integration.

Therefore, if we now take these definitions and sum (12.216) in the vertical direction, then

$$\Phi_{K+\frac{1}{2}} = \Phi_s + R \sum_{L=K+1}^{\mathfrak{K}} T_L \Delta \ln \sigma_L. \tag{12.245}$$

Thus if we sum (12.243) then we obtain

$$\ln p_s^{n+1} = \sum_{K=1}^{\mathfrak{K}} \frac{\left(\left(B_p\right)_{d,K}^n - \frac{\Delta t}{2} D_K^{n=1} \right) \Delta \sigma_K}{\left(1 - \Sigma_{top}\right)}, \tag{12.246}$$

$$\frac{\Delta t}{2} \dot{\sigma}_{k+\frac{1}{2}}^{n+1} = \sum_{L=1}^{K} \left(\left(B_p\right)_{d,L}^n - \frac{\Delta t}{2} D_L^{n+1} - \ln p_s^{n+1} \right) \Delta \sigma_L, \tag{12.247}$$

where G is defined by

$$G_K = \frac{\Phi_{K-\frac{1}{2}} + \Phi_{K+\frac{1}{2}}}{2} + RT_{0,K} \ln p_s. \tag{12.248}$$

If we now substitute (12.245)–(12.248) into (12.242), then we obtain

$$G_K^{n+1} + \frac{\Delta t}{2} \sum_L M_{KL} D_L^{n+1} = H_K, \tag{12.249}$$

where

$$H_K = \Phi_s + \frac{1}{2} R \left(B_T\right)_{d,K}^n \Delta \ln \sigma_K + R \sum_{L=K+1}^{\mathfrak{K}} \left(B_T\right)_{d,L}^n \Delta \ln \sigma_L + \sum_{L=1}^{\mathfrak{K}} M_{KL} \left(B_p\right)_{d,L}^n, \tag{12.250}$$

where $\mathbf{M}$ is a function of R, c_p, $T_{0,K}$, σ_{top}, and σ_K only.

If now assume that the matrix $\mathbf{M}$ is diagonalizable such that

$$\sum_{L,M} = E_{KL}^{-1} M_{LM} E_{MN} = \omega_{KN} \lambda_N, \tag{12.251}$$

then (12.240), (12.241), and (12.249) can be combined to yield

$$\frac{(\Delta t)^2 \lambda_K}{4} \nabla^2 \left(\mathbf{E}^{-1}\mathbf{G}\right)_K^{n+1} - \left(\mathbf{E}^{-1}\mathbf{G}\right)_K^{n+1} = -\left(\mathbf{E}^{-1}\mathbf{Y}\right)_K, \tag{12.252}$$

where

$$Y_K = H_K - \frac{\Delta t}{2} \sum_{L=1}^{\mathfrak{K}} M_{KL} \left(\nabla \cdot B_v\right)_{d,L}^n . \tag{12.253}$$

It is stated in [118] that when (12.252) has been solved, it yields $\left(\mathbf{E}^{-1}\mathbf{G}\right)_K^{n+1}$, and so D_K^{n+1} can be extracted from (12.249), and subsequently $\ln p_s^{+1}$ and $\dot{\sigma}_{K+\frac{1}{2}}^{n+1}$ from (12.246) and (12.247), respectively, and finally, we have u_K^{n+1}, v_K^{n+1}, and T_K^{n+1} calculated from (12.240), (12.241), and (12.242), respectively.

McDonald does go on to perform a stability analysis for the linearized equations of motion on an f-plane, and we refer the reader to [118] for those details.

We can see from the derivation that we have just worked through that it is quite difficult to implement the semi-Lagrangian approach in 3D spherical coordinates following this approach, there are some techniques to simplify some parts of the model.

12.9 Removing the Helmholtz equation

In [184] Tanguay and Robert devise a way to remove the Helmholtz equation in the model through using the modified form of the shallow water equations:

$$\frac{DU}{Dt} - fV + \overline{\frac{\partial \phi'}{\partial X}}^t + K\frac{\partial S}{\partial X} = 0, \tag{12.254}$$

$$\frac{DV}{Dt} + fU + \overline{\frac{\partial \phi'}{\partial Y}}^t + K\frac{\partial S}{\partial Y} = 0, \tag{12.255}$$

$$\frac{D}{Dt}\left(\phi + \mu\right) + \phi_0 \overline{\delta'}^t + \left(\phi - \phi_0\right)\delta = 0, \tag{12.256}$$

where

$$\mu = \phi_0^2 \Delta t^4 S \frac{\partial^2}{\partial X^2}\left(S\frac{\partial^2 \phi}{\partial Y^2}\right). \tag{12.257}$$

The equations above have been put into a conformal projection where we have

$$K = \frac{1}{2}\left(U^2 + V^2\right), \tag{12.258a}$$

$$\delta = S\left(\frac{\partial U}{\partial X} + \frac{\partial V}{\partial Y}\right), \tag{12.258b}$$

$$S = m^2, \tag{12.258c}$$

$$\left[U, V\right] = \left[\frac{u}{m}, \frac{v}{m}\right]. \tag{12.258d}$$

An important feature to note here is that the extra term that has been added to (12.256) is unphysical. The term is added so that the usual Helmholtz equation will factorize without any further approximations. The term is fourth-order in Δt so it should not introduce any significant errors in the corresponding integration.

The set of equations above are now approximated with a three-time-level scheme for the Lagrangian derivatives and time averaged along the trajectory for the time-mean terms. This then leads to

$$\left(U + \Delta t \frac{\partial \phi}{\partial X}\right)^{n+1} = Q_1, \tag{12.259}$$

$$\left(V + \Delta t \frac{\partial \phi}{\partial Y}\right)^{n+1} = Q_2, \tag{12.260}$$

$$\left(\phi + \phi_0^2 (\Delta t)^4 S \frac{\partial^2}{\partial X^2}\left(S \frac{\partial^2 \phi}{\partial Y^2}\right) + \Delta t \phi_0 \delta\right)^{n+1} = Q_3, \tag{12.261}$$

where

$$Q_1 = \left(U - \Delta t \frac{\partial \phi}{\partial X}\right)^{n-1} - 2\Delta t \left(S\left(U \frac{\partial U}{\partial X} + V \frac{\partial U}{\partial Y}\right) - fV + K \frac{\partial S}{\partial X}\right)^n, \tag{12.262}$$

$$Q_2 = \left(V - \Delta t \frac{\partial \phi}{\partial Y}\right)^{n-1} - 2\Delta t \left(S\left(U \frac{\partial V}{\partial X} + V \frac{\partial V}{\partial Y}\right) - fU + K \frac{\partial S}{\partial Y}\right)^n, \tag{12.263}$$

$$Q_3 = (\phi + \mu - \Delta t \phi_0 \delta)^{n-1} - 2\Delta t \left(S\left(U \frac{\partial}{\partial X}(\phi + \mu) + V \frac{\partial}{\partial Y}(\phi + \mu)\right) + (\phi - \phi_0)\delta\right)^n. \tag{12.264}$$

As we know the variables at times $t^n - \Delta t$ and at t^n, we can first evaluate Q_1, Q_2, and Q_3 from (12.262), (12.263), and (12.264), respectively. The next step consists of solving the implicit equations (12.259), (12.260), and (12.261) for the variables U, V, ϕ at time $t^n + \Delta t$. If we now use U and V from (12.259) and (12.260), then it is possible to remove the divergence term in (12.261), which results in

$$\phi - (\Delta t)^2 \phi_0 S \left(\frac{\partial^2 \phi}{\partial X^2} + \frac{\partial^2 \phi}{\partial Y^2}\right) + \phi_0^2 (\Delta t)^4 S \frac{\partial^2}{\partial X^2}\left(S \frac{\partial^2 \phi}{\partial Y^2}\right) = Q_3 - \Delta t \phi_0 S \left(\frac{\partial Q_1}{\partial X} + \frac{\partial Q_2}{\partial Y}\right). \tag{12.265}$$

If we did not have the fourth-order term in Δt here, then we would have the Helmholtz equation that we encountered with the other semi-implicit schemes. The reason that Tanguay and Robert introduce the unphysical parameter is because it allows (12.265) to be factorized into two simple tridiagonal systems:

$$\left(1 - (\Delta t)^2 \phi_0 S \frac{\partial^2}{\partial X^2}\right)\left(1 - (\Delta t)^2 \phi_0 S \frac{\partial^2}{\partial Y^2}\right)\phi = Q_3 - \Delta t \phi_0 S \left(\frac{\partial Q_1}{\partial X} + \frac{\partial Q_2}{\partial Y}\right). \tag{12.266}$$

It is stated in [184] that since the left-hand side of (12.266) factorizes into an operator in terms of the independent variables X coupled with an operator in terms of the variable Y, the authors find that this system is easier to solve for ϕ than the Helmholtz equation.

We now move on to consider a development that was done at the ECMWF involving the extrapolation that is needed in the two-time-level semi-Lagrangian approaches.

12.10 Stable extrapolation two-time-level scheme (SETTLS)

The European Centre for Medium-Range Weather Forecasting (ECMWF) implemented a two-time-level semi-Lagrangian scheme operationally in their numerical weather prediction model using a time step of 1800 s, with 31 hybrid levels in the vertical direction from December 1996 to April 1998. The reason for the short time in operational world for this scheme to be used was due to the detection of a weak instability in the form of noise in several forecasts fields. However, this noise is not present when a three-time-level scheme is employed for this same setup [82]. As a result of this feature, the time step had to be reduced to 990 s, which effectively removed the advantages of the semi-Lagrangian scheme.

We saw earlier, when using a spectral model, that the Gaussian grid is often used, where the computation of quadratic terms is alias-free if the number of points NF of the grid along each Gaussian latitude satisfies $NF > 3M + 1$, and the number of latitude points NG satisfies $NG \geq \dfrac{3M+1}{2}$, where M is the spectral truncation. Alias-free computation of quadratic terms is essential in the Eulerian methods because the Eulerian advection terms are large and quadratic, thus Hortal refers to this grid as the *quadratic grid*.

It was shown in [40] that it was possible to use a *reduced-resolution* Gaussian grid, in which $NF \geq 2M + 1$ and $NG \geq \dfrac{2M+1}{2}$, and this was the grid with the smallest number of points compatible with exact spectral transforms of a field, or a linear combination of fields. This is to say, if we start with a field in the spectral space, transform it to the gridpoint space using the reduced-resolution Gaussian grid, and transform it back to the spectral space, then the original field is recovered exactly. As such this grid is referred to as the *linear Gaussian grid*.

As we have seen throughout, the semi-Lagrangian approach does not require the evaluation of the advection terms and so it is seen that the linear grid is suitable for this form of integration. The advantage of the linear grid over the quadratic is that, given a spectral resolution, the number of grid points is smaller in the former by a factor of $\left(\frac{2}{3}\right)^2$. As most of the computations in a semi-Lagrangian model are performed in the gridpoint space, the use of the linear grid gives an efficiency increase of approximately 2.25 over the quadratic grid at the same spectral resolution.

If we assume a certain distribution of grid points corresponding to the quadratic Gaussian grid of a spectral resolution M, then the quadratic Gaussian grid will have $\frac{3M+1}{2}$ latitudes with $3M + 1$ points on each. When computing the direct spectral transform of a field represented on that set of points, the maximum number of spectral coefficients corresponds to a spectral truncation of $M' = \frac{3M}{2}$. Quadratic terms computed on that grid lead to aliasing for the wave numbers larger that $\frac{2M'}{3}$, and this is the reason why discarding the wavenumbers form M to M' avoids quadratic aliasing. The linear Gaussian grid corresponding to the spectral resolution M' is precisely the distribution of the grid points from which we started [82].

The important thing to notice here is that, given a Gaussian grid, it can be viewed as the quadratic grid for the spectral resolution M, or as the linear grid corresponding to the larger spectral resolution M', and so for the linear grid of resolution M' we keep all the spectral coefficients that are computed, whereas we discard the larger wavenumbers when it is considered as the quadratic grid of resolution M. Therefore, integrations that use the linear Gaussian grid are more sensitive to noise produced by the time-integration scheme, as the noise tends to concentrate in the shortest wavelengths. However, when using the quadratic Gaussian grid, the shortest wavelengths are eliminated, but they are retained in the linear grid, and it was this feature that was creating the noise in the ECMWF system. Thus Hortal developed an alternative two-time-level approach which we shall present now.

The problem that was causing the instability in the ECMWF original implementation of the semi-Lagrangian scheme was caused by the extrapolation in time. We shall show the derivation of the identification of this artifact, and then the new scheme to eliminate this noise.

As before we start with the general form of each equation in the framework of the semi-Lagrangian method which can be written as

$$\frac{DX}{Dt} \equiv N + L,$$

where the right-hand side is split into a linearized part L and the remainder, or nonlinear, term N. The original two-time-level ECMWF scheme discretizes this equation as

$$\frac{X_a^{n+1} - X_d^n}{\Delta t} = \frac{1}{2}\left(N_d^{n+\frac{1}{2}} + N_a^{t+\frac{1}{2}}\right) + \frac{1}{2}\left(L_a^{n+1} - L_d^n\right). \tag{12.267}$$

We should note here that the arrival point is a point on the Gaussian grid. The values of the quantities at $n+1$ are obviously unknown, whereas the nonlinear term at $n+\frac{1}{2}$ is found through a linear extrapolation in time from values at times n and $n-1$. The computation of the semi-Lagrangian trajectory is identical to the solution of every evolution equation. It is performed by solving the equation

$$\frac{D\mathbf{r}}{Dt} = \mathbf{u}, \tag{12.268}$$

where $\mathbf{r}$ is the position of a parcel of air and $\mathbf{u}$ is the three-dimensional velocity of that parcel.

In the original version of the ECMWF semi-Lagrangian scheme, they employed the second-order accurate scheme that is centered in time and space, which comes from [188], given by

$$\mathbf{r}_a^{n+1} \approx \mathbf{r}_d^n + \Delta t \cdot \mathbf{u}_m^{n+\frac{1}{2}}. \tag{12.269}$$

The positions $\mathbf{r}_a^{n+\frac{1}{2}}$ are the Gaussian arrival grid points, $\mathbf{r}_d^n$ are the departure points, and $\mathbf{u}_m^{t+\frac{1}{2}}$ are the velocities in the middle of the trajectory estimated at an instant midway between the present and the future time steps. This estimation is achieved by linear extrapolation in time exactly the same as with nonlinear term of the evolution equations. The interpolation to the middle of the trajectory can also be approximated by an average of the values at the arrival and departure points, as is done with the nonlinear term of the evolution equations.

It is stated in [82] that the cause of the noise was traced to the time extrapolation of the vertical velocity used in the semi-Lagrangian trajectory. Because the scheme was using a linear extrapolation in time of the velocity of an air parcel, it means that we have assumed that $\frac{\partial \mathbf{u}}{\partial t}$ is constant. According to [82], this procedure is unstable for CFL numbers larger than unity if the quantity $\mathbf{u}$ depends on unknown $\mathbf{r}$. However, it is then stated that an extrapolation *along the semi-Lagrangian trajectory* is stable according to [82]. A problem is that the this procedure cannot be applied to the nonlinear term of the momentum equations due to the semi-implicit treatment of these equations. The incompatibility of the extrapolation along the trajectory with the semi-implicit scheme comes from the requirement that the orographic part of the geopotential gradient term has to be in a very good equilibrium with the surface pressure part and this is only possible if the linear term is evaluated at the same points as the nonlinear term in the momentum equations. This is also the reason why in the ordinal ECMWF model the nonlinear term of the momentum equations is averaged along the semi-Lagrangian trajectory rather than interpolated to the middle of the trajectory.

12.10.1 SETTLS

In [82] Hortal introduces an alternative second-order accurate scheme to solve (12.268), which is derived by expanding the unknown quantity $\mathbf{r}$ in a Taylor series around the departure point of the semi-Lagrangian trajectory as

$$\mathbf{r}_a^{n+1} \approx \mathbf{r}_d^n + \Delta t \left[\frac{D\mathbf{r}}{D\mathbf{t}}\right]_d^n + \frac{(\Delta t)^2}{2}\left[\frac{D^2\mathbf{r}}{Dt^2}\right]_{\text{av}}, \tag{12.270}$$

where the subscript "av" represents some average value along the semi-Lagrangian trajectory. Substituting from (12.268) we obtain

$$\mathbf{r}_a^{n+1} \approx \mathbf{r}_b^n + \Delta t \mathbf{u}_d^t + \frac{(\Delta t)^2}{2}\left[\frac{D\mathbf{u}}{Dt}\right]_{\text{av}}. \tag{12.271}$$

It is stated in [82] that the equation above describes a uniformly accelerated movement, the trajectory that can no longer be considered as straight lies on a plane of arcs of a great circle in spherical geometry, which is the traditional case in semi-Lagrangian schemes, and the position of the middle point of the trajectory is no longer an average between the departure and arrival points.

We now have to consider how we are going to approximate the quantity

$$\left[\frac{D\mathbf{u}}{Dt}\right]_{\text{av}}. \tag{12.272}$$

In [82] the author states that the first possibility he explored was to use an average along the trajectory of the explicit estimate of the right-hand side of the momentum equations as the horizontal part of (12.272), and then the expression

$$\left[\frac{Dw}{Dt}\right]_{\text{av}} \approx \frac{w_a^n - w_{d_-}^{n-1}}{\Delta t}, \tag{12.273}$$

evaluated at the previous time step as the vertical part, where d_- means the position at time $n-1$ of the parcel of air arriving at gridpoint a at time t. We should note that (12.273) is not applicable to the nonlinear terms of the momentum equations, since it uses information from the departure point of the previous time step trajectory, which is not exactly the same as the point to which the linear term is interpolated. Hortal then says that this approach did remove the noise but did not produce the same level of forecasts as the original scheme when used in a 4DVAR data assimilation scheme.

The formulation that was eventually adopted comes form the following estimate:

$$\left[\frac{D\mathbf{u}}{Dt}\right]_{\text{av}} \approx \left[\frac{D\mathbf{u}}{Dt}\right]^{n-\frac{1}{2}} \approx \frac{\mathbf{u}_a^n - \mathbf{u}_d^{n-1}}{\Delta t}, \tag{12.274}$$

using the departure point of the semi-Lagrangian trajectory corresponding to the present time step instead of the departure point of the trajectory corresponding to the previous time as in (12.273). Here d means the position at time n of the parcel of air that will arrive at gridpoint a at time $n+1$.

The way to interpret this estimate is to assumes that the total time derivative of the velocity is constant in time and is also seen as some form of extrapolation along the trajectory, but the estimate only uses the arrival and departure points of the present trajectory and is therefore compatible with the semi-implicit treatment of the evolution equations. According to linear stability analysis, this scheme should be stable and was therefore named the **Stable Extrapolation Two-Time-Level Scheme, or SETTLS**. If we now substitute (12.274) into (12.271), then this yields

$$\mathbf{r}_a^{n+1} = r_d^n + \frac{\Delta t}{2}\left(\left[2\mathbf{u}^n - \mathbf{u}^{n-1}\right]_d + \mathbf{r}_a^n\right). \tag{12.275}$$

Stability analysis

The stability analysis for the original ECMWF scheme and the SETTLS scheme is derived as follows: Consider the forced one-dimensional semi-Lagrangian equation

$$\frac{D\psi}{Dt} = S(\psi),$$

where ψ is a generic variable, by applying the prototype linear problem in which $S(\psi) \equiv i\omega\psi + \lambda\psi$, where ω and λ are constants, ψ is complex, and the advecting velocity U is constant. We know that if the initial condition for ψ is given by the function $\psi(x,0) = f(x)$, then the solution is

$$\psi(x,t) = f(x - Ut)\, e^{(i\omega+\lambda)t},$$

which is nonamplifying for $\lambda \le 0$.

The original ECMWF two-time-level semi-Lagrangian scheme and the SETTLS schemes are given by

$$\frac{\psi\left(x_a,t^{n+1}\right)-\psi\left(x_d,t^n\right)}{\Delta t} = \frac{3}{2}S\left(x_m,t^n\right)-\frac{1}{2}S\left(x_a,t^{n-1}\right),$$

$$\frac{\psi\left(x_a,t^{n+1}\right)-\psi\left(x_d,t^n\right)}{\Delta t} = \frac{1}{2}\left(2S\left(x_d,t^n\right)-S\left(x_d,t^n\right)+S\left(x_m,t^{n-1}\right)\right).$$

If we now assume a single-wave solution of the form

$$\psi\left(x_i,t^n\right)\equiv\left(A_k\right)^n e^{ikj\Delta x},$$

and define $s=x_a-x_d$ and $\gamma=\lambda\Delta t+i\omega\Delta t$, then we obtain the following two quadratic equations for the stability zone of the original formulation and the SETTLS scheme:

$$A_k^2-A_k\left(\frac{3\gamma}{2}e^{-\frac{iks}{2}}+e^{-iks}\right)+\left(\frac{\gamma}{2}e^{-\frac{iks}{2}}\right) = 0,$$

$$A_k^2-A_k\left(\gamma e^{-iks}+\frac{\gamma}{2}+e^{-iks}\right)+\frac{\gamma}{2}e^{-iks} = 0,$$

respectively.

From the stability equations above, we can see that both stability regions are dependent on the distance between the arrival and the departure point through s, which is implying that the range of values of $\lambda\Delta t$ and $\omega\Delta t$ for which $|A_k|\leq 1$, which is proportional to the Courant number. The analysis of the stability domain for the original semi-Lagrangian formulation can be found in [46] where we have a copy of the stability figures from [82] in Fig. 12.21.

In the stability figures are regions of the $\omega\Delta t$–$\lambda\Delta t$ plane in which $|A_k|\leq 1$ for several values of ks between 0 and 2π. The most severe stability constraints are, as we have seen earlier, imposed by the $2\Delta x$ waves, for which these values of ks correspond to Courant numbers, $u\frac{\Delta t}{\Delta x}$, between 0 and 2. As the Courant number increases, the region of absolute stability rotates clockwise around the origin in the $\omega\Delta t$–$\lambda\Delta t$ plane. An important feature that is mentioned by Durran in [46] is that when $u\frac{\Delta t}{\Delta x}=2$, then all $2\Delta x$ waves that should properly be damped are amplified, and the only $2\Delta x$ waves that are damped are those that should be amplified.

An important remark that is made in [46] about the 1997 scheme, as it is referred to in [82], is that the instability that develops in the scheme as the Courant number increases may be qualitatively understood to result from a failure to match the numerical domain of dependence for the data used to compute $S(\psi)$ with the numerical domain of dependence for the true solution.

Returning to Hortal (2002), he states that for the three regions that are in the upper plot of Fig. 12.21 are those associated with $ks=0,\pi,2\pi$. The area of stability rotates and for $ks=2\pi$ does not overlap with the region corresponding to $ks=0$, but also to note is that the area of stability for $ks=2\pi$ is inside the region where the analytical solution is growing with time. Therefore, according to [82], all the prototype problems that have stable solutions for $ks=0$ are unstable for $ks=2\pi$ in the case of the original scheme.

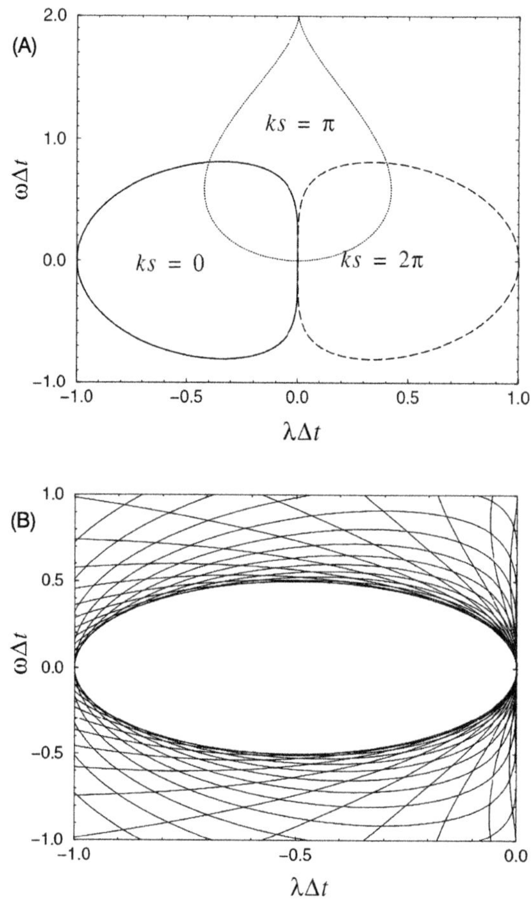

FIGURE 12.21 Copy of the stability regions for the 1997 method, and the new SETTLS scheme. Hortal, M., 2002: The development and testing of a new two-time-level semi-Lagrangian scheme(SETTLS) in the ECMWF forecast model, *Q. J. R. Meteor. Soc*, **128**, 1671–1687. ©Royal Meteorological Society.

Hortal then states the reason for the instability that was noticed in the ECMWF system, where if we consider the prototype problem that is stable for $ks = 0$, and the advection velocity is included in the evolution operator as in the case of weather forecast equations, and if the advection velocity is not everywhere zero, then there could be points in the solution domain and wave numbers for which $ks = 2\pi$. These wave numbers will become **unstable**, noise will be produced at that point in the velocity field, and the Courant number will change, rotating the area of stability. This rotation can prevent the model from blowing up, unlike the case of instability in an Eulerian scheme. One effect of the local instability will then be the production of noise in the forecast field, as observed in the ECMWF operational forecasts.

Now focusing on the new SETTLS scheme, we have that its stability region is in the lower plot in Fig. 12.21, which is plotted for the values of $ks = 0$ to 2π in intervals of $\frac{\pi}{10}$. For $ks = 0$ it coincides with the corresponding area of the original scheme. The area of stability does **not** rotate with the Courant number but only oscillates in shape, most of the values of $\lambda\Delta t$ and $\omega\Delta t$ for which the solution is stable for a Courant number of zero also being in the area of stability for larger values of the Courant number. The interior area of the second plot above is the area of unconditional stability where the intersection of the areas of stability for all the values of ks.

In 2004, Durran and Reinecke [49], provided a general measure to determine the stability for a class of explicit two-time-level semi-Lagrangian schemes for a forced advection equation. We start by considering that the two-time-level schemes can be in general be written in the form

$$\frac{\psi_a^{n+1}-\psi_d^n}{\Delta t}=\frac{1}{2}\left(L_a^{n+1}+L_d^n\right)+a_1N_a^n+a_2N_d^n+a_3N_a^{n-1}+a_4N_d^{-1}, \tag{12.276}$$

where the linear component of the forcing is being evaluated using a trapezoidal time difference and the nonlinear term is evaluated using some explicit time-integration formula. We should note here that in [67] they looked at the integration of the nonlinear term which, as we have seen earlier in this section, can cause instabilities. If we now wish to use a second-order in time accurate approximation to the nonlinear forcing in (12.276) that predicts $\psi\left(\mathbf{x}\left(t+\Delta t\right),t+\Delta t\right)$ from data at time levels t and $t-\Delta$ and spatial locations $\mathbf{x}\left(t\right)$ and $\mathbf{x}\left(t+\Delta t\right)$, then, according to [65], all of these schemes can be expressed in the form

$$\frac{\psi_a^{n+1}-\psi_d^n}{\Delta t}=\left(\frac{3}{4}-\kappa\right)N_a^n+\left(\frac{3}{4}+\kappa\right)N_d^n-\left(\frac{1}{4}-\kappa\right)N_a^{n-1}-\left(\frac{1}{4}+\kappa\right)N_d^{n-1}, \tag{12.277}$$

where κ is an undetermined parameter, and in [49] the authors investigate how the choice of κ influences the stability of the family of semi-Lagrangian schemes in (12.277). It is stated in [49] that if we set $\kappa=\frac{1}{4}$ in (12.277) then we obtain the SETTLS scheme.

The absolute stability of the family of the semi-Lagrangian approximations (12.277) follows when the unusual linear stability analysis is applied using the single Fourier mode, and if we let $\theta=u_0k\Delta t$ and $\gamma=\left(\lambda+i\omega\right)\Delta t$, then the general quadratic equation of the amplification factor A_k can be shown to be

$$A_k^2-\left(\gamma\left(\frac{3}{4}-\kappa+e^{-i\theta}\left(\frac{3}{4}+\kappa\right)\right)+e^{-i\theta}\right)+\gamma\left(\frac{1}{4}-\kappa+e^{-i\theta}\left(\frac{1}{4}+\alpha\right)\right)=0. \tag{12.278}$$

The amplification factor depends on the strength of the forcing per time step through γ and the Courant number through θ, where solutions to (12.278) are periodic over $[-\pi,\pi]$. The shortest wavelength resolved on the numerical mesh, as we have seen many times now, is the $2\Delta x$ wave, which here corresponds to $k=\dfrac{\pi}{\Delta x}$ and $\theta=\dfrac{\pi u_0\Delta t}{\Delta x}$. It is stated in [49] that the amplification factor for this wave completes one periodic cycle as $\dfrac{u_0\Delta t}{\Delta x}$ varies over the interval $[-1,1]$, and as such, except in the rare occasion that the semi-Lagrangian scheme is utilized with a very small time step, implies that a necessary condition for the absolute stability of any the family of schemes in (12.277) is that the magnitude of the roots of (12.278) be less than or equal to unity for all θ.

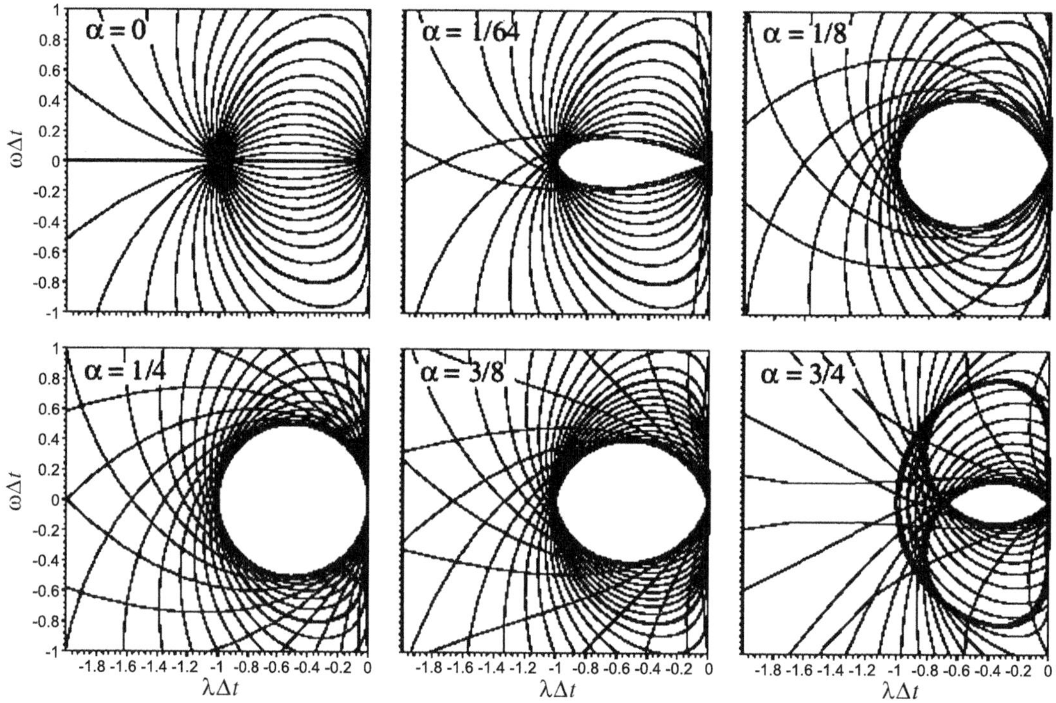

FIGURE 12.22 Copy of the absolute stability regions from [49]. Durran, D. R. and Reinecke, P. A., 2004: Instability in a class of explicit two-time-level semi-Lagrangian schemes, *Q. J. R. Meteor. Soc.*, **130**, 365–369. ©Royal Meteorological Society.

In Fig. 12.22 we have a copy of the regions of absolute stability for six choices for the κ parameter, referred to α in [49], where we labeled the parameter κ to avoid confusion with how we have been using α throughout the book. The six values of κ shown are $\kappa = 0, \frac{1}{64}, \frac{1}{8}, \frac{1}{4}, \frac{3}{8}$, and $\frac{3}{4}$. The important feature to notice about this plot, and which is highlighted in [49], is that the value for κ that has the largest absolute stability region is that associated with $\kappa = \frac{1}{4}$, which, as we indicated earlier, the SETTLS scheme of ECMWF uses.

12.11 Flux form on a sphere

Over the last two chapters we have considered the flux-form (FF) finite volume approach from Lin and Rood for the conservation of chemical species transport with a somewhat semi-Lagrangian (SL) approach. Now their theory has been taken further to be applied to the shallow water equations on the sphere to be applied to the conservation law for mass and indirectly to the conservation law for absolute vorticity via a reverse engineering procedure with a two-grid system, which is a combination of the C and D grids. The derivation that we show in the first part of this section in the horizontal comes from Lin and Rood (1997) [114].

12.11.1 Multidimensional flux-form semi-Lagrangian scheme

The multidimensional FFSL scheme, as we saw earlier, is derived from a direction-split perspective using strictly 1D operators. A procedure is then applied to remove the directional bias and the dominant first-order error resulting from the splitting approach. The conservation law which Lin and Rood refer to as the transport equation, for a density-like field Q, where in a shallow water context Q may represent the depth of the fluid or the absolute vorticity, is given by

$$\frac{\partial Q}{\partial t}+\nabla\cdot(\mathbf{u}Q)=0. \tag{12.279}$$

The two one-dimensional FF operators, F and G, for updating Q from (12.279) are for one time step in the zonal, λ-direction, and meridional, θ-direction, respectively. The 1D operators are assumed to be 1D finite volume schemes at this point, but no explicit form of the 1D operator is assumed a priori. We now recall the difference and average operators from Chapter 10:

$$\delta_{\sigma}q=q\left(\sigma+\frac{\sigma}{2}\right)-q\left(\sigma-\frac{\sigma}{2}\right), \tag{12.280a}$$

$$\overline{q}^{\sigma}=\frac{1}{2}\left(q\left(\sigma+\frac{\sigma}{2}\right)+q\left(\sigma-\frac{\sigma}{2}\right)\right). \tag{12.280b}$$

Thus the 1D FF operators can be written as

$$F\left(\overline{u}^{t},\Delta t;Q^{n}\right)=-\frac{\Delta t}{r\Delta\lambda\cos\theta}\delta_{\lambda}\left(\mathcal{X}\left(\overline{u}^{t},\Delta t;Q^{n}\right)\right), \tag{12.281a}$$

$$G\left(\overline{v}^{t},\Delta t;Q^{n}\right)=-\frac{\Delta t}{r\Delta\theta\cos\theta}\delta_{\theta}\left(\mathcal{Y}\left(\overline{v}^{t},\Delta t;Q^{n}\right)\right), \tag{12.281b}$$

where $\mathcal{X}$ and $\mathcal{Y}$ are time-averaged fluxes of Q in the zonal and meridional direction, respectively, and defined as

$$\mathcal{X}\left(\overline{u}^{t},\Delta t,Q^{n}\right)=\frac{1}{\Delta t}\int_{t}^{t+\Delta t}uQdt, \tag{12.282a}$$

$$\mathcal{Y}\left(\overline{v}^{t},\Delta t,Q^{n}\right)=\frac{1}{\Delta t}\int_{t}^{t+\Delta t}vQdt. \tag{12.282b}$$

To approximate the time-averaged flux $\mathcal{X}$, or $\mathcal{Y}$, across the boundaries of the grid cells, properly defined time-averaged or time-centered winds $\left(\overline{u}^{t},\overline{v}^{t}\right)$ and the cell-averaged field at time-level n, Q^{n}, are required. The fluxes are then used for updating the cell-averaged fields to the next time-level. We have a copy of the schematic of this grid in Fig. 12.23, where the advective winds and the cell-averaged Q field are defined on the C grid.

It is pointed out in [114] that the cross-derivative terms which are essential to a scheme's stability in multiple dimensions are automatically included by applying the 1D operators sequentially. However, the sequential splitting process introduces a *splitting error* term, which is usually directionally asymmetric. The first step to achieving the desired multidimensionality is to remove directional biases by averaging two antisymmetric operator-split algorithms.

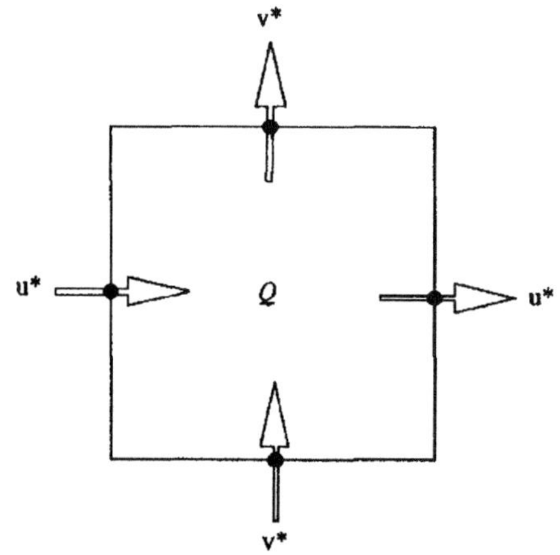

FIGURE 12.23 Copy of figure from [114] of the C grid arrangement for the transport of the cell-averaged density-like Q field by the 2D FFSL scheme. Lin, S.-J. and Rood, R. B., 1997: An explicit flux-form semi-Lagrangian shallow-water model on the sphere, *Q. J. R. Meteor. Soc.*, **123**, 2477–2498.

The resulting directional-bias-free algorithm as we saw earlier is

$$Q^{n+1}=Q^{n+1}+F\left[Q^n+\frac{1}{2}G\left(Q^n\right)\right]+G\left[Q^n+\frac{1}{2}F\left(Q^n\right)\right]. \tag{12.283}$$

To address a deformational error associated with this current scheme, Lin and Rood introduce a second step which is to replace F and G inside the square brackets in (12.283), the contributions from the cross-stream directions, with their advective-form counterparts f and g, respectively, to arrive at the following form of the 2D FFSL scheme:

$$Q^{n+1}=Q^n+F\left[Q^n+\frac{1}{2}g\left(\overline{v}_a^t,\Delta t,Q^n\right)\right]+G\left[Q^n+\frac{1}{2}f\left(\overline{u}_a^t,\Delta t,Q^n\right)\right], \tag{12.284}$$

where

$$\overline{u}_a^t\equiv\overline{\overline{u}^t}^{\lambda},\quad \overline{v}_a^t\equiv\overline{\overline{v}^t}^{\theta},$$

which is similar to before but now in spherical coordinates.

We now move on to implement this approach in a shallow water equations model on the sphere.

12.11.2 2D FFSL with the shallow water equations

In [114] the authors state that the mass conservation law for a shallow layer of water is given by

$$\frac{\partial h}{\partial t}+\nabla\cdot(\mathbf{u}h)=0, \tag{12.285}$$

where h represents the depth of the fluid. The vector invariant form of the momentum equations in spherical coordinates in component form is

$$\frac{\partial u}{\partial t} = \xi v - \frac{1}{a\cos\theta}\frac{\partial}{\partial \lambda}\left(\kappa + \Phi\right), \tag{12.286a}$$

$$\frac{\partial v}{\partial t} = -\xi u - \frac{1}{a}\frac{\partial}{\partial \theta}\left(\kappa + \Phi\right), \tag{12.286b}$$

where most variables have the same meaning as earlier, but now we have that $\xi = 2\omega\sin\theta + \nabla\times\mathbf{u}$ as the absolute vorticity, $\Phi \equiv \Phi_s + gh$ is the free surface geopotential, and Φ_s is the surface geopotential, finally, κ is the kinetic energy defined by $\kappa = \frac{1}{2}\mathbf{u}\cdot\mathbf{u}$.

Lin and Rood state that a significant advantage of the vector invariant form of the momentum equation is that the metric terms, which are singular at the poles, are absorbed into the definition of the relative vorticity, which is well defined and nonsingular at the poles. They also state that a disadvantage of this form is that the discretized form of the kinetic energy κ needs to be formulated carefully to minimize inconsistency between $\nabla\kappa$ and the discretized absolute vorticity fluxes, $u\xi$ and $v\xi$.

The next step in [114] is to form conservation law for the absolute vorticity which can readily be obtained by taking curl of the vector momentum equation, which yields

$$\frac{\partial \xi}{\partial t} + \nabla\cdot\left(\mathbf{u}\xi\right) = 0. \tag{12.287}$$

As we have seen before, the divergence equation $\delta = \nabla\cdot\mathbf{u}$ is obtained by applying the divergence operator to (12.285). Lin and Rood now make the statement that if the vorticity–divergence form of the equations is discretized, a way must be found to invert the pair (ξ, δ) back to (u, v) at each time-step for the time integration to proceed. They suggest that the spectral transform method is ideally suited to this purpose because the inversion is nearly trivial. Models based on a local discretization method, for example, finite difference or finite volume schemes, would need to solve an elliptic equation, which may in some cases be nonlinear and inefficient to solve.

As we shall see, the FFSL algorithm is designed to model the cell-averaged variables, and hence will only conserve "mass" (depth of the fluid, h) and the h-weighted potential vorticity, which is the absolute vorticity. Lin and Rood state that energy and potential enstrophy will, in general, decrease monotonically in time due to the subgrid mixing algorithm implicitly chosen with monotonic 1D finite-volume flux-form operators. The physical basis and interpretation of these subgrid mixing processes varies widely. When the van Leer or the PPM schemes are chosen, an implied subgrid distribution is also chosen. This distribution is piecewise linear for the van Leer scheme and piecewise parabolic with the PPM. When a monotonicity constraint is applied, local extrema are prohibited by readjusting the subgrid distribution. This adjustment process emulates the local mixing mechanisms. Since parabolic distributions allow more structure than the linear distributions, PPM schemes are generally less diffusive than van Leer schemes.

Lin and Rood now state that they have the goal of transporting h and ξ in a general divergent flow. They state that a requirement is that h and ξ be defined at the same point, or in the finite volume sense enclosed in the same cell. As the prognostic variables that Lin and Rood

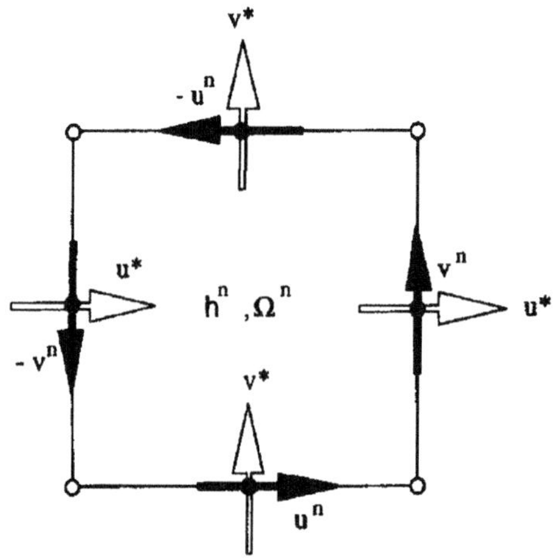

FIGURE 12.24 Copy of Fig. 2 from [114] of the two-grid system: the "CD grid". The time-centered advective winds $(\overline{u}^t, \overline{v}^t)$ (the hollow arrows) are staggered as in the C grid, whereas the prognostic winds (u^n, v^n) (the solid arrows) are staggered as in the D grid. The cell-averaged relative vorticity is computed by the Stokes theorem. Lin, S.-J. and Rood, R. B., 1997: An explicit flux-form semi-Lagrangian shallow-water model on the sphere, *Q. J. R. Meteor. Soc.*, **123**, 2477–2498.

are using are h and (u, v), rather than h and (ξ, δ), they state that the D grid would be the best choice. We have a copy of the CD grid from [114] in Fig. 12.24.

According to [114], as tangential winds are defined along the cell boundaries, the D grid is suited for computing the circulation (and hence, the cell-averaged relative vorticity, in the mean-value theorem sense). It is also the best grid on which to compute geostrophically balanced flow. It is known that any single-grid system, other than the C grid or the Z grid [155], generates two-grid-length gravity waves. This problem can be avoided by computing the time-centered advective winds $(\overline{u}^t, \overline{v}^t)$ on the C grid, as required by the multidimensional FFSL scheme.

Moving on to implementing this approach on the sphere, we have that if the first term on the right-hand side of (12.286a) and (12.286b) is interpreted as the time-averaged (from time t to time $t+\Delta t$) meridional and zonal flux of the absolute vorticity, then a consistently discretized absolute vorticity equation can be formed by taking the curl, numerically, of these two discretized component equations. Directly from (12.284), the discretized transport equations for h and ξ are:

$$h^{n+1} = h^n + F\left(\overline{u}^t, \Delta t; h^\theta\right) + G\left(\overline{v}^t, \Delta t; h^\lambda\right), \tag{12.288a}$$

$$\xi^{n+1} = \xi^n + F\left(\overline{u}^t, \Delta t; \xi^\theta\right) + G\left(\overline{v}^t, \Delta t; \xi^\lambda\right), \tag{12.288b}$$

where

$$(\cdot)^\theta \equiv (\cdot)^n + \frac{1}{2} g\left[\overline{\overline{v}^t}^\theta, \Delta t; (\cdot)^n\right], \qquad (\cdot)^\lambda \equiv (\cdot)^n + \frac{1}{2} f\left[\overline{\overline{u}^t}^\lambda, \Delta t : (\cdot)^n\right]. \tag{12.289}$$

The last part of the numerical approximation that we have to consider here is that for the pressure gradient terms, where Lin and Rood use either the *economical explicit* or the *forward–backward* scheme, which is conditionally stable and second-order accurate if it is combined with a forward in time advection scheme. Thus the final forms of the momentum equations are

$$u^{n+1} = u^n + \Delta t \left\{ \mathcal{Y}\left(\overline{v}^t, \Delta t; \xi^\lambda\right) - \frac{1}{a\Delta\lambda\cos\theta}\delta_\lambda\left(\overline{\kappa} + \overline{\overline{\Phi}^{n+1}}^\theta\right)\right\}, \tag{12.290a}$$

$$v^{n+1} = v^n - \Delta t \left\{ \mathcal{X}\left(\overline{u}^t, \Delta t; \xi^\theta\right) - \frac{1}{a\Delta\theta}\delta_\theta\left(\overline{\kappa} + \overline{\overline{\Phi}^{n+1}}^\theta\right)\right\}, \tag{12.290b}$$

where κ is the upstream biased kinetic energy defined at the four corners of the cell, which are represented by the hallow circles in Fig. 12.24, and formulated as

$$\overline{\kappa}^t = \frac{1}{2}\left(\mathcal{X}\left(\overline{\overline{u}^t}^\theta, \Delta t; u^n\right)\right). \tag{12.291}$$

The above form of κ minimizes the inconsistency in the momentum equation and thus avoids an instability that is referred to as a Hollingsworth–Kållberg instability [78]. It can be readily verified that (12.288b) can be recovered by taking "curl" of the two components, (12.290a) and (12.290b), of the vector momentum equation provided the cell-averaged relative vorticity is computed by the Stokes theorem as follows:

$$\nabla\times\mathbf{u}^n = \frac{1}{a^2\Delta\lambda\Delta\theta\cos\theta}\int_C \mathbf{u}^n\cdot d\mathbf{l} = \frac{1}{a\Delta\lambda\cos\theta}\delta_\lambda v^n - \frac{1}{a\Delta\theta\cos\theta}\delta_\theta\left(u^n\cos\theta\right), \tag{12.292}$$

where the contour integral is taken counterclockwise along the boundary of a cell and $d\mathbf{l}$ represents an element of the contour. At the two poles, the relative vorticity is similarly computed as the net circulation around the polar cap divided by the total area of the polar cap, which is equal to $2\pi a^2\left(1-\cos\left(\frac{\Delta\theta}{2}\right)\right)$. We have a copy of the polar cap figure from Lin and Rood (1997) in Fig. 12.25.

To advance the advective winds at time level n on the C grid, obtained by spatial averaging for a half-time-step to obtain the desired time-centered values. This is in effect employing a two-grid two-step time marching procedure, where the vorticity on the C grid is defined at the four cell corners, not the center of the cell, which is the mass point, as on the D grid. It is stated in [114] that the functional relation between h and ξ because the C-grid variables $\overline{h}^t$ and $\left(\overline{u}^t, \overline{v}^t\right)$ are not prognostic variables in the CD grid algorithm. According to Lin and Rood, these are discarded after the prognostic variables h^n and (u^n, v^n), which are defined on the D grid, are updated for a one time step.

To finish the description of the algorithm, we have the formulas of the complete time marching:

$$\overline{h}^t = h^n + F\left(\mathbf{u}_C^n, \frac{\Delta t}{2}; h^{\frac{\theta}{2}}\right) + G\left(v_C^n, \frac{\Delta t}{2}; h^{\frac{\lambda}{2}}; h^{\frac{\lambda}{2}}\right), \tag{12.293}$$

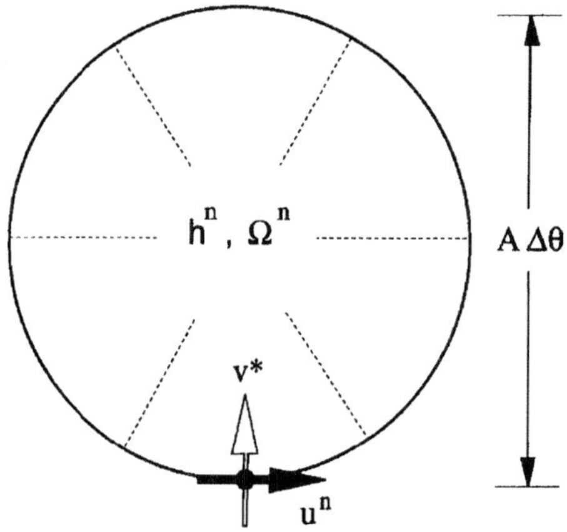

FIGURE 12.25 Copy of Fig. 3 from [114] of the North polar cap. The great circle distance from one side of the polar cap to the opposite side is $a\Delta\theta$. The edge of the polar cap is located at $\theta = \frac{\pi}{2} - \frac{\Delta\theta}{2}$. Lin, S.-J. and Rood, R. B., 1997: An explicit flux-form semi-Lagrangian shallow-water model on the sphere, *Q. J. R. Meteor. Soc.*, **123**, 2477–2498.

$$\overline{u}^t = u_c^n + \frac{\Delta t}{2}\left\{\mathcal{Y}\left(v^n, \frac{\Delta t}{2}, \xi_c^{\frac{\lambda}{2}}\right) - \frac{1}{a\Delta\lambda\cos\theta}\delta_\lambda\left[\kappa^{**} + \overline{\Phi}^t\right]\right\}, \tag{12.294}$$

$$\overline{v}^t = v_c^n + \frac{\Delta t}{2}\left\{\mathcal{X}\left(u^n, \frac{\Delta t}{2}, \xi_c^{\frac{\theta}{2}}\right) - \frac{1}{a\Delta\theta}\delta_\theta\left[\kappa^{**} + \overline{\Phi}^t\right]\right\}, \tag{12.295}$$

where

$$h^{\frac{\lambda}{2}} = h^n + \frac{1}{2}f\left(u_a^n, \frac{\Delta t}{2}; h^n\right), \qquad h^{\frac{\theta}{2}} = h^n + \frac{1}{2}g\left(v_a^n, \frac{\Delta t}{2}; h^n\right),$$

$$\xi_c^{\frac{\lambda}{2}} = \xi_c^n + \frac{1}{2}f\left(u_b^n, \frac{\Delta t}{2}; \xi_c^n\right), \qquad \xi_c^{\frac{\theta}{2}} = \xi_c^n + \frac{1}{2}g\left(v_b^n, \frac{\Delta t}{2}; \xi_c^n\right),$$

and

$$u_b^n \equiv \overline{u^n}^\lambda, \quad v_b^n \equiv \overline{v^n}^\theta, \quad u_c^n \equiv \overline{u_b^n}^\theta, \quad v_c^n \equiv \overline{v_b^n}^\lambda, \quad u_a^n \equiv \overline{u_c^n}^\lambda,$$
$$v_a^n \equiv \overline{v_c^n}^\theta, \quad \xi_v^n \equiv 2\omega\sin\theta + \nabla\times\mathbf{u}_c^n,$$

finally, κ^{**} is the upwind biased kinetic energy defined at the mass points as

$$\kappa^{**} = \frac{1}{2}\left[\mathcal{X}\left(u_a^n, \frac{\Delta t}{2}; u_c^n\right) + \mathcal{Y}\left(v_a^n, \frac{\Delta t}{2}; v_c^n\right)\right]. \tag{12.296}$$

After values for $(\overline{u}^t, \overline{v}^t)$ are obtained, the prognostic variables h^n and (u^n, v^n) are updated using (12.288a), (12.290a), and (12.290b).

12.11.3 Lagrangian control-volume

We now move on to briefly introduce the extension of the FFSL method to what is referred to as a vertically Lagrangian finite volume dynamical core for global models, which appeared in Lin (2004) [112], where Lin introduces the Lagrangian control volume vertical discretization and states that the introduction of the control-volume vertical discretization means that all prognostic equations in the global model are reduced to 2D, in the sense that they are vertically decoupled. The two-dimensional flux form is very similar to what we have just presented were the main difference now is the term $\overline{\pi}^t$ in our time-averaged notation, which is the time mean pseudodensity value of all material that passed through the cell edge. The important equation in [112] from a semi-Lagrangian point of view (labeled (6) there) is given by

$$\overline{\pi}^t\left(\overline{u}^t, \Delta t; \widetilde{\pi}\right) \approx \frac{1}{\Delta t}\int_t^{t+\Delta t} \pi \, d\tau, \tag{12.297}$$

where $\widetilde{\pi}$ is the finite volume representation of π such that

$$\widetilde{\pi} \equiv \frac{1}{a^2 \Delta\theta\, \Delta\lambda \cos\theta} \int_{\lambda-\frac{\Delta\lambda}{2}}^{\lambda+\frac{\Delta\lambda}{2}} \int_{\theta-\frac{\Delta\theta}{2}}^{\theta+\frac{\Delta\theta}{2}} \pi\left(t; \lambda\theta\right) a^2 \cos\theta \, d\theta \, d\lambda. \tag{12.298}$$

Lin describes (12.297) as needed to be carried out along **backward in time** trajectory of the **cell-edge** position from t^{n-1} to ^{n}. The essence of the 1D finite-volume algorithm is to construct here, based on the given initial cell-mean values of $\widetilde{\pi}$, an approximated subgrid distribution of the true π field, to enable an analytic integration of (12.297). Lin then assumes that there is no error in obtaining the time-mean wind $\overline{u}^t$, the only error produced by the 1D transport scheme would be solely due to the approximation to the true distribution of π using the assumed subgrid distribution. From this perspective, it can be said that the 1D finite volume transport algorithm combined the space–time discretization in the approximation of the time-mean cell-edge value π. Thus this scheme is not a true semi-Lagrangian finite volume scheme in that it is really an Eulerian flux form according to [99], where Lauritzen describes this approach that considers fluxes through the arrival cell walls rather than explicit integration over the departure area. The actual way to interpret this approach is to apply a sort of semi-Lagrangian approach in that it is not moving the whole arrival cell or area backward to a slightly warped Eulerian square but rather finding the swept flux areas that arrive at the regular area at time t^{n+1}. There is a very good figure highlighting this in [101], which we have a copy of Fig. 1 in Fig. 12.26.

To finish this section, we present a copy of a schematic of the vertical departure and arrival cells from [100] in Fig. 12.27 where we can see that each vertical level is separated from each other, and they can only arrive at the same pressure level, but the four volumes arrive into the uniform volume for *Eulerian* grid at the arrival time.

We now finish off this chapter with the definitions of a few initial conditions that can be used in spherical models, where these are from the suite of initial conditions from Williams et al. (1992) [206].

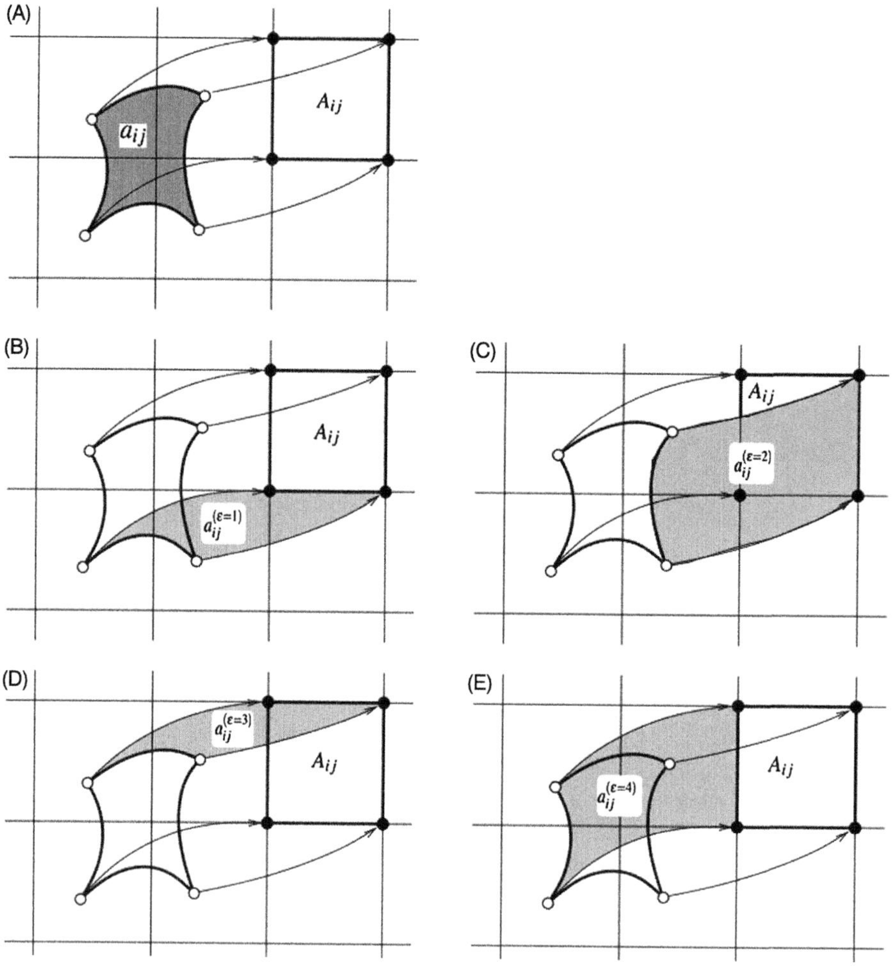

FIGURE 12.26 Copy of Fig. 1 from [101] where plot (A) is the true finite volume semi-Lagrangian illustration, while (B)–(E) illustrate the flux swept semi-Lagrangian advection. Lauritzen, P. H., 2017: CAM-SE–CSLAM: Consistent Coupling of a Conservative Semi-Lagrangian Finite-Volume Method with Spectral Element Dynamics, *Monthly Weather Review*, **145**, 833–855. ©American Meteorological Society. Used with permission.

12.12 Numerical test cases for the sphere

There is a very good paper in 1992 by Williamson et al. [206] where, as with the two-dimensional problem, we shall present the equations for initial conditions for these situations. We start by defining the parameters:

$$a = 6.37122 \times 10^6 \text{ m}, \qquad \Omega = 7.292 \times 10^{-5} \text{ s}^{-1}, \qquad g = 9.80616 \text{ ms}^{-2}.$$

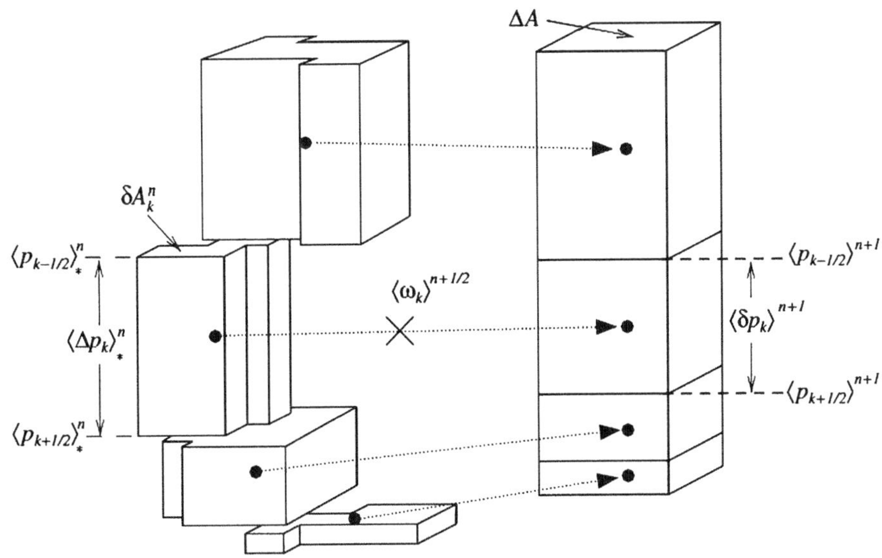

FIGURE 12.27 Copy of Fig. 4 from [100] of the departure and arrival cells that make up a deformed column on the left and a regular column on the right. Lauritzen, P. H., Kaas, E., Machenhauercand, B. and Lindberg, K., 2008: A mass-conservative version of the semi-implicit semi-Lagrangian HIRLAM, *Q. J. R. Meteor. Soc.*, **134**, 1583–1595. ©Royal Meteorological Society.

Advection of cosine bell over the pole

In [206] the authors state that the cosine bell is advected once around the sphere, where the present several orientations of the advecting winds are specified including around the equator, directly over the poles, and minor shifts from these two orientations. The advecting wind is given by

$$u = u_0\left(\cos\theta\cos\alpha + \sin\theta\cos\lambda\sin\alpha\right), \qquad v = -u_0\sin\lambda\sin\alpha.$$

In terms of stream function and velocity potential, this is

$$\psi = -au_0\left(\sin\theta\cos\alpha - \cos\lambda\cos\theta\sin\alpha\right), \qquad \chi = 0.$$

The parameter α is the angle between the axis of solid body rotation and the polar axis of the spherical coordinate system. The values of α that should be used as suggested by [206] are $\alpha = 0.0, 0.005, \frac{\pi}{2} - 0.05$, and $\frac{\pi}{2}$.

The initial cosine bell profile to be advected is given by

$$h\left(\lambda,\theta\right) = \begin{cases} \dfrac{h_0}{2}\left(1+\cos\dfrac{\pi r}{R}\right), & r < R, \\ 0, & r \geq R, \end{cases} \tag{12.299}$$

where $h_0 = 1000$ m and r is the great circle distance between (λ, θ) and the center, which can initially be taken as $(\lambda_c, \theta_c) = \left(\frac{3\pi}{2}, 0\right)$ and given by

$$r = a \arccos\left(\sin\theta_c \sin\theta + \cos\theta_c \cos\theta \cos(\lambda - \lambda_c)\right). \tag{12.300}$$

The radius $R = \frac{a}{3}$ and the advecting velocity $u_0 = \frac{2\pi a}{12 \text{ days}}$, where 12 days is $12 \times 24 \times 60 \times 60$ seconds, which is equivalent to 40 ms^{-1}. This solutions translate without changing shape.

Global steady state nonlinear zonal geostrophic flow

It is stated in [206] that this case is the steady state to the nonlinear shallow water equations. It consists of solid body rotation, or zonal flow, with the corresponding geostrophic height filled. The Coriolis parameter is a function of latitude and longitude, so the flow can be specified in spherical coordinates' poles not necessarily coincident with Earth's rotation axis.

The velocity field is initially, and for all time as it is time-independent, given as

$$u = u_0 \left(\cos\theta \cos\alpha + \cos\lambda \sin\theta \sin\alpha\right), \qquad v = -u_0 \sin\lambda \sin\alpha.$$

In terms of streamfunction and velocity potential, the velocity field is

$$\psi = -a u_0 \left(\sin\theta \cos\alpha - \cos\lambda \cos\theta \sin\alpha\right), \qquad \chi = 0.$$

The absolute vorticity is given by

$$\eta = \left(\frac{2u_0}{a} + 2\Omega\right)\left(-\cos\lambda \cos\theta \sin\alpha + \sin\theta \cos\alpha\right). \tag{12.301}$$

The analytical height field is given by

$$gh = gh_0 - \left(a\Omega u_0 + \frac{u_0^2}{2}\right)\left(-\cos\lambda \cos\theta \sin\alpha + \sin\theta \cos\alpha\right)^2. \tag{12.302}$$

It is stated in [206] that it may be desirable to modify the initial wind and height fields so that they satisfy a discrete nonlinear geostrophic relationship consistent with the scheme being tested. The Coriolis parameter associated with this solution is

$$f = 2\Omega\left(-\cos\lambda \cos\theta \sin\alpha + \sin\theta \cos\alpha\right). \tag{12.303}$$

Williamson et al. recommend the following values for the parameters: $u_0 = \frac{2\pi a}{12 \text{days}}$, $gh_0 = 2.94 \times 10^4 \text{ m}^2 \text{s}^{-2}$.

Zonal flow over an isolated mountain

The set of initial conditions provide insight into a zonal flow impinging on a mountain. The wind and height fields are as those for the test case above with $\alpha = 0$, $h_0 = 5960$ m, $u_0 = 20$ ms^{-1}. The surface of the mountain is given by

$$h_s = h_{s_0}\left(1 - \frac{r}{R}\right),$$

where $h_{s_0} = 2000$ m, $R = \frac{\pi}{9}$, and $r^2 = \min\left(R^2, (\lambda - \lambda_c)^2 + (\theta - \theta_c)\right)$. The center is taken as $(\lambda_c, \theta_c) = \left(\frac{3\pi}{2}, \frac{\pi}{6}\right)$.

Rossby–Haurwitz wave

Rossby–Haurwitz waves are analytical solutions of the nonlinear barotropic vorticity equation on the sphere. The parameters for these waves allow one to achieve different dynamical flows in the shallow water equations model even though these waves are not solutions of these equations. In Fletcher (2004), these waves were used to test a new form of control variable transform, as they allowed testing the transform with a geostrophic balance-like flow, as well as a fast deforming flow [60].

The initial velocity field is nondivergent and given by the streamfunction

$$\psi = -a^2\omega\sin\theta + a^2K\cos^R\theta\sin\theta\cos R\lambda, \tag{12.304}$$

where ω, K, and R are constants. This pattern moves from West to East without change of shape in a nondivergent barotropic model with angular velocity

$$\nu = \frac{R(3+R)\omega - 2\Omega}{(1+R)(2+R)}.$$

The velocity components are given by

$$u = a\omega\cos\theta + aK\cos^{R-1}\theta\left(R\sin^2\theta - \cos^2\theta\right)\cos R\lambda, \qquad v = -aKR\cos^{R-1}\theta\sin\theta\sin R\lambda,$$

and the vorticity is given by

$$\zeta = 2\omega\sin\theta - K\sin\theta\cos^R\theta\left(R^2 + 3R + 2\right)\cos R\lambda.$$

The height field is obtained from the streamfunction by solving the balance equation so that the initial tendency of the divergence is zero [206]:

$$\begin{aligned}
gh &= gh_0 + a^2A(\theta) + a^2B(\theta)\cos R\lambda + a^2C(\theta)\cos 2R\lambda,\\
A(\theta) &\equiv \frac{\omega}{2}(2\Omega + \omega)\cos^2\theta + \frac{1}{4}K^2\cos^{2R}\theta\left((R+1)\cos^2\theta + \left(2R^2 - R - 2\right) - 2R^2\cos^{-2}\theta\right),\\
B(\theta) &\equiv \frac{2(\Omega + \omega)K}{(R+1)(R+2)}\cos^R\theta\left(\left(R^2 + 2R + 2\right) - (R+1)^2\cos^2\theta\right),
\end{aligned}$$

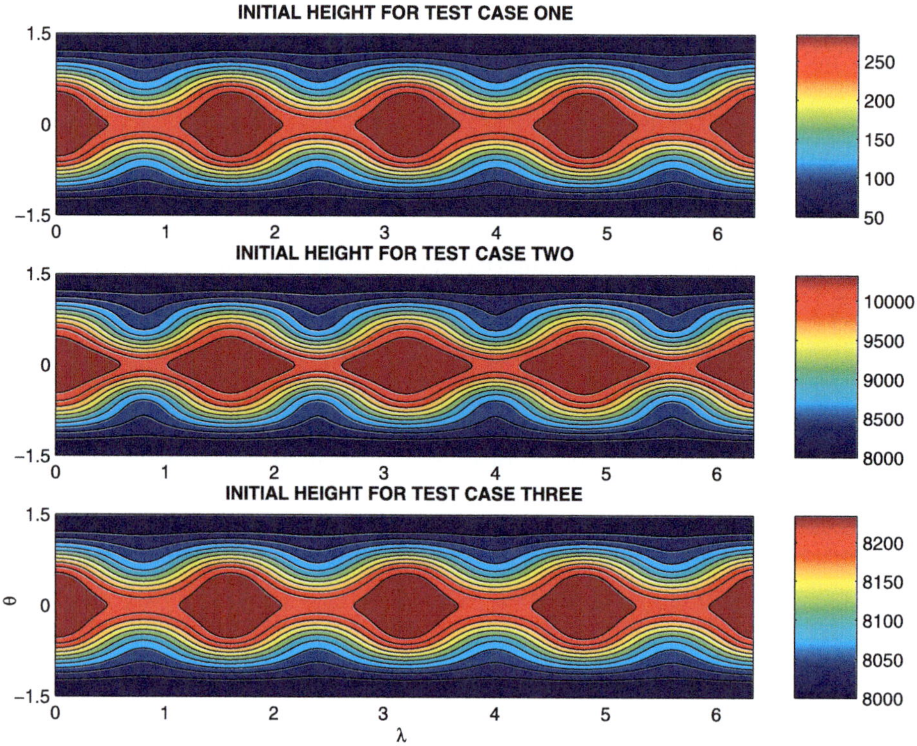

FIGURE 12.28 Plot of the three different Rossby–Haurwitz waves that describe different flows in the shallow water equations model.

$$C(\theta) = \frac{1}{4}K^2\cos^{2R}\theta\left((R+1)\cos^2\theta - (R+2)\right).$$

The parameters are $\omega = K = 7.848 \times 10^{-6}\ \mathrm{s}^{-1}$ and $h_0 = 8000$ m. The wave number R should be set to 4, as this is the highest stable wavenumber.

In Fig. 12.28 we have plotted the three initial profiles from [60] in a cylinder domain.

12.13 Summary

This chapter comprised two parts: in the first, we introduced how the vector differential operators transferred to spherical coordinates, then we introduced some different forms of projections, and then went on to the different forms of grid types that could be used to emulate the sphere. Given the basis for the numerical model, in the second part, we looked at different approaches that have been used to implement the semi-Lagrangian approach in the shallow water equations, in both limited area and global contexts, as well as in the primitive equations. We have looked at how to implement the semi-Lagrangian techniques in the spectral space, as well as with a finite volume/flux form. We have looked at a way to try and eliminate the Helmholtz equations that need to be solved in up to three dimensions for the height/geopotential fields, as well as introduced SETTLS which is the time interpolation scheme that the ECMWF uses in its semi-Lagrangian scheme. We finished this chapter by presenting some sets of initial conditions that could be used to test semi-Lagrangian advection techniques on the sphere.

Two things that we have not addressed yet in great detail about semi-Lagrangian methods are the needed shape-preserving and mass-conserving, and so we now move on to look at how these important properties of the continuous dynamical system can be achieved with the different forms of semi-Lagrangian integration.

CHAPTER

13

Shape-preserving and mass-conserving semi-Lagrangian approaches

In this chapter we consider properties that we require so that the semi-Lagrangian schemes are consistent with the continuous differential equations and introduce techniques that have been developed to address this. As we can see in the three plots in Fig. 13.1, which are for the linear, quadratic, and cubic Lagrange interpolation polynomial from the Eady model advection of the PV, the linear approach has almost complectly damped the PV, whereas the quadratic approach has quite extensive over- and undershoots which is clearly indicating that this scheme does not preserve the shape of the PV. We can also see that the cubic Lagrange interpolation approach, while not as severe as for the linear and quadratic cases, is still not quite preserving the shape of the PV.

Given this motivation, we now move on to consider techniques to improve the interpolation to ensure the shape of the advected quantity is preserved, before moving on to address the conservation of mass approaches.

13.1 Shape-preserving semi-Lagrangian advection

When we introduced the interpolation formulas earlier, we stated that we would like the interpolation scheme for finding the departure point to ensure that the associated advection remains monotonic when there is no diffusion present. However, there are many other properties that we would like for the semi-Lagrangian scheme to possess, where one of them is that the scheme be **shape-preserving**, and in [207] such a scheme is presented.

We summarize here the derivation of shape-preserving schemes from [207] and then show how it works with the 1D advection equations before considering a set of shape-preserving options for the 2D case. If we consider the one-dimensional problem and discretize with a grid of equally spaced points $x_1 < x_2 < \cdots < x_N$, where the associated data values at each point are $f_i = f(x_i)$, for $i = 1, 2, \ldots, N$, then we define a piecewise interpolant p on each of the grid intervals $\left[x_i, x_{i+1}\right]$ as

$$p(x) = p_i(\alpha)$$

https://doi.org/10.1016/B978-0-12-817222-3.00017-7

where

$$p(x_i) = f_i, \qquad \frac{dp}{dx}(x_i) = d_i,$$

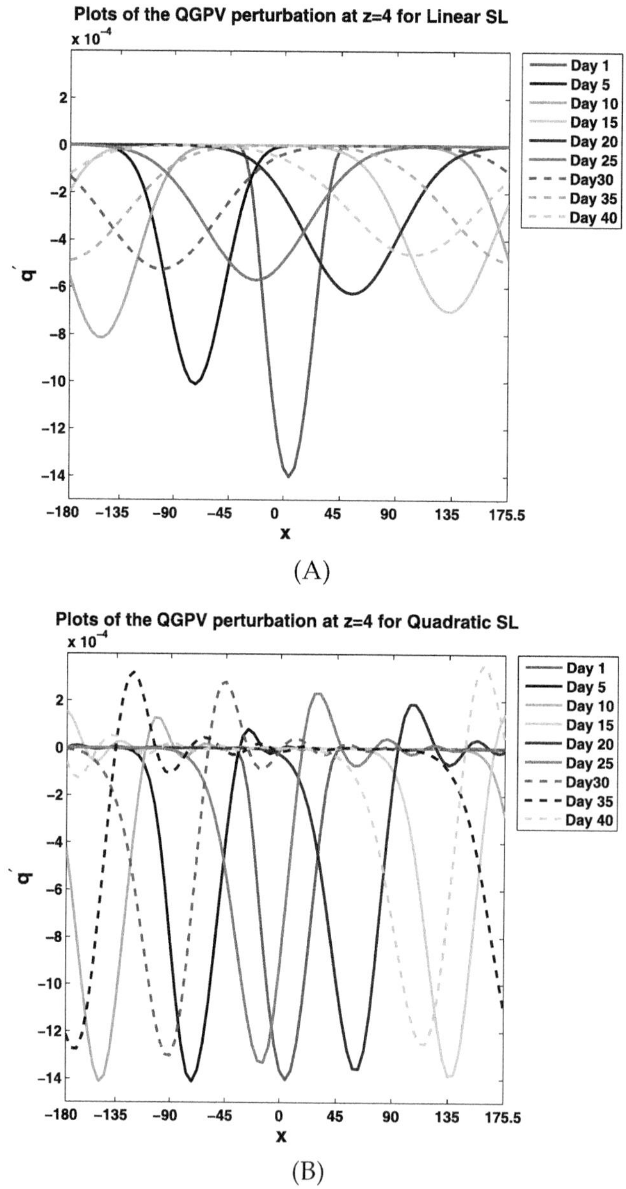

FIGURE 13.1 Plots of the cross-section of the PV for the linear, quadratic, and cubic Lagrange interpolation polynomial, where we can see that the shape of the PV is not preserved for any of these schemes.

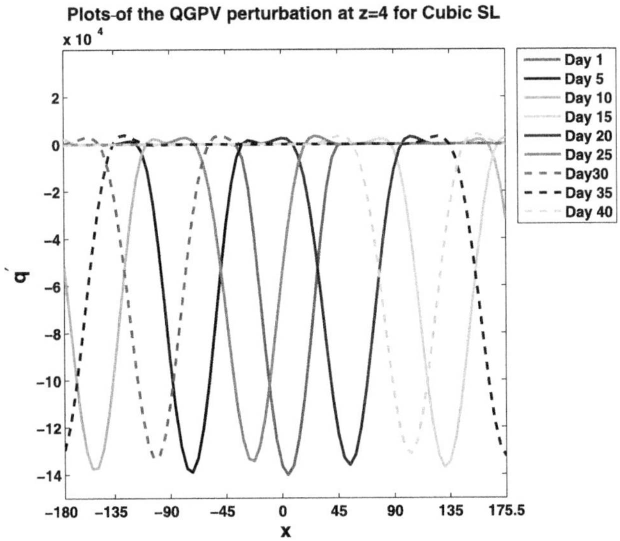

(C)

FIGURE 13.1 *(continued).*

for $i = 1, 2, \ldots, N$, and where d_i is some estimate of the derivative of the function f at the grid points. We should note that this is the definition of an Hermite polynomial. This then makes the interpolant a function of f_i and d_i at the end points of the intervals.

It is mentioned in [207] that it is possible to state two different forms of shape-preserving methods, which come from [85] and [44]. The shape preserving methods can be expressed in a general form [44] as

$$p_i \equiv \frac{P_i(\alpha)}{Q_i(\alpha)} \tag{13.1}$$

on the interval where $0 \leq \alpha \leq 1$, which is equivalent to $x_i \leq x \leq x_{i+1}$. The two functions of the parameter α in (13.1) are defined as

$$P_i(\alpha) = f_{i+1}\alpha^3 + (r_i f_{i+1} - \Delta x_i d_{i+1})\alpha^2(1-\alpha) + (r_i f_i + \Delta x_i d_i)\alpha(1-\alpha)^2 + f_i(1-\alpha)^3, \tag{13.2a}$$

$$Q_i(\alpha) = 1 + (r_i - 3)\alpha(1-\alpha), \tag{13.2b}$$

and, while the expressions above are a convenient way of expressing the interpolation polynomial, we should note that this may not be the most efficient way to apply them in practice.

If we substitute $r_i = 3$ into (13.2a) and (13.2b), then $Q_i = 1\ \forall i$, and $P_i(\alpha)$ would become the general form of the cubic spline. The second shape-preserving scheme from [44], which is referred to as the rational cubic interpolant, sets r_i to be

$$r_i = 1 + \frac{c_{i+1}}{c_i} + \frac{c_i}{c_{i+1}}, \tag{13.3}$$

where the c_is are differences between the discrete slope operator Δ and the continuous derivative as

$$c_i = \Delta_i - d_i, \qquad c_{i+1} = d_{i+1} - \Delta_i,$$

and, as a reminder, the discrete slope is defined as

$$\Delta_i = \frac{(f_{i+1} - f_i)}{(x_{i+1} - x_i)}.$$

In the companion paper to [207], the authors of [208] refer to two different constraints for preserving the shape of the field being advected. These two possible approaches are through imposing a monotonic constraint or, secondly, a convex/concave constraint. We shall consider the two constraints in the order just mentioned, starting with the monotonic constraint.

Monotonic shape-preserving constraints

For different interpolants to maintain monotonicity, the estimates of the derivative, d_i, must satisfy certain conditions. These conditions are enforced through constraints of the derivative estimates d_i and d_{i+1} at the endpoints of the grid interval as a function of the discrete slope, Δ_i, across the interval. However, we should note that the constraint on the derivative estimate d_i based upon the discrete slope Δ_i to the left may be different from the constraint that is based upon the discrete slope Δ_i of the interval to the right.

Given the possible differences in the discrete slopes from the two different intervals, one then has some choices to make. The first approach would be to constrain the derivatives differently for the interpolation over the two intervals; in doing so, the resulting interpolant would be C^0-continuous, which means that the interpolant is continuous, but that its first derivative contains jumps. A second approach would be to enforce that the constraints from both intervals are satisfied simultaneously; this would then make the interpolant C^1-continuous, implying that both the function and its first derivative are continuous.

In [207] there are two conditions that are defined to ensure that the interpolant satisfies one of the two continuity conditions just mentioned. The first condition is referred to as a **necessary condition for monotonicity $\mathbf{C^0}$ (NCM0)**. The NCM0 condition states that for the interpolating function to be monotonic in the grid interval, the derivative estimate at the end point must have the same sign as the discrete slope on the interval, which implies that

$$\begin{aligned}
&\operatorname{sign}(d_i) = \operatorname{sign}(\Delta_i) = \operatorname{sign}(\Delta_{i+1}), && \Delta_i \neq 0,\\
&d_i = d_{i+1} = 0, && \Delta_i = 0.
\end{aligned} \tag{13.4}$$

The second condition stated in [207] is referred to as the **necessary condition for monotonicity $\mathbf{C^1}$ (NCM1)**. For the interpolating function to be continuous between intervals, the derivative estimates at a gridpoint must have the same sign as the discrete slope of the adjacent grid intervals:

$$\begin{aligned}
&\operatorname{sign}(\Delta_{i-1}) = \operatorname{sign}(d_i) = \operatorname{sign}(\Delta_i), && \Delta_{i-1}\Delta_i > 0,\\
&d_i = 0, && \Delta_{i-1}\Delta_i \leq 0.
\end{aligned} \tag{13.5}$$

The reason for the choice of r in [44], which is mentioned in [207], is to ensure that the conditions NCM0 and NCM1 are also sufficient conditions for monotonicity for their resulting rational cubic interpolant. We should note here, again as stated in [207], that the two conditions, NCM0 and NCM1, are only necessary conditions for the cubic Hermite interpolant.

However, it is possible to arrive at a set of necessary and sufficient condition for monotonicity of the cubic Hermite interpolant which is derived in [62]. It is stated as follows: If $\Delta_i = 0$ and $d_i = d_{i+1} = 0$, then the necessary conditions (13.4) and (13.5) are also sufficient.

If the discrete slope Δ_i is not equal to zero, then we first define the ratio of the derivative estimates to the discrete slope by

$$\tau = \frac{d_i}{\Delta_i}, \tag{13.6a}$$

$$\eta = \frac{d_{i+1}}{\Delta_i}, \tag{13.6b}$$

then, for $\Delta_i \neq 0$, the Hermite cubic interpolant will be monotonic if and only if

$$0 \leq \tau \leq 3, \qquad 0 \leq \eta \leq 3, \tag{13.7}$$

or

$$(\tau - 1)^2 + (\tau - 1)(\eta - 1) + (\eta - 1)^2 - 3(\tau + \eta + 2) \leq 0. \tag{13.8}$$

The constraints above may appear a bit random in that the ratio of the derivative values to the discrete gradient slope at the various gridpoints needs to be constrained by the factor of 3. Therefore, as a quick aside we shall briefly summarize the derivation of the constraints in (13.7) and (13.8) from [62].

An aside to Fritsch and Carlson (1980) paper

We start by considering the general definition for a cubic polynomial over an interval $I_i = \left[x_{i-1}, x_i\right]$ where the point that we wish to interpolate to x is in this interval. Given the motivation that we have been using so far in this chapter, we wish to construct a piecewise cubic function $p(x)$ that is $C^1[\mathbf{I}]$, where $\mathbf{I} \equiv [x_1, x_N]$, and such that the polynomial is equal to the value of the fields at the gridpoints.

On each subinterval $I_i = \left[x_{i-1}, x_i\right]$ we define a cubic Hermite based polynomial as

$$p(x) \equiv f_{i-1} H_1(x) + f_i H_2(x) + d_{i-1} H_3(x) + d_i H_4(x), \tag{13.9}$$

where the d_js for $j = i-1, i$ are the derivatives of the polynomial at the relative gridpoints, and the $H_k(x)$s are the Hermite basis functions for the specific interval defined by

$$\begin{aligned} H_1 &\equiv 3\left(\left(\frac{x_i - \tilde{x}}{\Delta x_{i-1}}\right)^2\right) - 2\left(\left(\frac{x_i - \tilde{x}}{\Delta x_{i-1}}\right)^3\right), \\ H_2 &\equiv 3\left(\left(\frac{\tilde{x} - x_{i-1}}{\Delta x_{i-1}}\right)^2\right) - 2\left(\left(\frac{\tilde{x} - x_{i-1}}{\Delta x_{i-1}}\right)^3\right), \end{aligned}$$

$$
\begin{aligned}
H_3 &\equiv -\Delta x_i\left(\left(\frac{x_i-\tilde{x}}{\Delta x_{i-1}}\right)^3-\left(\frac{x_i-\tilde{x}}{\Delta x_{i-1}}\right)^2\right),\\
H_4 &\equiv \Delta x_i\left(\left(\frac{\tilde{x}-x_{i-1}}{\Delta x_{i-1}}\right)^2-\left(\frac{\tilde{x}-x_{i-1}}{\Delta x_{i-1}}\right)^2\right).
\end{aligned} \tag{13.10}
$$

The goal of [62] is to derive conditions to ensure that the derivative approximations needed for the cubic Hermite based approach are monotonic. Thus the monotonicity conditions, (13.6a) and (13.6b).

The next step is to assume that the discrete slope, $\Delta_{i-1}=\dfrac{f_i-f_{i-1}}{\Delta x_{i-1}}$, is not equal to zero. Now we expand the polynomial in (13.9) with the definitions from (13.10) in a Taylor series about $x=x_i$ which results in a cubic interpolation polynomial in terms of $(\tilde{x}-x_{i-1})$, f_{i-1}, d_i, d_{i-1}, and Δ_{i-1} as

$$
\begin{aligned}
p(\tilde{x}) &= \left(\frac{d_{i-1}+d_i-2\Delta_{i-1}}{\Delta x_{i-1}}\right)(\tilde{x}-x_{i-1})^3+\left(\frac{-2d_{i-1}-d_i+3\Delta_{i-1}}{\Delta x_{i-1}}\right)(\tilde{x}-x_{i-1})^2\\
&+ d_{i-1}(\tilde{x}-x_{i-1})+f_{i-1}.
\end{aligned} \tag{13.11}
$$

Exercise 13.1. *Verify the expression in* (13.11). ***Hint:*** *Expand* x_i *as* $x_i\equiv x_{i-1}+\Delta x_{i-1}$ *and* f_i *as* $f_i=f_{i-1}+\Delta_i\Delta x_{i-1}$.

Differentiating (13.11) with respect to $\tilde{x}$ results in

$$
p'(\tilde{x})=3\left(\frac{d_{i-1}+d_i-2\Delta_{i-1}}{\Delta x_{i-1}}\right)(\tilde{x}-x_{i-1})^2+2\left(\frac{-2d_{i-1}-d_i+3\Delta_{i-1}}{\Delta x_{i-1}}\right)(\tilde{x}-x_{i-1})+d_{i-1}. \tag{13.12}
$$

Differentiating the second time to obtain the Hessian results in

$$
p''(\tilde{x})=6\left(\frac{d_{i-1}+d_i-2\Delta_{i-1}}{\Delta x_{i-1}}\right)(\tilde{x}-x_{i-1})+2\left(\frac{-2d_{i-1}-d_i+3\Delta_{i-1}}{\Delta x_{i-1}}\right). \tag{13.13}
$$

We now consider two cases that can arise:

Case I: If we have $d_{i-1}+d_i-2\Delta_{i-1}=0$, then for this situation the interpolant is either quadratic or linear, and therefore the associated derivative is linear or constant, respectively. Since we have that $\min(d_{i-1},d_i)\leq p(\tilde{x})\leq\max(d_{i-1},d_i)$, and then the standard sign monotonicity condition is a sufficient condition.

Case II: If we have $d_{i-1}+d_i-2\Delta_i\neq 0$, then the derivative of the interpolant is quadratic. The derivative will be concave up if $d_{i-1}+d_i-2\Delta_{i-1}>0$ and concave down if $d_{i-1}+d_i-2\Delta_{i-1}<0$. In [62] the authors note here that if $f_{i-1}<f_i$ and $p'(\tilde{x})$ is concave down then the interpolant is monotone increasing, since $0\leq\min(d_{i-1},d_i)\leq p'(\tilde{x})$. By a similar argument, if the signs are reversed in the inequalities above and the min is replaced with the max operator, then the interpolant is monotone decreasing.

Given that it is possible for the interpolant to be monotone increasing or decreasing, we then need a condition that can accommodate both properties. Let τ_{i-1} and κ_{i-1} be as defined in (13.6a) and (13.6b), which are the respective ratios of the endpoint derivatives to the slope of the secant line. This then enables us to rewrite the condition for Case II as $(\tau_{i-1}+\kappa_{i-1}-2)\,\Delta_i$ and therefore the interpolant is monotone if $\tau_{i-1}+\kappa_{i-1}-2<0$, which is one of the conditions stated in [207].

If we now consider the other situation where $\tau_{i-1}+\kappa_{i-1}-2>0$, then it is possible to show that the derivative of the interpolant has a unique extremum at

$$\hat{x}=x_{i-1}+\frac{\Delta x_i}{3}\left(\frac{2\tau_{i-1}+\kappa_{i-1}-3}{\tau_{i-1}+\kappa_{i-1}-2}\right), \tag{13.14}$$

and

$$p'\left(\hat{x}\right)=\Psi\left(\tau_{i-1},\kappa_{i-1}\right)\Delta_{i-1}, \tag{13.15}$$

where

$$\Psi\left(\tau,\kappa\right)=\tau-\frac{1}{3}\frac{\left(2\tau+\kappa-3\right)^2}{\left(\tau+\kappa-2\right)}. \tag{13.16}$$

These expressions above are the conditions that are stated in [207] for concave monotonicity.

Returning to Williamson and Rasch (1989) and (1990) papers

In [207] it is highlighted that the elliptical component of the necessary and sufficient condition in (13.8) is difficult to apply because the derivative estimates at the two ends of a grid interval may depend on each other. The simpler condition in (13.7), which they refer to as a rectangular condition, does provide a sufficient, but not necessary, condition to ensure monotonicity of the Hermite cubic interpolation polynomial [43]. The sufficient condition in (13.7) is obviously easier to enforce than the more general necessary and sufficient conditions from (13.8) and (13.7).

Given an estimate of the derivatives, which we should note are not guaranteed to satisfy any of the conditions for monotonicity that have been stated, if a monotonic interpolant is the desired outcome, then the derivative estimates must be modified to satisfy the appropriate condition. It is stated in [207] that a C^0 and C^1 forms are defined based on whether the derivative estimates d are bounded only by Δ on the interval being interpolated or by the discrete slopes of the two adjacent intervals simultaneously.

The modifications just mentioned are defined in [207] for both the cubic rational interpolant and for the cubic Hermite interpolant as:

Cubic rational: If necessary conditions (13.6a) or (13.6b) are not met, depending on whether C^0 or C^1 continuity is the preferred outcome, then the appropriate derivative estimate is set to zero, as the sign of 0 can be considered either as $+$ or $-$.

Cubic Hermite: If the derivative estimate does not satisfy (13.6a) or (13.6b), again depending of which form of continuity is sought, then we follow the procedure set out for the cubic rational. As these conditions are not satisfied, then if the conditions in (13.7) are not met, that is to say, if $\tau \geq 3$ or $\eta \geq 3$, then these estimates are set to 3, which then gives both C^0 and C^1 continuity.

We now consider the Hyman derivative estimate as a candidate to help preserve shape, which is defined as

$$d_i = \frac{-f_{i+2} + 8f_{i+1} - 8f_{i-1} + f_{i-2}}{-x_{i+2} + 8x_{i+1} - 8x_{i-1} + x_{i-2}}, \tag{13.17}$$

and is a fourth-order estimate on a smooth varying grid. Another alternative is to consider a cubic derivative estimate, which arises from fitting a cubic Lagrange interpolant through the four point surrounding the interval. The derivatives at the end of the grid intervals are derived through differentiating the Lagrange estimate and evaluating it at the end of the grid interval.

An important thing to note here is that if the Lagrange estimate derivatives are not modified then an Hermite interpolation polynomial with these estimates is equivalent to the cubic Lagrange interpolation polynomial that we introduced earlier, which is not monotonic. However, one of the pleasant properties of the Hermite interpolants is that it is possible to modify their derivative estimates to ensure monotonicity of the interpolants.

Another important thing to note is that depending on the point where we wish to interpolate our grid point values of the function to, with respect to the side of the interval, if x_i is in the interval (x_{i-1}, x_{i+1}), then the associated 4 gridpoints are needed and the cubic approximation to the derivative will either use $x_{i-2}, x_{i-1}, x_i, x_{i+1}$ for $x \in (x_{i-1}, x_i)$, or $x_{i-1}, x_i, x_{i+1}, x_{i+2}$ if $x \in (x_i, x_{i+1})$. Therefore, the estimates for d_i in these two subintervals are given by

$$d_i = \begin{cases} f\left[x_i, x_{i-1}\right] + (x_i - x_{i-1})\, f\left[x_{i+1}, x_i, x_{i-1}\right] & \\ \quad + (x_i - x_{i-1})(x_i - x_{i+1})\, f\left[x_{i+2}, x_{i+1}, x_i, x_{i-1}\right], & x \in (x_i, x_{i+1}), \\ f\left[x_{i_1}, x_{i-2}\right] + ((x_i - x_{i-1}) + (x_i - x_{i-2}))\, f\left[x_i, x_{i-1}, x_{i-2}\right] & \\ \quad + (x_i - x_{i-2})(x_i - x_{i-1})\, f\left[x_{i+1}, x_i, x_{i-1}, x_{i-2}\right], & x \in (x_{i-1}, x_i), \end{cases} \tag{13.18}$$

where $f[\cdot]$ is the Newton divided difference function that we presented in Chapter 6. If using the difference described in (13.18), then this is a piecewise interpolant and will only be C^0-continuous.

The third approximation to the d_is that is considered in [207] is referred to as the **Akima estimate** [1], and is the weighted average of the discrete slope on the two sides of the point that we wish to interpolate the function to, in our case for semi-Lagrangian advection the departure point. The Akima derivative estimate is given by

$$d_i = \begin{cases} \dfrac{\iota \Delta_{i-1} + \kappa \Delta_i}{\iota + \kappa}, & \iota + \kappa \neq 0, \\ \dfrac{\Delta_{i-1} + \Delta_i}{2}, & \iota + \kappa = 0, \end{cases} \tag{13.19}$$

where we have used ι and κ instead of ν and η, because their definitions are different, and we do not want to be confused about which definition of a variable we are using at this specific time. This is also the reason why we have not use α and β which are the coefficients in [207] because α has already been considered in this chapter.

At first glance, (13.19) does not appear nonlinear, however, it is only when we consider the definitions of ι and κ, namely

$$\iota \equiv |\Delta_{i+1} - \Delta_i|, \quad \kappa \equiv |\Delta_{i-1} - \Delta_{i-2}|, \tag{13.20}$$

do we see how (13.19) is nonlinear. The weights in (13.19), $\frac{\iota}{\iota+\kappa}$ and $\frac{\kappa}{\iota+\kappa}$, are inversely proportional to an approximation to the curvature on that side. The definition in (13.19) is often referred to as a *geometric mean* estimation [62]. We now go on to consider this technique on the plane.

13.1.1 Advection on a plane

In [207], Williamson and Rasch extend the one-dimensional semi-Lagrangian advection with shape-preserving to two dimensions, and consider three different approaches: (1) splitting the two-dimensional advection operator into two one-dimensional operators via **explicit fractional** time steps, or rather what we know as time splitting, (2) using a monotonic piecewise bicubic interpolating, and (3) using a two-dimensional interpolation by tensor products of one-dimensional shape-preserving schemes. We present these three approaches now.

Fractional time steps

If we consider the one-dimensional advection equation again, then we have that

$$\frac{Df}{Dt} = \frac{\partial f}{\partial t} + u\frac{\partial f}{\partial x} = 0, \tag{13.21}$$

where the semi-Lagrangian solution is of the form

$$f\left(x_a, t^{n+1}\right) = f\left(x_d, t^n\right). \tag{13.22}$$

We know by now that the regular Eulerian grid is the same at both time steps and so we can consider the semi-Lagrangian solution as being expressed as an operator that maps from the regular grid at t^n to the same grid at t^{n+1}, such that

$$f^{n+1} = A_x\left(u, \Delta t\right) f^n. \tag{13.23}$$

We can write a similar expression for the y direction given by

$$f^{n+1} = A_y\left(v, \Delta t\right), \tag{13.24}$$

and the two-dimensional advection equation, as we know, is given by

$$\frac{Df}{Dt} = \frac{\partial f}{\partial t} + u\frac{\partial f}{\partial x} + v\frac{\partial f}{\partial y} = 0. \tag{13.25}$$

It is stated in [207] that Purnell [149], following Strang [180], suggested that an approximation to the two-dimensional problem above that is second-order accurate in time is of the form

$$f^{n+1} = A_x\left(u, \frac{\Delta t}{2}\right) A_y\left(v, \frac{\Delta t}{2}\right) A_y\left(v, \frac{\Delta t}{2}\right) A_x\left(u, \frac{\Delta t}{2}\right) f^n. \tag{13.26}$$

One time step of length Δt consists of a sequence of one-dimensional advection substeps, each of length $\frac{\Delta t}{2}$.

Williamson and Rasch show results from experiments with a solid body setup with a square block which should retain its shape where they compare the approach just described against the bicubic Lagrange interpolation polynomial which is equivalent to an unconstrained, or unlimited, form. They state that the time split cubic forms with both the monotonic and unconstrained forms have significant errors. They are in the form of diffusing the shape as well as introducing a position error by shifting the center of the feature away from the axis of rotation.

Monotonic bicubic interpolation

The monotonic piecewise cubic interpolation from [62] was extended to a bicubic version in [29]. The first step is to define monotonicity for bivariate functions. A bivariate function p is monotonic in the domain D_{ij}, which is defined by $x_i \leq x \leq x_{i+1}$ and $y_j \leq y \leq y_{j+1}$, if all univariate slices of p taken parallel to the coordinate axes are monotonic. Another condition is that the sign of the monotonicity must be the same for all x slices and for all y slices, although not necessarily of the same sign as for the x slices [207].

Bicubic polynomial interpolation with the domain D_{ij} can be written as

$$\begin{aligned} p(x,y) &= f(x_i,y_j)H_1(\alpha)G_1(\beta) + f(x_{i+1},y_j)H_2(\alpha)G_1(\beta) + f(x_i,y_{j+1})H_1(\alpha)G_2(\beta) \\ &+ f(x_{i+1},y_{j+1})H_2(\alpha)G_2(\beta) + f_x(x_i,y_j)H_3(\alpha)G_1(\beta) + f_x(x_{i+1},y_j)H_4(\alpha)G_1(\beta) \\ &+ f_x(x_i,y_{j+1})H_3(\alpha)G_2(\beta) + f_x(x_{i+1},y_{j+1})H_4(\alpha)G_2(\beta) \\ &+ f_y(x_i,y_j)H_1(\alpha)G_3(\beta) + f_y(x_{i+1},y_j)H_2(\alpha)G_3(\beta) \\ &+ f_y(x_i,y_{j+1})H_1(\alpha)G_4(\beta) + f_y(x_{i+1},y_{j+1})H_2(\alpha)G_4(\beta) \\ &+ f_{xy}(x_i,y_j)H_3(\alpha)G_3(\beta) + f_{xy}(x_{i+1},y_j)H_4(\alpha)G_3(\beta) \\ &+ f_{xy}(x_i,y_{j+1})H_3(\alpha)G_4(\beta) + f_{xy}(x_{i+1},y_{j+1})H_4(\alpha)G_4(\beta), \end{aligned} \tag{13.27}$$

where the derivative estimates are denoted by f_x, f_y, and f_{xy}, and the cubic basis functions are given by

$$\begin{aligned} H_1(\alpha) &= 2\alpha^3 - 3\theta^2 + 1, \\ H_2(\alpha) &= -2\alpha^3 + 3\alpha^2, \\ H_3(\alpha) &= \alpha^3 - 2\alpha^2 + \alpha, \\ H_4(\alpha) &= \alpha^3 - \alpha^2, \end{aligned} \tag{13.28}$$

with similar forms for the G_is.

A necessary condition that (13.27) is monotonic is

$$\operatorname{sgn}\left[\Delta_y\left(x_i\right)\right]=\operatorname{sgn}\left[\Delta_y\left(x_{i+1}\right)\right]=S_y, \tag{13.29a}$$
$$\operatorname{sgn}\left[\Delta_x\left(y_j\right)\right]=\operatorname{sgn}\left[\Delta_x\left(x_{j+1}\right)\right]=S_x, \tag{13.29b}$$

where

$$\Delta_y(x)=\frac{f\left(x, y_{j+1}\right)-f\left(x, y_j\right)}{y_{j+1}-y_j}, \tag{13.30a}$$
$$\Delta_x(y)=\frac{f\left(x_{i+1}, y_j\right)-f\left(x_i, y\right)}{x_{i+1}-x_i}. \tag{13.30b}$$

To ensure monotonicity at the corners, we must have

$$0 \leq S_y f_y(\xi, \eta) \leq S_y 3 \Delta_y(\xi), \tag{13.31a}$$
$$0 \leq S_x f_x(\xi, \eta) \leq S_x 3 \Delta_x(\eta), \tag{13.31b}$$

for $\xi=x_i$ and x_{i+1} and $\eta=y_j$ and y_{j+1}. These conditions are the univariate SCM0 applied along the edges of the rectangular domain D_{ij} and guaranteeing that the interpolant will be monotonic along the edges. Additional conditions on the first derivative for monotonicity within domain D_{ij} are

$$-S_y \frac{3 \Delta_y\left(x_i\right)}{x_{i+1}=x_i} \leq S_y \Delta_y^{\prime}\left(x_i\right), \tag{13.32}$$
$$S_y \frac{3 \Delta_y\left(x_{i+1}\right)}{x_{i+1}=x_i} \leq S_y \Delta_y^{\prime}\left(x_i\right), \tag{13.33}$$
$$-S_x \frac{3 \Delta_x\left(y_j\right)}{y_{j+1}=y_j} \leq S_x \Delta_x^{\prime}\left(y_j\right), \tag{13.34}$$
$$S_x \frac{3 \Delta_x\left(y_{j+1}\right)}{y_{j+1}=y_j} \leq S_x \Delta_x^{\prime}\left(y_j\right), \tag{13.35}$$

where

$$\Delta_y^{\prime}(x) = \frac{f_x\left(x, y_{j+1}\right)-f_x\left(x, y_j\right)}{y_{j+1}-y_j}, \tag{13.36}$$
$$\Delta_x^{\prime}(y) = \frac{f_y\left(x_{i+1}, y\right)-f_y\left(x_i, y\right)}{x_{i+1}-x_i}. \tag{13.37}$$

Finally, we have to consider the conditions on the cross-derivatives at the corners of the boxes. For $\eta=y_j$ and y_{j+1}, we have

$$-S_y \frac{3 f_y\left(x_i, \eta\right)}{x_{i+1}-x_i} \leq S_y f_{xy}\left(x_i, \eta\right) \leq S_y 3\left(\Delta_y^{\prime}\left(x_i\right)+\frac{3 \Delta_y\left(x_i\right)+f_y\left(x_i, \eta\right)}{x_{i+1}-x_i}\right), \tag{13.38}$$

$$S_y 3\left(\Delta_y'(x_{i+1}) - \frac{3\Delta_y(x_{i+1}) - f_y(x_{i+1},\eta)}{x_{i+1}-x_i}\right) \leq S_y f_{xy}(x_{i+1},\eta) \leq S_y \frac{3f_y(x_{i+1},\eta)}{x_{i+1}-x_i} \leq \frac{3S_y f_y(x_i,\eta)}{x_{i+1}-x_i}, \tag{13.39}$$

and for $\xi = x_i$ and x_{i+1}, we have

$$-S_x \frac{3f_x(\xi,y_j)}{y_{j+1}-y_j} \leq S_x f_{xy}(\xi,y_j) \leq S_x 3\left(\Delta_x'(y_j) + \frac{3\Delta_x(y_j) + f_x(\xi,y_j)}{y_{j+1}-y_j}\right), \tag{13.40}$$

$$S_x 3\left(\Delta_x'(y_{j+1}) - \frac{3\Delta_x(y_{j+1}) - f_x(\xi,y_{j+1})}{y_{j+1}-y_j}\right) \leq S_x f_{xy}(\xi,y_{j+1}) \leq S_x \frac{3f_x(\xi,y_{j+1})}{y_{j+1}-j_i} \leq S_x \frac{3f_x(\xi,y_{j+1})}{y_{j+1}-y_j}. \tag{13.41}$$

In [207], Williamson and Rasch provide a set of guidelines to ensure that the monotonicity is maintained, which is too much for us to put here, and the reader is referred to [207] for those details. They also show results from this scheme for the test case mentioned earlier; while it did not center the square feature correctly, it was suffering from slightly less straight sides than the unmodified version, but it did not suffer from over- and undershoots.

However, Williamson and Rasch decided against this approach as they ultimately wished to be able to preserve shape in 3D, and, as we saw above, the conditions on the boundaries can be quite complicated. Thus we move on to the last option that Williamson and Rasch present, which is the tensor product.

Tensor product

The essence of this approach is to obtain a two-dimensional interpolation at a point by first performing a series of interpolations in one dimension followed by one interpolation in the other dimensions. The number of interpolations done in the first series is given by the number of data values needed by the interpolation in the other direction, including those needed for the derivatives estimates.

For example, if the point (x, y) lies in the grid box $x_i \leq x \leq x_{i+1}$, $y_j \leq y \leq y_{j+1}$ and we use a cubic interpolant with the Hyman derivative estimate from earlier, then we would first perform six interpolations in x at the rows y_{j+k}, for $k = -2, \ldots, 3$ to obtain six interpolated values at (x, y_{j+k}). The next step is to calculate the required y derivatives at (x, y_j) and (x, y_{j+1}) and interpolate in y to obtain the desired values at (x, y). The x interpolated values (x, y_{j+k}) are used to calculate the Hyman derivative estimate in the y direction, after which these y derivative estimates can be modified for shape-preserving if desired.

13.2 Cascade interpolation

In the opening paragraph of Nair et al. (1999) [132], the authors state that at every time step of a semi-Lagrangian advection scheme employing backward trajectories, the advected fields need to be evaluated at the upstream points via interpolation of the known surrounding gridpoint values. A typical tensor-product interpolation in three dimensions (3D), based on a one-dimensional (1D) interpolator, requires $O\left(p^3\right)$ operations per gridpoint per field, where p here is the formal order of accuracy of the interpolator; as we know, for a cubic interpolation polynomial this would be $p=4$. A high-order interpolation formally improves the accuracy of the advection scheme, but unfortunately engenders additional computational expense. In Purser and Leslie (1991) [152], the authors proposed an alternative scheme which they termed **cascade** interpolation which will play an important part in the semi-Lagrangian scheme that we introduce in the next section.

The cascade technique is a sequence of 1D interpolations for the multidimensional interpolation problem associated with semi-Lagrangian advection, thereby reducing the cost of using high-order interpolation schemes in 2D and 3D to $O(p)$ operations per gridpoint per field, which we can clearly see is a significant saving. We now present the work from Purser and Leslie (1991) [152].

13.2.1 Introduction to cascade interpolation

In a paper that Purser and Leslie published in 1988 [151], they had argued that well-known deficiencies of the horizontally unstaggered grid, solution separation by grid splitting, nonlinear computational instability, and an inferior accuracy in the treatment of certain combinations of governing equations terms, that are crucial to geostrophic adjustment, are eradicated by the combined adoption of high-order differencing and the occasional application of selective low-pass spatial filtering to inhibit any tendency towards grid splitting or the noise symptomatic of nonlinear instability.

Leslie and Purser had also shown in [109] that the use of high-order accurate representations of the continuity and hydrostatic equations in the vertical direction, using Lagrange quadrature formulas, significantly improved the accuracy of forecasts, especially of precipitation. They then state that it is evident that there are considerable benefits to be gained from the adoption in a fully unstaggered grid, provided it is used in conjunction with high-order numerics and suitable horizontal filtering. They also make the comment that in a semi-Lagrangian model, it would be unsatisfactory to use high-order differencing and vertical quadrature for the adjustment terms while only a low-order technique to interpolate the dependent variables to each trajectory origin. As we know, the computational cost of using a higher-order interpolation in 3D is too high.

Therefore, Leslie and Purser draw our attention to a technique of interpolation between grids that, for high-order accurate schemes, provides significant computational savings, where this approach is based upon the Cartesian product of one-dimensional Lagrange interpolation polynomials. They state that this method dispenses with the Cartesian product interpolation in favor of a **cascade** of 1D interpolations through a succession of hybrid grids mixing the standard model coordinates with curvilinear Lagrangian coordinates.

Lagrange interpolation

We start by recalling that the general expression for the Lagrange interpolation between discrete values $\psi\left(\widehat{x}_i\right)$ of a generic variable ψ at node positions $\widehat{x}_i$ along a curve to a target location x is given by

$$\psi(x) \approx \sum_{i=m_1}^{m_2} L_i(x)\,\psi(\widehat{x}_i), \tag{13.42}$$

where

$$L_i(x) \equiv \prod_{\substack{j \neq i \\ j \in [m_1, m_2]}} \frac{x - \widehat{x}_j}{\widehat{x}_i - x_j}. \tag{13.43}$$

An interesting statement is then made in [152] that for best results m_1 and m_2 should be chosen to provide a span of grid points straddling the target point x as **symmetrically** as possible.

If we now consider that the interpolations from one **grid** to another are of **comparable** resolution, then it becomes necessary to evaluate a set of coefficients $L_i^{(g)}(x)$ of the form (13.43) for each grid interval $g = [x_g, x_{g+1}]$ in which target x is presumed to lie. In this case the denominator of the expression (13.43) for a complete grid line can be **precomputed** in a single step for a cost *per gridpoint* of only $O(N)$. Likewise the N numerators at each target x can be computed together for a cost of only $O(N)$.

Cartesian-product interpolation

We are now considering the interpolation of grided data $\psi(\widehat{x}_i, \widehat{y}_j)$ in two dimensions to a target point $\mathbf{x} = (x, y)$. This time the interpolation polynomial is formulated as a Cartesian product of basic Lagrange interpolations:

$$\psi(x, y) \approx \sum_{j=n_1}^{n_2} L_j(y) \left(\sum_{i=m_1}^{m_2} L_i(x)\,\psi(\widehat{x}_i, \widehat{y}_j) \right). \tag{13.44}$$

This was the approach that we introduced in the tricubic interpolation section in Chapter 11, where here we have $N^2 + N \approx O(N^2)$ operations when the Nth order polynomial is chosen in **both** directions.

Interpolation using hybrid grids

The problem, as stated so far, has been an interpolation to a target point in isolation. Purser and Leslie state that in practise the task a semi-Lagrangian model presents is to interpolate from the regular model grid $\widehat{\mathbf{x}} \equiv \{\widehat{x}_i, \widehat{y}_j, \widehat{z}_k\}$ to a nonstandard curvilinear grid of target points $\mathbf{x}(\widehat{\mathbf{X}}_{i',j',k'})$, where $\widehat{\mathbf{X}}_{i',j',k'} \equiv (\widehat{X}_{i'}, \widehat{Y}_{j'}, \widehat{Z}_{k'})$ are Lagrangian coordinates of the latter grid of points such that in the time step interval $[(n-1)\Delta t, n\Delta t]$ one has

$$\frac{D\mathbf{x}(\widehat{\mathbf{X}}, t)}{Dt} = \mathbf{u}(\mathbf{x}, t), \tag{13.45a}$$

$$\mathbf{x}(\widehat{\mathbf{X}}_{i',j',k'}, n\Delta t) = \mathbf{x}_{i',j',k'}. \tag{13.45b}$$

We now make an assumption that, provided the velocity field $\mathbf{u}$ is reasonably smooth at the grid scale and $\mathbf{x}(\mathbf{X})$ is a smooth function, the implicit function $\mathbf{X}(\mathbf{x})$ is also smooth. We also have to assume that there is not too much shearing of the Lagrangian coordinates; we will explain in the next section what this means. Given these assumptions, we see that the smoothness of the Lagrangian grid might be exploited by the fact that the values of $\psi\left(\widehat{X}_{i'}, \widehat{y}_j, \widehat{z}_k\right)$ at positions defined at time $t = (n-1)\Delta t$ by the intersection of the Lagrangian coordinate $\widehat{X}_{i'}$ with standard coordinates $\widehat{y}_j, \widehat{z}_k$ can be approximated by interpolation from the same variable ψ held at standard grid positions $\psi\left(\widehat{x}_i, \widehat{y}_j, \widehat{z}_k\right)$ simply by the one-dimensional application of the Lagrange interpolation along the standard grid lines

$$y = \widehat{y}_j, \quad j = 1, 2, \ldots, \qquad z = \widehat{z}_k, \quad k = 1, 2, \ldots,$$

provided that only the x coordinates of all the target points are known. Two further grid interpolations, which require only the one-dimensional Lagrange formula to be applied at each stage, suffice to yield the desired approximation $\psi\left(\widehat{X}_{i'}, \widehat{Y}'_j, \widehat{Z}_{k'}\right)$ of ψ at the Lagrangian points. We have a copy of the two-dimensional schematic of this approach from [152] in Fig. 13.2. Purser and Leslie go on to describe the procedure for the cascade interpolation, but they also have a summary in a table, which is reproduced in Table 13.1.

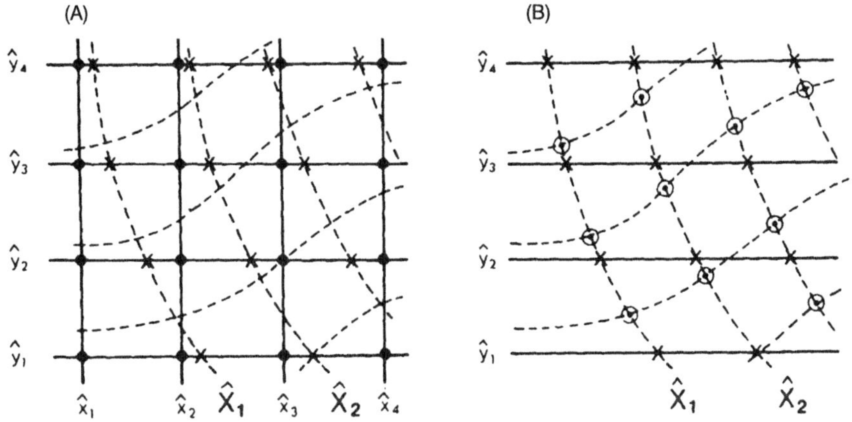

FIGURE 13.2 Copy of Fig. 2 from [152] of the illustration of a two-dimensional version of the cascade method of interpolating a generic variable ψ from model grid (solid lines) to Lagrangian grid (dashed lines). Purser, R. J. and Leslie, L. M., 1991: An efficient interpolation procedure for high-order three-dimensional semi-Lagrangian models, *Monthly Weather Review*, **119**, 2492–2498. ©American Meteorological Society. Used with permission.

TABLE 13.1 Summary of the five stages of the cascade method of grid interpolation to back trajectories.

Stage 1:	$x\left(\widehat{X}, \widehat{Y}, \widehat{Z}\right) \to x\left(\widehat{X}, \widehat{Y}, \widehat{z}\right)$ $y\left(\widehat{X}, \widehat{Y}, \widehat{Z}\right) \to y\left(\widehat{X}, \widehat{Y}, \widehat{z}\right)$	Using $z\left(\widehat{X}, \widehat{Y}, \widehat{Z}\right)$ to define nodes.
Stage 2:	$x\left(\widehat{X}, \widehat{Y}, \widehat{z}\right) \to x\left(\widehat{X}, \widehat{y}, \widehat{z}\right)$	Using $y\left(\widehat{X}, \widehat{Y}, \widehat{z}\right)$ to define nodes.
Stage 3:	$\psi\left(\widehat{x}, \widehat{y}, \widehat{z}\right) \to \psi\left(\widehat{X}, \widehat{y}, \widehat{z}\right)$	Using $x\left(\widehat{X}, \widehat{y}, \widehat{z}\right)$ to define nodes.
Step 4:	$\psi\left(\widehat{X}, \widehat{y}, \widehat{z}\right) \to \psi\left(\widehat{X}, \widehat{Y}, \widehat{z}\right)$	Using $y\left(\widehat{X}, \widehat{y}, \widehat{z}\right)$ to define nodes.
Step 5:	$\psi\left(\widehat{X}, \widehat{Y}, \widehat{z}\right) \to \psi\left(\widehat{X}, \widehat{Y}, \widehat{Z}\right)$	Using $z\left(\widehat{X}, \widehat{Y}, \widehat{Z}\right)$ to define nodes.

13.2.2 Monotonic cascade interpolation

The interpolation formula that we present now is used in the next section, and it is explaining the link between the Lagrangian and Eulerian frameworks to enable conservation of advected fields. This form of interpolation is referred to as **monotonic cascade interpolation** and is taken from [132].

If we let (X, Y) denote the curvilinear Lagrangian coordinate system corresponding to the Eulerian (x, y) coordinate system. A y-line of the regular Eulerian mesh is a straight line defined by $x = x_i$, whereas the associated mapped Lagrangian Y-curve is a curve corresponding to the mapped y-line that joins upstream points $(x_{i1}, y_{i1}), (x_{i2}, y_{i2}), \ldots, (x_{in} y_{in})$. We have a copy of the hand-drawn schematic from [132] in Fig. 13.3.

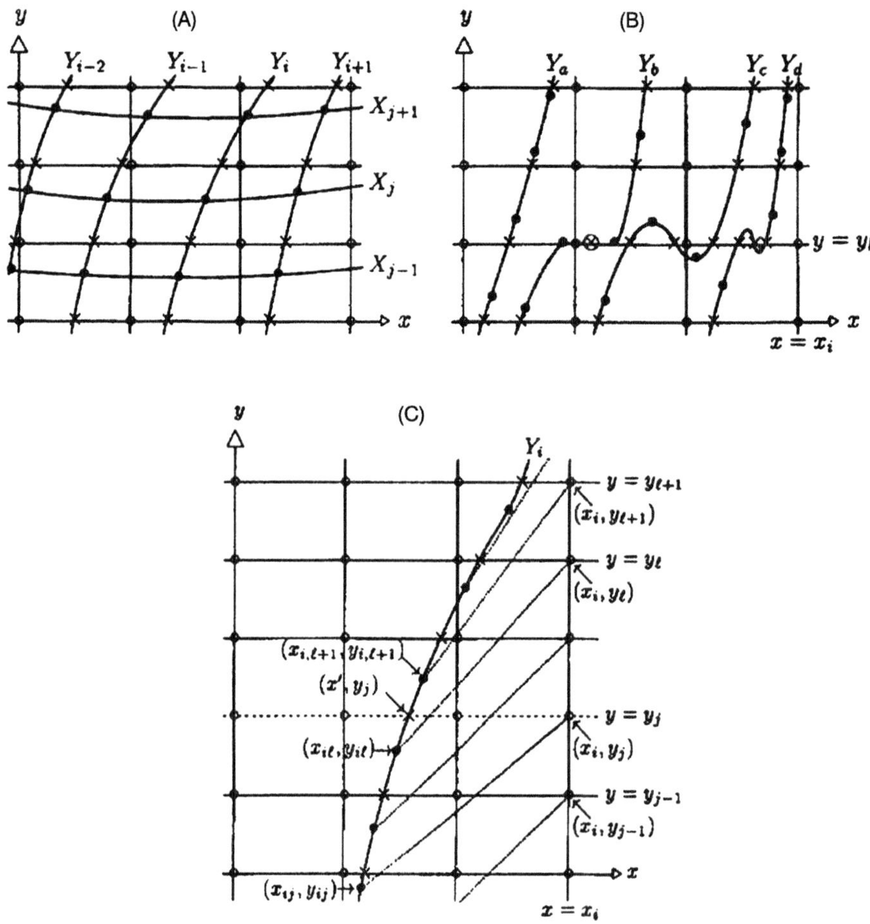

FIGURE 13.3 Copy of Fig. 1 from [132]. Nair, M., Côté, J. and Staniforth, A., 1999: Monotonic cascade interpolation for semi-Lagrangian advection, *Q. J. R. Meteor. Soc.*, **125**, 197–212. ©Royal Meteorological Society.

In [132] the authors state that a key component of the cascade-interpolation technique is the generation of a well-defined intermediate mesh, where an intermediate mesh can be constructed as the intersection of Y-curves with x-lines, or as the intersection of X-curves with y-lines. The simplest way of determining an intermediate mesh is by using linear interpolation to determine the points of intersection of each Y-curve with x-lines in the two-dimensional problem. A motivation to consider a linear interpolation comes from the work by McDonald in 1987 where his analysis, which we saw in Chapter 8, indicates that when the advected quantity is interpolated with a cubic interpolation, then it is sufficient to use linear interpolation when computing the trajectories that define the Lagrangian mesh, and is significantly more economical than the approach suggest by Purser and Leslie [152]. Therefore, although it is possible to use higher-order interpolation than linear, this is of little benefit since it does not reduce the order of the truncation error.

If we now consider plot 1(C) in Fig. 13.3 from [132], then if a specific Y-curve Y_i which joins the points $(x_{1i}, y_{i1}), (x_{i2}, y_{i2}), \ldots, (x_{in}, y_{in})$ intersects the x-line defined by $y = y_j$, then it will do so in a way that $y_{il} \leq y_j \leq y_{i,l+1}$ for some integer index l, where $1 \leq l \leq n-1$. By passing a straight line between the two points (x_{il}, y_{il}) and (x_{il+1}, y_{il+1}), its point of intersection with $y = y_j$ is simply (x', y_j) where

$$x' = x_{ii} + \frac{y_j - y_{il}}{y_{i,l+1} - y_{il}} \left(x_{i,l+1} - x_{il}\right). \tag{13.46}$$

In a 2D rectangular domain, the cascade interpolation procedure for semi-Lagrangian advection employing backward trajectories can be summarized as follows:

(i) For $i = 1, 2, \ldots, m$ and $j = 1, 2, \ldots, n$, obtain, in the usual way, the upstream departure points (x_{ij}, y_{ij}) that define the Lagrangian mesh.
(ii) Determine the location of the points that define the intermediate mesh using the linear interpolation algorithm (13.46).
(iii) For $j = 1, 2, \ldots, n$, interpolate, using 1D polynomial interpolation along the x-lines $y = y_j$, all advected fields from the points of the Eulerian mesh to the points of the intermediate mesh.
(iv) For $j = 1, 2, \ldots, n$, interpolate, using 1D polynomial interpolation along the Y-curves Y_{ii}, all advected fields from the points of the intermediate mesh to the target points of the Lagrangian mesh.

The interpolation in step (iv) above is performed using the distance s along the Y-curve as the independent variable. The distances s_l of the points (x_l, y_l) along the curve are recursively defined by

$$\begin{aligned} s_l &= 0, \\ s_{l+1} &= s_l + \sqrt{(x_{l+1} - x_l) + (y_{l+1} - y_l)}. \end{aligned} \tag{13.47}$$

A point (x', y_j) of the intermediate mesh that lies on the straight line joining (x_l, y_l) to (x_{l+1}, y_{i+1}), where x' is given by (13.46), is therefore located at

$$s' = s_l + \left(s_l + (s_{l+1} - s_l) \frac{(y_j - y_l)}{(y_{l+1} - y_l)} \right), \tag{13.48}$$

and the second term on the right-hand side of (13.48) is guaranteed to be positive.

As we saw in plot 1(B) in Fig. 13.3, there can be times where the Lagrangian Y curves intersect with specific x curves more than once, and to eliminate this possibility, it may be the case that the time step may have to be chosen sufficiently small.

In [132], Nair et al. state that cubic interpolation, with a monotonicity constraint, is adopted in the present work for the 1D polynomial interpolators of steps (iii) and (iv), since cubic interpolation of advected fields has been found to be a good compromise between accuracy and efficiency for semi-Lagrangian schemes [176]. Higher-order, for example, quintic, interpolators also merit serious consideration as pointed out in [152], provided that the advected fields have sufficient continuity, since this reduces the inherent damping caused by interpolation, and the incremental cost of increasing the polynomial order for cascade interpolation is much less than that for conventional Cartesian-product interpolation. Linear and quadratic interpolation are, however, not generally recommended since they can lead to an unacceptable amount of damping [177].

Nair et al. then move on to introduce what they refer to as *an improved monotonic filter* which is based upon the Quasi-Monotone Semi-Lagrangian (QMSL) filter from [20] which was modified in [197,196], and where this modified filter was applied immediately after each 1D interpolation of the interpolation cascade. In [197] their algorithm identifies the existence of a local extremum by examining slope variations of the function in the three-mesh-length interval surrounding the target evaluation-point: the QMSL filter is then turned off whenever an extremum is identified in order to capture it. This modification is designed to address a shortcoming of the QMSL filter, which is that under certain circumstances it clips some local extrema that we would ideally wish to retain. However, the filter in [197] is only partially successful, since it trades off an improvement in one circumstance against a deterioration in another. This is illustrated in Fig. 2 for a 1D interpolation function in a region with a strong Gibbs (over- and undershoots) phenomenon [132].

According to [132], the **QMSL filter** is defined as follows:

(a) Identify the local minimum and maximum surrounding the target point $x_i - \alpha_i$ as

$$f^- = \min\left(f_l, f_{l+1}\right), \tag{13.49}$$

$$f^+ = \max\left(f_l, f_{l+1}\right), \tag{13.50}$$

whenever $x_i - \alpha_i$ is in the interval $\left[x_l, x_{l+1}\right]$;

(b) Reset $f\left(x_i - \alpha_i\right)$ such that

$$f\left(x_i - \alpha_i\right) = f^- \quad \text{if } f\left(x_i - \alpha_i\right) < f^-, \tag{13.51}$$

$$f\left(x_i - \alpha_i\right) = f^+ \quad \text{if } f\left(x_i - \alpha_i\right) > f^+. \tag{13.52}$$

In [197] the modification that the authors do to the QMSL filter is to switch it off whenever an extremum is identified by the monotonicity condition

$$\left(f_l - f_{l+1}\right)\left(f_{l+2} - f_{l+1}\right) < 0, \tag{13.53}$$

provided $f_{\min} \leq f\left(x_i - \alpha_i\right) \leq f_{\max}$ where $f_{\min}$ and $f_{\max}$ are the global minimum and maximum.

Nair et al. present a simple, but effective example in [132] to illustrate what is wrong with the modification that was done to the QMSL filter in [197], in that the QMSL filter correctly enforces monotonicity and realistically controls the Gibbs behavior. However, although the modified filter mostly controls this behavior, it fails to do so at the point $x_i - \alpha_i$, resulting in a violation of monotonicity, an isolated small-scale perturbation being superimposed on the expected solution, and the introduction of a spurious mass-source.

Nair et al. state that by making a further small modification, it is possible to have the best of two worlds, that is to say, to have the robustness of the QMSL algorithm and the capture of local subgrid-scale extrema by the [197] modified one. The reason why the modified algorithm breaks down in a Gibbs-phenomenon situation is that in identifying local extrema, the local behavior of the function is considered only in the three-mesh-length-wide interval containing the extremum, which is insufficient to exclude the two-mesh-length noise signature of the Gibbs phenomenon. By extending the width to five mesh-lengths (or more if desired), and by insisting that the function has only a single extremum in this larger interval, the problem is avoided. The revised monotonic procedure can be expressed algorithmically as follows:

(i) If all the following four inequalities hold, do not apply the QMSL filter:

$$\begin{aligned} f_{\min} \leq f(x_i - \alpha_i) &\leq f_{\max}, && (13.54)\\ (f_{l-1} - f_{l-2})(f_l - f_{l-1}) &> 0, && (13.55)\\ (f_l - f_{l-1})(f_{l+2} - f_{l+1}) &> 0, && (13.56)\\ (f_{l+2} - f_{l+1})(f_{l+3} - f_{l+2}) &> 0, && (13.57) \end{aligned}$$

where the target meshpoint $x_i - \alpha_i$ is in the interval $\left[x_l, x_{l+1}\right]$.

(ii) If one or more of these inequalities is violated, then apply the QMSL filter.

Inequality (13.56) identifies $f(x_i - \alpha_i)$ as an extremum in the interval $\left[x_l, x_{l+1}\right]$, whereas inequalities (13.55) and (13.57) ensure monotonicity in the two-mesh-length-wide intervals on either side of this interval. Note that the new filter-formulation reduces to the modified filter from [197] by dropping conditions (13.55) and (13.57), and then further reduces to the Bermejo and Staniforth (1992) (QMSL) filter by dropping conditions (13.54)–(13.57) altogether.

13.2.3 Cascade interpolation on a sphere

In 1996 at the European Center for Medium-Range Weather Forecasts, it was recommended that because of the economic promise of the cascade interpolation it is should be developed further to allow for applications in spherical geometry [134]. It was mentioned in [152] that this is not straightforward due to the convergence of the meridians to the polar singularities of the spherical coordinate system. The method that we present here today is from [134], which is an extension of the work we presented in the last subsection from [132].

We start with the two-dimensional advection equation of a passive scalar ψ on a sphere, which mathematically is

$$\frac{\partial \psi}{\partial t} + \frac{u}{a\cos\theta}\frac{\partial \psi}{\partial \lambda} + \frac{v}{a}\frac{\partial \psi}{\partial \theta} = 0. \tag{13.58}$$

Now let (λ_i, θ_j) where $i = 1, 2, \ldots, m$ and $j = 1, 2, \ldots, n$ define the regularly spaced Eulerian meshpoints, where we assume that we have uniform resolution $(\Delta\lambda)$ in the λ direction, with m even. If the poles are meshpoints then $\theta_1 \equiv -\frac{\pi}{2}$ and $\theta_n \equiv \frac{\pi}{2}$. Therefore this formulation includes uniform-resolution latitude–longitude meshes, with or without poles, and the Gaussian grid of spectral models.

In [134] the authors now assume that for every meshpoint (λ_i, θ_j) at time $t + \Delta t$ there exists a unique upstream Lagrangian point $(\lambda_{ij}, \theta_{ij})$ at time t, and that the position of this latter point is determined by integrating backwards in the usual way along a trajectory of the flow. If we now let (Λ, Θ) denote the curvilinear Lagrangian coordinate system corresponding to the Eulerian (λ, θ) system, we can see this in the copy of Fig. 1 from [134] in Fig. 13.4. A θ circle of the regular Eulerian mesh is a longitude circle, defined by $\lambda = \lambda_i$ and its continuation $\lambda = \lambda_{i+\frac{m}{2}} = \lambda_i + \pi$. The associated Lagrangian θ-curve, corresponding to the mapping of this θ-circle, is simply a deformed circle that joins the upstream points

$$(\lambda_{i1}, \theta_{i1}), (\lambda_{i2}, \theta_{i2}), \ldots (\lambda_{in}, \theta_{in}) \left(\lambda_{i+\frac{m}{2},n-1}, \theta_{i+\frac{m}{2},n-1}\right) \left(\lambda_{i+\frac{m}{2},n-2}, \theta_{i+\frac{m}{2},n-2}\right), \ldots,$$
$$\left(\lambda_{i+\frac{m}{2},1}, \theta_{i+\frac{m}{2},1}\right).$$

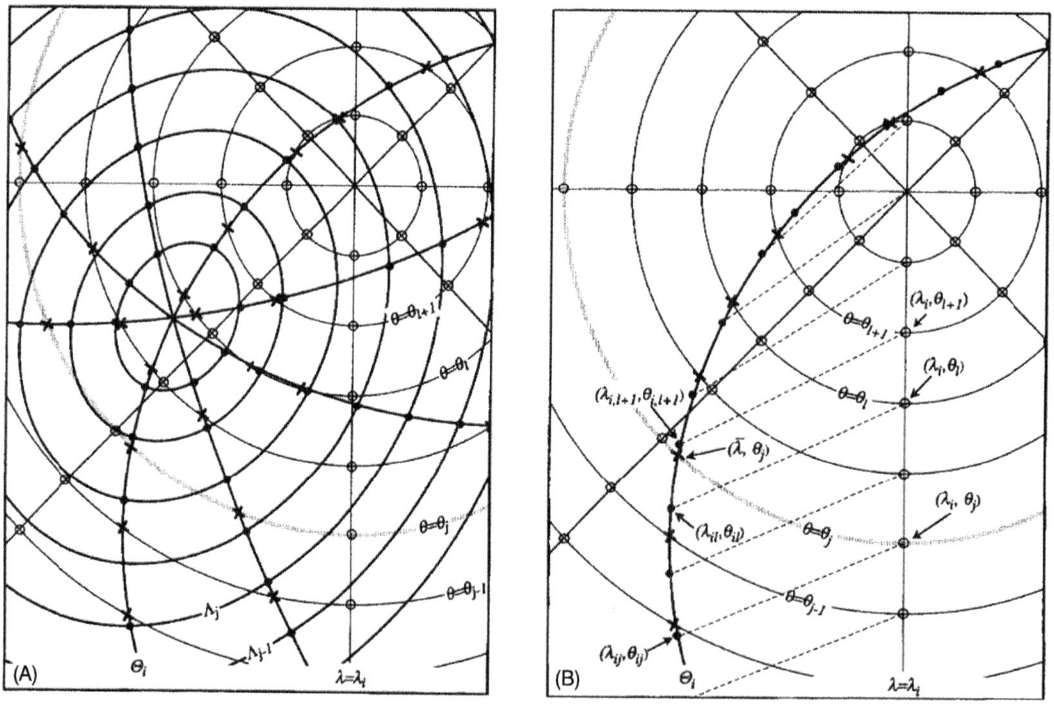

FIGURE 13.4 Copy of Fig. 1 from [134], Nair, M., Cfê, J. and Staniforth, A., 1999: Cascade interpolation for semi-Lagrangian advection over the sphere, *Q. J. R. Meteor. Soc*, **125**, 1445–1468. ©Royal Meteorological Society.

According to [134], there are only $\frac{m}{2}$ such closed curves Θ_i, $i = 1, 2, \ldots, \frac{m}{2}$, due to the convention that they adopt of *wrapping a longitude around the whole globe* to simplify the cascade procedure in polar regions, and make the remark that we should note that if the poles are meshpoints, then the South and North poles (λ_i, θ_1) and (λ_i, θ_n), where i is arbitrary, respectively map to the points $(\lambda_{i1}, \theta_{il})$ and $(\lambda_{in}, \theta_{in})$ which are common to all Θ_i-curves. We should be aware that the continuity and uniqueness of ψ at the South pole then imposes the constraints

$$\psi(\lambda_1, \theta_1) = \psi(\lambda_2, \theta_1) = \cdots = \psi(\lambda_m, \theta_1), \tag{13.59}$$

$$\psi\left(\lambda_{1,1}, \theta_{1,1}\right) = \psi\left(\lambda_{2,1}, \theta_{2,1}\right) = \cdots = \psi\left(\lambda_{m,1}, \theta_{m,1}\right), \tag{13.60}$$

and similar expressions can be imposed at the North pole.

In [134] the authors state that when the poles are the meshpoints, the upstream values at the corresponding Lagrangian poles can be determined by interpolating along any of the Θ_i-curves, where in current formulation there are $\frac{m}{2}$, and Nair et al. take an average of these values to perform their studies, but check to see what would happen if they picked an arbitrary curve. The authors found that the results were consistent and only differed by a truncation error.

Cascading the interpolation on a sphere

We start by defining an algorithm for an intermediate mesh which is the intersection of the Lagrangian Θ_i-curve with the Eulerian λ-lines:

(i) For $i = 1, 2, \ldots, m$ and $j = 1, 2, \ldots, n$, obtain in the usual way the upstream points $\left(\lambda_{ij}, \theta_{ij}\right)$ that define the Lagrangian mesh.

(ii) Determine the location of the intersection points of Θ-curves (Lagrangian longitudes) with λ-circles (Eulerian latitudes).

(iii) For $j = 1, 2, \ldots, n$, interpolate, using 1D polynomial interpolation along the λ-circles $\theta = \theta_j$, all advected fields from the points of the Eulerian mesh to the intersection meshpoints.

(iv) For $j = 1, 2, \ldots, n$, interpolate, using 1D polynomial interpolation along the Θ-curves (there are $\frac{m}{2}$ of them), all advected fields from the intersection meshpoints to the points of the Lagrangian mesh.

We now specify how to determine the location of the points of the intermediate mesh, and how to subsequently interpolate along the Lagrangian longitudes, that is, the Θ-curves.

Determining of mesh intersection points

In [134] the authors refer to Fig. 13.4B where we can see a specific Θ_i-curve. If this curve intersects the specific Eulerian latitude circle $\theta = \theta_j$, then it will in general do so such that $\theta_{ip} \leq \theta_j \theta_{i,p+1}$ for some value of the integer index p, where $\leq p \leq n - 1$. Nair et al. now state that it is natural to piecewise approximate the Θ-curve by great-circle segments since it is a deformed mapping of the great circle $\lambda = \lambda_i$. The intersection of the latitude circle with the segment of the great circle that joins the two points $\left(\lambda_{ip}, \theta_{ip}\right)$ and $\left(\lambda_{i,p+1}, \theta_{i,p+1}\right)$ of the Lagrangian mesh then defines a point $\left(\widetilde{\lambda}, \theta_j\right)$ of the intermediate mesh. It is convenient to perform the mesh navigation using a unit sphere ($a = 1$), and the Cartesian coordinates

$$\left.\begin{array}{l} x = \cos\theta\cos\theta \\ y = \sin\theta\cos\theta \\ z = \sin\theta \end{array}\right\} \Rightarrow x^2 + y^2 + z^2 = 1. \tag{13.61}$$

The equation of the θ_j-latitude circle is

$$x^2 + y^2 = \cos^2\theta_j. \tag{13.62}$$

The great-circle plane that passes through two points $\left(\lambda_{ip}, \theta_{ip}\right)$ and $\left(\lambda_{i,p+1}, \theta_{i,p+1}\right)$ is given by

$$Ax + By + Cz = 0, \tag{13.63}$$

where

$$(A, B, C) = \left(x_{ip}, y_{ip} z_{ip}\right) \times \left(x_{i,p+1}, y_{i,p+1}, z_{i,p+1}\right), \tag{13.64}$$

and $\left(x_{ip}, y_{ip}, z_{ip}\right)$ and $\left(x_{i,p+1}, y_{i,p+1}, z_{i,+1p}\right)$ are obtained through evaluating (13.61) at the latitude–longitude points $\left(\lambda_{ip}, \theta_{ip}\right)$ and $\left(\lambda_{i,p+1}, \theta_{i,p+1}\right)$. The intersection of this plane with the latitude plane $\theta = \theta_j$ is a straight line [134] given by

$$Ax + By = -C\sin\theta_j. \tag{13.65}$$

The coordinates of the intersection points of this line and the θ_j-latitude circle can be found through simultaneously solving (13.62) and (13.65), which leads to the quadratic equation

$$\left(A^2 + B^2\right)x^2 + \left(2AC\sin\theta_j\right)x + \left(\left(B^2 + C^2\right)\sin^2\theta_j - B^2\right) = 0. \tag{13.66}$$

The coordinates $\left(\widetilde{\lambda}, \theta_j\right)$ of the sought intersection point are found by solving (13.66) for x. Given our value of x, we can obtain y from (13.65). The values of $\widetilde{\lambda}$ is found through

$$\widetilde{\lambda} = \tan^{-1}\left(\frac{y}{x}\right). \tag{13.67}$$

If $\widetilde{\lambda}$ is in the interval $[-\pi, 0)$, then we add 2π to the result to locate roots in $[0, 2\pi)$.

Interpolating along the Lagrangian longitudes Θ_i

The interpolation along Lagrangian latitudes θ_i is performed using the distance s along the curve, which is normalized to the unit sphere, as an independent variable. The incremental distance Δs_{ip} along the great-circle segment joining together two Lagrangian points $\left(\lambda_{ip}, \theta_{ip}\right)$ and $\left(\lambda_{i,p+1}, \theta_{i,p+1}\right)$ of the Θ_i-curve, obtained using the dot product of the two position vectors from the origin, is

$$\Delta\widetilde{s} = \cos^{-1}\left(x_{ip}\widetilde{x} + y_{ip}\widetilde{y} + z_{ip}\widetilde{z}\right). \tag{13.68}$$

Of the two possible roots for $\widetilde{\lambda}$ from (13.66), the correct value, satisfies

$$\Delta\widetilde{s} \leq \Delta s_{ip}. \tag{13.69}$$

Now that the nonuniform relative position of all of the points along the piecewise defined Lagrangian latitude Θ_i is known, the periodic interpolation to the upstream departure points $(\lambda_{ij}, \theta_{ij})$ proceeds in the manner from [132,134].

We now move on to consider many different forms and applications of the cascade interpolating to be able to conserve mass of the flux that is being advected in the Semi-Lagrangian Inherently Conserving and Efficient Scheme, or SLICE for short.

13.3 Semi-Lagrangian inherently conserving and efficient scheme (SLICE)

The summary that we present here is from Zerroukat, Wood, and Staniforth (2002), ZWS hereafter [212]. The model that they are applying a form of semi-Lagrangian advection to is associated with a set of transport problems. We start with the 1D form of the passive advective transport of a scalar quantity ρ as we saw before, that is governed, when there are no sources and sinks, by

$$\frac{\partial \rho}{\partial t} + \frac{\partial}{\partial x}(u\rho) = 0, \tag{13.70}$$

where $\rho(x,t)$ is the density, which is the amount of scalar per unit length, of the transported quantity, and $u(x,t)$ is the transporting velocity as before.

The next step is different from what we have seen for the advection problem, ZWS now integrate (13.70) with respect to x between two arbitrary moving boundaries $x_1 = x_1(x,t)$ and $x_2(x,t)$ such that

$$\frac{d}{dt}\int_{x_1(x,t)}^{x_2(x,t)} \rho dx - \left(\rho(x_2,t)\frac{dx_2}{dt} - \rho(x_1,t)\frac{dx_1}{dt}\right) + \left(\rho(x_2,t)u(x_2,t) - \rho(x_1,t)u(x_1,t)\right) = 0. \tag{13.71}$$

If the boundaries x_1 and x_2 are moving with the fluid, that is to say,

$$\frac{dx_1}{dt} = u(x_1,t), \qquad \frac{dx_2}{dt} = u(x_2,t), \tag{13.72}$$

then (13.71) simplifies to the integral form of what is referred to as the **tracer conservation equation**

$$\frac{dM(x_1,x_2,t)}{dt} \equiv \frac{d}{dt}\left(\int_{x_1(t)}^{x_2(t)} \rho(x,t)\,dx\right) = 0, \tag{13.73}$$

where (13.73) states that the mass $M(x_1,x_2,t)$ contained between any two boundaries $x_1(t)$ and $x_2(t)$ moving with the fluid is invariant in time, which is the same as saying that the **mass M is conserved**.

Before we progress, we briefly define what Eulerian and Lagrangian control volumes are, as this approach is a finite-volume-based algorithm:

Definition 13.2. An **Eulerian Control Volume (ECV)** is fixed in space, and fluid can freely pass through the volume's boundary.

Definition 13.3. A **Lagrangian Control Volume (LCV)** is attached to the fluid where the volume is freely carried along with the fluid, and no fluid passes through its boundary.

Now given that $x_1(t)$ and $x_2(t)$ in (13.73) are any two points traveling with the fluid, from the copy of Fig. 1 from [212] in Fig. 13.5, it is possible to consider that these moving boundaries instantaneously coincide at time t^{n+1} with the boundaries of an Eulerian CV, while their upstream positions $x_1(t^n)$ and $x_2(t^n)$ at time t^n form the left and right boundaries of the associated upstream Lagrangian CV. This can be understood as saying that the fluid contained in the Lagrangian segment $[x_{d1}, x_{2d}] \equiv \left[x_1(t^n), x_2(t^n)\right]$ is transported to the Eulerian segment $\left[x_1\left(t^{n+1}\right), x_2\left(t^{n+1}\right)\right]$.

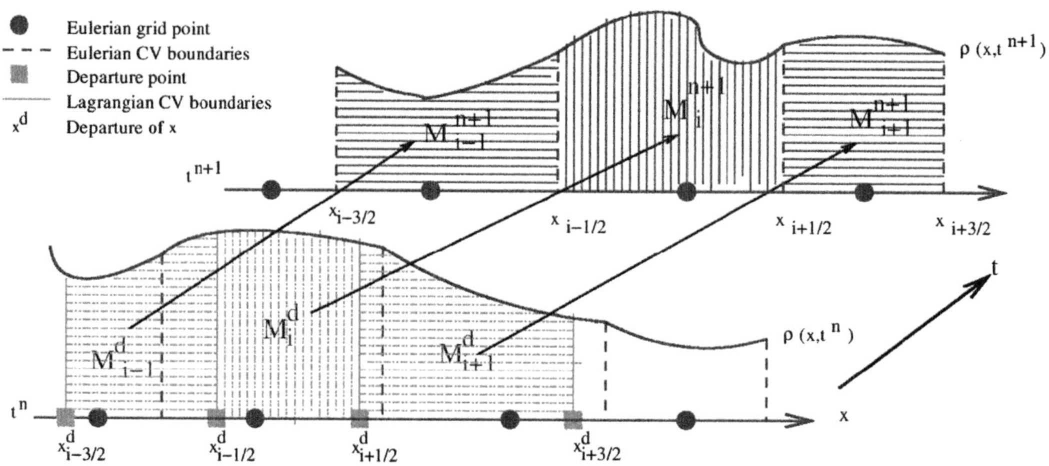

FIGURE 13.5 Copy of Fig. 1 from [212]. Zerroukat, M., Wood, N. and Staniforth, A., 2002: SLICE: A Semi-Lagrangian Inherently Conserving and Efficient scheme for transport problems, *Q. J. R. Meteor. Soc.*, **128**, 2801–2820. ©Royal Meteorological Society.

To discretize (13.73), we consider the case where we have a one-dimensional domain denoted by $\Omega = [x_{\min}, x_{\max}]$ and it is subdivided into N ECVs with an unequal spacing $\Delta x_i \equiv x_{i+\frac{1}{2}} - x_{i-\frac{1}{2}}$ for $i = 1, 2, \ldots, N$, where $x_{i-\frac{1}{2}}$ and $x_{i+\frac{1}{2}}$ are the left and right boundaries of ECV$_i$, respectively. Next there are N points associated with the grid point at the centers of ECV$_i$, namely $x_i \equiv \dfrac{x_{i+\frac{1}{2}} + x_{i-\frac{1}{2}}}{2}$, where the grid-box-averaged densities are stored, and which are defined at time t as

$$\overline{\rho_i}(t) \equiv \frac{1}{\Delta x_i} \int_{x_{i-\frac{1}{2}}}^{x_{i+\frac{1}{2}}} \rho(x,t)\,dx \equiv \frac{1}{\Delta x_i} M\left(x_{i-\frac{1}{2}}, x_{i+\frac{1}{2}}, t\right) \equiv \frac{M_i}{\Delta x_i}, \tag{13.74}$$

the time-discretization of (13.73) is written in a semi-Lagrangian form as

$$\overline{\rho}_i^{n+1} \equiv \overline{\rho}_i\left(t^{n+1}\right) \equiv \frac{M_i^{n+1}}{\Delta x_i} = \frac{M_{id}^n}{\Delta x_i}, \tag{13.75}$$

where

$$M_{id} \equiv \int_{x_{\left(i-\frac{1}{2}\right)d}}^{x_{\left(i+\frac{1}{2}\right)d}} \rho\,(x,t)\,dt, \tag{13.76}$$

with $x_{\left(i-\frac{1}{2}\right)d}$ and $x_{\left(i+\frac{1}{2}\right)d}$ being the left- and right-hand boundaries of the LCV_i at time t^n, respectively.

However, in general, the shape of $\rho\,(x,t^n)$ is not known a priori and therefore a piecewise representation that uses the given discrete grid-box-averaged values is usually adopted. In [212] the authors present a piecewise cubic method (PCM) which we shall briefly introduce here.

13.3.1 Piecewise cubic method

(i) *Definition of the cubic.* It is possible to construct a density profile over the whole domain using a series of local cubic polynomials. For any ECV_i, defined over the segment $\left[x_{i-\frac{1}{2}}, x_{i+\frac{1}{2}}\right]$, a cubic polynomial for density distribution at time t^n can be uniquely defined as

$$\rho_i\,(\xi) = a_i^{(0)} + a_i^{(1)}\xi + a_i^{(2)}\xi^2 + a_i^{(3)}\xi^3, \quad \xi \in [0,1], \tag{13.77}$$

where $\xi = \dfrac{x - x_{i-\frac{1}{2}}}{\Delta x_i}$ is a dimensionless local variable and the coefficients in the polynomial in (13.77) are such that the following four conditions are satisfied:

$$\rho_i\,(0) = \rho_i^L, \tag{13.78a}$$

$$\rho_i\,(1) = \rho_i^R, \tag{13.78b}$$

$$\int_0^1 \rho_i\,(\xi)\,d\xi = \overline{\rho_i}, \tag{13.78c}$$

$$\left.\frac{d\rho_i}{d\xi}\right|_{\xi=\frac{1}{2}} = \delta\rho_i, \tag{13.78d}$$

where ρ_i^L and ρ_i^R are local estimates of ρ at the left- and right-hand edges of ECV_i, and $\delta\rho_i$ is a scaled slope, $\Delta x_i \left.\dfrac{d\rho}{dx}\right|_{x_i}$, of the density distribution at its center. It is stated in [212] that the solutions to (13.78a)–(13.78d) are the following:

$$\begin{cases} a_i^{(0)} = \rho_i^L, \\ a_i^{(1)} = -6\rho_i^L + 6\overline{\rho}_i - 2\delta\rho_i, \\ a_i^{(2)} = 9\rho_i^L - 3\rho_i^R - 6\overline{\rho}_i + 6\delta\rho_i, \\ a_i^{(3)} = -4\rho_i^L + 4\rho_i^R - 4\delta\rho_i. \end{cases} \tag{13.79}$$

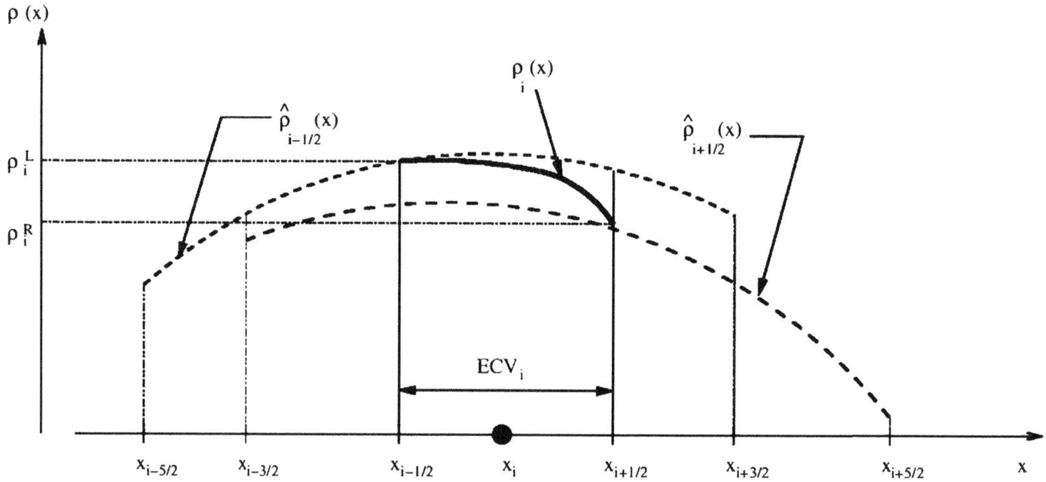

FIGURE 13.6 Copy of Fig. 2 from [212]. Zerroukat, M., Wood, N. and Staniforth, A., 2002: SLICE: A Semi-Lagrangian Inherently Conserving and Efficient scheme for transport problems, *Q. J. R. Meteor. Soc.*, **128**, 2801–2820. ©Royal Meteorological Society.

(ii) *Computation of the cubic parameters* ρ_i^L, ρ_i^R, *and* $\delta\rho$. The density values at the edges of the ECV_is can be computed through differentiating a locally defined **quartic cumulative mass function**, which is equivalent to evaluating a **cubic density function** $\widehat{\rho}_{i-\frac{1}{2}}$ at $x_{i-\frac{1}{2}}$, which is to say,

$$\rho_i^L = \widehat{\rho}_{i-\frac{1}{2}}\left(x_{i-\frac{1}{2}}\right). \tag{13.80}$$

We have a copy of Fig. 2 from [212] in Fig. 13.6, and the leftmost curve in that figure is centered about the point $x_{i-\frac{1}{2}}$. It is constructed to have specific masses $\Delta x_{i-2}\overline{\rho}_{i-2}$, $\Delta x_{i-1}\overline{\rho}_{i-1}$, $\Delta x_i\overline{\rho}_i$, and $\Delta x_{i+1}\overline{\rho}_{i+1}$ in the four immediately surrounding ECVs to satisfy the four conditions:

$$\left\{\begin{aligned}
&\int_{x_{i+\frac{1}{2}}}^{x_{i+\frac{3}{2}}} \widehat{\rho}_{i-\frac{1}{2}}(x)\,dx = \Delta x_{i+1}\overline{\rho}_{i+1},\\
&\int_{x_{i-\frac{1}{2}}}^{x_{i+\frac{1}{2}}} \widehat{\rho}_{i-\frac{1}{2}}(x)\,dx = \Delta x_{i}\overline{\rho}_{i},\\
&\int_{x_{i-\frac{3}{2}}}^{x_{i-\frac{1}{2}}} \widehat{\rho}_{i-\frac{1}{2}}(x)\,dx = \Delta x_{i-1}\overline{\rho}_{i-1},\\
&\int_{x_{i-\frac{5}{2}}}^{x_{i-\frac{3}{2}}} \widehat{\rho}_{i-\frac{1}{2}}(x)\,dx = \Delta x_{i-2}\overline{\rho}_{i-2}.
\end{aligned}\right. \tag{13.81}$$

To ensure continuity of $\rho(x)$ across control volume boundaries, ρ_i^R is obtained through

$$\rho_i^R = \rho_{i+1}^L = \widehat{\rho}_{i+\frac{1}{2}}\left(x_{i+\frac{1}{2}}\right), \tag{13.82}$$

where $\widehat{\rho}_{i+\frac{1}{2}}\left(x_{i+\frac{1}{2}}\right)$ satisfies (13.81) but with the index shift $i \to i+1$.

The slope $\left.\left(\frac{d\rho}{dx}\right)\right|_{x_i}$ at the center of ECV$_i$ is computed as the average of the two slopes $\left.\left(\frac{d\widehat{\rho}_{i-\frac{1}{2}}(x)}{dx}\right)\right|_{x_i}$ and $\left.\left(\frac{d\widehat{\rho}_{i+\frac{1}{2}}(x)}{dx}\right)\right|_{x_i}$, which yields

$$\delta\rho_i = \frac{\Delta x_i}{2}\left[\left.\left(\frac{d\widehat{\rho}_{i-\frac{1}{2}}(x)}{dx}\right)\right|_{x_i} + \left.\left(\frac{d\widehat{\rho}_{i+\frac{1}{2}}(x)}{dx}\right)\right|_{x_i}\right]. \tag{13.83}$$

(iii) *Computation of piecewise integrals.* This is concerned with the mass M_{id}^n contained in LCV$_i$ that extends over the segment $\left[x_{i-\frac{1}{2}d}, x_{i+\frac{1}{2}d}\right]$ at time t^n and is computed as

$$\begin{aligned} M_{id}^n &= \int_{x_{1-\frac{1}{2}d}}^{x_{1-\frac{1}{2}d}} \rho\left(x, t^n\right) dx, \\ &= \begin{cases} \Delta x_l \int_{\xi_{i-\frac{1}{2}d}}^{1} \rho_l^n(\xi)\, d\xi + \sum_{j=l+1}^{m-1} \Delta x_j \overline{\rho}_j^n + \Delta x_m \int_0^{\xi_{i+\frac{1}{2}d}} \rho_m^n(\xi)\, d\xi, & m \geq l+1, \\ \Delta x_l \int_{\xi_{i-\frac{1}{2}d}}^{\xi_{i+\frac{1}{2}d}} \rho_l^n(\xi)\, dx, & m = l. \end{cases} \end{aligned} \tag{13.84}$$

Here l and $m \geq l$ are the ECV indices associated with the segments in which $x_{i-\frac{1}{2}d}$ and $x_{i+\frac{1}{2}d}$ lie, that is to say, $x_{i-\frac{1}{2}d} \in \left[x_{l-\frac{1}{2}}, x_{l+\frac{1}{2}}\right]$ and $x_{i+\frac{1}{2}d} \in \left[x_{m-\frac{1}{2}}, x_{m+\frac{1}{2}}\right]$. The integrals on the right-hand side of (13.84) are evaluated analytically.

13.3.2 SLICE 2D

We start by assuming that the arrival points at time t^{n+1} collectively define an Eulerian (x, y) mesh that subdivides the entire computational domain into a finite number of contiguous and nonoverlapping regular 2D ECVs; similarity, the departure points at time t^n collectively define a Lagrangian (X, Y) mesh that subdivides the same domain into an equal number of contiguous an nonoverlapping irregular 2D LCVs.

Now let $\Omega = [x_{\min}, x_{\max}] \times \left[y_{\min}, y_{\max}\right]$ represent a 2D domain that subdivides into $N_x \times N_y$ ECVs with centers (x_i, y_i), $i = 1, 2, \ldots, N_x$, $j = 1, 2, \ldots, N_y$ and whose boundaries are the straight lines that join the four corner points $\left(x_{i\pm\frac{1}{2}}, y_{j\pm\frac{1}{2}}\right)$, $i = 1, 2, \ldots, N_x$, $j = 1, 2, \ldots, N_y$. The curvilinear Lagrangian coordinate system, which corresponds to the transform Eulerian (x, y) system by the action of the velocity field over the time interval $\left[t^n, t^{n+1}\right]$, is denoted by (X, Y). A y_j-line of the Eulerian grid is the horizontal line defined by the points

$\left\{\left(x_{i-\frac{1}{2}}, y_j\right), i = 1, 2, \ldots, N_x + 1\right\}$, and the Lagrangian Y_i is the curve defined by the set of points $\left\{\left(\widetilde{x}_{i,j-\frac{1}{2}}, \widetilde{y}_{i,j-\frac{1}{2}}\right), j = 1, 2, \ldots, N_y + 1\right\}$. These latter points are the center points of the bottom edges of $\text{LCV}_{i,j}$ along the Lagrangian $X_{j-\frac{1}{2}}$, which is equivalent to saying that

$$\left(\widetilde{x}_{i,j-\frac{1}{2}}, \widetilde{y}_{i,j-\frac{1}{2}}\right) = \frac{\left(x_{i-\frac{1}{2}d}, y_{j-\frac{1}{2}d}\right) + \left(x_{i+\frac{1}{2}d}, y_{j+\frac{1}{2}d}\right)}{2}.$$

To help understand what we have just presented, we have a copy of Fig. 3 from [212] in Fig. 13.7, where we can see the curvilinear Lagrangian curves overlaying the Eulerian grid underneath.

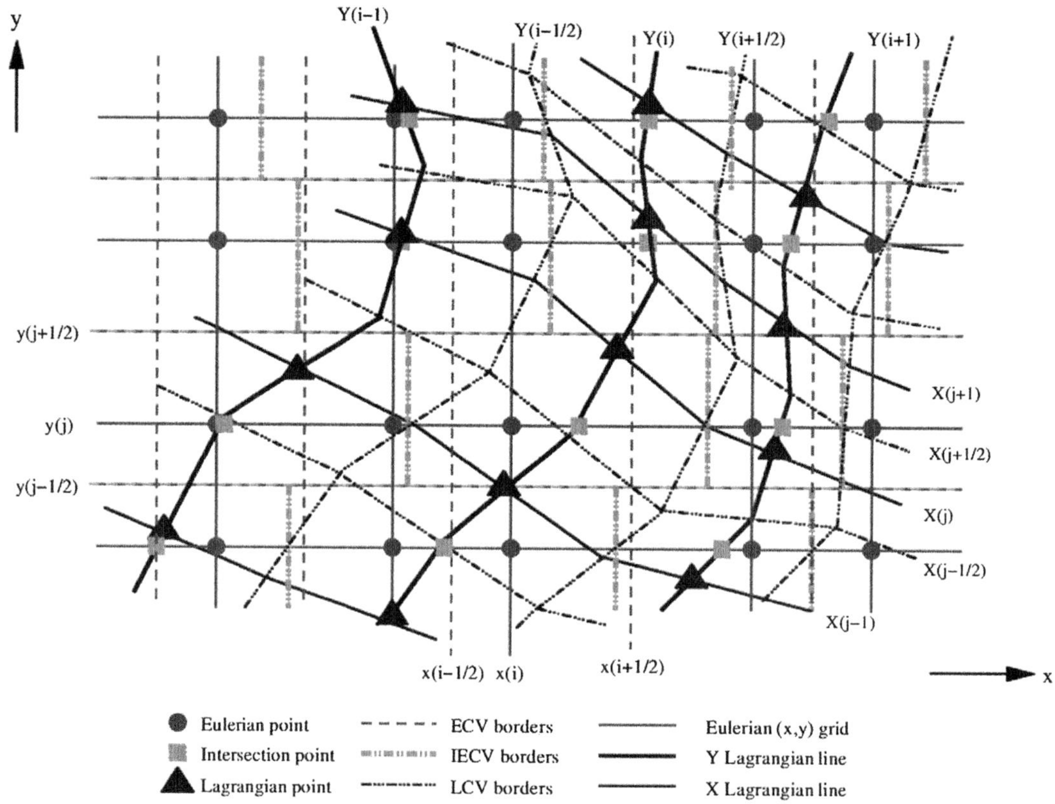

FIGURE 13.7 Copy of Fig. 3 from [212]. Zerroukat, M., Wood, N. and Staniforth, A., 2002: SLICE: A Semi-Lagrangian Inherently Conserving and Efficient scheme for transport problems, *Q. J. R. Meteor. Soc.*, **128**, 2801–2820. ©Royal Meteorological Society.

The technique that SLICE 2D here uses is to consider the problem of the evaluation of the time evolution of the density averaged over an Eulerian grid box, $\overline{\rho}(x, y, t^n)$, to compute $\overline{\rho}(x, y, t^{n+1})$. To achieve conservation, the scheme advectively transports the quantity conserved, mass here, then computes the grid-box-averaged density at the new time step from the mass that arrives at the ECV at time t^{n+1}. This is equivalent to computing the elementary masses of the LCVs, then translating them to their corresponding ECVs.

The first stage of the scheme is the computation of the masses $ME_{i,j}$ contained in $\text{ECV}_{i,j}$, associated with the Eulerian grid points (x_i, y_j), then their redistribution in a conservative way among the intersection Eulerian Control Volumes (IECV) using the SLICE 1D algorithm above, where the IECV are intermediate CVs associated with surrounding intersect points and aligned along the first cascade direction x, see Fig. 13.7. The second stage is to transfer the masses at $\text{IECV}_{i,j}$ to the Intersection Lagrangian Control Volume ILCV, $\text{ILCV}_{i,j}$, and then redistribute them conservatively to $\text{LCV}_{i,j}$ again using the SLICE 1D algorithm. We should note here that the ILCVs are intermediate CVs associated with surrounding intersection points but aligned along the second cascade direction Y.

There are several steps in this algorithm, and the reader is referred to [132] for more information about the cascade interpolation, and to [212] for the full derivation of the equations involved in this algorithm. However, we do here present the summary of the steps involved in this algorithm below.

1. Define the corner points of $\text{ECV}_{i,j}$ to be $\left(x_{i\pm\frac{1}{2}}, y_{j\pm\frac{1}{2}}\right)$, $i = 1, 2, \ldots, N_x$, $j = 1, 2, \ldots, N_y$.
2. Given $\overline{\rho}^n_{i,j} \equiv \overline{\rho}\left(x_i, y_j, t^n\right)$, compute the mass $ME^n_{i,j} = \overline{\rho}^n_{i,j} EX_{i,j}$ of $\text{ECV}_{i,j}$, where EA is the area of the ECV given by $EA_{i,j} = \left(x_{i+\frac{1}{2}} - x_{i-\frac{1}{2}}\right) \times \left(y_{j+\frac{1}{2}} - y_{j-\frac{1}{2}}\right)$.
3. Compute the coefficients of the cubic polynomial coefficients as with SLICE 1D for each $\text{ECV}_{i,j}$.
4. Locate the corners of $\text{LCV}_{i,j}$, which are the departure points for the corners of $\text{ECV}_{i,j}$, namely $\left(x_{i\pm\frac{1}{2}d}, y_{j\pm\frac{1}{2}d}\right)$.
5. Locate intersection points, see [212] for details.
6. Locate the $\text{IECV}_{i,j}$ boundaries $\left[x'_{i-\frac{1}{2},j}, x'_{i+\frac{1}{2},j}\right]$.
7. Compute the mass $\text{MI}^n_{i,j}$ of each $\text{IECV}_{i,j}$ using SLICE 1D.
8. Set the mass of each $\text{ILCV}_{i,j}$ equal to $\text{MI}^n_{i,j}$.
9. Compute the distances $s_{i,j}$ that define the boundaries of $\text{LCV}_{i,j}$ along Y_i.
10. Compute the distances $s'_{i,j}$ that define the boundaries of $\text{ILCV}_{i,j}$ along Y_i.
11. Compute the cubic coefficients for each $\text{ILCV}_{i,j}$.
12. Compute the mass $\text{ML}^n_{i,j}$ of each $\text{LCV}_{i,j}$ using SLICE-1D.
13. Transport $\text{ML}^n_{i,j}$ to its corresponding $\text{ECV}_{i,j}$ and compute $\overline{\rho}^{n+1}_{i,j} = \dfrac{ML^n_{i,j}}{EA_{i,j}}$.

We should note here that in future developments of SLICE in 1D and 2D, ZWS introduced a parabolic spline method (PSM) to replace what we have presented here, which they now refer to as the Piecewise Cubic Method (PCM) with an interpolation based upon the parabolic spline method (PSM) [214]. In [215] they implement the PSM into a two-dimensional problem both in Cartesian and spherical coordinates. We shall introduce SLICE-S in the next subsection. The reason for introducing the PSM approach is that of all piecewise parabolic functions that satisfy a given mass (average density) distribution, such as the one used by PPM, PSM is an optimal reconstruction since it possesses the minimum norm (or curvature) and best approximation properties [214]. Furthermore, an operation count shows that PSM is 60% more efficient than PPM, and its monotonic filter damps less than PPM's.

The PSM is formulated as follows: Let $\rho_i(x)$ be a local, specifically to ECV$_i$, parabola defined on the interval $\left[x_{i-\frac{1}{2}}, x_{i+\frac{1}{2}}\right]$; $\rho_i(x)$ may be uniquely defined in terms of two meshpoint values $\rho_{i-\frac{1}{2}} \equiv \rho_i\left(x_{i-\frac{1}{2}}\right)$, $\rho_{i+\frac{1}{2}} \equiv \rho_i\left(x_{i+\frac{1}{2}}\right)$ and the interval-averaged value

$$\overline{\rho}_i \equiv \int_{x-\frac{1}{2}}^{x+\frac{1}{2}} \rho_i(x)\,dx. \tag{13.85}$$

Thus

$$\rho_i(\zeta) = a_i^{(0)} + a_i^{(1)}\zeta + a_i^{(2)}\zeta^2, \quad \xi \in [0,1], \tag{13.86}$$

where $\zeta \equiv \dfrac{x - x_{i-\frac{1}{2}}}{\Delta x_i}$ is a dimensionless local coordinate and $\left(a_i^{(0)}, a_i^{(1)}, a_i^{(2)}\right)$ are the coefficients defined such that

$$a_i^{(0)} = \rho_{i-\frac{1}{2}}, \tag{13.87}$$

$$a_i^{(1)} = -4\rho_{i-\frac{1}{2}} - 2\rho_{i+\frac{1}{2}} + 6\overline{\rho}_i, \tag{13.88}$$

$$a_i^{(2)} = 3\rho_{i-\frac{1}{2}} + 3\rho_{i+\frac{1}{2}} - 6\overline{\rho}_i, \tag{13.89}$$

$$\overline{\rho}_i = \int_0^1 \rho_i(\zeta)\,d\zeta. \tag{13.90}$$

By construction, $\rho_i(x)$ as a piecewise parabola defined by (13.86) for $i = 1, 2, \ldots$ is continuous at all the meshpoints. However, only the interval-averaged quantities $\overline{\rho}_i$ are known, the meshpoint values are not. If we now impose the condition that its first derivatives are also continuous at these meshpoints, then this condition defines a parabolic spline, and yields the following equation:

$$\frac{1}{\Delta x_i}\rho_{i-\frac{1}{2}} + 2\left(\frac{1}{\Delta x_i} + \frac{1}{\Delta x_{i+1}}\right)\rho_{i+\frac{1}{2}} + \frac{1}{\Delta x_{i+1}}\rho_{i+\frac{3}{2}} = 3\left(\frac{1}{\Delta x_i}\overline{\rho}_i + \frac{1}{\Delta x}\overline{\rho}_{i+1}\right). \tag{13.91}$$

The reader is referred to both [214] and [215] for more details on how this approach performs with standard one- and two-dimensional test cases. We now move on to consider the derivation of SLICE on a two-dimensional spherical surface.

13.3.3 SLICE-S: SLICE on the sphere

As one can imagine, the name that was given to the 2D spherical version of SLICE by Zerroukat, Wood, and Staniforth was SLICE-S [213], where the motivation comes from the particular issue with conserving semi-Lagrangian schemes in their application to spherical geometry which is associated with considerable distortion of the control volumes close to the polar singularities. The scheme, SLICE-S, remaps the whole spherical field in a general approach using multiple sweeps of a 1D conservative remapping algorithm, and without any restriction on the meridional or zonal Courant numbers.

We start by considering passive 2D advective transport on a unit sphere of a scalar quantity ρ governed, in the absence of sources and sinks, by

$$\frac{\partial \rho}{\partial t} + \nabla \cdot (\rho \mathbf{u}) = 0, \tag{13.92}$$

where ρ is again the density of the transported quantity. Now let δV be a material volume. As it is 2D, it is really an area that moves with the field, and then an equivalent integral form of the conservation equation is

$$\frac{D}{Dt}\left(\int_{\delta v} \rho d V\right) = 0. \tag{13.93}$$

Integrating from time t_0 to t_1 yields

$$M_1 = M_0, \tag{13.94}$$

where

$$M_1 \equiv \int_{\delta V_1} \rho d V, \quad M_0 \equiv \int_{\delta V_0} \rho d V,$$

and δV_1 is the fluid volume at time t_1 that corresponds to the volume δv_0 at time t_0.

Now we let δV_1 be a known grid volume, fixed in time, with $\overline{\rho}_1$ being the associated value of the scalar averaged over this volume, so that $M_1 = \rho_1 \delta V_1$. Referring back to the Cartesian cases, δV_1 is then an Eulerian spherical volume (Eulerian Control Volume), whereas δV_0 is its corresponding Lagrangian volume (Lagrangian Control Volume). The problem simplifies to computing the discrete integral M_0, which is a remapping of a given average field $\overline{\rho}$ at time t_0 on a regular Eulerian grid to an irregular Lagrangian grid.

Let the surface of the sphere be subdivided into $N_\lambda \times N_\theta$ Eulerian Control Volumes denoted as before by $\mathrm{ECV}_{i,j}$, $i = 1, \ldots, N_\lambda$, $j = 1, \ldots, N_\theta$. Let the center of $\mathrm{ECV}_{i,j}$ have spherical polar coordinates $(\lambda_i, \theta_j) = \frac{1}{2}\left[\left(\lambda_{i-\frac{1}{2}}, \theta_{j-\frac{1}{2}}\right) + \left(\lambda_{i+\frac{1}{2}}, \theta_{j+\frac{1}{2}}\right)\right]$. Each $\mathrm{ECV}_{i,j}$ is delimited by the two lines of longitude $\lambda_{i\pm\frac{1}{2}}$ and by the two lines of latitude $\theta_{j\pm\frac{1}{2}}$, and has an Eulerian area $\mathrm{EA}_{i,j} = \left(\lambda_{i+\frac{1}{2}} - \lambda_{i-\frac{1}{2}}\right)\left(\sin\theta_{j+\frac{1}{2}} - \sin\theta_{j-\frac{1}{2}}\right)$. Thus the South and North poles correspond to the latitudes $\theta_{\frac{1}{2}}$ and $\theta_{N_\theta+\frac{1}{2}}$, respectively, and, by periodicity, $\lambda_{N_\lambda+\frac{1}{2}} \equiv \lambda_{\frac{1}{2}}$. The set of all corner points $\left(\lambda_{\pm\frac{1}{2}}, \theta_{j\pm\frac{1}{2}}\right)$ of these Eulerian control volumes forms the Eulerian latitude–longitude grid of the model. We have a copy of the cascade grids for a random part of the sphere and at the North pole in Fig. 13.8, which comes from [213].

We will not provide the full derivation here as there are a lot of indices involved, but we do provide the SLICE-S algorithm 15 steps that come from [213]:

1. Define the corner points of $ECV_{i,j}$ to be $\left(\lambda_{i\pm\frac{1}{2}}, \theta_{j\pm\frac{1}{2}}\right)$, $i = 1, \ldots, N_\lambda$, $j = 1, \ldots, N_\theta$.
2. Given $\overline{\rho}^n_{i,j} \equiv \rho\left(\lambda_i \theta_j, t^n\right)$, compute the mass $ME^n_{i,j} = \overline{\rho}^n_{i,j} \mathrm{EA}_{i,j}$ of $ECV_{i,j}$.

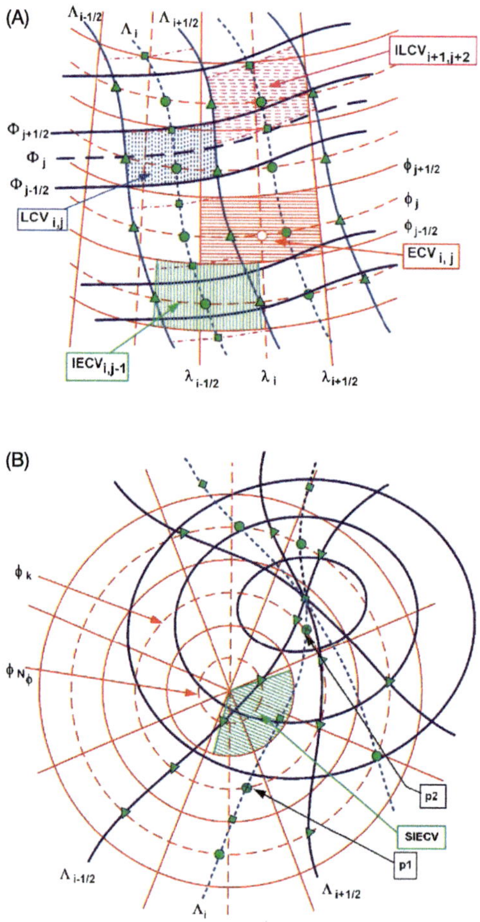

FIGURE 13.8 Copy of Fig. 1 from [213]: Superposition of Lagrangian and Eulerian grids (A) away from the poles and (B) in the vicinity of the North pole: Eulerian $\theta_{j-\frac{1}{2}}$ =red solid arc, Eulerian θ_j =red dashed arc, Eulerian $\lambda_{i-\frac{1}{2}}$ =red solid line, Eulerian λ_i =red dashed line, Lagrangian $\Phi_{j-\frac{1}{2}}$ =blue thick solid quasihorizontal curve, Lagrangian Φ_j =blue thick dashed quasihorizontal curve (only one shown), Lagrangian $\Lambda_{i-\frac{1}{2}}$ =blue thick solid quasivertical curve, Lagrangian Λ_i =blue thick dashed quasivertical curve, intersections of $\lambda_{i-\frac{1}{2}}$ with θ_j are shown by green filled triangles, intersections of λ_i with θ_j by green filled circles, boundaries of ILCVs along Λ_i by green filled squares, and the white open circle denotes the center of $\text{ECV}_{i,j}$. Zerroukat, M., Wood, N. and Staniforth, A., 2004: SLICE: A Semi-Lagrangian Inherently Conserving and Efficient scheme for transport problems on the sphere, *Q. J. R. Meteor. Soc.*, **130**, 2649–2664. ©Royal Meteorological Society.

3. Compute the piecewise coefficients for the first remapping of the SLICE-1D algorithm from earlier.
4. Locate the corners of $LCV_{i,j}$ – these are the departure points for the corners of $ECV_{i,j}$, that is, $\left(\lambda^d_{i\pm\frac{1}{2},j\pm\frac{1}{2}}, \theta^d_{i\pm\frac{1}{2},j\pm\frac{1}{2}}\right)$.

5. Locate intersection points $\left(\lambda'_{i-\frac{1}{2},j},\theta_j\right)$ and $\left(\lambda'_{i,j}\theta_j\right)$.
6. Locate the $IECV_{i,j}$ boundaries $\left(\lambda'_{i-\frac{1}{2},j},\lambda'_{i+\frac{1}{2},j}\right)$ with respect to ECV boundaries.
7. Compute the mass $MI^n_{i,j}$ of each $IECV_{i,j}$ using SLICE-1D.
8. Set the mass $MIL^n_{i,j}$ of each $ILCV_{i,j}$ equal to $MI^n_{i,j}$. There are some special case to this step, and the reader is referred to Sect. 2(a) of [213] for details.
9. Compute the distances $s_{i,j}$ that define the boundaries of $LCV_{i,j}$ along λ_i.
10. Compute the distances $s'_{i,j}$ that define the boundaries of $ILCV_{i,j}$ along Λ_i.
11. Compute the piecewise coefficients for the second remapping of the SLICE-1D algorithm from earlier.
12. Compute the mass $ML^n_{i,j}$ of each $LCV_{i,j}$ using SLICE-1D.
13. Transport $ML^n_{i,j}$ to its corresponding $ECV_{i,j}$ and compute $\overline{\rho}^*_{i,j} \equiv \dfrac{ML^n_{i,j}}{EA_{i,j}}$.
14. Compute the corrected density, $\overline{\rho}^{cc}_{i,j}$, in the extended polar cap regions, as a function of $\overline{\rho}^*_{i,j}$, where this is explained in Sect. 2(b) of [213].
15. Set $\overline{\rho}^{n+1}_{i,j} = \overline{\rho}^{cc}_{i,j}$ in the extended polar cap regions, and $\rho^n_{i,j} = \overline{\rho}_{i,j}$ elsewhere.

In [213] the authors state that noting the impact of spherical geometry on the above definitions, and with the exception of the calculation of special $IECV$s and a correction for the polar cap region, SLICE-S is identical to SLICE under the transformation from spherical coordinates to the Cartesian coordinates of [212]. In particular, the first cascade remapping of SLICE-1D shown earlier in this chapter is along the Eulerian θ_j-directions with the second remapping along the Lagrangian Λ_i-directions. It is also then mentioned in [214] that for clarity there is a change of notation: in the Cartesian scheme earlier, a Lagrangian Y_i denotes the set of departure points of an Eulerian line of constant x_i, but if the same convention were followed here, the equivalent of Y_i would be Φ_i as we showed earlier from [134]. However, in this paper the convention is changed to the more natural one, whereby Λ_i refers to the set of departure points associated with an Eulerian line of constant λ_i. Changing the order of the cascade remapping, however, is equally valid and has similar, albeit slightly different, challenges associated with polar regions. The intersection points between the $\Lambda_{i-\frac{1}{2}}$ and θ_j, and Λ_i and θ_j, are computed as detailed in Sect. 2(b) of Nair et al. (1999) [134]. As noted above, SLICE-S is identical to SLICE in all aspects except two, which we shall briefly outline here; the reader is referred to [213] for the exact details.

Special $IECV$ ($SIECV$s)

We let $SIECV_{i,j}$ be a special $IECV$ where the left boundary $\Lambda_{i-\frac{1}{2}}$ crosses θ_j, but neither Λ_i nor $\Lambda_{i+\frac{1}{2}}$ crosses θ_j; see Fig. 13.8 for details. The mass $MI_{i,j}$ of $SIECV_{i,j}$ is then simply allocated to the two nearest Lagrangian mass centers (p_1, p_2) (see Fig. 1(B) in Fig. 13.8) with a pair of weights (α_1, α_2), where $\alpha_1 + \alpha_2 = 1$. Hence, to each mass MIL_{pk} for $k = 1$ or 2, of the $ILCV$s centered at p_1 and p_2 is added a fraction α_k of the mass $MI_{i,j}$, that is to say, $MIL_{pk} \to MIL_{pk} + \alpha_k MI_{i,j}$, for $k = 1, 2$. ZWS state now that it is important to note that, for this aspect of the scheme, the main constraint is conservation (i.e., all the elementary masses have to be transferred to the second stage of cascade) and not accuracy. Accuracy aspects are dealt with by the correction stage detailed below. Therefore, the simple choice $\alpha_1 = \alpha_2 = \frac{1}{2}$ was found to be sufficient for this purpose in [213]. However, more complicated choices for the weights which might, for example, take into account relative distances, are of course possible and may conceivably lead to further improvements.

Correction of the polar cap region

As we can see from Fig. 13.8, the high distortion of the Lagrangian curves near the poles can lead to singular behavior in the grid intersections. However, it also means that the assumption (step 8 of the SLICE-S algorithm above), in which SLICE remaps mass from 2D CVs using two sweeps of 1D remapping along two cascade directions, can become nonnegligibly inaccurate because the first cascade direction, θ, is almost parallel to the second cascade direction, Λ. This distorts the distribution of mass along the cascade directions which, together with the conservative nature of the scheme and associated cumulative effects, can result in a distorted distribution around the poles [213].

To address this problem, ZWS start by letting k, which is less than or equal to N_θ, be the northern polar singular belt, that is to say, the latitude strip within which the Lagrangian North pole $\theta^d_{\forall i, N_\theta}$ is situated, $\theta_{i,k-\frac{1}{2}} \leq \theta^d_{\forall i, N_\theta} \leq \theta_{i,k+\frac{1}{2}}$, this is highlighted in Fig. 1(B) in Fig. 13.8. ZWS now define an *extended north polar cap* as the $\eta = N_\theta - k + 2$ latitudes surrounding the North pole. The extra latitude arises because the piecewise representation requires neighboring information. Thus the effect of inaccuracy in the subpolar cap of $N_\theta - k + 1$ latitudes has an impact on the neighboring $k+1$ latitude. ZWS now introduce a similar approach to that of Nair and Machenhauer in [135] which is used here to correct the density field for the η latitudes of the extended North polar cap while maintaining mass conservation, specifically

$$\overline{\rho}^{n+1}_{i,j} = \begin{cases} \overline{\rho}^{cc}_{i,j} = \dfrac{\sum\limits_{l=k-1}^{N_\theta}\sum\limits_{m=1}^{N_\lambda} \overline{\rho}^{*}_{l,m} EA_{l,m}}{\sum\limits_{l=k-1}^{N_\theta}\sum\limits_{m=1}^{N_\lambda} \rho^{SSL}_{l,m} EA_{l,m}} \rho^{SSL}_{i,j}, & i = 1, 2, \ldots, N_\lambda,\ j = k-1, k, \ldots, N_\theta, \\ \overline{\rho}^{*}_{i,j}, & \text{otherwise,} \end{cases} \tag{13.95}$$

where $\overline{\rho}^{SSL}_{i,j}$ is a the average density computed using a bicubic Lagrange nonconservative semi-Lagrangian scheme and $\overline{\rho}^{*}_{i,j}$ is the uncorrected density field obtained at step 13 of the SLICE-S algorithm. A similar correction formula is applied at the South pole.

In [213] ZWS go on to say that it is possible to use this correction scheme additionally after the first cascade remapping and that this approach would give a more accurate distribution for the mass of the $ILCV$s in the extended polar cap but at the cost of increased algorithmic complexity. However, the approach adopted here is cheap, simple, purely post-processing in nature, and improves the accuracy in the extended polar caps, while, by construction, retaining mass conservation. The scheme additionally has the desirable property that for zero advecting velocity, the exact result is trivially obtained. This is because for zero displacement, $ECV_{i,j}$, $IECV_{i,j}$, $ILCV_{i,j}$, and $LCV_{i,j}$ are all identical, and the remapping is also exact in both cascade directions. The piecewise density function is constructed in such a way that its 1D integral from one boundary to another of $ECV_{i,j}$ $(\equiv IECV_{i,j} \equiv ILCV_{i,j} \equiv LCV_{i,j})$ is exactly $\overline{\rho}_{i,j}$ as we saw earlier in this section.

13.3.4 C-SLICE

In [217] ZWS state that a C-grid staggering of mass and momentum variables is often used in atmospheric models to represent horizontal divergence accurately, and hence represent gravity-wave dispersion well. However, although the SLICE scheme intrinsically conserves mass, it is based effectively on a B-grid staggering of mass and momentum variables.

The original SLICE scheme (denoted B-SLICE) uses the corners of control volumes as the basis of a cascade remapping for both directions: this is similar to what would be achieved by a B-grid discretization of the continuity equation in the shallow water equations. C-SLICE instead uses the departure points associated with u-points as the basis of remapping in the first cascade direction, and those of v-points for the second cascade remapping. This has the advantage of less averaging, and a good correspondence with the computation of divergence on a C grid.

We have copies of Figs. 1 and 2 from [217] in Figs. 13.9–13.10, which highlight the difference between B-SLICE and C-SLICE. The first difference is associated with the departure points where we let $X_{i-\frac{1}{2}}$ be a curved Lagrangian line that connects the set of generic discrete departure points $\left\{P_{i-\frac{1}{2},j}, j=1,\ldots,N_y\right\}$ denoted by squares in Fig. 13.9 and 13.10. Depending upon the context, $P_{i-\frac{1}{2},j}$ is either a departure point associated with the corner of an ECV, Fig. 13.9, or a departure point associated with the center of the face of an ECV in the x-direction, located at a u-point, Fig. 13.10. Specifically, in B-SLICE, $P_{i-\frac{1}{2},j}$ is the departure point associated with a corner of an ECV, whereas in C-SLICE, it is the departure point associated with the *center of the face* of an ECV in the x-direction.

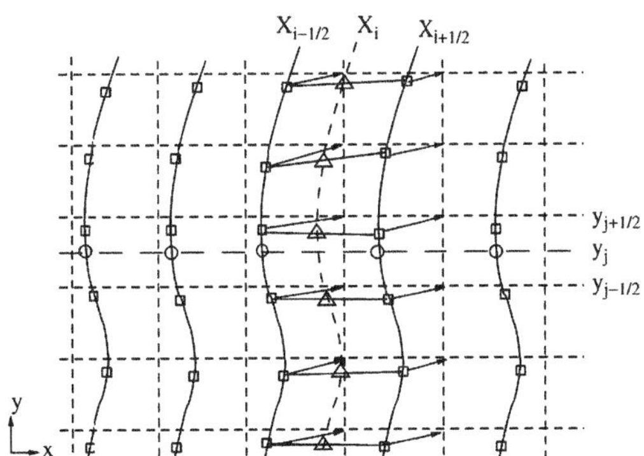

FIGURE 13.9 Superposition of Eulerian (x, y) and Lagrangian (X, Y) grids for B-SLICE. Boundaries of Eulerian control volumes (cells) are shown with horizontal and vertical dashed lines. Arrows symbolize the trajectory along which points/particles travel during a time step in a backward semi-Lagrangian context. Zerroukat, M., Wood, N. and Staniforth, A., 2009: An improved version of SLICE for conservative monotonic remapping on a C-grid, *Q. J. R. Meteor. Soc.*, **128**, 541–546. ©Royal Meteorological Society.

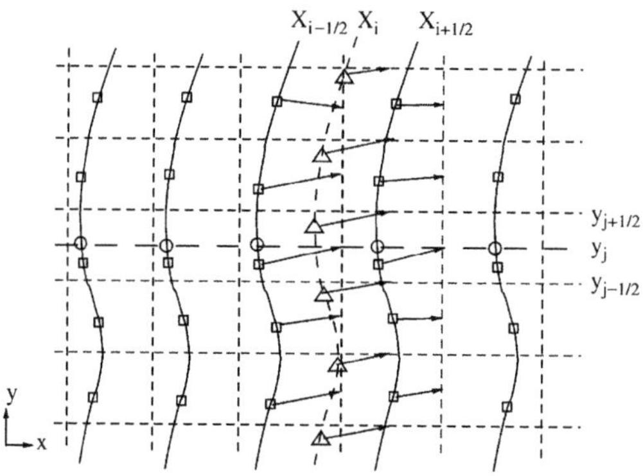

FIGURE 13.10 As Fig. 13.9 but for C-SLICE. Zerroukat, M., Wood, N. and Staniforth, A., 2009: An improved version of SLICE for conservative monotonic remapping on a C-grid, *Q. J. R. Meteor. Soc.*, **128**, 541–546. ©Royal Meteorological Society.

We now summarize the advantages of using the C-grid over the B-grid as set out in [217].

C-grid departure points

In [217] the authors provide the following description of the advantage of using the C-grid for the departure points in C-SLICE. Thus we now let X_i be a curved Lagrangian line shown as the quasivertical dashed line in Figs. 13.9 and 13.10, connecting the generic discrete departure points $\left\{P_{i,j-\frac{1}{2}}, j=1,\ldots,N_y+1\right\}$ which are at the triangles in the figures above. The departure points $P_{i,j-\frac{1}{2}}$ define the boundaries of the LCV in the y-direction. In B-SLICE the points $P_{i,j-\frac{1}{2}}$ are constructed from the $P_{i-\frac{1}{2},j}$ points, where the coordinates of $P_{i,j-\frac{1}{2}}$ computed as the average of the coordinates of the Lagrangian corner points $P_{i-\frac{1}{2},j}$, and $P_{i+\frac{1}{2},j}$, that is to say, the triangles are obtained by averaging the squares [212]; whereas for C-SLICE, $P_{i,j-\frac{1}{2}}$ are computed directly as the departure points associated with v-points $\left(i, j-\frac{1}{2}\right)$ (and independently from the computation of u-departure points).

The u and v points, which are denoted by squares and triangles in Figs. 13.9 and 13.10, are independently evaluated using appropriate staggered wind components, rather than averaged from other departure points.

Simplified computation of intersections

Following on, if we now let $\widetilde{x}_{i-\frac{1}{2},j}$ be the x-coordinate of the intersection of the Lagrangian line $X_{i-\frac{1}{2}}$ with the Eulerian straight line $y=y_j$, which are the circles in the provided schematics. In Cartesian geometry, $\widetilde{x}_{i-\frac{1}{2},j}$ is computed as the intersection of the straight line $y=y_j$ with the straight line that joins two points of $X_{i-\frac{1}{2}}$ located on either side of y_j, that it to say, the points $P_{i-\frac{1}{2},m}$ and $P_{i-\frac{1}{2},m+1}$, where m is such that $y^d_{i-\frac{1}{2},m} \le y_j \le y^d_{i-\frac{1}{2},m+1}$. In Cartesian geometry, this is a simple operation with a unique solution. However, for spherical geometry this is not the case [213].

In spherical geometry the points $P_{i-\frac{1}{2},m}$ and $P_{i-\frac{1}{2},m+1}$ are connected to one another by the arc of a unique great circle, which ZWS denote by $GC\left(P_{i-\frac{1}{2},m}, P_{i-\frac{1}{2},m+1}\right)$. Furthermore, $y = y_j$ is a latitude circle in the longitude–latitude coordinate system. The intersection of $GC\left(P_{i-\frac{1}{2},m}, P_{i-\frac{1}{2},m+1}\right)$ with $y = y_j$ then gives rise to a quadratic equation involving trigonometric functions and, in general, having two distinct solutions which we saw earlier [134]. This requires careful selection of the appropriate solution, adding to the cost of computing intersections. ZWS state that in C-SLICE they found that there is no significant loss of accuracy if the spherical aspects are neglected for the computation of $\widetilde{x}_{i-\frac{1}{2},j}$. ZWS say that this is consistent with the findings for cascade interpolation in [134].

Once the set of intersection points $\left\{\widetilde{x}_{i-\frac{1}{2},j}, i = 1, \ldots, N_x + 1\right\}$ has been computed, the first cascade remapping is a series of 1D-SLICE remappings, which remaps mass from the original Eulerian CVs

$$ECV_{i,j} = \left\{(x, y), x_{i-\frac{1}{2}} \le x \le x_{i+\frac{1}{2}}, y_{j-\frac{1}{2}} \le y \le y_{j+\frac{1}{2}}\right\},$$

to intermediate Eulerian CVs

$$IECV_{i,j} = \left\{(x, y), \widetilde{x}_{i-\frac{1}{2},j} \le x \le \widetilde{x}_{i+\frac{1}{2},j}, y_{j-\frac{1}{2}} \le y \le y_{j+\frac{1}{2}}\right\}.$$

Note that the boundaries of IECVs in their direction are the same as those of ECVs. An important feature to note here is that this first cascade remapping is the same for both B-SLICE and C-SLICE.

Simplified computation of distances

For the second cascade remapping, B-SLICE computes a cumulative physical distance $D_{i,j+\frac{1}{2}}$ along X_i, from a reference point $P_{i,0}$ to the point $P_{i,j+\frac{1}{2}}$. This is the sum of all the distances that connect intermediate points between $P_{i,0}$ and $P_{i,j+\frac{1}{2}}$, and so $D_{i,j+\frac{1}{2}} \equiv \sum_{k=0}^{j} f\left(P_{i,k-\frac{1}{2}}, P_{i,k+\frac{1}{2}}\right)$ (with $P_{i,-\frac{1}{2}} \equiv P_{i,0}$). In Cartesian geometry, $f\left(P_{i,k-\frac{1}{2}}, P_{i,k+\frac{1}{2}}\right) \equiv \left\|P_{i,k-\frac{1}{2}} - P_{i,k+\frac{1}{2}}\right\|$ is the Euclidean distance between the two points $P_{i,k-\frac{1}{2}}$ and $P_{i,k+\frac{1}{2}}$, and in spherical geometry $f\left(P_{i,k-\frac{1}{2}}, P_{i,k+\frac{1}{2}}\right) \equiv GC\left(P_{i,k-\frac{1}{2}}, P_{i,k+\frac{1}{2}}\right)$ is the distance along the great-circle arc between the two points. As a consequence, $D_{i,j+\frac{1}{2}}$ is both flow-dependent and (in general) time-dependent, and the set of distances $D_{i,j+\frac{1}{2}}$ has to be recomputed at every timestep; $D_{i,j+\frac{1}{2}}$ also determines the boundaries of $ILCV$s [213]. These boundaries are then used in the reconstruction part (i.e., finding the polynomial coefficients used in the remapping).

With C-SLICE, ZWS state that they had found that there is insignificant loss of accuracy if the second remapping is done along X_i, as originally, but using a simplified measure of distance (i.e., $DILCV_{i,j+\frac{1}{2}} = DILCV_{j+\frac{1}{2}} = \left\|y_{j+\frac{1}{2}} - y_0\right\|$ and $DLCV_{i,j+\frac{1}{2}} = \left\|y^d_{i,j+\frac{1}{2}} - y_0\right\|$, where $y_{j+\frac{1}{2}}$ are the boundaries in the y-direction of the original fixed ECVs, $y^d_{i,j+\frac{1}{2}}$ is the y-component of a departure point that arrives at a v-point $\left(i, j+\frac{1}{2}\right)$, and a reference $y_0 \leq \min\left(y_{j+\frac{1}{2}}, y^d_{i,j+\frac{1}{2}}\right)$ $\forall i, j$. Note that the flow-dependent nature of SLICE remains, because the cascade remapping is still along the Lagrangian lines X_i, which are flow-dependent; it is only the measure of distance between the centers of volumes along X_i that is redefined. This has the important advantage that the reconstruction part of both cascade remappings are time-independent, and can thus be precomputed once and for all.

Further aspects for spherical geometry

As seen earlier when we presented SLICE-S, there was a problem at the poles and so we had complicated computations of intersections and mass distribution there. However, in the C-SLICE algorithm there is no special treatment of the CV near the poles, which were referred to as the $SIECV$. The rationale for these special volumes was that the polar cap correction requires only the total mass of the polar cap region and, therefore, the details of how the mass is distributed within it are irrelevant before the correction step. Hence, the details of that mass redistribution can be computationally simplified without changing the total mass of the polar cap. For the polar cap region in C-SLICE, the original departure points associated with u and v arrival points are modified; see Fig. 13.11 which is a copy of Fig. 4 from [217].

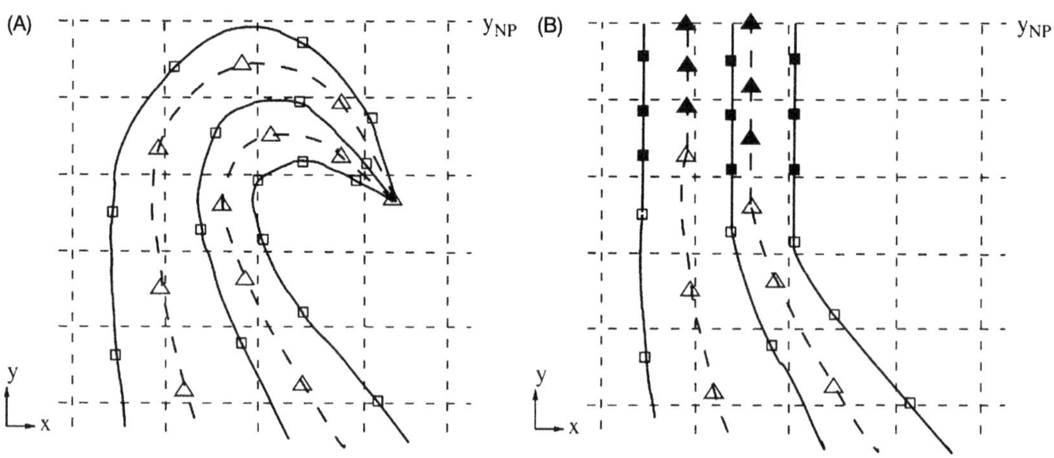

FIGURE 13.11 Illustrative schematic of the simplification of the Lagrangian grid inside the polar cap region (example with three points North-poleward modification). (A) Original departure points in (x, y) longitude – latitude grid with u-departure points (squares) and v-departure points (triangles); (B) three modified departure points are shown with u-departure points (filled squares) and v-departure points (filled triangles). Zerroukat, M., Wood, N. and Staniforth, A., 2009: An improved version of SLICE for conservative monotonic remapping on a C-grid, *Q. J. R. Meteor. Soc.*, **128**, 541–546. ©Royal Meteorological Society.

The departure points associated with the centers of ECVs (these are only needed for the polar cap regions) are left unchanged as these are used in the correction step. The departure points associated with u and v arrival points that are located poleward of the departure point associated with arrival at the North pole, which is to say, $y_{j>q} > y^d_{NP}$, where y^d_{NP} is the y-coordinate of the departure point of the North pole, and q is the latitude index for which no change is required, are modified as follows (with an analogous approach for the South polar region):

1. To preserve the monotonicity of $y^d_{i,j}$ with respect to index j, the relative distance of departure points from the North pole is made proportional to the relative distance of the corresponding arrival points from the North pole. The y-coordinate of a departure point is then modified as
$$y^d_{i,j>q} \to y_{NP} - \left(y_{NP} - y^d_{i,j}\right)\frac{\left(y_{NP} - y_j\right)}{\left(y_{NP} - y_q\right)},$$
where y_{NP} and y_j are the Eulerian y-coordinates of the North pole and the arrival point corresponding to a departure point $y^d_{i,j>q}$, respectively (for no motion, $y^d_{i,j} \to y_j$).
2. The x-coordinate of a departure point is modified as $x^d_{i,j} > q \to x^d_{i,q}$. This corresponds to placing all departure points $\left(x^d_{i,j>q}, y^d_{i,j>q}\right)$ poleward of $\left(x^d_{i,q}, y^d_{i,q}\right)$ on the meridian $\left(x^d_{i,q}, y^d_{i,q}\right)$. See Fig. 13.11 for an illustration of this.

Given the derivation of the spherical, as well as the C-grid, version of SLICE, we now move on to consider an implementation of the scheme with the spherical shallow water equations.

13.3.5 SLICE applied to global SISL SWE model

In [217] Zerroukat et al. introduce the SLICE formulation into a semi-implicit semi-Lagrangian (SISL) shallow water equations (SWE) model, and we present that implementation here now. We start with writing the SWE in a **mass-conservative Lagrangian form** as

$$\frac{D\mathbf{u}}{Dt} = -\Omega_r \times \mathbf{u} - \nabla\left(\Phi + \Phi^S\right) \equiv \Psi, \tag{13.96}$$
$$\frac{D}{Dt}\left(\int_{\delta A}\left(\Phi - \Phi^{\text{ref}}\right)dA\right) = -\Phi^{ref}\int_{\delta A}\nabla\cdot\mathbf{u}\,dA, \tag{13.97}$$

where $\mathbf{u}$ is the velocity vector, t is time, Ω_r is a radial vector with magnitude given by the radial component of the Earth's rotation vector Ω, Φ is the geopotential of the free surface relative to a rigid surface, Φ^S is the geopotential of the rigid surface relative to a mean sea level, and Φ^{ref} is a constant, representative, reference value for Φ, which is introduced as part of the SISL integration; δA is a finite material area that moves with the fluid, and where (13.97) is the integral form of the continuity equation

$$\frac{D}{Dt}\left(\Phi - \Phi^{\text{ref}}\right) + \Phi^{\text{ref}}\nabla\cdot\mathbf{u} = -\left(\Phi - \Phi^{\text{ref}}\right)\nabla\cdot\mathbf{u}. \tag{13.98}$$

As we are considering semi-Lagrangian methods in this book, we also have the trajectory, or kinematic equation, given by

$$\frac{D\mathbf{x}}{Dt} = \mathbf{u}, \tag{13.99}$$

where $\mathbf{x}$ is the position vector on the sphere relative to the center of the sphere.

The continuous equations (13.96)–(13.97) and (13.99) are discretized in a semi-implicit semi-Lagrangian approach as

$$(u-\alpha\Delta t\Psi)_a^{n+1} = (\mathbf{u}+\beta\Delta t\Psi)_d^n, \tag{13.100}$$

$$\left[\int_{\delta A}\left(\left(\Phi-\Phi^{\text{ref}}\right)+\alpha\Delta t\Phi^{\text{ref}}\nabla\cdot\mathbf{u}\right)dA\right]_a^{n+1} = \left[\int_{\delta A}\left(\left(\Phi-\Phi^{\text{ref}}\right)-\beta\Delta t\Phi^{\text{ref}}\nabla\cdot\mathbf{u}\right)dA\right]_d^n, \tag{13.101}$$

$$(\mathbf{x}-\beta_x\Delta t\mathbf{u})_d^n = (\mathbf{x}-\alpha_x\Delta t\mathbf{u})_a^{n+1}, \tag{13.102}$$

where α, α_x are time weights; $\beta\equiv 1-\alpha$ and $\beta_x\equiv 1-\alpha_x$, while a and d represent the arrival and departure points, respectively; when we are considering the integral, they are referring to a departure and arrival cell. To apply a centered second-order in time approach, we set the four parameters equal to 0.5.

If we now let (λ,θ) be the longitude and latitude spherical polar coordinates on a sphere of radius r, with velocity components

$$\mathbf{u}\equiv(u,v)^T\equiv\left(r\cos\theta\frac{D\lambda}{Dt},r\frac{D\theta}{Dt}\right)^T. \tag{13.103}$$

It is stated in [217] that the principal reason for not writing the momentum equation in component form in spherical geometry is the evaluation of a vector field at the departure point, that is to say, the right-hand side of (13.100). This is because in curvilinear geometries the unit base vectors at the departure point are not in general aligned with those at the arrival point. A full derivation of this can be found in [179] where the authors introduce a rotation matrix Λ whose elements depend on the unit basis vectors at both the departure and the arrival points. It is through applying this approach that we can rewrite (13.100) as

$$(\mathbf{u}-\alpha\Delta t\Psi)_a^{n+1}=\Lambda\left(\mathbf{u}+\beta\Delta t\Psi\right)_{dL}^n, \tag{13.104}$$

where the subscript dL denotes evaluation at the departure point in terms of the local basis vectors at the departure point, where Λ is defined as

$$\Lambda\equiv\begin{pmatrix}\Lambda_{11} & \Lambda_{12}\\ \Lambda_{21} & \Lambda_{22}\end{pmatrix},$$

where

$$\Lambda_{11}=\Lambda_{22}=\frac{Q_{11}+Q_{22}}{1+Q_{33}},\qquad \Lambda_{12}=\Lambda_{21}=\frac{Q_{12}-Q_{21}}{1+Q_{33}},$$

and

$$\begin{aligned}
Q_{11} &= \cos(\lambda_a-\lambda_d), \qquad Q_{12}=\sin\theta_d\sin(\lambda_a-\lambda_d),\\
Q_{21} &= -\sin\theta_a\sin(\lambda_a-\lambda_d), \qquad Q_{22}=\cos\theta_a\cos\theta_d+\sin\theta_a\sin\theta_d\cos(\lambda_a-\lambda_d),\\
Q_{33} &= \sin\theta_a\sin\theta_d+\cos\theta_a\cos\theta_d\cos(\lambda_a-\lambda_d).
\end{aligned}$$

Taking the components of (13.104) with respect to the unit vectors at the arrival point leads to

$$(u - \alpha \Delta t \Psi_u)_a^{n+1} = \Lambda_{11} (u + \beta \Delta t \Psi_u)_{dL}^n + \Lambda_{12} (v + \beta \Delta t \Psi_v)_{dL}^n, \tag{13.105}$$

$$(v - \alpha \Delta t \Psi_v)_a^{n+1} = \Lambda_{21} (u + \beta \Delta t \Psi_u)_{dL}^n + \Lambda_{22} (v + \beta \Delta t \Psi_v)_{dL}^n, \tag{13.106}$$

where

$$\Psi_u = -2 (\Omega_r \times \mathbf{u})_u - \frac{1}{r \cos\theta} \delta_\lambda \left(\Phi + \Phi^S \right), \tag{13.107}$$

$$\Psi_v = -2 (\Omega_r \times \mathbf{u})_v - \frac{1}{r} \delta_\theta \left(\Phi + \Phi^S \right), \tag{13.108}$$

where δ_λ and δ_θ represent the central differencing operators, defined here as

$$(\delta_\lambda F)_{i,j} \equiv \frac{F_{i+\frac{1}{2},j} - F_{i-\frac{1}{2},j}}{\Delta \lambda_i}, \qquad (\delta_\theta F)_{i,j} \equiv \frac{F_{i,j+\frac{1}{2}} - F_{i,j-\frac{1}{2}}}{\Delta \theta_j}.$$

The departure point (13.102) is solved iteratively using a local Cartesian transform approach which is derived in [199], where at each arrival point we define a local Cartesian coordinate system $Oxyz$, with its origin O at the center of the sphere and whose unit vectors $\mathbf{i}_c, \mathbf{j}_c, \mathbf{k}_c$ are parallel to the unite vectors $\mathbf{i}, \mathbf{j}, \mathbf{k}$ of the spherical coordinate systems there. If we now let U and V be the components of the wind in the local Cartesian coordinate system corresponding to the components (u, v) in the spherical system, then

$$\begin{pmatrix} U \\ V \end{pmatrix} = \Lambda \begin{pmatrix} u \\ v \end{pmatrix}.$$

The departure point (X_d, Y_d) in the local system at the kth iteration is computed as

$$X_d^{(k)} = -\gamma \Delta t \left(\alpha_x U_a^{n+1} + \beta_x U^n \left(\lambda_d^{(k-1)}, \theta_d^{(k-1)} \right) \right), \tag{13.109}$$

$$Y_d^{(k)} = -\gamma \Delta t \left(\alpha_x V_a^{n+1} + \beta_x V^n \left(\lambda_d^{(k-1)}, \theta_d^{(k-1)} \right) \right), \tag{13.110}$$

where $\gamma \equiv \frac{1 + Q_{33}}{2}$. It is stated in [217] that only two iterations are sufficient for a reasonable convergence, which is consistent with earlier results in [176]. The transform from (X_d, Y_d) to (λ_d, θ_d) is given by

$$(\lambda_d - \lambda_a) = \tan^{-1} \left(\frac{X_d}{Z_d \cos\theta_a - Y_d \sin\theta_a} \right), \tag{13.111}$$

$$\theta_d = \sin^{-1} \left(\frac{Y_d \cos\theta_a + Z_d \sin\theta_a}{\sqrt{X_d^2 + Y_d^2 + Z_d^2}} \right), \tag{13.112}$$

where $Z_d^2 = r^2 - X_d^2 - Y_d^2$.

Spatial discretization

The approach that is presented in [217] is based upon suggestions from [191] for the Arakawa C grid where v is placed at the poles in preference to u and Φ, and the discretization employs two-point differencing and averaging operators where for the domain

$$[\lambda_0, \lambda_N] \times [\theta_0, \theta_M] \equiv [0, 2\pi] \times \left[-\frac{\pi}{2}, \frac{\pi}{2}\right],$$

is divided into N intervals $\left[\lambda_{i-1}, \lambda_i\right]$ for $i = 1, 2, \ldots, N$ in the λ direction, and M intervals $\left[\theta_{j-1}, \theta_j\right]$ for $j = 1, 2, \ldots, M$ in the θ direction. Thus we have the u, v and Φ points located at $\left(\lambda_i, \theta_{j-\frac{1}{2}}\right)$, $\left(\lambda_{i-\frac{1}{2}}, \theta_j\right)$ and $\left(\lambda_{i-\frac{1}{2}}, \theta_{j-\frac{1}{2}}\right)$, respectively. In [217] the authors introduce the grid intervals as

$$\Delta\lambda_l \equiv \lambda_{l+\frac{1}{2}} - \lambda_{l-\frac{1}{2}}, \qquad \Delta\theta_l \equiv \theta_{l+\frac{1}{2}} - \theta_{l-\frac{1}{2}}, \tag{13.113}$$

where l is a positive integer multiple of $\frac{1}{2}$.

The averaging operators are defined as

$$\left.\overline{F}^{\lambda}\right|_{i,j} \equiv \left(\frac{\lambda_{i+\frac{1}{2}} - \lambda_i}{\Delta\lambda_i}\right) F_{i-\frac{1}{2},j} + \left(\frac{\lambda_i - \lambda_{i-\frac{1}{2}}}{\Delta\lambda_i}\right) F_{i+\frac{1}{2},j}, \tag{13.114}$$

$$\left.\overline{F}^{\theta}\right|_{i,j} \equiv \left(\frac{\theta_{j+\frac{1}{2}} - \theta_j}{\Delta\theta_j}\right) F_{i,j-\frac{1}{2}} + \left(\frac{\theta_j - \theta_{j-\frac{1}{2}}}{\Delta\theta_j}\right) F_{i,j+\frac{1}{2}}, \tag{13.115}$$

$$\left.\langle F\rangle^{\lambda}\right|_{i,j} \equiv \frac{1}{2}\left(F_{i-\frac{1}{2},j} + F_{i+\frac{1}{2},j}\right), \tag{13.116}$$

$$\left.\langle F\rangle^{\theta}\right|_{i,j} \equiv \frac{1}{2}\left(F_{i,j-\frac{1}{2}} + F_{i,j+\frac{1}{2},j}\right), \tag{13.117}$$

where i and j are the horizontal grid indices in the λ and θ directions, and are both positive integer multiples of $\frac{1}{2}$; λ_i denotes the value of λ at the ith grid point in the λ-direction, and θ_j denotes the value of θ at the jth grid point in the θ direction.

It is stated in [217], and shown in [191], that the discretization can be viewed as being based on either a finite differences or finite volumes. In [217] the authors adopt the finite volume approach, similar to the SLICE transport numerical scheme earlier.

Therefore, after applying the spatial discretization to (13.105) and (13.106), we obtain

$$\left(u - \alpha\Delta t\Psi_u\right)_a^{n+1} = \Lambda_{11}\left(u + \beta\Delta t\Psi_u\right)_{d_l^u}^n + \Lambda_{12}\left(v + \beta\Delta t\Psi_v\right)_{d_L^v}^n \equiv R_u^n, \tag{13.118}$$

$$\left(v - \alpha\Delta t\Psi_v\right)_a^{n+1} = \Lambda_{21}\left(u + \beta\Delta t\Psi_u\right)_{d_l^u}^n + \Lambda_{22}\left(v + \beta\Delta t\Psi_v\right)_{d_L^v}^n \equiv R_v^n, \tag{13.119}$$

where the subscript d_L indicates the gridpoint whose departure point is to be used, and (13.107) and (13.108) are redefined as

$$\Psi_u \equiv -2\left(\Omega_r \times \mathbf{u}\right)_u - \frac{1}{r\cos\theta}\delta_\lambda\left(\Phi + \Phi^S\right), \tag{13.120}$$
$$\Psi_v \equiv -2\left(\Omega_r \times \mathbf{u}\right)_v - \frac{1}{r}\delta_\theta\left(\Phi + \Phi^S\right). \tag{13.121}$$

The Coriolis terms in (13.120) and (13.121) are evaluated following [191], which leads for the linearized equations with v placed at the poles, to good Rossby mode dispersion properties, and also ensures that the Coriolis terms provide no source, or sink, of energy. Thus we have

$$-2\left(\Omega_r \times \mathbf{u}\right)_u = \frac{1}{r\cos\theta\Delta\lambda}\left\langle \langle\widetilde{v}\rangle^{\theta}\frac{f}{\Phi}\right\rangle^{\lambda}, \tag{13.122}$$
$$-2\left(\Omega_r \times \mathbf{u}\right)_v = -\frac{1}{r\Delta\theta}\left\langle \langle\widetilde{u}\rangle^{\lambda}\frac{f}{\Phi}\right\rangle^{\theta}, \tag{13.123}$$

where

$$\widetilde{u} \equiv r\Delta\theta\overline{\Phi}^{\lambda}u, \tag{13.124}$$
$$\widetilde{v} \equiv r\cos\theta\Delta\lambda\overline{\Phi}^{\theta}v \tag{13.125}$$

are mass flux variables [191], and

$$f \equiv 2\Omega\sin\theta \tag{13.126}$$

is assumed to be evaluated and stored at Φ points.

If we now turn our attention back to the discrete continuity equation (13.101), then this equation can be rewritten as

$$\left[\left(\Phi - \Phi^{\text{ref}}\right) + \alpha\Delta t\Phi^{\text{ref}}\nabla\cdot\mathbf{u}\right]_a^{n+1} = \frac{1}{\Delta A}\left[\int_{\delta A}\left(\left(\Phi - \Phi^{\text{ref}}\right) - \beta\Delta t\Phi^{\text{ref}}\nabla\cdot\mathbf{u}\right)dA\right]_d^n \equiv R_\theta^n, \tag{13.127}$$

which is computed using the C-SLICE conservative remapping algorithm that we summarized earlier from [216], and $\Delta A \equiv r^2\cos\theta\Delta\lambda\Delta\theta$ is the discrete element of area of the arrival cell. The divergence term is computed as

$$\nabla\cdot\mathbf{u}_a^{n+1} = \frac{1}{r\cos\theta}\left[\delta_\lambda u + \delta_\theta\left(v\cos\theta\right)\right]_a^{n+1}. \tag{13.128}$$

Helmholtz equation

As seen throughout all of the different applications of semi-Lagrangian techniques, at some point with the shallow water equations model, or the primitive equations, we have to solve a Helmholtz equation, and here the situation is no different. We start with the definitions

$$u' \equiv u_a^{n+1}, \qquad v' \equiv v_a^{n+1}, \qquad \Phi' \equiv \Phi_a^{n+1} - \Phi^{\text{ref}}, \tag{13.129}$$

and so the momentum equations (13.118)–(13.119) and the continuity equation (13.127) can be written in the form

$$u' + H_u \delta_\lambda \Phi' = R_u^* + R_u^n, \tag{13.130}$$

$$v' + H_v \delta_\theta \Phi' = R_v^* + R_v^n, \tag{13.131}$$

$$\Phi' + H_\Phi \left(\delta_\lambda \left(H_\lambda u'\right) + \delta_\theta \left(H_\theta v'\right)\right) = R_\Phi^n, \tag{13.132}$$

where

$$R_u^* \equiv \alpha \Delta t \left[-2\left(\Omega_r \times \mathbf{u}\right)_u - \frac{1}{r\cos\theta}\delta_\lambda \Phi^S\right]_a^{n+1}, \qquad R_v^* \equiv \alpha \Delta t \left[-2\left(\Omega_r \times \mathbf{u}\right)_v - \frac{1}{r}\delta_\theta \Phi^S\right]_a^{n+1},$$

and

$$H_u \equiv \frac{\alpha\Delta t}{r\cos\theta}, \qquad H_v \equiv \frac{\alpha\Delta t}{r}, \qquad H_\lambda \equiv r\Phi^{\text{ref}}, \qquad H_\theta \equiv r\cos\theta\,\Phi^{\text{ref}}, \qquad H_\Phi \equiv \frac{\alpha\Delta t}{r^2\cos\theta}.$$

If we now eliminate u' and v' from (13.130)–(13.132), then this leads to the Helmholtz problem:

$$H_\theta \delta_\lambda \left(H_\lambda \left(H_u \delta_\lambda \Phi'\right)\right) + H_\theta \delta_\theta \left(H_\theta \left(H_v \delta_\theta \Phi'\right)\right) - \Phi' = W^* + W^n, \tag{13.133}$$

where

$$W^* \equiv D\left(R_u^*, R_v^*\right),$$

$$W^n \equiv D\left(R_u^n, R_v^n\right) - R_\Phi^n,$$

$$D(F,G) \equiv H_\Phi \left(\delta_\lambda \left(H_\lambda F\right) + \delta_\theta \left(H_\theta G\right)\right).$$

The coefficients on the left-hand side of (13.133) are independent of time. The right-hand side of (13.133) contains both explicitly known terms, W^n, and an implicitly defined nonlinear term, W^*, leading to a nonlinear coupling with (13.130) and (13.131). The nonlinear set of coupled equations are solved using an iterative approach. Since the Coriolis terms are handled iteratively as they appear on the right-hand side of (13.130) and (13.131), this adds the constraint that $\frac{\Omega\Delta t}{2} \leq 1$.

The algorithm for the iterative solution of the nonlinear coupled set of Eqs. (13.130), (13.131), and (13.133) from [217] is summarized as follows:

- Do $n = 1 : T$ (time-step loop), where we have the solutions (u^n, v^n, Φ^n) at level $t = t^n$.
- Do $k = 1 : K$ (outer-loop iterations, departure-point loop) where we compute departure points using $(u, v)^n$ and the latest estimate for $(u, v)^{n+1}$. Evaluate $\left(R_u^n, R_v^n, R_\Phi^n\right)$.
- Do $l = 1 : L$ (inner-loop iteration, Coriolis loop) where we evaluate $\left(R_u^*, R_v^*\right)$ and the right-hand side of (13.133). Solve the Helmholtz problem (13.133) for Φ^{n+1}, and then update $(u, v)^{n+1}$ from (13.130) and (13.131), respectively.
- Enddo
- Enddo
- Enddo

13.3.6 SLICE-3D

The three-dimensional version of SLICE on the sphere was introduced in Zerroukat and Allen (2012) in [211], where they state that SLICE is a conservative remapping tool which inherently conserves mass due to being more mathematically based and inherently satisfies local and global mass conservation. Three-dimensional remappings can be relatively complex and computationally expensive, owing to the extra geometric computations needed. The complexity of three-dimensional remappings also makes their parallel implementation difficult. The use of a floating (Lagrangian) vertical coordinate can alleviate some of these issues by reducing the problem to two dimensions.

In [211] the authors state that SLICE-3D maintains the characteristic of achieving the full 3D remapping as multiple sweeps of one-dimensional conservative remappings. Unlike most other geometry-based schemes, SLICE-3D also does not compute any irregular Lagrangian volumes at any stage of the remapping process. The resulting scheme makes the possibility of higher-dimensional remapping with high-order reconstructions, without a prohibitive complexity and computational cost, a reality.

We start with the Lagrangian equivalent in volume form of the conservation law given by

$$\frac{D}{Dt}\left(\int_V \rho dV\right) = 0. \tag{13.134}$$

Integrating (13.134) from time t^n to t^{n+1}, using a semi-Lagrangian approach yields

$$\int_v \rho^{n+1} dV = M_d^n \equiv \int_{V_d} \rho^n dV, \tag{13.135}$$

where V_d is the volume from which the fluid ended up in V during the time step Δt. Now we let V be a known volume of a Eulerian, fixed in time, regular box, ECV, with ρ^n being the associated value of the scalar averaged over the ECV ($M^n = \rho^n V$). Let V_d be the volume of the corresponding upstream Lagrangian volume (LCV), where M_d^n is the mass contained in the Lagrangian volume V_d at time t^n. Any point $\mathbf{x}$ of LCV is linked to its Eulerian counterpart in ECV by the kinematic equation

$$\frac{D\mathbf{x}}{Dt} = \mathbf{u}. \tag{13.136}$$

Computing the evolution of the density distribution ρ^{n+1} from the known state ρ^n reduces to computing the discrete integrals M_d^n from M^n. This amounts to a remapping of a given field ρ at time t^n on a regular (Eulerian) grid to an irregular (Lagrangian) grid. Finally, ρ^{n+1} can be computed from

$$\rho^{n+1} = \frac{M_d^n}{V}. \tag{13.137}$$

We can see from (13.137) that there is no need to compute the complex Lagrangian volume V_d to compute the new density at time level $n+1$ as long as it is possible to calculate the mass M_d^n without knowing V_d, and this is what SLICE-3D does through simple remapping the mass from one grid to another without ever computing V_d, and then computed the density evolution through (13.137).

Given this setup, we know go into some of the theory of SLICE-3D.

Grid definitions

It is assumed in [211] that the we are using a C grid for the longitude and latitude spherical grid system (x, y, z) where z could be either the radial position r or the normalized terrain following height $\eta = \eta(r, r_{\min}, r_{\max})$, where $r_{\min}$ and $r_{\max}$ are the radii of the bottom surface and the top of the domain, respectively. We now let the three-dimensional spherical domain $\Omega \equiv [0, 2\pi] \times \left[-\frac{\pi}{2}, \frac{\pi}{2}\right] \times [z_{\min}, z_{\max}]$ be divided into $N_x \times N_y \times N_z$ spherical Eulerian Control Volumes $ECV_{i,j,k}$, $i = 1, \ldots, N_x$, $j = 1, \ldots, N_y$ and $k = 1, \ldots, N_z$. Each $ECV_{i,j,k}$ is defined as

$$ECV_{i,j,k} \equiv \left\{(x, y, z), x \in \left[x_i^u, x_{i+1}^u\right], y \in \left[y_j^v, y_{j+1}^v\right], z \in \left[z_{i,j,k}^w, z_{i,j,k+1}^w\right]\right\}, \qquad (13.138)$$

where the superscripts on the x, y, and z are referring to their C-grid location at that wind point, as well as the 3D vertical wind location of the z points. Each $ECV_{i,j,k}$ has a fixed volume, which is given by

$$V_{i,j,k} \equiv \frac{1}{3}\left[\left(r_{i,j,k+1}^w\right)^3 - \left(r_{i,j,k}^w\right)^3\right]\left(x_{i+1}^u - x_i^u\right)\left(\sin y_{j+1}^v - \sin y_j^v\right). \qquad (13.139)$$

Zerroukat and Allen now introduce a notation to simplify the locations of the different spatial coordinate with $\delta \equiv \{u, v, w, \rho\}$. The ρ-points are the centers of the ECVs and they are located at $\left(x_i^\rho, y_j^\rho, z_{i,j,k}^\rho\right)$, where $x_i^\rho \equiv \frac{1}{2}\left(x_i^u + x_{i+1}^u\right)$, $y_j^\rho \equiv \frac{1}{2}\left(y_j^v + y_{j+1}^v\right)$, and $z_{i,j,k}^\rho \equiv \frac{1}{2}\left(z_{i,j,k}^w + z_{i,j,k+1}^w\right)$, where we can see these points in the copies of Figs. 1 and 2 from [211] in Figs. 13.12 and 13.13.

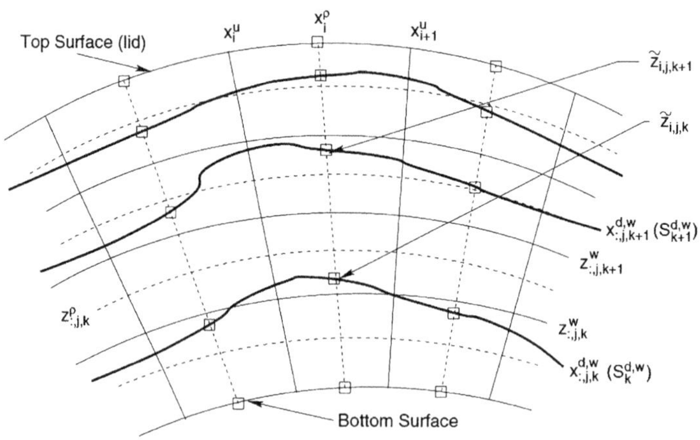

FIGURE 13.12 Copy of Fig. 1 from [211] of the vertical intersection schematic for a y-slice at fixed j. Zerroukat, M. and Allen, T., 2012: A three-dimensional monotone and conservative semi-Lagrangian scheme (SLICE-3D) for transport problems, *Q. J. R. Meteor. Soc.*, **138**, 1640–1651. ©Royal Meteorological Society.

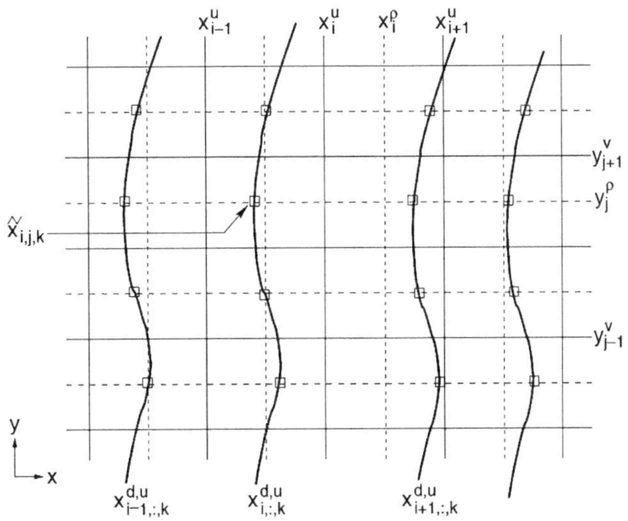

FIGURE 13.13 Copy of Fig. 2 from [211] of the horizontal intersection schematic for a z-slice at fixed k. A three-dimensional monotone and conservative semi-Lagrangian scheme (SLICE-3D) for transport problems, *Q. J. R. Meteor. Soc.*, **138**, 1640–1651. ©Royal Meteorological Society.

To perform the remapping, SLICE-3D requires the departure points of the u, v, w, and ρ points in 3D, but, as is mentioned in [211], the departure points are calculated outside of SLICE-3D and therefore any scheme can be used. Now given the density $\rho^n_{i,j,k}$ for each $ECV_{i,j,k}$, assumed to be located at the center of $ECV_{i,j,k}$ at time t^n, and the departure points $\mathbf{x}^{\delta}_{d}$, SLICE-3D computes the mass of the corresponding $LCV_{i,j,k}$, then the density $\rho^{n+1}_{i,j,k}$ for the new time level t^{n+1} using (13.137).

To simplify the precise description of the SLICE-3D algorithm, Zerroukat and Allen introduce the following notation: given a three-dimensional array $\mathbf{A} \equiv \{a_{i,j,k}; i = 1, \ldots, N_x; j = 1, \ldots, N_y; k = 1, \ldots, N_z\}$ then $\mathbf{A}_{:,j,k}$, $\mathbf{A}_{i,:,k}$ and $\mathbf{A}_{i,j,:}$ will be used to refer to the one-dimensional vectors $\{a_{i,j,k}; i = 1, \ldots, N_x\}$, $\{a_{i,j,k}; j = 1, \ldots, N_y\}$ and $\{a_{i,j,k}; k = 1, \ldots, N_z\}$, respectively.

Given this notation we say that SLICE-3D requires four major components, which are:

(i) vertical intersections,
(ii) horizontal intersections,
(iii) a one-dimensional remapping,
(iv) a polar cap correction.

An important thing to note here is that once the vertical intersections have been calculated, the rest of SLICE-3D is mostly same as the algorithm for SLICE-S.

Vertical intersections

In [211] the authors state that SLICE-3D requires $\tilde{z}_{i,j,k}$ to perform the first vertical remapping. Therefore, let $\tilde{z}_{i,j,k}$ be the z-coordinate of the intersection of the deformed spherical shell

$$S_k^{dw} \equiv \mathbf{x}_{:,:,k}^{d,w} \equiv \left\{ \left(x_{i,j,k}^{d,w}, y_{i,j,k}^{d,w}, z_{i,j,k}^{d,w} \right); i = 1, \ldots, N_x; j = 1, \ldots, N_y \right\},$$

which is the surface that is made up of all the departures of w-points for a fixed k, with the radial line that goes through the center of the Eulerian column $\left(x_i^\rho, y_j^\rho\right)$, which is implying that we are using $\widetilde{z}_{i,j,k} = \mathbf{x}_{:,:,k}^{d,w} \perp \left(x_i^\rho, y_j^\rho\right)$ which we can see in Fig. 13.12. This is computed using the triangle, which comprises the three points of $S_k^{d,w}$, that contains the point $\left(x_i^\rho, y_j^\rho\right)$. The search algorithm that they use in [211] identifies the quadrilateral, which is formed by the four points surrounding the target, then divides it into two triangles. Although $\widetilde{z}_{i,j,k}$ can be computed directly from the four surrounding points, through solving a quadratic equation, Zerroukat and Allen indicate that results show that the use of triangles achieves similar results and is simpler as it solves only a linear equation and avoids the extra cost for the selection of the appropriate solution if the quadratic option is used.

As mentioned earlier, once the vertical intersections are calculated, the SLICE-S, or SLICE-2D for the full Cartesian 3D implementations, is used. There is an illustration of the horizontal step from [211] in Fig. 13.13, where we are looking for the intersections of the Lagrangian line with the latitude circle.

Once all of the intersection points are known and the initial mass distribution, then the SLICE-3D algorithm is as follows:

Step 1: Compute the initial mass distribution $m_{i,j,k} = \rho_{i,j,k}^n V_{i,j,k}$.
Step 2: Compute the vertical intersection.
Step 3: Compute the horizontal intersections.
Step 4: For each vertical column, compute the mass through the one-dimensional remapping.
Step 5: For each East–West strip, compute the mass through the one-dimensional remapping.
Step 6: For each North–South strip, compute the mass through the one-dimensional remapping.
Step 7: Correct the mass for the polar cap regions.
Step 8: Compute the final density through $\rho_{i,j,k}^{n+1} = \dfrac{\overline{m}_{i,j,k}}{V_{i,j,k}}$, where $\overline{m}_{i,j,k}$ is the mass after being remapped through the three dimensions.

13.3.7 Applications of SLICE-3D: non-hydrostatic vertical-slice equations

We finish this section with a brief summary of two different applications of the implementation of SLICE-3D in some form. The implementation that we presented for the shallow water equations model was in a semi-implicit form. In Melvin et al. (2010) [129], the authors present the SLICE-3D in a semi-implicit implementation for the nonhydrostatic vertical-slice equations where, if we consider the continuity equation in flux form, we have

$$\frac{D}{Dt}\left(\int_{\delta A} \rho \, dA\right) = 0. \qquad (13.140)$$

However, in [129] the authors explain that the equation above is the integral equivalent of the differential form

$$\frac{D\rho}{Dt} + \rho\delta = 0 \tag{13.141}$$

of the continuity equations used in traditional semi-implicit semi-Lagrangian discretization, where δ is the divergence as presented earlier. The integral in (13.140) is over the finite material volume δA.

The Lagrangian integral form in (13.140) of the continuity equation is well suited for considering with a semi-Lagrangian discretization. However, in order to apply a semi-implicit scheme, and thereby couple the mass field stably to the wind field, the divergence field needs to be discretized implicitly, since it controls the prorogation of fast acoustic and gravity waves. In its present form (13.140), the divergence field does not explicitly appear. It is only implied through the trajectory equations, since they determine the departure area. In order for the divergence to appear, we follow a technique that was introduced in [217] and which we saw earlier with the shallow water equations implementation of SLICE-2D. To make the divergence field materialize [129], for the shallow water equations, [217] introduced a reference density profile ρ^{ref}. However, in the current model equations, this is more difficult because in the shallow water equations this reference state was a constant, but this is not the case here, where the reference state can vary spatially. This is overcome in [129] through the following derivation from the appendix of the said paper:

We start by adding and subtracting the reference state from the continuity equation (13.140), which yields the following:

$$\begin{aligned}
0 &= \frac{D}{Dt}\left[\int_{\delta A}\left(\rho - \rho^{\text{ref}}\right) dA\right] + \frac{D}{Dt}\left(\int_{\delta A}\rho^{\text{ref}} dA\right) \\
&= \frac{D}{Dt}\left[\int_{\delta A}\left(\rho - \rho^{\text{ref}}\right) dA\right] + \int_{\delta A}\frac{D\rho^{\text{ref}}}{Dt} dA + \int_{\delta A}\rho^{\text{ref}}\frac{D}{Dt}(dA) \\
&= \frac{D}{Dt}\left[\int_{\delta A}\left(\rho - \rho^{\text{ref}}\right) dA\right] + \int_{\delta A}\mathbf{u}\cdot\nabla\rho^{\text{ref}} dA + \int_{\delta A}\rho^{\text{ref}}(\nabla\cdot\mathbf{u})\, dA \\
&= \frac{D}{Dt}\left[\int_{\delta A}\left(\rho - \rho^{\text{ref}}\right) dA\right] + \int_{\delta A}\nabla\cdot\left(\rho^{\text{ref}}\mathbf{u}\right) dA,
\end{aligned}$$

where it has been assumed that ρ^{ref} is time-invariant, and that the following two identities hold:

$$\frac{D\rho^{\text{ref}}}{Dt} \equiv \mathbf{u}\cdot\nabla\rho^{\text{ref}}, \qquad \frac{D}{Dt}(dA) \equiv (\nabla\cdot\mathbf{u})\, dA.$$

This now means that the continuity equation is in a form where the theory from the shallow water equations implementation of SLICE can be applied. We should note that a similar reference technique is applied to the thermodynamical equation to obtain

$$\frac{D\left(\theta - \theta^{\text{ref}}\right)}{Dt} + \mathbf{u}\cdot\nabla\theta^{\text{ref}} = 0. \tag{13.142}$$

The reason for doing this is so that the semi-implicit method of rewriting the advection of thermodynamic variables can be applied to allow for the coupling between the thermodynamic variables and the vertical wind equations to be handled implicitly.

Next Melvin et al. introduce terrain-following vertical coordinates, denoted by $\eta = \eta(x, z)$, which are independent of time, so that orography could be incorporated. The transforms are assumed to be monotonic, and $\eta = 0$ corresponds to the lower boundary, $z = z_s(x)$, while $\eta = 1$ corresponds to the upper boundary $z = z_T =$ constant. The velocity components u, v, and w retain their original form. Additionally, the reference profiles are now assumed to be functions of η only. Thus the final set of continuous equations are:

$$
\begin{aligned}
\frac{Du}{Dt} &= -\frac{c_p\theta}{\frac{\partial z}{\partial \eta}}\left(\frac{\partial}{\partial x}\left(\pi\frac{\partial z}{\partial \eta}\right) - \frac{\partial}{\partial \eta}\left(\pi\frac{\partial z}{\partial x}\right)\right) + fv - Fw, && (13.143)\\
\frac{Dv}{Dt} &= -fu, && (13.144)\\
\delta_v\frac{Dw}{Dt} &= -\frac{c_p\theta}{\frac{\partial z}{\partial \eta}}\frac{\partial \pi}{\partial \eta} - f - Fu, && (13.145)\\
0 &= \frac{D}{Dt}\left[\int_{\delta A}\left(\rho - \rho^{\text{ref}}\right)dA\right] + \int_{\delta A}\nabla\cdot\left(\rho^{\text{ref}}\mathbf{u}\right)dA, && (13.146)\\
0 &= \frac{D\left(\theta - \theta^{\text{ref}}\right)}{Dt} + \mathbf{u}\cdot\nabla\theta^{\text{ref}}, && (13.147)
\end{aligned}
$$

with the kinematic equations

$$\frac{Dx}{Dt} \equiv u, \qquad \frac{D\eta}{Dt} \equiv \dot{\eta}, \tag{13.148}$$

where

$$\pi^{\frac{1-\kappa}{\kappa}} = \frac{R}{p_0}\rho\theta,$$

and the total derivative becomes

$$\frac{D}{Dt} \equiv \frac{\partial}{\partial t} + u\frac{\partial}{\partial x} + \dot{\eta}\frac{\partial}{\partial \eta}, \tag{13.149}$$

while now the divergence term is

$$\nabla\cdot\left(\rho^{\text{ref}}\mathbf{u}\right) = \left(\frac{\partial z}{\partial \eta}\right)^{-1}\left(\frac{\partial}{\partial x}\left(\rho^{\text{ref}}\frac{\partial z}{\partial \eta}u\right) + \frac{\partial}{\partial \eta}\left(\rho^{\text{ref}}\frac{\partial z}{\partial \eta}\dot{\eta}\right)\right), \tag{13.150}$$

and partial derivatives with respect to x are taken holding η, rather than z, constant [129]. The model above is an f–F plane approximation, where the equations are solved on a plane that is tangent to the sphere at a particular latitude. Then the Coriolis parameters f and F are fixed at these values at that latitude. In the final application of the model in [129] it is the

f-plane where F is set to zero. The term δ_v is a quasihydrostatic switch such that the equation set if fully compressible when $\delta_v = 1$, but reduces to the quasihydrostatic equation set when $\delta_v = 0$.

The equations above are discretized in the usual semi-implicit semi-Lagrangian fashion, as seen many times before, so what we present here is simply the flux-form discretization of the continuity equation given by

$$\left[\left(\rho - \rho^{ref}\right) + \alpha\Delta t\nabla\cdot\left(\rho^{\text{ref}}\mathbf{u}\right)\right]_a^{n+1} = \frac{1}{\Delta A}\left[\int_{\delta a}\left(\left(\rho - \rho^{\text{ref}}\right) - \beta\Delta t\nabla\cdot\left(\rho^{\text{ref}}\mathbf{u}\right)\right) dA\right]_d^n, \tag{13.151}$$

where

$$\nabla\cdot\left(\rho^{\text{ref}}\mathbf{u}\right) = \frac{\Delta x\Delta\eta}{\Delta A}\left\{\delta_x\left[\left(\delta_\eta z\right)\rho^{\text{ref}}u\right] + \delta_\eta\left[\left(\delta_\eta z\right)\overline{\rho^{\text{ref}}}^{\eta}\dot{\eta}\right]\right\}, \tag{13.152}$$

and

$$\Delta A = \left(\delta_\eta Z\right)\Delta x\Delta\eta, \tag{13.153}$$

and the averaging operator is given by

$$\left.\overline{\rho^{\text{ref}}}^{\eta}\right|_{i,k} \equiv \left(\frac{\eta_{k+\frac{1}{2}} - \eta_k}{\Delta\eta_k}\right)\rho^{\text{ref}}_{i,k-\frac{1}{2}} + \left(\frac{\eta_k - \eta_{k-\frac{1}{2}}}{\Delta\eta_k}\right)\rho^{\text{ref}}_{i,k+\frac{1}{2}}. \tag{13.154}$$

The reader is referred to [129] for the more detailed description of this implementation and the results obtained with this approach.

There has been more development of the SLICE algorithm where it was applied in a deep-atmosphere global nonhydrostatic model [209]. While we would have summarized the paper here, most of it is not concerned with the application of SLICE but rather the removal of the reference state to minimize its influence [192]. The main difference is that the continuity equation in semi-Lagrangian form is now written as

$$\rho_a^{n+1} - \rho^{\text{ref}} = \frac{1}{V_a}\left(\int_V \rho dV\right)_d^n - \rho^{\text{ref}}, \tag{13.155}$$

where V_a is the volume of the grid box centered at each arrival ρ point, and where the authors have used the version of SLICE from [211] combined with a piecewise parabolic method of Colellla and Woodward (1984) [36].

13.4 Flux-form semi-Lagrangian spectral element approach

This approach was introduced in Ullrich and Norman (2014) [193], where the motivation was to find a way to use spectral methods with tracer transport which was not as costly for this problem, but also did not require substantial parallel communications over a time step. Ullrich and Norman state that there are five properties that any semi-Lagrangian scheme should possess for it to be considered for an operational implementation for tracer transport:

1. Uses the native spectral element nodal grid.
2. Conserves mass without a posteriori mass fixers.
3. Allows for at least third-order accuracy.
4. Maintains stability up to a Courant number of 1.
5. Maintains strict positivity of the tracer field.

They also say that the Flux-Form Semi-Lagrangian Spectral Element (FF-SLSE) approach has several desirable properties. Since this scheme is formulated in flux-form, local and global conservation tracer mass is immediately guaranteed, and this approach admits both a variational and a finite-difference formulation. This being said, the scheme is described below.

The scheme is based upon 1D Gauss–Lobatto–Legedre (GLL) characteristic functions which are based upon the nodal Gauss–Lobatto grid. We now consider a 1D reference element on the closed interval $[-1, 1]$. For a given polynomial degree $n_p \geq 1$, the nodal Gauss–Lobatto points include both endpoints $\widetilde{x}_i = -1$ and $\widetilde{x}_{n_p+1}$, as well as $n_p - 1$ nodes in the interior, defined by the roots of $P'_{n_p-1}(\widetilde{x})$ of the Legendre polynomial. This then leads to an implied quadrature rule over $[-1, 1]$ with corresponding nodal weights

$$w_i = \frac{2}{n_p\left(n_p - 1\right)\left[P_{n_p-1}\left(\widetilde{x}_i\right)\right]^2}. \tag{13.156}$$

The ith GLL characteristic function on this grid is defined as the lowest-order polynomial which is 1 at $\widetilde{x}_i$ and 0 at all $\widetilde{x}_j$ with $j \neq i$. The schemes that the authors of [193] are interested in are third-degree (cubic) polynomial within each spectral element. When using quadrilateral (2D) and hexahedral (3D) spectral elements, the characteristic functions in 1D can be easily extended to higher dimensions via a tensor product. In the global domain, which consists of an appropriate tiling of spectral elements, nodal points are labeled using vector notation $\mathbf{x}_j$ where the index j spans all nodes.

In [193] they state that the spectral element method, and consequently the FF-SLSE method, further enforce continuity at element boundaries. The characteristic functions which are nonzero at element boundaries are mirrored in immediate neighbors to give an overlapping set of basis functions. This requirement also implies that all degrees of freedom at element boundaries are shared between spectral elements. Hence, each element contains one fewer degree of freedom than its number of basis functions (three degrees of freedom per spectral element for cubic basis functions). The basis function $\phi(\mathbf{x})$ is then defined as the characteristic function (or mirrored characteristic function) which is 1 at $\mathbf{x}_j$ and 0 at all other nodes.

We now assume a periodic domain $\mathcal{D}$ in all coordinate directions. As seen before, the conservative form of the advection equation is given by

$$\frac{\partial \psi}{\partial t} + \nabla \cdot \mathbf{F}(\psi; \mathbf{u}) = 0, \tag{13.157}$$

where $\psi = pq = \psi(\mathbf{x}, t)$ is the tracer density variable that is composed of the density ρ and tracer concentration q and $\mathbf{F}(\psi; \mathbf{u}) = \psi\mathbf{u}$. The tracer density ψ is approximated using a basis function expansion

$$\psi(\mathbf{x},t)=\sum_{j}\psi_j(t)\,\phi_j(\mathbf{x}), \tag{13.158}$$

where the sum is taken over the full set of basis functions and $\psi_j(t)=\psi\left(\mathbf{x}_j,t\right)$. The next step is to multiply the differential equation by $\phi_i(x)$ and integrate over $\mathcal{D}$:

$$\sum_{j=1}^{N}\frac{\partial\psi_j}{dt}\int_{\mathcal{D}}\phi_i(\mathbf{x})\,\phi_j(\mathbf{x})\,dV=-\int_{\mathcal{D}}\phi_i(\mathbf{x})\,\nabla\cdot\mathbf{F}(\psi;\mathbf{u})\,dV, \tag{13.159}$$

where the boundary integral term is dropped due to periodicity of equation image and dV is the volume element associated with equation image. Under the Galerkin formulation, this equation must be satisfied for all basis functions ϕ_i. Integration by parts is now applied to obtain the weak form,

$$\sum_{j=1}^{N}\frac{\partial\psi_j}{dt}\int_{\mathcal{D}}\phi_i(\mathbf{x})\,\phi_j(\mathbf{x})\,dV=\int_{\mathcal{D}}(\psi;\mathbf{u})\cdot\nabla\phi_i(\mathbf{x})\,dV. \tag{13.160}$$

In matrix form, this system of equations can be written as

$$\mathbf{M}\frac{D\boldsymbol{\psi}}{Dt}=\mathbf{G}, \tag{13.161}$$

where $\boldsymbol{\psi}=\boldsymbol{\psi}\,()=\left[\psi_j(t)\right]_{j=1}^{N}$ is the vector of nodal values, $\mathbf{M}$ is the symmetric mass matrix with components

$$(\mathbf{M})_{ij}=\int_{\mathcal{D}}\phi_i(\mathbf{x})\,\phi_j(\mathbf{x})\,dV, \tag{13.162}$$

and $\mathbf{G}$ is the vector of integrated fluxes

$$\mathbf{G}_i=\int_{\mathcal{D}}\mathbf{F}(\psi;\mathbf{u})\cdot\nabla\phi_i(\mathbf{x})\,dV. \tag{13.163}$$

All integrals are instead evaluated using the Gauss–Lobatto integration rule induced by the underlying grid. Using this approach, the mass matrix will be diagonal and the inversion will no longer require a global matrix–vector solving. This technique is also known as mass-lumping, since it is identical to replacing M with a diagonal matrix composed of the row sums of M. Another advantage of this technique is that fluxes must only be computed at GLL points, since each component of the flux vector will reduce to a weighted sum of pointwise fluxes.

We have to see here that (13.161) represents a semi-discrete system of equations. To integrate this equation a semi-Lagrangian integration is introduced here through integrating (13.161) in time over the interval $\mathcal{T}^n=\left[t^n,t^{n+1}\right]$, then we obtain the linear system

$$\left[\mathbf{M}\left(\widetilde{\boldsymbol{\psi}}^{n+1}-\boldsymbol{\psi}^{n}\right)\right]_i=\int_{\mathcal{D}}\overline{F}^{n}(\mathbf{x})\cdot\nabla\phi_i(\mathbf{x})\,dV, \tag{13.164}$$

with unknown advected tracer density $\widetilde{\boldsymbol{\psi}}^{n+1} = \widetilde{\boldsymbol{\psi}}\left(t^{n+1}\right)$ and time integrated pointwise flux

$$\overline{\mathbf{F}}^{n}(\mathbf{x}) = \int_{\mathcal{T}^{n}} \mathbf{F}(\psi; \mathbf{u})\, dt. \tag{13.165}$$

The semi-Lagrangian component here, according to [193], is in specifying the time-integrated flux (13.165) using **known** information. For consistency with the induced quadrature rule on the spatial grid, Gauss–Lobatto quadrature is used to approximate the integral (13.165), although any other appropriately chosen quadrature rule can also be used:

$$\overline{\mathbf{F}}^{n}(\mathbf{x}) = \frac{\Delta}{2} \sum_{m=1}^{M_q} w_m \mathbf{F}\left[\psi\left(\mathbf{x}, t^{n,m}\right); \mathbf{u}\left(\mathbf{x}, t^{n,m}\right)\right], \tag{13.166}$$

where $\Delta t = t^{n+1} - t^{n}$, M_q denotes the number of points in the corresponding quadrature rule, and

$$t^{n,m} = t^{n} + \frac{\Delta t}{2}\left(1 + \widetilde{x}_m\right). \tag{13.167}$$

The FF-SLSE uses a fourth-order Simpson's rule for integration in time, which consists of nodal points and weights

$$\widetilde{x}_1 = -1, \qquad \widetilde{x}_2 = 0, \qquad \widetilde{x}_3 = 1, \tag{13.168}$$

$$w_1 = \frac{1}{6}, \qquad w_2 = \frac{2}{3}, \qquad w_3 = \frac{1}{6}. \tag{13.169}$$

The final stage of this algorithm is to determine the pointwise fluxes as functions of the known state vector $\boldsymbol{\psi}^{n}$. Using the fact that tracer concentrations are transported with the flow, for a given $\mathbf{x}_i$ and t^{n+1}, there exists some point $\mathbf{x}_i^{*}$ so that

$$q\left(\mathbf{x}_i, t^{n+1}\right) = q\left(\mathbf{x}_i^{*}, t^{n}\right). \tag{13.170}$$

As we know by now, the points $\mathbf{x}_i$ and $\mathbf{x}_i^{*}$ lie along the same fluid parcel trajectory, which is determined by the differential equation

$$\frac{D\mathbf{x}}{Dt} = \mathbf{u}(\mathbf{x}, t), \tag{13.171}$$

for a sufficiently smooth velocity field $\mathbf{u}(\mathbf{x}, t)$. In [193] the authors employ a fourth order Runge–Kutta scheme given by

$$\mathbf{x}_i^{(1)} = \mathbf{x}_i - \frac{\left(t - t^{n}\right)}{2} \mathbf{u}\left(\mathbf{x}_i, t\right), \tag{13.172}$$

$$\mathbf{x}_i^{(2)} = \mathbf{x}_i - \frac{\left(t - t^{n}\right)}{2} \mathbf{u}\left(\mathbf{x}_i^{(1)}, \frac{t + t^{n}}{2}\right), \tag{13.173}$$

$$\mathbf{x}_i^{(3)} = \mathbf{x}_i - \left(t - t^{n}\right) \mathbf{u}\left(\mathbf{x}_i^{(2)}, \frac{t + t^{n}}{2}\right), \tag{13.174}$$

$$\mathbf{x}_i^* = \frac{1}{3}\left(-\mathbf{x}_i + \mathbf{x}_i^{(1)} + 2\mathbf{x}_i^{(2)} + \mathbf{x}_i^{(3)}\right) - \frac{(t-t^n)}{6}\mathbf{u}\left(\mathbf{x}_i^{(3)}, t^n\right). \tag{13.175}$$

Once $\mathbf{x}_i^*$ is known, the pointwise flux is evaluated via

$$\begin{aligned}\mathbf{F}\left[\psi\left(\mathbf{x}_i, t^{n,m}\right); \mathbf{u}\left(\mathbf{x}, t^{n,m}\right)\right] &= \mathbf{F}\left[q\left(\mathbf{x}_i^*, t^n\right)\rho\left(\mathbf{x}_i, t^{n,m}\right); \mathbf{u}\left(\mathbf{x}, t^{n,m}\right)\right] \\ &= q\left(\mathbf{x}_i^*, t^n\right)\rho\left(\mathbf{x}_i, t^{n,m}\right)\mathbf{u}\left(\mathbf{x}_i, t^{n,m}\right).\end{aligned} \tag{13.176}$$

It is then stated in [193] that the density comes from the dynamical core of the operational model to ensure that there is consistency between the tracers and the dynamics.

Time-integrated formulation

In [193] the authors present an alternative formulation that is based upon spectral finite-differencing where now if we let the flux $\mathbf{F}(\mathbf{x}, t)$ be approximated via an expansion of basis functions ϕ_i as

$$\mathbf{F}(\mathbf{x}, t) = \sum_i \mathbf{F}\left[\psi\left(\mathbf{x}_i, t\right); \mathbf{u}\left(\mathbf{x}_i, t\right)\right]\phi_i(\mathbf{x}), \tag{13.177}$$

then the continuity equation becomes

$$\frac{\partial \psi}{\partial t} = -\sum_i \mathbf{F}\left[\psi\left(\mathbf{x}_i, t\right); \mathbf{u}\left(\mathbf{x}_i, t\right)\right] \cdot \nabla\phi_i(\mathbf{x}). \tag{13.178}$$

The update equation is then obtained by integrating this expression over the time interval $\mathcal{T}^n$ and evaluating the expression at $\mathbf{x} = \mathbf{x}_j$ such that

$$\widetilde{\psi}_j^{n+1} = \psi_j^n - \sum_i \overline{\mathbf{F}}^n\left(\mathbf{x}_i\right) \cdot \nabla\phi_i\left(\mathbf{x}_j\right). \tag{13.179}$$

At element boundaries where the basis functions are not differentiable, the derivatives $\nabla\phi_i\left(\mathbf{x}_j\right)$ are averaged among the one-sided derivatives on all intersecting elements, which then makes the two approaches equivalent.

In [193] Ullrich and Norman mention that the approach just described is consistent and ensures accuracy to high order, but exhibits instability for certain values of the Courant number. They indicate that this instability is removed via application of an explicit high-order numerical diffusion operator to filter out the unstable computational modes. The application of diffusion is generally necessary for spectral approximations of advection–diffusion operators so as to remove spurious noise associated with the high-order basis functions. The diffusion operator is given by

$$\mathcal{H}^k\psi = \alpha\underbrace{(\nabla\cdot\nabla)\cdots(\nabla\cdot\nabla)}_{k}\psi = \alpha\nabla^{2k}\psi, \tag{13.180}$$

where α is the coefficient of diffusion and $2k$ specifies the order of the diffusion operator. The operator is applied using a forward Euler step, which is formulated as

$$\boldsymbol{\psi}^{n+1} = \widetilde{\boldsymbol{\psi}} + \Delta t\alpha\mathcal{H}^k\widetilde{\boldsymbol{\psi}}^{n+1}, \tag{13.181}$$

where $\widetilde{\psi}$ is the updated tracer density following the application of the advection. The high-order Laplacian $\mathcal{H}^k$ is applied recursively by defining $\mathcal{H}^0\psi = \widetilde{\psi}$ and

$$\mathcal{H}^k\psi = \mathcal{H}\left(\mathcal{H}^{k-1}\psi\right) = \nabla\cdot\nabla\left(\mathcal{H}^{k-1}\psi\right). \tag{13.182}$$

Applying the spectral-element discretization from earlier, along with the expansion

$$\mathcal{H}^k\psi(\mathbf{x},t) = \sum_j \left(\mathcal{H}^k\psi\right)_j(t)\,\phi_j(\mathbf{x}), \tag{13.183}$$

we obtain a matrix form as

$$\mathcal{H}^k\psi = \mathbf{M}^{-1}\mathbf{H}\left(\mathcal{H}^{k-1}\psi\right), \qquad (\mathbf{H})_{ij} \equiv \int_{\mathcal{D}} \nabla\phi_i\cdot\nabla\phi_j dV. \tag{13.184}$$

In summary, the scheme presented here can be viewed as the following algorithmic steps:

1. Backtrack trajectories from $\mathbf{x}_j$ and $t^{n,m}$ to obtain origin points $\mathbf{x}_j^*$.
2. Interpolate values of the tracer field via

$$q\left(\mathbf{x}_j^*, t^n\right) = \sum_i q_j^n\phi_i\left(\mathbf{x}_j^*\right).$$

3. Calculate the time-integrated fluxes $\overline{\mathbf{F}}^n\left(\mathbf{x}_j^*\right)$.
4. If flux-corrected transport is enabled, limit pointwise fluxes $\overline{\mathbf{F}}^n\left(\mathbf{x}_j\right)$, to ensure positivity in the spectral elements (see [193] for more details and associated references).
5. Calculate predicted pointwise updates $\widetilde{\psi}_j^{n+1}$ via either approach presented.

Once the advection step is completed, hyperdiffusion is then applied so as to ensure stability of the underlying scheme. This process proceeds as follows:

1. Calculate the Laplacian of the field through

$$\mathcal{H}\psi = \mathbf{M}^{-1}\mathbf{H}\widetilde{\psi}^{n+1}.$$

2. Compute hyperdiffusion fluxes

$$\mathbf{F}(\psi;\mathbf{u}) = \mathbf{F}_h(\psi) = \alpha\nabla\left(\mathcal{H}\psi\right).$$

3. If flux-corrected transport is enabled, limit pointwise fluxes $\mathbf{F}\left(\mathbf{x}_j\right)$.
4. Calculate pointwise updates ψ_j^{n+1}.

In [193] Ullrich and Norman go on to provide a stability analysis for the schemes, as well as show results with the three standard test cases that we presented in Chapter 10, and the reader is referred to [193] for the details of that analysis.

We now move on to consider a completely different use of the semi-Lagrangian method, which is important in the field of the geosciences, namely, the family of **Vlasov equations**.

13.5 Conservative semi-Lagrangian HWENO method for the Vlasov equations

We now move on to consider two new aspects of semi-Lagrangian methods. The first is the class of polynomials that are referred to as the **weighted essentially nonoscillatory** methods, where in the method that we present here, the authors use an Hermite based approach, hence the acronym HMENO, as well as introduce the Vlasov family of equations. The Vlasov–Poisson (VP) simulations introduced in [28] is based on dimensional splitting. The VP system, arising from collisionless plasma applications, is defined by

$$\frac{\partial f}{\partial t}+\mathbf{u}\cdot\nabla_{\mathbf{x}}f+E\left(t,\mathbf{x}\right)\cdot\nabla_{\mathbf{u}}f=0, \tag{13.185}$$

and

$$E\left(t,\mathbf{x}\right)-\nabla\phi\left(t,\mathbf{x}\right),\qquad -\nabla_{\mathbf{x}}\phi\left(t,\mathbf{x}\right)=\rho\left(t,\mathbf{x}\right), \tag{13.186}$$

where $\mathbf{x}$ and $\mathbf{u}$ are coordinates in phase space $(\mathbf{x},\mathbf{u})\in\mathbb{R}^3\times\mathbb{R}^3$, E is the electric field, ϕ here is the self-consistent electrostatic potential, and $f\left(t,\mathbf{x},\mathbf{u}\right)$ is a probability distribution function that describes the probability of finding a particle with velocity $\mathbf{u}$ at position $\mathbf{x}$ at time t. The probability distribution function couples to the long-range fields via the charge density such that

$$\rho\left(t,\mathbf{x}\right)=\int_{\mathbb{R}^3}f\left(t,\mathbf{x},\mathbf{u}\right)-1,$$

where in [28] the authors are taking the limit of uniformly distributed infinitely massive ions in the background.

Cai et al. introduce the SL Hermite interpolation in a flux-difference form for 1D transport problems and then go on to incorporate the HWENO mechanism into the flux function reconstruction procedure to realize a nonoscillatory capturing of solutions with sharp gradients. They start with the 1D transport problem

$$\frac{\partial f}{\partial t}+u\frac{\partial f}{\partial x}=0,\qquad f\left(x,t=0\right)=f_0\left(x\right), \tag{13.187}$$

where the domain of f is defined as $[a,b]$, and where u is a constant. For the Hermite method, we have to consider the evolution equation for the solution's derivative $g=\frac{\partial f}{\partial x}$. For the linear transport problem (13.187), g satisfies the same linear transport equation

$$\frac{\partial g}{\partial t}+u\frac{\partial g}{\partial x}=0. \tag{13.188}$$

The next step in [28] is to discretize the domain $[a,b]$ as

$$a=x_{\frac{1}{2}}<x_{\frac{3}{2}}<\cdots<x_{N+\frac{1}{2}}=b,$$

with the uniform grid points $x_i=a+\left(i-\frac{1}{2}\right)\Delta x$ and the cell size $\Delta x=x_{i+\frac{1}{2}}-x_{i-\frac{1}{2}}$. We now introduce the intervals denoted by $I_i=\left[x_{i-\frac{1}{2}},x_{i+\frac{1}{2}}\right]$ and $I_{i+\frac{1}{2}}=\left[x_i,x_{i+1}\right]$. In the HWENO approach, the numerical solutions associated with each grid point are point values f_i^n and derivatives g_i^n. In [28] the authors state that to design a semi-Lagrangian HWENO scheme, the functions $\left\{f_i^{n+1}, g_i^{n+1}\right\}_{i=1}^{N}$ are updated from the corresponding solutions at time t^n. We move on to introduce the interpolating schemes from [28]: For the linear problem with constant characteristic speed, as we know, the solutions f_i^{n+1} and g_i^{n+1} can be obtained by what is referred to in [28] as **shifting** the information at t^n in the semi-Lagrangian framework. The amount of shift here is what we have been referring to as α.

There are three cases that Cai et al. consider: (1) α shift to the right by some amount less than half a cell $\left(\alpha \in \left[0, \frac{1}{2}\right]\right)$, (2) shift to the left by some amount less than half a cell $\left(\alpha \in \left[-\frac{1}{2}, 0\right]\right)$, and (3) shift a distance greater than half a cell $\left(|\alpha| > \frac{1}{2}\right)$. To illustrate the idea, we consider Hermite interpolations for $\alpha \in \left[0, \frac{1}{2}\right]$, while that for $\alpha \in \left[-\frac{1}{2}, 0\right]$ is said to be mirror symmetric with respect to x_i of the previous interpolations. In the case when $|\alpha| > \frac{1}{2}$, the whole grid shifting is carried out and followed by a final update based on the procedure for $\alpha \in \left[-\frac{1}{2}, \frac{1}{2}\right]$. In the example that follows, Cai et al. present the Hermite interpolation with cubic polynomials. Higher-order schemes will be discussed later.

1. The underlying function at t^n can be approximated by an Hermite-type reconstruction, based on the stencil $\left\{f_{i-1}^n, f_{i-1}^n, g_i^n, g_{i-1}^n\right\}$ as

$$\begin{aligned}\widetilde{f}_{i-\frac{1}{2}}^n(\xi) &= f_i^n g_i^n \Delta x \xi + 2\left(g_i^n \Delta x - 3 f_i^n + 3 f_{i-1}^n + g_{i-1}^n \Delta x\right)\xi^2 \\ &+ \left(2 f_i^n - g_i^n \Delta x - 2 f_{i-1}^n - g_{i-1}^n \Delta x\right)\xi^3,\end{aligned}$$

   where $\xi(x) \equiv \dfrac{x - x_i}{x_{i-1} - x_i}$ is such that $\xi \in [0, 1]$, for $x \in I_{i-\frac{1}{2}}$.

2. Values f_i^{n+1} and g_i^{n+1}, as we know, can be obtained by tracing the characteristic back to time $t = t^n$ and evaluating the interpolant $\widetilde{f}_{i-\frac{1}{2}}^n(\xi)$ at the foot of the characteristics $x_d = x_i - u\Delta t$,

$$\begin{aligned}f_i^{n+1} &= f_i^n - \xi_0\left(\left(f_i^n \xi_0 - 2 f_i^n \xi_0^2\right) - 3\left(f_{i-1}^n \xi_0 - 2 f_{i-1}^n \xi_0^2\right)\right) \\ &- g_i^n \Delta x \xi_0 + \left(2 g_i^n \Delta x + g_{i-1}^n \Delta x\right)\xi_0^2 - \left(g_i^n \Delta x + g_{i-1}^n \Delta x\right)\xi_0^3, \qquad (13.189)\\ g_i^{n+1} &= g_i^n - \left(4 g_i^n - 6\frac{f_i^n - f_{i-1}^n}{\Delta x} + 2 g_{i-1}^n\right)\xi_0 - \left(6\frac{f_i^n - f_{i-1}^n}{\Delta x} - 3\left(g_i^n + g_{i-1}^n\right)\right)\xi_0^2, \\ & \qquad (13.190)\end{aligned}$$

   where $\xi_0 \equiv \dfrac{x_d - x_i}{x_{i-1} - x_i}$. We will point out here that we have corrected a mistake, or just a mistyped factor of 6 that was missing in the second set of terms in (13.190).

In [28] the authors then state for the above linear scheme (13.189) and (13.190), that they have the following mass conservation result.

Proposition 13.4. *If* $\sum_{i=1}^{N} g_i^n \equiv 0$ *and with periodic boundary condition, the scheme* (13.189)–(13.190) *conserves the total mass, which is to say that*

$$\sum_{i=1}^{n} f_i^{n+1} = \sum_{i=1}^{N} f_i^n \quad \text{and} \quad \sum_{i=1}^{N} g_i^{n+1} \equiv 0.$$

The proof is quite simple and is omitted here.

In order to guarantee the assumption in the proposition above, Cai et al. introduce a sliding average function, $h(x)$, which satisfies

$$f(x) = \frac{1}{\Delta x} \int_{x-\frac{\Delta x}{2}}^{x+\frac{\Delta x}{2}} h(\xi)\, d\xi, \tag{13.191}$$

which then implies

$$g(x) = f_x(x) = \frac{1}{\Delta x}\left(h\left(x + \frac{\Delta x}{2}\right) - h\left(x - \frac{\Delta x}{2}\right)\right).$$

Thus $\sum_{i=1}^{N} g_i^0 = \sum_{i=1}^{N}\left(h_{i+\frac{1}{2}}^0 - h_{i-\frac{1}{2}}^0\right) \equiv 0$, where $h_{i\pm\frac{1}{2}}^0 \approx h\left(x \pm \frac{\Delta x}{2}\right)$ can be obtained by what is referred to as **reconstruction** from $\left\{f_j^0\right\}_j$. In [28] they adopt a fifth-order upwind-biased WENO reconstruction for obtaining initial conditions of h. If $u > 0$, they use $\{f_{i-2}^0, f_{i-1}^0, f_i^0, f_{i+1}^0, f_{i+2}^0\}$, however, if $v < 0$ they use $\{f_{i-1}^0, f_i^0, f_{i+1}^0, f_{i+2}^0, f_{i+3}^0\}$.

The next step is to rewrite (13.189) and (13.190) in a flux-difference form, in order to ensure local mass conservation, especially when the nonlinear HWENO mechanism is applied. To achieve this, Cai et al. propose to update $\left\{f_i^n, h_{i+\frac{1}{2}}^n\right\}_i$ instead of $\{f_i^n, g_i^n\}_i$ observing that g_i^n can be recovered from $\left\{h_{i+fh}^n\right\}_i$ by $g_i^n = \dfrac{h_{i+fh}^n - h_{i-\frac{1}{2}}^n}{\Delta x}$. Thus for (13.189) we obtain

$$f_i^{n+1} - \xi_0\left(\widehat{f}_{i+\frac{1}{2}}^n(\xi_0) - \widehat{f}_{i-\frac{1}{2}}^n(\xi_0)\right), \tag{13.192}$$

where

$$\widehat{f}_{i-\frac{1}{2}}^n(\xi) \equiv \begin{pmatrix} f_{i-1}^n & h_{i-\frac{1}{2}}^n & h_{i+\frac{1}{2}}^n \end{pmatrix} \begin{pmatrix} 0 & 3 & -2 \\ 1 & -2 & 1 \\ 0 & -1 & 1 \end{pmatrix} \begin{pmatrix} 1 \\ \xi_0 \\ \xi_0^2 \end{pmatrix}. \tag{13.193}$$

See [28] for the full derivation of the expressions above. Given these values, it is possible to update the gradients

$$g_i^{n+1} = \frac{h_{i+\frac{1}{2}}^{n+1} - h_{i-fh}^{n+1}}{\Delta x}, \tag{13.194}$$

where

$$h_{i-\frac{1}{2}}^{n+1} \equiv \begin{pmatrix} f_{i-1}^n & h_{i-\frac{1}{2}}^n & h_{i+\frac{1}{2}}^n \end{pmatrix} \begin{pmatrix} 0 & 6 & -6 \\ 1 & -4 & 3 \\ 0 & -2 & 3 \end{pmatrix} \begin{pmatrix} 1 \\ \xi_0 \\ \xi_0^2 \end{pmatrix}.$$

It is stated in [28] that there are two main advantages to work with the flux-difference form:

1. The flux-difference form can ensure local mass conservation.
2. We can design a nonlinear HWENO mechanism for the flux reconstructions.

That being said, we now go on to summarize the HWENO nonlinear mechanism.

13.5.1 HWENO reconstruction for flux functions

In general, high-order fixed stencil reconstruction of numerical fluxes performs well when the solution is smooth. However, around discontinuities, oscillations will be introduced. So we summarize a nonlinear SL HWENO procedure for reconstructing the flux $\widehat{f}_{i-\frac{1}{2}}^n(\xi)$. Cai et al. choose to adaptively assigning nonlinear weights to neighboring candidate stencils, the nonlinear HWENO reconstruction preserves high-order accuracy of the linear scheme around smooth regions of the solution, while producing a sharp and essentially nonoscillatory capture of discontinuities.

The scheme that Cai et al. introduce is a **fifth-order** HWENO scheme based upon the reconstruction stencil $\left\{f_{i-2}^n, f_{i-1}^n, f_i^n, f_{i+1}^n, g_{i-2}^n, g_{i+1}^n\right\}$, which in matrix form is given by

$$f_i^{n+1} = f_i^n - \xi_0 \begin{pmatrix} f_{i-2}^n & f_{i-1}^n & f_i^n & f_{i+1}^n & h_{i-\frac{5}{2}}^n & h_{i-\frac{3}{2}}^n & h_{i+\frac{1}{2}}^n & h_{i+\frac{3}{2}}^n \end{pmatrix} \mathbf{B}_5^L \begin{pmatrix} 1 \\ \xi_0 \\ \xi_0^2 \\ \xi_0^3 \\ \xi_0^4 \end{pmatrix}, \quad (13.195)$$

where

$$\mathbf{B}_5^L \equiv \begin{pmatrix} \frac{8}{27} & \frac{19}{108} & -\frac{5}{12} & -\frac{19}{108} & \frac{13}{108} \\ -1 & -1 & \frac{3}{4} & \frac{1}{2} & -\frac{1}{4} \\ 0 & \frac{7}{4} & -\frac{1}{4} & -\frac{3}{4} & \frac{1}{4} \\ \frac{19}{27} & -\frac{25}{27} & -\frac{1}{12} & \frac{23}{54} & -\frac{13}{108} \\ -\frac{1}{9} & -\frac{1}{18} & \frac{1}{6} & \frac{1}{18} & -\frac{1}{18} \\ \frac{1}{9} & \frac{1}{18} & -\frac{1}{6} & -\frac{1}{18} & \frac{1}{18} \\ \frac{2}{9} & -\frac{2}{9} & -\frac{1}{6} & \frac{2}{9} & -\frac{1}{18} \\ -\frac{2}{9} & \frac{2}{9} & \frac{1}{6} & -\frac{2}{9} & \frac{1}{18} \end{pmatrix}.$$

Then f_i^{n+1} can be written in the flux-difference form as

$$f_i^{n+1} = f_i^n - \xi_0\left(\left(\begin{array}{ccccc} f_{i-1}^n & f_i^n & f_{i+1}^n & h_{i-\frac{3}{2}}^n & h_{i+\frac{3}{2}}^n \end{array}\right)\mathbf{C}_5^L - \left(\begin{array}{ccccc} f_{i-2}^n & f_{i-1}^n & f_i^n & h_{i-\frac{5}{2}}^n & h_{i+\frac{1}{2}}^n \end{array}\right)\mathbf{C}_5^L\right)\begin{pmatrix} 1 \\ \xi_0 \\ \xi_0^2 \\ \xi_0^3 \\ \xi_0^4 \end{pmatrix}, \tag{13.196}$$

where

$$\mathbf{C}_5^L \equiv \begin{pmatrix} -\frac{8}{27} & -\frac{19}{108} & \frac{5}{12} & \frac{19}{108} & -\frac{13}{108} \\ \frac{19}{27} & \frac{89}{108} & -\frac{1}{3} & -\frac{35}{108} & \frac{7}{54} \\ \frac{19}{27} & -\frac{25}{27} & -\frac{1}{12} & \frac{23}{54} & -\frac{13}{108} \\ \frac{1}{9} & \frac{1}{18} & -\frac{1}{6} & -\frac{1}{18} & \frac{1}{18} \\ -\frac{2}{9} & \frac{2}{9} & \frac{1}{6} & -\frac{2}{9} & \frac{1}{18} \end{pmatrix}$$

and

$$h_{i-\frac{1}{2}}^{n+1} = \left(\begin{array}{ccccc} f_{i-1}^n & f_i^n & f_{i+1}^n & h_{i-\frac{3}{2}}^n & h_{i+\frac{3}{2}}^n \end{array}\right)\mathbf{D}_5^L\begin{pmatrix} 1 \\ \xi_0 \\ \xi_0^2 \\ \xi_0^3 \\ \xi_0^4 \end{pmatrix},$$

with $\mathbf{D}_5^L$ satisfying $\mathbf{D}_5^L(:,k) = k\mathbf{C}_5^L(:,k)$, $k = 1, 2, \ldots, 5$.

We now move on to show how to reconstruct the flux-function, where this is for the flux $\widehat{f}_{i-\frac{1}{2}}^n$ and $h_{i-\frac{1}{2}}^{n+1}$, for which the stencil $\left\{f_{i-2}^n, f_{i-1}^n, f_i^n, h_{i-\frac{5}{2}}, h_{i+\frac{1}{2}}^n\right\}$ is used. It is composed of the information from three stencils:

$$S_1 = \left\{h_{i-\frac{5}{2}}^n, f_{i-2}^n, f_{i-1}^n\right\}, \quad S_2 = \left\{f_{i-2}^n, f_{i-1}^n, f_i^n\right\}, \quad S_3 = \left\{f_{i-1}^n, f_i^n, h_{i-\frac{1}{2}}^n\right\}, \tag{13.197}$$

Cai et al. state that, intuitively, in regions where the function is smooth, we want to use information from S_1, S_2, and S_3 in an optimal way to obtain a fifth-order approximation. On the other hand, around a big jump, we only want to use the information from the relatively smooth stencil, thus we only use the HWENO mechanism in adaptively reconstructing the coefficients in front of the constant 1 in the equation for $\widehat{f}_{i-\frac{1}{2}}^n$ and h_{i-fh}^{n+1}, while leaving the coefficients for $\xi_0, \xi_0^2, \xi_0^3, \xi_0^4$ unchanged. We can observe that the first column of matrix $\mathbf{C}_5^L$ is the same as that of $\mathbf{D}_5^L$. Thus we only consider the HWENO procedure for constructing $\widehat{f}_{i-\frac{1}{2}}^n$:

1. Compute the linear weights γ_1, γ_2, and γ_3 such that

$$\begin{pmatrix} f_{i-2}^n & f_{i-1}^n & f_i^n & h_{i-\frac{5}{2}}^n & h_{i+\frac{1}{2}}^n \end{pmatrix} \mathbf{C}_5^L(:,1) = \gamma_1 \begin{pmatrix} f_{i-2}^n & f_{i-1}^n & h_{i-\frac{5}{2}}^n \end{pmatrix} \begin{pmatrix} -2 \\ 2 \\ 1 \end{pmatrix}$$
$$+ \gamma_2 \begin{pmatrix} f_{i-2}^n & f_{i-1}^n & f_i^n \end{pmatrix} \begin{pmatrix} -\frac{1}{6} \\ \frac{5}{6} \\ \frac{1}{3} \end{pmatrix} + \gamma_3 \begin{pmatrix} f_{i-1}^n & f_i^n & h_{i+\frac{1}{2}}^n \end{pmatrix} \begin{pmatrix} \frac{1}{4} \\ \frac{5}{4} \\ -\frac{1}{2} \end{pmatrix}, \tag{13.198}$$

where the terms multiplying the γ_is are third-order reconstructions of fluxes from the three stencils S_1, S_2, and S_3, where for the situation here we have $\gamma_1 = \frac{1}{9}$, $\gamma_2 = \frac{4}{9}$, and $\gamma_3 = \frac{4}{9}$.

2. Next we compute the smoothness indicators, denoted by β_j for each stencil S_j, which measure how smooth the function $p_j(x)$ is in the target cell I_i. The smaller this smoothness indicator β_j, the smoother the function $p_j(x)$ is in the target cell. Thus we have

$$\beta_j = \sum_{l=1}^{2} \int_{I_i} \Delta x^{2l-1} \left(\frac{\partial}{\partial x^l} p_j(x) \right)^2 dx. \tag{13.199}$$

The smoothness indicators β_j are written out explicitly as quadratic forms of the points $\left\{ f_i^n, h_{i+\frac{1}{2}}^n \right\}_i$ in the stencil,

$$\begin{aligned}
\beta_1 &= \frac{13}{3}\left(-\frac{9}{4} f_{i-2}^n + \frac{3}{2} h_{i-\frac{5}{2}}^n + \frac{3}{4} f_{i-1}^n \right)^2 + \left(\frac{31}{4} f_{i-2}^n - \frac{9}{2} h_{i-\frac{5}{2}}^n - \frac{13}{4} f_{i-1}^n \right)^2, \\
\beta_2 &= \frac{13}{12}\left(-f_{i-2}^n + 2 f_{i-1}^n - f_i^n \right)^2 + \left(-\frac{3}{2} f_i^n + 2 f_{i-1}^n - \frac{1}{2} f_{i-2}^n \right)^2, \\
\beta_3 &= \frac{13}{3}\left(-\frac{9}{4} f_i^n + \frac{3}{4} f_{i-1}^n + \frac{3}{2} h_{i+fh}^n \right)^2 + \left(\frac{5}{4} f_i^n + \frac{1}{4} f_{i-1}^n - \frac{3}{2} h_{i+\frac{1}{2}}^n \right)^2.
\end{aligned}$$

3. The next step is to compute the nonlinear weights based on the smoothness indicators:

$$\omega_j = \frac{\overline{\omega}_j}{\sum_{k=1}^{3} \overline{\omega}_k}, \quad j = 1, 2, 3, \quad \overline{\omega}_k = \frac{\gamma_k}{(\epsilon + \beta_k)^2}, \tag{13.200}$$

where in [28] the authors take $\epsilon = 10^{-6}$ to prevent the denominator becoming zero.

4. Compute numerical fluxes constructed in via HWENO. Define matrices $\widetilde{\mathbf{C}}_5^L$ and $\widetilde{\mathbf{D}}_5^L$ as

$$\begin{aligned}
\widetilde{\mathbf{D}}_5^L(:,1) &= \widetilde{\mathbf{C}}_5^L(:,1) = \omega_1 \begin{pmatrix} -2 & 2 & 0 & 1 & 0 \end{pmatrix} + \omega_2 \begin{pmatrix} -\frac{1}{6} & \frac{5}{6} & \frac{1}{3} & 0 & 0 \end{pmatrix} \\
&+ \omega_3 \begin{pmatrix} 0 & \frac{1}{4} & \frac{5}{4} & 0 & -\frac{1}{2} \end{pmatrix},
\end{aligned}$$

$$\begin{aligned}
\widetilde{\mathbf{C}}_5^L(:,k) &= \mathbf{C}_5^L(:,k), \\
\widetilde{\mathbf{D}}_5^L(:,k) &= k\mathbf{C}_5^L(:,k), \quad k = 2, \ldots, 5.
\end{aligned}$$

The updated numerical flux is computed using $\widetilde{\mathbf{C}}_5^L$ and $\widetilde{\mathbf{D}}_5^L$ as

$$\widehat{f}^n_{i-\frac{1}{2}}(\xi_0) = \begin{pmatrix} f^n_{i-2} & f^n_{i-1} & f^n_i & h^n_{i-\frac{5}{2}} & h^n_{i+\frac{1}{2}} \end{pmatrix} \widetilde{\mathbf{C}}_5^L \begin{pmatrix} 1 \\ \xi_0 \\ \xi_0^2 \\ \xi_0^3 \\ \xi_0^4 \end{pmatrix}, \tag{13.201}$$

$$h^{n+1}_{i-\frac{1}{2}} = \begin{pmatrix} f^n_{i-2} & f^n_{i-1} & f^n_i & h^n_{i-\frac{5}{2}} & h^n_{i+\frac{1}{2}} \end{pmatrix} \widetilde{\mathbf{D}}_5^L \begin{pmatrix} 1 \\ \xi_0 \\ \xi_0^2 \\ \xi_0^3 \\ \xi_0^4 \end{pmatrix}. \tag{13.202}$$

Given this setup, we now move on to a process referred to as **Strang splitting** to enable the solution to the Vlasov equations.

13.5.2 Strang splitting SL HWENO scheme for the Vlasov–Poisson system

We start by denoting the function f by $f(t,x,u) \geq 0$ to be the distribution function of electrons in phase space and $E(t,x)$ is the self-consistent electric field, the dimensionless VP system then is defined as

$$\frac{\partial f}{\partial t} + u\frac{\partial f}{\partial x} + E(t,x)\frac{\partial f}{\partial u} = 0, \tag{13.203}$$

$$\frac{DE}{Dt}(t,x) = \rho(t,x) = \int_{-\infty}^{\infty} f(t,x)\,dv - 1, \tag{13.204}$$

on the domain $[a,b] \times [-l,l]$ with periodic boundary conditions and zero boundary conditions for the velocity domain. For the Hermite method, we have that not only the distribution function f but also its derivatives in x and u directions are advected. This then implies the following equations for f_x and f_u:

$$\begin{aligned}
\frac{\partial f_x}{\partial t} + u\frac{\partial f_x}{\partial x} + \frac{\partial (E(x,t) f_v)}{\partial x} &= 0, & (13.205)\\
\frac{\partial f_v}{\partial t} + u\frac{\partial (u f_x)}{\partial x} + E(t,x)\frac{\partial f_v}{\partial v} &= 0. & (13.206)
\end{aligned}$$

When Strang splitting is applied to the set of governing equations above, we obtain

$$(SL_x)\quad \begin{cases} \dfrac{\partial f}{\partial t} + u\dfrac{\partial f}{\partial x} = 0 & (SL_x^0), \\[2ex] \dfrac{\partial f_x}{\partial t} + u\dfrac{\partial f_x}{\partial x} = 0 & (SL_x^1), \\[2ex] \dfrac{\partial f_u}{\partial t} + \dfrac{\partial u f_x}{\partial u} = 0 & (FD_x), \end{cases} \tag{13.207}$$

and

$$(SL_u)\quad \begin{cases} \dfrac{\partial f}{\partial t} + E(t,x)\dfrac{\partial f}{\partial v} = 0 & (SL_u^0), \\[2ex] \dfrac{\partial f_u}{\partial t} + E(t,x)\dfrac{\partial f_v}{\partial u} = 0 & (SL_u^1), \\[2ex] \dfrac{\partial f_x}{\partial t} + \dfrac{\partial E(t,x) f_v}{\partial x} = 0 & (FD_u). \end{cases} \tag{13.208}$$

In what follows, $f_{ij}^n \equiv f(x_i, u_j)$, $(f_x)_{ij}^n \equiv f_x(x_i, u_j)$, and $(f_u)_{i,j}^n \equiv f_u(x_i, u_j)$ at time $t = t^n$. We also have a second interval denoted by $J_j \equiv \left[u_{j-\frac{1}{2}}, u_{j+\frac{1}{2}}\right]$, and $T_{ij} \equiv \left[x_{i-\frac{1}{2}}, x_{i+f\frac{1}{2}}\right] \times \left[u_{j-\frac{1}{2}}, u_{j+\frac{1}{2}}\right]$. We now introduce two new sets of variables $\left\{\Phi_{i-\frac{1}{2},j}^n\right\}$ and $\left\{\Psi_{i,j-\frac{1}{2}}^n\right\}_{ij}$ such that

$$(f_x)_{ij}^n \equiv \frac{\Phi_{i+\frac{1}{2},j}^n - \Phi_{i-\frac{1}{2},j}^n}{\Delta x}, \qquad (f_u)_{ij}^n \equiv \frac{\Psi_{i,j+\frac{1}{2}}^n - \Psi_{ij-\frac{1}{2}}^n}{\Delta u}. \tag{13.209}$$

To solve the original partial differential equations, we first reconstruct $\Phi_{i+\frac{1}{2},j}^n$ and $\Psi_{i,j+\frac{1}{2}}^n$ in the x and u directions, respectively, as described earlier for the 1D SL HWENO. Next we have to update $\left\{f_{ij}^n, \Phi_{i+\frac{1}{2},j}^n\right\}$ by the 1D SL HWENO scheme: $\Psi_{i,j+\frac{1}{2}}^n$ is updated by the (FD_x) equation above via treating the derivative term as a coarse term, and applying a central difference scheme coupled with a trapezoid rule:

$$\begin{aligned} (f_u)_{ij}^{n+1} &= (f_u)_{ij}^n \\ &- \frac{\Delta t}{2}\left(\frac{u_{i,j-2}(f_x)_{i,j-2}^n - 8u_{i,j-1}(f_x)_{i,j-1}^n + 8u_{i,j+1}(f_x)_{i,j+1}^n - u_{i,j+2}(f_x)_{i,j+2}^n}{24\Delta u}\right. \\ &- \left.\frac{u_{i,j-2}(f_x)_{i,j-2}^{n+1} - 8u_{i,j-1}(f_x)_{i,j-1}^{n+1} + 8u_{i,j+1}(f_x)_{i,j+1}^{n+1} - u_{i,j+2}(f_x)_{i,j+2}^{n+1}}{24\Delta u}\right). \end{aligned} \tag{13.210}$$

This can be summarized in the following algorithm:

Step 1. Apply WENO limiter as a preprocessing procedure to reconstruct $\Phi^n_{i+\frac{1}{2},j}$.

First, the TVB limiter is used to identify the *troubled cells*, where these are the cells which might need the limiting procedure. Let

$$\tilde{f}_{ij} = \Phi^n_{i+\frac{1}{2},j} - f^n_{i,j}, \quad \tilde{\tilde{f}}_{i,j} = -\Phi^n_{ij} + f^n_{ij}. \tag{13.211}$$

These are modified by the modified minmod function:

$$\tilde{f}^{(\mathrm{mod})}_{ij} = \tilde{m}\left(\tilde{f}^n_{ij}, f^n_{i+1,j} - f^n_{i,j}, f^n_{i,j} - f^n_{i-1,j}\right), \tag{13.212}$$

$$\tilde{\tilde{f}}^{(\mathrm{mod})}_{ij} = \tilde{m}\left(\tilde{\tilde{f}}^n_{ij}, f^n_{i+1,j} - f^n_{i,j}, f^n_{i,j} - f^n_{i-1,j}\right), \tag{13.213}$$

where $\tilde{m}$ is given by

$$\tilde{m}(a_1, a_2, \ldots, a_n) \equiv \begin{cases} a_1, & |a_1| \le M_x (\Delta x)^2, \\ m(a_1, a_2, \ldots, a_n), & \text{otherwise,} \end{cases} \tag{13.214}$$

and the minmod function m is given by

$$m(a_1, a_2, \ldots, a_n) = \begin{cases} s \cdot \min\limits_{1 \le j \le n} |a_j|, & \text{if } \operatorname{sign}(a_1) = \operatorname{sign}(a_2) = \cdots = \operatorname{sign}(a_n) = s, \\ 0, & \text{otherwise.} \end{cases} \tag{13.215}$$

The limiter parameter $M_x > 0$ is a constant. If $\tilde{f}^{(\mathrm{mod})}_{ij} \neq \tilde{f}^n_{ij}$ or $\tilde{\tilde{f}}^{(\mathrm{mod})}_{ij} \neq \tilde{\tilde{f}}^n_{ij}$, cell T_{ij} is declared a troubled cell. If this is the case, then $\Phi^n_{i+\frac{1}{2},j}$ and $\Phi^n_{i-\frac{1}{2},j}$ in those troubled cells for WENO reconstructions.

Step 2. Perform a half-time step advection in the physical space, i.e., use equations in (13.207).

Step 3. Compute the electric field at the half-step by substituting f into Eq. (13.204) and solve for $E^*(x)$ where $*$ represents n or $n+1$.

Step 4. Similar to Step 1, apply WENO limiter as a preprocessing procedure to reconstruct $\Psi^*_{i,j+\frac{1}{2}}$.

Step 5. Perform a full time step advection in velocity space, using equations in (13.208).

Step 6. Use the same preprocessing procedure as in Step 1 to reconstruct $\Phi^{**}_{i+\frac{1}{2},j}$.

Step 7. Perform a half-time step advection in the physical space.

Cai et al. go on to provide results for both the 1D linear advection and the VP system for the following initial conditions:

1. 1D transport with initial conditions

$$f(x,0) = \begin{cases} \frac{1}{6}\left(G(x,\beta,z-\delta) + G(x,\beta,z+\delta) + 4G(x,\beta,z)\right), & -0.8 \le x \le -0.5, \\ 1, & -0.4 \le x \le -0.2, \\ 1 - |10(x-0.1)|, & 0 \le x \le 0.2, \\ \frac{1}{6}\left(F(x,\beta,a-\delta) + F(x,\beta,a+\delta) + 4F(x,\beta,z)\right), & 0.4 \le x \le 0.6, \\ 0, & \text{otherwise}, \end{cases}$$

where $G(x,\beta,z) \equiv \exp\left\{-\beta(x-z)^2\right\}$, $F(x,\beta,a) \equiv \sqrt{\max\left(1-\alpha^2(x-a)^2, 0\right)}$, $a = 0.5$, $z = -0.7$, $\delta = 0.005$, $\alpha = 10$, and $\beta = \frac{\log 2}{36\delta^2}$. Cai et al. state that for these mixture of discontinuities and smooth profiles, nonoscillatory numerical capture of discontinuities was observed.
2. 2D rigid body rotation.
3. The VP system: Two-stream instability starts with the initial distribution function:

$$f(x,u,t-0) = \frac{2}{7\sqrt{2\pi}}\left(1+5u^2\right)\left(1+\alpha\left(\frac{(\cos 2kx + \cos 3kx)}{1.2} + \cos kx\right)\right)\exp\left\{-\frac{u^2}{2}\right\}, \tag{13.216}$$

where $\alpha = 0.01$, $k = 0.5$, and the domain is $l = \frac{2\pi}{k}$.
4. Weak and strong Landau damping: These initial conditions are given by

$$f(x,u,t=0) = \frac{1}{\sqrt{2\pi}}\left(1+\alpha\cos(kx)\right)\exp\left\{-\frac{u^2}{2}\right\}, \tag{13.217}$$

where for the weak damping we have $\alpha = 0.01$ and $k = 0.5$, whereas for the strong damping we have $\alpha = 5$ and $k = 0.5$.
5. Two-stream instability, with

$$f(x,u,t=0) = \frac{1}{2u_{th}\sqrt{2\pi}}\left[\exp\left\{-\frac{(u-q)^2}{2u_{th}^2}\right\} + \exp\left\{-\frac{(u+q)^2}{2u_{th}^2}\right\}\right](1+0.05\cos kx), \tag{13.218}$$

where $q = 0.99$, $u_{th} = 0.3$, and $k = \frac{2}{13}$.

Because of the periodicity in space, Cai et al. use a fast Fourier transform to solve the 1D Poisson equation; $\rho(x,t)$ is computed by the rectangular rule, which is spectrally accurate, when the underlying function is smooth enough. The following norms in the VP system below should remain constant in time:

1. L^p-norm for $1 \le p \le \infty$,

$$\|f\|_p \equiv \left(\int_u \int_x |f(x,u,t)|^p \, dx du\right)^{\frac{1}{p}}; \tag{13.219}$$

2. Energy,

$$\text{Energy} = \int_u \int_x f(x,u,t)\,u^2 dx du + \int_x E^2(x,t)\,dx, \tag{13.220}$$

where $E(x,t)$ is the electric field;

3. Entropy,

$$\text{Entrophy} = \int_u \int_x f(x,u,t)\log\left(f(x,u,t)\right) dx dv. \tag{13.221}$$

Cai et al. state that it is expected that the scheme will conserve mass; however, the positivity of f, as stated, will not be preserved and so when numerically computing the entropy, Cai et al. use

$$\int_u \int_x f(x,u,t)\log|f(x,u,t)|\,dx dv.$$

The time step in the experiments is set to $\Delta t = \dfrac{CFL}{\dfrac{u_{\max}}{\Delta x} + \max\left(\dfrac{E(x)}{\Delta u}\right)}$, where $u_{\max}$ is the maximum velocity of the phase space mesh, with CFL specified in each experiment.

In some of the plots that are shown in [28], Cai et al. have also plotted the troubled cells. In Fig. 13.14 we have a copy of the Fig. 4.3 from [28] which shows the results from using a

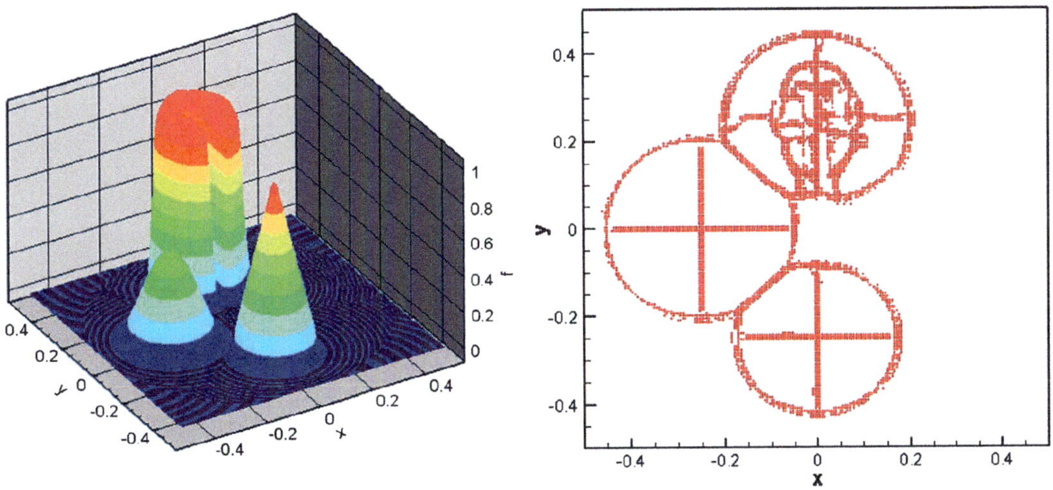

FIGURE 13.14 Copy of Fig. 4.3 from [28] which shows plots of the results where the authors have used the CSL-HWENO-W, which is the Conservative Semi-Lagrangian HWENO with limiter, and the troubled cells. Reprinted from Journal of Computational Physics. Cai, X., Qiu, J. and Qiu, J.-M., 2016: A conservative semi-Lagrangian HWENO method for the Vlasov equation, **323**, 95–114. ©(2016), with permission from Elsevier.

limiter, when solving the nonlinear partial differential equations

$$f_t - 2\pi y f_x + 2\pi x f_y = 0, \qquad (x, y) \in [-0.5, 0.5] \times [-0.5, 0.5], \tag{13.222}$$

with a prescribed $CFL = 2.2$ and the initial conditions being a slotted disk, cone, and smooth bump.

Cai et al. state in [28] that nonoscillatory capturing of discontinuities is observed, yet we can see faint ripples and wakes around the three bodies.

The second figure we show here is for the two-stream instability plots in Fig. 13.15, where we can see that the troubled cells are different for the two Courant numbers, which will

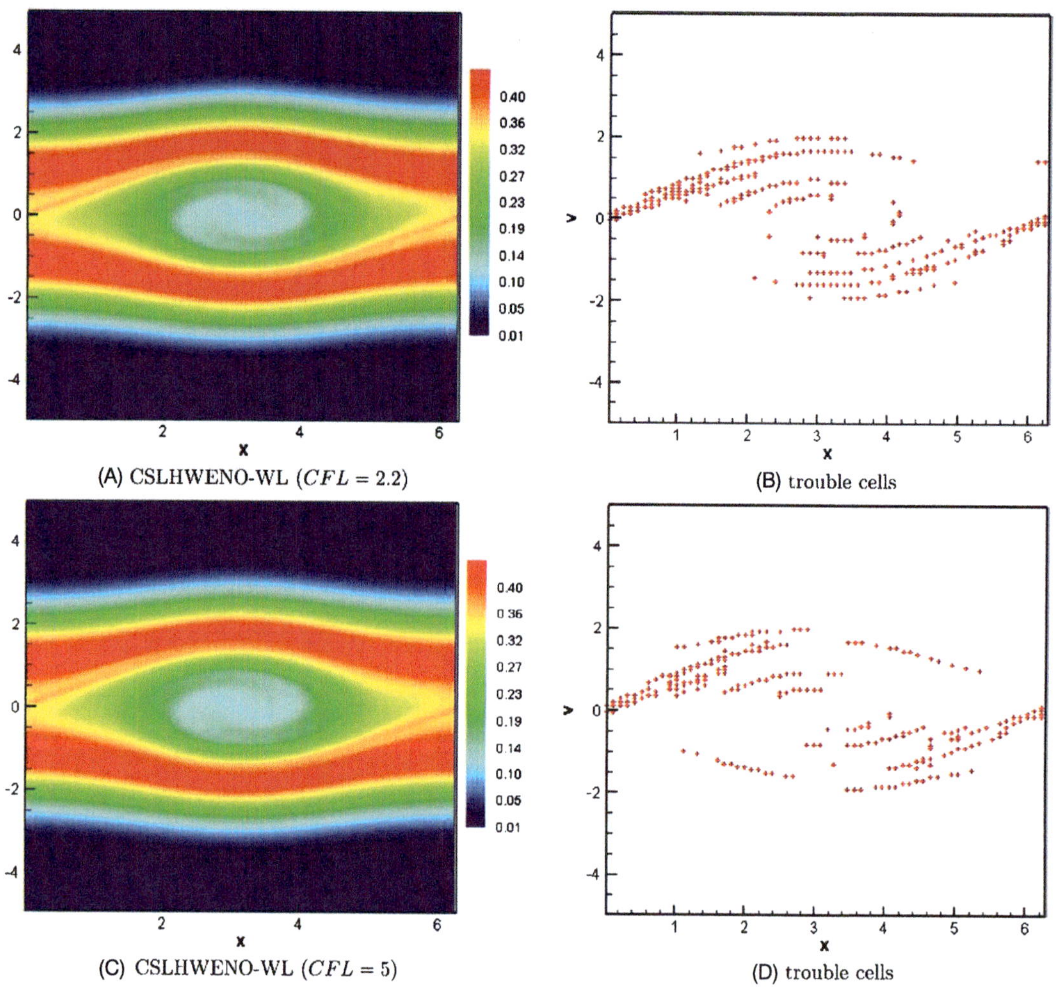

FIGURE 13.15 Copy of Fig. 4.7 from [28] of the phase space plots of the two-stream instability test case with the troubled cells. Reprinted from Journal of Computational Physics. Cai, X., Qiu, J. and Qiu, J.-M., 2016: A conservative semi-Lagrangian HWENO method for the Vlasov equation, **323**, 95–114. ©(2016), with permission from Elsevier.

change the u_{max} for the grids, and that the location of the troubled cells has moved around the central body.

The last figure we show is for the two-stream instability test case, Fig. 13.16, for the CSLWENO, CSLHWENO-WO, and CSLHWENO-WL, where the first acronym is for a similar situation to what we have presented here but not using Hermite interpolation. We can see an interesting feature in the troubled cells in that they are not confined to the main structure, but appear to also be up- and downstream of the main feature. The reason for presenting this plot is to show that semi-Lagrangian methods in this form are able to capture a feature and move around it with quite good accuracy.

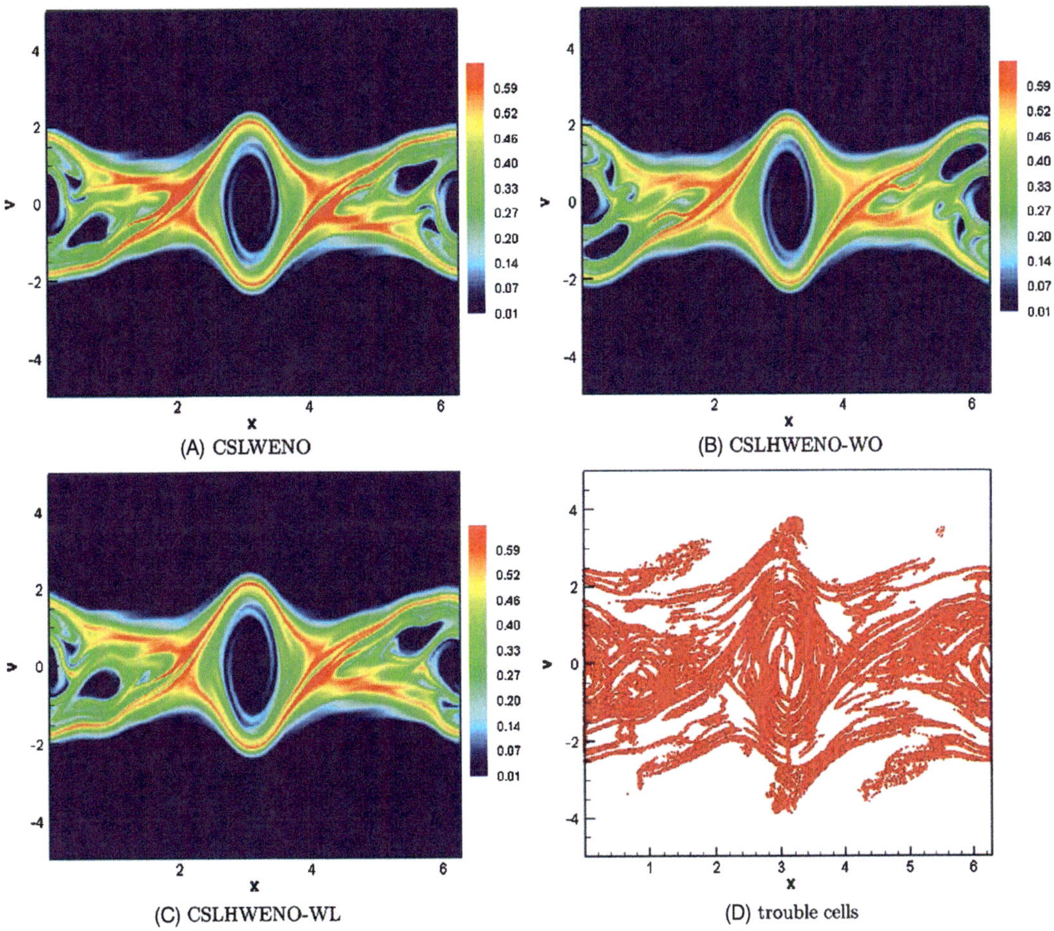

FIGURE 13.16 Figure 4.12 from [28] here: Copy of Fig. 4.12 from [28] of the plots of the phase space for the two-stream instability test case for the three different schemes mentioned in the text, along with the plot of the troubled cells. Reprinted from Journal of Computational Physics. Cai, X., Qiu, J. and Qiu, J.-M., 2016: A conservative semi-Lagrangian HWENO method for the Vlasov equation, **323**, 95–114. ©(2016), with permission from Elsevier.

13.6 Summary

In this chapter we have introduced many different approaches that have been developed to try and reduce the errors associated with the semi-Lagrangian approaches of not being mass preserving and shape preserving. We have introduced different interpolation polynomials to address these issues and recast the problem into integral form to apply a finite volume based approach in flux form for the transport problem, the shallow water equations model in spherical coordinates, and Cartesian coordinates for a transport problem where we presented the SLICE scheme. We have introduced the powerful and useful cascade interpolation methods and seen them applied in varying complex models up to a deep atmosphere nonhydrostatic configuration. We have also introduced a new set of differential equations that have a form of highly nonlinear advection, referred to as the Vlasov family, and shown that it is possible to use a finite volume based approach with Hermite weighted essentially nonoscillatory (HWENO) scheme.

We had finish this chapter somewhere, and we acknowledge that there are many other techniques we could have included, but it was starting to head towards a book in itself.

CHAPTER

14

Tangent linear modeling and adjoints of semi-Lagrangian methods

An important part of four-dimensional variational and hybrid data assimilation methods is the ability to run a tangent linear, and its adjoints, backward-in-time integration. As tangent linear model involves linearization and small perturbations, the need to avoid discontinuities, as well as sharp gradients, is quite important. We will find a condition that will be quite restrictive on the ability of the tangent linear model and its adjoints of semi-Lagrangian schemes to remain stable. Adjoints are also an important tool for singular vectors to identify sensitive areas and error propagation, as well as for observation impact studies and OSSEs, [61], and so the ability to form some form of approximation to the tangent linear model and the adjoint of a semi-Lagrangian approach is critical.

We start this chapter with an introduction to forming the tangent linear approximation and then move on to its adjoint, before looking at how these approaches are applicable to the semi-Lagrangian method.

14.1 Derivation of the linearized model

If we consider a general nonlinear initial value problem denoted by

$$y_i = \mathcal{M}\left(x_1, x_2, \ldots, x_N\right), \tag{14.1}$$

where n represents the time index, N is the total number of grid points in the numerical grid, x_js are the discrete model variables, and the y_is are the outputs from the numerical model, then what we may be seeking, for example, in applications of data assimilation, are answers to the questions of how $x\left(t^{n+1}\right)$ changes with respect to $x_i\left(t^n\right)$ and, when we are analyzing the output y_i, what features in x_j caused this. It is possible to attempt to quantify answers to these two questions through considering tangent linear models and the adjoint model.

To derive the linearized model, we start by expressing the x_js as either a background, or reference state, and a perturbation such that

$$x_j = \overline{x}_j + \delta x_j. \tag{14.2}$$

We are trying to ascertain how the nonlinear model is affected by the perturbation to the state, x_j, which can be approximated through considering the difference between the outputs from

https://doi.org/10.1016/B978-0-12-817222-3.00018-9

$y(\overline{x}+\delta x)$ and $y(\overline{x})$. We have plotted the situation just described in Fig. 14.1, where we see that we can approximate the change in the output by δy which is approximated through the tangent approximation.

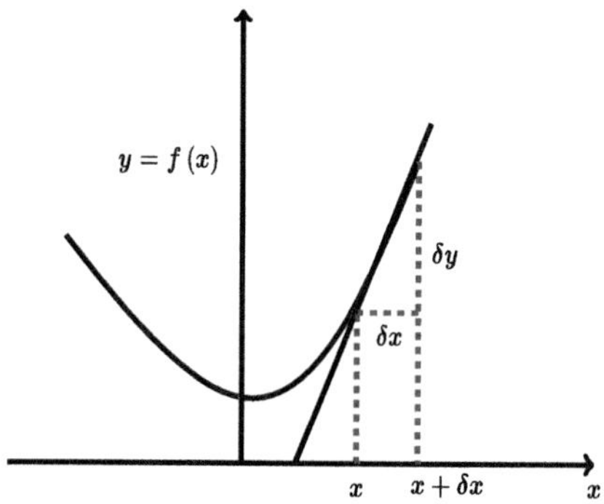

FIGURE 14.1 Schematic of the tangent linear approximation to the function $f(x)$.

Mathematically we can express the gradient as

$$\frac{dy}{dx} \approx \frac{\delta y_i}{\delta x_i} \Longrightarrow \delta y_i \approx \left(y_i\left(\overline{x}+\delta x\right) - y_i\left(\overline{x}\right)\right)\delta x_i. \tag{14.3}$$

We now substitute the nonlinear model in for y in (14.3) and expand the nonlinear model about the reference state to obtain

$$\begin{aligned} \delta y_i &= y_i\left(\overline{x}+\delta x\right) - y_i\left(\overline{x}\right) = \mathcal{M}\left(\overline{x}_1+\delta x_1, \overline{x}_2+\delta x_2, \ldots, \overline{x}_N+\delta x_N\right) - \mathcal{M}\left(\overline{x}_1, \overline{x}_2, \ldots, \overline{x}_N\right) \\ &= \mathcal{M}\left(\overline{x}_1, \overline{x}_2, \ldots, \overline{x}_N\right) + \frac{\partial \overline{y}_i}{\partial x_1}\delta x_1 + \frac{\partial \overline{y}_i}{\partial x_2}\delta x_2 + \cdots - \mathcal{M}\left(\overline{x}_1, \overline{x}_2, \ldots, \overline{x}_N\right), \\ \delta y_i &= \sum_j^N \overline{\frac{\partial y_i}{\partial x_j}}\delta x_j. \end{aligned} \tag{14.4}$$

The expression in (14.4) is the **tangent linear model**, which is quite often abbreviated to TLM, and it gives an approximation for the growth of the perturbations, δx_j, $j = 1, 2, \ldots, N$. The tangent linear model enables us to address the question of how y_i changes with respect to x_j, for $j = 1, 2, \ldots, N$ and $i = 1, 2, \ldots, M$, where M is the number of time steps the numerical model has taken. Given the answer to the first question, we now consider the second question posed earlier, which we do through introducing the adjoint operator.

14.2 Adjoints

As just mentioned, we now consider how to address the second question of linking the behavior in y_i to x_j. We start by introducing a scalar measure of the outputs

$$J = J\left(\mathbf{y}\left(\mathbf{x}\right)\right). \tag{14.5}$$

Next we expand (14.5) in a Taylor series, which leads to a change in the measure, δJ, as

$$\delta J = \sum_{i=1}^{M} \frac{\partial \overline{J}}{\partial y_i} \delta y_i, \tag{14.6}$$

but we can also consider variations with respect to the x_js which results in

$$\delta J = \sum_{j=1}^{N} \frac{\partial \overline{J}}{\partial x_j} \delta x_j, \tag{14.7}$$

where (14.6) and (14.7) are **equal when both functions are linear**.

It is (14.7) which is of interest here as it depends on the perturbation inputs. Given the way that we defined J, it is possible to differentiate using the chain rule to obtain estimates for $\frac{\partial \overline{J}}{\partial x_j}$ as

$$\frac{\partial \overline{J}}{\partial x_j} = \sum_{i=1}^{M} \overline{\frac{\partial y_i}{\partial x_j} \frac{\partial \overline{J}}{\partial y_i}}. \tag{14.8}$$

An important, and quite often seen as being very powerful, property of (14.8) is that it enables us to integrate the future time gradient, or sensitivities, **backwards** to the initial time.

The first feature to notice here is that the tangent linear model sums the terms $\frac{\partial y_i}{\partial x_j}$ over the js, while the summation in (14.8) is over the is. To help keep track of the different terms here, we will now use matrix–vector notation. First let the matrix $\mathbf{M}$ be defined as having entries

$$\mathbf{M}_{i,j} \equiv \frac{\partial y_i}{\partial x_j}, \tag{14.9}$$

then for the tangent linear model we see that we are summing over the columns of $\mathbf{M}$, while for the model given by (14.8) we are summing over the rows of $\mathbf{M}$. From the rules of matrix–vector multiplication and introducing the vectors $\delta \boldsymbol{y}$ and $\delta \boldsymbol{x}$, we can write (14.4) as

$$\delta \boldsymbol{y} = \mathbf{M} \delta \boldsymbol{x}. \tag{14.10}$$

We can then write (14.8) as

$$\frac{\partial J}{\partial \boldsymbol{x}} = \mathbf{M}^T \frac{\partial J}{\partial \boldsymbol{y}}. \tag{14.11}$$

The matrix $\mathbf{M}$ as defined in (14.9) is referred to as either the **resolvent** of the tangent linear model, or as the **Jacobian** of the nonlinear model. Given the definition for the resolvent, it is possible to write the perturbation to the nonlinear model as

$$\mathcal{M}(\overline{\boldsymbol{x}} + \delta \boldsymbol{x}) = \mathcal{M}(\overline{\boldsymbol{x}}) + \mathbf{M} \delta \boldsymbol{x}, \tag{14.12}$$

which is equivalent to the expression in the introduction to this chapter. As an aside, depending on the papers or textbooks one reads, the resolvent is also denoted by **L**. The model that maps sensitivities backwards is referred to as the **adjoint model** as it uses the adjoint of the tangent linear model.

We now consider the derivation of the tangent linear model and its adjoint for the one-dimensional tracer advection problem in its Eulerian form, for the tracer $q\left(x,u,t\right)$, which is given by

$$\frac{\partial q}{\partial t}+u\frac{\partial q}{\partial x}=0. \tag{14.13}$$

The first step in deriving the tangent linear and adjoint models is to introduce perturbations for q and u as $\overline{q}+\delta q$ and $\overline{u}+\delta u$, and then substitute these perturbations into (14.13). This then results in the continuous tangent linear equations as

$$\frac{\partial \delta q}{\partial t}+\delta u\frac{\overline{q}}{\partial x}+\overline{u}\frac{\partial \delta q}{\partial x}+\delta u\frac{\partial \delta q}{\partial x}=0. \tag{14.14}$$

The second order term in δ is ignored in the tangent linear approximation. If we now discretize (14.14) with a forward upwind scheme in both space and time, then we have

$$\delta q_j^{n+1}=\delta q_j^n-\Delta t\left(\delta u_j^n\frac{\overline{q}_{j+1}^n-\overline{q}_j^n}{\Delta x}-\overline{u}_j^n\frac{\delta q_{j+1}^n-\delta q_j^n}{\Delta x}\right)=0. \tag{14.15}$$

Before we progress any further, we introduce the notion of the **active** variables for the derivation of the adjoint. The definition of the active variables comes from Kalnay (2003) [91], where active variables are those who's values are used to **modify** a set, or all, of the active variables. An important feature to ensure when deriving the adjoint is that the active variables are **not** modified. One approach to ensure that unmodified variables remain such is achieved through including the condition $q_j^n=q_j^n$ as an example into the numerical model.

Returning to the advection example, we can identify that we have four active variables in (14.15), δq_{j+1}^n, δq_j^n, δu_j^n, and δq_j^{n+1}, but only one of them is modified, namely q_j^{n+1}. Therefore, we need to ensure that the remaining three are not modified in the tangent linear model. Thus, we can write (14.15) as a matrix–vector multiplication, which is given by

$$\begin{pmatrix}\delta q_{j+1}^n\\ \delta q_j^n\\ \delta u_j^n\\ \delta q_j^{n+1}\end{pmatrix}=\begin{pmatrix}1 & 0 & 0 & 0\\ 0 & 1 & 0 & 0\\ 0 & 0 & 0 & 1\\ 1+b_j^n & -b_j^n & a_j^n & 0\end{pmatrix}\begin{pmatrix}\delta q_{j+1}^n\\ \delta q_j^n\\ \delta u_j^n\\ \delta q_j^{n+1}\end{pmatrix}, \tag{14.16}$$

where

$$a_j^n=-\frac{\Delta t}{\Delta x}\left(\overline{q}_{j+1}^n-\overline{q}_j\right),\quad b_j^n\equiv\overline{u}_j^n\frac{\Delta t}{\Delta x}.$$

Introducing the adjoint variables $\begin{pmatrix} \widehat{\delta q_{j+1}^{n}} \\ \widehat{\delta q_{j}^{n}} \\ \widehat{\delta u_{j}^{n}} \\ \widehat{\delta q_{j}^{n+1}} \end{pmatrix}$, enables us to obtain the adjoint equations through taking the transpose of the matrix in (14.16) and multiplying out the matrix–vector products, which results in

$$
\begin{aligned}
\widehat{\delta q_{j+1}^{n}} &= \widehat{\delta q_{j+1}^{n}} - b_j^n \widehat{\delta q_{j+1}^{n+!}}, \\
\widehat{\delta q_{j}^{n}} &= \widehat{\delta q_{j}^{n}} + \left(1 + b_j^n\right) \widehat{\delta q_{j+1}^{n+1}}, \\
\widehat{\delta u_{j}^{n}} &= \widehat{\delta u_{j}^{n}} + a_j^n \widehat{\delta q_{j+1}^{n+1}}, \\
\widehat{\delta q_{j+1}^{n+1}} &= 0.
\end{aligned}
$$

Exercise 14.1. *Derive the tangent linear model and the adjoint for a centered-time, centered-space scheme applied to* (14.13).

Exercise 14.2. *Derive the tangent linear model and the adjoint for the Lax–Wendroff scheme applied to* (14.13).

14.3 Test of the tangent linear and adjoint models

There are a series of tests that can be applied to verify if the derivation and the coding of the tangent linear and adjoint models are correct. If we consider the tangent linear model first, then we have shown that if we have a nonlinear $\mathcal{M}(\boldsymbol{x})$, and $\mathbf{M}$ is the tangent linear model, then for small perturbations $\delta \boldsymbol{x}$, we have

$$\mathcal{M}(\boldsymbol{x} + \delta \boldsymbol{x}) - \mathcal{M}(\boldsymbol{x}) \approx \mathbf{M} \gamma \delta \boldsymbol{x}.$$

If we now define the relative error as

$$E_R = \frac{\mathcal{M}(\boldsymbol{x} + \gamma \delta \boldsymbol{x}) - \mathcal{M}(\boldsymbol{x})}{\mathbf{M} \gamma \delta \boldsymbol{x}}, \tag{14.17}$$

then, provided the tangent linear approximation is a good approximation, as $\gamma \to 0$ we should have that the relative error tends to zero.

Now if we consider the adjoint model, then we have the following identity for the tangent linear and adjoint model:

$$\langle \mathbf{M} \delta \boldsymbol{x}, \mathbf{M} \delta \boldsymbol{x} \rangle = \left\langle \delta \boldsymbol{x}, \mathbf{M}^T \mathbf{M} \right\rangle, \tag{14.18}$$

for an inner product $\langle \cdot, \cdot \rangle$ and any $\delta \boldsymbol{x}$. This is a mechanism to test the accuracy of the adjoint calculation.

The third measure to ascertain the accuracy of the adjoint derivation and coding is through what is referred to as the **gradient test**. If we recall that we have a functional that is associated with the adjoint given by J, then the gradient of J is ∇J, and so it is possible to check that the gradient of the functional has been coded correctly through the identity

$$\Psi(\alpha) = \frac{J(\boldsymbol{x} + \alpha \boldsymbol{h}) - J(\boldsymbol{x})}{\alpha \boldsymbol{h}^T \nabla J(\boldsymbol{x})} = 1 + O(\alpha), \tag{14.19}$$

where $\boldsymbol{h}$ is a vector of unit length, often taken to be $\frac{\nabla J(\boldsymbol{x})}{||\nabla J(\boldsymbol{x})||_2}$. Therefore, for some values of α away from the accuracy of the machine precision, if the gradient is accurate then we would expect the function $\Psi(\alpha)$ to be approximately 1.

We now move on to consider an alternative to deriving the adjoint directly, namely to using the computer code of the tangent linear model to obtain the adjoint, which is referred to as **differentiating the code**.

14.4 Differentiating the code to derive the adjoint

In the derivations of the adjoints that we have presented so far, we have used either the continuous forms to derive the tangent linear equations and then derived the continuous form of the adjoint equations, or we have taken the discrete equations for the linear advection equations into a matrix, and then transposed that matrix to form the adjoint. It is quite often the case that we have a set of computer codes of the tangent linear model, and it is shown in [136] that it is possible to *differentiate the computer code* rather than having to derive the equations themselves. We now present a summary of the appendix from Navon et al. (1992) where the authors show the equivalency of differentiating the code to transposing the tangent linear matrix.

If we consider a linear model which consists of a DO or FOR loop, depending on the computing language one is using, then we have

$$\begin{aligned} &\text{DO } I = 1 : N-1 \\ &X(I) = aY(I+1) \\ &\text{END DO} \end{aligned}$$

where X and Y are vectors. This DO loop is equivalent to the following algebraic matrix equations:

$$\begin{pmatrix} X(1) \\ X(2) \\ \vdots \\ X(N-1) \end{pmatrix} = \begin{pmatrix} 0 & a & 0 & \dots & 0 \\ 0 & 0 & a & \dots & 0 \\ \vdots & \vdots & \vdots & \ddots & \vdots \\ 0 & 0 & 0 & \dots & a \end{pmatrix} \begin{pmatrix} Y(1) \\ Y(2) \\ \vdots \\ Y(N) \end{pmatrix}, \tag{14.20}$$

if the $Y(I)$s are not reused outside of the loop, or alternatively, if the variables are used outside of the loop then the matrix equation becomes

$$\begin{pmatrix} Y(1) \\ Y(2) \\ \vdots \\ Y(N) \\ X(1) \\ X(2) \\ \vdots \\ X(N) \end{pmatrix} = \begin{pmatrix} 1 & 0 & 0 & \dots & 0 \\ 0 & 1 & 0 & \dots & 0 \\ 0 & 0 & 1 & \dots & 0 \\ \vdots & \vdots & \vdots & \ddots & \vdots \\ 0 & 0 & 0 & \dots & 1 \\ 0 & a & 0 & \dots & 0 \\ 0 & 0 & a & \dots & 0 \\ \vdots & \vdots & \vdots & \ddots & \vdots \\ 0 & 0 & 0 & \dots & a \\ 0 & 0 & 0 & \dots & 0 \end{pmatrix} \begin{pmatrix} Y(1) \\ Y(2) \\ \vdots \\ Y(N) \end{pmatrix}. \tag{14.21}$$

We know that the adjoints of (14.20) and (14.21) are given by the transposes of the matrices in these equations which yields

$$\begin{pmatrix} \widehat{Y(1)} \\ \widehat{Y(2)} \\ \vdots \\ \widehat{Y(N)} \end{pmatrix} = \begin{pmatrix} 0 & 0 & 0 & \dots & 0 \\ a & 0 & 0 & \dots & 0 \\ 0 & a & 0 & \dots & 0 \\ \vdots & \vdots & \vdots & \ddots & \vdots \\ 0 & 0 & 0 & \dots & 0 \end{pmatrix} \begin{pmatrix} \widehat{X(1)} \\ \widehat{X(2)} \\ \vdots \\ \widehat{X(N-1)} \end{pmatrix} \tag{14.22}$$

for (14.20), and

$$\begin{pmatrix} \widehat{Y(1)} \\ \widehat{Y(2)} \\ \vdots \\ \widehat{Y(N)} \end{pmatrix} = \begin{pmatrix} 1 & 0 & 0 & \dots & 0 & 0 & 0 & \dots & 0 & 0 \\ 0 & 1 & 0 & \dots & 0 & a & 0 & \dots & 0 & 0 \\ 0 & 0 & 1 & \dots & 0 & 0 & a & \dots & 0 & 0 \\ \vdots & \vdots & \vdots & \ddots & \vdots & \vdots & \vdots & \ddots & \vdots & \vdots \\ 0 & 0 & 0 & \dots & 1 & 0 & 0 & \dots & a & 0 \end{pmatrix} \begin{pmatrix} \widehat{Y(1)} \\ \vdots \\ \widehat{Y(N)} \\ \widehat{X(1)} \\ \vdots \\ \widehat{X(N)} \end{pmatrix} \tag{14.23}$$

for (14.21). Therefore the adjoint of the DO loop will be in one of two forms, either

$$\begin{aligned} &\text{DO } I = 1 : N - 1 \\ &YHAT(I+1) = aXHAT(I) \\ &\text{END DO} \end{aligned}$$

or

$$\begin{aligned} &\text{DO } I = 1 : N - 1 \\ &YHAT(I+1) = YHAT(I+1) + aXHAT(I) \\ &\text{END DO} \end{aligned}$$

The general rules for coding adjoints are:

1. Make a copy of the original tangent linear program.
2. Remove all of the lines in the program that are not acting upon the perturbations.
3. Reverse the order of **all** the operations, including the loops!
4. For each line inside of the loop:
   - (i) For each perturbation that occurs on the right-hand side of the equal sign in the tangent linear model, create a new line with that perturbation being the left-hand side variable now.
   - (ii) Set these perturbations equal to themselves and then with a "+".
   - (iii) Take the original coefficient that was multiplying the perturbation, multiply it by the old left-hand side tangent linear variable, and then add it on to the perturbation.
5. Set the old left-hand side variable to zero.

Given the theory that we have presented here, we now go on to consider how we can apply it to some semi-Lagrangian integration schemes.

14.5 Tangent linear approximations to semi-Lagrangian schemes

The paper by Polavarapu et al. (1996) [144] provides a rigorous derivation of a consistent approach in deriving the tangent linear model for semi-Lagrangian schemes. In [144] the authors state the initial problem associated with a tangent linear model of a semi-Lagrangian advection scheme through considering Burger's equation with no viscosity, which is given by

$$\frac{\partial u}{\partial t}+u\frac{\partial u}{\partial x}=0.$$

They consider a two-time-step semi-Lagrangian scheme of the equation above given by

$$u\left(x_i,t^{n+1}\right)=u\left(x_i-\alpha_i,t^n-\Delta t\right), \tag{14.24}$$

where the subscript i refers to the ith grid point and α_i is determined through a second-order in time approximation to

$$\frac{dx}{dt}=u\left(x,t\right). \tag{14.25}$$

The velocity, as we know for the nonconstant case, needs to be interpolated between grid points, if it does not coincide with a grid point. In [144] the authors present an example where there is regular grid spacing and use linear interpolation so that

$$u\left(x_i-\alpha_i,t^n-\Delta t\right)=au\left(x_{j-1},t^n-\Delta t\right)+bu\left(x_j,t^n-\Delta t\right), \tag{14.26}$$

where

$$b=\frac{x_i-\alpha_i-x_j-1}{\Delta x}, \qquad a=1-b.$$

As we have seen many times now, we need to determine the index of the nearest gridpoint to the apparent departure, or upstream interpolation, point. Using FORTRAN, they state in [144], that this is equivalent to the step

$$j = \text{INT}\left(\frac{x_i - \alpha_i}{\Delta x}\right) + 1.$$

However, INT is an intrinsic FORTRAN function, there is an equivalent function in MATLAB®, *floor*. Unfortunately, these steps are not differentiable; also note that j is a function of the control variable α_i, while j is not. Therefore, the tangent linear code is linearized about the determined grid interval, but it should be noted that a perturbed interpolation point **could lie outside this interval**.

The **tangent linear model**, as we have seen, is derived though considering a model that is denoted by $\mathcal{M}(\boldsymbol{x})$ which could be linear, or nonlinear, that updates the state variable $\boldsymbol{x}$ as

$$\boldsymbol{x}^n = \mathcal{M}\left(\boldsymbol{x}^{n-1}\right), \tag{14.27}$$

where n is the time index, $\mathcal{M}$ is a vector-valued function of dimension N, with N being the number of the time steps we require to move the initial conditions $\boldsymbol{x}_0$ forward in time. The next step is to denote a specific trajectory through time and space by $\overline{\boldsymbol{x}}^n$, for $n = 0, 1, \ldots, N$, that satisfy (14.27); we consider perturbations that are also referred to as variations about this reference trajectory that evolve according to

$$\delta\boldsymbol{x}^n = \mathcal{M}\left(\overline{\boldsymbol{x}}^{n-1} + \delta\boldsymbol{x}^{n-1}\right) - \mathcal{M}\left(\overline{\boldsymbol{x}}^{n-1}\right). \tag{14.28}$$

We now introduce the *tangent linear hypothesis* which states that for small perturbations, where by small we mean that $|\delta\boldsymbol{x}| \ll |\overline{\boldsymbol{x}}|$, an associated good approximation to (14.28) is given by the linear part of the nonlinear variation

$$\frac{d\mathcal{M}}{d\boldsymbol{x}}\left(\overline{\boldsymbol{x}}^{n-1}\right)\delta\boldsymbol{x}^{n-1}. \tag{14.29}$$

Therefore, by definition, the tangent linear model describes the evolution of the linear, sometimes referred to as the first-order variation, which is tangent to the reference trajectory in phase space. Thus the tangent linear model should contain all of the linear parts of the nonlinear variation and is therefore equal to the expression in (14.29). It is stated in [144] that the hypotheses is true for continuous, or theoretical, models. However, in numerical geophysical modeling, we have discrete numerical models where the tangent linear model is obtained through differentiating the codes line by line, which is what is done in [136] and [111], and due to the nondifferentiability of discrete programming structures, the tangent linear model is not necessarily equal to the linear variation. This is an important point that needs to be kept in mind when deriving the tangent linear model.

In [144] the authors introduce the concept of **correctness**, which they state refers to the ability of the model to be asymptotic to the linear part of the nonlinear variation as the size of the perturbation tends to zero. Given the property of correctness, the authors of [144] devise a measure to distinguish between the linear variation predicted by the tangent linear model, $(\delta \boldsymbol{x}^n)^L$, defined as

$$(\delta \boldsymbol{x}^n)^L = \boldsymbol{L}\left(\boldsymbol{x}^{n-1}\right)\delta \boldsymbol{x}^{n-1}, \tag{14.30}$$

where $\boldsymbol{L}$ is the linear operator, and the linear part of the nonlinear variation which is given by (14.29). The tangent linear model is correct when it is equal to the linear variation, that is to say, when $\boldsymbol{L} = \frac{d\mathcal{M}}{dx}$. In [144] they next define the **linearization error** which is given by

$$E \equiv \delta \boldsymbol{x}^n - (\delta \boldsymbol{x}^n)^L = \mathcal{M}\left(\overline{\boldsymbol{x}}^{n-1} + \delta \boldsymbol{x}^{n-1}\right) - \mathcal{M}\left(\overline{\boldsymbol{x}}^{n-1}\right) - (\delta \boldsymbol{x}^n)^L. \tag{14.31}$$

Expanding the nonlinear model in (14.31) in a Taylor series and considering the ith component of $\boldsymbol{x}$, it can be shown that this expansion is equal to

$$\begin{aligned} E_i &= \sum_{j=1}^{N}\left(\frac{\partial \mathcal{M}_i}{\partial x_j}\left(\overline{\boldsymbol{x}}^{n-1}\right) - l_{ij}\left(\overline{\boldsymbol{x}}^{n-1}\right)\right)\delta x_j^{n-1} \\ &+ \frac{1}{2}\sum_{j=1}^{N}\sum_{k=1}^{N}\frac{\partial^2 \mathcal{M}_i}{\partial x_j \partial x_k}\left(\overline{\boldsymbol{x}}^{n-1}\right)\delta x_j^{n-1}\delta x_k^{n-1} + o\left(\left(\delta \boldsymbol{x}^{n-1}\right)^3\right). \end{aligned} \tag{14.32}$$

The l_{ij} term in (14.32) refers to the (i,j)th element of $\boldsymbol{L}$. Thus the tangent linear model is correct if and only if for all i and all $\delta \boldsymbol{x}^{n-1}$, E_i is second-order in terms of the perturbed variables as $\delta \boldsymbol{x}^{n-1}$ tends to zero. The reader is recommended to read [144] for more details about the interpretation of the linearization error. What is important for us here is how this measure is used to ascertain the tangent linear approximation, and therefore by association, the adjoint of the semi-Lagrangian schemes.

14.5.1 Linearizing semi-Lagrangian interpolation

Again in this subsection we carry on following the theory set out in [144] as we did in the previous subsection. We start by considering a continuously differentiable function $y(x)$ with independent variable x. The interpolation of y at a given point ξ could be described as

$$\eta = y(\xi). \tag{14.33}$$

As a result of $y(x)$ being known everywhere, the interpolation has simply become a function evaluation here. However, it is more likely the case that y is only known at a finite number of points so that $y(x)$ would be approximated by some interpolating function. The next step is to assume $y(x)$ and ξ consist of two parts, $y(x) = \overline{y(x)} + \delta y(x)$ and $\xi = \overline{\xi} + \delta\xi$, where the two perturbations just mentioned are small. The tangent linear variation of η due to the variation in the function itself, $\delta y(x)$, and the point of interpolation, $\delta\xi$, is

$$\delta\eta^L = \delta y\left(\overline{\xi}\right) + \left.\frac{d\overline{y}}{dx}\right|_{x=\overline{\xi}}\delta\xi. \tag{14.34}$$

In [144], the expression in (14.34) is labeled the *tangent linear equation* as it is not a model. From (14.33) we can see that η is a linear function of $y(x)$ but in general is a nonlinear function of ξ. If we had the case were there was no variation in ξ, then the interpolation would be a linear process. Therefore in [144] the authors consider the case where there are only variations in ξ and set $\delta y(x) = 0$ to isolate the nonlinearity of the interpolation and to identify conditions for the linearization. Given these assumptions, it can be shown that the linearization error for this situation is given by

$$E = \overline{y}\left(\overline{\xi} + \delta\xi\right) - \overline{y}\left(\overline{\xi}\right) - \frac{d\overline{y}}{dx}\left(\overline{\xi}\right)\delta\xi. \tag{14.35}$$

Given the expression in (14.35), we now turn our attention to deriving the tangent linear equation for the interpolation operator in the semi-Lagrangian methods where we have a grid with $N+1$ grid points that are denoted by $x_0, x_1, \ldots, x_N$, and at these grid points we have values of the function y denoted by $y_0, y_1, \ldots, y_N$ and denote a piecewise interpolation function by $P(x)$ that takes the values of y_j at x_j. We now assume that the original interpolation point, $\overline{\xi}$, lies in the half-open interval $\left[x_{j-1}, x_j\right)$ and that the perturbed interpolation point lies in the half-open interval

$$\overline{\xi} + \delta\xi \in \left[x_{i-1}, x_i\right). \tag{14.36}$$

We have presented a copy of the figure from [144] that provides an illustration of where the perturbed departure point could end up in, relative to the departure point in Fig. 14.2.

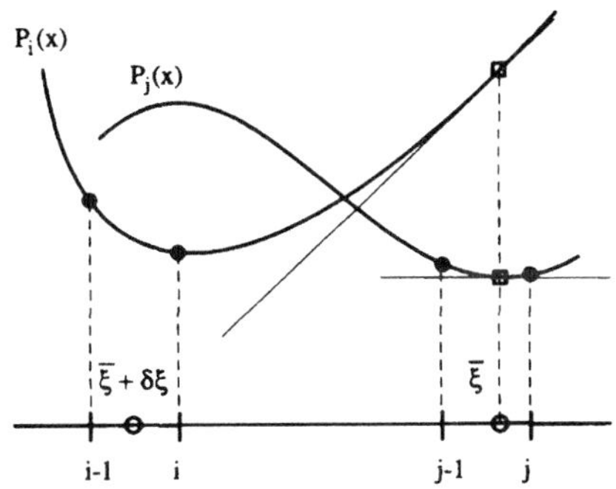

FIGURE 14.2 Copy of Fig. 1 from Polavarapu, S. and Tanguay, M. and Mànard and Staniforth, A, 1996: The tangent linear model for semi-Lagrangian schemes: linearizing the process of interpolation, *Tellus*, **48A**, 74–96. https://www.tandfonline.com/doi/abs/10.3402/tellusa.v48i1.11633.

We now consider the nonlinear variation which is determined by

$$P_i\left(\overline{\xi} + \delta\xi\right) - P_j\left(\overline{\xi}\right), \tag{14.37}$$

where the subscripts refer to the interval for which the interpolation functions is appropriate. Therefore, the tangent linear equation for this situation is

$$\delta P_j\left(\overline{\xi}\right) = \frac{dP_j}{d\xi}\left(\overline{\xi}\right)\delta\xi \tag{14.38}$$

and, according to (14.35), the linearization error is

$$E = P_i\left(\overline{\xi} + \delta\xi\right) - P_j\left(\xi\right) - \frac{dp_j}{d\xi}\left(\overline{\xi}\right)\delta\xi, \tag{14.39}$$

which can be expanded in a Taylor series to show that

$$E = \left[P_i\left(\overline{\xi}\right) - P_j\left(\xi\right)\right] + \left[\frac{dP_i}{d\xi}\left(\overline{\xi}\right) - \frac{dP_j}{d\xi}\left(\overline{\xi}\right)\right]\delta\xi + \frac{1}{2}\frac{d^2P_i}{d\xi^2}\left(\overline{\xi}\right)\left(\delta\xi\right)^2 + O\left(\left(\delta\xi\right)^3\right). \tag{14.40}$$

As pointed out in [144], we have to note here that the evaluation of P_i and its derivatives at $\overline{\xi}$ involves **extrapolation** of this function beyond the interval for which it is defined. Therefore, for the remainder of this derivation, all of the interpolating functions must be defined over the whole domain, such as polynomials. Given this definition of the linearization error, Polavarapu et al. point out that there are several points to note regarding the linearization error for piecewise continuous interpolating functions:

- If the functions P_i and P_j are not equal at $\overline{\xi}$, then there is a zeroth-order term in $\delta\xi$.
- If the first derivatives of P_i and P_j are not equal at $\overline{\xi}$, then there is a first-order term in $\delta\xi$.
- If the functions P_i and P_j, as well as their first derivatives, are equal at $\overline{\xi}$, then there is no linearization error even if higher-order derivatives are not equal at $\overline{\xi}$.
- The second- and higher-order terms involve extrapolating the function appropriate for the interval containing $\overline{\xi} + \delta\xi$ to the interval containing $\overline{\xi}$. Thus the better this extrapolation, the smaller the linearization error and the larger the range of validity of the tangent linear hypothesis.

Therefore, it easy to see that the TLE is correct for all $\overline{\xi}$ if and only if $P_i = P_j$, for all i and j. An important remark not highlighted in [144] is that the perturbed interpolation/departure point is determined by the variation of the velocity, which is continuous. Thus the perturbation departure point is likely to remain within one or two grid intervals of the unperturbed departure point. We now consider the following two cases to determine the effect on the linearization error:

Case 1: If the departure point and the perturbed departure point lie in the same grid interval, then $i = j$ and

$$E = \frac{1}{2}\frac{d^2Pj}{d\xi^2}\left(\xi\right) + O\left(\left(\delta\xi\right)^3\right). \tag{14.41}$$

This then implies that the TLE is correct since the interpolating function and its first derivative are continuous within a grid interval.

Test 2: If the departure point lands on a grid point, $\overline{\xi} \equiv \overline{\xi}_{j-1}$, and since polynomials are continuous at x_{j-1}, then the error is

$$E = \left[\frac{dP_{j-1}}{d\xi}\left(x_{j-1}\right) - \frac{dP_j}{d\xi}\left(x_{j-1}\right)\right]\left(\delta\xi\right) + \frac{1}{2}\frac{d^2P_{j-1}}{d\xi^2}\left(x_{j-1}\right)\left(\delta\xi\right)^2 + O\left(\left(\delta\xi\right)^3\right). \tag{14.42}$$

For this situation the TLE is correct if and only if the interpolating function has continuous first derivatives at gridpoints.

Given the expression of the tangent linear model for the semi-Lagrangian interpolation operator, we now consider the case from [144] where a general interpolation scheme is given by

$$P(x) = a_j y_{j-1} + b_j y_j + \frac{h_j^2}{6}\left(\left(a_j^3 - a_j\right) y''_{j-1} + \left(b_j^3 - b_j\right) y''_j\right), \tag{14.43}$$

where

$$a_j = \frac{x_j - x}{h_j}, \quad b_j = \frac{x - x_{j-1}}{h_j}, \quad h_j = x_j - x_{j-1}, \tag{14.44}$$

and the expressions for how the second derivatives of y are approximated determine the order and type of interpolation.

The next step in the derivation of the tangent linear model of the semi-Lagrangian scheme is to differentiate the general interpolation formula in (14.43) with respect to $\overline{\xi}$. The differentiation of the polynomial is achieved through the chain rule and by evaluating (14.43) and (14.44) at $x = \overline{\xi}$. This then results in the tangent linear model being

$$\delta P_j\left(\overline{\xi}\right) = \left(\frac{\overline{y}_j - \overline{y}_{j-1}}{h_j} - \frac{h_j}{6}\left(3\overline{a}_j^2 - 1\right)\overline{y}''_{j-1} - \left(3\overline{b}_j^2 - 1\right) y''_j\right)\delta\xi, \tag{14.45}$$

where

$$a_j = \frac{x_j - \overline{\xi}}{h_j}, \quad b_j = \frac{\overline{\xi} - x_{j-1}}{h_j}. \tag{14.46}$$

If we now consider the cubic Lagrange interpolation, then we now move on to consider the work in Tanguay and Polavarapu (1999) [182], which is where the tangent linear and the adjoint for the cubic Lagrange interpolation are derived.

Tanguay and Polavarapu (1999) paper

In [182] the authors apply the theory from [144] to a 1D passive tracer equation on the sphere in the longitudinal direction, which is defined by

$$\frac{\partial F}{\partial t} + \omega(\lambda)\frac{\partial F}{\partial \lambda} = 0, \tag{14.47}$$

where $F = F(\lambda, t)$ represents the tracer field at time t over the interval $\lambda \in [0, 2\pi]$, $\omega(\lambda)$ is the angular velocity which is greater than zero. Since the equation in (14.47) is linear with respect to the tracer and we do not allow any variation in the wind, the tangent linear and nonlinear models are the same equation.

As we have seen many times, the Lagrangian property of (14.47) states that if (14.47) is integrated forward one time step, then the field at the arrival point λ at time $t + \Delta t$ is equal to the value of the tracer at the departure point $\xi(\lambda)$ at time t, through

$$F(\lambda, t + \delta t) = F(\xi(\lambda), t), \tag{14.48}$$

where

$$\int_{\xi(\lambda)}^{\lambda} \frac{d\overline{\lambda}}{\omega\left(\overline{\lambda}\right)} = \Delta t. \tag{14.49}$$

It is also very highly likely that the departure, or upstream, point does not coincide with a gridpoint and so there will be a need for an interpolation of the function to that point.

From the theory that we presented earlier in this chapter, we require an inner product for the adjoint. In [182] the authors define their inner product to be

$$\langle F, G\rangle = \int_0^{2\pi} F(\lambda)\, G(\lambda)\, d\lambda. \tag{14.50}$$

As the formulation of the adjoint in [182] is associated with 4D VAR, see Fletcher (2017) for details on data assimilation [61], they refer to a forcing term $\widehat{F_T}(\lambda)$ at the end of the assimilation period of length T, which then makes the adjoint equation associated with (14.47) as

$$\frac{\partial \widehat{F}}{\partial t} + \frac{\partial}{\partial \lambda}\left(\omega(\lambda)\,\widehat{F}\right) = \frac{\partial \widehat{F}}{\partial t} + \frac{d\omega(\lambda)}{dx}\widehat{F} + \omega(\lambda)\frac{\partial \widehat{F}}{\partial x} = 0, \tag{14.51}$$

where $\widehat{F} \equiv \widehat{F}(\lambda, t)$ represents the adjoint variable with $\widehat{F_T}(\lambda, T) = \widehat{F_T}(\lambda)$. This then gives rise to the conservative form of the passive tracer equation. Now if (14.51) is integrated backward for one time step, then the adjoint field $\widehat{F}$ at the departure point λ at time t is given in terms of the adjoint variable $\widehat{F}$ at the arrival point $\widehat{\xi}(\lambda)$ at time $t + \Delta t$ as

$$\widehat{F}(\lambda, t) = \widehat{F}\left(\widehat{\xi}(\lambda), t + \Delta t\right)\frac{\omega\left(\widehat{\xi}(\lambda)\right)}{\omega(\lambda)}, \tag{14.52}$$

where

$$\int_{\lambda}^{2\pi} \frac{d\overline{\lambda}}{\omega\left(\overline{\lambda}\right)} = \Delta t. \tag{14.53}$$

According to [182], this expression can be deduced from a series of inner product relationships between the forward and adjoint models:

$$\begin{aligned}
\int_0^{2\pi} F(\lambda, t+\Delta t)\,\widehat{F}(\lambda, t+\Delta t)\, d\lambda &= \int_0^{2\pi} F(\xi(\lambda), t)\,\widehat{F}(\lambda, t+\Delta t)\, d\lambda \\
&= \int_0^{2\pi} F(\lambda, t)\,\widehat{F}\left(\widehat{\xi}, t+\Delta t\right)\frac{D\widehat{\xi}(\lambda)}{d\lambda} d\lambda \\
&= \int_0^{2\pi} F(\lambda, t)\,\widehat{F}\left(\widehat{\xi}, t+\Delta t\right)\frac{\omega\left(\widehat{\xi}(\lambda)\right)}{\omega(\lambda)} d\lambda \\
&= \int_0^{2\pi} F(\lambda, t)\,\widehat{F}(\lambda, t+\Delta t).
\end{aligned} \tag{14.54}$$

Generally, the upstream point position $\xi = \xi(\lambda)$ defined in (14.49) will lie between gridpoints and the evaluation of $F(\xi(\lambda), t)$ in (14.48) will require an interpolation as we only have values of F at gridpoints when we are considering the numerical approximation to (14.47). The piecewise continuous interpolation to obtain the value of F in the semi-Lagrangian formulation at the point $\lambda_i = i\Delta\lambda$ is given by

$$F_i(t+\Delta t) = P_{i-p}(\xi), \tag{14.55}$$

where P_{i-p} are the interpolating functions over the interval $(i-p, i-p-1)$ with $F_i = F(\lambda_i)$. If we recall the general definition for the interpolation scheme, then we have

$$P_{i-p}(\xi) = aF_{i-p-1} + (1-a)F_{i-p} + \frac{\Delta\lambda^2}{6}\left(xF''_{i-p-1} + yF''_{i-p}\right),$$

where for the cubic Lagrange interpolation we have

$$F''_i = \frac{F_{i-1} - 2F_i + f_{i+1}}{\Delta\lambda_{i-1}}\Delta\lambda^2, \tag{14.56}$$

and as before, the index $i-p$ refers to the interval $\left[(i-p-1)\Delta x, (i-p)\Delta x\right]$. We recall that the Courant number is calculated through $\alpha = \alpha(\lambda) = \dfrac{\lambda - \xi(\lambda)}{\Delta\lambda}$, and p is the largest integer value such that $p = \int(K)$ where $p \leq K$. The displacement between the two gridpoints is denoted as $a = a(\lambda)$ such that

$$\xi = (i-p)\Delta\lambda - a\Delta\lambda, \tag{14.57}$$

where $x = \left(a^3 - a\right)$ and $y = (1-a)^3 - (1-a)$.

It is possible to write all of these equations for the cubic Lagrange interpolation in matrix–vector form as

$$\mathbf{F}(t+\delta t) = \mathbf{M}\mathbf{F}(t), \tag{14.58}$$

where the matrix $\mathbf{M}$ is defined as

$$\mathbf{M} = \mathbf{I}_a\mathbf{S}_{p+1} + \mathbf{I}_{(1-a)}\mathbf{S}_p + \frac{\Delta\lambda^2}{6}\left(\mathbf{I}_x\mathbf{S}_{p+1} + \mathbf{I}_y\mathbf{S}_p\right)\mathbf{D}_2, \tag{14.59}$$

and the $\mathbf{I}_z$s, for $z = a, (1-a), x$, and y, are diagonal matrices with the definitions for the values of z on the diagonal entries, $\mathbf{S}$ is the p-shift operator such that the entries at $(i, i-p(\lambda_i))$ are equal to 1, while all the other entries are equal to 0, and finally, $\mathbf{D}_2$ is the numerical approximation to the second derivative operator.

The adjoint of (14.58) with respect to a discretized form of (14.49), given by

$$\langle F, G\rangle = \Delta\lambda\sum_{i=1}^{N} F_i G_i, \tag{14.60}$$

is

$$\widehat{\mathbf{F}}(t) = \mathbf{M}^T\widehat{\mathbf{F}}(t+\Delta t), \tag{14.61}$$

where

$$\mathbf{M}^T = \mathbf{S}_{p+1}^T \mathbf{I}_a + \mathbf{S}_p^T \mathbf{I}_{(1-a)} + \frac{\Delta\lambda^2}{6} \mathbf{D}_2 \left(\mathbf{S}_{p+1}^T \mathbf{I}_x + \mathbf{S}_p^T \mathbf{I}_y \right). \tag{14.62}$$

The transposed shift operator matrix, $\mathbf{S}_p^T$, depends upon the variation in the integer value p of the Courant number as a function of the gridpoint. The simplest case is when p is the same for all gridpoints. Under these circumstances, $\mathbf{S}_p^T = \mathbf{S}_{-p}$, which shares the eigenvalue basis of $\mathbf{D}_2$ [182]. If this is the case, then from (14.60) and (14.61), we have that the numerical adjoint at the gridpoint λ_i is given by

$$\widehat{F}_i(t) = \left((1-a) \widehat{F} \right)_{i+p} + \left(a\widehat{F} \right)_{i+p+1} + \frac{\Delta\lambda^2}{6} \left(\left(y\widehat{F} \right)''_{i+p} + \left(x\widehat{F} \right)'' \right), \tag{14.63}$$

where $\widehat{F}$ on the right-hand side of the equation in (14.63) is evaluated at $t + \Delta t$. An important feature to note here is that if the angular velocity is constant, then $\left(a\widehat{F}\right)_{i+p+1} = a \left(\widehat{F}\right)_{i+p+1}$. Therefore, from (14.63) we can say that the numerical adjoint of the 1D passive tracer equation is a backward integration of a passive tracer problem starting from the forcing $\widehat{F}_T(\lambda)$ at time T and moving downstream to the flow using a semi-Lagrangian formulation using the same interpolation scheme as the forward integration.

There are many different versions of the adjoint of the semi-Lagrangian method that can occur due to variations to many different parts of the equation, and in [182] Tanguay and Polavarapu consider three cases, A, B, and C. The authors presented them in an illustration of where they only allow one discontinuity in p over the grid. We now assume that $p_K = p_J$ for $K \geq J$ and $p_K = p_{J-1}$ for $k \leq J-1$, where $\lambda_J = J\Delta\lambda$ is a specific gridpoint with its upstream position at ξ_J in the interval $\left[(J - p_j - 1)\Delta\lambda, (J - p_J)\Delta\lambda \right]$. We have a copy of the schematic from Tanguay and Polavarapu (1999) in Fig. 14.3, which shows the points associated with the three test cases, where they are varying the locations of the upstream position ξ_{J-1} with respect to the interval containing ξ_J. In test case A, we have that the integer value for p is the same for all grid points. For case B, a local increase in the wind leads to $p_{J-1} = P_j - 1$. whereas for case C we experience a deceleration pattern leading to $p_{J-1} = p_j + 1$.

Adjoints of cases A, B, and C

For the three test cases A, B, and C just mentioned, Tanguay and Polavarapu state that they will recover an explicit formula for $\widehat{F}$ at the gridpoint λ_I with the index I defined as $I = J - p_J - 1$ and represented in bold in Fig. 14.3. We present this work for the linear and cubic Lagrange interpolations by applying the direct conversion method. We now introduce a general index K such that for the linear interpolation we have

$$F_K\left(t^n + \Delta t\right) = a_K F_{K-p_K-1} + (1 - a_K) F_{K-p_K} + HOT, \tag{14.64}$$

where F on the right-hand side is evaluated at $t = t^n$. The higher order terms (HOT) are equal to zero for the linear interpolation, while for the cubic Lagrange interpolation these are

$$\frac{1}{6} x_K \left(F_{K-p_k-2} - 2F_{K-p_k-1} + F_{K-p_k} \right) + \frac{1}{6} y_K \left(F_{K-p_k-1} - 2F_{K-p_k} + F_{K-p_k+1} \right). \tag{14.65}$$

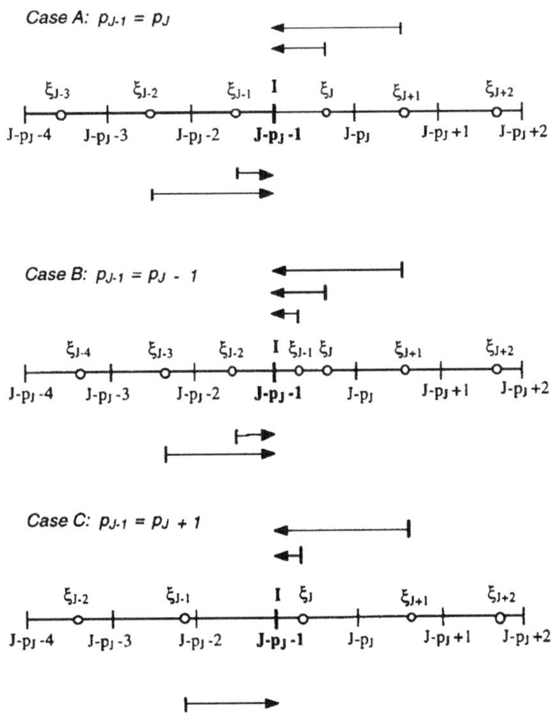

FIGURE 14.3 Copy of Fig. 1 which is an illustration of the cases A, B, and C, obtained by varying the relative location of the upstream position. From Tanguay, M. and Polavarapu, S., 1999: The Adjoint of the Semi-Lagrangian Treatment of the Passive Tracer Equation, *Monthly Weather Review*, **127**, 551–564. ©American Meteorological Society. Used with permission.

The direct conversion method referred to above requires us to find the K indices for which the field value F_I appears during the forward integration on the right-hand side of (14.64) and in (14.65).

For case A, the field value F_I appears during the forward integration on the right-hand side of (14.64) for $K = J - 1$ and J. Also, F_I appears in (14.65) for $K = J - 2$ to $J + 1$. Tanguay and Polavarapu indicate that the arrows drawn in Fig. 14.3 with the head at the bold index I and starting from each upstream position for which the interpolation scheme uses F_I to estimate F at that upstream position. Using the direct conversion method, the adjoint $\widehat{F}(t)$ at the index I is the sum of the adjoints $\widehat{F}(t + \Delta t)$ at the indices $J - 2$ to $J + 1$, each multiplied with the associated coefficient of F_I appearing on the right-hand side of (14.64) or in (14.65). For example, $(1 - a_J)\widehat{F}_{J-1}(t + \Delta t)$ contributes to $\widehat{F}_I(t^n)$ because $(1 - a_{J-1})F_I(t^n)$ appears on the right-hand side of (14.64) for $K = J - 1$. Adding all these contributions, the numerical adjoint at the gridpoint λ_I becomes

$$\widehat{F}_I(t^n) = \left[(1 - a)\widehat{F}\right]_{J-1} + \left[a\widehat{F}\right]_J + \frac{(\Delta\lambda)^2}{6}\left[\left(y\widehat{F}\right)''_{J-1} + \left(x\widehat{F}\right)''_J\right], \tag{14.66}$$

where $\widehat{F}$ on the right-hand side is evaluated at time $t^n + \Delta t$.

For case B, Tanquay and Polavarapu state that an explicit formula for $\widehat{F}$ with the index I is obtained the same way. Here F_I appears on the right-hand side of (14.64) or $K = J - 2$ to J, in (14.65) for $K = J - 3$ to $J + 1$, and the numerical adjoint at λ_I is

$$\begin{aligned}\widehat{F_i}\left(t^n\right) &= \left[(1-a)\widehat{F}\right]_{J-2} + \left[a\widehat{F}\right]_{J-1} + \left[a\widehat{F}\right]_J \\ &+ \frac{1}{6}\left(\left[y\widehat{F}\right]_{J-3} - 2\left[y\widehat{F}\right]_{J-2} + \left[y\widehat{F}\right]_{J-1} + \left[y\widehat{F}\right]_J \right.\\ &+ \left.\left[x\widehat{F}\right]_{J-2} - 2\left[x\widehat{F}\right]_{J-1} - 2\left[x\widehat{F}\right]_J + \left[x\widehat{F}\right]_{J+1}\right), \end{aligned} \tag{14.67}$$

for the cubic Lagrange interpolation.

For case C, F_I appears on the right-hand side of (14.64) only for $K = J$, and in (14.65) for $K = J - 1$ to $J - 1$. The numerical adjoint at λ_I becomes

$$\widehat{F_I}\left(t^n\right) = \left[a\widehat{F}\right]_J + \frac{1}{6}\left(\left[y\widehat{F}\right]_{J-1} + \left[y\widehat{F}\right]_J - 2\left[x\widehat{F}\right]_J + \left[x\widehat{F}\right]_{J+1}\right). \tag{14.68}$$

Tanguay and Polavarapu state that by comparing (14.66), (14.67), and (14.68), we find that the gridpoint formula for $\widehat{F}$ at the index I varies when there is a change in the relative distribution of the upstream position ξ_{J-1} generated by a discontinuity in the integral value p_J of the Courant number at the gridpoint λ_J, and that this will affect the accuracy of the adjoint at the gridpoint λ_I. A detailed analysis of the accuracy of the adjoints derived above can be found in [182], and the reader is referred to that paper for the details.

14.5.2 Adjoint of 2D tracer on the sphere

In longitude–latitude (λ, θ) coordinates, the 2D advection of a passive tracer on the sphere can be written as

$$\frac{\partial F}{\partial t} + \frac{u}{a\cos\theta}\frac{\partial F}{\partial \lambda} + \frac{v}{a}\frac{\partial F}{\partial \theta} = 0. \tag{14.69}$$

Based upon the its Lagrangian property, if (14.69) is integrated forward one time step, the field F at the arrival point $\mathbf{g}$ at time $t^n + \Delta t$ is given in terms of F at the departure point $\xi(\mathbf{g})$ at time t^n by

$$F\left(\mathbf{g}, t^n + \Delta t\right) = F\left(\xi\left(\mathbf{g}\right), t^n\right), \tag{14.70}$$

where

$$\frac{d\lambda}{dt}\left(t^n\right) = \frac{u\left(\lambda\left(t^n\right), \theta\left(t^n\right), t^n\right)}{a\cos\theta\left(t^n\right)}, \qquad \frac{d\theta}{dt}\left(t^n\right) = \frac{v\left(\theta\left(t^n\right), \theta\left(t^n\right), t^n\right)}{a} \tag{14.71}$$

are integrated from $\xi(\mathbf{g})$ at time t^n to $\mathbf{g}$ at $t^n + \Delta t$. Tanguay and Polavarapu state that generally the upstream position $\xi(\boldsymbol{\gamma})$ will lie between gridpoints and the evaluation of $F(\xi(\mathbf{g}))$ in (14.70) will require an interpolation since, as we know, only gridpoint values of F are predicted, where now $\langle\cdot,\cdot\rangle$ is defined as

$$\langle F, G\rangle = \int_0^{2\pi}\int_{-\frac{\pi}{2}}^{\frac{\pi}{2}} F(\lambda,\theta)\, G(\lambda,\theta)\, a^2\cos\theta\, d\lambda\, d\theta, \tag{14.72}$$

the associated adjoint of (14.69), assuming a forcing $\widehat{F_T}(\lambda,\theta)$ at the end of a 4D VAR assimilation window [182], is

$$\frac{\partial \widehat{F}}{\partial t}+\frac{1}{a\cos\theta}\frac{\partial}{\partial\lambda}\left[u\widehat{F}\right]+\frac{1}{a\cos\theta}\frac{1}{a\cos\theta}\frac{\partial}{\partial\theta}\left[v\widehat{F}\cos\theta\right]=0, \tag{14.73}$$

where $\widehat{F}=\widehat{F}(\lambda,\theta,t^n)$ represents the adjoint variable with $\widehat{F}(\lambda,\theta,t^n)=\widehat{F_T}(\lambda,\theta)$. This gives rise to the conservative form of the 2D passive tracer equation. It is said in [182] that it is verified that if (14.73) is integrated backward for one time step, the field $\widehat{F}$ at the departure point $\mathbf{g}=(\lambda,\theta)$ at time t^n is given in terms of $\widehat{F}$ at the arrival point $\widehat{\xi}(\mathbf{g})=\left(\widehat{\lambda},\widehat{\theta}\right)$ at time $t^n+\Delta t$ by

$$\widehat{F}\left(\mathbf{g},t^n\right)=\widehat{F}\left(\widehat{\xi}\left(\mathbf{g}\right),t^n+\Delta t\right)\frac{\dfrac{\partial(x,y)}{\partial(\lambda,\theta)}\left(\widehat{\xi}\right)}{\dfrac{\partial(x,y)}{\partial(\lambda,\theta)}\left(\mathbf{g}\right)}\frac{\partial\left(\widehat{\lambda},\widehat{\theta}\right)}{\partial(\lambda,\theta)}. \tag{14.74}$$

A very detailed study of the linearization errors associated with the tangent linear model can be found in [183].

14.6 Perturbation forecast modeling

So far in this chapter we have presented the theory and examples for tangent linear modeling and adjoints through the linearization of the discrete version of the equations. In [106] the authors refer to obtaining the linear model through linearizing the discrete linear model as the *tangent linear model*, where the adjoint is found through the transposition of the matrix that represents the discrete tangent linear model. As we have seen, this approach had the advantage that the tangent linear model can be found by directly linearizing the nonlinear model source code, which is achieved through transposing the tangent linear source code [106]. This process is known as *automatic differentiation.*

An alternative approach that was developed at the UK Meteorological Office to be used with their operational 4D incremental VAR data assimilation system involved taking the continuous equations of the nonlinear model and to first linearize these to form a set of linear equations. It was these linearized equations that were then discretized with some suitable numerical scheme, to form the discrete linear model which is referred to as the **perturbation forecast model**. Given this approach, the adjoint model was obtained from the perturbation forecast model by a transposition of the perturbation forecast models' source code. It is shown in [104] and [115] that a 4D incremental VAR scheme that uses this approach still contains the exact adjoint of the discrete linear model. In [106] the authors refer to this process of deriving the adjoint as a *semi-continuous* method. The reason for this designation was to differentiate this approach from the *continuous approach* of obtaining an adjoint model, where the adjoint model is coded directly from the continuous adjoint equations.

In [106] the authors state two advantages of the perturbation forecast model over the discrete tangent linear model approach. We summaries these advantages here:

1. The first advantage is based upon the premise that, although the tangent linear model is valid for infinitesimal perturbations, it is the finite perturbations that are comparable to the size of uncertainties in the initial conditions [54]. A perturbation forecast model can be designed to be based upon physical principles and can make some small approximations to the true tangent linear model. The semi-continuous approach to developing linear model allows such approximations to be made to both the equations of the linear model, before any discretization, which could be achieved through scale analysis, and in the implementation of the numerical scheme. As a result of this approach, there can be savings made in the execution costs of the linear model and its adjoint. As numerical models become more complex, such approximations are likely to become more important, since a direct linearization of the discrete nonlinear model is likely to lead to many small terms that are costly to evaluate but also do not add much information to the data assimilation process.
2. A second advantage of the semi-continuous approach is that it is possible to avoid some of the problems that occur when linearizing complex systems. We have stated in the summary of the work in [144] and [182] that problems occur in the direct linearization of the interpolation with the semi-Lagrangian advection schemes. By forming the linear model from the continuous linear equations many of these difficulties are avoided.

14.6.1 Example with a 1D shallow water equations model

In [106] the authors present an example of the differences in the derivation and performance of the perturbation forecast model and the tangent linear approximation, where this example is the one-dimensional shallow water equations model. A more detailed study of the two techniques can be found in Lawless (2001) [105].

The continuous nonlinear equations used in [106] were those that describe a one-dimensional shallow water system, where the flow is of a single layer fluid over an obstacle in the absence of rotation. The associated nonlinear partial differential equations are given by

$$\frac{Du}{dt} + \frac{\partial \phi}{\partial x} = -g\frac{\partial H}{\partial x}, \tag{14.75a}$$

$$\frac{D(\ln \phi)}{Dt} + \frac{\partial u}{\partial x} = 0, \tag{14.75b}$$

where

$$\frac{D}{Dt} \equiv \frac{\partial}{\partial t} + u\frac{\partial}{\partial x}, \tag{14.76}$$

is the material derivative, $H = H(x)$ is the height of the bottom orography, u is the velocity of the fluid, $\phi = gh$ is the geopotential, where g is the gravitational constant, h, being greater than zero, is the depth of the fluid above the orography. The problem is defined on the domain of $x \in [0, L]$ and $t \in [0, T]$. The spatial boundary condition are taken to be periodic for the three fields,

$$u(0,t) = u(L,t), \qquad \phi(0,t) = \phi(L,t), \qquad H(0) = H(L).$$

The initial conditions for the wind and the geopotential fields across the whole domain are given by

$$u(x,0) = u_0(x), \qquad \phi(x,0) = \phi(x).$$

Linearized equations

The linearized versions of (14.75a) and (14.75b) are formed through considering the fields u and ϕ and perturbations δu and $\delta\phi$ about a spatially and temporally varying basis state $\overline{u}$ and $\overline{\phi}$ that satisfy the nonlinear equations. Thus we have

$$u(x,t) = \overline{u}(x,t) + \delta u(x,t), \tag{14.77a}$$

$$\phi(x,t) = \overline{\phi}(x,t) + \delta\phi(x,t). \tag{14.77b}$$

Substituting (14.77a) and (14.77b) into (14.75a), the momentum equation, and (14.75b), the continuity equation, yields

$$\frac{D\delta u}{Dt} + \delta u \frac{\partial \overline{u}}{dx} + \frac{\partial \delta\phi}{\partial x} = 0, \tag{14.78a}$$

$$\frac{D}{DT}\left(\frac{\delta\phi}{\overline{\phi}}\right) + \delta u \frac{\partial \ln \overline{\phi}}{\partial x} + \frac{(\delta u)}{\partial x} = 0, \tag{14.78b}$$

where the material derivative $\frac{D}{Dt}$ is defined in (14.77a), whereas the velocity is the linearization state $\overline{u}$.

Numerical models

The numerical approximation used in [106] is a two-time-level semi-implicit semi-Lagrangian scheme [188], but with an off-centered time averaging of the forcing term along the trajectory. An Arakawa C grid was used for the numerical modeling.

Nonlinear model. The arrival and departure points for the wind and geopotential height fields are denoted by au, $a\phi$, du, and $d\phi$, respectively. The time discretization of the nonlinear model is then

$$\frac{u_{au}^{n+1} - u_{du}}{\Delta t} + (1-\alpha_1)\left(\frac{\partial\phi}{\partial x} + g\frac{\partial H}{\partial x}\right)_{du}^{n} + \alpha_1\left(\frac{\partial\phi}{\partial x} + g\frac{\partial H}{\partial x}\right)_{au}^{n+1} = 0, \tag{14.79a}$$

$$\frac{(\ln\phi)_{a\phi}^{n+1} - (\ln\phi)_{d\phi}^{n}}{\Delta t} + (1-\alpha_2)\left.\frac{\partial u}{\partial x}\right|_{d\phi}^{n} + \alpha_2\left.\frac{\partial u}{\partial x}\right|_{a\phi}^{n+1} = 0, \tag{14.79b}$$

where the superscript is the discrete time index, and the coefficients α_1 and α_2 are time weighting parameters chosen to lie in the interval [0.5, 1]. When $\alpha_1 = \alpha_2 = 0.5$, the scheme is second-order in time. The procedure for implementing this numerical scheme can be found in [106].

Tangent linear model. The tangent linear model used in [106] is obtained through differentiating the nonlinear model's source code. The nondifferentiable procedures within the semi-Lagrangian scheme are treated by assuming that the perturbations do not move the departure point outside of the grid interval defined by the linearization state. The interpolation scheme used in all three models is the cubic Lagrange interpolation scheme.

Perturbation forecast model. The perturbation forecast model is developed through the continuous linear equations (14.78a) and (14.78b). A comparison of these equations with the nonlinear model equations in (14.75a) and (14.75b) indicates that they have the same structure, except that the linear equations have an extra term in which the wind field perturbation multiplies the gradient of the linearization state. These terms are discretized as a semi-implicit semi-Lagrangian with the off-centered averages along the trajectory. The remaining terms in (14.78a) and (14.78a) are discretized as the corresponding terms in the nonlinear model. Therefore, the numerical equations for the perturbation forecast model for this example are

$$\frac{\delta u_{au}^{n+1}-\delta u_{du}^{n}}{\Delta t}+(1-\alpha_1)\left.\frac{\partial \delta\phi}{\partial x}\right|_{du}^{n}+\alpha_1\left.\frac{\partial \delta\phi}{\partial x}\right|_{au}^{n+1}+(1-\alpha_3)\left(\delta u\frac{\partial \overline{u}}{\partial x}\right)_{du}^{n}+\alpha_3\left(\delta u\frac{\partial \overline{u}}{\partial x}\right)_{au}^{n+1}=0, \tag{14.80a}$$

$$\frac{1}{\Delta t}\left(\left(\frac{\delta\phi}{\overline{\phi}}\right)_{au}^{n+1}-\left(\frac{\delta\phi}{\overline{\phi}}\right)_{d\phi}^{n}\right)+(1-\alpha_2)\left.\frac{\partial \delta u}{\partial x}\right|_{d\phi}^{n}+\alpha_2\left.\frac{\partial \delta u}{\partial x}\right|_{a\phi}^{n+1}+(1-\alpha_4)\left(\delta u\frac{\partial\left(\ln\overline{\phi}\right)}{\partial x}\right)_{d\phi}^{n}$$
$$+\alpha_4\left(\delta u\frac{\partial\left(\ln\overline{\phi}\right)}{\partial x}\right)_{a\phi}^{n+1}=0, \tag{14.80b}$$

where α_i, for $i=1,2,3,4$, are time weighting coefficients. As with the discrete nonlinear model and the tangent linear model, if the time weights are all set to 0.5 then the numerical approximations is second-order in time.

We now finish this chapter with a caveat about the dynamically based departure point iterations.

14.7 Sensitivity of adjoint of semi-Lagrangian integration to departure point iterations

In Chapter 11 we summarized a paper by Diamantakis and Magnusson about the sensitivity that had been detected in the ECMWF operational forecasting system to the number of the departure point (d.p.) iterations. While it improved the performance of the nonlinear model, when this was introduced to the operational hybrid 4D VAR system, there appeared to be a slight problem. In [45] it is stated that

> *IFS uses a 4D VAR system to compute the initial state of the atmosphere [154]. A central component of this system is the tangent linear model (TLM) and its adjoint. These two models are used in a minimization procedure to compute the atmospheric state that provides the best fit of a short range background forecast to available observations within a 12-h assimilation window. The TLM is derived from the nonlinear forecast model (NLM) linearized around the best available estimate of the*

atmospheric state (the nonlinear trajectory) and is used to evolve analysis increments over the assimilation window. The TLM code for the semi-Lagrangian dynamics is derived directly from the NLM code using a line-by-line approach; that is, every line of the discrete NLM code is converted to the discrete TLM code, and from this we derive the adjoint code.

Tests of the 4D VAR system with a large number of iterations revealed occasional instabilities that may result in forecast failures. Separate stand-alone tests of the TLM exposed these instabilities, showing that in some cases of strong stratospheric cross-polar flow, fast-amplifying noise emerges and spreads in three dimensions, "contaminating" the solution in a large region of the atmosphere. Although the precise mechanism of this instability is not fully understood and it is difficult to analyze it given that a full nonlinear problem must be considered for this (see appendix for a discussion), experimental evidence suggests that in IFS this flow-generated noise occurs at regions of strong wind shear near the poles. Li et al. (1993) [111] explain that ***the TLM is more likely to break down in regions where sharp gradients prevail in the interpolated fields coupled with strong advective wind (p. 1761).*** *This is the situation encountered here. The special geometry near the poles seems to contribute further to this problem. Furthermore, the zonal resolution near the poles is still higher than the equatorial zonal resolution even with the reduced Gaussian grids used by IFS. As shown in the appendix, the linearization of the SL scheme for the transport of a variable includes both the transport of the perturbed variable and the perturbation to the d.p. It can thus be expected that the condition that governs stability of the TLM will not be the same with the semi-Lagrangian model, which is unconditionally stable. Furthermore, as discussed in the appendix, additional iterations may amplify instability.*

This is a very important statement. Throughout the book we have been focused on the forward modeling with semi-Lagrangian methods, but only now have we introduced the backward models (adjoint). In [45] this identified problem is investigated as follows:

Given that IFS experimental evidence suggests that multiple d.p. iterations of the TLM excite this instability, it would be interesting to investigate the impact of the dynamic DPI approach if this is applied to the TLM and adjoint model as well. Indeed, this technique seems to effectively control this instability and the system can be run stably with an increased number of iterations. The implementation for the TLM scheme can be summarized in two steps and resembles a master–slave process:

1. *The NLM computes the linearization state (trajectory) and during this computational process determines dynamically the number of required DPI per grid point and saves this number;*
2. *The precomputed number of DPI is used for the TL and adjoint models at corresponding grid points. For example, if maximum number of iterations is $\nu_{\max} = 10$ but the NLM has computed that 3 iterations are required for the trajectory on a grid point, the TLM/adjoint at this grid point will be restricted to do 3 iterations.*

We have a copy of Fig. 7 from [45] in Fig. 14.4 where the TLM instability and the impact of DPI control are presented, where the difference [111] is given by

$$D[\delta x(t_0)] = M^T(t_n, t_0)\,\delta x(t_0) - (M(t_n, t_0)[x_0(t_0) + \delta x(t_0)] - M(t_n, t_0)[x_0(t_0)]). \tag{14.81}$$

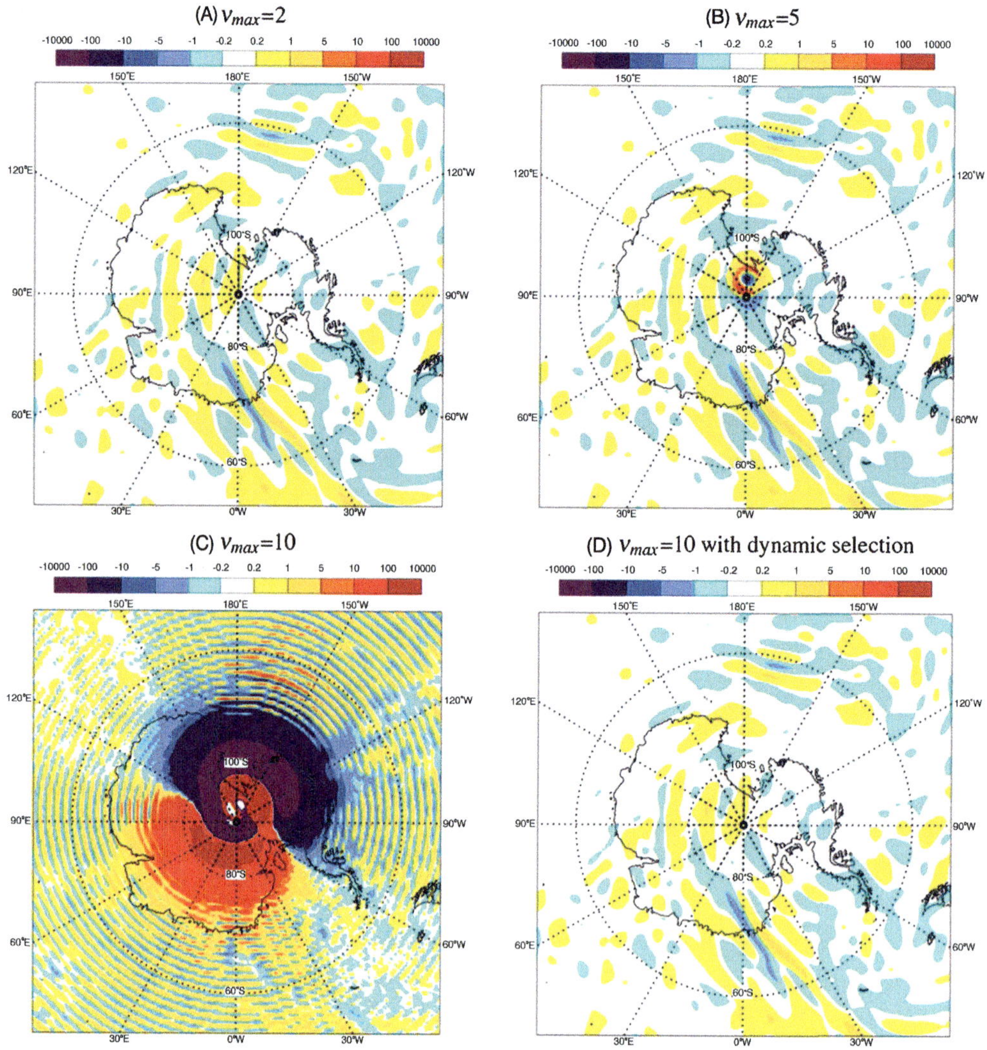

FIGURE 14.4 Copy of Fig. 7 from [45] of the tangent linear–nonlinear model perturbation difference for temperature $D\left[\delta T\left(t_0\right)\right]$ at time $t+12h$ and vertical level 20 (3 hPa): (A)–(C) using different numbers of DPI ($\upsilon_{\max}$); (D) including the convergence control algorithm. Diamantakis, M. and Magnusson, L., 2016: Sensitivity of the ECMWF model to semi-Lagrangian departure point iterations, *Monthly Weather Review*, **144**, 3233–3250. ©American Meteorological Society. Used with permission.

Here we shall briefly summarize the work from the appendix of [45] to highlight the mathematical linkages to the instability. As always we start with the one-dimensional nonlinear wind advection equation

$$\frac{Du}{Dt} \equiv \frac{\partial u}{\partial t} + u\frac{\partial u}{\partial x} = 0.$$

The semi-Lagrangian discretization of the equation above is on a grid x_i with N points and uniform Δx resolution, $x_i = (i-1)\,\Delta x$, $i = 1, 2, \ldots, N$ is obtained by interpolating to the departure points $x^n_{d_i}$:

$$\begin{aligned} u_i^{n+1} &= \left(1-\widehat{\alpha}_i^n\right) u_{i-p}^n + \widehat{\alpha}_i^n u_{i-p-1}^n, \\ \widehat{\alpha}_i^n &= \frac{x_{i-p} - x_{d(i)}^n}{\Delta x}, \quad p = \lfloor \frac{u_i^n \Delta t}{\Delta x} \rfloor, \end{aligned} \tag{14.82}$$

where

$$x_{d(i)}^n = x_i - u_i^n \Delta t.$$

In the analysis in [45] the authors have used a simple Euler method for computing the departure point, and a linear interpolating scheme. The latter, they say, is the method commonly used to interpolate the wind fields during the departure point iterations.

Taking a linear perturbation of the discretization above results in

$$\delta u_i^{n+1} = \left(1-\widehat{\alpha}_i^n\right) \delta u_{i-p}^n + \alpha_i^n \delta u_{i-p-1}^n + \delta \widehat{\alpha}_i^n \left(u_{i-p-1}^n - u_{i-p}^n\right), \tag{14.83}$$

where δ is the tangent linear perturbations. In [45] the authors state that Eq. (14.83) shows that the tangent linear of the SL advection of u includes both the transport of the perturbed variable δu and the perturbation to the d.p. (contained in $\delta\widehat{\alpha}_i$). The presence of the last term in (14.83) suggests that there is no guarantee that the tangent linear perturbation is monotonic, even if linear interpolation is used for SL advection, which is monotonic by nature. The perturbation δ_i^{n+1} may be outside the range of time level n input data $\left[\min\left(\delta u_{i-p}^n, \delta u_{i-p-1}^n\right), \max\left(\delta u_{i-p}^n, \delta u_{i-p-1}^n\right)\right]$ and therefore there is more damping in the nonlinear state compared with the perturbed state.

In both [111] and [144], the authors demonstrate that the perturbed d.p. $x_{d(i)} + \delta x_{d(i)}$ should lie in the same interval with $x_{d(i)}$ otherwise a large error may be introduced and the tangent linear approximation would become invalid. In the following paragraph we discuss how this condition may be violated when having multiple iterations. To do this, we use the ECMWF scheme with the SETTLS extrapolation technique; however, a similar result may be obtained using another extrapolation formula for the midpoint of the SL trajectory.

The tangent linear of the SETTLS iterative procedure for finding the d.p. would be

$$\delta x_{d(i)}^{[\nu]} - \frac{\Delta t}{2}\left(\delta u_i^n + \delta u_{d(i)}^{*[\nu-1]}\right), \quad \nu = 2, 3, \ldots, \tag{14.84}$$

where $\delta u_{d(i)}^{*[\nu-1]}$ denotes the perturbation of the quantity $\delta u_{d(i)}^{*[\nu-1]} = \left[2u^n - u^{n-1}\right]_{d(i)}$, which is interpolated to the d.p. estimate. Thus for u^* we have

$$\delta x_{d(i)}^{[\nu]} = -\frac{\Delta t}{2}\left(\delta u_i^n + \left(1-\widehat{\alpha}_i^{[\nu-1]}\right)\delta u_{i-p}^* + \widehat{\alpha}_i^{[\nu-1]}\delta u_{i-p-1}^* + \delta\widehat{\alpha}_i^{[\nu-1]}\left(u_{i-p-1}^* - u_{i-p}^*\right)\right), \tag{14.85}$$

for $\nu = 2, 3, \ldots,$ where

$$\widehat{\alpha}_i^{[\nu]} \equiv \frac{x_{i-p} - x_{d(i)}^{[\nu]}}{\Delta x}, \quad \delta\widehat{\alpha}_i^{[\nu]} \equiv -\frac{\delta x_{d(i)}^{[\nu]}}{\Delta x}, \tag{14.86}$$

and $\left[x_{i-p-1}, x_{i-p}\right]$ is the interval containing the d.p. estimate at iteration ν. We can see that both linear interpolation weight $\widehat{\alpha}_i$ and the number p in (14.86) depend on iteration number ν. For $\nu = 1$ we have the simple starting formula derived from

$$\delta x_{d(i)}^{[1]} = -\Delta t \delta u_i^n. \tag{14.87}$$

Next Diamantakis and Magnusson split (14.86) so as to be able to express it in terms of the Lipschitz number of the extrapolated field u^*:

$$\delta x_{d(i)}^{[\nu]} = -\frac{\Delta t}{2}\left(\delta u_i^n + \left(1 - \widehat{\alpha}_i^{[\nu-1]}\right)\delta u_{i-p}^* + \alpha_i^{[\nu-1]}\delta u_{i-p-1}^*\right) - \frac{1}{2}\delta x_{d(i)}^{[\nu-1]} L_{i-p-\frac{1}{2}}^*, \tag{14.88}$$

where

$$L_{i-p-\frac{1}{2}}^* \equiv \Delta t \frac{u_{i-p}^* - u_{i-p-1}^*}{\Delta x}. \tag{14.89}$$

The number above is an estimate of the Lipschitz number centered at the interval containing the d.p. of grid point i. Diamantakis and Magnussen then go on to analyze the equations as follows: For a long time step Δt, (14.88) implies that the likelihood that the perturbed d.p. lies in a different interval than the d.p. increases as the perturbation $\delta x_{d(i)}$ may become large. This is a highly unlikely situation for an infinitesimal perturbation and small time step; however, semi-Lagrangian NWP models use long time steps and in certain weather regimes the perturbations can be large, which raises the probability of this situation. Time extrapolation used to derive u^* may generate noise and jumpiness in the discrete Lipschitz number, which directly influences the magnitude of $\delta x_{d(i)}^{[\nu]}$. Considering that (14.88) is a recursive formula, having a multiplicative effect on the magnitude of the d.p. perturbation, we may conclude that the likelihood of producing a large perturbation $\delta x_{d(i)}^{[n]}$ increases with multiple iterations when time steps are long and the Lipschitz number is large. The latter implies existence of large CFL number in the vicinity, otherwise no regions with sufficiently large wind gradients would exist to give rise to large Lipschitz numbers.

14.8 Summary

In this chapter we have given a very brief introduction to tangent linear models and their adjoints. We have shown how to apply this technique to the semi-Lagrangian in one and two dimensions, and on the sphere. We have seen that in the tangent linear models we are perturbing not just the wind field or the tracer, but also the departure point. We have seen that we do not really want the perturbed departure point to be outside the current interval as this could cause serious problems. We have also introduced the perturbation forecast model and applied it to the semi-Lagrangian shallow water equations model to derive the associated model equations. We have also introduce d the concept of differentiating the code, and another example of this can be found in [111]. Finally, we look at how the departure point iterations for the nonlinear model could have a detrimental effect on the tangent linear model.

That's it for the theory chapters, there is still a lot more we could have put into the book, but we can not capture everything, no matter how hard we tried. We now move on to the final chapter which sets of examples where semi-Lagrangian integration has been used in some form of modeling in the geosciences.

CHAPTER

15

Applications of semi-Lagrangian methods in the geosciences

In this chapter we will show examples of how semi-Lagrangian methods are used in different disciplines in the geosciences. We shall summarize the models that the authors have used, and show a few plots of the performance of the schemes. We will begin this summary with the discipline that started it all – atmospheric sciences.

15.1 Atmospheric sciences

In this section we summarize some of the implementations of semi-Lagrangian techniques associated with different disciplines of atmospheric sciences. As we have seen throughout the book, the Canadian scientists at ECMWF and the scientists at the Irish meteorological service, or Met Éireann as it is know since the 1990s, have been very instrumental in enabling the concept of semi-Lagrangian approach to be considered operationally viable.

Starting with ECMWF, this center introduced a version of the two-time-level semi-implicit semi-Lagrangian schemas in the IFS in 1995. This was with a spectral model using a Gaussian grid. Not far behind was the Met Office, who went operational with a semi-implicit semi-Lagrangian nonhydrostatic 3D primitive equations model in the gridpoint space in 2002 [42]. In their scheme, the Met Office applied semi-Lagrangian approaches for all prognostic variables, except density, with conservative and monotone treatment of tracers, predictor–corrector implementation of a two-time-level semi-implicit time integration scheme, with a three dimensional iterative solution of a variable coefficient elliptic equation for the pressure increment at each time step. A very detailed explanation of the implementation, with the manipulation of the partial differential equations involved, can be found in Davies et al. (2005) [42].

In [42] the authors show that the during winter trial periods the new scheme was seen to be consistently better in forecasting major storms. The two depressions shown in Fig. 15.1 produced severe weather across France and neighboring countries on December 26–27, 1999. The MSLP analysis (lower panels) show the first storm over Germany and the second approaching northwest France. The middle panels show the then-operational 4- and 5-day forecasts which fail to capture either storm adequately. The upper panels show the same forecasts for the new model which are considerably better.

Semi-Lagrangian Advection Methods and Their Applications in Geoscience
https://doi.org/10.1016/B978-0-12-817222-3.00019-0

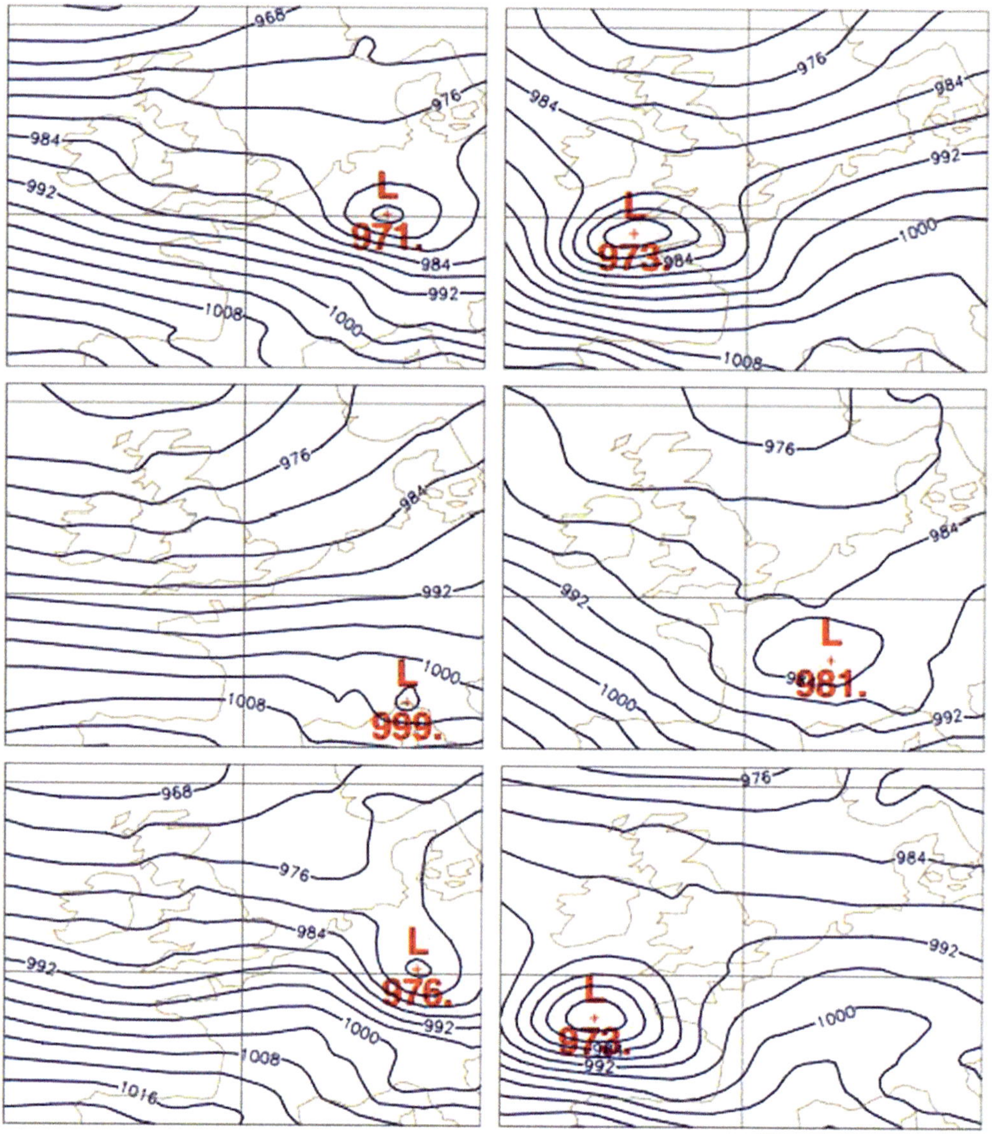

FIGURE 15.1 Copy of Fig. 2 from [42] presenting plots of the situations described in the text, Davies, T., Cullen, M. J. P., Malcolm, A. J., Mawson, M., Staniforth, A., White, A. A., and Wood, N., 2005: A new dynamical core for the Met Office's global and regional modelling of the atmosphere. *Q. J. R. Meteor. Soc.*, **131**, 1759–1782. ©Royal Meteorological Society.

Recently the dynamical core at the Met Office was updated, which is now called Even Newer Dynamics for General Atmospheric Modelling of the Environment (ENDGame). We have met ENDGame already; the new core is SLICE-based and is described in [209]. ENDGame builds on the previous New Dynamics core [42] and improves its accuracy, scalability, and stability. Another positive effect of ENDGame is an increase in synoptic variability because it is less diffusive, and this leads to a more energetic model. This also leads to a beneficial increase in the spread between members of the MOGREPS ensemble. ENDGame was first implemented in the operational global NWP models on July 15, 2014, with most limited-area models (including MOGREPS-UK) following on February 3, 2015.

Most of the description above is from [69] where MOGREPS stands for the Met Office Global and Regional Ensemble Prediction System, where the UK part is referring to the high resolution 2.2 km model, where the operational high resolution hybrid 4D VAR system just over the United Kingdom (UK) runs at 1.5 km. The reason for mentioning this it to highlight that all of the new models run with a semi-implicit semi-Lagrangian scheme, similar to the scheme introduced in Chapter 13.

Probabilistic forecasts are routinely generated from MOGREPS-UK for the Operational Meteorologists via the Gridded Post-Processing (GPP) system. Products include probabilities of exceeding, or being below, predefined thresholds for rainfall rates and accumulations, visibility, temperature, and wind gusts. Products showing the probability of exceeding the 10th, 50th, and 90th percentiles for visibility, as well as the mean, spread, and percentiles for temperature are also produced. One of the most important applications for MOGREPS-UK is for flood forecasting. This was motivated by a series of severe flooding events during 2007.

The high resolution of the MOGREPS-UK model allows precipitation to be forecast on the scale of river catchments, and, with the help of hydrological models, the forecasts provide valuable guidance to flood forecasters. Probabilistic forecasts of precipitation amounts exceeding thresholds that could lead to surface-water flooding are routinely provided to Operational Meteorologists (particularly those in the Flood Forecasting Centre) via the GPP. These are generated as the probability of a threshold being exceeded somewhere within a 30×30 km square rather than for each gridpoint. This is useful to indicate that heavy rain is a possibility within an area even if the chance at any particular location is very small. In addition, MOGREPS-UK forecasts, with statistical postprocessing to provide a wider range of scenarios, are fed in to the grid-to-grid (G2G) UK-wide hydrological model which provides forecast scenarios for river flows and levels [69]. A copy of Fig. 3 from [69] showing an example of the 1 hour rainfall amounts from the deterministic model, along with the probability of rainfall exceeding 4 mm per hour probability forecast from MOGREPS-UK is in Fig. 15.2.

The reason to show the plots in Fig. 15.2 is because at the heart of the numerical model that lead to those figures is a semi-Lagrangian approach, and to highlight that the semi-Lagrangian approach is applicable at 2.2 km.

A version of semi-Lagrangian advection has been implemented in many of the world's global numerical weather prediction systems, e.g., NAVGEM at the Naval Research Laboratory [76], but we should note here that a next generation of global model is due to go operational soon at Fleet Numeric, Météo-France, Korean Meteorological Agency, Bureau of Meteorological in Australia, to name but a few.

15.1.1 Moisture transport

In a study presented by Molina et al. in 2019 [131], the authors are interested in how the Amazon forests and climatological precipitation patterns in South America are interrelated. A fundamental question is how these patterns depend on the presence of forests. Molina

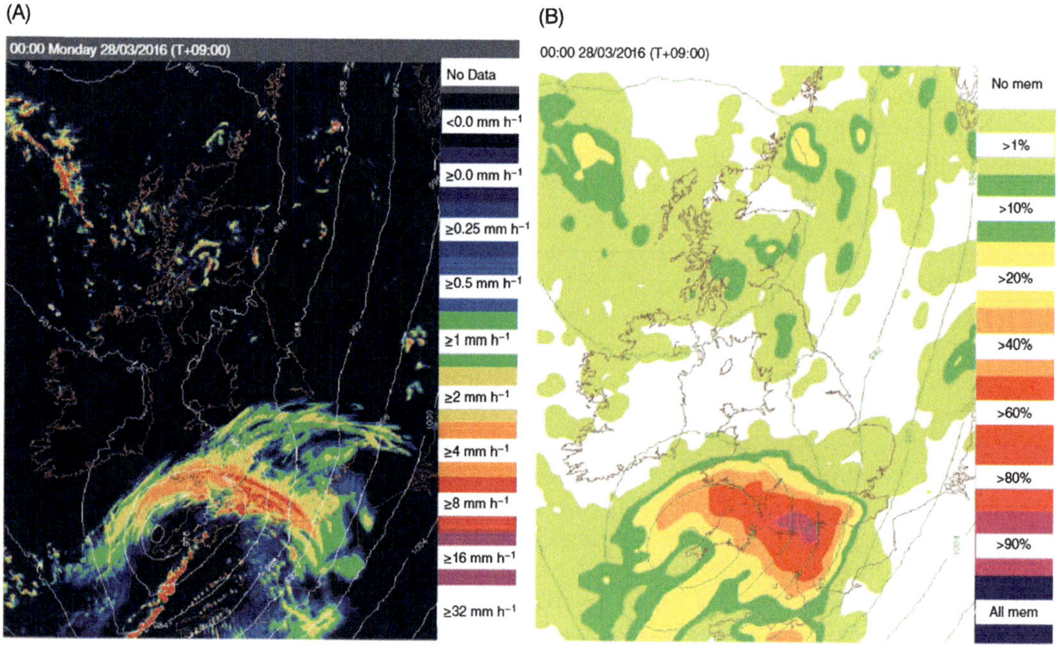

FIGURE 15.2 Copy of Fig. 3 from [69] of examples of rainfall forecasts: (A) 1 hour accumulations from UKV deterministic model and (B) probability of rainfall exceeding 4 mm per hour from MOGREPS-UK within an 18 × 18 km neighborhood. Both forecasts are T+9 h forecasts valid at 0000 UTC March 28, 2016 (storm Katie). Hagelin, S., Son, J., Swinbank, R., McCabe, A., Roberts, N. and Tennant, W., 2017: The Met Office Ensemble convective scale ensemble, MOGREPS-UK, *Q. J. R. Meteor. Soc.*, **143**, 2846–2861. ©Royal Meteorological Society.

et al. investigate this relationship by studying how precipitation varies with distance from the ocean along wind streamlines linking the Atlantic Ocean to northwestern and southern South America through the Amazon forests. Through a robust observation-based analysis, they found that precipitation exponentially increases with distance from the ocean along wind streamlines flowing over forests, while it exponentially decreases downwind of the forests. These patterns are consistent with multiple mechanisms through which forests influence the transport of atmospheric moisture and precipitation production over the continent.

The semi-Lagrangian component comes in through what is referred to as the **Dynamic Recycling Model (DRM)**, where the DRM is a two-dimensional semi-Lagrangian model for estimating exchanges of atmospheric moisture between different regions. Atmospheric moisture can reach a given target region after being evaporated from the surface from a given source region. Molina et al. are interested in the atmospheric moisture contributed by source areas within the forest or the ocean to precipitable water over target areas representing pre-forest, forest, and postforest regions exchange of moisture estimated by the DRM based on the equation of conservation of atmospheric moisture integrated in the vertical direction, assuming that the atmospheric column is well mixed.

The experiments that Molina et al. show required inputs for the DRM which include evaporation, precipitable water, and the vertical integral of horizontal moisture flux vector. For the

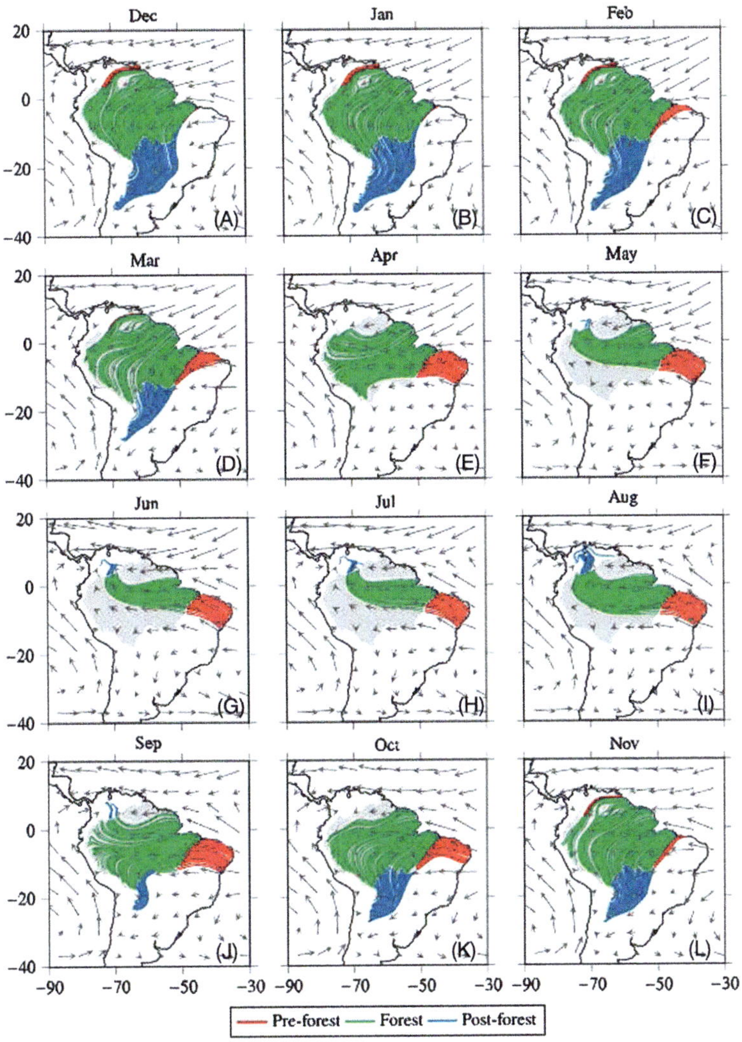

FIGURE 15.3 Plot of the monthly variation in precipitation. Fig. 1 from Molina, R. D., Salazar, J. F., Martinez, J. A., Villegas, J. C., Arias, P. A., 2019: Forest-induced exponential growth of precipitation along climatological wind streamlines over the Amazon, *J. Geoph. Res. – Atmos.*, **124**, 2589–2599. ©American Geophysical Union.

DRM experiments, they focus on the seasons of December–February (DJF) and June–August (JJA), where they do this because of the moisture transport patterns that are characteristic for each of these seasons. We have a copy of their Fig. 1 in Fig. 15.3, where (i) presents northeasterly winds with preforest-to-forest transition during DJF and (ii) shows southeasterly winds with preforest-to-forest transition during JJA. They use daily and 6-hourly fields over a 0.75 grid; thus, the estimated transport (as they say) includes the effect of variations at the synoptic scale. In the results from this work, the authors refer to P as the amount of precipitation and X the distance inland, or along the streamlines, from the coast. The results from the DRM are shown in Fig. 15.4.

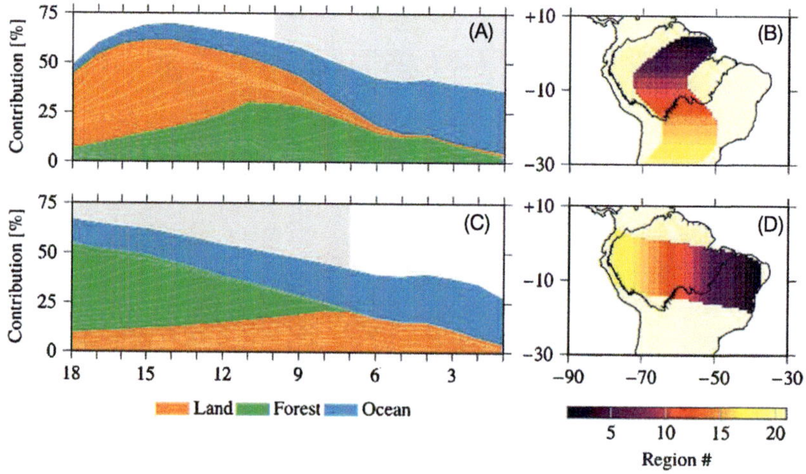

FIGURE 15.4 Results from the Dynamic Recycling model, Fig. 6 from Molina, R. D., Salazar, J. F., Martinez, J. A., Villegas, J. C., Arias, P. A., 2019: Forest-induced exponential growth of precipitation along climatological wind streamlines over the Amazon, *J. Geoph. Res. – Atmos.*, **124**, 2589–2599. ©American Geophysical Union.

The conclusions that Molina et al. provide for the DCM results were:

Model results show that the contribution of Amazon forests to precipitable water progressively increases with X *as the streamlines pass over the forest cover, progressively decreases along the postforest region, and is absent in the preforest region. In contrast to the behavior over forests, ocean contribution to precipitation progressively decreases with distance to the Atlantic coast. For trajectories over nonforest areas, there is also an increase in the moisture contributed by nonforest cover but not as pronounced as over the Amazon forests.*

15.1.2 Convection scale

The following description of the organizations comes from [18]. There is a strong history of active collaboration between European meteorological institutes on numerical weather prediction (NWP) in order to develop and maintain numerical short-range weather forecasting systems for operational use. The international research program High Resolution Limited Area Model (HIRLAM) was initiated in 1985 and consists today of the National Meteorological Services (NMSs) from 10 countries: Denmark, Estonia, Finland, Iceland, Ireland, Lithuania, the Netherlands, Norway, Spain, and Sweden, with France as an associate member. Similarly, the collaboration among the NMSs of central Europe, Aire Limitée Adaptation Dynamique Développement International (ALADIN) started in 1991 and consists today of 16 member countries: Algeria, Austria, Belgium, Bulgaria, Croatia, the Czech Republic, France, Hungary, Morocco, Poland, Portugal, Romania, Slovakia, Slovenia, Tunisia, and Turkey.

The ALADIN NWP system is being developed within the frameworks of Action de Recherche Petite Échelle Grande Èchelle (ARPEGE) and Integrated Forecasting System (IFS) software, developed jointly by the European Centre for Medium-Range Weather Forecasts (ECMWF) and Météo-France. A more detailed explanation of the ALADIN code architecture and its canonical model configurations, Applications of Research to Operations at Mesoscale (AROME), and Aire Limitee Adaptation/Application de la Recherche a l'Operationnel (ALARO),

HARMONIE-AROME uses the same nonhydrostatic (NH) dynamical core as AROME-France, which has been developed by ALADIN. It is based on the fully compressible Euler equations. The evolution of the equations is discretized in time and space using a semi-Lagrangian advection scheme on an A grid and a semi-implicit two-time-level scheme, with spectral representation of most prognostic variables based on a double Fourier decomposition. The spectral SISL scheme originates from the global IFS used operationally at ECMWF. Horizontal diffusion is applied both by linear spectral diffusion and nonlinear flow-dependent diffusion which acts through SL advection and, thus, was given the name semi-Lagrangian horizontal diffusion. Quasimonotonic operators in the interpolation process are used in order to remove the appearance of negative values for positive definite fields, as well as an unrealistic increase of eddy kinetic energy during the forecast.

The stable extrapolation two-time-level scheme (SETTLS), specific to the HARMONIE-AROME configuration, is used as the second-order two-time-level scheme in order to avoid extrapolation in time of the velocities used for the computation of the trajectories, and for the nonlinear terms, of the evolution equations. Furthermore, in order to assure stability of the integrations, a new method for treating the upper boundary conditions was implemented, using the same relaxation scheme as for the horizontal. This method makes it possible to use SETTLS also for horizontal resolutions below 1 km, so there is no need to use the predictor–corrector.

An important statement that is made about HARMONIE-AROME is that the horizontal advection for all of the cloud hydrometeors is done by a semi-Lagrangian approach.

Semi-Lagrangian techniques have also been applied in the **advection–condensation** problem as stated in the paper *"The performance of a semi-Lagrangian transport scheme for the advection–condensation problem."* We should note here that this paper is dedicated in the memory of to André Robert [141]. The problem considered here is for heat and water substances in two phases, gaseous and condensed, which in continuous form is given by

$$\frac{D\theta}{Dt} = \frac{L\theta_0}{C_p T_0} C_d, \qquad \frac{Dq_v}{Dt} = -C_d, \qquad \frac{Dq_c}{Dt} = +C_d,$$

where T is temperature, θ is potential temperature, q_v is the water vapor mixing ratio, q_c is the cloud water content, L is the latent heat of condensation of water, and C_p is the specific heat of air at constant pressure. The condensation rate C_d is determined so as to convert supersaturation into cloud water when $q_s > q_{vs}$, ($C_d > 0$), and to evaporate cloud water into subsaturated air when $q_c > 0$ and $q_v < q_{vs}$ ($C_d < 0$).

The set of equations above are solved numerically by a process-splitting method, which implies a fractional time step. In the first step, an adiabatic transport is calculated with a three-time-level semi-Lagrangian transport scheme with zero forcing. Next the vertical displacement in one time step are computed. In the second step, the interim forecast fields are affected by the condensation process C_p. An important thing to note here, which is pointed out in [141], is that other methods are possible that would take into account the fact that condensation is occurring during the displacement.

The reader is referred to [141] to see the results of this approach, but in their conclusions Pellerin et al. state that the semi-Lagrangian transport scheme has been applied to a set of coupled field equations for internal energy, water vapor, and cloud vapor, the coupling being provided by phase change of water substance. It was shown in [141] that the SL approach with unconstrained cubic interpolation does generate some ripples in fields that exhibit rapid variations. These ripples are said to be substantially weaker than those in most Eulerian leap-frog schemes. These spurious ripples maybe amplified by the condensation process. To overcome these ripples, a constraint was devised that is effective in preventing their occurrences.

The scheme was also applied in a full SISL area model solving the fully elastic equations for a moist bubble convection experiment, and it was shown that the SL approach produced less noise than the Eulerian based approach to the same problem. Another meso-scale application can be found in [66] for precipitating clouds, where here the authors are using a two-time-level semi-Lagrangian scheme.

15.2 Atmospheric chemistry

We now move onto consider applications of semi-Lagrangian approaches with chemical transport. We shall consider applications of pollutant transport, as well as volcanic ash modeling.

15.2.1 Pollutant transport

In a paper by Hansen et al. (2011) [75], the authors provide a detailed summary of semi-Lagrangian methods that are used in pollution models. The study that they perform compares the flowing schemes: first, classical semi-Lagrangian cubic interpolation, second, semi-Lagrangian cubic cascade interpolation [133], third, semi-Lagrangian cubic interpolation with the modified interpolation weights, called Locally Mass Conserving Semi-Lagrangian [87], and last, semi-Lagrangian cubic interpolation with a locally mass conserving monotonic filter [88].

The opening paragraph of this paper mentions that a main disadvantage of traditional SL methods is that they are not mass-conservative when applied to the volume density continuity equation. The SLICE scheme proposed by [212,213] and the LMCSL scheme by [87] are examples of new SL schemes that are mass conserving. There are two aspects of mass conservation, global and local. Global conservation can be imposed by correcting the advected field after determining the global mass loss. However, this does not ensure local conservation of mass. Local mass conservation can be achieved by making the scheme inherently conserving. Both the SLICE and LMCSL are examples of locally mass-conserving schemes.

In [75] the authors state that in air pollution modeling, accurate methods are important to be able to model steep gradients in the concentration fields caused by steep gradients in the emission fields and by nonlinear atmospheric chemistry. In chemical transport models, it is crucial that no negative values are generated. Not only are they "unphysical", they also cause the chemical part of the model to break down. The present advection scheme in the Danish Eulerian Hemispheric Model (DEHM) applied to the mixing ratio version of the continuity equation is the ASD, which is very accurate compared to other schemes The ASD scheme can, however, create problems near sharp gradients. Spurious waves known as the Gibbs phenomenon can occur when such features are present, and they can result in unphysical negative values.

Hansen et al. then go on to provide testing of the different schemes in 2D using some of the standard test cases we presented in Chapter 10 but that the profiles are now different chemical species and pollutants in rural and urban environments. In this study the current Eulerian scheme did perform quite well, not being outpoured by the SL schemes, however, in the situations where there were sharp gradients, it appeared that the SL approaches performed slightly better, which is likely to be the case in an urban setting.

15.2.2 Volcanic ash transport

Those of us who remember the trouble that the eruption of Eyjafjallajökull caused in 2010, where all flights across the North Atlantic were grounded for over a week due to not knowing if the particulate that could become glass in the heat of the airplane's engines was present in the air. After this event, and when it was realized that the models at the time were not sufficient to handle such a major disruptive event, a lot of time and effort was put in to developing more advanced models, as well as data assimilation systems, along with better observations.

We now summarize a recent paper that describes a new adaptive semi-Lagrangian approach for transport of volcanic emissions in the atmosphere by Gerwing et al. published in 2018 [63].

Gerwing et al. start this paper in their abstract with the comment that the dispersion of volcanic emissions in the Earth atmosphere is of interest for climate research, air traffic control, and human wellbeing. Current volcanic emission dispersion models rely on fixed-grid structures that often are not able to resolve the fine filamented structure of volcanic emissions being transported in the atmosphere. In [63] they extend an existing adaptive semi-Lagrangian advection model for volcanic emissions including the sedimentation of volcanic ash. The advection of volcanic emissions is driven by a precalculated wind field. For evaluation of the model, the explosive eruption of Mount Pinatubo in June 1991 is chosen, which was one of the largest eruptions in the 20th century. They compare simulations of the climactic eruption on June 15, 1991 to satellite data of the Pinatubo ash cloud and evaluate different sets of input parameters. The authors could reproduce the general advection of the Pinatubo ash cloud, and, owing to the adaptive mesh, simulations could be performed at a high local resolution while minimizing computational cost.

Volcanic SO_2 injected into the stratosphere has a global impact due to its conversion to sulfate aerosol which disturbs the Earth's radiation balance. Tropical volcanic eruptions thereby lead to warmer winters and colder summers on the Northern Hemisphere continents through dynamical feedbacks and radiative forcing, respectively. In addition, volcanic aerosols lead to an increase in stratospheric particle surface area, enhancing the ozone destruction, especially in high latitudes. The amplitude of the diurnal cycle of the surface air temperature is reduced by volcanic tephra remaining in the atmosphere on timescales from minutes to weeks. Thus a good model of the transport and dispersion of volcanic emissions is essential.

A difference in the way the semi-Lagrangian advection is implemented in this work is in the displacement calculation where the concentration at the departure point is scaled as

$$C(\mathbf{x}, t) = \frac{\left|\overline{V}(\mathbf{x} - \boldsymbol{\alpha}, t - \Delta t)\right|}{\left|\overline{V}(\mathbf{x}, t)\right|} C(\mathbf{x} - \mathbf{x}, t - \Delta t) + \Delta t R_d, \tag{15.1}$$

where $V(t)$ is a time dependent reference volume, R is the right-hand side forcing terms integrated over this volume, and $\overline{V}(\mathbf{x}, t)$ is the dual cell area. The scaling factor is to ensure that the scheme is mass-conserving.

The semi-Lagrangian method described above employs an adaptive mesh: in regions where a high spatial resolution is required, the mesh is refined, whereas the mesh size in other parts of the model domain is kept relatively coarse. Thereby, memory requirements can potentially be decreased by orders of magnitude without losing accuracy. In our case the refinement criterion is based on the concentration gradient in a mesh element, which is a set of tetrahedra representing the computational domain.

We have a copy of the initial mesh from (Fig. 2 in [63]) in Fig. 15.5.

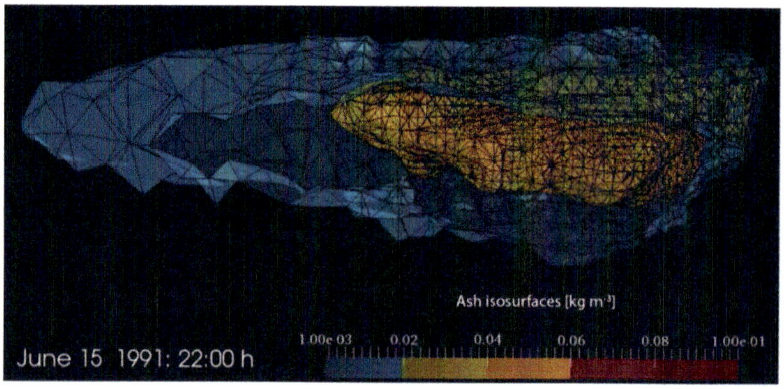

FIGURE 15.5 Triangulated mesh on the surface of the ash cloud. Fig. 2 from Gerwing, E., Hort, M., Behrens, J., and Langmann, B.: An adaptive semi-Lagrangian advection model for transport of volcanic emissions in the atmosphere, Nat. Hazards Earth Syst. Sci., 18, 1517–1534, https://doi.org/10.5194/nhess-18-1517-2018, 2018.

In Fig. 15.6 we have a copy of Fig. 6 from [63] which is of the modeled ash cloud, showing the evolution and the concentration on the surface of the ash cloud.

The conclusions from [63] were that in this study the authors have demonstrated the versatility of adaptive meshing algorithms for modeling the dispersion of volcanic emissions. More specifically, the high performance of this code would allow, if implemented into operational ash dispersion models, a significant improvement of dispersion predictions as model runs

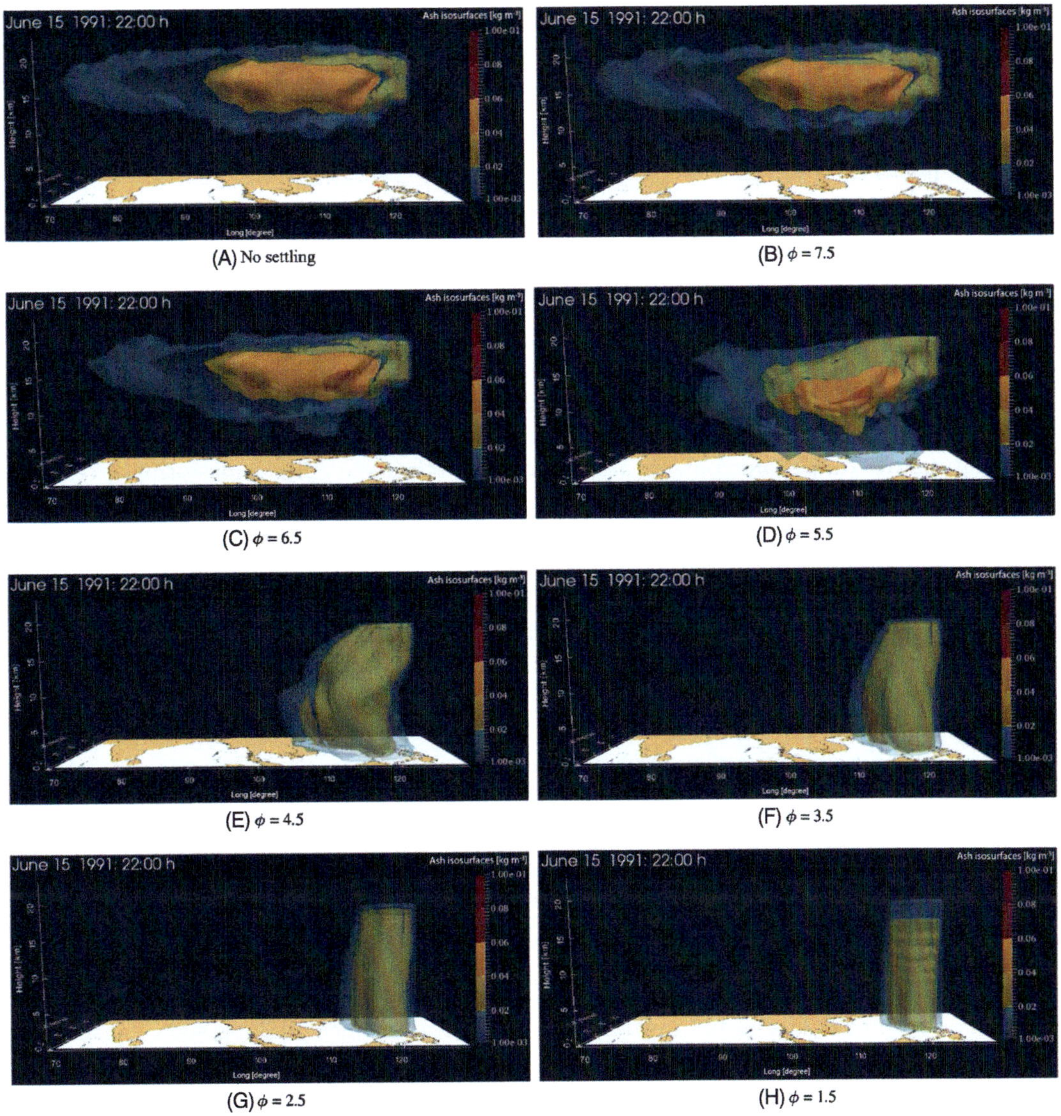

FIGURE 15.6 Concentration on the surface of the ash cloud. Fig. 6 from Gerwing, E., Hort, M., Behrens, J., and Langmann, B.: An adaptive semi-Lagrangian advection model for transport of volcanic emissions in the atmosphere, Nat. Hazards Earth Syst. Sci., 18, 1517–1534, https://doi.org/10.5194/nhess-18-1517-2018, 2018.

could be carried out significantly faster compared to codes using a fixed grid. The research community would benefit from such a faster code by being able to resolve the fine filamented structure of volcanic emissions during their transport, as well as test more boundary conditions, newly developed sedimentation models, and complex chemical reactions which could occur between different trace gases in the atmosphere while being transported.

15.3 Hydrological and ocean applications

In this section we shall present applications of semi-Lagrangian methods for water temperature and glacier movements, as well as for coastal ocean model, tidal flows, and a free surface simulation.

15.3.1 Water temperature model

The summary that we present here is the work from Yearsley (2009) [210], which is on a one-dimensional stream temperature model that is computationally efficient and highly scalable in both time and space. The model is developed within the framework of state space structure. The time-dependent equations for the conservation of thermal energy in a flowing stream or river are solved using a semi-Lagrangian numerical scheme. Solutions are obtained by tracking individual water parcels along their flow characteristics and storing the simulated results at discrete points on a fixed grid. Computational efficiency and accuracy of the numerical scheme are demonstrated by comparison of model estimates with observations of stream temperatures from rivers in the Pacific Northwest, as well as with results from a closed-form solution of the energy equation. A preliminary analysis of the impact of climate changes on stream temperature in the Columbia River system illustrates the strengths of the semi-Lagrangian method for addressing water quality issues of regional, national, and, ultimately, global scale. Further development of the semi-Lagrangian method has the potential to improve the ability of water quality planners to perform uncertainty analysis, risk analysis, and forecasting for large, complex river systems.

We have a copy of Fig. 3 from [210], in Fig. 15.7, which is showing how well the new scheme that Yearsley is introducing here is doing against established Eulerian models.

We can clearly see that the new semi-Lagrangian approach is able to cope with the discontinuity very well compared to the other schemes.

15.3.2 Tidal flow simulations

In [51] El-Amrani and Seaïd introduce an essentially nonoscillatory semi-Lagrangian method for tidal flow simulations. The scheme is designed to solve the incompressible Navier–Stokes equations with assumptions of shallow water flows including bed frictions, eddy viscosity, wind sheer stresses, and Coriolis forces. The method that is introduced in [51] is stated as employing the modified method of characteristics to discretize the convective term in a finite element framework. Limiters are incorporated in the method to reconstruct an essentially nonoscillatory algorithm at minor additional cost. The central idea consists in combining linear and quadratic interpolation procedures using nodes of the finite element where departure points are localized. The resulting semi-discretized system is then solved by an explicit Runge–Kutta Chebyshev scheme with extended stages. This scheme adds in a natural way a stabilizing stage to the conventional Runge–Kutta method using the Chebyshev polynomials. The proposed method is verified for the recirculation tidal flow in a channel with forward-facing step. We refer the reader to the paper for the mathematical details, but

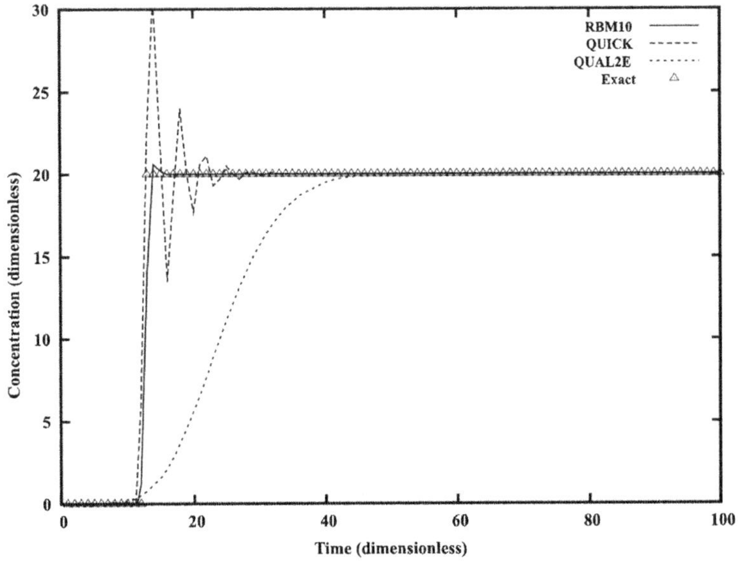

FIGURE 15.7 Copy of Fig. 3 from [210] of a plot of the advection of a step initial conditions for the semi-Lagrangian approach gainst QUICK, and an implicit scheme. See [210] for more details. Yearsley, J. R., 2009: A semi-Lagrangian water temperature model for advection-dominated river systems, *Water Res. Res.*, **45**, W12405. ©American Geophysical Society.

we will say that the authors refer to the semi-Lagrangian method as **SLAG** throughout the paper, and that their version is **ENOSLAG**.

We have a copy of Fig. 3 from [51], Fig. 15.8, which is showing the performance of the proposed scheme with one of the standard test cases we introduced earlier for the deformational flow, where we can see that the performance of ENOSLAG is a lot more consistent with the nonoscillatory property than without, compared to the true solution in the bottom panel.

It should be noted that the new semi-Lagrangian scheme was tested with the strait of Gibraltar for a more rigorous test, and the reader is referred to [51] for more details.

15.3.3 Displacement of free surface

We now consider a very different form of semi-Lagrangian application. We will not go into the mathematics here, but the reader is referred to [102] for the exact details. The summary of the paper is that the authors are presenting a numerical method based on an adaptive octree space discretization for the simulation of displacement of free surfaces, proposed and applied to 3D free surface flow problems. AVOF approach is combined with a mass-conserving semi-Lagrangian time-stepping scheme. An interface prediction algorithm is used to refine the octree at the predicted location of the interface in order to ensure detail preservation. Subsequently, the fluid is advected, and a coarsening algorithm adapts the mesh to avoid excess refinement in noninterfacial regions. SLIC and decompression algorithms are used for postprocessing to limit numerical diffusion and correct numerical compression of the VOF

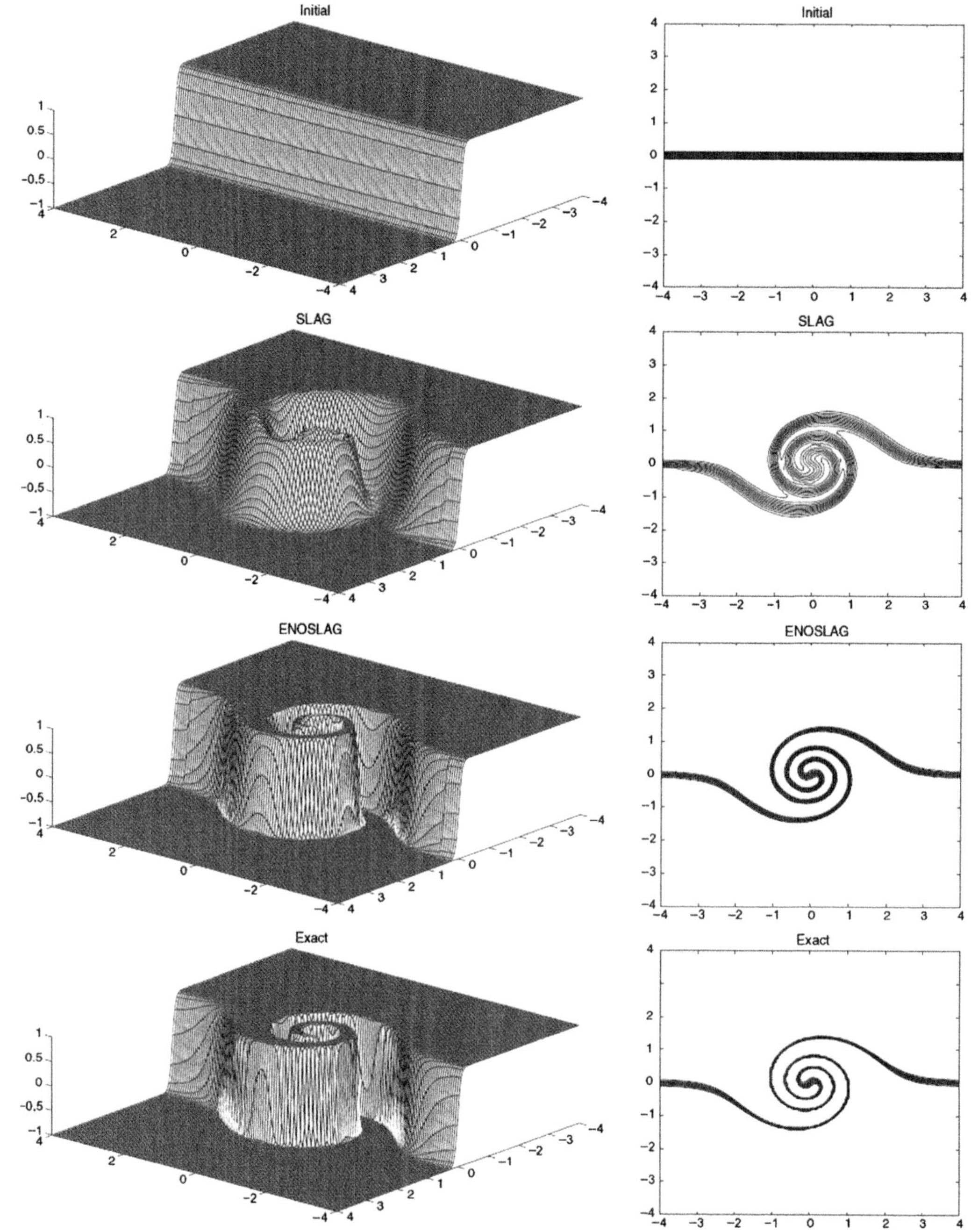

FIGURE 15.8 Results from the deformational flow test case, Fig. 3 from El-Amrani, M. and Seaïd, M., 2009: An essentially non-oscillatory semi-Lagrangian method for tidal flow simulations, *Int. J. Numer. Meth. Eng.*, **81**, 805–834. ©John Wiley and Sons, Ltd.

function. The scheme is unconditionally stable with respect to the CFL number and does not require solving of a linear system. The octree scheme allows an isotropy and refinement of interfacial cells to an arbitrary level. It does not require a 2 : 1 cell size ratio condition between neighboring cells. The authors validate their work on benchmark test cases. The scheme is coupled with a Stokes solver on a tetrahedral grid for solving of time-dependent Navier–Stokes equations.

We have copies of Figs. 20 and 21 from [102] in Figs. 15.9 and 15.10, which are showing a wave profile cross-section, and then the output with using an adaptive mesh to show that, as the refinement increases, the scheme is approximating the true wave profile, which is from sensors in a wave tank.

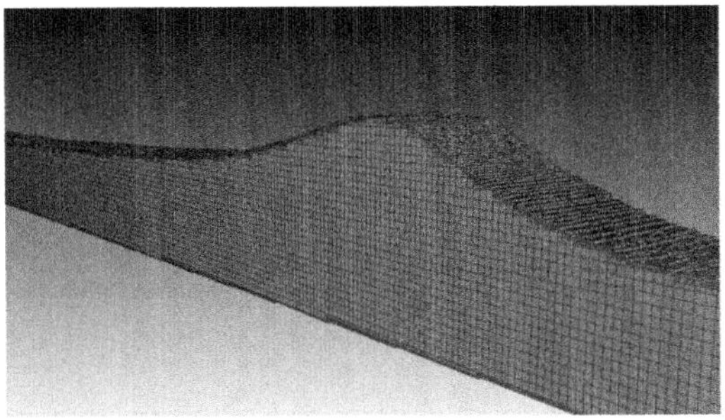

FIGURE 15.9 Copy of Fig. 20 of a wave profile from Laurma, V., Picasso, M. and Steiner, G., 2016: An octree-based adaptive semi-Lagrangian VOF approach for simulating the displacement of free surfaces, *Computer and Fluids*, **131**, 190–204. https://doi.org/10.1016/j.compfluid.2016.03.005. https://creativecommons.org/licenses/by/4.0/.

15.3.4 Glacier surges

The work that we summarize here is from two papers by Hambrey and Clarke in 2019 [72,35] where they are modeling the glacier movement on the Trapridge Glacier in Yukon, Canada. The authors are using a thermomechanical numerical ice dynamics model to simulate the evolution of glacier geometry. The flow solver assumes the shallow ice approximation. The semi-Lagrangian application comes in when they need to compute particle trajectories. From the point of ice deposition in the accumulation zone $\mathbf{X}_d$ to the point of emergence in the ablation zone $\mathbf{X}_e$ the particle follows a trajectory $\mathbf{X}(\mathbf{t})$. There are also semi-Lagrangian tracers that are modeled here, which are for the ice particles, and this is used to model the ice deposition time.

A semi-Lagrangian approach is also applied to the transport of medial moraines where these are referred to as debris trajectories having an englacial component which starts at the surface despoliation site and ends where the debris emerges and is deposited by surface meting, on a ice-free bed.

15.3.5 Coastal ocean models

In [64] Gillibrand and Herzfeld introduce a flux-form semi-Lagrangian model that is mass conserving for scalar transport in a coastal ocean model. They are testing to see if

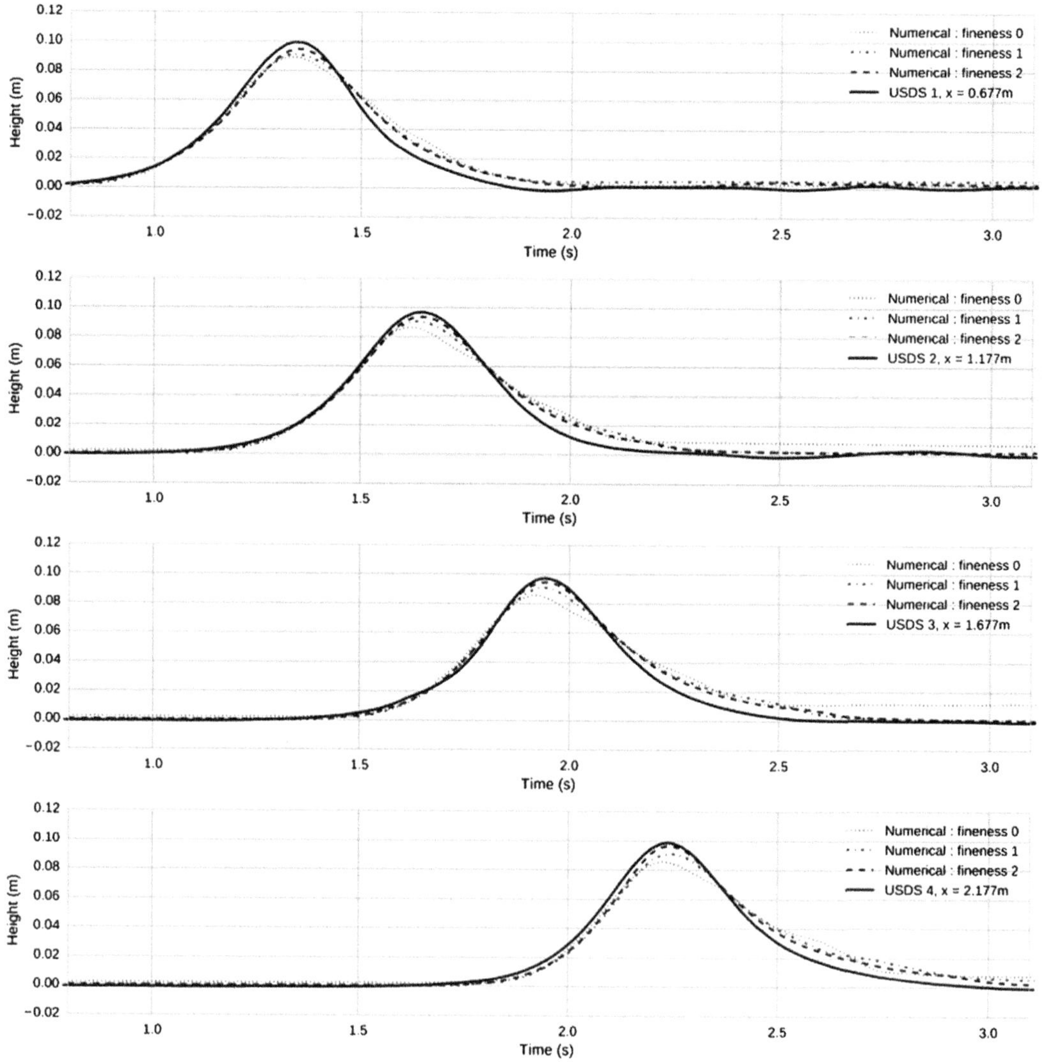

FIGURE 15.10 Copy of Fig. 21 of the pointwise height of the free surface from Laurma, V., Picasso, M. and Steiner, G., 2016: An octree-based adaptive semi-Lagrangian VOF approach for simulating the displacement of free surfaces, *Computer and Fluids*, **131**, 190–204. https://doi.org/10.1016/j.compfluid.2016.03.005. https://creativecommons.org/licenses/by/4.0/.

1. The transport model advection scheme produces comparable tracer distributions to those simulated by the full hydrodynamic model.
2. The tracer mass is conserved by the transport model.
3. The advection scheme meets constancy requirements, i.e., a domain-wide uniform tracer concentration of unity remains constant.
4. The scheme is more computationally efficient.

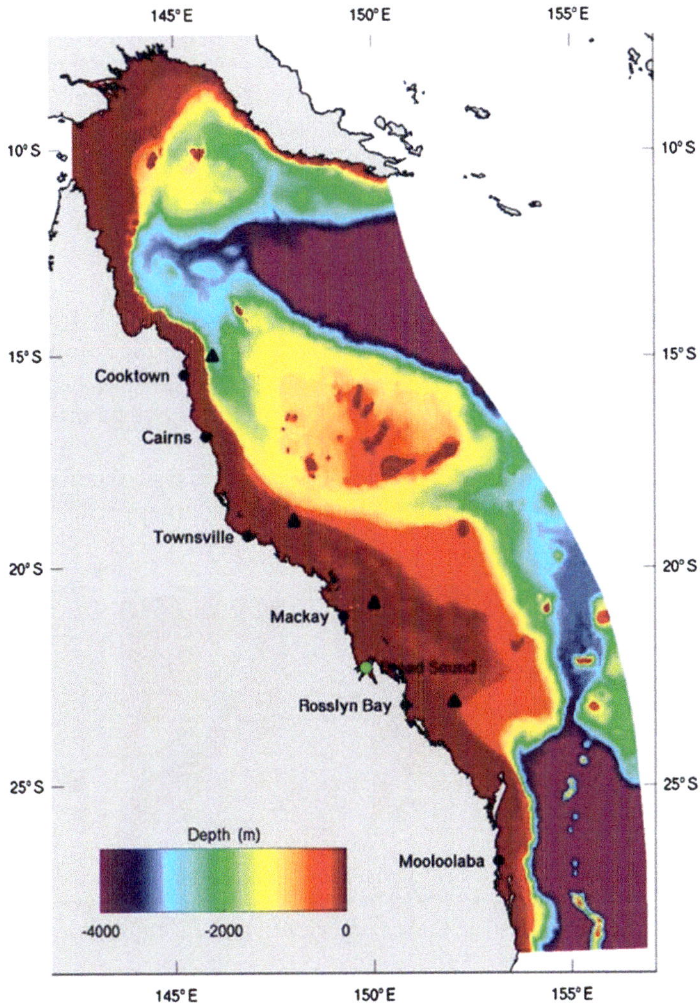

FIGURE 15.11 Copy of Fig. 9 from [64] showing the model bathymetry and the source of the passive tracer. Reprinted from Ocean-Modelling. Gillibrand, P. A. and Herzfeld, M., 2016: A mass conserving advection scheme for offline simulation of scalar transport in coastal ocean models, *Ocean Modelling*, **101**, 1–16. ©(2016) with permission from Elsevier.

They run their form of semi-Lagrangian integration with an estuary basin closed to the sea. The basin had a depth of 20 m, shoaling along the estuary at the western side to 5 m depth at the head. The eastern boundary was closed, which is effectively the same as having a wall there. Two experiments were performed for this setup. In the first, the initial tracer concentration throughout the whole domain was zero and a point source of tracer was discharged in the center of the basin. As there is no open boundary to the sea, there should be no tracer loss. A second tracer, with an initial concentration of 1.0 throughout the domain and a riverine concentration also of 1.0, was simulated to test constancy. Over the course of the simulation, concentrations of this second tracer should remain constant at exactly 1.0. The second set of experiments was for an open estuary where the setup was as for the first experiment, but now the eastern boundary was open. The same two passive tracer experiments were performed as previously. The first tracer had an initial concentration of zero throughout the domain, with a point source of tracer discharging in the center of the basin, at a rate of 1000 $mg\,s^{-1}$. At the open eastern boundary, a no-gradient condition was used during outflow and an oceanic tracer value of zero was specified during inflow. The second tracer had an initial concentration of 1.0 throughout the domain and riverine and open boundary concentrations also of 1.0, designed to test constancy.

The realistic case chosen to assess the advection algorithm is that of the Great Barrier Reef (GBR) developed through the eReefs initiative. The hydrodynamic component of this initiative represents the first time that a three-dimensional model has been applied to the whole reef, from Papua New Guinea to the south Queensland border; we have a copy of Fig. 9 from [64] in Fig. 15.11.

We have a copy of Fig. 11 from [64] in Fig. 15.12 that shows how the flux form does against the full model, as well as against a traditional semi-Lagrangian approach.

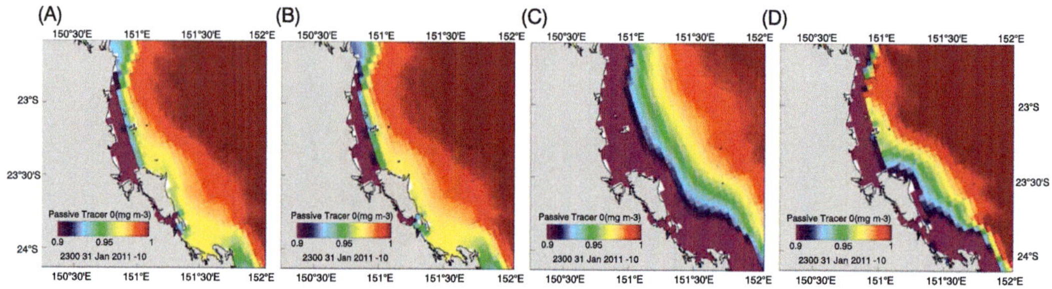

FIGURE 15.12 Copy of Fig. 11 from [64] showing the surface distributions of passive tracers. Reprinted from Ocean-Modelling. Gillibrand, P. A. and Herzfeld, M., 2016: A mass conserving advection scheme for offline simulation of scalar transport in coastal ocean models, *Ocean Modelling*, **101**, 1–16. ©(2016) with permission from Elsevier.

The conclusions from [64] were that the FFSL scheme presented here provides an accurate and mass-conserving advection scheme that may be used to simulate the transport of tracers in offline coastal ocean models. The scheme is comparable in terms of computational efficiency to traditional semi-Lagrangian advection schemes, but offers more accurate solutions relative to the source hydrodynamic model results. As such, the scheme should offer an opportunity for marine biogeochemical, sediment transport, and water quality modelers to move their models offline without suffering from loss of mass or unacceptable inaccuracy as a consequence.

15.4 Earth's mantle and interior

In this section we shall look at a few applications of semi-Lagrangian methods for modeling processes at the Earth's crust and below.

15.4.1 Mantle plumes

The first situation we consider is an application of semi-Lagrangian methods with the three-dimensional modeling of mantle plume evolution [86]. In [86], the authors investigate the effects of thermal diffusion on the evolution of mantle plumes by means of three-dimensional numerical modeling, forward and backward in time. Mantle plumes are fed by a hot, low-viscosity material from the thermal boundary layer. The material of the plumes is mainly advected toward the Earth's surface with some effects of thermal diffusion. However, the feeding can become weaker with time, and then thermal diffusion can take over and control the evolution of the plumes. Numerical experiments forward in time show that a week feeding of mantle plumes by the hot material from the boundary layer results in the diffusive disappearance of plume tails first and plume heads later. This is the most likely explanation for the seismically detected low-velocity mantle structures (mantle plumes) with prominent heads and almost invisible tails at mid-mantle depths. We develop restoration models (backward in time) tore cover strong features of mantle plumes in the geological past after they have dissipated due to thermal diffusion and analyze effects of thermal diffusion and temperature-dependent viscosity on the reconstruction of the mantle plumes. We investigate the impact of thermal diffusion on the performance of our restoration (variational data assimilation) algorithm. For a given range of Rayleigh number Ra and two values of the viscosity ratio r (between the upper and lower boundaries of the model domain), we show that (1) the residuals between the temperature predicted by the forward model and that reconstructed by the backward modeling become larger, and (2) the restoration process becomes poorer as Ra decreases and r increases. We assimilate the temperature obtained from high-resolution seismic tomography data for the southeastern Carpathians and show that present diffused mantle structures can be restored to their prominent state in the Miocene times.

The semi-Lagrangian integration is applied to temperature in this application. We have a copy of Fig. 2 from [86] in Fig. 15.13 just to illustrate that semi-Lagrangian approaches can be used for this type of problem. Please see [86] for more details.

15.4.2 Crustal accretion

In [189] the authors present the results from a coupled model that solves simultaneously for crustal accretion and hydrothermal cooling. Their approach resolves both processes within one 2D finite-element model that self-consistently solves for crustal, mantle, and hydrothermal flow. The formation of new oceanic crust is approximated as a gabbro glacier, in which the entire lower crust crystallizes in one shallow melt lens.

They find that the depth of the melt lens and the shape of hot (potentially molten) lower crust are highly dependent on the ridge permeability structure. The predicted depth of the melt lens is primarily controlled by the permeability at the ridge axis, whereas the off-axis permeability determines the width of hot lower crust. We are not sure what exactly the semi-Lagrangian advection is applied here to, but in a throwaway line in [189] the authors say that the equations for Stokes and Darcy flows and thermal diffusion are solved with a 2D finite element method implemented in the MATLAB® code MILAMIN. Later in the next paragraph they state that "*Advection is resolved with a semi-Lagrangian scheme.*"

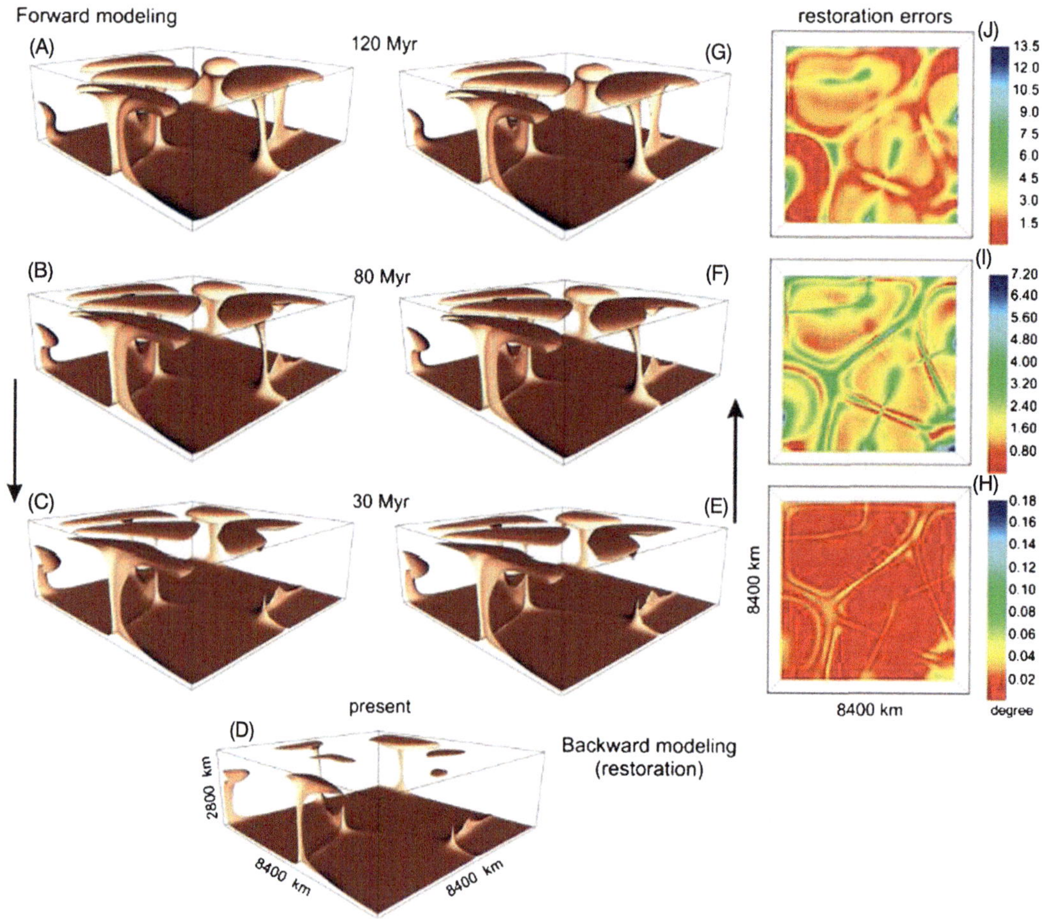

FIGURE 15.13 Mantle plume diffusion, Fig. 2 from Ismail-Zadeh, A., Schbert, G., Tsepelev, I. and Korotkii, A., 2009: Three-dimensional forward and backward numerical modeling of mantle plumes evolution: Effects of thermal diffusion, *J. Geo. Res. B.*, **111**, B06401. ©American Geophysical Union.

However, we have a copy of Fig. 8 from [189] in Fig. 15.14 which shows no instability from using a semi-Lagrangian approach for this problem.

15.4.3 Hydrothermal circulation

The paper that we summarize here is [8] by Bani-Hassan et al. who state that they present quantitative modeling results for the effects of surface relief on hydrothermal convection at ocean-spreading centers, investigating how vent site locations and subsurface flow patterns are affected by bathymetry induced sub-seafloor pressure variations. The model is based on a 2D FEM solver for a fluid flow in porous media and used to simulate hydrothermal convection systematically in 375 synthetic studies. The results of these studies show that bathymetric

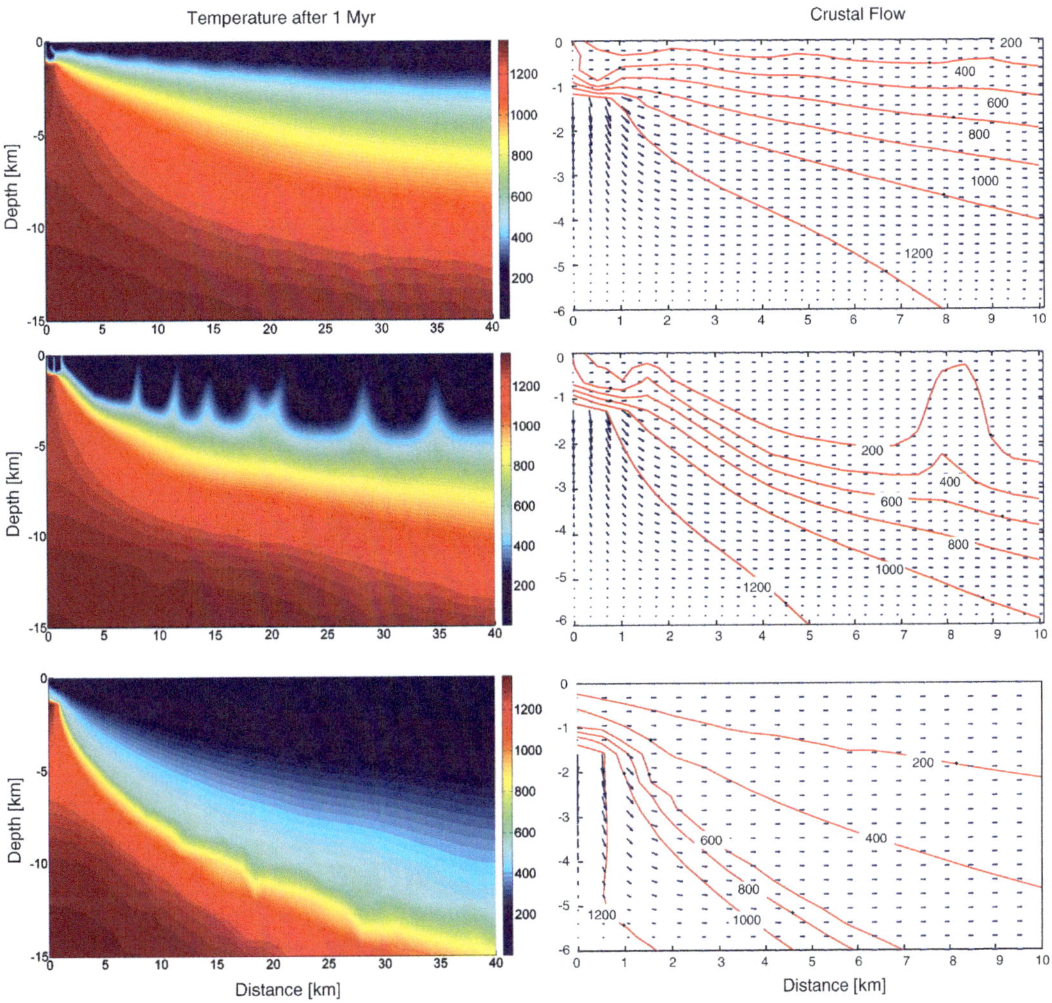

FIGURE 15.14 Copy of Fig. 8 showing influence of off axis-permeability, reprinted from Earth and Planetary Science Letters: Theissen-Krah, S., Iyer, K., Rüpke, L. H., Phipps Morgan, J., 2011: Coupled mechanical and hydrothermal modeling of crustal accretion at intermediate to fast spreading ridges, *Earth and Planetary Sci. Lett.*, **311**, 275–286. ©(2011) with permission from Elsevier.

relief has a profound effect on hydrothermal flow: bathymetric highs induce subsurface pressure variations that can deviate upwelling zones and favor venting at structural highs. The deviation angle from vertical upwelling can be expressed by a single linear dependence relating deviation angle to bathymetric slope and depth of the heat source. These findings are confirmed in two case studies for the East Pacific Rise and Lucky Strike hydrothermal fields. In both cases, it is possible to predict the observed vent field locations only if bathymetry is taken into account. Our results thereby show that bathymetric relief should be considered in simulations of submarine hydrothermal systems and plays a key role, especially in focusing venting of across axis hydrothermal flow onto the ridge axis of fast spreading ridges.

Again not many details are provided about what aspects of semi-Lagrangian theory are being used, but the authors do say that it is in the advective part of the energy equation. Again the figures that we show here are to illustrate that the scheme stays stable and produces what appear to be realistic results. In Figs. 15.15–15.17 we have reproduced Figs. 1, 4, and 7, respectively. Fig. 15.15 is of a single plume simulation, Fig. 15.16 is of simulations of temperature fields for a selected number of bathymetric models, and Fig. 15.17 is of 2D temperature field for the hydrothermal vent.

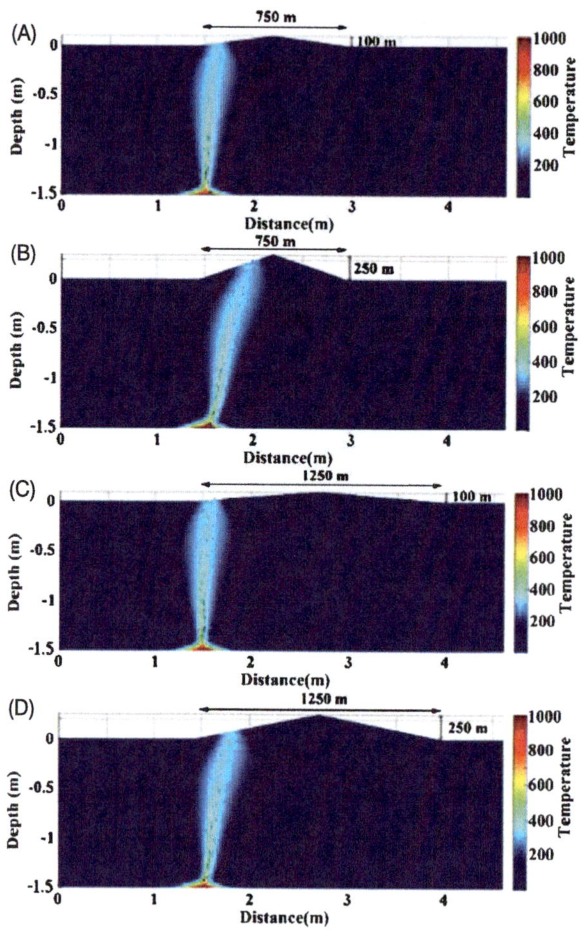

FIGURE 15.15 Copy of Fig. 1 showing a single plume from Bani-Hassan, N., Iyer, K., Rüpke, L. H. and Borgia, A., 2012: Controls of bathymetric relief on hydrothermal fluid flow at mid-ocean ridge, *GeoChem. Geophy. Geosys.*, **13**, Q05002. ©American Geophysical Union.

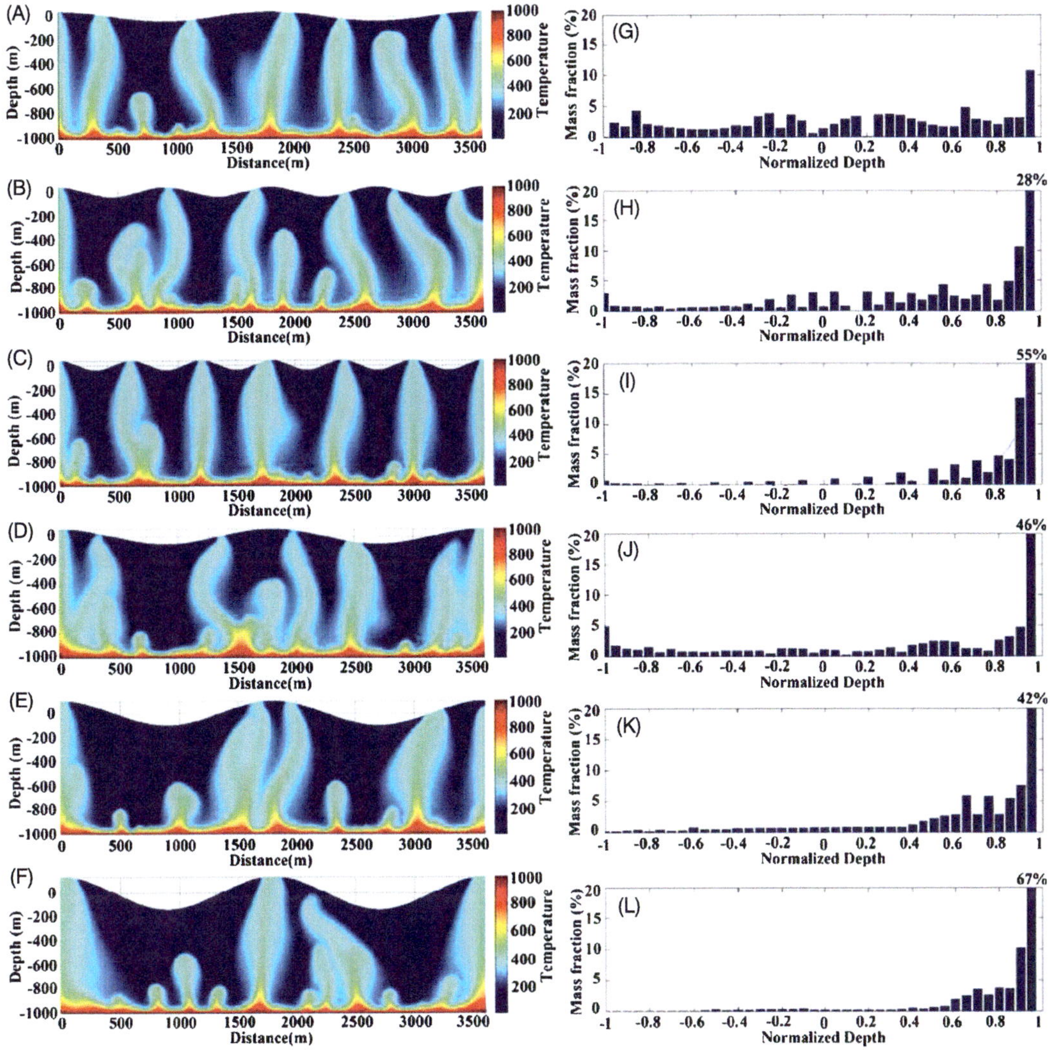

FIGURE 15.16 Copy of Fig. 4 for variable bathymetric models from Bani-Hassan, N., Iyer, K., Rüpke, L. H. and Borgia, A., 2012: Controls of bathymetric relief on hydrothermal fluid flow at mid-ocean ridge, *GeoChem. Geophy. Geosys.*, **13**, Q05002. ©American Geophysical Union.

15.5 Other applications

We have found many other different applications for the semi-Lagrangian approaches in the geosciences, and we list a few below with references for further reading

- Ionospheric disturbances [181]
- Ion Cyclotron motion [194]

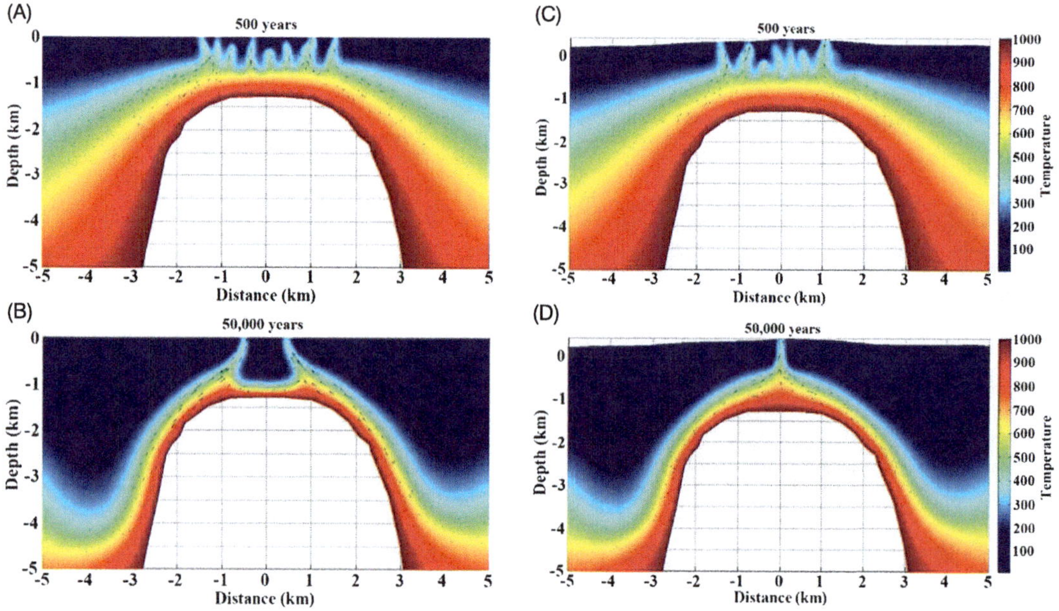

FIGURE 15.17 Copy of Fig. 7 of 2D temperature field for the Lucky strike vent field from Bani-Hassan, N., Iyer, K., Rüpke, L. H. and Borgia, A., 2012: Controls of bathymetric relief on hydrothermal fluid flow at mid-ocean ridge, *GeoChem. Geophy. Geosys.*, **13**, Q05002. ©American Geophysical Union.

- Earth's magnetosheath [77]
- Nonlinear conservative laws with WENO [83]
- Multiphase flows [139]
- Nonlinear ship-wave [130]
- Accretion disc modeling in a microquasar [98]

to name but a few applications outside of atmospheric sciences.

15.6 Summary

In this chapter we have shown that semi-Lagrangian methods are used in a multitude of applications, not just in the geosciences, and definably not just for the modeling of flows in the atmosphere. We have seen applications throughout the Earth's system from the mantle to the atmosphere, and indicated that semi-Lagrangian methods are used in the magnetosphere and the ionosphere.

That being said, semi-Lagrangian methods continue to adapt to the problem at hand with the changing of the number of departure point iterations, to ensuring that the same order polynomial used in high resolution simulations for the velocity field is the same as that of the field itself. We finish with one last figure which comes from [83], Fig. 15.18, where we are con-

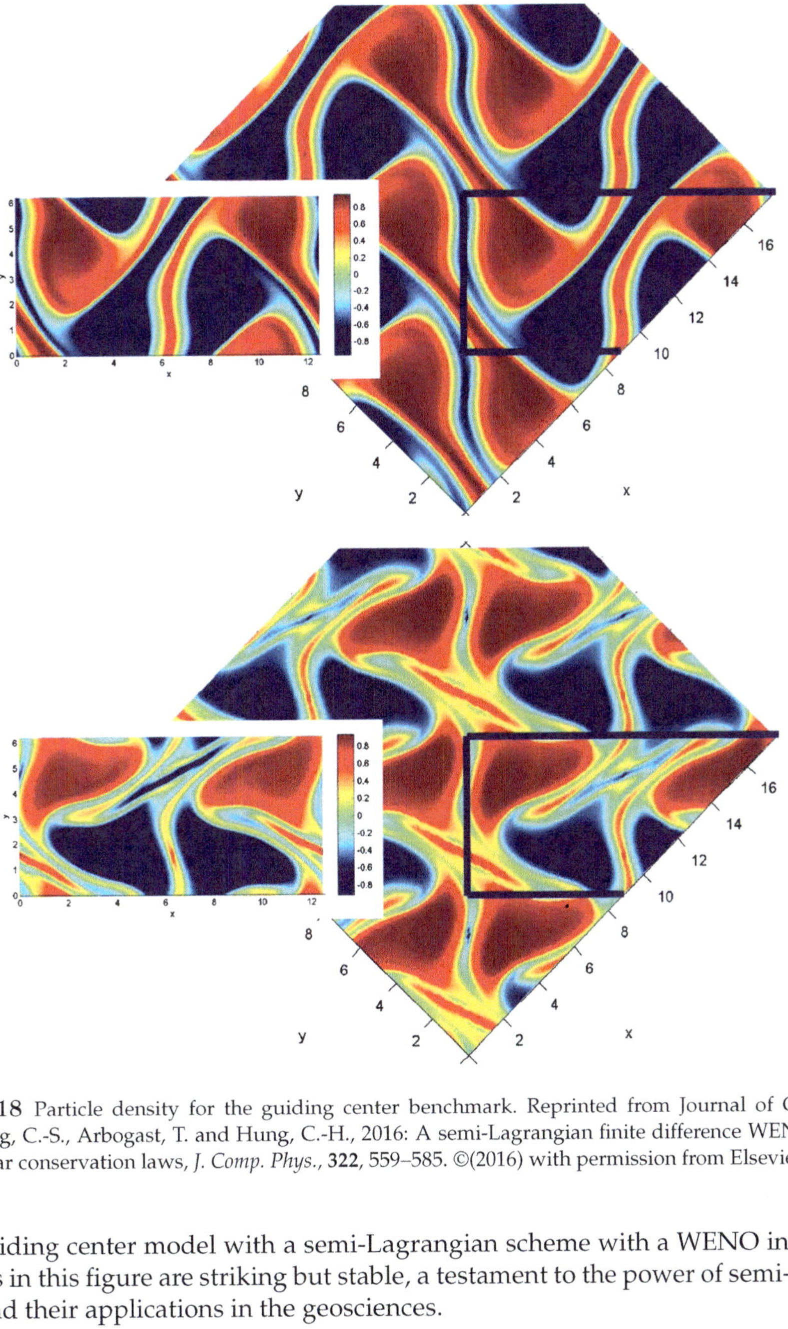

FIGURE 15.18 Particle density for the guiding center benchmark. Reprinted from Journal of Computational Physics. Huang, C.-S., Arbogast, T. and Hung, C.-H., 2016: A semi-Lagrangian finite difference WENO scheme for scalar nonlinear conservation laws, *J. Comp. Phys.*, **322**, 559–585. ©(2016) with permission from Elsevier.

sidering guiding center model with a semi-Lagrangian scheme with a WENO interpolation, the patterns in this figure are striking but stable, a testament to the power of semi-Lagrangian methods and their applications in the geosciences.

CHAPTER

16

Solutions to select exercises

Solution to Exercise 2.2

We will only show this for the first term; the approach is the same for the other two. We start by expanding the brackets and take one term at a time: For the first term, the one multiplying $f\left(t^{n-1},\psi^{n-1}\right)$, we have

$$\frac{\left(t-t^{n+1}\right)\left(t-t^{n}\right)}{\left(t^{n-1}-t^{n+1}\right)\left(t^{n-1}-t^{n}\right)}. \tag{16.1}$$

The denominator is simply $(-2\Delta t)(-\Delta t)=2\Delta t^2$. Focusing on the numerator, we have to expand the multiplication with the brackets which results in

$$t^2-tt_{n+1}-t_nt+t_nt_{n+1}, \tag{16.2}$$

where we have moved the superscript to subscript to make the powers easier to follow in the next step. Integrating (16.2) with respect to dt yields

$$\left[\frac{t^3}{3}-\frac{t^2t_{n+1}}{2}-\frac{t^2t_n}{2}-t_nt_{n+1}t\right]_{t_n}^{t_{n+1}}, \tag{16.3}$$

which results in

$$\frac{t_{n+1}^3}{3}-\frac{t_n^3}{3}-\frac{t_{n+1}^3}{2}+\frac{t_n^2t_{n+1}}{2}-\frac{t_nt_{n+1}^2}{2}+\frac{t_n^3}{2}+t_nt_{n+1}^2-t_n^2t_{n+1}. \tag{16.4}$$

We now substitute $t_{n+1}\equiv t_n+\Delta t$ in to (16.4) and multiply out. We do not do that step here, but, needless to say, all the terms cancel except for

$$\frac{\Delta t^3}{3}-\frac{\Delta t^3}{2}=-\frac{\Delta t^3}{2}.$$

The final step is to now divide by $2\Delta t^2$, which yields

$$-\frac{\Delta t}{12},$$

i.e., the coefficient in the text.

https://doi.org/10.1016/B978-0-12-817222-3.00020-7

Solution to Exercise 3.14

Here we recall that the local truncation error is defined as

$$\tau_L \equiv \psi\left(t^{n+1}\right) - \psi_i^{n+1},$$

which implies that

$$\begin{aligned}
\tau_L &= \psi\left(t^{n+1}\right) - \psi_i^n + \Delta t\left(\frac{\psi_{i+1}^n - \psi_{i-1}^n}{2\Delta x}\right) \\
&= \psi_i^n + \Delta t \psi_i^{n'} + \frac{\Delta t^2}{2}\psi_i^{n''} + \frac{\Delta t^3}{6}\psi_i^{n'''} + \cdots - \psi_i^n \\
&- \frac{\Delta t}{2\Delta x}\left(\psi_i^n + \Delta x \psi_{i,x}^n + \frac{\Delta x^2}{2}\psi_{i,xx}^n + \frac{\Delta x^3}{6}\psi_{i,xxx}^n + \cdots\right. \\
&- \left.\psi_i^n + \Delta x \psi_{i,x}^n - \frac{\Delta x^2}{2}\psi_{i,xx}^n + \frac{\Delta x^3}{6}\psi_{i,xxx}^n + \cdots\right).
\end{aligned}$$

And as we have seen for the example, terms cancel, then there is the term which is the differential equation which cancels and the leading order terms are $\left((\Delta x)^2, (\Delta t)^2\right)$. You would follow a similar procedure for the other exercises similar to this one.

Solution to Exercise 6.4

As we know, the distance from the departure point if we have it located in the grid box $i-p$ and $i-p-1$ implies that $\alpha \equiv x_i - \widetilde{x}$. Therefore the distance from x_{p-3} is $3-\alpha$. The distance to x_{p+2} is $2+\alpha$, finally the distance to x_{p+3} is $3+\alpha$. We will only show the term multiplying x_{p-3} in the centered approach, which is given by

$$\begin{aligned}
P_{4,1}(\alpha) &= \frac{\left(\widetilde{x} - x_{p-2}\right)\left(\widetilde{x} - x_{p-1}\right)\left(\widetilde{x} - x_p\right)\left(\widetilde{x} - x_{p+1}\right)}{\left(x_{p-3} - x_{p-2}\right)\left(x_{p-3} - x_{p-1}\right)\left(x_{p-3} - x_p\right)\left(x_{p-3} - x_{p+1}\right)} \\
&= \frac{(2-\alpha)(1-\alpha)\alpha(1+\alpha)}{(-1)(-2)(-3)(-4)} \\
&= \frac{\alpha(1+\alpha)(1-\alpha)(2-\alpha)}{24}.
\end{aligned}$$

You would follow the procedure above for the other three points substituting x_{p-2}, x_{p-1}, x_p, and x_{P-1} in the denominator where we have x_{p-3} above, and in the numerator, you would form the differences but omitting each point in turn to obtain 5 factors for the five points. You would follow a similar derivation for the quintic interpolation polynomial as well.

Bibliography

[1] H. Akima, A new method of interpolation and smooth curve fitting based on local procedures, J. Assoc. Comput. Mach. 17 (1970) 589–602.

[2] R. Alexander, Diagonally implicit Runge–Kutta methods for stiff O.D.E.'s, SIAM J. Numer. Anal. 14 (1977) 1006–1021.

[3] A. Arakawa, V.R. Lamb, Computational design of the basic dynamical processes of the UCLA general circulation model, Methods Comput. Phys. 17 (1977) 173–265.

[4] A. Arakawa, S. Moorthi, Baroclinic instability in vertically discrete systems, J. Atmos. Sci. 45 (1988) 1688–1707.

[5] R. Asselin, Frequency filter for time integration, Mon. Weather Rev. 100 (1972) 487–490.

[6] K.E. Atkinson, An Introduction to Numerical Analysis, John Wiley and Sons, 1988.

[7] J. Badger, B.J. Hoskins, Simple initial value problems and mechanisms for baroclinic growth, J. Atmos. Sci. 58 (2001) 39–49.

[8] N. Bani-Hassan, K. Iyer, L.H. Rüpke, A. Borgia, Controls of bathymetric relief on hydrothermal fluid flow at mid-ocean ridge, Geochem. Geophys. Geosyst. 13 (2012) Q05002.

[9] A.M. Baptista, Y.L. Zhang, A. Chawla, M. Zulauf, C. Seaton, E.P. Myers III, J. Kindle, M. Wilkin, M. Burla, P.J. Turner, A cross-scale model for 3D baroclinic circulation in estuary-plume-shelf systems II: application to the Columbia River, Cont. Shelf Res. 25 (2005) 935–972.

[10] P. Bartello, S.J. Thomas, The cost-effectiveness of semi-Lagrangian advection, Mon. Weather Rev. 124 (1996) 2883–2897.

[11] G.K. Batchelor, An Introduction to Fluid Dynamics, Cambridge University Press, 2000.

[12] J.R. Bates, An efficient semi-Lagrangian ad alternating direction implicit method for integrating the shallow water equations, Mon. Weather Rev. 112 (1984) 2033–2047.

[13] J.R. Bates, Y. Li, A. Brandt, S.F. McCormick, A global barotropic primitive equation model based on the semi-Lagrangian advection of potential vorticity, Q. J. R. Meteorol. Soc. 121 (1995) 1981–2005.

[14] J.R. Bates, A. McDonald, Multiply-upstream, semi-Lagrangian advective schemes: analysis and application to a multi-level primitive equation model, Mon. Weather Rev. 110 (1982) 1831–1842.

[15] J.R. Bates, S. Moorthi, R.W. Higgins, A global multilevel atmospheric model using a vector semi-Lagrangian finite difference scheme. Part I: adiabatic formulation, Mon. Weather Rev. 121 (1993) 244–263.

[16] J.R. Bates, F.H.M. Semazzi, R.W. Higgins, S.R.M. Barros, Integration of the shallow-water equations on the sphere using a vector semi-Lagrangian scheme with multigrid solver, Mon. Weather Rev. 118 (1990) 1615–1627.

[17] A. Beljadid, A. Mohammadin, M. Charron, C. Girard, Theoretical and numerical analysis of a class of semi-implicit semi-Lagrangian schemes potentially applicable to atmospheric models, Mon. Weather Rev. 142 (2014) 4458–4476.

[18] L. Bengtsson, U. Andrae, T. Aspelien, Y. Batrak, J. Calvo, W. de Rooy, E. Gleeson, B. Hansen-Sass, M. Homleid, M. Hortal, K.-I. Ivarsson, S. Lenderink, G. ad Niemela, K. Pagh Nielsen, J. Onvlee, P. Rontu, L. Samuelsson, D. Santos Munoz, A. Subias, S. Tijm, V. Toll, X. Yang, M. Ødegaard Køltzow, The HARMONIE-AROME model configuration in the ALADIN-HIRLAM NWP system, Mon. Weather Rev. 145 (2017) 1919–1935.

[19] R. Bermejo, On the equivalence of semi-Lagrangian schemes and particle-in-cell finite element methods, Mon. Weather Rev. 118 (1990) 979–987.

[20] R. Bermejo, A. Staniforth, The conversion of semi-Lagrangian advection schemes to quasi-monotone schemes, Mon. Weather Rev. 120 (1992) 2622–2632.

[21] V. Bjerknes, Das problem der wettervorhersage, betrachtet vom stanpunkt der mechanik and der physik, Meteorol. Z. 21 (1904) 1–7.

[22] G. Brassington, B. Sanderson, Semi-Lagrangian and COSMIC advection in flows with rotation or deformation, Atmos.-Ocean 37 (1999) 369–388.

[23] J.C. Butcher, Coefficients for the study of Runge–Kutta integration processes, J. Aust. Math. Soc. 3 (1963) 185–201.

[24] J.C. Butcher, Implicit Runge–Kutta processes, Math. Comput. 18 (1964) 50–64.
[25] J.C. Butcher, Integration processes based on Radau quadrature formulas, Math. Comput. 18 (1964) 233–244.
[26] J.C. Butcher, An algebraic theory of integration methods, Math. Comput. 26 (1972) 79–106.
[27] J.C. Butcher, Numerical Methods for Ordinary Differential Equations, Wiley, 2016.
[28] X. Cai, J. Qiu, J.-M. Qiu, A conservative semi-Lagrangian HWENO method for the Vlasov equation, J. Comput. Phys. 323 (2016) 95–114.
[29] R.E. Carlson, F.N. Fritsch, Monotone piecewise bicubic interpolation, SIAM J. Numer. Anal. 22 (1985) 386–400.
[30] V. Casulli, R.T. Cheng, Stability, accuracy and efficiency of a semi-implicit method for 3D shallow water flow, Comput. Math. Appl. 27 (1994) 99–112.
[31] A. Caya, P. Laprise, R. an Zwack, Consequences of using the splitting method for implementing physical forecasting in a semi-implicit semi-Lagrangian model, Mon. Weather Rev. 126 (1998) 1707–1713.
[32] D. Caya, R. Laprise, A semi-implicit semi-Lagrangian regional climate model: the Canadian RCM, Mon. Weather Rev. 127 (1999) 341–362.
[33] J.G. Charney, N.A. Phillips, Numerical integration of the quasi-geostrophic equations for barotropic and simple baroclinic, J. Meteorol. 10 (1953) 71–99.
[34] P.E. Chase, Stability properties of predictor-corrector methods for ordinary differential equations, J. Assoc. Comput. Mach. 9 (1962) 457–468.
[35] G.K.C. Clarke, M.J. Hambrey, Structural evolution during cyclic glacier surges: 2. Numerical modeling, J. Geophys. Res., Earth Surf. 124 (2019) 495–525.
[36] P. Colella, P.R. Woodward, Piecewise parabolic method for gas-dynamical simulations, J. Comput. Phys. 54 (1984) 174201.
[37] J.W. Cooley, O.W. Tukey, An algorithm for the machine calculation of complex Fourier series, Math. Comput. 19 (1966) 297–301.
[38] E. Cordero, N. Wood, A. Staniforth, Impact of semi-Lagrangian trajectories on the discrete modes of a non-hydrostatic vertical-column model, Q. J. R. Meteorol. Soc. 131 (2005) 93–108.
[39] J. Côté, A Lagrange multiplier approach for the metric terms of semi-Lagrangian models on the sphere, Q. J. R. Meteorol. Soc. 114 (1988) 1347–1352.
[40] J. Côté, A. Staniforth, A two-time-level semi-Lagrangian semi-implicit scheme for spectral models, Mon. Weather Rev. 116 (1988) 2003–2012.
[41] J. Crank, P. Nicolson, A practical method for numerical evaluation of solutions of partial differential equations of the heat conduction type, Proc. Camb. Philos. Soc. 43 (1) (1947) 50–67.
[42] T. Davies, M.J.P. Cullen, A.J. Malcolm, M. Mawson, A. Staniforth, A.A. White, N. Wood, A new dynamical core for the Met Office's global and regional modelling of the atmosphere, Q. J. R. Meteorol. Soc. 131 (2005) 1759–1782.
[43] C. de Boor, B. Swatrz, Piecewise monotone interpolation, J. Approx. Theory 21 (1977) 411–416.
[44] R. Delbourgo, J.A. Gregory, Shape preserving piecewise rational interpolation, SIAM J. Sci. Stat. Comput. 6 (1985) 967–976.
[45] M. Diamantakis, L. Magnusson, Sensitivity of the ECMWF model to semi-Lagrangian departure point iterations, Mon. Weather Rev. 144 (2016) 3233–3250.
[46] D.R. Durran, Numerical Methods for Wave Equations in Geophysical Fluid Dynamics, Springer, 1999.
[47] D.R. Durran, Numerical Methods for Fluid Dynamics With Applications to Geophysics, Springer, 2010.
[48] D.R. Durran, P.N. Blossey, Implicit-explicit multistep methods for fast-wave-slow-wave problems, Mon. Weather Rev. 140 (2012) 1307–1325.
[49] D.R. Durran, P.A. Reinecke, Instability in a class of explicit two-time-level semi-Lagrangian schemes, Q. J. R. Meteorol. Soc. 130 (2004) 365–369.
[50] E. Eady, Long waves and cyclone waves, Tellus 1 (1949) 33–52.
[51] M. El-Amrani, M. Seaïd, An essentially non-oscillatory semi-Lagrangian method for tidal flow simulations, Int. J. Numer. Methods Eng. 81 (2009) 805–834.
[52] A. Eliassen, The quasi-static equations of motion with pressure as independent variable, Geofys. Publ. 17 (1949) 3.
[53] G. Erbes, A semi-Lagrangian method of characteristics for the shallow-water equations, Mon. Weather Rev. 121 (1993) 3443–3452.
[54] R.M. Errico, K.D. Reader, An examination of the accuracy of the linearization of a mesoscale model with moist physics, Q. J. R. Meteorol. Soc. 125 (1999) 169–195.

[55] B. Farrell, The initial growth of disturbances in a baroclinic flow, J. Atmos. Sci. 39 (1982) 1663–1686.
[56] R. Fjørtoft, On a numerical method of integrating the barotropic vorticity equation, Tellus 4 (1952) 179–194.
[57] R. Fjørtoft, On the changes in spectral distribution of kinetic energy for two dimensional, nondivergent flow, Tellus 5 (1953) 225–230.
[58] R. Fjørtoft, On the use of space-smoothing in physical weather forecasting, Tellus 7 (1955) 462–480.
[59] S.J. Fletcher, Numerical Approximations to Buoyancy Advection in the Eady Model, Master's thesis, University of Reading, Department of Mathematics, 1999.
[60] S.J. Fletcher, Higher Order Balance Conditions Using Hamiltonian Dynamics for Numerical Weather Prediction, Ph.D. thesis, University of Reading, Department of Mathematics, 2004.
[61] S.J. Fletcher, Data Assimilation for the Geosciences: From Theory to Applications, Elsevier, 2017.
[62] F.N. Fritsch, R.E. Carlson, Monotone piecewise cubic interpolation, SIAM J. Numer. Anal. 17 (1980) 238–246.
[63] E. Gerwing, M. Hort, J. Behrens, B. Langmann, An adaptive semi-Lagrangian advection model for transport of volcanic emissions in the atmosphere, Nat. Hazards Earth Syst. Sci. 18 (2018) 1517–1534.
[64] P.A. Gillibrand, M. Herzfeld, A mass conserving advection scheme for offline simulation of scalar transport in coastal ocean models, Ocean Model. 101 (2016) 1–16.
[65] I.G. Gospodinov, V.G. Spiridonov, J.F. Geleyn, Second-order accuracy of two-time-level semi-Lagrangian schemes, Q. J. R. Meteorol. Soc. 127 (2001) 1017–1033.
[66] W.W. Grabowski, P.K. Smolarkiewicz, Two-time-level semi-Lagrangian modeling of precipitating clouds, Mon. Weather Rev. 124 (1996) 487–497.
[67] S. Gravel, A. Staniforth, J. Côté, A stability analysis of a family of baroclinic semi-Lagrangian forecast models, Mon. Weather Rev. 117 (1989) 130–137.
[68] D.X. Guo, J.B. Drake, A global semi-Lagrangian spectral model of the shallow water equations with variable resolution, J. Comput. Phys. 206 (2005) 559–577.
[69] S. Hagelin, J. Son, R. Swinbank, A. McCabe, N. Roberts, W. Tennant, The Met Office ensemble convective scale ensemble, MOGREPS-UK, Q. J. R. Meteorol. Soc. 143 (2017) 2846–2861.
[70] E. Hairer, G. Wanner, Solving Ordinary Differential Equations II: Stiff and Differential Algebraic Problems, Springer, 1996.
[71] G.J. Haltiner, R.T. Williams, Numerical Prediction and Dynamic Meteorology, 2nd ed., John Wiley and Sons, New York, 1980.
[72] M.J. Hambrey, G.K.C. Clarke, Structural evolution during cyclic glacier surges: 1. Structural glaciology of Trapridge Glacier, Yukon, Canada, J. Geophys. Res., Earth Surf. 124 (2019) 464–494.
[73] R. Heikes, D.A. Randall, Numerical integration of the shallow water equations on a twisted icosahedral grid. Part I: basic design and results of tests, Mon. Weather Rev. 123 (1995) 1862–1880.
[74] R. Heikes, D.A. Randall, Numerical integration of the shallow water equations on a twisted icosahedral grid. Part II: a detailed description of the grid and an analysis of numerical accuracy, Mon. Weather Rev. 123 (1995) 1881–1887.
[75] A.B. Hensen, J. Brandt, J.H. Christensen, E. Kaas, Semi-Lagrangian methods in air pollution models, Geosci. Model Dev. 4 (2011) 511–541.
[76] T.F. Hogan, L. Ming, J.A. Ridout, M.S. Peng, T.R. Whitcomb, B.C. Ruston, C.A. Reynolds, S.D. Eckermann, J.R. Moskaitis, N.L. Baker, J.P. McCormack, K.C. Viner, J.G. McLay, M.A. Flatau, L. Xu, C. Chen, S.W. Chang, The Navy global environmental model, Oceanography 127 (2014) 116–125.
[77] S. Hoilijoki, M. Palmroth, B.M. Walsh, Y. Pfau-Kempf, S. von Alfthan, U. Ganse, O. Hannuksela, R. Vainio, Mirror modes in the Earth's magnetosheath: results from a global hybrid-Vlasov simulation, J. Geophys. Res. Space Phys. 121 (2016) 4191–4204.
[78] A. Hollingsworth, P. Kållberg, V. Renner, D.M. Burridge, An internal symmetric computational instability, Q. J. R. Meteorol. Soc. 109 (1983) 417–428.
[79] J.R. Holton, An Introduction to Dynamic Meteorology, Academic Press, 1992.
[80] R.B. Hood, Numerical advection algorithms and their role in atmospheric transport and chemistry models, Rev. Geophys. 20 (1987) 71–100.
[81] M. Hortal, Some recent advances at ECMWF, LAM News 27 (1998) 32–36.
[82] M. Hortal, The development and testing of a new two-time-level semi-Lagrangian scheme (SETTLS) in the ECMWF forecast model, Q. J. R. Meteorol. Soc. 128 (2002) 1671–1687.
[83] C.-S. Huang, T. Arbogast, C.-H. Hung, A semi-Lagrangian finite difference WENO scheme for scalar nonlinear conservation laws, J. Comput. Phys. 322 (2016) 559–585.

[84] S.Z. Husain, C. Girard, Impact of consistent semi-Lagrangian trajectory calculations on numerical weather prediction performance, Mon. Weather Rev. 145 (2017) 4127–4150.
[85] J.M. Hyman, Accurate monotonicity preserving cubic interpolation, SIAM J. Sci. Stat. Comput. 4 (1983) 645–654.
[86] A. Ismail-Zadeh, G. Schbert, I. Tsepelev, A. Korotkii, Three-dimensional forward and backward numerical modeling of mantle plumes evolution: effects of thermal diffusion, J. Geophys. Res. B 111 (2006) B06401.
[87] E. Kaas, A simple and efficient locally mass conserving semi-Lagrangian transport scheme, Tellus 60A (2008) 305–320.
[88] E. Kaas, J. Nielsen, A mass conserving quasi-monotonic filter for use in semi-Lagrangian models, Mon. Weather Rev. 138 (2010) 1588–1876.
[89] A. Kageyama, T. Sato, Yin-Yang grid: an overset grid in spherical geometry, Geochem. Geophys. Geosyst. 5 (2004) 1–15.
[90] K. Kageyama, Dissection of a sphere and Yin-Yang grids, J. Earth Sim. 3 (2005) 20–28.
[91] E. Kalnay, Atmospheric Modeling, Data Assimilation and Predictability, Cambridge University Press, 2003.
[92] S.K. Kar, An explicit time difference scheme with an Adams–Bashforth predictor and a trapezoidal corrector, Mon. Weather Rev. 140 (2012) 307–322.
[93] S. Kumar Das, A.J. Weaver, Semi-Lagrangian advection algorithms for ocean circulation models, J. Atmos. Ocean. Technol. 12 (1995) 935–950.
[94] P.K. Kundu, I.M. Cohen, Fluid Mechanics, Academic Press, 2002.
[95] H.-C. Kuo, R.T. Williams, Semi-Lagrangian solutions to the inviscid Burgers equation, Mon. Weather Rev. 118 (1990) 1278–1288.
[96] Y. Kurihara, Numerical integration of the primitive equations on primitive grids, Mon. Weather Rev. 93 (1965) 399–415.
[97] M. Kwizak, A.J. Robert, A semi-implicit scheme for grid point atmospheric models of the primitive equations, Mon. Weather Rev. 99 (1972) 32–36.
[98] G. Lanzafame, Implicit integrations for SPH in semi-Lagrangian approach: application to the accretion disc modeling in a microquasar, Comput. Phys. Commun. 184 (2013) 671–688.
[99] P.H. Lauritzen, A stability analysis of finite-volume advection schemes permitting long time steps, Mon. Weather Rev. 135 (2007) 2658–2673.
[100] P.H. Lauritzen, E. Kaas, B. Machenhauer, K. Lindberg, A mass-conserving version of the semi-implicit semi-Lagrangian HIRLAM, Q. J. R. Meteorol. Soc. 134 (2008) 1583–1595.
[101] P.H. Lauritzen, M.A. Taylor, J. Overfelt, P.A. Ullrich, R.D. Nair, S. Goldhaber, R. Kelly, CAM-SE-CSLAM: consistent coupling of a conservative semi-Lagrangian finite-volume method with spectral element dynamics, Mon. Weather Rev. 145 (2017) 833–855.
[102] V. Laurma, M. Picasso, G. Steiner, An octree-based adaptive semi-Lagrangian VOF approach for simulating the displacement of free surfaces, Comput. Fluids 131 (2016) 190–204.
[103] A. Lavrov, Stability and dispersion analysis of semi-Lagrangian methods with Hermite interpolation, Numer. Heat Transf., Part B, Fundam. 155 (3) (2009) 177–195.
[104] A.S. Lawless, A perturbation forecast model and its adjoint, in: Preprints of the 11th Conference on Numerical Weather Prediction, American Meteorological Society, 1996.
[105] A.S. Lawless, Development of Linear Models for Data Assimilation in Numerical Weather Prediction, Ph.D. thesis, Department of Mathematics, University of Reading, 2001.
[106] A.S. Lawless, N.K. Nichols, S.P. Ballard, A comparison of two methods for developing the linearization of a shallow-water model, Q. J. R. Meteorol. Soc. 129 (2003) 1237–1254.
[107] D.Y. Le Roux, C.A. Lin, A. Staniforth, An accurate interpolating scheme for semi-Lagrangian advection on an unstructured mesh for ocean models, Tellus 49A (1997) 119–138.
[108] D.Y. Le Roux, C.A. Lin, A. Staniforth, A semi-implicit semi-Lagrangian finite-element shallow water ocean model, Mon. Weather Rev. 128 (2000) 1384–1401.
[109] L.M. Leslie, R.J. Purser, High-order numerics in an unstaggered three-dimensional time-split semi-Lagrangian forecast model, Mon. Weather Rev. 119 (1991) 1612–1623.
[110] L.M. Leslie, R.J. Purser, Three-dimensional mass conserving semi-Lagrangian scheme employing forward trajectories, Mon. Weather Rev. 123 (1995) 2551–2566.
[111] Y. Li, I.M. Navon, P. Courtier, P. Gauthier, Variational data assimilation with a semi-Lagrangian semi-implicit global shallow water equation model and its adjoint, Mon. Weather Rev. 121 (1993) 1759–1769.

[112] S.-J. Lin, A vertically Lagrangian finite volume dynamical core for global models, Mon. Weather Rev. 132 (2004) 2293–2307.
[113] S.-J. Lin, R.B. Rood, Multidimensional flux-form semi-Lagrangian transport schemes, Mon. Weather Rev. 124 (1996) 2046–2070.
[114] S.-J. Lin, R.B. Rood, A explicit flux-form semi-Lagrangian shallow water model on the sphere, Q. J. R. Meteorol. Soc. 123 (1997) 2477–2498.
[115] A.C. Lorenc, et al., The Met. Office global three dimensional variational data assimilation scheme, Q. J. R. Meteorol. Soc. 126 (2000) 2991–3012.
[116] E.N. Lorenz, Energy and numerical weather prediction, Tellus 12A (1960) 364–373.
[117] A. McDonald, Accuracy of multiply-upstream, semi-Lagrangian advective schemes, Mon. Weather Rev. 112 (1984) 1267–1275.
[118] A. McDonald, A semi-Lagrangian and semi-implicit two time-level integration scheme, Mon. Weather Rev. 114 (1986) 824–830.
[119] A. McDonald, Accuracy of multiply-upstream semi-Lagrangian advective scheme II, Mon. Weather Rev. 115 (1987) 1446–1450.
[120] A. McDonald, Semi-Lagrangian methods, in: ECMWF Seminar Series on Numerical Methods in Atmospheric Modelling, 1991, pp. 257–277.
[121] A. McDonald, An examination of alternative extrapolation to find the departure point position in a two-time-level semi-Lagrangian integration, Mon. Weather Rev. 127 (1999) 1985–1993.
[122] A. McDonald, Boundary conditions for semi-Lagrangian schemes: testing some alternatives in one-dimensional models, Mon. Weather Rev. 128 (2000) 4084–4096.
[123] A. McDonald, Well-Posed Boundary Conditions for Semi-Lagrangian Schemes: the One Dimensional Case, Part II, Tech. report, 44, HIRLAM, 2000.
[124] A. McDonald, A step towards transparent boundary conditions for meteorological models, Mon. Weather Rev. 130 (2002) 140–151.
[125] A. McDonald, Transparent boundary conditions for the shallow water equations: testing in a nested environment, Mon. Weather Rev. 131 (2003) 698–705.
[126] A. McDonald, J.R. Bates, Improving the estimate of the departure point position in a two-time level semi-Lagrangian and semi-implicit scheme, Mon. Weather Rev. 115 (1987) 737–739.
[127] A. McDonald, J.R. Bates, Semi-Lagrangian integration of a gridpoint shallow water model on the sphere, Mon. Weather Rev. 117 (1989) 130–137.
[128] A. McDonald, J. Haugen, A two-time-level, three-dimensional semi-Lagrangian, semi-implicit limited-area gridpoint model of the primitive equations, Mon. Weather Rev. 120 (1992) 2603–2621.
[129] T. Melvin, M. Dubal, N. wood, A. Staniforth, M. Zerroukat, An inherently mass-conserving iterative semi-implicit semi-Lagrangian discretization of the non-hydrostatic vertical-slice equations, Q. J. R. Meteorol. Soc. 136 (2010) 799–814.
[130] A. Mola, L. Heltai, A. DeSimone, A stable and adaptive semi-Lagrangian potential model for unsteady and nonlinear ship-wave interactions, Eng. Anal. Bound. Elem. 37 (2013) 128–143.
[131] R.D. Molina, J.F. Salazar, J.A. Martinez, J.C. Villegas, P.A. Arias, Forest-induced exponential growth of precipitation along climatological wind streamlines over the Amazon, J. Geophys. Res., Atmos. 124 (2019) 2589–2599.
[132] R. Nair, J. Côté, A. Staniforth, Monotonic cascade interpolation for semi-Lagrangian advection, Q. J. R. Meteorol. Soc. 125 (1999) 197–212.
[133] R. Nair, J.S. Scroggs, F.H.M. Semazzi, Efficient conservative global transport schemes for climate and atmospheric chemistry models, Mon. Weather Rev. 130 (2002) 2059–2073.
[134] R.D. Nair, J. Côté, A. Staniforth, Cascade interpolation for semi-Lagrangian advection over the sphere, Q. J. R. Meteorol. Soc. 125 (1995) 1445–1468.
[135] R.D. Nair, B. Manchenhauer, The mass-conservative cell-integrated semi-Lagrangian advection scheme on the sphere, Mon. Weather Rev. 130 (2002) 649–667.
[136] I.M. Navon, X. Zou, J. Derber, J. Sela, Variational data assimilation with an adiabatic version of the NMC spectral model, Mon. Weather Rev. 120 (1992) 1433–1446.
[137] A. Oliveria, A.M. Baptista, A comparison of integration and interpolation Eulerian–Lagrangian methods, Int. J. Numer. Methods Fluids 21 (1995) 183–204.
[138] A. Oliveria, A.M. Baptista, On the role of tracking on Eulerian–Lagrangian solutions of the transport methods, Adv. Water Resour. 21 (1998) 539–554.

[139] M. Owkes, O. Desjardins, A mass and momentum conserving unsplit semi-Lagrangian framework for simulating multiphase, J. Comput. Phys. 332 (2017) 21–46.
[140] B.P. Leonard, A.P. Lock, M.K. MacVean, Conservative explicit unrestricted-time-step multidimensional consistency-preserving advection schemes, Mon. Weather Rev. 124 (1996) 2558–2606.
[141] P. Pellerin, R. Laprise, I. Zawadzki, The performance of a semi-Lagrangian transport scheme for the advection-condensation problem, Mon. Weather Rev. 123 (1995) 3318–3330.
[142] N.A. Phillips, A coordinate system having some special advantages for numerical forecasting, J. Meteorol. 14 (1957) 184–185.
[143] N.A. Phillips, An example of non-linear computational instability, in: Bert Bolin (Ed.), The Atmosphere and the Sea in Motion, Oxford University Press, 1959, pp. 501–504.
[144] S. Polavarapu, M. Tanguay, R. Mánard, A. Staniforth, The tangent linear model for semi-Lagrangian schemes: linearizing the process of interpolation, Tellus 48A (1996) 74–95.
[145] W.H. Press, S.A. Teukolsky, W.T. Vetterling, B.P. Flannery, Numerical Recipes, 2nd. ed., Cambridge University Press, 1992.
[146] A. Priestly, A quasi-conservative version of the semi-Lagrangian scheme, Mon. Weather Rev. 121 (1993) 621–629.
[147] J. Pudykiewicz, R. Benoit, A. Staniforth, Preliminary results from a partial LARP model based on existing meteorological forecast model, Atmos.-Ocean 23 (1985) 267–303.
[148] J. Pudykiewicz, A. Staniforth, Some properties and comparative performance of the semi-Lagrangian methods of Robert in the solution of the advection-diffusion equation, Atmos.-Ocean 22 (1984) 283–308.
[149] D.K. Purnell, Solution of the advective equation by upstream interpolation with a cubic spline, Mon. Weather Rev. 104 (1976) 42–48.
[150] R.J. Purser, Accurate numerical differencing near a polar singularity of a skipped grid, Mon. Weather Rev. 116 (1988) 1067–1076.
[151] R.J. Purser, L.M. Leslie, A semi-implicit semi-Lagrangian finite difference scheme using high-order spatial differencing on a non-staggered grid, Mon. Weather Rev. 116 (1988) 2069–2080.
[152] R.J. Purser, L.M. Leslie, An efficient interpolation procedure for high-order three dimensional semi-Lagrangian models, Mon. Weather Rev. 119 (1991) 2492–2498.
[153] R.J. Purser, L.M. Leslie, An efficient semi-Lagrangian scheme using third-order semi-implicit time integration and forward trajectories, Mon. Weather Rev. 122 (1994) 745–756.
[154] F. Rabier, H. Jarvinen, E. Klinker, J.-F. Mahfouf, A. Simmons, The ECMWF implementation of four dimensional variational assimilation. Part I: experimental results with simplified physics, Q. J. R. Meteorol. Soc. 126A (2000) 1143–1170.
[155] D.A. Randall, Geostrophic adjustment and the finite difference shallow water equations, Mon. Weather Rev. 122 (1994) 1371–1377.
[156] L. Richardson, Weather Prediction by Numerical Process, Cambridge University Press, 1922.
[157] H. Ritchie, Eliminating the interpolation associated with the semi-Lagrangian scheme, Mon. Weather Rev. 114 (1986) 135–146.
[158] H. Ritchie, Semi-Lagrangian advection on a Gaussian grid, Mon. Weather Rev. 115 (1987) 608–619.
[159] H. Ritchie, Applications of the semi-Lagrangian method to a spectral model of the shallow water equations, Mon. Weather Rev. 116 (1988) 1587–1598.
[160] H. Ritchie, C. Temperton, A. Simmons, M. Hortal, T. Davies, D. Dent, M. Hamrud, Implementation of the semi-Lagrangian method in a high-resolution version of the ECMWF forecast model, Mon. Weather Rev. 123 (1995) 489–514.
[161] A. Robert, A stable numerical integration scheme for the primitive meteorological equations, Atmos.-Ocean 19 (1981) 35–46.
[162] A. Robert, A semi-Lagrangian and semi-implicit numerical integration scheme for the primitive meteorological equations, J. Meteorol. Soc. Jpn. 60 (1982) 319–325.
[163] A.J. Robert, The integration of a low order spectral form of the privative meteorological equations, J. Meteorol. Soc. Jpn. 44 (1966) 237–244.
[164] L.P. Rushøjgaard, S.E. Cohn, Y. Li, R. Ménard, The use of spline interpolation in semi-Lagrangian transport models, Mon. Weather Rev. 126 (1998) 2008–2016.
[165] R. Sadourny, Conservative finite difference approximations of the primitive equations on quasi uniform spherical grids, Mon. Weather Rev. 100 (1972) 136–144.

[166] R. Sadourny, A. Arakawa, A. Mintz, Integration of the non-divergent barotropic vorticity equation with an icosahedral-hexagonal grid for the sphere, Mon. Weather Rev. 96 (1968) 351–356.
[167] J.S. Sawyer, A semi-Lagrangian method of solving the vorticity advection equation, Tellus A 15 (1963) 336–342.
[168] F. Schmidt, Variable fine mesh in a spectral global model, Beitr. Phys. Atmos. 50 (1977) 211–217.
[169] F.H.M. Semazzi, P. Dekker, Optimal accuracy semi-Lagrangian model, Mon. Weather Rev. 122 (1994) 2139–2159.
[170] R. Shapiro, Smoothing filtering and boundary effects, Rev. Geophys. Space Phys. 8 (1970) 359–387.
[171] I. Silberman, Planetary waves in the atmosphere, J. Meteorol. 11 (1954) 27–34.
[172] C.J. Smith, The Semi-Lagrangian Method for Atmospheric Modelling, Ph.D. thesis, University of Reading, Department of Mathematics, 2000.
[173] P.K. Smolarkiewicz, J.A. Pudykiewicz, A class of semi-Lagrangian approximations for fluids, J. Atmos. Sci. 49 (1992) 2082–2096.
[174] M.R. Spiegel, Vector Analysis, Schaum's Outline, McGraw–Hill, 1959.
[175] A. Staniforth, Review - formulating efficient finite-element codes for flows in regular domains, Int. J. Numer. Methods Fluids 7 (1987) 1–16.
[176] A. Staniforth, J. Côté, Semi-Lagrangian integration schemes for atmospheric models - a review, Mon. Weather Rev. 119 (1991) 2206–2223.
[177] A. Staniforth, J. Pudykiewicz, Reply to comments on and addenda to "Some properties and comparative performance of the semi-Lagrangian method of Robert in the solution of the advection-diffusion equation", Atmos.-Ocean 23 (1985) 195–200.
[178] A. Staniforth, A. White, N. Wood, Analysis of semi-Lagrangian trajectory calculations, Q. J. R. Meteorol. Soc. 129 (2003) 2065–2085.
[179] A. Staniforth, N. White, A.A. Wood, Treatment of vector equations in deep-atmosphere, semi-Lagrangian models. I: momentum equations, Q. J. R. Meteorol. Soc. 136 (2010) 497–506.
[180] G. Strang, On the construction and comparison of difference schemes, SIAM J. Numer. Anal. 5 (1968) 506–517.
[181] D. Sydorenko, R. Rankin, Simulation of ionospheric disturbances created by Alfvén waves, J. Geophys. Res. 117 (2012) A09229.
[182] M. Tanguay, S. Polavarapu, The adjoint of the semi-Lagrangian treatment of the passive tracer equation, Mon. Weather Rev. 127 (1999) 551–564.
[183] M. Tanguay, S. Polavarapu, P. Gauthier, Temporal accumulation of first-order linearization error for semi-Lagrangian passive advection, Mon. Weather Rev. 125 (1996) 1296–1311.
[184] M. Tanguay, A. Robert, Elimination of the Helmholtz equation associated with the semi-implicit scheme in a grid point model of the shallow water equations, Mon. Weather Rev. 114 (1986) 2154–2162.
[185] M. Tanguay, A. Simard, A. Staniforth, A three dimensional semi-Lagrangian scheme for the Candian regional finite-element forecast model, Mon. Weather Rev. 117 (1989) 1861–1871.
[186] M. Tanguay, E. Yakimiw, H. Ritchie, A. Robert, Advantages of spatial averaging in semi-implicit and semi-Lagrangian schemes, Mon. Weather Rev. 120 (1992) 113–123.
[187] C. Temperton, Treatment of the Coriolis terms in semi-Lagrangian spectral models, Atmos.-Ocean 35 (1997) 293–302.
[188] C. Temperton, A. Staniforth, An efficient two-time-level semi-Lagrangian semi-implicit integration scheme, Q. J. R. Meteorol. Soc. 115 (1987) 1025–1039.
[189] S. Theissen-Krah, K. Iyer, L.H. Rüpke, J. Phipps Morgan, Coupled mechanical and hydrothermal modeling of crustal accretion at intermediate to fast spreading ridges, Earth Planet. Sci. Lett. 311 (2011) 275–286.
[190] J. Thuburn, A PV-based shallow-water model on a hexagonal-icosahedral grid, Mon. Weather Rev. 125 (1997) 2328–2347.
[191] J. Thuburn, A. Staniforth, Conservation and linear Rossby-mode dispersion on the spherical C-grid, Mon. Weather Rev. 132 (2004) 641–653.
[192] J. Thuburn, M. Zerroukat, N. Wood, A. Staniforth, A.A. White, Coupling a mass-conserving semi-Lagrangian scheme (SLICE) to a semi-implicit discretisation of the shallow-water equations: minimizing the dependence on a reference atmosphere, Q. J. R. Meteorol. Soc. 136 (2010) 146–154.
[193] P.A. Ullrich, M.R. Norman, The flux-form semi-Lagrangian spectral element (FF-SLSE) method for tracer transport, Q. J. R. Meteorol. Soc. 140 (2014) 1069–1085.
[194] T. Umeda, Effect of ion cyclotron motion on the structure of wakes: a Vlasov simulation, Earth Planets Space 64 (2012) 231–236.

[195] F. Voitus, P. Termonla, P. Bénard, Well-posed lateral boundary conditions for spectral semi-implicit semi-Lagrangian schemes: test in a one-dimensional model, Mon. Weather Rev. 137 (2009) 315–330.
[196] W.-Y. Sun, K.-S. Yeh, A general semi-Lagrangian advection scheme employing forward trajectories, Q. J. R. Meteorol. Soc. 123 (1997) 2463–2476.
[197] W.-Y. Sun, K.-S. Yeh, R.-Y. Sun, A simple semi-Lagrangian scheme for advection equations, Q. J. R. Meteorol. Soc. 122 (1996) 1211–1226.
[198] P. Welander, Studies on the general development of motion in a two-dimensional ideal fluid, Tellus 17 (1955) 141–156.
[199] N. White, A.A. Wood, A. Staniforth, Treatment of vector equations in deep-atmosphere, semi-Lagrangian models. II: kinematic equation, Q. J. R. Meteorol. Soc. 136 (2010) 507–516.
[200] A. Wiin-Nielsen, On the application of trajectory methods in numerical forecasting, Tellus 11 (1959) 180–196.
[201] P.D. Williams, A proposed modification to the Robert–Asselin filter, Mon. Weather Rev. 137 (2009) 2538–2546.
[202] P.D. Williams, The RAW filter: an improvement to the Robert–Asselin filter in semi-implicit integrations, Mon. Weather Rev. 139 (2011) 1996–2007.
[203] P.D. Williams, The RAW filter: an improvement to the Robert–Asselin filter in semi-implicit integrations, Mon. Weather Rev. 139 (2011) 1996–2007.
[204] R.T. Williams, R.P. Kurth, Formation of discontinuities in a stratified rotating atmosphere, J. Geophys. Res. 81 (1976) 1133–1140.
[205] D.L. Williamson, Integration of barotropic vorticity equations on a spherical geodesic grid, Tellus 20A (1968) 642–653.
[206] D.L. Williamson, J.B. Drake, J.J. Hack, R. Jakob, P.N. Swartrauber, A standard test set for numerical approximations to the shallow water equations in spherical geometry, J. Comput. Phys. 102 (1992) 211–224.
[207] D.L. Williamson, P.J. Rasch, Two-dimensional semi-Lagrangian transport with shape-preserving interpolation, Mon. Weather Rev. 117 (1989) 117–129.
[208] D.L. Williamson, P.J. Rasch, On shape preserving interpolation and semi-Lagrangian transport, SIAM J. Sci. Stat. Comput. 11 (1990) 656–687.
[209] N. Wood, A. Staniforth, A. White, T. Allen, M. Diamantakis, M. Gross, T. Melvin, C. Smith, S. Vosper, M. Zerroukat, J. Thuburn, An inherently mass-conserving semi-implicit semi-Lagrangian discretization of the deep-atmosphere global non-hydrostatic equations, Q. J. R. Meteorol. Soc. 140 (2014) 1505–1520.
[210] J.R. Yearsley, A semi-Lagrangian water temperature model for advection-dominated river systems, Water Resour. Res. 45 (2009) W12405.
[211] M. Zerroukat, T. Allen, A three-dimensional monotone and conservative semi-Lagrangian scheme (SLICE-3D) for transport problems, Q. J. R. Meteorol. Soc. 138 (2012) 1640–1651.
[212] M. Zerroukat, N. Wood, A. Staniforth, SLICE: a semi-Lagrangian inherently conserving and efficient scheme for transport problems, Q. J. R. Meteorol. Soc. 128 (2002) 2801–2820.
[213] M. Zerroukat, N. Wood, A. Staniforth, SLICE-S: a semi-Lagrangian inherently conserving and efficient scheme for transport problems on the sphere, Q. J. R. Meteorol. Soc. 130 (2004) 2649–2664.
[214] M. Zerroukat, N. Wood, A. Staniforth, The parabolic spline method (PSM) for conservative transport problems, Int. J. Numer. Methods Fluids 11 (2006) 1297–1318.
[215] M. Zerroukat, N. Wood, A. Staniforth, Application of the parabolic spline method (PSM) to multi-dimensional conservative semi-Lagrangian transport scheme (SLIC), J. Comput. Phys. 225 (2007) 935–948.
[216] M. Zerroukat, N. Wood, A. Staniforth, An improved version of SLICE for conservative monotonic remapping on a C-grid, Q. J. R. Meteorol. Soc. 135 (2009) 541–546.
[217] M. Zerroukat, N. Wood, A. Staniforth, A.A. White, J. Thuburn, An inherently mass-conserving semi-implicit semi-Lagrangian discretisation of the shallow-water equations on the sphere, Q. J. R. Meteorol. Soc. 135 (2009) 1104–1116.
[218] Y. Zhang, A.M. Baptista, SELFE: a semi-implicit Eulerian–Lagrangian finite-element model for cross-scale ocean circulation, Ocean Model. 21 (2008) 71–96.

Index

C

D

T

CPI Antony Rowe
Eastbourne, UK
November 23, 2019